DIFFERENTIAL EQUATIONS WITH BOUNDARY VALUE PROBLEMS
A Systems Approach

DIFFERENTIAL EQUATIONS WITH BOUNDARY VALUE PROBLEMS

A Systems Approach

Bruce P. Conrad
Temple University

Prentice Hall

Pearson Education, Inc.
Upper Saddle River, NJ 07458

Library of Congress Cataloging in Publication Data

Conrad, Bruce P.
    Differential equations with boundary value problems: a systems approach / Bruce P. Conrad.
      p. cm.
    Includes index.
    ISBN 0-13-093419-4
    1. Differential equations. 2. Boundary value problems. I. Title.

QA372.C6417 2003
515'.35--dc21

2002074985

Acquisitions Editor: George Lobell
Editor-in-Chief: Sally Yagan
Vice President/Director of Production and Manufacturing: David W. Riccardi
Executive Managing Editor: Kathleen Schiaparelli
Senior Managing Editor: Linda Mihatov Behrens
Assistant Managing Editor: Bayani de Leon
Production Editor: Steven S. Pawlowski
Manufacturing Buyer: Alan Fischer
Manufacturing Manager: Trudy Pisciotti
Marketing Assistant: Rachel Beckman
Editorial Assistant: Jennifer Brady
Art Director: John Christiana
Cover Designer/Interior Designer: Michelle White
Cover Image: Burj Al Arab Hotel, Dubai, United Arab Emirates (front); Burj Al Arab Hotel at night (back);
The Burj Al Arab (Arabian Tower) Hotel is one of the tallest hotels ever built, with an atrium over 300 feet in
the air, and a total height of 1053 feet. The hotel was built on an island in the Persian Gulf off of Dubai.
Cover Image Credit: Peter Ginter/ Bilderberg/AURORA (front); Simon Warren/ CORBIS (back)
Image Resource Center:
        Director: Melinda Reo
        Manager, Rights and Permissions: Zina Arabia
        Cover Image Specialist: Karen Sanatar
        Image Permission Coordinator/Photo Researcher: Kim Marsden
Art Studio: Laserwords

©2003 by Pearson Education, Inc.
Pearson Education, Inc.
Upper Saddle River, New Jersey 07458

Printed in the United States of America

10 9 8 7 6 5 4 3 2

ISBN 0-13-093419-4

Pearson Education LTD., *London*
Pearson Education Australia PTY, Limited, *Sydney*
Pearson Education Singapore, Pte. Ltd
Pearson Education North Asia Ltd, *Hong Kong*
Pearson Education Canada, Ltd., *Toronto*
Pearson Educacion de Mexico, S.A. de C.V.
Pearson Education – Japan, *Tokyo*
Pearson Education Malaysia, Pte. Ltd

For Becky

Contents

Preface

If an application of mathematics has a component that varies continuously as a function of time, then it probably involves a differential equation. For this reason, ordinary differential equations are of great importance in engineering, applied mathematics and the sciences. This has been recognized since the founders of calculus, Newton and Leibniz, made their contributions to the subject in the late seventeenth century.

A differential equations text has to address an audience with diverse interests. Science and engineering majors are required to take a differential equations course because it provides them with valuable mathematical tools. Mathematics majors may take courses in differential equations because the subject is interesting; because it is an essential component of applied mathematics; or because it is prerequisite for the study of differential geometry, dynamical systems, and mathematical modeling. All students who take a differential equations course will gain a deeper understanding of the concepts and applications of calculus.

There are numerous differential equations texts on the market, and it is reasonable to ask why I chose to write another one. I had taught many differential equations courses before I decided to write this text, and what tipped the scale was a conversation overheard in the mathematics library between two students. The gist of the conversation was that the differential equations course was trivial, which may be true if the course focuses only on the formalities of solving equations for which algorithms are available. If viewed as an application of the quadratic formula, solving second-order linear equations with constant coefficients, for example, *is* easy. A differential equations course, like any mathematics course, needs to offer more intellectual challenge than that. I wanted to write a text that will enable students to visualize a differential equation as a direction or vector field, and to use the standard formal solution procedures with a full understanding of their limitations.

This text maintains a moderate level of rigor. Proofs are included if they are accessible and have the potential to enhance the reader's understanding of the subject. For example, the existence theorem for solutions of initial value problems is not actually proved—its prerequisite, Ascoli's theorem, would not ring a bell for most readers—but I allude to Peano's proof, which uses Euler's method to show that a solution exists. On the other hand, the proof of the uniqueness theorem is included, as a special case of Proposition 2.4.3, which specifies an upper bound for the rate at which solutions of a differential equation can diverge from one another.

Although applications usually involve systems of differential equations, the emphasis in most differential equations texts is second-order equations. When faced with a system, there is a rather complicated algorithm that finds an equivalent higher-order equation. This approach doesn't carry one very far, and it stems from a desire to avoid linear algebra in general and characteristic roots (eigenvalues) in particular. In fact, there is no result presented in the introductory linear algebra course that is not useful in differential equations, and linear algebra ought to be a prerequisite for the differential equations course. In spite of this, many universities, including the one that employs me, require only two semesters of calculus as a prerequisite for their differential equations courses. There are texts that present differential equations and linear algebra as a combined course. This is an acceptable approach, to which this text is an alternative. I have attempted to accommodate the needs of readers who have not had a linear algebra course, without wasting the time of those who have had the course. Thus, Chapter 4 presents linear systems of differential equations in matrix form but is limited to systems of two equations—and thus involves only 2×2 matrices. The time spent making this accessible to those who have not had a linear algebra course is not excessive. Linear operators are defined and discussed in Chapter 5, along with further concepts from linear algebra: linear combinations and linear independence.

We should tell students about IVP (initial value problem) solvers. These are computer programs that can calculate and plot an approximate solution of an initial value problem. People who work with differential equations find them indispensable; yet many students complete a differential equations course without ever using one. It is common to find the IVP solver algorithms in differential equations texts. The Runge-Kutta algorithm is often found, for example. These algorithms should be studied in numerical analysis courses, but our job is to get the students to use them. *A solution of a differential equation that was not obtained by symbol manipulation is for many people a first encounter with a function that is not presented as a formula.* While no student should pass a differential equations course without learning how to solve certain differential equations by analytic means, students must be trained to use numerical methods as well. My goal has been to induce an intuitive understanding of the concept of a solution of an initial value problem in order to resolve potential confusion about what we are approximating when we call a numerical method. This is not a numerical analysis text, and the discussion of numerics is confined to Euler's method—which advances the understanding of what a differential equation is—and a brief user's guide to more advanced methods. I do not hesitate to call upon effective numerical methods when symbolic methods can't be used.

Technology and Supplements

It is widely believed that computer algebra software (CAS) can make short work of the routine calculations that have bedeviled generations of students in introductory differential equations courses. I have found that this software is sometimes beneficial as a labor-saving device, and that it is definitely useful for producing illustrations. There are three admirable CAS programs available: *Maple*, *Mathematica*, and MATLAB. Most of the illustrations in this text were produced with *Mathematica*. Examples using CAS programs can be found on the text Web site, **http://www.prenhall.com/conrad**. This text can be used effectively without the benefit of CAS, but an IVP solver that will display graphs of solutions of differential equations on a computer or calculator screen is required. Every CAS can function as an IVP solver, and special-purpose IVP solvers may be downloaded by following links on the text Web site. The Web site also has a list of currently available hand-held graphing calculators that include IVP solvers. There is a general discussion of IVP solvers in Section 2.3.

For those who wish to use Maple with this text, there is a new manual, Maple Projects for Differential Equations (013-047974-8), by Gilbert and Hsiao, that can be shrink wrapped with the text for one half the manual's normal price.

The same is also true for Polking and Arnold's Ordinary Differential Equations Using MATLAB (013-011381-6). This text also is accompanied by a Student Solutions Manual and an Instructor Solutions Manual.

Organization

To narrow the field when selecting a differential equations text, an instructor may ask if the applications are realistic. This text does not favor applications involving "real data." While there is much to be said for studying realistic applications, the many complications may obscure the differential equation that should be the center of attention. It is preferable to consider a simple problem that exemplifies the differential equations component of an actual application.

The existence and uniqueness theorems for initial value problems have implications about the structure of the set of solutions of a differential equation or system. This text explores that structure, as well as the additional structural properties of special kinds of equations, such as linear equations or autonomous systems. The properties of linearity and of being autonomous are important for many applications, and by associating these properties with particular applications, we can bring our physical experience to the task of learning about differential equations.

This text is presented in four parts: The first introduces both linear differential equations and nonlinear systems and provides a foundation. The second part is devoted to linear differential equations, including systems of first-order equations, the single second-order equation, Laplace transform methods, and equations with variable coefficients. The third part focuses on nonlinear differential equations and dynamical systems. The fourth part of the text, on boundary value problems, focuses on applications of the technique of separation of variables for partial differential equations. The dependency graph that follows the table of contents will be of assistance in navigating the text and in planning class syllabi. In this graph, the Sections marked with asterisks in the table of contents are treated as separate nodes. They can be skipped without interruption of the sequence within the Chapters where they reside. In a recent course at Temple University, given over a semester of fourteen weeks (not including the final examination), I covered Chapters 1–6, and Section 8.1, skipping Sections 1.5, 2.5, 3.4, 3.5, 5.9, and 6.8. The class met three hours per week, with an additional weekly *Maple*-based computer laboratory.

I have used an approach to definitions that is more common in lower-level mathematics texts. Instead of placing formal, numbered statements of definitions in the text, I have provided each Chapter with a glossary that contains all of the definitions, in alphabetical order. This allows for a more informal discussion of a term when it is introduced. The first use of a term that is defined in the glossary is in boldface. I have reserved the bold typeface for that purpose—except when it is used in Section headings and mathematical expressions. My hope is that readers, who are accustomed to hypertext will find this approach to be convenient.

Part I is divided into three Chapters, covering linear first-order equations, nonlinear first-order equations, and systems of first-order equations, respectively. The goal of Chapter 1 is to present linear equations and a few applications (linear growth and mixture problems) that illuminate the meaning of linearity. The method of solution is variation of constants, rather than the Leibniz integrating factor, because it leads to a discussion of the structure of the general solution as the sum of a particular solution and the associated homogeneous solution. The Chapter includes an introduction to initial value problems, with a careful explanation of the existence and uniqueness theorems.

Chapter 2 opens by visualizing differential equations with the aid of a direction field, and this leads to Euler's method. We then turn to the symbolic algorithms for solving separable and exact equations. We notice that these algorithms do not actually produce an explicit solution as the algorithm for solving linear equations did; instead, the output is an integral: a function $F(x, y)$ whose level curves define solutions. The theoretical component of the Chapter is the existence and uniqueness theorems, and the applications are to nonlinear growth and falling bodies with air drag.

Systems of differential equations traditionally make their appearance later in a text than they do here. Since systems do play a prominent role in applications, and there are many elementary things to be said about them, I have introduced them in Part I. A solution of a system of two first-order differential equations is viewed as a pair of parametric equations for a curve in the phase plane. If the system is autonomous, it can be visualized as a vector field on the phase plane. The method for finding an integral of a system of two autonomous equations is presented, and to make accurate graphs of solutions, there is an informal discussion of IVP solvers— the user's guide that was mentioned previously. Applications are to the van der Pol equation and to population models for two interacting species.

Part II is devoted to linear differential equations, starting with systems. Practically all of the systems in Chapter 4 are of two equations with constant coefficients, stated in explicit form. The solution algorithm is invariably to compute the characteristic roots and vectors of the coefficient matrix, determine a fundamental matrix solution, and in the inhomogeneous case, use variation of constants to determine a particular solution. This algorithm elucidates the structure of the solution set by enabling us to sketch phase portraits for homogeneous systems and to identify centers, nodes, and saddles readily (these geometric aspects of systems are found in Section 8.1, which can be read just after Section 4.3). Readers are expected to learn to find characteristic roots and vectors of 2×2 matrices, as well as the definition of characteristic roots and vectors of larger matrices. Chapter 4 contains an introduction to matrices (concentrating on the 2×2 case) and a review of the complex number system.

The centerpiece of the traditional first course in ordinary differential equations is analysis of second-order linear equations. By placing this topic after linear systems, we realize an economy, because second-order equations are easily transformed into equivalent systems, and the results of Chapter 4 can be applied. The connection with systems is exploited to streamline the presentation of the theoretical aspects of second-order equations, but the solution algorithm for equations with constant coefficients is presented in the usual way. In Chapter 5, I have emphasized the vector space structure of the set of solutions of a linear homogeneous differential equation and the concept of a linear differential operator. About half of the Chapter is devoted to an application, damped mechanical systems with one degree of freedom. I have not used this application to convince readers that differential equations is applicable mathematics, but to enable visualization of the concepts presented. The important properties of second-order linear differential equations can be better presented in terms of mechanical systems than in any other context. The

Chapter also includes a discussion of mechanical systems with two degrees of freedom, which are modeled by systems of two second-order equations. The emphasis is on finding fundamental frequencies and modes of vibration.

The Laplace transform is presented in Chapter 6 as an alternate means of solving inhomogeneous linear equations or systems with constant coefficients. The goal is to understand how transform methods work. It is natural to use the Laplace transform to motivate and explore the convolution, and this leads to a different interpretation of the variation of constants formula that was studied in Chapter 5. A second goal of this Chapter is to introduce the Heaviside unit step function and its "derivative," the Dirac delta function. The applications in Chapter 6 are primarily to electrical circuits.

Chapter 7 is devoted to series solutions of linear differential equations with variable coefficients. The reason to study series solutions is to express solutions of linear differential equations with variable coefficients in terms of elementary and special functions. The expansion of a solution near a regular singular point is considered in some detail, as a generalization of the solution of Cauchy-Euler equations. Thus, Bessel functions can be introduced, and we use them in Chapter 11 to explore the solution of the diffusion equation in a cylindrical domain.

Nonlinear differential equations are the subject of Part III, which can be read at any time after Chapter 4. Chapter 8 is about the stability and asymptotic stability of stationary points of autonomous nonlinear systems. Stability is analyzed by means of linearization and by the method of Lyapunov. A detailed understanding of phase portraits and the stability of linear systems is an essential prerequisite for comprehending linearization, and these topics are covered in the first two Sections of Chapter 8. Instructors who emphasize the geometric aspect of systems will want to cover these Sections in the midst of Chapter 4.

Chapter 9 asks why two-dimensional autonomous systems are so simple and three-dimensional systems so complicated? Our analysis of this question leads us to consider transversals and the first return mapping, and thus to dynamical systems. The first return mapping associated with a system of two autonomous differential equations is a one-dimensional dynamical system, while a three-dimensional system of differential equations reduces to a dynamical system on the plane. To answer the question about the two-dimensional case, we include a proof of the Poincaré-Bendixson theorem, which explains how the topology of the plane constrains solutions: The first return mapping turns out to be monotone. In three dimensions, the simplest nontrivial example of a first return mapping occurs in the case of systems of two linear equations with periodic coefficients, and the theorem of Floquet. In general, it would not even make sense to say the first return mapping for any three-dimensional system is monotone, since a plane is not naturally presented as an ordered set. Chapter 9 presents the Lorenz equations as an example of the chaos phenomenon, and, in a final exercise, considers the forced Duffing eqation.

Part IV starts with the pure two-point boundary value problem. The variation of constants formula reappears in the guise of Green's function, which is constructed as a way of exploring existence and uniqueness of solutions of two-point boundary value problems. The application here is to a mechanical system, for which data on position or velocity are obtained at two different times. After Section 10.1, our attention will be focused on boundary problems that do *not* have unique solutions. The initial motivation for such problems is to understand why the existence and uniqueness of solutions, which we take for granted with initial value problems, are not available in other venues. What is more surprising is that with boundary value problems, the situations where uniqueness fails are of more interest in applications than the situations where uniqueness holds. We turn nonuniqueness to our advantage by studying the classical partial differential equations of mathematical physics—the initial boundary value problems for the diffusion and wave equations, and the three boundary value problems for the Laplace equation—by the method of separation of variables. The basic technique, separation of variables, is presented in Chapter 10 within the context of the diffusion equation. This leads to a brief introduction to Fourier sine series. The other boundary value problems are postponed for coverage in Chapter 11. Chapter 10 concludes with the Sturm oscillation theorems, providing yet another opportunity for phase plane analysis of second-order linear differential equations.

Chapter 11 brings the results of Chapter 7 on series solutions, and Chapter 10, on separation of variables, to bear upon the classical partial differential equations of mathematical physics: the diffusion, wave, and Laplace equations. Bessel functions and Legendre polynomials, which were introduced in Chapter 7, are used to express solutions on domains with particular kinds of symmetry.

Acknowledgments

I have had the benefit of many helpful suggestions in writing this text, and I would like to thank the many reviewers who made them: Eduardo Cattani, Alan Genz, Tomas Gideon, Peter Wolfe, Esteban Tabak, Baruch Cahlon, Darrell Schmidt, and Richard Tangeman. I would also like to thank Robert Stern for his encouragement at an early stage of this project, and the Prentice Hall staff, in particular George Lobell, Dennis Kletzing, Steven Pawlowski, and Melanie VanBenthuysen who have been very patient with me. I had assistance with the answer manual from Xiangdong Wen, which is much appreciated. I have had productive discussions about differential equations with Eric Grinberg, David Hill, Omar Hijab, Nicholas Macri, colleagues at Temple University. Finally, I thank my family for putting up with me while I was engaged in writing this text.

Bruce P. Conrad
conrad@euclid.math.temple.edu

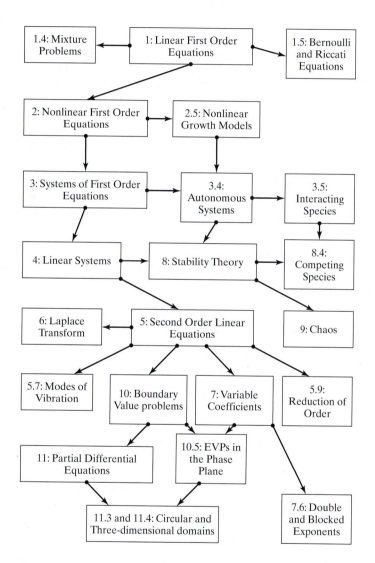

Chapter Dependency Graph

PART I

FIRST-ORDER EQUATIONS

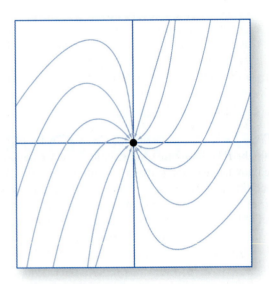

CHAPTER 1

Linear Equations

1.1 Introduction

A differential equation is a relation involving an unknown function and some of its derivatives. For example,

$$\frac{dy}{dt} = y + e^t$$

is a differential equation that asks for a function, $y = f(t)$, whose derivative is equal to the function plus e^t. By differentiating, you can verify that a function such as $y = te^t$ meets this specification.

Differential equations are a source of fascinating mathematical problems, but the main reason that people study them is that they have numerous applications. Consider the problem of determining the velocity of a falling object. Newton's second law of motion tells us that the net force on the object is equal to the mass times the acceleration (the derivative of the velocity with respect to time). This law can be restated in the form of a differential equation,

$$m\frac{dv}{dt} = F,$$

where the unknown function is the velocity, v. The mass of the object is m, and F is the net force. If we are ignoring air resistance, and the object is close to the earth's surface, then $F = mg$ is a constant force due to gravitation, directed downward. In the meter-second system of units, the gravitational acceleration g is approximately $9.80\text{m}/\text{sec}^2$.

Taking the downward direction as positive, we have a differential equation:

$$m\,v' = mg.$$

Let us simplify and solve this differential equation as follows: Cancel the mass to obtain $v' = g$. This means that v is an antiderivative of the constant g; thus $v = gt + C$, where C is an arbitrary constant.

This solution is not a single function, but a family of functions depending on an arbitrary constant C. To determine the equation of motion of this particular falling object, more information is needed.

For example, an **initial condition**, specifying the velocity when $t = 0$, will determine the equation of motion. If we assume that the object was falling from rest, so that $v = 0$ when $t = 0$, then we know that the solution that we seek has the property $v(0) = 0$. We have knowledge of the starting, or initial value of v. It is easy to infer from this initial condition that the constant C is equal to 0, and $v = gt$.

What if we *don't* ignore air resistance? It is known that air resistance is a force whose magnitude is proportional to the square of the velocity, provided that the velocity does not approach the speed of sound. Thus the air resistance component of the force on the falling object has magnitude $k\,v^2$, where k is a constant of proportionality (called the **drag coefficient**). If we combine air resistance and gravitation, the force on the falling object becomes $F = mg - kv^2$ The differential equation now takes the form

$$\frac{dv}{dt} = g - \frac{k}{m}v^2.$$

Now the unknown function v appears on the right side of the equation. We may be tempted to integrate as we did before, but then we need to calculate

$$v = \int \left(g - \frac{k}{m}v^2 \right) dt.$$

Since the "unknown function" v appears in the integrand here, there is no way to calculate this integral without first knowing the answer! The purpose of a differential equations text is to develop ways to study solutions in spite of this quandary.

The **order** of a differential equation is the order of the highest derivative of the unknown function occurring in the equation. The differential equations describing the velocity of a falling object that we just considered were first order. In the related second-order equation, $y'' = g$, the unknown function represented by the variable y represents the distance the object has fallen. The velocity would be $v = y'$, so if we include air resistance, we have another second-order equation, $y'' = g - k(y')^2/m$.

A differential equation involving only derivatives with respect to one independent variable is called an **ordinary differential equation**, or **ODE**. The falling-body differential equations that we just considered are ODEs, because they only involve derivatives with respect to the independent variable t. If partial derivatives with respect to two or more independent variables appear in a differential equation, then it is a **partial differential equation,** or **PDE**. As an example, here is the second-order PDE that describes the vibration of a guitar string:

$$\frac{\partial^2 y}{\partial t^2} - c^2 \frac{\partial^2 y}{\partial x^2} = 0.$$

The unknown function y represents the displacement of a point on the string x centimeters from the bridge at time t, and c is a constant related to the tension and density of the string.

In any differential equation, a **dependent variable** is a variable that represents an unknown function. A **solution** of a differential equation is a function that can be substituted for the dependent variable to produce an identity. Thus, we found that the solutions of the falling object ODE $v' = g$ are $v = gt + C$, where C is an arbitrary constant.

Example 1.1.1   The ODE

$$\frac{dy}{dt} = ky \tag{1.1}$$

can be used to compute growth of populations, compound interest on savings accounts, and so on (see Section 1.2). Show that $y = Ce^{kt}$, where C is an arbitrary constant, satisfies this ODE.

Solution. If $y = Ce^{kt}$, then $\frac{dy}{dt} = Cke^{kt}$ and $ky = Cke^{kt}$ as well. Hence equation (1.1) becomes an identity. ❑

Differential equations typically have infinite families of solutions, but we often need just one solution from the family. We refer to a single solution of a differential equation as a **particular solution** to emphasize that it is one of a family.

The **general solution** of a differential equation is the family of all solutions that the differential equation has. The general solution of an ODE on an interval (a, b) is a family of solutions, all defined at every point of the interval (a, b), such that every solution that is defined on that interval belongs to the family. Finding the general solution of any ODE usually involves two steps: *calculation* and *verification*. The calculation step is exemplified by our solution of the falling-body equation $v' = g$. Unless a mistake was made in the integration, the family of solutions we found will satisfy the ODE.

The verification step is to show that there are no "extra" solutions of the ODE that don't belong to this family. The following theorem from calculus will be useful:

Theorem 1.1   Equal derivatives theorem

Let $f_1(t)$ and $f_2(t)$ be defined and differentiable on an interval (a, b) (infinite endpoints are permitted), and assume that $f_1'(t) = f_2'(t)$ for all $t \in (a, b)$. Then there is a constant C such that $f_1(t) = f_2(t) + C$ for all $t \in (a, b)$.

Example 1.1.2   Verify that the family of solutions $v = gt + C$ for $y' = g$ is the general solution.

Solution. Let $y_0(t) = gt$. If $y_1(t)$ is another solution of the differential equation, then $y_1'(t) = y_0'(t) = g$. It follows from the equal derivatives theorem that $y_1(t) \equiv y_0(t) + C$. In other words, *every solution of the ODE belongs to the family* $y = gt + C$. ❑

Since ODEs typically have families of solutions, they are frequently coupled with a constraint in order to single out the one solution that is of interest in a given application. In other words, problems are frequently stated as "Find the solution of _____ (a given ODE) such that _____ (a constraint) is satisfied." The type of constraint that we have already encountered, and shall frequently encounter in the future, is an initial condition, where the value of the solution function is specified at some initial time. Thus, when we were told that an object was falling from rest, that could be translated into an initial condition for the velocity: $v(0) = 0$. An ODE coupled with an initial condition is called an **initial value problem**, or **IVP**. Thus, the motion of a body falling from rest with air resistance would be modeled by the IVP,

$$\frac{dv}{dt} = g - kv^2; \ v(0) = 0.$$

Example 1.1.3   The IVP

$$y' = 0.05 \, y; \ y(0) = 1000$$

is a mathematical model[1] for the balance of a savings account that earns 5% interest, when the principal is initially \$1000. Solve this IVP to find the balance as a function of time.

Solution.   In Example 1.1.1, we found that the solution of the ODE, $y' = k \, y$, is $y = Ce^{kt}$, where C is constant. For this bank account, $k = 0.05$, so $y = C \, e^{0.05 \, t}$. When we substitute $t = 0$ and $y = 1000$ to incorporate the initial condition, we find $C = 1000$. Hence $y = 1000e^{0.05 \, t}$. ◻

In this text, we will study techniques for solving several specific kinds of ODEs. We also have to consider ODEs that have solutions that can't be expressed as formulas in terms of standard mathematical notation. To work with an ODE without expressing its solutions as a formula, a computer is useful. In fact, since the invention of the first computer, solutions of intractable ODEs and PDEs have been calculated by computers.

These calculations are done by *numerical* means; that is, the computer does not work with a formula for a solution but calculates a table of values that give a close approximation of a solution of the differential equation from the differential equation itself. As a simple example, consider the ODE $y' = f(t)$. The solution is

$$y = \int f(t) \, dt + C,$$

which is helpful if we can determine a formula for the antiderivative of $f(t)$. If no such formula is available, we would have to turn to a numerical method of evaluating the integral, such as the rectangle rule, the trapezoidal rule, or perhaps Simpson's rule. Numerical methods for solving ODEs of the form $y' = f(t, y)$, where the right side involves y as well as t, are generalizations of these "rules."

A numerical method cannot compute a family of solutions; it can only approximate one solution at a time. The user is expected to input constraints specifying the

---

[1] This model is known as the continuous compounding model.

solution to be approximated. A program that is designed to approximate solutions of ODEs with initial conditions as the constraints is called an **IVP solver**.

While there are programs that are primarily IVP solvers, general-purpose mathematical software systems—that is, *Macsyma, Maple, Mathematica*, and MATLAB—include subroutines that are IVP solvers and are also capable of finding a formula for the general solution of practically any differential equation for which there is an established method of solution. These programs, which we call computer algebra software (CAS), follow rules for manipulating formulas instead of performing numerical calculations. There are advanced calculators that incorporate IVP solvers and CAS as well.

Figure 1.1 displays the graphs of several solutions of the ODE $v' = g - kv^2$ representing the motion of a falling object with air resistance proportional to the square of the velocity. Displaying graphs of several solutions may suggest features that all solutions of a given ODE have in common (in this case they all have a common asymptote). Any explanation of the behavior of solutions of an ODE that is based on an analysis of the equation itself, and not on a formula for the general solution, is called a **qualitative study**. A qualitative study may result in a better understanding of the physical phenomenon represented by the ODE than a formula for the general solution could. For example, we see in Figure 1.1 that if the velocity starts with $v = 0$, it will increase but not without bound. There is a limiting velocity, known as the **terminal velocity**. The figure also shows what happens if an object is induced to fall at a velocity faster than its terminal velocity: It will slow down.

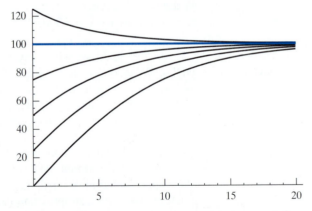

FIGURE 1.1   Several solutions of the ODE $v' = g - kv^2$. Each graph represents the velocity of a falling object subject to air resistance. The values $g = 9.8$ meters/second^2 and $k = 9.8 \times 10^{-4}$ kilograms /meter were used.

To see how a qualitative study might proceed, let's look more closely at the ODE $v' = g - kv^2$. Define a constant v_∞ by $v_\infty = \sqrt{g/k}$, so that $g = kv_\infty^2$. Recast the model as

$$v' = k(v_\infty^2 - v^2).$$

For $v < v_\infty$ we see that $v' > 0$. This means that v is increasing. When $v > v_\infty$, $v' < 0$, and v is decreasing. In both cases, v tends toward v_∞, as in Figure 1.1.

This qualitative study correlates well with the physics. For $v < v_\infty$, the gravitational force is greater than the air resistance, so the net force is directed downward and causes the object to accelerate. If $v > v_\infty$ the air resistance is dominant, and the object decelerates. Finally, if $v = v_\infty$ gravity and air resistance balance each other, and the speed is constant.

EXERCISES

For each of the differential equations in Exercises 1–13,

   (**i**) *determine the* order,

   (**ii**) *determine whether the differential equation is an ODE or a PDE.*

1. $y'' + y = 0$

2. $y' = (x + y)^4$

3. $(y')^2 + xy = e^x$

4. $\dfrac{\partial y}{\partial t} + 6y\dfrac{\partial y}{\partial t} = \dfrac{\partial^3 y}{\partial x^3}$

5. $y' = t^2 + y^2$

6. $y^{(100)} - y = 0$

7. $y' = \sin y$

8. $y^2 + (y')^2 = 1$

9. $y^3 + (y')^3 = 1$

10. $t^2 y'' + ty' + (t^2 - 1)y = 0$

11. $\dfrac{\partial^2 u}{\partial x \partial y} = 0$

12. $\dfrac{\partial u}{\partial t} = \dfrac{\partial^2 u}{\partial x^2}$

13. $\dfrac{d^3 u}{dt^3} + 3t\dfrac{du}{dt} - u\cosh t = \dfrac{1}{t^2 + 1}$

14. Show that $y = C\sec(t)$, is a family of solutions of

$$y' - \tan(t)y = 0$$

and find solutions that satisfy the following initial conditions:

  a. $y(0) = 0$ b. $y(\pi/4) = 1$

15. Show that $y = C\sin(t) + D\cos(t)$ is a family of solutions of

$$y'' + y = 0,$$

and find all solutions that satisfy the following constraints:

  a. $y(0) = 0$

  b. $y(0) = 0$, $y(\pi) = 0$

  c. $y(0) = 0$, $y(\pi/6) = 1$

  d. $y(0) = 1$, $y'(0) = -1$

16. Show that $y = 2(\sqrt{t} - 1) + Ce^{-\sqrt{t}}$ is a solution of

$$y' + \frac{y}{2\sqrt{t}} = 1,$$

and use it to solve the IVP that couples this ODE with the initial condition $y(1) = 0$.

17. Show that $y = Ce^{t^2}$ is a solution of the ODE $y' = 2ty$, and solve the IVP $y' = 2ty$; $y(0) = 3$.

18. Let $y(t) = \begin{cases} 0 & \text{if } t \le 0 \\ t^2 & \text{if } t > 0. \end{cases}$ Show that $y(t)$ is a solution of the IVP $y' = 2\sqrt{y}$; $y(1) = 1$. Also show that

$$y = t^2 \text{ for all real } t$$

is *not* a solution!

  ◆ ***Hint:*** Remember that $\sqrt{t^2} = |t|$.

19. Let $y(t)$ be a differentiable function defined on the interval $(-1, 1)$ with the property that for all t,

$$t^2 + y(t)^2 = 1$$

(there are two such functions). Show that $y(t)$ is a solution of the ODE

$$yy' + t = 0.$$

The following problems require showing a given family of functions to be the general solution of a given differential equation. This entails verifying that every member of the family is a solution and that every solution belongs to the family. In each case, the equal derivatives theorem can be used.

20. Verify that $y(t, C) = (t + C)e^{-t}$ is the general solution of the ODE

$$\frac{d}{dt}(e^t y) = 1$$

by showing that for every value of the constant C, $y(t, C)$ satisfies the ODE, and that every solution has the form $y(t, C)$ for some C.

21. Verify that $y = t\ln t + Ct$ is the general solution of

$$\frac{ty' - y}{t^2} = \frac{1}{t}$$

on the interval $(0, \infty)$.

  ◆ ***Hint:*** $(ty' - y)/t^2 = \dfrac{d}{dt}\left(\dfrac{y}{t}\right).$

1.2 Linear Models for Growth and Decay

An ODE that can be used to simulate a scientific or engineering phenomenon is a type of **mathematical model** of the phenomenon. Developing mathematical models involves theorizing, computation, and testing. Theorizing starts by identifying a variable (or variables) that describes the phenomenon (for a falling body, the variables might be the velocity, the distance fallen, or both). The ODE then describes how the variable or variables change with time. The ODE may include parameters such as masses, interest rates, and friction constants. These are constants in a given system but can be varied to make the model applicable to other systems.

Since nature is complicated, it may be necessary to ignore some of the smaller influences to make the model comprehensible. For example, a model used to compute the trajectory of a projectile might be based on ignoring air resistance, so that the only force to be considered is gravitational.

The computation and testing stage of the modeling process involves finding a solution of the ODE and comparing it with observed data. If the computed solution and the data don't agree, then either the theorizing oversimplified by ignoring a factor that significantly influences the data, the theory was wrong, or the computation was faulty.

The derivative of a quantity y with respect to time gives its absolute rate of change. It often happens that the **relative rate of change**, rather than the absolute rate of change, fits naturally into a mathematical model. The relative rate of change of y is defined to be the absolute rate of change, expressed as a fraction of y. The distinction between absolute and relative rates of change can be readily seen by comparing simple and compound interest. Simple interest is calculated as a percentage of the original deposit and does not change as long as no further deposits are made. The absolute rate of change of value of this account is the dollar amount of interest earned per unit time. For example, if the account pays 4% interest and $4000 is initially deposited, then interest payments of $0.04(\$4000) = \$160$ will be paid each year for the life of the account. The value y of the account is determined by the ODE

$$\frac{dy}{dt} = 160,$$

with $y(0) = 4000$. This indicates a constant *absolute* rate of change. After t years, the value will be

$$4000 + \int_0^t 160\, dt = \$(4000 + 160t).$$

With compound interest, the interest previously accumulated in the account is included in calculating the interest to be paid. The relative rate of change is the absolute growth rate expressed as a percentage of the account's current balance. Banks quote this rate as the annual percentage rate and often use the abbreviation APR. The formula for the relative rate of change of the balance y is

$$\frac{1}{y}\frac{dy}{dt}.$$

Assuming that the bank uses continuous compounding, as most do, the relative rate of change of the account balance is equal to the APR, which is constant. If the APR is k, then the ODE

$$\frac{dy}{dt} = k \cdot y \tag{1.2}$$

is an appropriate model. Equation (1.2) is called the **linear growth equation**. The word *linear* refers to the ODE, not its solution.

The linear growth equation can be used as a model for the balance of a bank account with continuously compounded interest (as noted previously), as well as population growth and radioactive decay. We will study these, as well as a number of other phenomena to which equation (1.2) applies.

Example 1.2.1   The family of solutions $y = Ce^{kt}$ of equation (1.2) was found in Example 1.1.1 on page 4. Verify that it is the general solution.

Solution. We must show that any solution $y = \phi(t)$ of equation (1.2) belongs to the family $y = Ce^{kt}$. We can achieve this objective by showing that the product $e^{-kt}\phi(t)$ is constant.

By the product rule for differentiation,

$$\frac{d}{dt}[e^{-kt}\phi(t)] = -ke^{-kt}\phi(t) + e^{-kt}\phi'(t).$$

Since $\phi(t)$ is a solution of equation (1.2), we can replace $\phi'(t)$ with $k\phi(t)$, and then

$$\frac{d}{dt}[e^{-kt}\phi(t)] = -ke^{-kt}\phi(t) + e^{-kt}k\phi(t) = 0.$$

By the equal derivatives theorem, $e^{-kt}\phi(t) = C$, where C is constant. Thus

$$e^{kt}e^{-kt}\phi(t) = e^{kt}C.$$

Since $e^{kt}e^{-kt} = 1$, it follows that $\phi(t) = Ce^{kt}$. ❑

Every solution of equation (1.2) will display *exponential growth* if $k > 0$, and *exponential decay* if $k < 0$. Figures 1.2 and 1.3 show graphs of $y = e^{kt}$ for $k > 0$ and $k < 0$, respectively.

For exponential growth, the **doubling time** is the time it takes for a solution of (1.2) whose initial value is $y(0)$ to reach the value $2y(0)$. To find the doubling time, substitute $t = 0$ in the formula $y = Ce^{kt}$:

$$y(0) = Ce^0 = C.$$

Now substitute $t = T$, where T denotes the doubling time, and $y = 2C$. We find that

$$2C = Ce^{kT},$$

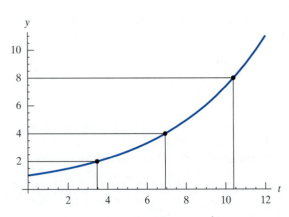

FIGURE 1.2 Exponential growth: $y = e^{kt}$ $(k > 0)$.
Vertical lines mark the doubling time.

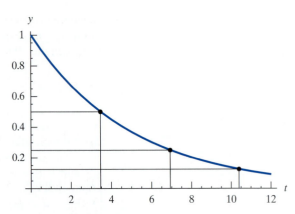

FIGURE 1.3 Exponential decay: $y = e^{kt}$ $(k < 0)$.
Vertical lines are at intervals of one half-life.

and hence $e^{kT} = 2$. Taking logarithms, $kT = \ln(2)$, and therefore

$$T = \frac{\ln(2)}{k}.$$

To summarize, the doubling time T is inversely proportional to the constant k and does not depend on $y(0)$. Furthermore,

$$y(t + T) = Ce^{k(t+T)} = Ce^{kt}e^{kT} = 2Ce^{kt} = 2y(t).$$

It follows that $y(2T) = 2y(T) = 4y(0)$, $y(3T) = 8y(0)$, and so on.

Population growth. Let $y(t)$ represent the population, at time t years after a reference time. Although the population at any time is an integer, we will be using a continuous variable to represent it. This *continuous approximation* of a discrete variable restricts this model to large populations. Under the assumption that the relative birth and death rates, b and d, respectively (expressed per thousand population), are constant, the relative rate of change for the population is $\frac{b-d}{1000}$:

$$y'(t) = \frac{b - d}{1000} \cdot y(t),$$

which is equation (1.2), with constant $k = \frac{b-d}{1000}$. In the past, birth and death rates for human populations have not remained constant for long periods of time. Therefore, predictions based on the linear model are only reliable over short time spans.

Example 1.2.2   According to the 1980 census, the population of the United States was then 227 million. The birth rate in 1980 was 15.9 per thousand, and the death rate was 8.7 per thousand. Use these data with the linear growth model to estimate the population in 2000.

Solution. The relative growth rate is $k = \frac{15.9-8.7}{1000} = 0.0072$. Let y represent the population, in millions. Our model takes the form of an IVP, $y' = 0.0072y$;

$y(1980) = 227$. The general solution of the ODE is $y = Ce^{0.0072t}$. If we set $t = 1980$, and $y = 227$, we find

$$227 = Ce^{(0.0072)(1980)}$$

and hence $C = 227e^{-(0.0072)(1980)}$. The solution of the IVP is therefore

$$y = 227e^{-(0.0072)(1980)}\, e^{0.0072t} = 227e^{0.0072(t-1980)}.$$

We therefore estimate that in 2000, the population will be

$$y(2000) = 227e^{0.0072(20)} = 262 \text{ million.}$$

(The 2000 census recorded a considerably larger population, 281 million. The discrepancy, which is about the same as the population of the state of New York, shows that the linear growth model is an oversimplification.) ❑

Compound interest. A bank deposit earns **continuous compound interest** if interest is continuously paid into the account at a rate proportional to the sum of the original principal and all interest accumulated. As long as the interest rate is constant, the linear growth model accurately represents compound interest, where the constant k is the interest rate.

The following example shows how the k can be calculated from two data points, $y(0)$ and $y(t_1)$, for some $t_1 > 0$.

Example 1.2.3   An investment of $10,000, made on June 30, 1993 in the Standard and Poor's 500 Stock Index would have been worth $17,987 on December 31, 1996, assuming all dividends were reinvested when received. What rate of compound interest would be required to match this performance? What would the doubling time be?

Solution. Let k be the interest rate. Then if $y(t)$ denotes the balance at time t, the IVP $y' = ky$; $y(0) = 10,000$ holds. The general solution of the ODE is $y = Ce^{kt}$, and the initial condition specifies $C = y(0) = 10,000$. The time elapsed is $t = 3.5$ years. Hence $17,987 = 10,000e^{3.5k}$, so $k = \frac{1}{3.5}\ln(1.7987) = 0.1677 = 16.77\%$. The doubling time is $\ln(2)/.1677 = 4.133$ years. ❑

Radioactive decay. The linear model for the decay of a radioactive isotope agrees extremely well with experimental evidence. In this model, $y(t)$ denotes the mass (in grams) of the isotope that has not yet decayed at time t. It is known that the rate at which atoms of the isotope disintegrate is proportional to the number of atoms present. Furthermore, all atoms of the isotope have the same mass, so we can say that the rate of disintegration is proportional to $y(t)$. Thus $y(t)$ decreases at a rate proportional to $y(t)$, and the linear decay model applies:

$$\frac{dy}{dt} = -ky.$$

The general solution of the linear decay equation is $y = Ce^{-kt}$, so we can say that the mass of the isotope is undergoing *exponential decay*.

The analog of the doubling time in this case is the **half-life,** which is the time it takes for a solution of equation (1.2) whose initial value is y_0 to reach the value $\frac{1}{2}y_0$.

Example 1.2.4   The half-life of the isotope I^{131} of iodine is 8 days. A nuclear accident releases 100 kilograms of I^{131} into the environment. How much I^{131} will remain in the environment after a period of one year? After two years?

Solution. A rough estimate is made by noting that a year is made up of slightly more that 45 eight-day periods. At the end of each eight-day period, there is half as much I^{131} present as there was at the beginning. Thus, after the first eight days, there are 50 kg of the isotope left; after the first 16 days, 25 kg will be left; and so on. After a year, the amount will be about $100 \cdot 2^{-45} = 2.84 \times 10^{-12}$ kilograms left. Two years make up more than 91 eight-day periods, so the amount left after two years will be about $100 \cdot 2^{-91} = 4.04 \times 10^{-26}$ kilograms left.   ❑

It is prudent to view answers involving extremely large or extremely small numbers critically. An atom of I^{131} has mass 2.17×10^{-25} kilograms, and it is impossible to have less than an atom of any substance. Therefore, the estimate we have made of the amount of I^{131} left after two years is wrong. This is not surprising since the model that we are using is not accurate when only a few atoms of isotope are present.

Homogeneous linear equations

A first-order **linear ODE** is an equation of the form

$$a_1(t)y' + a_0(t)y = b(t).$$

The functions $a_1(t)$ and $a_0(t)$ are called the **coefficient functions**, and $b(t)$ is the **source term**.

For example, equation (1.2), $y' - k\,y = 0$ is linear, and the falling-body equation with air drag, $v' + kv^2 = g$, is not linear because of the quadratic term in the variable v.

A linear ODE is **homogeneous** if the source term is 0. Equation (1.2) is homogeneous, and every homogeneous first-order linear ODE can be put in the form

$$y' = k(t)\,y, \tag{1.3}$$

where $k(t) = -a_0(t)/a_1(t)$. Thus a homogeneous linear equation with variable coefficients can be used as a model for growth or decay when the relative growth rate is a function of time.

To solve a homogeneous first-order linear ODE we can mimic the derivation of the solution of equation (1.2) as follows: Substitute $y = Ce^{K(t)}$, where $K(t)$ is a function that we will attempt to determine and C is an arbitrary constant. Using the chain rule, we find that

$$\frac{dy}{dt} = C\,K'(t)\,e^{K(t)}.$$

With this substitution, equation (1.3) is converted to

$$C\,K'(t)\,e^{K(t)} = k(t)\,C\,e^{K(t)}.$$

We can now divide through by $Ce^{K(t)}$ to obtain $K'(t) = k(t)$. In other words, if $K(t)$ is an antiderivative of $k(t)$, then $y = Ce^{K(t)}$ is a family of solutions of equation (1.3). This generalizes what we found in the constant coefficient case, since then $k(t) = k$ is constant, and $K(t) = k\,t$.

Proposition 1.2.1

Let $k(t)$ be a function that is continuous on an interval (c, d), and let $K(t)$ be an antiderivative of $k(t)$ on that interval. Then the family of solutions $y = Ce^{K(t)}$ of $y' = k(t)y$ is the general solution.

Proof We will rely on the equal derivatives theorem, just as we did when proving e^{kt} is the general solution of $y' = ky$ when k is constant. Let $y = \phi(t)$ be any solution of equation (1.3). By the product rule for differentiation,

$$\frac{d}{dt}[e^{-K(t)}\phi(t)] = -K'(t)\,e^{-K(t)}\,\phi(t) + e^{-K(t)}\,\phi'(t)$$

$$= -k(t)\,e^{-K(t)}\,\phi(t) + e^{-K(t)}\,k(t)\,\phi(t)$$

$$= 0.$$

It follows from the equal derivatives theorem that $e^{-K(t)}\phi(t)$ is constant, and hence that $\phi(t) = Ce^{K(t)}$. Thus, ϕ belongs to the family of solutions. ∎

The Homogeneous Case

An ODE that can be put into the form $a_1(t)y' + a_0(t)y = 0$ is called a homogeneous first-order linear equation. If $k(t) = -a_0(t)/a_1(t)$ is continuous on an interval (c, d) then the general solution is $y = Ce^{K(t)}$, where $K(t)$ is an antiderivative of $k(t)$ and C is a constant. Thus, the following formula for the general solution holds:

$$y = Ce^{-\int (a_0(t)/a_1(t))\,dt}.$$

A linear ODE with variable coefficients may have **singular points**. These are points where $k(t) = -a_0(t)/a_1(t)$ is discontinuous. Our algorithm for solving these equations fails at singular points, and we can only expect the expression $y = Ce^{K(t)}$ to represent the general solution on an interval that does not contain any singular points. This comment is applicable to the next example, since the ODE there has a singular point at 0. In this example we will substitute $y = Ce^{K(t)}$ into the equation without first dividing through by the coefficient of y'.

Example 1.2.5   Find the general solution of

$$ty' + my = 0,$$

where m is a constant, on the interval $(0, \infty)$, and find the solution that satisfies the initial condition $y(1) = -1$.

Solution. Substitute $y = Ce^{K(t)}$ and $y' = CK'(t)e^{K(t)}$ to obtain

$$tCK'(t)e^{K(t)} + mCe^{K(t)} = 0.$$

Solving for $K'(t)$, we obtain $K'(t) = -m/t$. Hence

$$K(t) = -\int \frac{m}{t}\,dt = -m\ln|t|.$$

(We don't have to worry about an integration constant here, because we don't need *the most general antiderivative of m/t.*) It follows that $y = Ce^{-m \ln |t|}$ is a family of solutions. Since $-m \ln |t| = \ln |t^{-m}|$, $e^{-m \ln |t|} = \pm t^{-m}$ and the general solution of this ODE is

$$y = Ct^{-m}.$$

(The constant C has been modified to absorb the $\pm$ sign.) You should check this result by differentiation.

To satisfy the initial condition, set $t = 1$ and $y = -1$ to obtain $-1 = C$. The solution of the IVP is $y = -t^m$.   ❑

EXERCISES

1. The doubling time for a certain population is 35 years. How long does it take for this population to triple?

✔**Answer:** 55.47 years

2. The U.S. national debt amounted to $1 trillion in 1980. By 1988 it had grown to $2 trillion, and by 1992, $4 trillion. Does this information indicate that the debt was growing exponentially between 1980 and 1992, or does it indicate otherwise?

3. Thomas Robert Malthus[2] was one of the first to recognize that the population will increase exponentially if it is not controlled by limitations of resources or governmental regulation. In his words,

> In the United States of America, where the means of subsistence have been more ample, the manners of the people more pure, and consequently the checks to early marriages fewer than in any of the modern states of Europe, the population has been found to double itself in twenty-five years.

By how much did the birth rate (per thousand) exceed the death rate in the United States at the time of Malthus?

4. If the birth and death rates of 1980 (see Example 1.2.2) were sustained indefinitely, how long would it take for the population to double?

✔**Answer:** 96 years

5. Two countries, A and B, each had a population of 1 million in 1900. The population of A obeys the exponential growth equation (1.2) with relative growth rate $k = .04$, and it is observed that the population of B, in millions, is always equal to the square root of the population of A, in millions. Thus, for example, when the population of A reached 4 million, the population of B was 2 million. Show that the population of B also obeys equation (1.2), and find its relative growth rate.

6. An investor sold her stock in two companies, A and B, for $44,200 and $2800, respectively. Each stock had been bought for an initial purchase price of $1000, with all dividends reinvested. The A stock had been held for 13.26 years, while the B stock had been held for 5.75 years. For each stock, compute the compound interest rate which would yield an equivalent return over the same time period. Which was the more lucrative investment?

7. A bond, dated 1700, promises to pay the bearer £1 sterling plus accumulated interest, compounded continuously at 6% per annum. What is the bond worth in 2000?

✔**Answer:** £65,659,969.14

In Exercises 8–15, find the general solution of the ODE and sketch the graph of the solution that satisfies the given initial condition.

8. a. $y' - 2y = 0$; $y(0) = 1$
   b. $y' + 2y = 0$; $y(0) = 1$

9. $ty' - 12y = 0$; $y(2) = 1$

10. $y' + \sin(t)y = 0$; $y(0) = 1$

11. $\cos(t)y' - \sin(t)y = 0$; $y(0) = -1$

12. $y' + ty = 0$; $y(1) = 1$

13. $ty' + y = 0$; $y(1) = 1$

14. a. $ty' + |t|y = 0$; $y(0) = 1$
   b. $ty' - |t|y = 0$; $y(0) = 1$

15. $(t^2 + 1)y' - ty = 0$; $y(3) = 0$

Carbon dating *A radioactive isotope of carbon, C^{14}, has a half-life of 5730 years. The fraction of C^{14} found in atmospheric carbon dioxide is constant at about one part per billion (abbreviated 1 ppb), because the decay rate is matched by the*

[2] *Essay on the Principle of Population,* 1798. The quotation was taken from page 22 of Norton Critical Edition, edited by Philip Appleman, Norton, New York 1976.

rate at which new C^{14} *is created by the effect of cosmic radiation. A living plant acquires all of its carbon by respiration of atmospheric carbon dioxide; hence the fraction of a living plant's carbon, which is the isotope* C^{14}*, is also 1 ppb. When the plant dies, decaying* C^{14} *is not replaced, so the time elapsed since the death of the plant can be estimated by measuring the fraction of* C^{14} *that remains. This technique, known as* C^{14} ***radiometric dating***, *was discovered in 1947 by Willard F. Libby. Before his work, no one had noticed that atmospheric carbon dioxide contained* C^{14}*. Libby received the 1960 Nobel Prize for Chemistry as a result of this discovery.*

16. Suppose that a tree dies at $t = 0$. Show that the fraction of C^{14} in wood from the tree, in ppb, is modeled by the IVP,

$$y' = -0.000121y; \quad y(0) = 1.$$

17. King Tutankhamen died in 1325 B.C.E., and his tomb was discovered in 1922 by Howard Carter. What would the proportion of C^{14} in a wood sample taken from the tomb? (Of course, this wouldn't have been measured until after 1947.)

18. A sample of wood from an archaeological investigation contains 0.7 ppb C^{14}. Assuming that the wood, when part of a living tree, had 1 ppb C^{14}, estimate the age of the wood.

✔**Answer:** 2950 years

19. A sample of wood taken from a bristlecone pine tree on White Mountain is determined to be 4850 years old by counting growth rings. What would the expected proportion of C^{14} in the sample be?

20. C^{14} radiometric dating is considered useful to determine the age of objects between 500 and 50,000 years old. Use this information to estimate the following measures of precision of the device used to determine the number of ppb C^{14}, on the basis of this statement.

Sensitivity The least concentration that can be detected.

Relative error The measurement error, expressed as a percentage of the true concentration.

✔**Answer:** sensitivity: .002 ppb, relative error, 6%

1.3   **Linear First-Order Equations**

If an object is hotter or colder than the environment in which it is situated, its temperature will approach the temperature of the environment, which we call the *ambient temperature*. Sir Isaac Newton devised a mathematical model of this phenomenon in the form of an ODE called **Newton's law of cooling**, based on the principle that the rate of change of the temperature of an object is negatively proportional to the difference between its temperature and that of the surrounding environment (the **ambient temperature**). If the object is colder than the environment, its temperature will increase, and if it is colder, its temperature will increase. To express this law as an ODE, let $T(t)$ denote the temperature of the object at time t, and let $A(t)$ be the ambient temperature. Then

$$T'(t) = -k[T(t) - A(t)]. \tag{1.4}$$

The coefficient k in equation (1.4), called the **transmission coefficient**, measures how well heat is transmitted between the object and its environment. If the object is insulated, k is usually inversely proportional to the thickness of the insulating material.

For Newton's law of cooling to be valid, the temperature of the object must be uniform at all times. We can consider this hypothesis to be satisfied if the object is small and conducts heat easily, or if the object is filled with a fluid that is in motion. Thus, we could expect the model to predict the temperatures of a red hot coin as it cools, a well-stirred flask of hot water, or a convection oven, but it might fail as a model for the temperature of a large baked potato.

Equation (1.4) can be written in the form

$$T' + kT = kA(t),$$

and this reveals that it is linear, but not homogeneous, because of the presence of the source term, $kA(t)$. A linear differential equation that is not homogeneous is called **inhomogeneous**.

To solve an inhomogeneous equation, it is necessary to solve the **associated homogeneous equation** first. This equation is obtained by removing the source term. Thus, the associated homogeneous equation for

$$a_1(t)y' + a_0(t)y = b(t), \tag{1.5}$$

is

$$a_1(t)y' + a_0(t)y = 0. \tag{1.6}$$

By Proposition 1.2.1 on page 13, the general solution of (1.6) has the form $y = Ce^{K(t)}$, where C is a constant, and $K(t) = -\int a_0(t)/a_1(t)\,dt$. Let us use $y_h(t)$ to denote the particular solution $e^{K(t)}$. The subscript h stands for "homogeneous"; thus the general solution of the homogeneous equation (1.6) is $y = C\,y_h(t)$. The following theorem tells how to express the general solution of an inhomogeneous linear ODE.

Theorem 1.2

Suppose that the coefficient functions $a_0(t)$ and $a_1(t)$, and the source term $b(t)$ in the inhomogeneous linear ODE (1.5) are continuous on an interval (c, d), and that $a_1(t) \neq 0$ on that interval. Let $y_p(t)$ be a particular solution of (1.5), defined on (c, d), and let $y_h(t)$ be a particular solution of the associated homogeneous equation (1.6). Then the general solution of (1.5) on (c, d) is

$$y = y_p(t) + C\,y_h(t). \tag{1.7}$$

Proof To verify that every function of the form (1.7) is a solution, just substitute $y = y_p(t) + C\,y_h(t)$ in (1.5):

$$a_1(t)[y_p(t) + C\,y_h(t)]' + a_0(t)[y_p(t) + C\,y_h(t)]$$
$$= \underbrace{a_1(t)y'_p(t) + a_0(t)y_p(t)}_{=b(t)} + C\underbrace{[a_1(t)y'_h(t) + a_0(t)y_h(t)]}_{=0}$$
$$= b(t).$$

Now let's verify that every solution of (1.5) belongs to the family (1.7). Let $y = \phi(t)$ be any solution of (1.5) that is defined on (c, d), and put $\psi(t) = \phi(t) - y_p(t)$. Then

$$a_1(t)\psi'(t) + a_0(t)\psi(t)$$
$$= \underbrace{a_1(t)\phi'(t) + a_0(t)\phi(t)}_{=b(t)} - \underbrace{[a_1(t)y'_p(t) + a_0(t)y_p(t)]}_{=b(t)}$$
$$= 0.$$

Therefore, $\psi(t)$ is a solution of the associated homogeneous equation (1.6). By Proposition 1.2.1, there is a constant C such that $\psi(t) = C\,y_h(t)$. It follows that

$$\phi(t) = y_p(t) + C\,y_h(t). \qquad \blacksquare$$

The technique that we will use to determine a particular solution $y_p(t)$ of an inhomogeneous linear ODE is known as the method of **variation of constants**.[3] With this method, we replace the constant C that appears in the solution of the associated homogeneous equation with a new dependent variable v, and substitute the resulting expression in the inhomogeneous equation. Thus, substitute $y = vy_h(t)$ in (1.5). By the product rule for differentiation,

$$y' = v'y_h(t) + vy'_h(t).$$

When we substitute this expression into equation (1.5), the result is

$$a_1(t)(v'y_h(t) + vy'_h(t)) + a_0(t)vy_h(t) = b(t)$$

or, more simply,

$$a_1(t)v'(t)y_h(t) + v(t)[a_1(t)y'_h(t) + a_0(t)y_h(t)] = b(t). \tag{1.8}$$

Since $y_h(t)$ is a solution of the homogeneous equation,

$$a_1(t)y'_h(t) + a_0(t)y_h(t) = 0;$$

thus the expression in square brackets in equation (1.8) drops out, leaving

$$a_1(t)v'(t)y_h(t) = b(t).$$

Therefore,

$$v'(t) = \frac{b(t)}{a_1(t)y_h(t)},$$

and hence

$$v(t) = \int \frac{b(t)}{a_1(t)y_h(t)} \, dt + C,$$

where C is constant. Upon multiplying this $v(t)$ by $y_h(t)$, we obtain the following expression for the general solution of equation (1.5) on the interval (c, d):

$$y(t) = y_h(t) \int \frac{b(t)}{a_1(t)y_h(t)} \, dt + Cy_h(t) \tag{1.9}$$

where

$$y_h = Ce^{-\int (a_0(t)/a_1(t)) \, dt}$$

is the solution of the homogeneous equation

$$a_1(t)y' + a_0(t)y = 0.$$

We will revisit the method of variation of constants when solving higher-order linear equations and systems of linear equations. In these cases, more than one constant will be "varied"—that is the reason for the plural "constants" in "variation of constants."

---

[3]Some texts refer to the method as "variation of parameters."

Example 1.3.1   Find the general solution of

$$y' + 3y = 2e^{-t}$$

on the interval $(-\infty, \infty)$, and determine the solution that satisfies the initial condition $y(0) = 0$.

Solution. The associated homogeneous equation is $y' + 3y = 0$, and we can take $y_h = e^{-3t}$.

Substitute $y = ve^{-3t}$ and $y' = v'e^{-3t} - 3ve^{-3t}$ in the in the inhomogeneous ODE to get

$$v'e^{-3t} - 3ve^{-3t} + 3ve^{-3t} = 2e^{-t}.$$

When simplified, this reduces to $v'e^{-3t} = 2e^{-t}$. Multiplying through by e^{3t} yields $v' = 2e^{2t}$. Thus

$$v = \int 2e^{2t}\, dt = e^{2t} + C.$$

Since $y = ve^{-3t}$, the general solution is the family $y = e^{-t} + Ce^{-3t}$.

To satisfy the initial condition, set $y = 0$ and $t = 0$. This yields $0 = 1 + C$, so $C = -1$, and

$$y = e^{-t} - e^{-3t}.$$

As expected, the general solution found in Example 1.3.1 splits as the sum of a particular solution, $y_p = e^{-t}$, and the general homogenous solution, $Cy_h = Ce^{-3t}$. Figure 1.4 displays the graphs of several solutions.

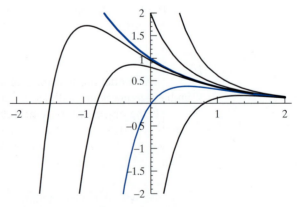

FIGURE 1.4   Solutions of $y' + 3y = 2e^{-t}$. See Example 1.3.1.

In the following example, we apply the method of variation of constants to a cooling problem.

Example 1.3.2   The temperature in an oven is 200°C when the oven is turned off. After 10 minutes, the temperature is 175°C. The temperature in the kitchen is 20°C. Find an expression for the temperature of the oven as a function of time.

Solution. We will use the IVP

$$\frac{dT}{dt} = -k(T - A(t)); \quad T(0) = 200$$

as our model, where T is the temperature in °C, and $A(t) = 20$ is the ambient temperature. The first step is to find the general solution of the ODE,

$$\frac{dT}{dt} = -k(T - 20),$$

or $T' + kT = 20k$. The associated homogeneous equation is

$$T' + kT = 0,$$

and we find $T_h = e^{-kt}$. Substitute $T = ve^{-kt}$ in the inhomogeneous equation. After simplifying, we obtain $e^{-kt}v' = 20k$, or $v' = 20ke^{kt}$. Integration yields

$$v = 20e^{kt} + C,$$

and hence $T = e^{-kt}v = 20 + Ce^{-kt}$. To evaluate the parameters C and k, we need to use the data. Substituting the initial condition $T(0) = 200$,

$$200 = 20 + Ce^{0},$$

and $C = 180$. Now substitute the other data point: When $t = 10$, $T = 175$. This yields

$$175 = 20 + 180e^{-10k},$$

and it follows that $e^{-10k} = 155/180$. Taking logarithms, $k = -\frac{1}{10}\ln(155/180) \approx 0.015$. Thus

$$T(t) = 20 + 180e^{-0.015t}. \qquad \Box$$

The model predicts that the difference between the oven temperature and the ambient temperature will decay exponentially.

If the object whose temperature is being modeled contains a source of heat, equation (1.4) must be modified by inserting another source term. Let $H(t)$ denote the rate that heat is generated within the object [$H(t)$ would be negative in some cases, such as air conditioning]. Then

$$T'(t) = -k[T(t) - A(t)] + mH(t), \tag{1.10}$$

where m is a positive constant, inversely proportional to the heat capacity of the object. For a small object, such as a toaster oven, m would be relatively large, while for a large object, such as a domed sports stadium, m would be small. In practice, we do not need to know the individual values of m or $H(t)$; all we need is the product $mH(t)$.

Example 1.3.3   The insulation of a building has transmission coefficient $k = 0.05$ hour^{-1}. When the furnace of the building is operating continuously and the outdoor temperature is $0°C$, the indoor temperature will be maintained at a constant $20°C$. Find the temperature inside the building as a function of time, with the furnace operating continuously, assuming that the outdoor temperature varies between $-7°C$ and $13°C$ each day, and that at time $t = 0$ the initial temperature indoors is $20°C$, and outdoors it is $3°C$ and getting warmer.

Solution. We are given that when the ambient temperature is 0, the solution is $T(t) \equiv 20$. This specifies a constant solution of equation (1.10), which can be used to determine the magnitude of the heat source term.

Substitute $T(t) = 20$, $T'(t) = 0$, and $A(t) = 0$ to get

$$0 = -0.05(20 - 0) + mH(t).$$

Hence $mH(t) \equiv 4$ degrees per hour.

We assume the ambient temperature $A(t)$ varies sinusoidally with a period of 24 hours, average temperature $T_a = \frac{-7+13}{2} = 3$ degrees centigrade, and amplitude $V = 13 - T_a = 10$ degrees. Hence

$$A(t) = 3 + 10 \sin\left(\frac{2\pi t}{24}\right),$$

where $t = 0$ is the time in the morning that $A(t) = T_a$. Hence the temperature satisfies the IVP,

$$T'(t) = -0.05\left\{T(t) - \left[3 + 10\sin\left(\frac{\pi t}{12}\right)\right]\right\} + 4; \quad T(0) = 20.$$

The ODE can be written more simply as

$$T' + 0.05T = 4.6 + 2\sin\left(\frac{\pi t}{12}\right).$$

The homogeneous solution is $e^{-0.05t}$, so we substitute $T = e^{-0.05t}v$ and simplify to find

$$v' = e^{0.05t}[4.6 + 2\sin(.26t)]$$

($\frac{\pi}{12} \approx 0.26$). Now integrate (it is fair to use a table of integrals here) to find

$$v = 23e^{0.05t} + 18.6e^{0.05t}(0.2\sin(0.26t) - 0.26\cos(0.26t)) + C.$$

Thus

$$T(t) = e^{-0.05t}v = 23 + 3.7\sin(0.26t) - 4.8\cos(0.26t) + Ce^{-0.05t}.$$

When $t = 0$ and $T = 20$, we have $20 = 23 - 4.8 + C$. It follows that $C \approx 1.8$. ❑

Figure 1.5 displays a graph of the temperature function found in Example 1.3.3. In our solution, we took the trouble to evaluate the constant C, using the initial condition. It is clear from Figure 1.5 that this was unnecessary for most purposes, because the temperature settles into a periodic regime after some time has elapsed. The term $Ce^{-0.2t}$, which is the homogeneous solution, is called a **transient** because it decays to 0 with increasing time. The periodic solution

$$T = 23 + 3.7\sin(0.26t) - 4.8\cos(0.26t))$$

of the inhomogeneous ODE is said to be **stable** because every solution is the sum of this and a transient term. Thus, the key information that we can extract from this example is that there is a stable periodic temperature with a mean of 23°C, ranging between 17°C and 29°C (the temperature range was determined from the graph). It is likely that the occupants of the building (or a thermostat) will turn off the heat to prevent it from getting as warm as 29°C, but that would invalidate our model: The ODE would no longer be linear (see Exercise 22 at the end of this section).

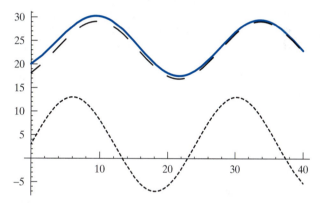

FIGURE 1.5   Temperature in a heated building (solid curve), and the stable periodic temperature (dashed curve). The outdoor temperature is indicated by the lower curve. See Example 1.3.3.

Example 1.3.4   Find the solution of the IVP,

$$ty' - y = t^2 \ln(t); \quad y(1) = 2, \tag{1.11}$$

on the interval $(0, \infty)$.

Solution. The general solution $y = Ct^{-m}$ of the homogeneous equation, $ty' + my = 0$, was found in Example 1.2.5. For the associated homogeneous equation in this example, $m = -1$, so we will use the homogeneous solution $y = Ct$ of $ty' - y = 0$.

Substitute $y = tv$, and $y' = v + tv'$ in equation (1.11). After simplifying, we have $t^2 v' = t^2 \ln(t)$, so $v' = \ln(t)$. Thus

$$v = \int \ln(t)\, dt = t\ln(t) - t + C,$$

and the general solution of the ODE is

$$y = t^2 \ln(t) - t^2 + Ct.$$

Now substitute $y = 2$ and $t = 1$ to obtain $2 = 0 - 1 + C$, and $C = 3$. The solution of the IVP is

$$y = t^2(\ln(t) - 1) + 3t. \qquad \blacksquare$$

The general solution in Example 1.3.4 again has the form $y_p + Cy_h$, where $y_p = t^2(\ln(t) - 1)$ and $y_h = t$.

EXERCISES

You are encouraged to use CAS or a table to calculate the integrals. You may also check your answers with a CAS.

In Exercises 1–12, find the general solutions of the ODEs on the intervals indicated (if no interval is indicated, the solution should be valid for all real t). Then find the particular solutions that satisfy the initial conditions. In some of these problems, a pair of differential equations that differ only in the sign of the coefficient is given. Use a CAS (or graphing calculator) to compare the graphs of several solutions of each of these paired equations. In cases where the solutions are defined for all real t, do the solutions seem to converge to some value as $t \to \infty$, or as $t \to -\infty$?

1. $y' = 3t - 4y$ on $(-\infty, \infty)$; $y(0) = 0$
2. $ty' - y = t^3 - 2t$ on $(0, \infty)$; $y(1) = 0$
3. $2y' + y = t^{-1}e^{-t/2}$ on $(0, \infty)$; $y(2) = 0$
4. $ty' + y = 1$ on $(0, \infty)$; $y(1) = 0$
5. $y' + 2ty = e^{-t^2}$ on $(-\infty, \infty)$; $y(0) = 0$
6. $y' + ty = t$ on $(-\infty, \infty)$; $y(0) = 0$
7. a. $y' + 10y = 1$ on $(-\infty, \infty)$
   b. $y' - 10y = 1$ on $(-\infty, \infty)$
   Initial condition for both: $y(0) = 1$
8. a. $ty' + 12y = t^3$ on $(0, \infty)$
   b. $ty' - 12y = t^3$ on $(0, \infty)$
   Initial condition for both: $y(1) = 0$
9. a. $y' + 4y = 2e^{-4t}\sin(2t)$ on $(-\infty, \infty)$
   b. $y' - 4y = 2e^{-4t}\sin(2t)$ on $(-\infty, \infty)$
   Initial condition for both: $y(0) = 0$
10. a. $y' + \tan(t)y = \sec^3(t)$ on $(-\pi/2, \pi/2)$
    b. $y' - \tan(t)y = \sec^3(t)$ on $(-\pi/2, \pi/2)$
    Initial condition for both: $y(0) = 1$
11. $\cos(t)y' = (y - 1)\sin(t)$; $y(0) = 0$ on $(-\pi/2, \pi/2)$

12. $\dfrac{y'}{y - 1} = t$; $y(0) = 1 - 10^{-6}$ on $(-\infty, \infty)$

13. A penny is heated to $800°$C and is then allowed to cool. The temperature after a minute is $600°$C, and the room temperature is $20°$C. When will it be safe to pocket the coin (the temperature should be less than $50°$C)?

    ✔**Answer:** 11 minutes

14. A roast has an internal temperature of $15°$C when it is put in a $175°$C oven. After 1 hour, a meat thermometer placed in the roast registers $50°$C. How much longer will it take for the roast to reach the state of medium rare ($65°$C)?

15. A turkey is at room temperature ($20°$C) when it is put into the oven, and it is removed after 5 hours, when its internal temperature has reached $85°$C. The oven temperature is $160°$C. The turkey is allowed to stand outside the oven for 30 minutes prior to carving. Estimate the internal temperature of the turkey when it is carved. What assumptions are necessary?

    ✔**Answer:** $81°$C

16. The building described in Example 1.3.3 is given additional insulation, reducing the transmission coefficient to $k = 0.1$ hour^{-1}. The furnace output is reduced to half of its capacity to compensate for the additional insulation. Find the new mean temperature of the building and the amplitude of its variation. Ignore transients.

17. At 3:00 A.M., the temperature inside a house is $15°$C, and the heat is turned off by a timer. When the heat is turned on again at 6:00 A.M., the temperature has fallen to $10°$C. Throughout this period, the temperature outdoors is $-20°$C. Find the rate constant for heat loss in the building.

    ✔**Answer:** 0.0514 hour^{-1}

18. (Continuation of Exercise 17) The outdoor temperature remains at $-20°$C. Assume that the furnace produces heat at a constant rate and remains on until 9:00 A.M., when the temperature inside the house reaches $20°$C. How long

would it take to warm the house from $10°C$ to $20°C$ if the temperature outdoors were $0°C$?

✔**Answer:** 2 hours 17.5 minutes

19. A cabin has two identical wood stoves; only one is in operation. The temperature outside is $-15°C$, and the indoor temperature has stabilized at an uncomfortable $10°C$. Therefore, the second stove will be put into use. At what temperature will the temperature now stabilize?

20. When the stove is lighted in a mountain cabin, the temperature in the cabin is the same as the temperature outside: $-20°C$. After an hour the temperature inside has reached $0°C$, and after 2 hours, the temperature is $10°C$. Find the temperature as a function of time, assuming that the stove continues to operate. How warm will the cabin eventually get?

✔**Answer:** $20°C$

21. The insulation of a house has a transmission coefficient of $k = .04$ hour^{-1}. The furnace is capable of heating the house at $1°C$ per hour (neglecting heat loss). Assume that the outdoor temperature is $0°C$, and that the heat is alternately turned on for 4 hours, then off for 1 hour, by an automatic timer. What will be the average temperature inside the house, if this continues indefinitely?

✔**Answer:** $20°C$

22. The building described in Example 1.3.3 is equipped with a thermostat that turns off the furnace when the interior temperature $T > 20°C$ and turns it on again when $T \le 20°C$.

  a. Let

$$mH(T) = \begin{cases} 4 & \text{if } T \le 20 \\ 0 & \text{if } T > 20. \end{cases}$$

  Show that the IVP $T' + 0.2T = 0.6 + 2\sin\left(\frac{\pi t}{12}\right) + mH(T)$; $T(0) = 20$ is an appropriate model for the temperature in the building.

  b. Explain why the ODE in part (a) is not linear.

  c. Use an IVP solver to solve the IVP in part (a). Explain the dips in the graph and the jagged appearance of the horizontal segments.

23. For each of the following ODEs, find the periodic solution, if there is one, and decide whether or not it is stable. In other words, express the general solution as the sum of the periodic solution and a family of exponential functions. If the exponential functions decay to 0 as $t \to \infty$, they represent transients and the periodic solution is stable.

  a. $y' + 5y = 5\cos(2t)$ b. $y' - y = 7\cos(4t)$

  c. $y' + 2y = \cos(t) - 3\sin(t)$

  d. $y' - 5y = 4\cos(t) + 3\sin(t)$

  e. $y' + y = e^{-t}\sin(t)$

24. Find the periodic solution of each of the following ODEs and show that it is stable.

  a. $y' + y = \sin(2t)$ b. $y' + y = \sin(2000t)$

  c. $y' + 5y = \cos(\pi t)$ d. $y' + 10000y = \sin(t)$

  e. $y' + .0006y = \sin(.0008t)$

  f. $y' + py = \cos(\omega t)$, where $p > 0$ is a constant. What would happen if $p < 0$?

The remaining problems in this section are designed to be done with a CAS.

25. Find the general solution of each of the following:

  a. $y' + 0.1y = \sin(2t)$

  b. $y' + 2ty = 1$. Attempt to solve this equation first without the computer's help. How does the computer get around the problem of evaluating $\int e^{t^2}\, dt$?

  c. $ty' + 12y = [\ln(t)]^3$ d. $\sqrt{t}y' + y = t$

26. For each of the following linear differential equations, find the general solution, and graph several solutions by substituting values for the constant, using the domain $-10 \le t \le 10$ and range $-10 \le y \le 10$. Make note of any properties that are common to all solutions.

  a. $y' + y = 1$ b. $y' - y = 1$

  c. $y' + y = t$ d. $y' - y = -t$

  e. $y' + y = \sin(t)$ f. $y' + 0.05y = \sin(t)$

  g. $\frac{1}{\pi}\sin(\pi t)y' + \cos(\pi t)y = 1$

  h. $\cos\left(\frac{\pi}{20}t\right)y' + \left(\frac{\pi}{20}\right)\sin\left(\frac{\pi}{20}t\right)y = 1$

27. Using the plot range $-\pi \le t \le \pi$, $-10 \le y \le 10$, graph the solution of the differential equation

$$ty' + t\cot(t)y = 1$$

that has a finite value at $t = 0$. What is that value?

✔**Answer:** $y(0) = 1$

28. Let $g(t)$ denote the solution of

$$ty' - 2y = e^{-t}$$

that has the property $g(1) = g'(1)$. Draw the graph of $g(t)$ on the interval $0 \le t \le 2$.

29. Plot the solution of $y' + 2y = ty^2$ that has a relative minimum at $t = \frac{1}{4}$. What is the value of $y(1/4)$?

✔**Answer:** $y(1/4) = 8$

30. Let $\phi(t)$ be the solution of $y' + 2ty = y^2$ that has a maximum at $t = 1$. Plot the graph of $\phi(t)$ on the interval $0 \le t \le 2$, and calculate $\phi(0)$.

✔**Answer:** $\phi(0) = 1.074386372$

*1.4 Mixture Problems

A typical mixture problem involves a tank that initially contains V_0 liters of a salt solution, with concentration C_0 grams per liter. A salt solution containing K grams of salt per liter is being poured into the tank at J liters per second; simultaneously, solution is pumped out of the tank at L liters per second. The problem is to find the concentration of salt in the tank as a function of time. See Figure 1.6.

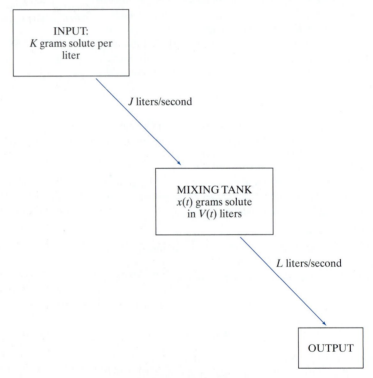

FIGURE 1.6   A typical mixture problem.

The mathematical model of the mixture problem depends on a simplifying assumption, which we will call the **uniformity hypothesis**: The solution in the tank is instantly and perfectly mixed, so that at any time, its concentration is uniform throughout the tank. Let $C(t)$ denote the salt concentration in the tank, in grams per liter.

We will actually use the *amount* of salt in the tank, denoted $x(t)$, rather than the concentration $C(t)$, as the dependent variable. We can determine the rate of change of $x(t)$ by subtracting the rate at which the salt removed (the output rate) from the rate that salt enters the tank (the input rate). The input rate is JK grams per minute, since J liters of solution, each containing K grams of salt, enter the tank each minute. The output rate is $LC(t)$ grams per minute, where $C(t)$ is the concentration of salt in the tank. Thus

$$x'(t) = JK - LC(t).$$

To complete this model, we have to express the concentration in terms of $x(t)$. Let $V(t)$ be the volume of solution in the tank at that time. If V_0 is the initial volume in the tank, then

$$V(t) = V_0 + (J - L)t$$

since the tank gains $J - L$ liters per minute. Therefore,

$$C(t) = \frac{x(t)}{V(t)} = \frac{x(t)}{V_0 + (J - L)t}.$$

It follows that $x(t)$ satisfies the linear ODE

$$\frac{dx}{dt} = JK - L\frac{x}{V_0 + (J - L)t}, \tag{1.12}$$

with the initial condition $x(0) = C_0 V_0$.

In our first example, the volume V is constant, because the rate J at which fluid flows in is the same as the rate L at which fluid is pumped out.

Example 1.4.1   An accident has caused 10 kg of potassium permanganate to spill into a mountain lake, turning its water purple in color. The lake covers an area of 100 square meters, and its average depth is 5 meters. A stream feeds the lake at 1000 liters per minute, and another stream takes water from the lake at the same rate.

Find the concentration of potassium permanganate in the lake as a function of time, assuming that the uniformity hypothesis holds in this situation. How much of the substance will remain in the lake after 24 hours have elapsed?

Solution. Let $x(t)$ denote the amount of potassium permanganate dissolved in the lake water. To obtain the concentration $C(t)$, divide $x(t)$ by the volume V of water in the lake. In cubic meters, V is equal to the product of the lake's area by its average depth: $V = 100 \times 5$. To convert to liters, remember that one cubic meter is equal to 1000 liters; thus $V = 5 \times 10^5$ liters.

In this lake, the input of potassium permanganate zero, for the stream feeding the lake was not affected by the spill. The output is the product of the concentration and the rate at which the stream that drains the lake is flowing: $C(t)$ kg/L $\times$ 1000 liters per minute. Thus

$$x'(t) = -\frac{x(t)}{5 \times 10^5} \times 1000 = -0.002\, x(t) \text{ kg/min.}$$

Thus, the amount of potassium permanganate in the lake satisfies the linear decay equation, $x' = -0.002\, x$. It follows that $x(t) = Ae^{-0.002t}$, where A is a constant equal to the initial value 10 of x. Thus $x(t) = 10e^{-0.002t}$. The half-life of $x(t)$ is $\frac{1}{60}(\ln(2)/0.002) \approx 6$ hours. Since 24 hours amounts to 4 half-lives, we expect that $10 \times 2^{-4} = \frac{5}{8}$ kilogram will remain in the lake after a day.   ❑

When the volume of solution is not constant, we encounter variable coefficients in the input-output equation, as in the following example.

Example 1.4.2   A tank initially contains 1000 liters of salt solution with 70 grams of salt per liter. A solution containing 120 grams of salt per liter enters the tank at the rate of 9 liters per minute, and the well-mixed solution is pumped out at 10 liters per minute. Find the concentration of salt as a function of time.

Solution. The salt input rate is

$$(120 \text{ grams/liter}) \times (9 \text{ liters/min}) = 1080 \text{ grams/min}.$$

The volume of solution in the tank is decreasing at a rate of 1 liter/min, so $V(t) = 1000 - t$ liters. Thus, the output rate is

$$\left(\frac{x}{1000 - t} \text{ grams/liter} \right) \times \left(10 \text{ liters/min} \right).$$

The input-output model results in the ODE

$$\frac{dx}{dt} = 1080 - 10 \left(\frac{x}{1000 - t} \right),$$

or, in standard form,

$$x' + \left(\frac{10}{1000 - t} \right) x = 1080.$$

The homogeneous solution is

$$\exp \left(- \int \frac{10}{1000 - t} \, dt \right) = e^{10 \ln(1000 - t)} = (1000 - t)^{10}.$$

Now we substitute $x = v(1000 - t)^{10}$ in the differential equation and simplify to get

$$v' = 1080(1000 - t)^{-10}.$$

Integration yields $v = 120(1000 - t)^{-9} + A$, where A is a constant. Since $x(t) = v(1000 - t)^{10}$, it follows that

$$x(t) = 120(1000 - t) + A(1000 - t)^{10}. \tag{1.13}$$

Since $x(0) = (70 \text{ grams/liter}) \times (1000 \text{ liters}) = 70{,}000$ grams, we can substitute $t = 0$, and $x = 70{,}000$ in equation (1.13) to obtain $70{,}000 = 120{,}000 + A \cdot 1000^{10}$, or $A = -50{,}000 \times 1000^{-10} = -50 \times 1000^{-9}$. If we substitute this value of A into equation (1.13) and simplify a little, we will have

$$x(t) = 120(1000 - t) - 50(1000 - t) \left(1 - \frac{t}{1000} \right)^9.$$

To obtain the concentration $C(t)$, divide by $V(t) = 1000 - t$:

$$C(t) = 120 - 50(1 - .001t)^9 \text{ grams per liter}.$$

This formula is valid for $0 \le t < 1000$. When $t = 1000$ the tank becomes empty, and the model is no longer applicable.

Input-output problems in personal finance

Many investment plans consider factors such as inflation and the effect of regular withdrawals. Suppose, for example, that a person has a pension account of P dollars upon retirement and withdraws money at a rate of $E(t)$ dollars per year to meet living expenses. If $y(t)$ denotes the account balance t years after retirement, then the rate of change of $y(t)$ is the rate at which interest income is received minus $E(t)$.

If r denotes the interest rate, then our model is an ODE,

$$y' = ry - E(t).$$

Assuming that the expenses grow with inflation according to the exponential growth law $E' = kE$, where k is the annual rate of inflation, the expenses after t years will be $E(t) = E_0 e^{kt}$, where E_0 is the initial rate of withdrawal.

Example 1.4.3   Ms. Doe retired yesterday at age 65. Her IRA account has a principal of $450,000, which was invested with a guaranteed interest rate of 5.25%, compounded continuously. Her budget calls for annual expenses of $20,000, with projected inflation of 2.5%. Determine Ms. Doe's savings account balance t years after her retirement. How long will her money last?

Solution. Let $y(t)$ denote the balance t years after retirement, in dollars. The input in this problem is interest amounting to $0.0525y(t)$, and the output is expenses of $20,000e^{0.025t}$ dollars per year. Thus

$$y' = 0.0525y - 20000e^{0.025t}.$$

This can be put into the standard form for a linear ODE:

$$y' - 0.0525y = -20000e^{0.025t}.$$

The homogeneous solution is $y = Ce^{0.0525t}$, so we substitute $y = ve^{0.0525t}$ and simplify to obtain $v' = -20,000e^{-0.0275t}$. Therefore, $v = 20,000e^{-0.017t}/0.0275 + C$, and since $y = ve^{0.0525t}$,

$$y = \frac{20000}{0.0275}e^{0.025t} + Ce^{0.0525t}.$$

Since $y(0) = 450,000$, and $20,000/0.0275 \approx 730,000$, $450,000 = 730,000 + C$, and thus $C \approx -280,000$. Ms. Doe's account balance in t years will be about

$$y(t) = 730,000e^{0.025t} - 280,000e^{0.0525t}.$$

This function is decreasing, and $y(t) = 0$ when $730,000 = 280,000e^{0.0275t}$; that is, when $e^{0.0275t} = 2.6$. This will be when Ms. Doe reaches the age of 100.   ❑

EXERCISES

1. A 16% salt solution enters a tank, initially containing 100 liters of pure water, at 2 liters per second. The solution is instantly and perfectly mixed, and brine is pumped out of the tank at 2 liters per second. How long does it take for the concentration of salt in the tank to reach: 8%? 12%? 15%?

2. A 4% salt solution is poured into a large tank at 2 liters per second. The solution is stirred to be always homogeneous. Brine is removed from the tank at 1.9 liters per second. Given that the tank initially contains 50 liters of 2% salt solution, what is the concentration of salt in the tank after one minute has elapsed?

✔ *Answer:* 3.79%

3. A tank contains 16 liters of 100 ppm potassium iodide, in which the iodide is the unstable isotope I^{131}, with a half-life of 8 days. Solution is drained from the tank at the rate of 1 liter per day, and fresh 100 ppm solution is added at the rate of 1 liter per day. Find the steady-state concentration of I^{131} in the tank.

✔ *Answer:* 42 ppm

4. (Continuation of Exercise 3) Suppose fresh 100 ppm solution of KI^{131} is added at the rate of 2 liters per day, while the well-mixed solution drains out at 1 liter per day, as before. If the initial volume is 16 liters and the initial concentration is 100 ppm, find the concentration after 8 days have elapsed.

✔ *Answer:* 63 ppm

5. A beverage bottling plant has had an accident. Someone poured quinine (intended for the tonic) in the orange drink mixing tank, and the resulting liquid is unpalatable. No one knows (or will say) when this happened, but we must find out, because the orange drink bottled since the mishap must not leave the plant. The concentration of quinine in the tank is now 0.01%. The tank now holds 10000 liters, and orange drink is being transferred to the carbonation tank at 1000 liters per hour. Fresh ingredients have been added to the tank at the rate of 900 liters per hour. It is known that the amount of quinine that was put into the tank was not more than 4 kilograms, and that it was introduced into the tank within the last 24 hours. Estimate how long ago, at worst, that the quinine was introduced.

6. A tank contains 4 kg of salt dissolved in 100 liters of water. A salt solution with concentration 0.01 kilograms per liter enters the tank at the rate of 2 liters per day. Evaporation removes 1 liter of water per day (salt does not evaporate), and an additional liter of solution is drained from the tank each day. Find the concentration of salt in the tank as a function of time.

7. When brought to the emergency room, an accident victim has 3 liters of blood and is still losing blood at 0.25 liter per hour. He is immediately given continuous blood transfusions at 0.5 liter per hour, and an antibiotic drug is administered intravenously at 0.5 gram per hour. Four hours later, the bleeding is stopped, and the transfusions stop in another two hours. Determine the concentration of the antibiotic in the patient's blood at the time that the transfusions stop.

✔ *Answer:* 0.55 grams per liter

8. Two mixing tanks each initially contain 2 liters of pure water. A 10% salt solution enters the first tank at 0.1 liter per minute. The well-mixed solution is pumped out of this tank at the same rate, and into the second tank. Instantaneous and perfect mixing occurs in the second tank, and the resulting solution is removed from it, again at the rate of 0.1 liters per minute. Determine the concentration of salt in the second tank as a function of time.

9. Consider two mixing tanks, labeled A and B. Initially, tank A contains 3 liters of 12% salt solution, and tank B contains 1 liter of water. Both tanks are stirred constantly, and solution is pumped from tank A to tank B at 0.2 liters per hour. Solution is also pumped from tank B to tank A at the same rate. Find the salt concentration of each tank as a function of time.

◆ *Hint:* Since this is a closed system, the total amount of salt held by the tanks is constant.

10. A solar water heater with a 200 liter capacity absorbs heat at a rate proportional to the sine of the angle of the sun over the horizon. The sun rises at 0600 and sets at 1800 on the 24 hour scale. Assume that the heat absorption rate is $1500 \sin\left(\frac{\pi}{12}(t - 6)\right)$ kilocalories per hour, where t is hours after midnight of the equinox. Of course, no heat is absorbed at night—the absorption rate is 0 for $0 < t < 6$, $18 < t < 30$, and so on.

The water heater loses heat by conduction to the environment and by hot water usage. The ambient temperature is 20°C, and the rate constant for heat conduction to the rest of the house is $k = 0.01$ hour^{-1}. Hot water is removed from the tank at an average rate of 5 liters per hour and is replaced at the same rate by cold water with a temperature of 10°C with instant and perfect mixing. If the water in the tank is at 60°C at 0600, find its temperature 24 hours later.

◆ *Hint:* Treat this as a mixture problem, and keep track of heat input and output. One kilocalorie of heat will raise the temperature of one liter by one centigrade degree.

11. A retired citizen's living expenses are $30,000 per year now, and he wants to allow for inflation of 3% per year.

How much money should he invest, at 6% interest, so that he can meet his living expenses and not touch the principal, no matter how long he lives?

✔ **Answer:** $1 million

12. An endowed chair in mathematics is to be funded with one million dollars. The endowment committee believes it can realize at least an 8% return on the funds, and expects to offer the occupant of the chair a 5% salary increase each year to keep pace with inflation. What should the initial salary be if the principal of the endowment fund is never to be drawn upon?

13. The *age-specific death rate* $d(t)$ is the number of deaths of individuals who are t years old per 1000 individuals of that age. Assume that $d(20) = 2.8$, $d(50) = 17.6$, and $d(t)$ satisfies the linear growth equation $d'(t) = kd(t)$. Of a sample of 100,000 twenty-year-olds, how many will survive to age 50?

✔ **Answer:** about 79,000

*1.5   Bernoulli and Riccati Equations

It is sometimes possible to convert a nonlinear first-order ODE to a linear one by a clever change of variables, known as a *linearizing substitution*. For example, $v = y^m$ is a linearizing substitution for the nonlinear ODE,

$$y^{m-1}y' + p(t)y^m = q(t). \tag{1.14}$$

By the chain rule, $dv/dt = my^{m-1}y'$. Multiply equation (1.14) by m and make the substitution. The resulting ODE,

$$v' + mp(t)v = mq(t),$$

is linear, and thus we can find its general solution. It is then a simple matter to replace v with y^m and obtain the general solution of (1.14).

A *Bernoulli equation* is a first-order nonlinear ODE of the form

$$y' + p(t)y = q(t)y^n, \tag{1.15}$$

where $n \neq 0, 1$. The power n to which y is raised on the right side is the **exponent**. Observe that by dividing equation (1.15) by y^n, it can be made to assume the form of equation (1.14), in which $m = 1 - n$. It is therefore possible to linearize a Bernoulli equation and thus to find its general solution.

The first Bernoulli equation was encountered by Johann Bernoulli in 1695. Bernoulli was trying to determine the trajectory of a projectile subject to the forces of gravity and air resistance. He knew that the force of air resistance would be proportional to the square of the speed v of the projectile, and he derived the ODE

$$\frac{dv}{dt} = -kv^2 - g\sin(\theta), \tag{1.16}$$

where g is the gravitational acceleration and θ denotes the inclination of the trajectory from the horizontal. Of course, θ is not constant, and Bernoulli derived an ODE that it would satisfy:

$$\frac{d\theta}{dt} = -\frac{g}{v}\cos(\theta). \tag{1.17}$$

The two ODEs, (1.16) and (1.17), form a coupled system, a topic that we will take up in Chapter 3. By the chain rule,

$$\frac{dv}{d\theta} = \frac{v'(t)}{\theta'(t)};$$

thus Bernoulli divided (1.16) by (1.17) to obtain

$$\frac{dv}{d\theta} = \frac{k\,v^3}{g\,\cos(\theta)} + \tan(\theta)\,v. \tag{1.18}$$

Example 1.5.1   Show that (1.18) is a Bernoulli equation and linearize it.

Solution. Rearrange (1.18) in the form

$$\frac{dv}{d\theta} - \tan(\theta)\,v = \frac{k}{g}\sec(\theta)\,v^3. \tag{1.19}$$

This has the form of (1.15), where v is the dependent variable, θ is the independent variable, $p(\theta) = -\tan(\theta)$, and $q(\theta) = (k/g)\sec(\theta)$. Thus it is a Bernoulli equation, with exponent $n = 3$. To linearize, define a new variable $u = v^{1-n} = v^{-2}$. Then

$$\frac{du}{d\theta} = -2v^{-3}\frac{dv}{d\theta};$$

or

$$\frac{dv}{d\theta} = -\frac{v^3}{2}\frac{du}{d\theta}.$$

Substitute this expression in (1.19) to obtain

$$-\frac{v^3}{2}\frac{du}{d\theta} - \tan(\theta)\,v = \frac{k}{g}\sec(\theta)\,v^3.$$

Divide through by v^3 to obtain

$$-\frac{1}{2}\frac{du}{d\theta} - \tan(\theta)\,v^{-2} = \frac{k}{g}\sec(\theta).$$

Finally, replace v^{-2} with u, and multiply through by -2 to obtain

$$\frac{du}{d\theta} + 2\tan(\theta)\,u. = -\frac{2k}{g}\sec(\theta). \qquad \square$$

Riccati equations. Riccati's equations is $y' + py^2 = ct^n$. This ODE was first studied by an Italian mathematician of the eighteenth century, J. F. Riccati. We will be interested here in a related equation, called the *generalized Riccati equation*. This is an ODE of the form

$$y' + p(t)y^2 + q(t)y = r(t). \tag{1.20}$$

For the (not generalized) Riccati equation, $p(t)$ is constant, $q(t) \equiv 0$, and $r(t) = ct^n$.

Solving generalized Riccati equations is not always possible. If, however, a particular solution $y_1(t)$ of equation (1.20) is available, then we can linearize it. The method is as follows: Let v be a new dependent variable defined by $y = v + y_1(t)$. Make this substitution in (1.20) to obtain

$$v' + y_1'(t) + p(t)\, v^2 + 2p(t)y_1(t)\, v + p(t)(y_1(t))^2 + q(t)\, v + q(t)y_1(t) = r(t).$$

Since $y_1(t)$ is a solution of (1.20), the sum of the second, fifth, and seventh terms on the left side of this equation is $r(t)$. We can therefore simplify and obtain

$$v' + p(t)\, v^2 + 2p(t)y_1(t)\, v + q(t)\, v = 0,$$

which can be rearranged to get a Bernoulli equation:

$$v' + (2p(t)y_1(t) + q(t))\, v = -p(t)\, v^2.$$

The process for linearizing the generalized Riccati equation is then completed by linearizing the Bernoulli equation, using the method outlined previously.

Example 1.5.2   Find the general solution of the generalized Riccati equation,

$$y' - \frac{1}{t^2(t+1)}\, y^2 - \frac{1}{t}\, y = -1, \tag{1.21}$$

given that $y = t^2 + t$ is a particular solution.

Solution. Let $y = v + t^2 + t$. Then

$$y' = v' + 2t + 1,$$

$$-\frac{1}{t^2(t+1)}\, y^2 = -t - 1 - \frac{2}{t}\, v - \frac{1}{t^2(t+1)}\, v^2$$

$$-\frac{1}{t}\, y = -t - 1 - \frac{1}{t}\, v.$$

Substituting these in (1.21) and simplifying, we obtain

$$v' - \frac{3}{t}\, v - \frac{1}{t^2(t+1)}\, v^2 = 0,$$

which is a Bernoulli equation with exponent 2. Thus we substitute $w = v^{-1}$. Since $w' = -v^{-2}v'$, we can substitute $v' = -v^2 w'$. Thus

$$-v^2\, w' - \frac{3}{t}\, v = \frac{1}{t^2(t+1)}\, v^2.$$

Now we can divide through by $-v^2$ and obtain

$$w' + \frac{3}{t}\, v^{-1} = -\frac{1}{t^2(t+1)}.$$

Since $w = v^{-1}$, we have a linear ODE with w as the dependent variable:

$$w' + \frac{3}{t} w = -\frac{1}{t^2(t+1)}. \tag{1.22}$$

The solution of the associated homogeneous equation, $w' + 3\,w/t = 0$, is $w = Ct^{-3}$. To find the general solution of (1.22), substitute $w = u\,t^{-3}$, where u is a new dependent variable. This yields

$$t^{-3} u' - 3t^{-4} u + \frac{3}{t} (t^{-3}u) = -\frac{1}{t^2(t+1)}.$$

After simplifying, we find that

$$u' = -\frac{t}{t+1} = \frac{1}{t+1} - 1.$$

Thus, integration yields $u = \ln|t + 1| - t + C$. We can now unwind our sequence of substitutions:

$$w = u\,t^{-3} = \frac{\ln|t + 1| - t + C}{t^3}$$

$$v = w^{-1} = \frac{t^3}{\ln|t + 1| - t + C}, \quad \text{and}$$

$$y = v + t^2 + t = t^2 + t + \frac{t^3}{\ln|t + 1| - t + C}. \qquad \square$$

EXERCISES

Find the general solution of the following equations:

1. $ty^2 y' + y^3 = 1$

2. $y' + (\tan t)y = y^2$

3. $y' + 3y = y^3 \sin t$

4. $e^{2y} y' + e^{2y} = e^{-2t}$

5. $y' + \dfrac{y}{2t + 1} = 12[(2t + 1)/y]^3$

6. $y' + \left(1 - \dfrac{3}{t}\right) y = \dfrac{y^{4/3}}{t}$

7. Can your CAS handle Bernoulli equations? Attempt to find the general solution of each of these equations with your computer.

Find the general solutions of the following generalized Riccati equations. A particular solution $y_1(t)$ is given to get the process started.

8. $y' + 2y^2 = t^{-2}; \; y_1 = t^{-1}$

9. $y' + y^2 + y = 2; \; y_1 \equiv 1$

10. $y' + y^2 = t^{-4}; \; y_1 = t^{-1} - t^{-2}$

11. $y' + ty^2 - \dfrac{1}{t}y = t^3; \; y_1 = t$

12. Solve the preceding Riccati equations using a CAS, and display the graphs of several solutions.

1.6 Glossary

Associated homogeneous equation The homogeneous linear ODE obtained from a given linear ODE by replacing the source function with 0.

Bernoulli equation A nonlinear ODE of the form

$$y' + p(t)y = q(t)y^n,$$

where $n \neq 0, 1$. The substitution $v = y^{1-n}$ converts a Bernoulli equation into a linear ODE, with v as the dependent variable.

CAS Computer **a**lgebra **s**ystem: A computer program or calculator that can differentiate, integrate, and perform algebraic manipulations. Examples include *Maple*, *Mathematica*, and MATLAB.

Coefficient function One of the functions $a_i(t)$ in the linear ODE

$$a_1(t)y' + a_0(t)y = b(t).$$

Continuous compound interest Interest that is computed using the linear growth equation to determine the balance of a bank account, where the growth constant is the interest rate.

Dependent variable A variable representing an unknown function in a differential equation.

Differential equation An equation involving an unknown function and its derivatives. See **ODE** and **PDE**.

Doubling time The time taken for the value of the function $f(t) = Ce^{kt}$ to increase from $y_1 = f(t_1)$ to $y_2 = 2y_1$. The doubling time is independent of the value of y_1 and is inversely proportional to k.

Equal derivatives theorem Let $f(t)$ and $g(t)$ be functions that are differentiable on an interval (a, b). If $f'(t) = g'(t)$ for all $t \in (a, b)$, then there is a constant C such that $f(t) = g(t) + C$ for all $t \in (a, b)$.

General solution A family of functions that represents the entire set of solutions that a differential equation has.

Half-life The time taken for the value of the function $f(t) = Ce^{-kt}$ to decrease from $y_1 = f(t_1)$ to $y_2 = \frac{1}{2}y_1$. The half-life is independent of the value of y_1 and is inversely proportional to k.

Homogeneous equation A linear ODE in which the source function is $b(t) \equiv 0$.

IVP Acronym for **i**nitial **v**alue **p**roblem. This is an ODE, provided with an initial condition.

Inhomogeneous equation A linear ODE with a nonzero source term.

Initial condition An equality $y(t_0) = y_0$ that determines a particular solution of a differential equation.

Input-output problem To determine the amount of substance present in a container, given the rate at which the substance enters (the input rate), and the rate at which it is removed (the output rate).

Linear growth equation The linear ODE, $y' = ky$ that serves as a model for population growth, etc. The word *linear* refers to the ODE, not the solution.

Linear first-order ODE An ODE that can be put in the form

$$a_1(t)y' + a_0(t)y = b(t).$$

Mathematical model A mathematical construction, such as an ODE, that simulates a natural or engineering phenomenon.

Mixture problem To determine the concentration of a solute in a container, given the rate at which a solution of known concentration enters the container, and the rate at which the thoroughly mixed solution is removed from the container.

Newton's law of cooling The difference in temperature between an object and its environment decreases at a rate proportional to that difference.

ODE Acronym for **o**rdinary **d**ifferential **e**quation, a differential equation involving derivatives with respect to only one independent variable.

Order The degree of the highest derivative in a differential equation.

Particular solution A function that is a solution of a differential equation.

PDE Acronym for **p**artial **d**ifferential **e**quation, a differential equation involving partial derivatives with respect to more than one independent variable.

Qualitative study An analysis of an ODE that does not depend on calculating solutions. Properties of solutions are inferred by examining the ODE itself.

Radiometric dating Determination of the age of an object by measuring the concentration of an unstable isotope.

Relative rate of change The rate of change of a quantity, expressed aas a fraction of the quantity. In symbols, the relative rate of change of $f(t)$ is equal to $f'(t)/f(t)$.

Riccati equation   A nonlinear ODE of the form

$$y' + py^2 = ct^n,$$

where p and c are constants. If a particular solution $y = y_1(t)$ is known, the substitution $y = v + y_1(t)$ converts a Riccati equation into a linear ODE, with v as the dependent variable.

Singular point   A point t_1 where either $a_0(t)/a_1(t)$ or $b(t)/a_1(t)$ is discontinuous is a singular point of the linear ODE $a_1(t)y' + a_0(t)y = b(t)$.

Solution   A function or family of functions that, when substituted for the dependent variable of a differential equation, produces an identity.

Source term   The function $b(t)$ in the linear ODE

$$a_1(t)y' + a_0(t)y = b(t).$$

Stable periodic solution   A solution $y_p(t)$ of a linear ODE that is periodic, and such that the general solution is $y = y_p(t) + y_h(t)$, where $y_h(t)$ is a transient.

Terminal velocity   The limiting velocity of an object that is falling subject to air resistance.

Transient   A term in the solution of a linear ODE that becomes negligible as $t \to \infty$.

Transmission coefficient   The rate of temperature change for an object, expressed as a fraction of the difference of the object and its surroundings. In symbols, $k = T'/(A - T)$, where k is the transmission coefficient, T is the temperature, and A is the ambient temperature.

Uniformity hypothesis   The assumption in a input-output problem that the contents of the container are instantly and perfectly mixed.

Variation of constants   A method for finding a solution of an inhomogeneous linear ODE. It proceeds by substituting for y in the given ODE the expression $y = ve^{-P(t)}$, where $y = Ce^{-P(t)}$ is the general solution of the associated homogeneous equation. The constant, C, is replaced with a new dependent variable, v.

1.7 Chapter Review

Find the general solution of each of the ODEs in Exercises 1–12.

1. $y' = ty + t + y + 1$ **2.** $y' = y - t^2$

3. $y' = y + e^t$ **4.** $y' + \tan(t)y = \cos^2(t)$

5. $y' = e^t - y$ **6.** $ty' + y = e^t \sin(2t)$

7. $ty' + 12y = 5t^2 + 3t - 2$ **8.** $y' + \tan(t)y = 1$

9. $y' = ty + 1$ **10.** $y' = \dfrac{5y}{t}$

11. $y' + y = e^{-t}$ **12.** $y' = y \cos(t)$

In Exercises 13–22, solve the initial value problem.

13. $y' = \dfrac{t+y}{t}$; $y(1) = 2$

14. $y' + y = e^{-t}$; $y(0) = 0$

15. $y' = 2t$; $y(2) = 4$

16. $y' + 4y = 3e^{-4t}\sin(3t)$; $y(0) = 0$

17. $ty' - 3y = 5t^3$; $y(1) = 1$

18. $y' = \sqrt[3]{y}$; $y(1) = 1$

19. $y' = ty + 2t + 3y + 6$; $y(0) = 0$

20. $y' = ty + 2t + 3y + 6$; $y(0) = -2$

21. $y' - \dfrac{2y}{4 - t^2} = \dfrac{1}{\sqrt{2 - t}}$; $y(0) = 3\sqrt{2}$

22. $y' + ty = 2t + t^3$; $y(0) = 1$

23. The balance of a bank account bearing continuous compound interest at a constant rate triples in 30 years, with no deposits or withdrawals. How long did it take to double?

24. If the world's population follows the linear growth model with a relative growth rate of 1.3% (0.013), how long will it take the population to double?

25. A piece of wood contains 0.8 ppb C^{14}, and when it was part of a living tree, it contained 1 ppb. How old is it? The half-life of C^{14} is 5730 years.

26. Find the general solution of $y' + y = \cos(4x)$. Is there a stable periodic solution?

27. A tank initially contains 1000 liters of brine with a concentration of 100 grams per liter. Brine, containing 50 grams salt per liter, is pumped into the tank at 200 liters per hour, and the well-mixed solution is pumped out at 100 liters per hour. After 10 hours, the tank overflows. What is the concentration of salt in the tank at that time?

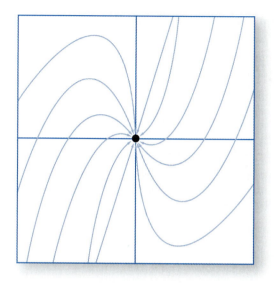

CHAPTER 2

Nonlinear Differential Equations

2.1 Separable ODEs

Let $v(t)$ denote the velocity of an object that is moving along a straight line. In addition to the forces (such as gravity) that affect the body's motion regardless of its speed, there are forces—usually labeled as friction—whose magnitude and direction are determined by the velocity of the object. Several mathematical models of friction exist. In the **linear model** the magnitude of the friction is assumed to be proportional to the velocity; thus

$$F = -b\,v,$$

where b is the **friction constant**. In the meter-kilogram-second-ampere, or MKSA, system of units, force is measured in **newtons**, and the friction constant b is measured in kilograms per second. Newton's second law of motion leads to the linear ODE,

$$mv' = -b\,v + f(t), \tag{2.1}$$

where $f(t)$ is the component of the force that does not depend on v, and m is the mass.

The linear model has the dubious virtue of being approximately correct most of the time. Although friction is inherently nonlinear, its nonlinear effects are often not apparent at low velocities.

Example 2.1.1   An object with mass 2 kilograms falls from rest, subject to the forces of gravity and linear friction. Given that the friction constant is 0.4 kilograms per second, describe the motion as a function of time.

35

Solution. The object's motion is governed by (2.1), in which the parameters are $m = 2$, $b = 0.4$, and $f(t) = mg$ newtons, where g denotes the acceleration due to gravity (9.8 meters/second^2). Thus,

$$2\,v' = -0.4\,v + 19.6. \tag{2.2}$$

The associated homogeneous equation is $2v' = -0.4\,v$, so we put $v_h = Ce^{-.2t}$. Set $v = w\,v_h$ in (2.2) and simplify to get

$$w' = 9.8\,e^{.2t}$$

so $w = 49e^{.2t} + C$. It follows that the general solution of (2.2) is

$$v = w\,v_h = 49 + Ce^{-.2t}.$$

To determine the value of C, recall that the object was said to fall from rest. This means that (2.2) is to be solved with the initial condition $v(0) = 0$. Setting $v = 0$ and $t = 0$ yields $C = -49$ and

$$v = 49(1 - e^{-.2t}).$$

The graph of this solution is shown in Figure 2.1. The horizontal asymptote $v = 49$ is a prominent feature of the graph, indicating that the velocity stabilizes at a value of 49 meters per second. At this velocity, the force of friction, $kv = 19.6$ newtons, exactly matches the force of gravity, $mg = 19.6$ newtons. This limiting velocity is called the **terminal velocity** of the object. We will use v_∞ to denote the terminal velocity. ❑

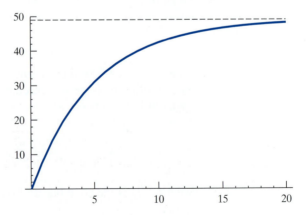

FIGURE 2.1   The velocity of a falling object.

A nonlinear model

Air resistance on a falling body is not primarily due to friction. As an object falls through the atmosphere, it has to move the air out of its path, and the force required to do this is in the opposite direction of the motion, with magnitude $k\,v^2$ proportional to the square of the object's speed. This force is called the **drag force**. For a body falling from rest, the IVP,

$$m \frac{dv}{dt} = -k\, v^2 + mg; \quad v(0) = 0, \tag{2.3}$$

where m is the mass of the object and k is a constant, will serve as a mathematical model. The sign of the term $-k\, v^2$ is chosen to be negative because as long as the object falls in the downward (positive) direction, the drag force will be directed upward (the negative direction). The ODE (2.3) is not linear, and of course, this means we can't use the method of Chapter 1 to solve it. We will soon see that there is a way to solve this IVP, but before we do, let us note that its solution does have something in common with the solution of the linear friction model in Example 2.1.1: There is a terminal velocity. The forces on the object are due to gravity, directed downward with magnitude $m\, g$; and air resistance, directed upward with magnitude $k\, v^2$. If

$$m\, g = k\, v^2,$$

the two forces will be in balance and the object will fall with the constant speed $v_\infty = \sqrt{m\, g / k}$. If $v < v_\infty$, then the gravitational force will dominate and the net force will be downward, causing the object to accelerate and the velocity to approach v_∞. Thus we would expect the velocity of a falling object to approach v_∞ as $t \to \infty$, as our notation is intended to suggest.

Equation (2.3) falls into a class of ODEs called **separable equations**. In general, a separable ODE has the form

$$\frac{dy}{dt} = g(t)\, h(y), \tag{2.4}$$

in which the right side can be expressed as the product of two single-variable functions. For example, the ODEs

$$\frac{dy}{dt} = t^2 \sin y$$

and

$$\frac{dy}{dt} = e^{t+3y} \ (= e^t\, e^{3y})$$

are separable. The ODE (2.4) includes as special cases ODEs of the form $\frac{dy}{dt} = g'(t)$, where $h(y) \equiv 1$, and $\frac{dy}{dt} = h(y)$, where $g(t) \equiv 1$.

By contrast, the ODE

$$\frac{dy}{dt} = 1 + t\, y^2$$

is not separable, because $1 + ty^2$ can't be factored as the product of a function of t and a function of y.

It is convenient to write separable ODEs in terms of *differentials,* rather than derivatives. If there is a functional relationship $y = \phi(t)$ between a variable t and a variable y, then by definition, $dy = \phi'(t)\, dt$. If $y = \phi(t)$ is a solution of the

ODE (2.4), then $\phi'(t) = f(t, y)$, so this ODE can be written in terms of differentials as

$$dy = g(t)h(y)\,dt. \qquad (2.5)$$

To solve this separable ODE, divide by $h(y)$ [provided that $h(y) \neq 0$] to get the following **separated equation:**

$$\frac{dy}{h(y)} = g(t)\,dt. \qquad (2.6)$$

To say $y = \phi(t)$ is a solution of the separated equation (2.6) means that

$$\frac{\phi'(t)\,dt}{h(\phi(t))} = g(t)\,dt, \qquad (2.7)$$

or $\phi'(t) = g(t)h(\phi(t))$. In other words, every solution of (2.6) is also a solution of the original ODE (2.5).

There may be solutions of (2.5) that do not satisfy the separated equation (2.6): the constant solutions. Let y_1 be a number such that $h(y_1) = 0$. Then the constant function $y \equiv y_1$ satisfies (2.5) because then $dy \equiv 0 \equiv h(y_1)g(t)\,dt$. Equation (2.6) is meaningless for $y \equiv y_1$, since the denominator is 0.

Integrating the separated equation   Suppose that $y = \phi(t)$ is a solution of the separated equation. If $H(y)$ and $G(t)$ denote antiderivatives of $[h(y)]^{-1}$ and $g(t)$, respectively, then in (2.7) we can replace $1/h(\phi(t))$ with $H'(\phi(t))$, dy with $\phi'(t)\,dt$, and $g(t)$ with $G'(t)$, to obtain

$$H'(\phi(t))\,\phi'(t)\,dt = G'(t)\,dt. \qquad (2.8)$$

By the chain rule,

$$\frac{d}{dt}H(\phi(t)) = H'(\phi(t))\,\phi'(t).$$

Thus, (2.8) tells us that $H(\phi(t))$ and $G(t)$ have equal derivatives. It follows from the equal derivatives theorem (see page 4) that on any interval upon which $H(\phi(t))$ and $G(t)$ are both defined, there is a constant C such that

$$H(\phi(t)) = G(t) + C. \qquad (2.9)$$

If $H(y)$ happens to be strictly increasing or strictly decreasing, there is a unique inverse function H^{-1}, and

$$\phi(t) = H^{-1}(G(t) + C)$$

is the general solution of the separated equation (2.6); otherwise the solution is defined implicitly by (2.9).

Solving a separable ODE

To solve the ODE

$$dy = g(t)\,h(y)\,dt$$

start by finding the constant solutions $y \equiv y_i$. This is done by determining the roots $y_1, y_2, \ldots$ of $h(y) = 0$. Then find the nonconstant solutions by integrating both sides of the separated equation

$$\frac{dy}{h(y)} = g(t)\,dt.$$

Example 2.1.2   Solve the initial value problem

$$y' = \frac{t}{y}; \quad y(0) = -1.$$

Solution. Since $h(y) = 1/y \neq 0$ for all y, there are no constant solutions. The separated equation is $y\,dy = t\,dt$, and integration of both sides yields $\frac{1}{2}y^2 = \frac{1}{2}t^2 + C$. To determine the value of C, substitute $t = 0$ and $y = -1$:

$$\frac{1}{2}(-1)^2 = \frac{1}{2}(0)^2 + C$$

so $C = \frac{1}{2}$. Thus $\frac{1}{2}y^2 = \frac{1}{2}t^2 + \frac{1}{2}$. Solving for y yields two solutions:

$$y = \pm\sqrt{t^2 + 1}.$$

Only one of these solutions,

$$y = -\sqrt{t^2 + 1},$$

is valid, since the initial condition specifies that $y(0)$ is negative. Figure 2.2 shows several solutions of the differential equation $y' = t/y$.

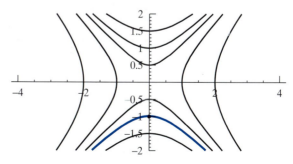

FIGURE 2.2   Solutions of the differential equation $y' = t/y$. The solution with initial value $y(0) = -1$ is shown as a heavy curve.

It is sometimes impractical to find a formula for y as an explicit function of t and the constant C, and even if it is possible to do this, it may be unnecessary. We will say that a function $F(t, y)$ is an **integral** of an ODE $y' = f(t, y)$ if, for all solutions $y = \phi(t)$ of the ODE, $F(t, \phi(t))$ is a constant function of t. Technically speaking, a constant function will be an integral for any ODE, but it does not define a solution implicitly. Integrals are usually obtained by integrating the ODE in some way, as we have done in the process of solving separable equations.

If $F(t, y)$ is a function of two variables, its level curves are the curves defined by the equations $F(t, y) = C$, where C is constant. If we have found an integral $F(t, y)$ of an ODE, then the graphs of all solutions will be subsets of level curves of F. For example, in the process of working out the solution of Example 2.1.2, we found that

$$F(t, y) = \frac{1}{2}y^2 - \frac{1}{2}t^2 \text{ is an integral of } y' = \frac{t}{y}.$$

It follows that if $y = \phi(t)$ is a particular solution of $y' = t/y$, then the graph of $\phi(t)$ is a subset of a hyperbola $F(t, y) = C$.

Example 2.1.3   Find the general solution of

$$y' = \frac{2y}{t}. \tag{2.10}$$

Solution. Since $h(y) = y$ has a zero at $y = 0$, there is a constant solution, $y \equiv 0$. The separated equation is

$$\frac{dy}{y} = \frac{2\,dt}{t}.$$

Integrating both sides, we have

$$\ln|y| = 2\ln|t| + C,$$

so $F(t, y) = \ln|y| - 2\ln|t|$ is an integral. Simplifying, we have

$$F(t, y) = \ln\left|\frac{y}{t^2}\right|.$$

Since $F(t, y)$ is an integral, so are the functions $\pm e^{F(t,y)} = y/t^2$. Thus, solutions take the form $y/t^2 = C$, or $y = C\,t^2$, for some constant C. It happens that the constant solution $y \equiv 0$ that we found at the outset belongs to this family (put $C = 0$). ❑

We will now solve an ODE that determines the velocity of a falling body influenced by gravitational and drag forces.

Example 2.1.4   An object is falling from a rest position, and its velocity (measured in meters per second) satisfies the initial value problem

$$v' = 9.8 - 9.8 \times 10^{-4}v^2; \quad v(0) = 0.$$

Find the velocity as a function of time.

Solution. The ODE is separable because the variable t does not appear on the right side. Notice that the right side can be factored as

$$9.8 \times 10^{-4}(100^2 - v^2) = 9.8 \times 10^{-4}(100 + v)(100 - v).$$

Thus there are two constant solutions, $v = \pm 100$. The solution $v = 100$ represents the terminal velocity, and $v = -100$ is extraneous, as it indicates a constant upward velocity. In fact, the ODE is wrong for $v < 0$ since then it has the motion and air resistance in the same direction. The separated equation is obtained by dividing through by $9.8 - 9.8 \times 10^{-4}v^2$, and it can be written

$$\frac{1}{9.8 - 9.8 \times 10^{-4}v^2}\, dv = dt.$$

We can simplify this slightly by multiplying through by 9.8×10^{-4} to get

$$\frac{1}{100^2 - v^2}\, dv = 9.8 \times 10^{-4}\, dt.$$

The partial fractions expansion for $1/(100^2 - v^2)$ is

$$\frac{1}{100^2 - v^2} = \frac{1}{200}\left(\frac{1}{100 - v} + \frac{1}{100 + v}\right).$$

Integrating both sides of the separated equation yields

$$\frac{1}{200}\left(\ln|100 + v| - \ln|100 - v|\right) = 9.8 \times 10^{-4}t + C.$$

The value of C can be determined at this point by referring to the initial condition: $t = 0$ and $v = 0$. It is easily seen that this gives $0 = C$. Setting $C = 0$, multiplying through by 200, and combining the logarithms yields

$$\ln\left|\frac{100 + v}{100 - v}\right| = 0.0196t.$$

Taking the exponential of both sides, we have

$$\frac{100 + v}{100 - v} = \pm e^{0.0196t}$$

and we choose the plus sign because it is in accord with the initial condition. It only remains to solve for v and obtain

$$v = 100\frac{e^{0.0196t} - 1}{e^{0.0196t} + 1}. \qquad \square$$

The solution of Example 2.1.4 has the expected properties: Starting from a velocity of 0, the object will accelerate and its velocity will approach a terminal velocity of 100 meters per second.

Example 2.1.5   Find a family of solutions of the ODE

$$y' = 3y^{2/3}. \tag{2.11}$$

Solution. Notice the constant solution $y \equiv 0$. Now, let's turn to the separated equation

$$\frac{1}{3}y^{-2/3}\,dy = dt$$

for $y \neq 0$. Integrating, we have $y^{1/3} = t + C$ and this leads to a family of solutions

$$y = (t + C)^3.$$

Although the constant solution $y \equiv 0$ does not belong to the family, the graph of each solution in the family is tangent to its graph, the t-axis, see Figure 2.3.   ❑

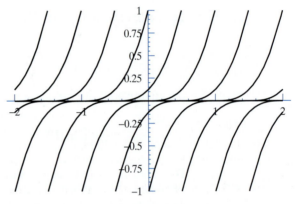

FIGURE 2.3   Solutions of $y' = 3y^{2/3}$. See Example 2.1.5.

The solution $y \equiv 0$ in Example 2.1.5 called a **singular solution** because it does not belong to the family of solutions that we derived.

EXERCISES

1. Which of the following ODEs is separable?

a. $\dfrac{dy}{dt} = \dfrac{t^2 + 5}{ty}$

b. $y\,dt = t\,dy$

c. $\dfrac{dy}{dt} = e^{ty}$

d. $\dfrac{dy}{dt} = e^{-(t^2 + y^2)}$

e. $\dfrac{dy}{dt} = \cot y$

f. $y' = \dfrac{t + y}{t - y}$

2. Find an integral for each of the following ODEs, *if it is separable*. The nonseparable equations should be skipped!

a. $y' = \tan(ty)$

b. $y' = \ln(t^y)$

c. $y' = \sec y$

d. $y' = \ln(t + y)$

3. A ball whose mass is 1 kilogram rolls on a level surface, subject only to the force of friction. The initial velocity is 1 meter per second; after 1 second, the velocity is 0.8

meters per second. Assuming that the magnitude of the frictional force is proportional to the velocity,

a. What is the friction constant?

b. When will the velocity be .5 meters per second?

c. How far did the ball roll in the first second?

d. Evaluate an improper integral to determine how far the ball will roll if given an infinite amount of time.

✔ *Answer:* (a) 0.223 kg/sec, (c) 0.896 m

4. Repeat problem 3, assuming the magnitude of the frictional force is proportional to the square of the velocity. The friction constant is now the ratio between the magnitude of the friction force and v^2.

✔ *Answer:* (a) 0.25 kg/m, (c) 0.893 m

5. An object in free fall is subject to gravitational force of mg and frictional force of $-bv$. Show that the terminal velocity v_∞ satisfies the equation $mg - bv_\infty = 0$, and therefore $v_\infty = mg/b$.

6. An object falls from rest. The terminal velocity is observed to be 98 meters per second. Assuming that the force of friction is

a. proportional to the velocity

b. proportional to the square of the velocity

how long does it take for the object to reach a speed of 49 meters per second?

✔ *Answer:* (a) 6.93 sec (b) 5.49 sec

7. An object with mass 1 kilogram is driven by a force of $30 \sin 3t$ newtons. The motion of the object is opposed by a frictional force that is proportional to the velocity. As $t \to \infty$, the velocity of the object approaches the function $2 \sin(3t - \theta)$, where $\theta = \sin^{-1}\left(\frac{1}{5}\right)$. Calculate the friction constant.

✔ *Answer:* $\sqrt{216}$ newtons per meter

In Exercises 8–15, find a family of solutions for the ODE. Then find the particular solution that satisfies the given initial condition (be sure to specify its domain).

8. $y' = 3t^2 y^2$; $y(0) = -1$ 9. $\dfrac{dy}{dt} = \dfrac{y}{2t}$; $y(1) = 2$

10. $tyy' = 1$; $y(e) = 2$ 11. $y' = ye^t$; $y(0) = 1$

12. $dy = (y^2 + 1)\, dt$; $y(0) = 1$

13. $\dfrac{dy}{dt} = y^2$; $y(0) = 1$ 14. $\dfrac{dy}{dt} = \dfrac{1+y}{\sqrt{t}}$; $y(4) = 0$

15. $y' = t(y^2 - 1)$; $y(0) = 0$

16. If an object is initially moving upward (we consider upward velocity to be negative), then the force of air resistance will be downward (positive), and the ODE

$$v' = g + \frac{k}{m}v^2$$

will apply as long as $v \le 0$, but when $v > 0$, the sign of the quadratic term should be changed to minus.

a. Show that the ODE

$$v' = g - \frac{k}{m}v\,|v|$$

is applicable to both upward and downward motion.

b. The object in Example 2.1.4 is thrown upward with an initial velocity of 100 meters/second ($v(0) = -100$). Give its velocity as a function of time for the duration of its upward motion.

c. Adapt the solution of Example 2.1.4 to obtain the velocity as a function of time for downward motion.

d. Sketch the graph of $v(t)$ for $t > 0$.

17. A basketball is dropped from a tower. There is no wind, and the drag force is proportional to the square of the velocity. After 2 seconds of fall, its velocity is 14.7 meters per second. Calculate its terminal velocity. It will be necessary to use Newton's method.

✔ *Answer:* about 19 meters per second

18. A rocket has a mass of 6 kilograms, including 2 kilograms of fuel. The fuel burns uniformly for 20 seconds, leaving no residue, and produces a constant thrust of 120 newtons. After 20 seconds, the fuel is spent, and the only forces are gravity and air drag. The drag force is proportional to the square of the velocity, with drag coefficient 0.02 kilograms per meter. Write initial value problems to determine the velocity of the rocket as a function of time for

a. $0 \le t \le 20$

b. $20 \le t \le T_1$, where T_1 is the time that the rocket reaches its maximum altitude

c. $T_1 \le t \le T_2$, where T_2 is the time that the rocket hits the ground

Be sure to take the variable mass of the rocket into account. Solving these ODEs is optional and is recommended only for readers who are using a CAS. Use the convention that upward motion has positive velocity.

19. A ball is thrown upward with a velocity v_0 (negative). Its terminal velocity in free fall is v_∞, and the drag force is proportional to the square of the velocity. Find formulas for the following as as functions of v_0 and v_∞.

a. the maximum height attained by the ball

b. the time taken to reach that height

c. the time taken to return to the ground

d. the velocity when the ball hits the ground

✔ *Answer:* $\dfrac{|v_0|v_\infty}{\sqrt{v_0^2 + v_\infty^2}}$

What happens if the drag force is negligible ($v_\infty = \infty$)?

20. Does the ball in Exercise 19 spend more time on the way up or on the way down?

✔ *Answer:* on the way down

21. A ball falls from rest from a high tower, with drag force proportional to the square of its velocity. How long does it take for the ball to attain a speed of half its terminal velocity?

✔ *Answer:* $t = (|v_\infty| \ln(3))/(2g)$

22. A ball is thrown downward from a high tower, with $v(0) = 4v_\infty$. How long does it take for the ball to reach a speed of $2v_\infty$? Assume the drag force is proportional to the square of the velocity.

✔ *Answer:* $t = (|v_\infty| \ln(1.8))/(2g)$

2.2   Exact Form and Integrating Factors

Suppose that $F(x, y)$ is a function that has continuous partial derivatives. If $y = \phi(x)$ is a function whose graph is part or all of the level curve

$$F(x, y) = C, \tag{2.12}$$

so that $F(x, \phi(x)) \equiv C$, we say that $\phi(x)$ is *defined implicitly* by equation (2.12). The derivative of $\phi(x)$ can then be determined[1] by differentiating the identity $F(x, \phi(x)) = C$ with respect to x and using the chain rule, with the following result:

$$\frac{\partial F}{\partial x}(x, \phi(x)) + \frac{\partial F}{\partial y}(x, \phi(x)) \cdot \phi'(x) = 0.$$

Therefore, $y = \phi(x)$ satisfies the ODE

$$\frac{\partial F}{\partial x} + \frac{\partial F}{\partial y} \cdot \frac{dy}{dx} = 0. \tag{2.13}$$

Equivalently, $F(x, y)$ is an integral of equation (2.13).

The **total differential** of $F(x, y)$ is defined as

$$dF = \frac{\partial F}{\partial x} dx + \frac{\partial F}{\partial y} dy.$$

The ODE (2.13) is equivalent to the equation $dF = 0$.

One approach to integrating an ODE

$$P(x, y)\, dx + Q(x, y)\, dy = 0 \tag{2.14}$$

would be to try to find a function $F(x, y)$ whose total differential is

$$dF = P(x, y)\, dx + Q(x, y)\, dy.$$

---

[1] In calculus courses, this process is called implicit differentiation.

This function $F(x, y)$ would have to satisfy the requirements

$$\frac{\partial F}{\partial x}(x, y) = P(x, y) \quad \text{and} \quad \frac{\partial F}{\partial y}(x, y) = Q(x, y). \tag{2.15}$$

Equations (2.15) overdetermine F—that is, usually there is no function that satisfies *both* equations. If there exists a function $F(x, y)$ that does satisfy (2.15) for all (x, y) in a domain $\mathcal{D}$ in the x, y-plane, the ODE (2.14) is said to be **exact** on $\mathcal{D}$. The remainder of this section will be devoted to explaining how to recognize exact ODEs, and find an integral for an exact ODE. We will also see that—in some circumstances—we can replace an ODE that is not exact with an equivalent ODE that is exact.

The following theorem tells us how to recognize an exact ODE.

Theorem 2.1

Assume that the first partial derivatives of $P(x, y)$ and $Q(x, y)$ are continuous on a rectangular domain $\mathcal{D}$ in the plane. Then the ODE (2.14) is exact on $\mathcal{D}$ if and only if

$$\frac{\partial P}{\partial y}(x, y) = \frac{\partial Q}{\partial x}(x, y) \tag{2.16}$$

for all $(x, y) \in \mathcal{D}$.

We will call condition (2.16) the **exactness condition.**

Proof First we will prove that if the ODE (2.14) is exact, then the exactness condition holds (this establishes that the exactness condition is necessary). Suppose that there is a function $F(x, y)$ such that $P(x, y) = \frac{\partial F}{\partial x}(x, y)$ and $Q(x, y) = \frac{\partial F}{\partial y}(x, y)$. Since P and Q have continuous first partial derivatives, F has continuous second partial derivatives on the domain $\mathcal{D}$.

We can now refer to a theorem on partial derivatives that says if a function $F(x, y)$ has continuous second partial derivatives on a domain in the plane, then

$$\frac{\partial^2 F}{\partial x \partial y} = \frac{\partial^2 F}{\partial y \partial x}.$$

Thus

$$\frac{\partial P}{\partial y} = \frac{\partial}{\partial y}\left(\frac{\partial F}{\partial x}\right) = \frac{\partial^2 F}{\partial y \partial x}$$

and

$$\frac{\partial Q}{\partial x} = \frac{\partial}{\partial x}\left(\frac{\partial F}{\partial y}\right) = \frac{\partial^2 F}{\partial x \partial y}$$

are equal: The exactness condition holds.

Now we will show that if $P(x, y)$ and $Q(x, y)$ satisfy the exactness condition, then there is a differentiable function $F(x, y)$ defined on $\mathcal{D}$ such that $dF = P(x, y)\,dx + Q(x, y)\,dy$.

Let (x_0, y_0) be a point in $\mathcal{D}$. Define a function

$$H(y) = \int_{y_0}^{y} Q(x_0, s) \, ds$$

and

$$F(x, y) = \int_{x_0}^{x} P(t, y) \, dt + H(y).$$

By the fundamental theorem of calculus,

$$\frac{\partial F}{\partial x} = P(x, y) + \frac{\partial}{\partial x} H(y) = P(x, y).$$

Since P has continuous partial derivatives, we can apply the Leibniz rule when differentiating F with respect to y:

$$\frac{\partial F}{\partial y} = \frac{\partial}{\partial y} \int_{x_0}^{x} P(t, y) \, dt + H'(y)$$

$$= \int_{x_0}^{x} \frac{\partial P}{\partial y}(t, y) \, dt + H'(y).$$

Now we will use the exactness condition to replace $\frac{\partial P}{\partial y}(t, y)$ with $\frac{\partial Q}{\partial t}(t, y)$ in the integral. This enables us to use another form of the fundamental theorem of calculus:

$$\int_{x_0}^{x} \frac{\partial Q}{\partial t}(t, y) \, dt = Q(x, y) - Q(x_0, y).$$

Thus we have shown that

$$\frac{\partial F}{\partial y} = Q(x, y) - Q(x_0, y) + H'(y). \tag{2.17}$$

Since $H(y)$ is by definition an antiderivative of $Q(x_0, y)$, the last two terms in (2.17) cancel, and the proof is complete. ■

Example 2.2.1   Which of the following ODEs are exact?

  i. $(y^2 + x^2 - 2x + 3) \, dx + (2xy - y^2 + 10) \, dy = 0$
  ii. $x(y^2 + x^2 - 2x + 3) \, dx + x(2xy - y^2 + 10) \, dy = 0$

Solution.

  i. $\frac{\partial P}{\partial y}(x, y) = 2y$ and $\frac{\partial Q}{\partial x}(x, y) = 2y$ so the equation is exact.

  ii. $\frac{\partial P}{\partial y}(x, y) = 2xy$ and $\frac{\partial Q}{\partial x}(x, y) = 4xy - y^2 + 10x$; therefore, the exactness condition does not hold. ❑

Observe that equation (ii) in Example 2.2.1 was obtained by multiplying equation (i) by x, but only equation (i) is exact. This shows that even though two equations may be *equivalent*, one may be exact while the other is not.

To integrate an exact equation, we will use a simplified version of the method used in the proof of Theorem 2.1. Start by computing *one* of the following indefinite integrals, whichever is easier:

$$M(x, y) = \int P(x, y)\, dx \quad \text{or} \quad N(x, y) = \int Q(x, y)\, dy.$$

When integrating with respect to x to find $M(x, y)$, treat y as a constant, and when computing $N(x, y)$, treat x as a constant. Let us assume that we have computed $M(x, y)$. Then any function of the form

$$F(x, y) = M(x, y) + H(y)$$

will satisfy the requirement $\frac{\partial F}{\partial x} = P(x, y)$. We can view the $H(y)$ term here as a "constant of integration," which depends on y since y was a constant when the antiderivative $M(x, y)$ was computed.

Now set $\frac{\partial F}{\partial y} = Q(x, y)$. In other words, solve the equation

$$\frac{\partial M}{\partial y} + H'(y) = Q(x, y)$$

to determine $H(y)$. This yields

$$H(y) = \int \left(Q(x, y) - \frac{\partial M}{\partial y} \right) dy.$$

Of course, $H(y)$ is not allowed to depend on x. The exactness condition ensures that all terms involving x in the difference $Q(x, y) - \frac{\partial M}{\partial y}$ cancel out, because

$$\frac{\partial}{\partial x} \left(Q(x, y) - \frac{\partial M}{\partial y} \right) = \frac{\partial Q}{\partial x} - \frac{\partial^2 M}{\partial x \partial y}$$

$$= \frac{\partial P}{\partial y} - \frac{\partial P}{\partial y} = 0.$$

Therefore, $F(x, y) = M(x, y) + H(y)$ is an integral of the ODE (2.14).

Example 2.2.2   Find an integral for the ODE

$$(y^2 + x^2 - 2x + 3)\, dx + (2xy - y^2 + 10)\, dy = 0.$$

Solution. According to Example 2.2.1, part (i), this equation is exact. The integral $F(x, y)$ will have the form

$$F(x, y) = \int (y^2 + x^2 - 2x + 3)\, dx + H(y)$$

$$= xy^2 + \frac{1}{3}x^3 - x^2 + 3x + H(y).$$

To determine $H(y)$, differentiate this expression with respect to y:

$$\frac{\partial F}{\partial y} = 2xy + \frac{dH}{dy}.$$

Since $\frac{\partial F}{\partial y} = Q(x, y) = 2xy - y^2 + 10$, it follows that

$$2xy + \frac{dH}{dy} = 2xy - y^2 + 10.$$

After canceling, we obtain $\frac{dH}{dy} = -y^2 + 10$. Therefore, $H(y) = -\frac{1}{3}y^3 + 10y$, and the integral is

$$F(x, y) = xy^2 + \frac{1}{3}x^3 - x^2 + 3x - \frac{1}{3}y^3 + 10y.$$

To find an explicit solution, we would have to solve the cubic equation $F(x, y) = C$ for y in terms of x.   ❑

Integrating factors

Equation (ii) of Example 2.2.1 isn't exact, but comparing it with the exact equation (i) of the example shows that it can be put into exact form by multiplying by an appropriate function, namely x^{-1}. This inspires the following definition: A function $m(x, y)$ is an **integrating factor** for the equation

$$P(x, y)\,dx + Q(x, y)\,dy = 0 \tag{2.18}$$

if

$$m(x, y)P(x, y)\,dx + m(x, y)Q(x, y)\,dy = 0$$

is exact.

If we can determine an integrating factor for a given ODE, then we have an equivalent ODE in exact form, and we can find an integral. In special circumstances, we can put this strategy into action.

One-variable integrating factors. Suppose that the ODE (2.18) has an integrating factor m. Then

$$m\, P(x, y)\,dx + m\, Q(x, y)\,dy = 0$$

is exact, so by the exactness condition,

$$\frac{\partial}{\partial y}[m\, P(x, y)] = \frac{\partial}{\partial x}[m\, Q(x, y)]. \tag{2.19}$$

Equation (2.19) is a PDE whose unknown function is the integrating factor m. It is not always easy to solve for m, but if it happens that there is an integrating factor $m(x)$ or $m(y)$ that depends on only one of the variables x or y, then equation (2.19) reduces to a separable ODE, and the integrating factor can be found.

Assume that there is an integrating factor of the form $m(x)$. By applying the product rule for differentiation to equation (2.19), we obtain

$$m(x)\frac{\partial P}{\partial y} = m'(x)\, Q(x, y) + m(x)\frac{\partial Q}{\partial x}.$$

This equation can be rearranged as

$$\frac{m'(x)}{m(x)} = \frac{1}{Q(x, y)}\left(\frac{\partial P}{\partial y} - \frac{\partial Q}{\partial x}\right). \tag{2.20}$$

Since its left side is independent of y, equation (2.20) only makes sense if its right side is also independent of y. Thus, if the right side is *not* independent of y, there is no integrating factor $m(x)$. If the right side is independent of y, it is straightforward to integrate both sides and calculate $m(x)$.

The same reasoning shows that there is an integrating factor $m(y)$ if and only if

$$\frac{1}{P(x, y)}\left(\frac{\partial Q}{\partial x} - \frac{\partial P}{\partial y}\right)$$

is a function of y alone, and that in this case

$$\frac{m'(y)}{m(y)} = \frac{1}{P(x, y)}\left(\frac{\partial Q}{\partial x} - \frac{\partial P}{\partial y}\right)$$

(see Exercise 21 at the end of this section).

Do not memorize the preceding material. It is easy to derive at a moment's notice from the exactness condition [in the form of equation (2.19)] by making the assumption that m is a function of x alone or of y alone.

Example 2.2.3   If possible, find an integrating factor $m(x)$ for the ODE

$$[2(x + y^2)\cos x + \sin x]\, dx + 2y \sin x\, dy = 0$$

and use it to determine an integral for the ODE.

Solution. An integrating factor $m(x)$ must satisfy

$$\frac{\partial}{\partial y}[m(x)(2(x + y^2)\cos(x) + \sin(x))] = \frac{\partial}{\partial x}[m(x)(2y \sin(x))].$$

If you do the differentiation (remember to use the product rule on the right side), you will obtain

$$m(x)(4y\,\cos(x)) = m'(x)(2y\,\sin(x)) + m(x)(2y\,\cos(x)).$$

We can cancel a factor of y and simplify this equation to the form

$$\frac{m'(x)}{m(x)} = \frac{\cos(x)}{\sin(x)}.$$

Now we can integrate: $\ln(|m(x)|) = \ln(|\sin(x)|) + C$. Therefore, $m = \sin(x)$ is an integrating factor. After multiplying equation (i) by this integrating factor, we have the exact equation

$$[2(x + y^2)\sin(x)\cos(x) + \sin^2(x)]\,dx + 2y\sin^2(x)\,dy = 0.$$

Since it's easier to integrate $Q(x, y) = 2y\sin^2(x)$ with respect to y than it is to integrate $P(x, y)$ with respect to x, set

$$F(x, y) = \int 2y\sin^2(x)\,dy + K(x) = y^2\sin^2(x) + K(x).$$

Then $\frac{\partial F}{\partial y} = 2y\sin^2(x)$. Since $\frac{\partial F}{\partial x} = 2y^2\sin(x)\cos(x) + K'(x)$,

$$2y^2\sin(x)\cos(x) + K'(x) = 2(x + y^2)\sin(x)\cos(x) + \sin^2(x),$$

and $K'(x) = 2x\sin(x)\cos(x) + \sin^2(x)$. A final integration shows that $K(x) = x\sin^2(x)$ and so $F(x, y) = (x + y^2)\sin^2(x)$ is the integral that we have sought. Graphs of all solutions of the ODE are subsets of level curves

$$(x + y^2)\sin^2(x) = C. \qquad \square$$

Example 2.2.4   Find, if possible, a one-variable integrating factor for the ODE

$$\left(2xy - \frac{2}{x^3}\right)dx + \left(4x^2 + \frac{3}{x^2 y}\right)dy = 0,$$

and use it to integrate the ODE.

Solution. We'll start by looking for an integrating factor $m(x)$. If there is one, it must satisfy

$$\frac{\partial}{\partial y}\left[m(x)\left(2xy - \frac{2}{x^3}\right)\right] = \frac{\partial}{\partial x}\left[m(x)\left(4x^2 + \frac{3}{x^2 y}\right)\right].$$

After you carry out the differentiation and simplify, you should obtain

$$m(x)\left(-6x + \frac{6}{x^3 y}\right) = m'(x)\left(4x^2 + \frac{3}{x^2 y}\right),$$

and then

$$\frac{m'(x)}{m(x)} = \frac{6 - 6x^4 y}{4x^5 y + 3}.$$

Since the right side depends on y, this equation has no solution, and hence there is no integrating factor $m(x)$.

To find if there is an integrating factor $m(y)$, we must find solutions (if any) of

$$\frac{\partial}{\partial y}\left[m(y)\left(2xy - \frac{2}{x^3}\right)\right] = \frac{\partial}{\partial x}\left[m(y)\left(4x^2 + \frac{3}{x^2 y}\right)\right].$$

Now we use the product rule in differentiating the left side of the equation:

$$m'(y)\left(2xy - \frac{2}{x^3}\right) + m(y)(2x) = m(y)\left(8x - \frac{6}{x^3 y}\right).$$

You can simplify this equation and obtain

$$\frac{m'(y)}{m(y)} = \frac{3}{y}.$$

Integrate both sides to get $\ln |m(y)|) = 3\ln|y| + C$. Thus $m(y) = y^3$ is an integrating factor. Multiplying equation (ii) by y^3 results in the exact equation

$$\left(2xy^4 - \frac{2y^3}{x^3}\right)dx + \left(4x^2 y^3 + \frac{3y^2}{x^2}\right)dy = 0.$$

Put

$$F(x, y) = \int (2xy^4 - 2x^{-3}y^3)\,dx + H(y)$$
$$= x^2 y^4 + x^{-2}y^3 + H(y).$$

Then $\frac{\partial F}{\partial y} = 4x^2 y^3 + 3x^{-2}y^2 + H'(y)$, and it follows that $H'(y) \equiv 0$. The integral is $F(x, y) = x^2 y^4 + x^{-2}y^3$. ❑

EXERCISES

General instruction for Exercises 1–14: Decide if the ODE is exact; if it is, find an integral, and if it isn't, do nothing. You are encouraged to use a computer algebra system to help find antiderivatives.

1. $(2x + 5y + 3)\,dx + (5x - 4y + 2)\,dy = 0$
2. $y\,dx + (x + y)\,dy = 0$
3. $(y^2 - 1)\,dx + (2xy - x)\,dy = 0$
4. $(y^2 - y)\,dx + (2xy - x)\,dy = 0$
5. $3e^{3x}y(\ln y - 1)\,dx + \ln y(e^{3x} - y)\,dy = 0$
6. $(x^2 + 2xy - y^2)\,dx + (x^2 - 2xy - y^2)\,dy = 0$
7. $x(x^2 + y^2 - 1)\,dx + y(x^2 + y^2 + 1)\,dy = 0$

8. $x(3\sqrt{x^2 + y^2} - 2)\,dx + y(3\sqrt{x^2 + y^2} + 2)\,dy = 0$
9. $\left[1 + \left(\frac{y}{x+y}\right)^2\right]dx + \left[\left(\frac{x}{x+y}\right)^2 - 1\right]dy = 0$
10. $\frac{y\,dx}{x} + \ln x\,dy = 0$
11. $(3x^2 + 6xy + 9y^2)\,dx + (3x^2 + 18xy + 51y^2)\,dy = 0$
12. $\frac{2x}{y}\,dx + \frac{y^2 - x^2}{y^2}\,dy = 0$
13. $2xy\,dx + (y^2 - x^2)\,dy = 0$
14. $2xy\,dx - (y^2 - x^2)\,dy = 0$

In Exercises 15–19, find an integrating factor for the given ODE, and use it to determine an integral.

15. $(x^2 + xy^2 + 1)\,dx + 2y\,dy = 0$

16. $x\,dy - (y - x)\,dx = 0$

17. $(x^2 + 2x + 2xy + 2y + 3y^2)\,dx + (2x + 6y)\,dy = 0$

18. $(5x + 20y + 28y^3)\,dx + (5x + 21xy^2)\,dy = 0$

19. $2xy\,dx + (3x^2 + 2y)\,dy = 0$

20. Show that the ODE $(5xy - 2y^2)\,dx + (5x^2 - 3xy)\,dy = 0$ does not have an integrating factor $m(x)$ or $m(y)$.

21. Show that there is an integrating factor $m(y)$ for the ODE $P(x, y)\,dx + Q(x, y)\,dy = 0$ if and only if the expression

$$\frac{1}{P(x, y)}\left(\frac{\partial Q}{\partial x} - \frac{\partial P}{\partial y}\right)$$

is independent of x.

22. Show that if $y' = f(x, y)$ is a separable ODE, with $f(x, y) = g(x)\,h(y)$, then $[h(y)]^{-1}$ is an integrating factor for

$$dy - f(x, y)\,dx = 0.$$

Thus, the method for solving separable ODEs is a special case of the integrating factor method.

23. Write the linear ODE, $y' + p(x)y = q(x)$, in the equivalent form

$$dy + [p(x)y - q(x)]\,dx = 0,$$

and find a one-variable integrating factor $m(x)$. (This integrating factor, discovered by Leibniz in 1692, has been memorized by ten generations of students in ODE courses.)

2.3 Graphical Analysis of ODEs

Nonlinear ODEs play an important role in mechanics, population studies, and countless other areas. There are two approaches to solving a nonlinear ODE: analytical and numerical. The goal of the analytical approach is to find a formula for the general solution, as we have done for the linear ODE. This is feasible for us only if the solution lies in the realm of **elementary functions**; that is, functions that can be obtained from the algebraic, exponential, and trigonometric functions by combining them through the operations of addition, multiplication, division, composition, and inversion. These are the functions that are studied in calculus courses and can be evaluated by all hand-held scientific calculators. The elementary functions form a small minority in the set of all functions, and many simple differential equations have solutions that are not elementary. Thus, we cannot expect that the analytical approach will always be successful.

The numerical approach works with IVPs, which usually have only one solution, and does not attempt to find the general solution of an ODE. Let $y = \phi(t)$ denote the solution of an IVP. We don't assume that a formula for $\phi(t)$ exists. All that we know about $\phi(t)$ is that $\phi(t_0) = y_0$ and that $y = \phi(t)$ satisfies the ODE. For any $t_1 \neq t_0$, either $\phi(t_1)$ does not exist, or $\phi(t_1)$ is a number, like π, that we can only approximate. A numerical solution of the IVP is a sequence of points $(t_0, y_0), (t_1, y_1), (t_2, y_2)\ldots$ in the plane such that (t_0, y_0) is the initial point of the IVP [that is, $\phi(t_0) = y_0$] and $y_i \approx \phi(t_i)$ for $i > 0$. The symbol $\approx$ means "is approximately equal to" and emphasizes that the numerical solution deals in approximation. The goal in calculating a numerical solution is to make the approximations sufficiently accurate for their intended purpose. While a numerical solution is defined only on a discrete set of points $\{t_0, t_1, \ldots t_n\}$, it may be presented as a graph by interpolating between the computed points.

It is possible to obtain valuable information about the solutions of an ODE by examining the ODE itself, and this analysis is a good complement to either of the above approaches to finding a solution. There are cases where we can avoid solving the ODE altogether. For example, consider the falling-body equation,

$$v' = g - \frac{k}{m}v^2.$$

We can solve this separable equation analytically—an unnecessary effort if our only goal is to find the terminal velocity, v_∞. When $v = v_\infty$, the acceleration is equal to zero; thus

$$g - \frac{k}{m}v_\infty^2 = 0$$

and hence

$$v_\infty = \pm\sqrt{\frac{mg}{k}}.$$

As usual, the positive direction is downward, so $v_\infty = +\sqrt{mg/k}$.

Direction fields

Imagine that the section of the t, y-plane extending from $t = A$ on the left to $t = B$ on the right, and from $y = C$ at the bottom to $y = D$ at the top is represented on a sheet of graph paper. We want to sketch some solutions of an ODE $y' = f(t, y)$ on this paper. The paper is marked, as all graph paper is, with vertical grid lines $t = t_i$ and horizontal grid lines $y = y_j$, where i runs from 0 at the left edge to N at the right, and j runs from 0 at the bottom to M at the right. The points (t_i, y_j) where the grid lines intersect are called **grid points**.

A **direction field** for the ODE $y' = f(t, y)$ can be drawn by placing at each grid point (t_i, y_j) a short line segment, called an **element**. The element is centered at the grid point, and its slope is equal to $f(t_i, y_j)$. Figure 2.4 shows a direction field for the ODE $y' = t - y^2$ on graph paper covering the region $-1.5 \le t \le 1.5$, $-1 \le y \le 1$.

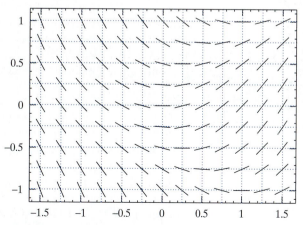

FIGURE 2.4   Direction field for $y' = t - y^2$.

Although the elements of the direction field can be drawn at only a finite set of points (we have chosen the grid points), we should imagine that there is an element located at each point of the plane. If $y = \phi(t)$ is a solution of the ODE, then at each point $(t, \phi(t))$ on the graph of ϕ the slope of the graph of ϕ is equal to $f(t, \phi(t))$, which is, by definition, the slope of the direction field at that point. Thus the graph of ϕ is tangent to the element at $(t, \phi(t))$. Perfect knowledge of the direction field would enable us to draw a perfect graph of $\phi(t)$.

Although its representation on the graph paper is discrete, the direction field still assists us in sketching solutions of the ODE that it represents. The graph of a solution $y = \phi(t)$ is unlikely to pass through very many grid points, but at each point the slope will be approximately equal to the slope of the nearest element of the direction field.

It is tedious to draw direction fields by hand, so we prefer to relegate the task to the computer. Any CAS is capable of drawing a direction field, and there are also limited-purpose programs, such as *MDEP* (for PCs) and *Macmath*, that will do the job with less fuss. Figure 2.5 shows two typical direction fields, as drawn by a CAS. They belong to the ODEs,

$$v' = 10 - .01v^2, \tag{2.21}$$

and

$$v' = 10 - .01e^{-.01t}v^2, \tag{2.22}$$

respectively. Equation (2.21) represents the velocity of a falling body with a constant drag coefficient, while in (2.22), the drag coefficient decreases with time. Compare the two: They indicate that solutions of (2.21) tend to a terminal velocity while solutions of (2.22) do not have a terminal velocity, since there is evidently no constant solution, where the direction field elements are always horizontal. Indeed, we can prove that (2.22) has no constant solution $v(t) \equiv v_\infty$. If such a solution existed, then $v' \equiv 0$. Thus, substituting $v' = 0$ and $v = v_\infty$ in (2.22) yields $0 = 10 - .01e^{-.01t}v_\infty^2$, which we solve for v_∞ to obtain

$$v_\infty = \sqrt{1000e^{.01t}},$$

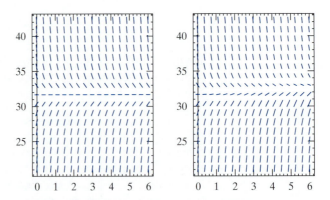

FIGURE 2.5   Direction fields corresponding to models for the velocity of a falling object. On the left, the drag coefficient is constant, and on the right, the drag coefficient decreases with time.

which is *not* constant. Since a terminal velocity must be constant, there is no terminal velocity.

Difference equations

For any sequence u_n of numbers, we can form a new sequence, called the sequence of **forward differences**, as

$$u_1 - u_0, u_2 - u_1, u_3 - u_2, \ldots .$$

The mth term of this sequence is $\Delta u_m = u_{m+1} - u_m$. In this notation, the sequence of forward differences is

$$\Delta u_0, \Delta u_1, \Delta u_2, \ldots .$$

Example 2.3.1   Find the sequence of forward differences for the squares 0, 1, 4, 9, 16, 25, 36, ...

Solution. In the given sequence, $u_m = m^2$. The first few entries of the sequence of forward differences suggest a familiar pattern: $\Delta u_0 = 1 - 0 = 1$, $\Delta u_1 = 4 - 1 = 3$, $\Delta u_2 = 5$, and we can easily see that the next few forward differences are 7, 9, 11, ... You can verify that $\Delta u_m = 2m + 1$ by factoring the difference of two squares. ❑

Example 2.3.2   Let y_m be the value of a bank account m years after an initial deposit in an account bearing an interest rate r, with annual compounding. Find a formula for the value of an account in which there is one initial deposit of P, after t years.

Solution. Let us compute the sequence of forward differences in the value of the account. Δy_m is the amount that the account's value increased or decreased during the $(m + 1)$st year. Since there were no deposits or withdrawals in that year, Δy_m is the interest earned:

$$\Delta y_m = r\, y_m.$$

It follows that $y_{m+1} = y_m + \Delta y_m = (1 + r)y_m$. The year-end balances form a geometric sequence y_0, $y_0(1 + r)$, $y_0(1 + r)^2$, ... Since y_0 is equal to the initial deposit, P, we have $y_m = P(1 + r)^m$. ❑

To make a rough sketch of the solution of an IVP, $y' = f(t, y)$; $y(t_0) = y_0$, we could take the following approach, which is a modification of the direction field procedure. Draw the direction field element centered at (t_0, y_0). Let (t_1, y_1) be the point at the right end of that element, and draw the right half of the direction field element at (t_1, y_1), extending to a point (t_2, y_2). Continuing this process, we would obtain a **polygonal curve** (also known as a broken line graph) that would approximate the graph of the solution of the IVP. The accuracy of the approximation would depend on the length of the direction field elements being used: With shorter elements, the graph will be more accurate and will require more effort to draw.

Let us focus on the endpoints (t_0, y_0), (t_1, y_1), (t_2, y_2), ... of the direction field elements. These yield two sequences t_n and y_n. We will stipulate that the direction field elements are drawn so that $\Delta t_n = h$ is a constant sequence. This means that the t_n are evenly spaced, with $t_n = t_0 + nh$. The slope of the direction field element at (t_n, y_n) is therefore

$$\frac{\Delta y_n}{\Delta t_n} = \frac{1}{h} \Delta y_n.$$

Since the slope of this direction field element is equal to $f(t_n, y_n)$, we have

$$\Delta y_n = h \, f(t_n, y_n). \tag{2.23}$$

Equation (2.23) is a **difference equation.** Difference equations are like ODEs, and we can exploit them to learn more about ODEs (and vice versa). To see how this analogy works, let us return to the sequence $y_0, y_1, y_2 \ldots$ of bank balances in Example 2.3.2. The sequence y_m is the solution of an initial value problem consisting of the difference equation $\Delta y_m = r \, y_m$ and $y_0 = P$.

The ODE analog of the compound interest difference equation,

$$\frac{dy}{dt} = r \, y,$$

is used to compute *continuously compounded interest*. The solution, with initial balance P, is $y = P e^{rt}$, and the year-end balances also form a geometric sequence,

$$P, \, P e^r, \, P(e^r)^2, \ldots, P(e^r)^m, \ldots.$$

Since e^r is slightly greater than $1 + r$, continuous compounding is more rewarding to the depositor.

In the context of difference equations, an IVP takes the form

$$\Delta y_m = F(m, y_m); \quad y_0 = A.$$

Provided that $F(m, y)$ is defined for all (m, y) such that m is a nonnegative integer, this IVP is guaranteed to have a unique solution. We don't even have to assume F is continuous. To see why this is so, rewrite the difference equation as

$$y_{m+1} = y_m + F(m, y_m). \tag{2.24}$$

Equation (2.24) is *recursive*, in the sense it gives a formula for computing the next term in the sequence. Since we have a starting value for y_0 (the initial condition), y_1, y_2, and so on can be computed, one after the other. Many well-known sequences are defined recursively—for example, the factorial sequence $n!$ is defined by $0! = 1$ and $n! = n \cdot (n-1)!$.

Euler's method

Leonhard Euler, one of the brightest stars of eighteenth-century mathematics, discovered a way to use a difference equation IVP to approximate the solution of a differential equation IVP. Let $y = \phi(t)$ denote the solution of the IVP,

$$y' = f(t, y); \tag{2.25}$$
$$y(t_0) = A. \tag{2.26}$$

Choose a **time step** h (h must be nonzero but is allowed to be negative) and set $t_n = nh + t_0$. Euler's method uses the difference equation (2.23) with initial condition $y_0 = A$ to determine a sequence $y_0, y_1, y_2, \ldots$. In this sequence, y_n serves as the approximation of $\phi(t_n)$.

We have pictured the approximate solution of an IVP given by Euler's method by plotting the sequence of points (t_m, y_m) and connecting adjacent points with straight line segments to form a polygonal curve.

Example 2.3.3 Use Euler's method with time step $h = \pm 0.25$ to approximate the solution of the IVP $y' = t - y^2$, $y(0) = 0$, for $-1.5 \le t \le 1.5$.

Solution. The solution must be propagated to the left and to the right from the initial point $(0, 0)$. To propagate leftward, we use $h = -0.25$; and to propagate rightward, $h = 0.25$. We will start with $h = -0.25$. Euler's method then generates a sequence y_m defined by the initial condition $y(0) = 0$ and the difference equation $\Delta y_m = -0.25(t_m - y_m^2)$, where $t_0 = 0$, $t_1 = -0.25$, $t_2 = -0.5$, and so on. Since $t_6 = -1.5$, we will need to compute y_m for $1 \le m \le 6$.

Write the difference equation in recursive form as

$$y_{m+1} = y_m - 0.25(t_m - y_m^2).$$

Starting with $y_0 = 0$, we have

$$y_1 = 0 - 0.25(0 - 0^2) = 0$$
$$y_2 = 0 - 0.25(-0.25 - 0^2) = 0.0625$$
$$y_3 = 0.0625 - 0.25(-0.5 - 0.0625^2) \approx 0.1885$$
$$y_4 \approx 0.1885 - 0.25(-0.75 - 0.1885^2) \approx 0.3849$$
$$y_5 \approx 0.3849 - 0.25(-1 - 0.3849^2) \approx 0.6719$$
$$y_6 \approx 0.6719 - 0.25(-1.25 - 0.6719^2) \approx 1.0972.$$

The point of this calculation is that if $y = \phi(t)$ denotes the solution of the IVP, then $\phi(-0.25) \approx 0$, $\phi(-0.5) \approx 0.0625$, and so on up until $\phi(-1.5) \approx 1.0972$. It does not tell us how good these approximations are, and in fact they are not very good. This ODE is a Riccati equation that can be solved analytically by a CAS (the solution involves *special functions* and will not be shown here). The CAS tells us that $\phi(-0.5) = 0.1266$ and $\phi(-1.5) = 1.7857$ (both rounded to 4 decimal places). To get this level of precision with Euler's method, it would be necessary to reduce the step size to $h = 0.0001$; then $\phi(-1.5)$ would be approximated by $y_{15,000}$. Obviously you would need a computer.

To propagate to the right, we start over with $h = +0.25$. The numbers $t_0, t_1, t_2 \ldots$ are $0, 0.25, 0.5 \ldots$, and the calculation proceeds as follows:

$$y_1 = 0 + 0.25(0 - 0^2)$$
$$y_2 = 0 + 0.25(0.25 - 0^2) = 0.0625$$
$$y_3 = 0.0625 + 0.25(0.5 - 0.0625^2) \approx 0.1865$$
$$y_4 \approx 0.1865 + 0.25(0.75 - 0.1885^2) \approx 0.3653$$
$$y_5 \approx 0.3653 + 0.25(1 - 0.3653^2) \approx 0.5820$$
$$y_6 \approx 0.5820 + 0.25(1.25 - 0.5820^2) \approx 0.8098.$$

We find that $\phi(0.25) \approx 0$, $\phi(0.5) \approx 0.0625$, $\phi(0.75) \approx 0.1865$, and so on up to $\phi(1.5) \approx 0.8098$. Due to the large step size, the approximations are again pretty sloppy. The correct values, rounded to 4 decimal places, are $\phi(0.25) = .0312$, $\phi(0.5) = 0.1235$, $\phi(0.75) = 0.2700$, and $\phi(1.5) = 0.8574$.

Figure 2.6 shows the graph of the approximate solution that we have produced, with the direction field for the ODE in the background.   ❑

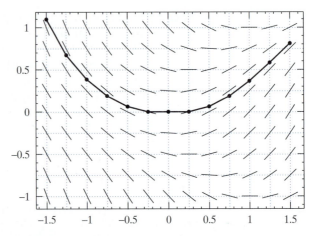

FIGURE 2.6   Approximate solution by Euler's method and direction field for $y' = t - y^2$.

To make a convincing argument that Euler's method really does produce an approximation of $\phi(t)$, we need to refer to the second mean value theorem, a special case of Taylor's theorem. The proof can be found in any calculus text.

Theorem 2.2   Second mean value theorem

Suppose that a function $\phi(t)$ is twice differentiable on an interval (A, B), and that t_0 and h are numbers such that $t_0, t_0 + h \in (A, B)$. Then there is a number c_1 between t_0 and $t_0 + h$, such that

$$\phi(t_0 + h) = \phi(t_0) + h\,\phi'(t_0) + \frac{1}{2}h^2\,\phi''(c_1). \tag{2.27}$$

If $\phi(t)$ is the solution of the IVP (2.25), (2.26), then we know that $\phi(t_0) = y_0$ and

$$h\,\phi'(t_0) = h\,f(t_0, y_0) = \Delta y_0$$

in the notation of the difference equation (2.23). Let's assume that ϕ is sufficiently differentiable so that we can use Theorem 2.2. Then we can rephrase equation (2.27) as

$$\phi(t_0 + h) = \underbrace{y_0 + \Delta y_0}_{y_1} + \frac{1}{2}h^2\,\phi''(c_1)$$

and hence $\phi(t_0 + h) - y_1 = \frac{1}{2}h^2\,\phi''(c_1)$, where c_1 is between t_0 and t_1. As $h \to 0$ $\frac{1}{2}h^2\phi''(c_1) \to 0$ at a faster rate, and that is why y_1 is a good approximation of $\phi(t_0 + h)$.

We have seen why Euler's method approximates the solution of an IVP over one step. We still need to justify the approximation as it proceeds through many steps. In the second step, Euler's method produces a line segment starting at (t_1, y_1). Let $\phi_1(t)$ be the solution of the IVP

$$y' = f(t, y) \quad y(t_1) = y_1.$$

Applying Theorem 2.2 to $\phi_1(t)$, we find that

$$\phi_1(t_1 + h) = \phi_1(t_2) = \underbrace{y_1 + \Delta y_1}_{y_2} + \frac{1}{2}h^2\,\phi_1''(c_2)$$

with c_2 between t_1 and t_2. The approximation error after the second step is

$$\phi(t_2) - y_2 = [\phi(t_2) - \phi_1(t_2)] + [\phi_1(t_2) - y_2]$$

$$= [\phi(t_2) - \phi_1(t_2)] + \frac{1}{2}h^2\,\phi_1''(c_2).$$

We can separate this expression into two terms, the **local error** and the **accumulated error**. The local error is defined to be the error due to *one step* of the approximation. In the step where y_2 is computed, the local error, which we denote LE_2, is equal to $\phi_1(t_2) - y_2$. The purpose of referring to the second mean value theorem is to estimate local error, and it tells us that $LE_2 = \frac{1}{2}h^2\,\phi_1''(c_2)$.

The accumulated error, denoted AE_m, is the error due to all previous steps taken in propagating the solution by Euler's method. Thus the first accumulated error is equal to

$$AE_1 = \phi(t_2) - \phi_1(t_2).$$

This is the difference between two solutions of the same ODE at $t = t_2$, given that the two solutions are close when $t = t_1$.

As we continue, each step approximates the solution $\phi_m(t)$ of a new IVP, whose initial point is the point (t_m, y_m) reached in the previous step. There is a new local error $LE_{m+1} = \frac{1}{2}\phi''(c_{m+1})$, and a new accumulated error,

$$AE_m = \phi(t_{m+1}) - \phi_m(t_{m+1}).$$

We can't expect that AE_m will be equal to the sum of the local errors LE_1, LE_2, ..., LE_m. Solutions of ODEs with nearby initial conditions may diverge away from each other (if this happens, the accumulated error will be larger than the sum of the previous local errors), or they may converge toward each other, causing the accumulated error to be less than the sum of the local errors.

Example 2.3.4   Determine the solution of the initial value problem

$$y' = y + \frac{1}{4}; \ y(0) = 0.$$

Using a time step $h = 0.5$, use Euler's method to calculate

a. y_1, y_2, y_3, and y_4
b. LE_1, LE_2, LE_3, and LE_4
c. AE_1, AE_2, and AE_3

Solution. We will start by computing $y_1 \ldots y_4$. The large time step will make the approximation unreliable. This suits the purpose of this example, because we are studying errors, but in any practical application it would be unacceptable.

The difference equation prescribed by Euler's method is $\Delta y_m = 0.5(y_m + 0.25)$, with initial condition $y_0 = 0$. Thus

$$y_1 = y_0 + \Delta y_0 = 0 + 0.5(0 + .25) = 0.125$$
$$y_2 = y_1 + \Delta y_1 = 0.125 + 0.5(0.125 + 0.25) = 0.3125$$
$$y_3 = y_2 + \Delta y_2 = 0.3125 + 0.5(0.3125 + .25) = 0.59375$$
$$y_4 = y_3 + \Delta y_3 = 0.65625 + 0.5(0.65625 + .25) = 1.015625.$$

To determine the errors precisely, we need to have the general solution of the ODE (in practice, we will never have access to this, since the purpose of Euler's method is to approximate a solution that can't be obtained analytically). Since $y' = y + \frac{1}{4}$ is a linear ODE, you can derive the solution $y = Ce^t - \frac{1}{4}$. The solution that we are trying to approximate is satisfies the initial condition $\phi(0) = 0$, so $C = \frac{1}{4}$.

Each solution $\phi_m(t)$, which we will use to quantify the errors, satisfies initial conditions $\phi(t_m) = y_m$, so $\phi_m(t) = C_m e^t - \frac{1}{4}$, where the constant C_m is determined by

$$y_m = C_m e^{t_m} - \frac{1}{4}.$$

Thus $C_m = (y_m + 0.25)e^{-t_m}$ and

$$\phi_m(t) = (y_m + 0.25)e^{t - t_m} - 0.25.$$

In particular,

$$\phi_1(t) = 0.375e^{t - 0.5} - 0.25$$
$$\phi_2(t) = 0.5625e^{t - 1} - 0.25$$
$$\phi_3(t) = 0.84375e^{t - 1.5} - 0.25.$$

Figure 2.7 displays the graphs of $\phi(t)$, and $\phi_m(t)$ for $m = 1, 2, 3$. It also shows the broken line graph connecting (t_0, y_0), (t_1, y_1), (t_2, y_2), (t_3, y_3), and (t_4, y_4). Notice that the graph of $\phi(t)$ is tangent to the first segment of the broken line graph, $\phi_1(t)$ is tangent to the second segment, and so on.

The local errors are as follows:

$$LE_1 = \phi(0.5) - y_1 = (0.25e^{0.5} - 0.25) - 0.125 \approx 0.037$$

$$LE_2 = \phi_1(1) - y_2 = (0.375e^{0.5} - 0.25) - 0.3125 \approx 0.056$$

$$LE_3 = \phi_2(1.5) - y_3 = (0.5625e^{0.5} - 0.25) - 0.59375 \approx 0.084$$

$$LE_4 = \phi_3(2) - y_4 = (0.84375e^{0.5} - 0.25) - 1.015625 \approx 0.125.$$

The accumulated errors are

$$AE_1 = \phi(1) - \phi_1(1) = (0.25e^1 - 0.25) - (0.375e^{0.5} - 0.25) \approx 0.061$$

$$AE_2 = \phi(1.5) - \phi_2(1.5) = (0.25e^{1.5} - 0.25) - (0.5625e^{0.5} - 0.25) \approx 0.193$$

$$AE_3 = \phi(2) - \phi_3(2) = (0.25e^2 - 0.25) - (0.84375e^{0.5} - 0.25) \approx 0.456.$$

In Figure 2.7 we can picture LE_4 as the vertical distance between the bottom two curves at $t = 2$ (at the right edge of the graph). AE_3 is the distance from the top curve to the second-to-the-bottom curve and represents the consequences of the local errors LE_1, LE_2, and LE_3.   ❑

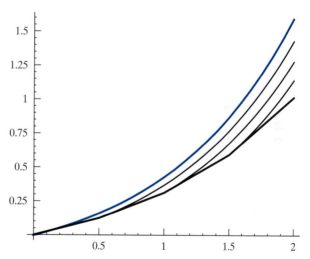

FIGURE 2.7   Approximate solution of $y' = y + 0.25$; $y(0) = 0$ by Euler's method. The actual solution $\phi(t)$ and the graphs of $\phi_m(t)$ for $m = 1, 2, 3$ are also shown.

Example 2.3.5   The velocity $y(t)$ (in meters per second) of a rock sinking in sea water is given by the differential equation

$$y' = 8 - 2y^2. \tag{2.28}$$

Given that $y(0) = 0$, use Euler's method with $h = 0.1$ to approximate the solution for $0 \le t \le 2$ seconds.

Solution. The difference equation given by Euler's method with $h = 0.1$ is

$$\Delta y_m = 0.1(8 - 2y_m^2).$$

The solution of this difference equation is the sequence beginning 0, 0.8, 1.472, ... By connecting successive points (t_m, y_m), (t_{m+1}, y_{m+1}) with line segments, we obtain a broken line graph that serves as an approximation of the solution of our initial value problem. This graph is shown in Figure 2.8. Although Euler's method can't be expected to give a very accurate approximation with such a large time step, it should be noted that the graph shown levels off at a velocity of 2 meters per second, which is the correct terminal velocity. This difference equation, like the differential equation it approximates, has a constant solution, $y_m = 2$, and our solution converges to it. In other words, the qualitative behavior of the Euler difference equation matches that of the differential equation. Unfortunately, this does not always happen—see Exercise 25 at the end of this section.   ❑

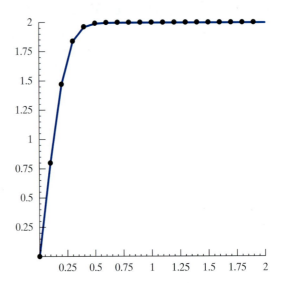

FIGURE 2.8   Velocity of a rock sinking in seawater, determined by Euler's method. See Example 2.3.5.

How to implement Euler's method with a spreadsheet

A spreadsheet program is designed to manipulate tables of numbers. Each entry in a table occupies a **cell** of the spreadsheet, whose location is identified by specifying its row and column. Columns are labeled with letters A, B, C, and so on. If there are more than 26 columns, symbols AA, AB, AC, ... would be used for columns 27, 28, 29 .... Rows are numbered. Thus C6 specifies the sixth cell from the top of column C, and AB99 is the 99th cell in column AB. After moving the cursor to a given cell, we can type an entry. This may be a text label, a number, or a formula. Formulas may involve arithmetic functions ($+$, $-$, $\times$, $\div$) or a wide variety of elementary functions. In some spreadsheets, all function names begin with @; thus @SIN stands for the sine function.

Euler's method will use three columns of the spreadsheet. Column A lists $t_0, t_1, t_2, t_3, \ldots$, Column B is for $y_0, y_1, y_2, y_3, \ldots$, and we put $\Delta y_0, \Delta y_1, \Delta y_2, \Delta y_3, \ldots$ in column C. The spreadsheet has the capability of copying a cell onto a range of cells, and this feature minimizes the typing necessary. To see how this works, let us fill in column A. Type the initial time in cell A1. Cell A2 will contain the formula for t_1, $(h + \text{A1})$ [type the value of the time step being used, not the letter h; thus if $h = 0.1$ type $(0.1 + \text{A1})$]. Now copy cell A2 to fill the remainder of column A. Spreadsheets use a relative notion of copying, so you will find that cell A3 contains the formula $(0.1 + \text{A2})$, A4 contains $(0.1 + \text{A3})$, and so on. Turning to column B, enter the initial value y_0 in cell B1. Cell B2 is to contain $y_1 = y_0 + \Delta y_0$. Since we plan to put Δy_0 in cell C1, enter the formula $(\text{B1} + \text{C1})$ in cell B2. The value appearing in this cell won't be correct until we have made an entry in cell C1. Copy cell B2 to fill the rest of column B.

In cell C1 we place the formula for Δy_0. If the differential equation is $y' = f(t, y)$, enter $h * f(\text{A1}, \text{B1})$, because the values of t_0 and y_0 are found in cells A1 and B1, respectively. Don't simply type $f(\text{A1}, \text{B1})$; the function f must be translated to a spreadsheet formula. For example, if $f(t, y) = t - y^2$ and $h = 0.1$, we would type $0.1 * (\text{A1} - \text{B1} \wedge 2)$. Finally, copy the contents of cell C1 to the remainder of column C.

Spreadsheets are capable of producing a graph of the approximate solution. The menu of graph types contains many choices; the only appropriate one is the xy-graph, with x-axis values from column A and "first series" values from column B. Although a polygonal graph is appropriate, commercial spreadsheets now draw smooth curves by employing sophisticated interpolation techniques. Smooth curves are no more accurate than the polygonal graphs, but they look nicer.

EXERCISES

1. In Exercise 16 (page 43), the ODE

$$v' = g - \frac{k}{m} v \, |v|$$

was introduced as a model for the velocity of a ball that was initially thrown upward and is subject to gravitational and drag forces (in the model, the positive direction is downward). Draw a direction field for this ODE. Use as parameters $g = 9.8$ m/sec^2 and $\frac{k}{m} = 0.002$ m^{-1}. Remember that in this model, the positive direction is downward.

In Exercises 2–12, sketch graphs of solutions of the IVPs on the given direction fields. You may wish to obtain sheets of clear acetate to draw upon, or to photocopy the direction fields and draw on the copies. If a computer with software that can draw and print out direction fields is available, you are encouraged to print out your own direction fields.

2. $y' = ty$; initial conditions $y(0) = 0$, $y(0) = 1$, and $y(0) = -1$

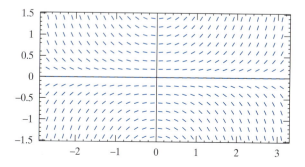

3. $y' = 3t - 2y$; initial conditions $y(0) = -1$, $-\frac{3}{4}$, and 0

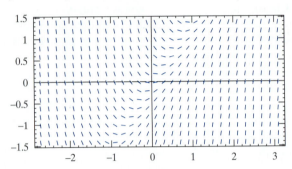

4. $y' = y - y^2$; initial conditions $y(0) = 0$ and $y(0) = \pm 1$

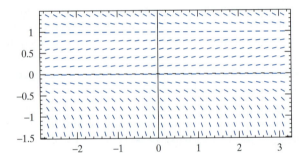

5. $y' = y^2 - 2ty + t^2$; initial conditions $y(0) = 0$, $y(1) = 1$, $y(0) = 1$, and $y(0) = -1$

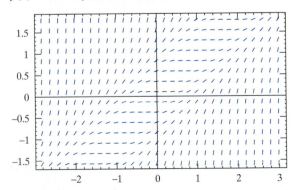

6. $y' = 2te^{-y}$; initial conditions $y(0) = 0$, $y(1) = 0$, and $y(-1) = 0$

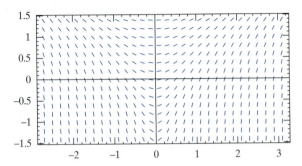

7. $y' = -2y$; $y(0) = -1$

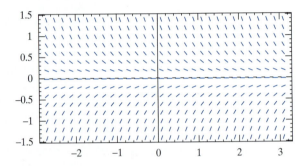

8. $y' = 2t$; $y(0) = 0$

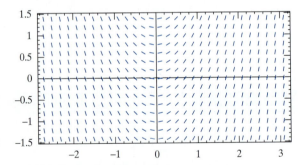

9. $y' = t^2 - y$; $y(0) = 0$

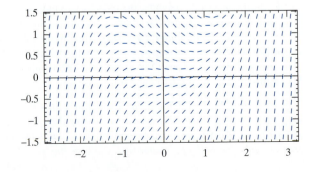

10. $y' = \dfrac{1}{t - 3y}$; $y(0) = 1$

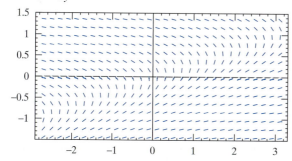

11. $y' = \dfrac{1}{t^2 + y^2}$; $y(0) = -0.2$

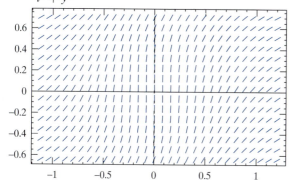

12. $y' = \sin y$; $y(0) = \dfrac{\pi}{6}$

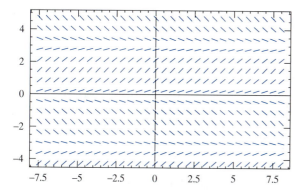

13. Show that the graph of the solution $y(t)$ of the IVP $y' = t - 3y$; $y(0) = 1$ crosses the line $t - 3y = 0$ exactly once and has a relative minimum at the crossing point. Do *not* use the explicit solution, $y = \frac{1}{9}(3t - 1 + 10e^{-3t})$. Instead, consider what the slope of the solution will be at a crossing point.

14. Find the sequence of forward differences for each of the following sequences.

a. $1, 2, 4, 8, 16, 32, \ldots, 2^m, \ldots$

b. $1, 3, 9, 27, 81, 243, \ldots, 3^m, \ldots$

c. $0, 1, 8, 27, 64, 125, \ldots, m^3, \ldots$

d. $1, \frac{1}{2}, \frac{1}{3}, \frac{1}{4}, \frac{1}{5}, \ldots, \frac{1}{m}, \ldots$

15. For each difference equation, determine the first three or four terms of the sequence y_m and then find an expression for y_m.

a. $\Delta y_m = 1$, $y_1 = 0$

b. $\Delta y_m = y_m$, $y_1 = 1$

c. $\Delta y_m = k y_m$, $y_1 = C$, where k and C are constants

d. $\Delta y_m = y_m + 1$, $y_1 = 0$

16. The bank offers you a choice of an account bearing an interest rate of 6%, compounded continuously, and an account with 6.25% interest, compounded annually. Both rates are guaranteed for five years, and you don't intend to make withdrawals before that time is up. Which account should you choose?

17. Let a_m and b_m be sequences. The difference equation

$$\Delta y_m = a_m y_m + b_m$$

is called the **linear difference equation** with coefficients a_m and source b_m. You will want to compare this definition with that of a linear ODE. If $b_m = 0$ for all m, the difference equation is homogeneous.

Define a new sequence A_m by $A_1 = 1$, and

$$A_m = (1 + a_1)(1 + a_2) \cdots (1 + a_{m-1}) \text{ for } m > 1.$$

a. Show that every solution of a homogeneous difference equation $\Delta y_m = a_m y_m$ has the form $y_m = C A_m$, where C is constant.

b. Explain how to solve an initial value problem involving an inhomogeneous linear difference equation.

◆ *Hint:* The procedure resembles the variation of constants method for solving an inhomogeneous linear ODE.

c. Find the general solution of

$$\Delta y_m = \frac{1}{m} y_m.$$

d. Use the method of part (b) to solve

$$\Delta y_m = \frac{1}{m} y_m + m + 1, \quad y_1 = 0.$$

18. a. Show that the solution of the IVP

$$y' = f(x); \quad y(a) = b, \qquad (2.29)$$

is

$$y(x) = b + \int_a^x f(u) \, du. \qquad (2.30)$$

b. Show that approximating the solution of the IVP 2.29 by Euler's method amounts to calculating a Riemann sum for the integral in equation (2.30).

Use Euler's method and a spreadsheet or programmable calculator to find approximate solutions of the differential equations in Exercises 19 and 20, over the intervals indicated, and with the prescribed step sizes. Draw broken line graphs of the solutions.

19. $y' = y^2$; $y(0) = 1$. $h = .05$, $0 \le t \le 1$.

✔ **Answer:** $y_{20} = 9.55$. In this case, the final local error (LE_{20}) is infinite, since the analytic solution, $y = 1/(1 - t)$, has a singularity at $t = 1$.

20. $y' = \dfrac{t}{\sqrt{t^2 + y^2}}$; $y(0) = 1$. $h = \pm 0.1$ for $-1 \le t \le 1$. Can you save time by exploiting the symmetry in the difference equations for propagating to the left and to the right?

21. Let $y = \phi(t)$ denote the solution of the IVP,

$$y' = t + y; \quad y(0) = 0.$$

Since the differential equation is linear, you can solve this IVP and find a formula for $\phi(t)$. This problem asks you to calculate $\phi(1)$ and to approximate $\phi(1)$ by using Euler's method. The purpose is to see how fast the approximation improves when the time step is reduced.

  Use Euler's method with time steps $h = 1$, 0.5, 0.25, 0.1, 0.05, 0.02, and 0.01 to approximate $\phi(1)$, and let $E(h) = |\phi(1) - y_N|$ (where $Nh = 1$) denote the approximation error obtained with time step h. Plot a graph of $E(h)$ as a function of h.

22. The purpose of this problem is to trace the consequence of one local error in Euler's method. Calculate the approximation y_1 of $e^{-0.1}$ given by solving the IVP $y' = -y$; $y(0) = 1$ by Euler's method with $h = 0.1$. Let $y = \phi_1(t)$ be the solution of $y' = -y$ with initial condition $y(0.1) = y_1$. Calculate $e^{-t} - \phi_1(t)$ for $t = 10$ and for $t = 20$.
  Repeat this calculation for the IVP $y' = y$, $y(0) = 1$ and its solution, $y = e^t$.

23. Solve the initial value problem $y' = y + \sin(\pi t)$; $y(0) = 0$, and approximate the solution by calculating y_1, y_2, y_3, and y_4 by Euler's method, using time step $h = 0.25$. Determine all of the local and accumulated errors.

24. Given that the general solution of the ODE $y' = 1 + y^2$ is $y = \tan(t - C)$, where C is an arbitrary constant, calculate the local and accumulated errors in approximating the solution of the IVP with initial condition $y(0) = 0$ with time step $h = 0.5$ for y_1, y_2, and y_3. Remember to use radians, and draw a graph showing the solution $\phi(t) = \tan(t)$ as well as the functions $\phi_1(t)$ and $\phi_2(t)$.

Exercises 25–27 demonstrate that Euler's method can give an answer that is drastically wrong if too large a time step is used. A modification of Euler's method that avoids this problem is suggested.

25. The solution $y = e^{-100t}$ of the initial value problem

$$y' = -100y; \quad y(0) = 1 \qquad (2.31)$$

converges to 0 very rapidly. This is apparent even if we restrict our attention to $0 \le t \le 1$, since $e^{-100 \times 1}$ is of the order of magnitude of 10^{-43}.

a. Use Euler's method with $h = 0.1$ to approximate the solution of the initial value problem (2.31) for $0 \le t \le 1$. Does your computed solution appear to converge to 0?

b. How small should h be to ensure that the solution converges to 0?

26. The backward Euler method. One approach that avoids erratic behavior observed in the solution computed in Exercise 25 is to use *backward* differences instead of forward differences with Euler's Method. Thus, we use the difference equation[2]

$$\nabla y_{m+1} = hf(t_{m+1}, y_{m+1}) \ (m \ge 0),$$

where $\nabla y_{m+1} = y_{m+1} - y_m$, as a model for the differential equation $y' = f(t, y)$. This difference scheme is said to be *implicit* because it defines y_{m+1} implicitly in terms of y_m by means of the relation

$$y_{m+1} - y_m - hf(t_{m+1}, y_{m+1}) = 0.$$

Test the backward Euler method by approximating the solution of the initial value problem (2.31), again with $h = 0.1$, for $0 \le t \le 1$.

27. This problem uses the forward and backward versions of Euler's method with time step $h = 0.1$ to approximate the solution of $y' = \sqrt{1 - y^2}$ with $y(0) = 0$ for $0 \le t \le 2$.

a. Explain why the forward version fails.

b. The difference equation for the backward version is

$$\nabla y_{m+1} = 0.1\sqrt{1 - y_{m+1}^2}.$$

Square both sides and use the quadratic formula to derive the following recursive equation:

$$y_{m+1} = \frac{y_m \pm 0.1\sqrt{1.01 - y_m^2}}{1.01}.$$

Decide which of the signs is correct, and which is extraneous.

[2]∇ is pronounced "nabla."

c. Calculate y_m for $1 \le m \le 20$.

d. The solution of the IVP is

$$y = \begin{cases} \sin(t) & \text{if } 0 \le t \le \pi/2 \\ 1 & \text{if } t > \pi/2. \end{cases}$$

Determine the errors in both the forward and backward versions. Which is the more accurate?

In Exercises 28–31, a separable equation and a corresponding direction field are given. Find a family of solutions and note any singular solutions. Sketch several solutions (including singular solutions) on the direction field.

28. $y' = \sqrt{ty}$

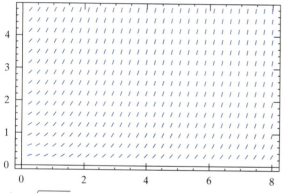

29. $y' = \sqrt{9 - y^2}$

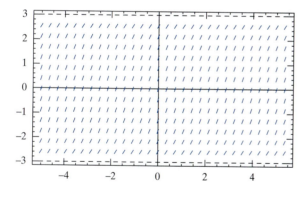

30. $y' = \cos(t)\tan(y)$

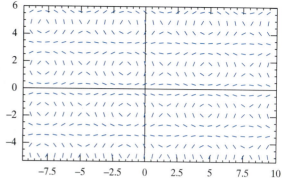

31. $(2 - t)\,dy = (1 - y)\,dt$

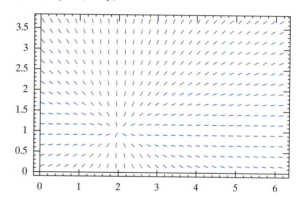

2.4   **Initial Value Problems**

We have learned how to solve IVPs involving linear first-order ODEs, and how to approximate solutions of IVPs involving nonlinear first-order ODEs. All of this has been done without explaining why we should expect an IVP to have one and only one solution. Before addressing this question, we need to say precisely what a solution of an IVP is.

Let us consider an IVP involving a first-order ODE with an initial condition:

$$y' = f(t, y); \tag{2.32}$$
$$y(t_0) = y_0. \tag{2.33}$$

The point in the t, y-plane with coordinates (t_0, y_0) is called the **initial point**. For the IVP (2.32), (2.33) to be correctly posed, (t_0, y_0) must be in the domain of $f(t, y)$.

A function $\phi(t)$ is a solution of the IVP (2.32), (2.33) if

- $y = \phi(t)$ satisfies the ODE (2.32),
- the domain on which $\phi(t)$ is defined is an open interval that contains t_0, and
- $\phi(t_0) = y_0$.

If we can find the general solution of the ODE (2.32), we can solve the IVP by substituting the initial condition (2.33) to determine the value of the constant. Since the domain of the solution of an IVP must be an interval, it may be necessary to specify the domain of the solution.

Example 2.4.1   Solve the IVP

$$\frac{dy}{dt} = -\frac{y}{t+1}; \quad y(0) = 1. \tag{2.34}$$

Solution. The ODE is linear and homogeneous, and so its general solution is of the form $y = Ce^{K(t)}$. Substituting this and $y' = CK'(t)e^{K(t)}$ yields

$$CK'(t)e^{K(t)} = -\frac{Ce^{K(t)}}{t+1},$$

so that $K'(t) = -1/(t+1)$. Hence we will put $K(t) = -\ln(t+1)$ and $e^{K(t)} = 1/(t+1)$. The general solution of the ODE is

$$y = \frac{C}{t+1}.$$

Substituting $t = 0$ and $y = 1$ yields $C = 1$. The solution of the IVP is

$$y = \frac{1}{t+1},$$

defined on the largest interval containing $t = 0$ and *not* containing $t = -1$: $(-1, \infty)$.   ❑

The graph of the solution of Example 2.4.1 is the solid curve in Figure 2.9. In the figure, the portion of the graph shown as a dashed curve is not part of the solution, because the solution must be defined on an interval. Intuitively speaking, it should be possible to trace the graph of a solution of an IVP without lifting one's pencil from the paper.

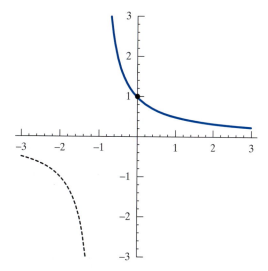

FIGURE 2.9   The solid curve is the solution of the IVP 2.34.

Existence and uniqueness theorems

Is there some property of a two-variable function $f(t, y)$ that we can easily verify and that guarantees the IVP (2.32), (2.33) has a solution? This question is answered by the following theorem, which is one of the foundations of our subject. Briefly, it says that the property we are looking for is continuity of f (as a function of two variables). Like most theorems, it must be stated carefully.

Theorem 2.3   Existence theorem

Let $f(t, y)$ be a function that is continuous at every point of some rectangle

$$\mathcal{D} = \{(t, y) : a < t < b, c < y < d\},$$

and let (t_0, y_0) be a point[3] in $\mathcal{D}$.

Then there is a function $\phi(t)$, defined on an open interval (h, k) that contains t_0, such that $\phi(t_0) = y_0$ and for all $t \in (h, k)$,

$$\phi'(t) = f(t, \phi(t)).$$

The existence theorem does not tell us how to find a solution of the IVP. Furthermore, it says nothing about the extent of the interval on which the solution is defined. The following example shows that even if $f(t, y)$ is continuous on the entire t, y-plane, the solution may only exist on an extremely short interval.

---

[3]To ensure that (t_0, y_0) is not on an edge of the domain where $f(t, y)$ is continuous, our rectangle is *open*: $\mathcal{D}$ contains the interior points, but not the edges.

Example 2.4.2   Show that

$$y = \frac{100}{1 - 100t^2} \tag{2.35}$$

is a solution of the IVP $y' = 2ty^2$; $y(0) = 100$. What is the domain of this solution?

Solution. Setting $t = 0$ in $100/(1 - 100t^2)$ yields $y(0) = 100$. This is the initial condition. Differentiating both sides of (2.35),

$$\begin{aligned} \frac{dy}{dt} &= \frac{-100(-200\,t)}{(1 - 100t^2)^2} \\ &= \frac{2t\,(100)^2}{(1 - 100t^2)^2} \\ &= 2t\,y^2. \end{aligned}$$

Therefore, (2.35) satisfies the ODE. This solution is undefined at $t = \pm\frac{1}{10}$, and the domain is the largest interval containing $t = 0$ but not $t = \pm\frac{1}{10}$: $\left(-\frac{1}{10}, \frac{1}{10}\right)$.   ❑

Figure 2.10 displays several graphs of solutions of the ODE that was the focus of Example 2.4.2.

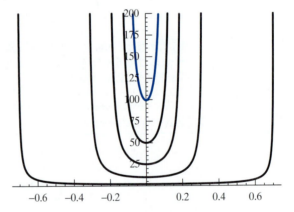

FIGURE 2.10   Solutions of $y' = ty^2$.

There are many proofs of the existence theorem, the most direct being the proof given by the Italian mathematician Giuseppe Peano in 1890. This proof shows that a solution of the IVP can be obtained as a limit of approximations given by Euler's method as the time step $h \to 0$. The details of the proof require a good dose of advanced calculus and are therefore omitted.

Uniqueness of solutions

It is easy to verify that $y \equiv 0$ and $y = t^3$ are solutions of the IVP

$$y' = 3y^{2/3}; \quad y(0) = 0$$

(see Figure 2.11). If $y = t^3$, then $y' = 3t^2$ and $y^{2/3} = t^2$; hence the first solution is valid. If $y \equiv 0$, then the ODE is satisfied because both sides are identically 0.

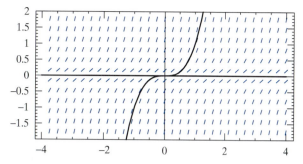

FIGURE 2.11 Two solutions of the IVP $y' = 3y^{2/3}$; $y(0) = 0$.

IVPs often serve as mathematical models for physical, biological, social, or engineering phenomena. Predictions will not be reliable unless we know that the IVP doesn't have multiple solutions. Fortunately, if the function $f(t, y)$ is nice enough, there will be only one solution. In rough terms, a nice function should not increase or decrease too rapidly. Requiring a function to be continuous rules out functions that have sudden jumps (discontinuities), but not functions that increase infinitely fast as the function $f(y) = 3y^{2/3}$ does at $y = 0$: $f'(0) = 2(0^{-1/3}) = \infty$. We will show that multiple solutions are impossible if the function $f(t, y)$ satisfies the following condition, known as a **Lipschitz condition**:

Lipschitz condition

A function $f(t, y)$ satisfies a Lipschitz condition with respect to the variable y in a rectangle $\mathcal{D}$ in the plane if there is a constant K (called a *Lipschitz constant*) such that for any (t, y_1) and $(t, y_2) \in \mathcal{D}$,

$$|f(t, y_2) - f(t, y_1)| \leq K |y_2 - y_1|.$$

Example 2.4.3   Show that $f(t, y) = 3\, y^{2/3}$ does not satisfy a Lipschitz condition on any domain that intersects the t-axis.

Solution. If we take $y_1 = 0$, then

$$|f(t, y_2) - f(t, y_1)| = 3\, y_2^{2/3},$$

while $K|y_2 - y_1| = K|y_2|$. If the Lipschitz condition holds, then for all y_2

$$3\, y_2^{2/3} \leq K\, |y_2|.$$

However, this can only hold for $|y_2| > (3/K)^3$. No matter how large we choose K to be, there will be values of y_2 close to 0 for which the inequality does not hold. ❑

The following theorem provides an easy way to check if a function $f(t, y)$ satisfies a Lipschitz condition.

Theorem 2.4

Suppose that the a function $f(t, y)$ is defined of a rectangle $\mathcal{D}$ in the t, y-plane. Then f satisfies a Lipschitz condition with respect to y on $\mathcal{D}$ if $\frac{\partial f}{\partial y}$ is continuous at each point (t, y) in $\mathcal{D}$ and at each point on the line segments that form its boundary.

Proof Since $\frac{\partial f}{\partial y}$ is continuous on $\mathcal{D}$ and on its boundary, there is a finite maximum value of $\left| \frac{\partial f}{\partial y} \right|$ in $\mathcal{D}$. Let K denote this maximum.

Suppose that (t_0, y_1) and (t_0, y_2) are points in $\mathcal{D}$, and for convenience, assume $y_1 < y_2$. Let $g(y) = f(t_0, y)$; then g is differentiable on $[y_1, y_2]$—in fact,

$$g'(y) = \frac{\partial f}{\partial y}(t_0, y).$$

Thus, by the mean value theorem, there is a number $c \in (y_1, y_2)$ such that

$$g(y_2) - g(y_1) = g'(c)(y_2 - y_1).$$

Let's take absolute values of both sides, and put this equation in terms of f:

$$|f(t_0, y_2) - f(t_0, y_1)| = \left| \frac{\partial f}{\partial y}(t_0, c) \right| |y_2 - y_1|.$$

Since $\left| \frac{\partial f}{\partial y}(t_0, c) \right| \leq K$, it follows that $|f(t_0, y_2) - f(t_0, y_1)| \leq K|y_2 - y_1|$.   ∎

There are functions $f(t, y)$ that satisfy Lipschitz conditions even though they are not differentiable. See Exercise 22 at the end of this section for an example.

Theorem 2.5   Uniqueness theorem

Suppose that $y = \phi_1(t)$ and $y = \phi_2(t)$ are two solutions of the IVP

$$y' = f(t, y); \ y(t_0) = y_0.$$

If the function $f(t, y)$ is continuous and satisfies a Lipschitz condition with respect to y on a rectangle

$$\mathcal{D} = \{(t, y) : a < t < b, c < y < d\}$$

that contains the initial point (t_0, y_0), then there is an open interval (h, k) with $t_0 \in (h, k)$, such that $\phi_1(t) = \phi_2(t)$ for all $t \in (h, k)$.

Suppose that $y = \phi_1(t)$ and $y = \phi_2(t)$ are solutions of an ODE, $y' = f(t, y)$, that satisfy initial conditions $\phi_1(t_0) = y_1$ and $\phi_2(t_0) = y_2$, respectively. If $y_1 = y_2$, the uniqueness theorem says that $\phi_1(t) \equiv \phi_2(t)$, provided that $f(t, y)$ is continuous and satisfies a Lipschitz condition with respect to y. The more general problem is to estimate how fast the graphs of $\phi_1(t)$ and $\phi_2(t)$ can diverge from one another if

$y_1 \neq y_2$. This problem is of practical importance, since initial conditions are often derived from measurements that can be subject to error. The true initial value might be y_1 and the measured value might be y_2. It is necessary to estimate how far the true solution can diverge from the one based on the measurement.

In the case of a linear ODE,

$$y' - k(t)y = q(t),$$

we can be precise, because we can compute $d(t) = \phi_1(t) - \phi_2(t)$. Since

$$
\begin{aligned}
d'(t) &= \phi_1'(t) - \phi_2'(t) \\
&= (k(t)\phi_1(t) + q(t)) - (k(t)\phi_2(t) + q(t)) \\
&= k(t)d(t),
\end{aligned}
$$

$d(t)$ satisfies the homogeneous linear ODE $y' - k(t)y = 0$. It follows that $d(t) = C\,e^{K(t)}$, where C is a constant and $K(t)$ is an antiderivative of $k(t)$. Since $d(t_0) = y_1 - y_2$, we can calculate that $C = (y_1 - y_2)e^{-K(t_0)}$ and hence

$$d(t) = (y_1 - y_2)\,e^{K(t) - K(t_0)}.$$

This proves the uniqueness theorem in the linear case, because clearly if $y_1 = y_2$, $d(t) \equiv 0$. When $y_1 \neq y_2$ we can also draw interesting conclusions. For example, suppose it is known that $K(t) \to \infty$ as $t \to \infty$. Then solutions with unequal initial conditions will diverge from one another. On the other hand, if $K(t) \to -\infty$ as $t \to \infty$, solutions will approach each other and become indistinguishable as $t \to \infty$.

Proof of the uniqueness theorem[4]

We will prove the uniqueness theorem by examining how fast solutions of IVPs with different initial conditions can diverge from each other. We will thus be extending the proof of the uniqueness theorem for linear ODEs that was just given. To put this program into effect, we need two lemmas. The first allows us to replace an IVP with an **integral equation**.

Lemma 2.4.1

The function $\phi(t)$ satisfies the IVP (2.32), (2.33) if and only if

$$\phi(t) = y_0 + \int_{t_0}^{t} f(u, \phi(u))\,du. \tag{2.36}$$

Proof By the fundamental theorem of calculus, for any differentiable function ϕ,

$$\phi(t) = \phi(t_0) + \int_{t_0}^{t} \phi'(u)\,du.$$

If $\phi(t)$ satisfies the IVP (2.32), (2.33), then we can replace $\phi(t_0)$ with y_0 and $\phi'(u)$ with $f(u, \phi(u))$. This shows that $\phi(t)$ satisfies equation (2.36).

If $\phi(t)$ is a solution of equation (2.36), then $\phi(t_0) = y_0$ because when $t = t_0$ the integral vanishes. Furthermore, the rule for differentiating integrals shows that $\phi'(t) = f(t, \phi(t))$; that is, $\phi(t)$ satisfies the ODE (2.32). ∎

---

[4]This material can be skipped if necessary.

The second piece of the puzzle is the following lemma, due to a Swedish-American mathematician, Thomas Gronwall.

Lemma 2.4.2   Gronwall's inequality

Suppose $g(t)$ is a function that is continuous on an interval $[a, b)$ and that $g(t) \geq 0$ for all $t \in (a, b)$. Suppose also that there are constants $M \geq 0$ and $C \geq 0$ such that the inequality

$$g(t) \leq C + M \int_a^t g(u)\, du \tag{2.37}$$

holds for all $t \in [a, b)$. Then $g(t) \leq C\, e^{M\,(t-a)}$ for all $t \in (a, b)$.

Proof Put

$$F(t) = e^{-Mt} \int_a^t g(u)\, du.$$

By the Leibniz rule for differentiating integrals,

$$F'(t) = -M\, e^{-Mt} \int_a^t g(u)\, du + e^{-Mt} g(t)$$

$$= e^{-Mt} \left(g(t) - M \int_a^t g(u)\, du \right).$$

Referring to inequality (2.37), it follows that $F'(t) \leq C\, e^{-Mt}$ for all $t \in [a, b)$. Since $F(a) = 0$,

$$F(t) = \int_a^t F'(u)\, du$$

$$\leq \int_a^t C e^{-Mu}\, du$$

$$= -\frac{C}{M}(e^{-Mt} - e^{-Ma})$$

$$= e^{-Mt} \frac{C}{M}(e^{M(t-a)} - 1).$$

Hence

$$\int_a^t g(u)\, du = e^{Mt} F(t)$$

$$\leq \frac{C}{M}(e^{M(t-a)} - 1).$$

Referring to inequality (2.37) again, it follows that

$$g(t) \leq C + M \frac{C}{M}(e^{M(t-a)} - 1) = C\, e^{M(t-a)}. \qquad \blacksquare$$

The uniqueness theorem now follows from the following proposition.

Proposition 2.4.3

Let $f(t, y)$ be continuous and satisfy a Lipschitz condition with respect to y on a rectangular domain

$$\mathcal{D} = \{(t, y) : a < t < b, c < y < d\}.$$

Let $\phi_1(t)$ and $\phi_2(t)$ be solutions of the ODE (2.32) that are defined on an interval $(p, q) \subset (a, b)$, and let $t_0 \in (p, q)$. Put $y_1 = \phi_1(t_0)$ and $y_2 = \phi_2(t_0)$. If $c < y_1, y_2 < d$, then there is an interval (h, k) such that for all $t \in (h, k)$,

$$|\phi_1(t) - \phi_2(t)| \leq |y_1 - y_2| e^{K|t - t_0|},$$

where K is a Lipschitz constant for f on $\mathcal{D}$.

Proof Since ϕ_1 and ϕ_2 are continuous and $c < \phi_1(t_0), \phi_2(t_0) < d$, there is an interval (h, k) such that for all $t \in (h, k)$ the inequalities $c < \phi_1(t), \phi_2(t) < d$ hold. By Lemma 2.4.1,

$$\phi_i(t) = y_i + \int_{t_0}^{t} f(u, \phi_i(u)) \, du$$

for $i = 1, 2$. By subtracting we find

$$\phi_1(t) - \phi_2(t) = y_1 - y_2 + \int_{t_0}^{t} [f(u, \phi_1(u)) - f(u, \phi_2(u))] \, du.$$

Assume $t > t_0$. By the triangle inequality,

$$|\phi_1(t) - \phi_2(t)| \leq |y_1 - y_2| + \int_{t_0}^{t} |f(u, \phi_1(u)) - f(u, \phi_2(u))| \, du.$$

Since $(u, \phi_i(u)) \in \mathcal{D}$ for $u \in [t_0, t]$, the Lipschitz condition allows us to substitute $K|\phi_1(u) - \phi_2(u)|$ for $|f(u, \phi_1(u)) - f(u, \phi_2(u))|$ in the integral to get

$$|\phi_1(t) - \phi_2(t)| \leq |y_1 - y_2| + K \int_{t_0}^{t} |\phi_1(u) - \phi_2(u)| \, du.$$

By Gronwall's inequality (Lemma 2.4.2) it follows that for $t_0 \leq t < h$,

$$|\phi_1(t) - \phi_2(t)| \leq |y_1 - y_2| e^{K(t - t_0)}.$$

For $t < t_0$ there is a simple trick that saves us from having to repeat this proof. Substitute $\tau = -t$ in the ODE $y' = f(t, y)$ to obtain

$$\frac{dy}{d\tau} = \frac{dy}{dt} \cdot \frac{dt}{d\tau} = f(t, y) \cdot (-1) = -f(-\tau, y),$$

with solutions $\phi_i(t) = \phi_i(-\tau)$. Since $g(\tau, y) = -f(-\tau, y)$ satisfies a Lipschitz condition with respect to y on

$$\mathcal{D}' = \{(\tau, y) - b < \tau < -a, c < y < d\}$$

(the Lipschitz constant is K), the preceding proof goes through to show that for $-t_0 \leq \tau < -h$,

$$|\phi_1(-\tau) - \phi_2(-\tau)| \leq |y_1 - y_2| e^{K(\tau - (-t_0))}.$$

Substituting $t = -\tau$, this yields

$$|\phi_1(t) - \phi_2(t)| \leq |y_1 - y_2| e^{K(-t + t_0)}$$

for $h < t \leq t_0$.   ■

To complete the proof of the uniqueness theorem we only have to apply Proposition 2.4.3 with $y_1 = y_2$ since the ϕ_i satisfy the same initial condition. The conclusion is that $|\phi_1(t) - \phi_2(t)| = 0$ for all $t \in (h, k)$.

Proposition 2.4.3 also proves that the solution of an IVP depends continuously on the initial conditions.

Theorem 2.6

Suppose that $f(t, y)$ is continuous and satisfies a Lipschitz condition with respect to y on a rectangular domain

$$\mathcal{D} = \{(t, y) : a < t < b, c < y < d\}.$$

Let $t_0 \in (a, b)$, and denote by $\phi(t, v)$ the solution of the IVP

$$y' = f(t, y); \quad y(t_0) = v.$$

Given an interval (C, D) with $c < C < D < d$, there is an interval (h, k) containing t_0 such that $\phi(t, v)$ is defined and continuous on the rectangle

$$\mathcal{D}' = \{(t, v) : h < t < k, C < y < D\} \subset \mathcal{D}.$$

While the proof of Theorem 2.6 will not be given in its entirety, it is worth noting where Proposition 2.4.3 fits in. To show $\phi(t, v)$ is continuous, we need to show that the difference $|\phi(t, v) - \phi(s, w)|$ can be made arbitrarily small by choosing (s, w) sufficiently close to (t, v). By the triangle inequality,

$$|\phi(t, v) - \phi(s, w)| \leq P + Q$$

where $P = |\phi(t, v) - \phi(t, w)|$ and $Q = |\phi(t, w) - \phi(s, w)|$. We can make Q small by choosing s close enough to t because $\phi(t, w)$, as a solution of an IVP [with initial condition $y(t_0) = w$] is continuous. Proposition 2.4.3 is used to show that P can be made arbitrarily small by choosing w close enough to v.

EXERCISES

In Exercises 1–8, solve the IVP, and find the largest interval on which the solution is defined.

1. $y' = 0$; $y(1) = -2$

2. $y' - 5y = 25t$; $y(0) = 10$

3. $y' + 2y = \sin(5t)$; $y(0) = 0$

4. $ty' + y = e^{t-1}$; $y(1) = 1$

5. $ty' - 2y = t^3$; $y(1) = 0$

6. $\sqrt{t}y' - y = -t$; $y(1) = 2$

7. $y' - 2ty = t$; $y(0) = \frac{1}{2}$

8. $y' - \tan(t)y = \sec^3 t$; $y(\frac{\pi}{6}) = 5$

9. Suppose that $f(t, y)$ is continuous and satisfies a Lipschitz condition with respect to y on a rectangular domain $\mathcal{D}$. Show that if $y = \phi_1(t)$ and $y = \phi_2(t)$ are solutions of the ODE $y' = f(t, y)$, then their graphs do not intersect each other in $\mathcal{D}$, unless they are identical in $\mathcal{D}$.

◆ **Hint:** Use the uniqueness theorem.

10. Solve the the ODE $\cos(t)y' + y = 1$, with initial condition $y(0) = 1$.

11. Let $f(t)$ be a function that is continuous on $(-\infty, \infty)$, with $f(t) \neq 0$ for all t. Solve the IVP $f(t)y' + y = 1$; $y(0) = 1$.

◆ **Hint:** The IVP in Exercise 10 is of this type.

In Exercises 12–21, find the set of initial points (t_0, y_0) for which the ODE, with initial value $y(t_0) = y_0$ satisfies the hypotheses of

    (i) the existence theorem
    (ii) the uniqueness theorem

12. $y' = t - \sqrt{y}$

13. $y' = \dfrac{t}{y^2 + 1}$

14. $yy' = t^2 + y^2$

15. $\sin(t)y' + y = 0$

16. $(y')^3 = t + 2y$

17. $y = ty' + \dfrac{1}{y'}$

18. $y = ty' + (y')^2$

19. $y' = \dfrac{\sqrt[3]{y - 1}}{\sqrt{ty}}$

20. $y' = \ln(t^2 + y^2 - 1)\ln(9 - t^2 - y^2)$

21. $y' = \sqrt[3]{t - y}$

22. Determine which of the following functions satisfy a Lipschitz condition with respect to y on the domain

$$\mathcal{D} = \{(t, y) : -1 < t < 1, -1 < y < 1\},$$

and find a Lipschitz constant for those that do.

  a. $f(t, y) = t - y^2$ b. $f(t, y) = 4t\,y$

  c. $f(t, y) = |y|$

  d. $f(t, y) = \lfloor y \rfloor$ ($\lfloor y \rfloor$ is the greatest integer $\leq y$)

  e. $f(t, y) = \lfloor t \rfloor$

23. Show that $f(t, y) = \sqrt{1 - y^2}$ satisfies a Lipschitz condition on the domain

$$\mathcal{D} = \{(t, y) : a < t < b, c < y < d\}$$

if $-1 < c$ and $d < 1$, but not if $c = -1$ or $d = 1$.

24. Use the uniqueness theorem to prove the equal derivatives theorem quoted on page 4.

◆ **Hint:** Consider the IVP satisfied by $f_1(t) - f_2(t)$.

25. *CAS exercise.* Find an initial value y_0 such that the solution of the IVP

$$\cos(x)y' + y = \sin(x) + 2\sin(2x); \quad y(0) = y_0$$

satisfies $y(1) = 1$.

✔**Answer:** $y_0 = -2.001386626$

26. *CAS exercise.* Find an initial value y_0 such that the solution of the IVP

$$y' + 2xy = 1, \quad y(0) = y_0$$

satisfies $y(1) = 0$.

✔**Answer:** $y_0 = -1.462651746$

*2.5   Nonlinear Growth Models

The linear model for population growth presented in Section 1.2 was based on an assumption that the relative growth rate of the population is constant. It leads to the linear homogeneous ODE,

$$y' = k\,y.$$

The linear model does not account for the effect that limited resources may have on the relative growth rate, although it is reasonable to expect the relative growth rate to decrease as resources become scarce. We can include resource limits in our model by setting the relative growth rate,

$$\frac{1}{y}\frac{dy}{dt},$$

equal to a function $f(y)$ that decreases as the population increases. The model thus leads to an ODE,

$$\frac{dy}{dt} = y\,f(y). \tag{2.38}$$

Since $f(y)$ is not a constant function, this ODE is not linear.

Both the linear and the nonlinear models can be generalized by allowing the relative growth rate to depend on time. This is useful if it is necessary to model a population whose reproduction is seasonal. In this section, we will only consider models that are nonlinear and independent of time.

The simplest way to impose resource limitations on a growth model is to stipulate that the environment has a **carrying capacity** M, representing the largest population that it can support. We expect that if for some reason the population exceeds the carrying capacity, it will decrease; and if the population is less than the carrying capacity, it will increase.

In 1838, P. F. Verhulst presented such a model for population growth. In his model, the relative growth rate decreases linearly from a value k when the population is a negligible fraction of the carrying capacity to 0 when the carrying capacity M is reached. The relative growth rate in this model is

$$f(y) = k\left(1 - \frac{y}{M}\right).$$

When we substitute this $f(y)$ in equation (2.38), the following ODE, known as the **logistic equation**, results:

$$y' = k\,y\left(1 - \frac{y}{M}\right). \tag{2.39}$$

The logistic equation has two constant solutions, $y = 0$, and $y = M$. The graphs of these constant solutions divide the t, y-plane into three regions. Below the line $y = 0$ and above the line $y = M$ the right side of equation (2.39) is negative. Therefore, any solution of the logistic equation with initial conditions in one of these two regions will be decreasing. The same reasoning shows that a solution with initial condition between the two constant solutions must be increasing. Figure 2.12 shows one such solution. A nonconstant solution cannot cross either of the lines $y = 0$ or $y = M$, because at the crossing point, the uniqueness theorem for initial value problems would be violated.

The population values $y = 0$ and $y = M$ are called **stationary points** of the logistic equation, because when the initial population has either of these values, it will remain constant (that is, stationary). The population $y = M$ is a **stable**

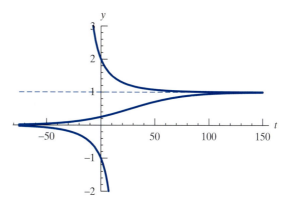

FIGURE 2.12   Solutions of the logistic equation
satisfying the initial conditions $y(0) = 2$, $y(0) = \frac{1}{4}$,
and $y(0) = -1$. The parameters k and M were 0.04
and 1, respectively.

stationary point because every solution whose initial value is sufficiently close to
M (any positive initial value will do) will converge to M as $t \to \infty$. On the other
hand, $y = 0$ is an **unstable** stationary point, because there are solutions with initial
values $y(0)$ arbitrarily close to 0 that diverge away from 0 as $t \to \infty$.

In the following example, we will derive the general solution of the logistic
equation. The solution involves three parameters: M, k, and the initial population;
hence three data points are required to determine the solution. To make the calculation as easy as possible, the data have been taken at evenly spaced time intervals.

Example 2.5.1   Find the general solution of the logistic equation (2.39), and
find values of the parameters to fit the data $y(0) = 1$, $y(1) = 2$, $y(2) = 3$.

Solution. The logistic equation is separable, and in separated form it appears as
follows:

$$\frac{1}{y(M - y)} \, dy = \frac{k}{M} \, dt.$$

Integration by partial fractions yields

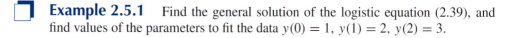

$$\frac{1}{M} \int \left(\frac{1}{y} + \frac{1}{M - y} \right) dy = \frac{1}{M} \ln \left(\frac{y}{M - y} \right) = \frac{k}{M} t + B,$$

where B denotes the integration constant. Thus

$$\frac{y}{M - y} = e^{kt + MB}.$$

This equation can be solved for y to obtain

$$y = \frac{M}{1 + Ae^{-kt}},$$

where $A = e^{-MB}$. To simplify further, set $v = e^{-k}$; then

$$y = \frac{M}{1 + Av^t}.$$

Since $v < 1$, we have $\lim_{t \to \infty} y(t) = M$. This confirms that the population will converge to the carrying capacity as $t \to \infty$. The parameters to be evaluated are M, A, and v. The three data points yield the following equations for these parameters:

$$\frac{M}{1 + A} = 1$$

$$\frac{M}{1 + Av} = 2$$

$$\frac{M}{1 + Av^2} = 3.$$

By the first equation, $M = 1 + A$, which we put in the second equation and solve for A to get $A = (1 - 2v)^{-1}$. Now substitute this and

$$M = 1 + A = 1 + \frac{1}{1 - 2v}$$

in the third equation. After simplifying, the result is a quadratic equation,

$$3v^2 - 4v + 1 = 0,$$

whose solutions are $v = \frac{1}{3}$ and $v = 1$. Since $v < 1$, the second solution is extraneous and $v = \frac{1}{3}$. The remaining parameters are readily evaluated—$A = 3$ and $M = 4$—and the solution is

$$y = \frac{4}{1 + 3^{1-t}}.$$

The graph of this solution is shown in Figure 2.13. ❑

The classic application of the logistic equation to demography is a 1920 study by R. Pearl and L. Reed,[5] who used the censuses of 1790, 1850, and 1910 to determine parameters for the logistic equation. They calculated that the carrying capacity of the United States was 197,274,000 (in fact, the actual population exceeded this value by 1970). Their solution of the logistic equation did not display a significant deviation from the census data of the years 1800–1840 and 1860–1900 and was considered to be a great success. Their predictions continued to be amazingly accurate until 1950. Figure 2.14 shows the Pearl and Reed logistic curve, and the census data covering the years 1790–1990. It is interesting to consider the causes of failure for this model in the years after 1950.

---

[5] Raymond Pearl and Lowell Reed, "On the rate of growth of the population of the United States since 1790 and its mathematical representation," *Proceedings of the National Academy of Sciences, Washington* vol. 6 (1920), pp. 275–288.

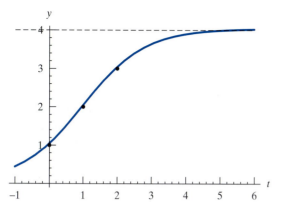

FIGURE 2.13   Solution of the logistic equation with parameters chosen to fit the data given in Example 2.5.1.

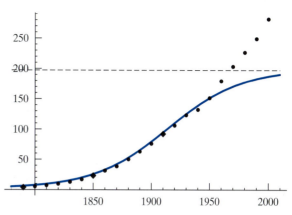

FIGURE 2.14   Population of the United States in millions, censuses of 1790–2000. The logistic curve resulting from Pearl and Reed's study is superimposed, and the data points used in their study are shown with diamond-shaped marks. The dashed line represents the carrying capacity that Pearl and Reed projected.

General nonlinear growth models

The key property of the growth models where the relative growth rate depends on the population is that they are based on an ODE

$$\frac{dy}{dt} = y \, f(y)$$

that is **autonomous**; that is, the independent variable t does not appear on the right side. In general, an autonomous first-order ODE will have the form

$$\frac{dy}{dt} = g(y). \qquad (2.40)$$

Every autonomous ODE is separable, but we should not rush to integrate. Properties of solutions such as whether they are increasing or decreasing functions, the location of any asymptotes, and so on, can be determined without calculating an integral. We will be assuming that

  i. $g'(y)$ is continuous.
 ii.   The zeros of $g(y)$ are *isolated*. This means that if $g(y_j) = 0$, then there is an open interval $(y_j - h, y_j + h)$ that contains no other zeros of g.

The purpose of assumption (i) is to enable us to use the existence and uniqueness Theorems [recall that if $g'(y)$ is continuous, then $g(y)$ is also continuous and satisfies a Lipschitz condition].

The zeros y_i of $g(y)$ are the stationary points of the ODE (2.40). Each stationary point represents a constant solution $y \equiv y_i$. It follows from assumption (ii) that the stationary points separate the line into a collection of open intervals of the form $(-\infty, y_1)$, (y_i, y_{i+1}), or (y_n, ∞), and on each of these intervals, g is nonzero, with no sign changes. If $g(y)$ has no zeros, there is just one interval, $(-\infty, \infty)$. We will call an interval on which $g(y)$ is positive an **up interval**, and an interval where $g(y)$ is negative is a **down interval**. It will be seen that the range of any solution of the ODE (2.40) will be one of these intervals. If the range is an up interval, the solution will be increasing—because in this case $y' = g(y)$ is positive—and if the range is a down interval, then the solution is decreasing.

The stationary points of the logistic equation, $g(y) = k\,y\left(1 - \frac{y}{M}\right)$, are $y_1 = 0$ and $y_2 = M$. There is one up interval, $(0, M)$, and two down intervals, $(-\infty, 0)$ and (M, ∞). Figure 2.12 displays typical solutions whose ranges are these intervals: Note that the solution $y = \phi(t)$ with $\phi(0) \in (0, M)$ is strictly increasing, with $\phi(t) \to M$ as $t \to \infty$ and $\phi(t) \to 0$ as $t \to -\infty$. If $\phi(0)$ lies in one of the two down intervals, then $\phi(t)$ is strictly decreasing, but its graph has only one asymptote.

Proposition 2.5.1

Let $y = \phi(t)$ be a solution of an autonomous ODE of the form (2.40). Then $\phi(t)$ is either a strictly increasing function, a strictly decreasing function, or a constant function.

Proof If $\phi(t)$ is not strictly increasing or decreasing, then it must have a relative minimum or maximum. Thus there is a number t_0 such that $\phi'(t_0) = 0$.

Set $y_0 = \phi(t_0)$. Since $g(y_0) = \phi'(t_0) = 0$, y_0 is a stationary point. It follows that $y \equiv y_0$ and $\phi(t)$ are solutions of the same initial value problem, $y'(t) = g(y)$; $y(t_0) = y_0$. By the uniqueness theorem, $\phi(t) \equiv y_0$. We have thus shown that if ϕ is neither an increasing function nor a decreasing function, then it must be constant. ■

To emphasize that Proposition 2.5.1 is only applicable to autonomous equations whose right sides $g(y)$ are nice, let us consider the following example.

Example 2.5.2   Explain why the following two examples are consistent with Proposition 2.5.1.

i. The function

$$y(t) = \begin{cases} t^2 & \text{for } t > 0, \\ 0 & \text{for } t \le 0 \end{cases}$$

is a solution of the ODE $y' = 2\sqrt{y}$, and it is not strictly monotone.

ii. The function $y(t) = e^{t^2}$ is a solution of the ODE $y' = 2ty$, and it is not monotone.

Solution.

i. $g(y) = 2\sqrt{y}$ is not differentiable at $y = 0$.
ii. The ODE $y' = 2ty$ is not autonomous.   ❑

Our next observation will be that if $\phi(t)$ is a nonconstant solution of equation (2.40), then the range of $\phi(t)$ will be an entire up interval (if ϕ is increasing) or an entire down interval (if ϕ is decreasing). We reason first that the range cannot overlap into two such intervals. If it did so, then the graph of $\phi(t)$ would cross the graph of a constant solution corresponding to a stationary point that separates the two intervals—a violation of the uniqueness theorem. To see that the range of ϕ cannot be a *proper* subset of an up or down interval, we will use the following proposition.

Proposition 2.5.2

Let $y(t)$ denote a solution of equation (2.40). If

$$\lim_{t \to \infty} y(t) = L,$$

where L is finite, then L is a stationary point of equation (2.40); and if $N = \lim_{t \to -\infty} y(t)$ is finite, then N is a stationary point.

Proof If L is finite,

$$\lim_{t \to \infty} y'(t) = \lim_{t \to \infty} g(y(t))$$
$$= \lim_{y \to L} g(y)$$
$$= g(L).$$

Thus, if $g(L) \neq 0$, then $y'(t)$ would have a nonzero limit as $t \to \infty$. For example, suppose $g(L) = c > 0$. Then for some number m, $y'(t) > c/2$ for all $t > m$. Integrating, we would have $y(t) > ct/2 + d$ (where d is an integration constant) for $t > m$. This contradicts the fact that $y(t)$ converges to a finite number L as $t \to \infty$. We can reach the same sort of contradiction if we assume that $g(L) < 0$; hence $g(L) = 0$.

The proof that N is a stationary point if it is finite is the same. ∎

Propositions 2.5.1 and 2.5.2 are the key ingredients of the proof of the following theorem.

Theorem 2.7

Let $y = \phi(t)$ be a solution of the autonomous equation $y' = g(y)$, where $g(y)$ has isolated zeros and $g'(y)$ is continuous everywhere. Then either ϕ is constant, or ϕ is a strictly increasing function whose range is an entire up interval, or ϕ is strictly decreasing, with range an entire down interval.

With the aid of Theorem 2.7, we can sketch graphs of solutions of an autonomous ODE like equation (2.40) without integrating. We start by drawing on the y-axis a **phase diagram** for the ODE, as follows. Locate the stationary points y_i by solving the equation $g(y) = 0$. These will be the endpoints of the up and down intervals. Mark the stationary points on the y-axis and draw horizontal lines to show the constant solutions. Test the sign of $g(y)$ by evaluating g at a convenient point in each interval (y_i, y_{i+1}) on the y-axis to see if it is an up or down interval. Finally, mark each up interval with an arrow directed upward, and each down interval with a downward arrow.

When the phase diagram and stationary solutions have been drawn, it is easy to sketch graphs of the nonconstant solutions. All solutions with initial points in a given up or down interval will be strictly increasing or decreasing, with ranges covering the entire interval. Stationary points at the ends of the up and down intervals will be horizontal asymptotes of the solution.

Example 2.5.3   Draw the phase diagram for the ODE

$$y' = 4y - y^3, \tag{2.41}$$

and make a sketch of the graphs of the solutions of the four initial value problems with $y(0) = \pm 1, \pm 3$.

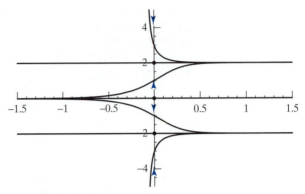

FIGURE 2.15   Solutions of the ODE $y' = 4y - y^3$. The phase diagram is superimposed on the y-axis.

Solution. The stationary points are determined by solving $4y - y^3 = 0$ to obtain $y_1 = -2$, $y_2 = 0$, and $y_3 = 2$. The up and down intervals are therefore $(-\infty, -2)$, $(-2, 0)$, $(0, 2)$, and $(2, \infty)$. To determine the direction of $(-\infty, -2)$, evaluate $g(y)$ at an arbitrarily chosen point in the interval: $f(-3) = 15$. Since we have found a positive value, this is an up interval and is to be marked accordingly. The other three intervals are marked as follows. Since $f(-1) < 0$, $(-2, 0)$ is a down interval; $f(1) > 0$, so $(0, 2)$ is an up interval; and, $f(3) < 0$, indicating that $(2, \infty)$ is a down interval. The phase diagram is imposed on the y-axis of Figure 2.15.

The stationary solutions are indicated by the horizontal lines in Figure 2.15. Each nonconstant solution is asymptotic to the constant solutions corresponding to the boundary of its range. Hence the solution of the initial value problem with $y(0) = 1$ is asymptotic to the t-axis as $t \to -\infty$ and to the line $y = 2$ as $t \to \infty$. In addition, the solution is strictly increasing and has a slope of 3 at its crossing of the y-axis. With this information, it is not difficult to make a sketch of the solution.

Similarly, if $y(0) = -1$, then $y(t)$ is decreasing, asymptotic to the t-axis as $t \to -\infty$, and to $y = -2$ as $t \to \infty$. If $y(0) = 3$, $y(t)$ is again decreasing, and asymptotic to $y = 2$ as $t \to \infty$; if $y(0) = -3$, then $y(t)$ is increasing and asymptotic to $y = -2$.   ❑

Stability

Consider an isolated stationary point y_1 of an autonomous ODE $y' = f(y)$. Since y_1 is isolated, there is an interval (a, b) containing y_1 and no other stationary points. We will say that y_1 is a **stable stationary point** if all solutions $\phi(t)$ with initial values in (a, b) converge to y_1 as $t \to \infty$. If there are solutions with initial values in (a, b) that do not converge to y_1, then we would say that y_1 is an **unstable stationary point**.

Figure 2.16 displays three phase diagrams in which the points y_i are stationary points of autonomous ODEs. The intervals immediately above and below the stationary point y_1 are both directed toward it, so that stationary point is stable. In fact, the interval I consisting of the point y_1, the up interval below it, and the down interval above it has the required property that every solution with initial point in I converges to y_1 as $t \to \infty$. If one or both of the intervals adjacent to a stationary point y_i are directed away from y_i, then y_i is unstable. In these cases, if we only know that the initial point y_0 is near to an unstable stationary point y_i, but do not know whether $y_0 < y_1$ or $y_0 > y_1$, it is impossible to predict the limit of the function $y(t)$.

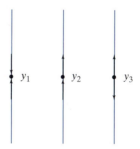

FIGURE 2.16   Stationary points: On the left, y_1 is stable. In the middle and right diagrams, y_2 and y_3 are unstable.

The following proposition offers an alternative way to determine the stability of a stationary point.

Proposition 2.5.3

Let y_1 be a stationary point of the autonomous ODE $y' = g(y)$ such that $g'(y_1) \neq 0$. Then y_1 is stable if $g'(y_1) < 0$ and is unstable if $g'(y_1) > 0$.

The proof of Proposition 2.5.3 is left as an exercise. (See Exercise 25 at the end of this section.) In the following example, we will use the proposition to determine the stability of each stationary point of an ODE, and then use the stability information to draw a phase diagram.

Example 2.5.4   Determine the stability status of each stationary point of

$$y' = \sin(y) - \frac{2}{\pi}y,$$

and draw the phase diagram.

Solution. Figure 2.17 shows the graphs of $u = \sin(v)$ and $u = 2v/\pi$. They cross at $v = 0$, and $v = \pm\pi/2$; therefore, the set of stationary points is $\{-\pi/2, 0, \pi/2\}$. Since $f'(0) = \cos(0) - 2/\pi$ is positive, 0 is unstable. $f'(\pm\pi/2) = \cos(\pi/2) - 2/\pi < 0$, so the other two stationary points are stable.

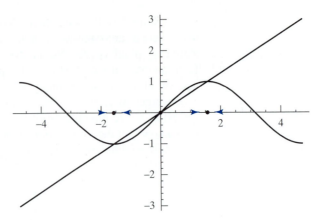

FIGURE 2.17   Graphs of $u = \sin(v)$ and $u = 2v/\pi$.

It follows that the intervals $(0, \pi/2)$ and $(-\pi/2, 0)$ are directed away from the stationary point at 0: $(0, \pi/2)$ is up and $(-\pi/2, 0)$ is down. Since the equilibria at $\pm\pi/2$ are stable, adjacent intervals are directed toward these points. Therefore, $(-\infty, -\pi/2)$ is an up interval and $(\pi/2, \infty)$ is a down interval. The phase diagram is drawn on the y-axis of Figure 2.17 (that is the horizontal axis in this case). When sketching graphs of solutions of $y' = \sin(t) - 2y/\pi$, the y-axis would be vertical, and the phase diagram in Figure 2.17 would be rotated counterclockwise 90° before sketching.   ❑

Proposition 2.5.3 does not apply to any stationary point y_1 where $f'(y_1) = 0$. Exercise 26 shows that in this case, it is possible for a stationary point to be either stable or unstable.

Dependence on parameters

The population P of codfish in a certain marine fishery is modeled by a modified logistic equation,

$$P' = kP(1 - P/M) - H. \tag{2.42}$$

The growth parameter k and the carrying capacity M are taken from the logistic equation, while H is the rate at which fish are harvested. We are interested in how the fate of the fish population depends on the parameter H.

Let $f(P) = -kP^2/M + kP - H$ be the quadratic expression on the right side of equation (2.42). Figure 2.18 shows graphs of $f(P)$ representing four different harvest rates.

Solving $f(P) = 0$ with the quadratic formula, we find that the stationary points are

$$P_1, P_2 = \frac{1}{2}M\left[1 \pm \sqrt{1 - \frac{4H}{kC}}\right], \tag{2.43}$$

provided that these are real numbers. If there are no real stationary points, then $f(P)$ is negative for all P, as in the graph corresponding to $H = 12000$ in Figure 2.18. Thus $(-\infty, \infty)$ is a down interval, and the codfish will be extinct when

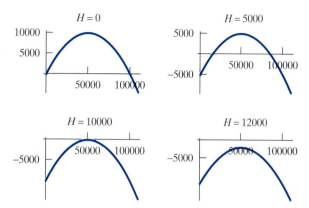

FIGURE 2.18   Population growth rates, as functions of the population, for various rates of harvest. In each graph, the growth parameter has been set at $k = 0.4$ per individual organism per year, and the carrying capacity at $M = 10^5$.

the population reaches 0. If there are two distinct stationary points, as in the graphs with $H = 0$ and $H = 5000$ of the figure, P_1 will be located between 0 and $\frac{1}{2}M$, and P_2 will be between $\frac{1}{2}M$ and M. The interval (P_1, P_2) is an up interval, and $(-\infty, P_1)$ and (P_2, ∞) are down intervals. Hence P_2 is stable, and will represent the limiting population.

We can identify a critical situation when P_1 and P_2 merge together as one stationary point, as in the graph corresponding to $H = 10000$ of Figure 2.18. The intervals above and below are both down intervals, and the stationary point is unstable, with extinction looming. The quadratic equation $f(P) = 0$ has a double root when

$$1 - \frac{4H}{kC} = 0;$$

that is, when $H = \frac{k}{4}M$. This is the critical harvest rate.

Equation (2.42) is actually an infinite family of ODEs depending on a parameter H. The solutions undergo a fundamental change, called a **bifurcation** at a critical value H_0 of the parameter. It is best to describe this situation with a two-dimensional diagram called a **bifurcation diagram**, as shown in Figure 2.19. The horizontal axis represents the value of the H, which is called the *bifurcation parameter* in this context. The vertical axis represents the population. The stationary points are considered as functions $P_1(H)$ and $P_2(H)$ and plotted on the diagram. By convention, stable stationary points are depicted by solid curves; unstable stationary points by dashed curves. Vertical arrows indicate whether the population is increasing or decreasing.

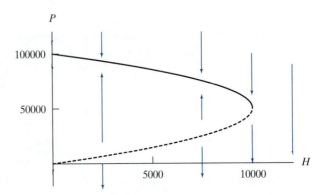

FIGURE 2.19 Bifurcation for the logistic equation with harvesting. The harvesting rate is the bifurcation parameter; the growth parameter and the carrying capacity are the same as in Figure 2.18.

EXERCISES

1. A lake can support a population of 1000 fish. There are now 600 fish in the lake, and on the same date last year there were 300. Assuming that the logistic model determines the fish population in the lake, how many fish will be in the lake a year from now?

   ✔ **Answer:** 840

2. Suppose that the lake in Exercise 1 has been stocked, so that now it contains 1200 fish. How many will be in the lake a year from now?

3. Solve the initial value problem

$$y' = .02y(200 - y); \quad y(0) = 10.$$

4. The populations of two communities are governed by the logistic equation,

$$u' = k u \left(1 - \frac{u}{M}\right).$$

   The second community has a carrying capacity twice as large as the first, and the growth constant k for the each community is the same. Show that if the initial population of the second community is twice the initial population of the first, then the second community will always have twice the population of the first.

Exercises 5–8 are based on the following data concerning the population of the United States.

Year	Population (millions)
1790	4
1880	50
1970	203

5. Find a solution of the logistic equation to fit the data.

6. What is the predicted carrying capacity, according to the logistic model?

7. When will the population reach a level of 210 million?

   ✔ **Answer:** In mid-1975

8. Use the logistic model to predict the population in 2050.

9. A rumor that a top psychic has predicted the sky is falling is spreading in Gossipville. Each citizen of that town calls ten others per day at random to chat; if one of the parties has heard the rumor and the other hasn't, the rumor spreads. The population of Gossipville is 10,000, and at this point 1000 have already heard the rumor. How long will it be before 90% of the people in Gossipville have heard the rumor?

   ◆ **Hint:** Let $N(t)$ be the number of people who have been warned about the sky. Show that N can be approximated by a continuous variable which satisfies the logistic equation. Of course, it is necessary to figure out the values of M and k.

   ✔ **Answer:** about $10\frac{1}{2}$ hours, assuming they chatter 24 hours per day

10. The growth of a certain population is limited by two resources: food and water. The food resources will support a population of M, and there is enough water for a population of $2M$. A researcher postulates the following nonlinear growth model, which accounts for both limitations:

$$y' = k(M - y)(2M - y)y.$$

Draw a phase diagram and show that the population has three stationary values but that only one is stable. Do all solutions converge to the same stable population? What happens if the population initially exceeds $2M$? Does this make sense?

Exercises 11–14 refer to an alternative to the logistic model for population growth, the Gompertz model. It is based on the ODE

$$y' = ky \ln\left(\frac{M}{y}\right),$$

where M is the carrying capacity and k is constant.

11. Show that the Gompertz model has, like the logistic model, two stationary populations, $y \equiv 0$ and $y \equiv M$, where $y \equiv 0$ is unstable, and that $y \equiv M$ is stable.

12. Show that the substitution $v = \ln(y)$ produces a linear ODE with v as the dependent variable.

13. Given the data $y(0) = 1$, $y(1) = 2$, and $y(2) = 3$, determine the carrying capacity according to the Gompertz model. It is interesting to compare this result with that of Example 2.5.1.

    ✔**Answer:** 5.31261

14. Use the following data to find the carrying capacity of the United States, according to the Gompertz model (these are the data used by Pearl and Reed in the study cited previously).

    ✔**Answer:** 18.35 billion

Year	Population (millions)
1790	4
1850	23
1910	92

A chemical reaction proceeds at a rate determined by the concentration of the reactants and catalysts. Specifically, if $y(t)$ denotes the concentration of one of the products of the reaction, and $u(t)$, $v(t)$, and so on. denote the concentrations of the reactants and catalysts, then

$$\frac{dy}{dt} = k[u(t)]^n[v(t)]^m \cdots,$$

where the exponents n, m, etc. are called the orders *of the reaction with respect to the corresponding reactant or catalyst. The number k is called the rate constant.*

15. In an *autocatalytic reaction,* the product acts as a catalyst to further the reaction. Assume that an autocatalytic reaction has order 1 with respect to the reactant and the catalyst, and that for each molecule of the reactant that is consumed, one molecule of the catalyst is produced. If the reaction is allowed to proceed in a closed system, show that its rate is governed by the logistic equation.

16. The reaction of gases

$$2NOBr \rightarrow 2NO + Br_2$$

is second order with respect to the concentration of NOBr. In an experiment, a 1-liter flask initially contains 2.7×10^{22} molecules of NOBr. After 2 minutes, 10^{22} molecules of the reactant are left. Determine the rate constant for this reaction. When was exactly half of the NOBr that was originally present consumed?

    ✔**Answer:** 1.2 minutes

17. Which of the following equations are autonomous?
    a. $y' = 1 - y$ b. $y' = \sin(t) + y^2$
    c. $y' = \begin{cases} y & \text{for } 0 < t < 1 \\ 0, & \text{otherwise} \end{cases}$
    d. $y' = \begin{cases} y & \text{for } 0 < y < 1 \\ 0, & \text{otherwise} \end{cases}$

Find the stationary points of the ODEs in Exercises 18–22 and draw their phase diagrams.

18. $y' = 4 + 3y - y^2$ 19. $y' = y^2$
20. $y' = \sin(y) - \frac{1}{2}$ 21. $y' = e^y$
22. $y' = y - \cos(y)$

23. Which of the following functions are solutions of first-order autonomous ODEs? In the case of those that are solutions, where would the stationary points be located?
    a. $y = 1 + t$. Domain, $(-\infty, \infty)$.
    b. $y = 1 + t^2$. Domain, $(-\infty, \infty)$.
    c. $y = e^{-2t}$. Domain, $(-\infty, \infty)$.
    d. $y = \tan(t)$. Domain, $(-\frac{\pi}{2}, \frac{\pi}{2})$.
    e. $y = \sin(t)$. Domain, $(-\infty, \infty)$.

24. The set of stationary points for an ODE $y' = f(y)$ is $\{0, 1, 2, 3, 4\}$. Suppose that $f'(0) = 2$, $f'(1) = -5$, $f'(2) = 0$, $f'(3) = -1$, and $f'(4) = 12$. Determine the stability of each stationary point, and draw the phase diagram.

25. Prove Proposition 2.5.3.
    ◆ **Hint:** Show that the hypothesis implies $g(y)$ changes sign at y_1.

26. An stationary point y_1 of the ODE $y' = g(y)$ is said to be *degenerate* if $f'(y_1) = 0$. For example, every point is a degenerate stationary point of the equation $y' = 0$. Give an example of a ODE with a degenerate, isolated, stable stationary point, and an example of a ODE with a degenerate, isolated, unstable stationary point.

27. Show that a ODE whose phase diagram is identical to the phase diagram for $y' = y^2$ must have a degenerate stationary point at 0.

28. Use Proposition 2.5.3 to determine the stability of 0 as an stationary point of

$$y' = \sin(y) - ky,$$

in the case where the parameter $k > 1$. Also determine the stability when $k < 1$. What can be said when $k = 1$? Draw phase diagrams for $k = 1$, for k slightly less than 1, and for k slightly greater than 1.

29. For the growth equation with harvesting,

$$P' = P(1 - P) - H.$$

a. What is the critical harvest rate?

b. Suppose $H = 0.1$, and $P(0) = 1$. What will be the limiting population?

c. Draw the bifurcation diagram.

30. The following ODE depends on a parameter m:

$$y' = y^2 - m.$$

a. For which values of m are there stationary points?

b. Draw the bifurcation diagram.

31. **Fitting a logistic curve to data.** In Example 2.5.1 we found that we could find a logistic curve passing through three points in the plane. What if there are more than three points? In this case, we cannot expect the logistic curve to pass through all of the points, but we can try to find parameters that will give the best logistic curve for the data. The following method works well in cases where the logistic equation is an appropriate model for population growth, such as the U.S. population, 1790–1950.

Write the logistic equation $P' = kP(1 - P/M)$ in the form

$$\frac{1}{P}\frac{dP}{dt} = mP + b,$$

where $m = -k/M$ and $b = k$. If we know the population at times $t_0 - h$, t_0, and $t_0 + h$, then we can approximate the relative growth rate as follows:

$$\left.\frac{1}{P}\frac{dP}{dt}\right|_{t=t_0} \approx \frac{P(t_0 + h) - P(t_0 - h)}{2hP(t_0)}. \qquad (2.44)$$

a. Let $P(t)$ denote the U.S. population (see Table 2.1) in year t. Tabulate, for $t = 1800, 1810, 1820, \ldots, 1940$, $P(t)$ and $(P(t + 10) - P(t - 10))/(20P(t))$.

b. Using a statistical calculator or a spreadsheet,[6] find the least squares line that best fits the data, with the population as the independent variable and the approximate relative growth rate, as calculated by equation (2.44), as the dependent variable.

c. The result of part (b) will be the slope m and the intercept b of the least squares line. Use these to determine the carrying capacity.

d. Use a CAS (or a pencil and paper) to find the solution of

$$P' = P(mP + b)$$

with initial condition $P(1900) = 76.21$ million. Plot a graph of the solution curve extending from the years 1790 to 2000. Include the census data, as was done in Figure 2.14.

Table 2.1 U.S. population in millions, censuses of 1790–2000.

t	$P(t)$	t	$P(t)$	t	$P(t)$
1790	3.9	1870	38.6	1950	151.3
1800	5.3	1880	50.2	1960	179.3
1810	7.2	1890	63.0	1970	203.3
1820	9.6	1900	76.2	1980	225.6
1830	12.9	1910	92.2	1990	248.7
1840	17.1	1920	106.0	2000	281.4
1850	23.2	1930	123.2		
1860	31.4	1940	132.2		

2.6 Glossary

$\approx$ Approximately equal to.

Δy_m The sequence of forward differences of a sequence y_m.

Accumulated error   The portion of the approximation error in propagating a solution an IVP by numerical means that is attributable to all previous steps. At each step, the error is equal to the sum of the accumulated error and the local error for that step.

[6]Statistical calculators and spreadsheets have least squares functions that you may use.

Ambient temperature   The temperature of the surrounding environment.

Autonomous ODE   An ODE of the form $y' = f(y)$, where the independent variable does not appear on the right side.

Backward Euler method   A modification of the Euler method for solving an IVP, $y' = f(t, y)$; $y(t_0) = y_0$, that uses backward rather than forward differences. Given y_m, solve the equation

$$y_{m+1} - y_m = h f(t_{m+1}, y_{m+1}),$$

for y_{m+1}. As in the forward Euler method, h denotes the time step and $t_m = t_0 + m h$.

Bifurcation   A change in stability of constant solutions of an autonomous ODE $y' = f(y, \mu)$ resulting from a change in the value of the parameter μ.

Bifurcation diagram   A diagram in the y, μ-plane (y-axis vertical) showing the level set $f(y, \mu) = 0$, where $y' = f(y, \mu)$ is an autonomous ODE depending on a parameter μ. The diagram is marked to indicate the stability of the stationary points (y, μ).

Carrying capacity   The maximum population that can be supported by the environment.

Cell   A position in a spreadsheet where an entry can be made.

Cooling, Newton's law of   The rate of change of the temperature T of an object is proportional to the difference $A - T$, where A is the ambient temperature.

Difference equation   An equation of the form $\Delta y_m = f(m, y_m)$. Given an initial condition $y_0 = A$, a difference equation determines a unique sequence y_m.

Direction field   A diagram for illustrating a ODE $y' = f(t, y)$ on graph paper. At each grid point (t_i, y_j) a short line segment with slope $m_{ij} = f(t_i, y_j)$ is shown.

Down interval   An interval, containing no stationary points, but bounded by stationary points of an autonomous ODE $y' = f(y)$, on which $f(y) < 0$. Any solution with initial value in an down interval will be a decreasing function; its range will be the entire down interval.

Drag coefficient   The ratio of the magnitude of the drag force on a moving object to the square of its velocity.

Drag force   The force required for an object moving through air, water, or other fluid to clear its path.

Element   A line segment appearing in a direction field.

Elementary functions   Functions that can be obtained from the algebraic, exponential, and trigonometric functions by combining them through the operations of addition, multiplication, division, composition, and inversion.

Euler method   A method for propagating the solution of an IVP, $y' = f(t, y)$, $y(t_0) = y_0$, by replacing the ODE with a difference equation

$$\Delta y_m = h f(t_m, y_m),$$

where h denotes the time step and $t_m = t_0 + m h$.

Exact equation   An ODE $P(x, y) dx + Q(x, y) dy = 0$, such that there is a function $F(x, y)$ whose total differential $dF = P(x, y) dx + Q(x, y) dy$. The function $F(x, y)$ is then an integral of the ODE. It is unusual for an equation to be exact; see the entry for *Exactness condition*.

Exactness condition   An ODE, $P(x, y) dx + Q(x, y) dy = 0$, where $P(x, y)$ and $Q(x, y)$ have continuous partial derivatives on a rectangular domain $\mathcal{D}$, is exact if and only if for all $(x, y) \in \mathcal{D}$ the equality $\frac{\partial P}{\partial y} = \frac{\partial Q}{\partial x}$ holds.

Existence theorem   Gives conditions that guarantee that an IVP has a solution. It is stated as follows. If $f(t, y)$ is continuous as a function of two variables on a rectangular domain $\mathcal{D}$ containing the initial point (t_0, y_0), then the IVP $y' = f(t, y)$; $y(t_0) = y_0$ has a solution $y = \phi(t)$ defined for t in some open interval (a, b) containing t_0.

Forward differences   Given any sequence $y_0, y_1, y_2 \ldots$ the sequence of forward differences is

$$(y_1 - y_0), (y_2 - y_1), (y_3 - y_2), \ldots .$$

Friction constant   In the linear model for friction, the resistive force of friction is negatively proportional to the velocity, and the constant of proportionality is $-b$. The number b is the friction constant.

Grid points   The points on graph paper where grid lines intersect.

Gronwall's inequality   An inequality that is used to estimate how fast solutions of a given ODE diverge from one another. It is stated on page 74.

Initial point   The point (t_0, y_0) in the t, y-plane corresponding to the initial condition of an IVP.

Integral   Given an ODE, $P(x, y) dx + Q(x, y) dy = 0$, an integral is a function $F(x, y)$ with the property that for every solution $y = \phi(x)$ of the ODE, $F(x, \phi(x))$ is constant.

Integral equation   An equation involving an unknown function and its integrals. The IVP $y' = f(t, y)$; $y(t_0) = y_0$ can be restated as the following integral equation:

$$y(t) = y_0 + \int_{t_0}^{t} f(u, y(u)) \, du.$$

Integrating factor (for an ODE $P(x, y)\, dx + Q(x, y)\, dy = 0$). A function $m(x, y)$ with the property that the equivalent ODE,

$$m(x, y)\, P(x, y)\, dx + m(x, y)\, Q(x, y)\, dy = 0,$$

is exact.

Level curve A level curve of a function $F(t, y)$ is any curve that is defined by an equation $F(t, y) = C$, where C is some constant.

Linear difference equation A difference equation of the form

$$\Delta y_m = a_m y_m + b_m,$$

where a_m and b_m are given sequences.

Linear model A linear ODE that serves as a mathematical model for some process. The linear model for friction is based on an assumption that the force due to friction is proportional to the velocity.

Lipschitz condition A condition that has an important role in the uniqueness theorem. A function $f(t, y)$ satisfies a Lipschitz condition with respect to y on a rectangular domain $\mathcal{D}$ if there is a constant K such that for all $(t, y_1), (t, y_2) \in \mathcal{D}$,

$$|f(t, y_1) - f(t, y_2)| \le K\, |y_1 - y_2|.$$

Local error The error introduced in carrying the numerical solution of an ODE forward one step.

Logistic equation A nonlinear population growth model that arises from the assumption that the relative growth rate decreases linearly as a function of the population.

Newton The unit of force in the meter-kilogram-second system. A force of 1 newton will cause a 1 kilogram mass to accelerate at the rate of 1 meter per second^2.

Periodic function A function $f(t)$ that satisfies an identity $f(t+T) \equiv f(t)$ for some fixed number T (called the period). For example, the sine and cosine functions are periodic (the period is 2π), and the tangent function is periodic with period π.

Phase diagram The y-axis, marked with the stationary points of an autonomous ODE $y' = f(y)$. All up intervals and down intervals are marked with arrows, directed toward increasing y and decreasing y, respectively.

Polygonal graph A graph that is composed of straight line segments.

Separable ODE An ODE of the form

$$\frac{dy}{dt} = g(t)\, h(y)$$

in which the right side can be factored as a product of a function of the independent variable and a function of the dependent variable.

Singular solution A solution of an ODE that is not a member of a family of solutions that was derived by analytic means. For example, $y \equiv 1$ is a singular solution of $y' = (1 - y^2)^{1/2}$ since it doesn't fit into the family $y = \sin(t - C)$.

Stable stationary point An stationary point y_0 of an autonomous ODE such that all solutions whose initial values are sufficiently close to y_0 converge to y_0 as $t \to \infty$. To recognize a stable stationary point, look at the phase diagram: y_0 is stable if and only if it is *both* the lower endpoint of a down interval *and* the upper endpoint of an up interval, as shown in the margin.

Stable periodic solution A periodic solution $\phi(t)$ of a linear ODE whose general solution can be described as the sum of $\phi(t)$ and a transient.

Stationary point A solution of an autonomous ODE $y' = f(y)$ that is a constant function. The stationary points are the same points as the zeros of $f(y)$.

Terminal velocity The velocity of an object falling in air (or some other fluid) at which the net force, considering gravity and air resistance, is 0.

Time step Euler's method approximates the solution $\phi(t)$ of an IVP as a sequence of numbers y_i, where $y_i \approx \phi(t_i)$. The times $t_0, t_1, t_2, \ldots$ are evenly spaced, and the time step is the difference $h = t_{i+1} - t_i$. The accuracy of the approximation given by Euler's method usually improves, and the computational effort increases, as the time step gets smaller in magnitude.

Total differential $dF = \frac{\partial F}{\partial x}\, dx + \frac{\partial F}{\partial y}\, dy$ is the total differential of $F(x, y)$.

Transient A term in a solution of an ODE that converges rapidly to 0 with increasing time, so that it can be neglected if the purpose is to predict what will happen after a long time.

Transmission coefficient The ratio of $T'(t)$ to $A(t) - T(t)$, where T is the temperature of an object and A is the ambient temperature.

Uniqueness theorem Gives conditions that guarantee that an IVP will have only one solution. It is stated as follows. Suppose $f(t, y)$ satisfies a Lipschitz condition with respect to y in a rectangular domain $\mathcal{D}$ containing the initial point (t_0, y_0). Then if $y = \phi_1(t)$ and $y = \phi_2(t)$ are solutions of the IVP $y' = f(t, y)$; $y(t_0) = y_0$, there is an open interval (a, b) containing t_0 such that $\phi_1(t) = \phi_2(t)$ for all $t \in (a, b)$.

Unstable stationary point   An stationary point y_0 of an autonomous ODE such that some nonconstant solution converges to y_0 as $t \to -\infty$. To recognize an unstable stationary point, look at the phase diagram: y_0 is unstable if and only if *either* it is the lower endpoint of an up interval (as shown on the right) *or* the upper endpoint of an down interval, *or both*.

Up interval   An interval, containing no stationary points, but bounded by stationary points of an autonomous ODE $y' = f(y)$, on which $f(y) > 0$. Any solution with initial value in an up interval will be an increasing function and its range will be the entire up interval.

2.7 Chapter Review

1. Sketch the graphs of the solutions of $y' = t^2 - y^2$, with initial conditions $y(-2) = 0$ and $y(2) = 0$, respectively, on the direction field shown.

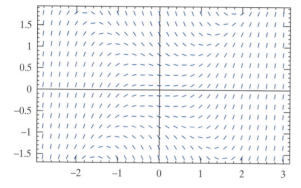

2. Use Euler's method with time step $h = 0.25$ to compute an approximate solution of the IVP, $y' = t^2 - y^2$; $y(0) = 0$. Draw the polygonal graph of the approximate solution on the direction field accompanying Exercise 1.

3. Sketch the graph of the solution of the IVP, $y' = t - \sqrt{y}$; $y(0) = 2$ on the direction field shown.

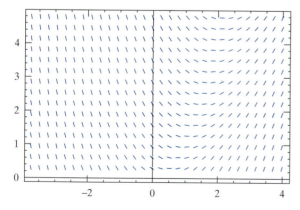

4. Use Euler's method with time step $h = 0.5$ to approximate the solution of the IVP $y' = t - \sqrt{y}$; $y(-2) = 4$. Draw the polygonal graph of this solution on the direction field accompanying Exercise 3.

5. Match the direction fields shown below with the ODEs.

   a. $y' = 0.03(5t - y^2)$ b. $y' = \sin(y)$

   c. $y' = \sin(t + y)$ d. $y' = -0.01y$

   e. $y' = 0.05y(\pi - y)$ f. $y' = 0.02(t^2 + y^2)$

   I

   II

   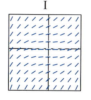

   III

   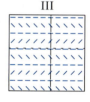

   IV

   V

   VI

   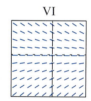

In Exercises 6–13, find a family of solutions and note any singular solutions.

6. $y' = ty^2$ 7. $y' = e^y$

8. $y' = y(1 - y)$ 9. $y' = \sqrt{\dfrac{1 - y^2}{1 - t^2}}$; $-1 < t < 1$

10. $y' = \sqrt{\dfrac{1 - y^2}{1 - t^2}}$; $|t| > 1$ 11. $y' = t(1 + y^2)/(1 + t^2)$

12. $y' = \sqrt{\dfrac{y}{t}}$ 13. $y' = e^{4y} \tan(3t)$

14. An object that is sinking in water is subject to forces due to gravity (downward), and buoyancy and drag (both upward). By the principle of Archimedes, the buoyancy force is equal to the weight of water displaced by the object. Thus, if σ denotes the specific gravity of the object, the buoyancy force is equal to $\frac{mg}{\sigma}$. The drag force is proportional to the square of the velocity. A diamond ($\sigma = 3.5$)

has a terminal velocity of 21 meters per second when sinking in water. If you toss it gently into the middle of Lake Tanganyika (1463 meters deep),

a. How far will the diamond travel in the first 10 seconds?

b. Estimate how long it will take for the diamond to get to the bottom of the lake.

c. Suppose that you throw the diamond vertically into the lake so that its initial velocity after it is immersed in water is 42 meters per second. Answer questions (a) and (b).

15. A certain population is growing according to the logistic equation. In year A, the population is 100,000 and growing at 4500 per year, and in year B the population is 200,000, growing at 8000 per year.

a. Calculate the relative growth rate in each of the years.

b. Find the carrying capacity.

c. If year A is 1980, when was year B?

✔ *Answer:* 1996

In Exercises 16–21, draw phase diagrams for the ODEs, identify all stable equilibrium points, and describe the limiting behavior of the solution of the IVP with increasing time. **Do not solve these ODEs.**

16. $y' = y(2y+1); \; y(0) = -1$

17. $y' = -y \sin^2(y); \; y(0) = 30$

18. $y' = y^2 + 2y + 2; \; y(0) = 0$

19. $y' = 5 - 3y; \; y(0) = -2$

20. $y' = y^{-1} - 1; \; y(0) = 3$

21. $y' = y^{-1} - 1; \; y(0) = -3$

In Exercises 22–24, it will be useful to differentiate both sides of the ODE to obtain an expression for y''. You don't need formulas for the solutions to do these problems!

22. Show that the graph of any solution of $y' = \frac{1}{y}$ is either increasing and concave down, or decreasing and concave up.

23. Imagine that the set of points in the plane satisfying the inequality $t + y + 1 > 0$ is colored red, and the set of points satisfying $t + y + 1 < 0$ is colored blue. Show that if $y(t)$ is a solution of the ODE $y' = t + y$, then the graph of $y(t)$ is concave up at red points, and concave down at blue points.

24. Show that the graph of any solution of $y' = e^y$ is increasing and concave up. What can be said about the graph of a solution of $y' = e^{-y}$?

25. Richardson's extrapolation. Let $\phi(t)$ be the solution of the IVP $y' = t - y^2; \; y(0) = 0$.

a. Using a spreadsheet or other computer version of Euler's method with $h = 0.1, 0.05,$ and 0.025, compute approximate values of $\phi(1)$.

b. Richardson's extrapolation works with the assumption that the accumulated error in each of these approximations is approximately equal to Ch, where C is an unknown constant. If Y_{h_1} and Y_{h_2} are approximations of $\phi(1)$ obtained using steps h_1 and $h_2 = h_1/2$, respectively,

$$\phi(1) \approx Y_{h_1} + Ch_1 \qquad (2.45)$$

$$\phi(1) \approx Y_{h_2} + \frac{1}{2}Ch_1. \qquad (2.46)$$

Eliminate the constant C from (2.45) and (2.46), to find a better approximation of $\phi(1)$. Let Z_{h_1} denote the approximation of $\phi(1)$ thus obtained. This Z_{h_1} is called a first-order extrapolated approximation.

c. Calculate $Z_{0.1}$ and $Z_{0.05}$.

d. Further analysis shows that there is a constant D such that

$$\phi(1) \approx Z_h + Dh^2.$$

Use the values of $Z(0.1)$ and $Z_{0.05}$ just calculated to eliminate the constant D, thus obtaining a second-order extrapolated approximation.

✔ *Answer:* The second-order extrapolation yields $\phi(1) \approx 0.4555383$. This is considerably closer to the correct value (found by solving the initial value problem using a CAS), $\phi(1) = 0.4555445$ than $Y_{0.025} = 0.4477666$.

26. For which values of y_0 does the IVP

$$y' = \sqrt[3]{\frac{1+y+t}{1-y+t}}; \; y(0) = y_0$$

have a solution? A unique solution?

27. The IVP $y' = y \sin(t - y), \; y(0) = 0$ has the solution $y \equiv 0$. How can you be sure that it's the only solution?

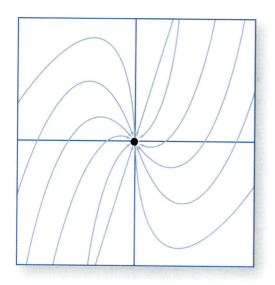

CHAPTER 3

Systems of Differential Equations

3.1 Introduction

A **system** of ODEs is a set of two or more ODEs to be treated simultaneously. Each ODE involves the same independent variable (we will always use t, for elapsed time, as the independent variable), but there are as many dependent variables as there are equations. The systems that we will study have the form

$$\left. \begin{array}{l} x' = f(t, x, y) \\ y' = g(t, x, y) \end{array} \right\}, \tag{3.1}$$

where x' and y' denote the derivatives of x and y with respect to t.

Most applications of ODEs involve systems. We have considered the velocity of a falling object in some detail in the previous chapters. Since the object was assumed to fall in a vertical direction, the velocity could be represented by a single ODE. In a situation where the motion is not in a straight line, it is necessary to represent the velocity as a vector. Let $u(t)$ and $v(t)$ represent the horizontal and vertical components of the velocity, respectively. The drag force is a vector whose direction is opposite to the velocity vector, and we will denote its horizontal and vertical components by $-h(u, v)$ and $-k(u, v)$, respectively. The gravitational force, mg, is vertical. By Newton's second law of motion, u and v must satisfy the system of ODEs,

$$m u' = -h(u, v)$$
$$m v' = mg - k(u, v).$$

In an IVP involving a system, the values of all dependent variables are specified at a starting time t_0. For example, an IVP involving a system of two ODEs would appear as follows:

$$\left. \begin{array}{l} x' = f(t, x, y); \ x(t_0) = x_0 \\ y' = g(t, x, y); \ y(t_0) = y_0 \end{array} \right\}. \tag{3.2}$$

☐ **Example 3.1.1**   Show that $x = \cos(t)$, $y = -\sin(t)$ is a solution of the initial value problem

$$x' = y; \quad x(0) = 1$$
$$y' = -x; \ y(0) = 0.$$

Solution. Substituting $x = \cos(t)$ and $y = -\sin(t)$ in the first equation yields the identity $-\sin(t) = -\sin(t)$. The same substitution in the second equation results in $\cos(t) = \cos(t)$. Finally, $\cos(0) = 1 = x_0$ and $-\sin(0) = 0 = y_0$.   ☐

The system in the preceding example occurs in the following application to mechanics. An object with a mass of 1 unit is suspended from a spring that satisfies Hooke's law. This means that when the spring is stretched x units past its equilibrium position (if the spring is compressed, x is negative), the net force on the object, including the spring and gravity, is $F = -kx$. As usual, the positive direction is downward. By Newton's second law of motion, the product of the mass and the acceleration is equal to the net force. Since the acceleration is equal to $\frac{d^2x}{dt^2}$, it follows that x satisfies the second-order ODE,

$$\frac{d^2x}{dt^2} = -kx. \tag{3.3}$$

Let y denote the velocity of the object, so that

$$\left. \begin{array}{l} x' = y \\ y' = -kx \end{array} \right\}. \tag{3.4}$$

The first equation of the system (3.4) simply repeats that y represents the rate of change of x (i.e., y is the velocity). The second is a rephrasing of equation (3.3), since the second derivative of x with respect to t is the same as the first derivative of y.

The statement that the object was pulled one unit below its equilibrium position and released can be interpreted as initial conditions $x(0) = 1$, $y(0) = 0$. If $k = 1$, the initial value problem of Example 3.1.1 is the same as the system (3.4), with these initial conditions.

More generally, consider a second-order ODE,

$$\frac{d^2x}{dt^2} = f\left(t, x, \frac{dx}{dt}\right). \tag{3.5}$$

Let $y = \frac{dx}{dt}$ denote the rate of change of x with respect to time. Then we can form the system of first-order ODEs,

$$\left. \begin{array}{l} x' = y \\ y' = f(t, x, y) \end{array} \right\}. \tag{3.6}$$

If $x = \phi(t)$, $y = \psi(t)$ is a solution of the system (3.6), then $\phi'(t) = \psi(t)$ (by the first equation). Differentiating and using the second equation, we have

$$\phi''(t) = \psi'(t) = f(t, \phi(t), \phi'(t));$$

in other words, $x = \phi(t)$ is a solution of equation (3.5).

Conversely, suppose that $x = \phi(t)$ is a solution of equation (3.5), and put $\psi(t) = \phi'(t)$. Then $x = \phi(t)$, $y = \psi(t)$ is a solution of the system (3.6). We will say that the system (3.6) **replaces** the ODE (3.5).

Example 3.1.2   Find a system of first-order ODEs that replaces the ODE

$$y'' + t\, y' + \sin(y) = 0.$$

Solution. The ODE can be written as

$$\frac{d^2 y}{dt^2} = -\left(t\frac{dy}{dt} + \sin(y) \right).$$

Set $v = \frac{dy}{dt}$ and $v' = \frac{d^2 y}{dt^2}$. Then the system

$$y' = v$$
$$v' = -(t\, v + \sin(y))$$

replaces the given ODE. ❑

Uncoupled equations

A system of two ODEs is **uncoupled** if one of the differential equations does not involve one of the dependent variables. Thus, an uncoupled system has the form

$$x' = f(t, x)$$
$$y' = g(t, x, y).$$

To find the general solution of this system, we start by solving the first equation. This gives a formula $x = \phi(t, C)$, where C is a constant. Substituting this formula for x in the second equation yields another first-order equation with one dependent variable, $y' = g(t, \phi(t, C), y)$. Thus, solving an uncoupled system of two differential equations is accomplished by solving two single first-order equations.

Example 3.1.3   Find the general solution of the system

$$x' = 2t\, x^2$$
$$y' = 2t\, x\, y.$$

In addition, find the solutions that satisfy the following sets of initial conditions.

a. $x(0) = 1$, $y(0) = 1$
b. $x(0) = 0$, $y(0) = 1$
c. $x(0) = 1$, $y(0) = 0$

Solution. The equation $x' = 2x^2 t$ is separable and can be written

$$\frac{dx}{x^2} = 2t\,dt.$$

Integration of both sides yields $-x^{-1} = t^2 + C$, or

$$x = \frac{-1}{t^2 + C}. \tag{3.7}$$

There is also a singular solution, $x \equiv 0$. We now substitute formula (3.7) for x in the second equation to get

$$y' = 2t \left(\frac{-1}{t^2 + C} \right) y,$$

which is also separable and can be written in the equivalent form

$$\frac{dy}{y} = -\frac{2t\,dt}{t^2 + C}.$$

Integrating again, we have $\ln |y| = -\ln |t^2 + C| + D$, where D is a second constant. Taking the exponential of each side of this equation, we obtain

$$|y| = e^{-\ln |t^2 + C| + D}$$

$$= \frac{e^D}{t^2 + C}.$$

Thus

$$y = \frac{K}{t^2 + C},$$

where $K = \pm e^D$. It is also possible that $K = 0$ (why?).

If we substitute the singular solution $x \equiv 0$ in the second equation [instead of formula (3.7)], then $y' \equiv 0$, so y would be constant.

The solution of the system will thus be one of the following function pairs:

$$\left\{ \begin{array}{l} x = -\dfrac{1}{t^2 + C} \\[2mm] y = \dfrac{K}{t^2 + C} \end{array} \right\} \quad \text{or} \quad \left\{ \begin{array}{l} x = 0 \\ y = K \end{array} \right\},$$

where C and K are constants.

We now solve the initial value problems.

a. Set $t = 0$ and $x = 1$ in $x = \frac{-1}{t^2 + C}$. This yields $1 = -1/C$, so $C = -1$. Now we set $t = 0$, $y = 1$, and $C = -1$ in $y = \frac{K}{t^2 + C}$, to get $1 = K/(-1)$. Hence $K = -1$ and the solution is

$$x = \frac{-1}{t^2 - 1}$$

$$y = \frac{-1}{t^2 - 1}.$$

b. Since $x(0) = 0$, the solution of the first equation is the singular solution $x \equiv 0$. It follows that y is constant and must be equal to its initial value of 1. The solution is

$$x \equiv 0$$
$$y \equiv 1.$$

c. As in part (a), $x = \frac{-1}{t^2-1}$. Since $y(0) = 0$, $y \equiv 0$. The solution is

$$x \equiv \frac{-1}{t^2 - 1}$$
$$y \equiv 0.$$ ❑

Example 3.1.4   Find the general solution of the second-order ODE,

$$\frac{d^2x}{dt^2} = \frac{dx}{dt} + 1.$$

Solution. Put $x' = y$ so that the ODE is replaced by the system

$$\left. \begin{array}{l} x' = y \\ y' = y + 1 \end{array} \right\}.$$ (3.8)

The second equation does not involve the variable x, and thus (3.8) is uncoupled. That equation is linear, and its solution is $y = C e^t - 1$.

Referring to the first equation of (3.8), we have $x' = C e^t - 1$, which can be solved by antidifferentiating. The solution is

$$x = C e^t - t + D,$$
$$y = C e^t - 1,$$

where C and D are constants.   ❑

EXERCISES

1. Show that for all real C, $x = C e^t$, $y = C e^t$ is a solution of the system

$$x' = y$$
$$y' = x,$$

and find all solutions of the form $x = A e^{-t}$, $y = B e^{-t}$, where A and B are constants.

2. Show that $x = C e^{-2t}$, $y = -3C e^{-2t}$ is a solution of

$$x' = x + y$$
$$y' = 3x - y,$$

and find all solutions of the form $x = A e^{2t}$, $y = B e^{2t}$, where A and B are constants.

3. Show that $x = 2C \cos(3t)$, $y = C[\cos(3t) + 3\sin(3t)]$ is a solution of

$$x' = x - 2y$$
$$y' = 5x - y,$$

and find all solutions where $x = 2A \sin(3t)$.

4. Show that $x = \sin(2t) + e^t$, $y = 2\cos(2t) - 4e^t$ is a solution of

$$x' = 5e^t + y$$
$$y' = -4x,$$

and find a family of solutions of this system.

5. Show that $x = e^{2t}(t + 1)$, $y = e^{2t}(t - 1)$ is a solution of

$$x' = x + y + 3e^{2t}$$
$$y' = 2x - 3e^{2t}.$$

Furthermore, show that $x = y = C e^{2t}$ is a family of solutions of the associated homogeneous system

$$x' = x + y$$
$$y' = 2x,$$

and find a family of solutions of the inhomogeneous system.

In Exercises 6–9 find a system of two first-order equations that replaces the given second-order ODE.

6. $\dfrac{d^2 x}{dt^2} = x \dfrac{dx}{dt} + 1.$

7. $y'' + 3y' + 4y = t^2$

8. $z'' + (z^2 - 1) z' + z = 0$

9. $u'' u' u = 1 + t^2$

In Exercises 10–18 determine if the system is uncoupled. If the system is uncoupled, find its general solution.

10. $\begin{cases} x' = 4x + 1 \\ y' = y \end{cases}$

11. $\begin{cases} x' = x^2 \\ y' = 1/x \end{cases}$

12. $\begin{cases} x' = xy \\ y' = -y \end{cases}$

13. $\begin{cases} x' = y^2 \\ y' = -x^2 \end{cases}$

14. $\begin{cases} x' = x - e^t \\ y' = x + y \end{cases}$

15. $\begin{cases} x' = t/x \\ y' = t x y \end{cases}$

16. $\begin{cases} x' = x + y^2 \\ y' = y/2 \end{cases}$

17. $\begin{cases} x' = x + y \\ y' = x - y \end{cases}$

18. $\begin{cases} x' = e^t x \\ y' = e^{2t} x + y \end{cases}$

3.2   The Phase Plane

There are several ways to present a solution $x = \phi(t)$, $y = \psi(t)$ of a system of two ODEs graphically, the simplest of which is to draw separate graphs of $\phi(t)$ and $\psi(t)$. This is useful, but it does not show any relationship between x and y. For example, it is easily verified that $x = \cos(t)$, $y = \sin(t)$ is a solution of the system

$$x' = -y$$
$$y' = x,$$

but it is not obvious from the graphs of the sine and cosine functions that $x^2 + y^2 = 1$.

The x, y-plane, whose coordinates represent the *dependent variables* of the system of ODEs, is called the **phase plane**. Let $x = \phi(t)$, $y = \psi(t)$ be a solution of a system of two differential equations. The curve in the phase plane described by the parametric equations $x = \phi(t)$, $y = \psi(t)$ is called the **orbit** of the solution. It is usually marked to show the direction of increasing t. For example, the orbit of the solution $x = \cos(t)$, $y = \sin(t)$ is the circle $x^2 + y^2 = 1$, marked with an arrow in the counterclockwise direction, since as t increases from 0 to $\frac{\pi}{2}$, $(\cos(t), \sin(t))$ moves along the circle from $(1, 0)$ to $(0, 1)$.

Example 3.2.1   Describe the orbits of the system

$$x' = x$$
$$y' = y.$$

Solution. This system is readily solved because it is uncoupled. The general solution of $x' = x$ is $x = C\,e^t$ and the solution of $y' = y$ is $y = D\,e^t$. (The constants C and D are independent of each other.)

If $C = D = 0$, the orbit is simply the origin itself (no arrow is attached to this "stationary" orbit). If $C = 0$, the orbit is the positive y-axis, directed upward if $D > 0$; or the negative y-axis, directed downward, if $D < 0$.

Assuming $C \neq 0$, we can substitute $e^t = x/C$ in the formula for y to obtain $y = (D/C)\,x$. The orbit consists of all points on this straight line that lie on the same side of the origin as (C, D) and is directed away from the origin. See Figure 3.1.
❑

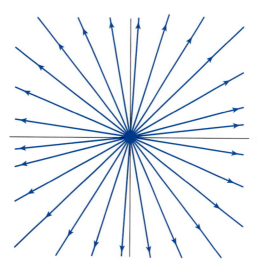

FIGURE 3.1   Orbits of the system $x' = x$, $y' = y$.

Example 3.2.2   Find a system of ODEs that replaces

$$y'' = y' + 2y. \tag{3.9}$$

Given that $y = e^{-t}$, $y = e^{2t}$ and $y = e^{2t} - e^{-t}$ are solutions of equation (3.9), find the corresponding solutions of the system and describe their orbits in the phase plane.

Solution. Define a new variable v as $v = y'$. Then $y = \phi(t)$ is a solution of equation (3.9) if and only if $(y, v) = (\phi(t), \phi'(t))$ is a solution of the system

$$y' = v$$
$$v' = v + 2y.$$

This is the system that replaces equation (3.9). The solution corresponding to $y = e^{-t}$ is $y = e^{-t}$, $v = -e^{-t}$. Notice that $v = -y$ for all t, and that as $t \to \infty$, $(y, v) \to (0, 0)$. Furthermore, y is positive and v is negative, which puts the orbit in the fourth quadrant of the y, v-plane. Therefore, the orbit is the part of the straight line $v = -y$ in the fourth quadrant and is directed toward the origin.

The orbit of the solution corresponding to $y = e^{2t}$ is $(y, v) = (e^{2t}, 2\,e^{2t})$. Here we have $v = 2y$, and as $t \to \infty$, both y and $v \to +\infty$. This orbit is therefore the part of the line $v = 2y$ in the first quadrant, directed away from the origin.

The parametric equation of the orbit of the solution corresponding to $y = e^{2t} - e^{-t}$ is $(y, v) = (e^{2t} - e^{-t}, 2\,e^{2t} + e^{-t})$. This will be left in parametric form, since it is not easy to eliminate t to obtain a relation between v and y. However, we can note that $(y(0), v(0)) = (0, 3)$, and that as $t \to -\infty$, $(y(t), v(t)) \to (-\infty, \infty)$, while $(y(t), v(t)) \to (\infty, \infty)$ as $t \to \infty$. The three orbits are displayed in Figure 3.2. ❑

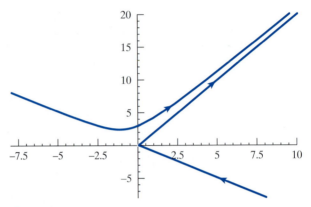

FIGURE 3.2   Three orbits of the system that replaces the ODE $y'' = y' + 2y$. The horizontal axis is the y-axis; the vertical axis represents y'.

A system

$$\left. \begin{array}{l} x' = f(x, y) \\ y' = g(x, y) \end{array} \right\} \tag{3.10}$$

with the property that the independent variable does not appear on the right side of either equation is said to be **autonomous**. An autonomous system describes a **vector field** that assigns to each point (x, y) a vector

$$\mathbf{v}(x, y) = f(x, y)\mathbf{i} + g(x, y)\mathbf{j}.$$

(Here, $\mathbf{i}$ and $\mathbf{j}$ denote the usual horizontal and vertical unit vectors in the plane.) A vector field can be represented graphically by drawing the vectors $\mathbf{v}(x_m, y_n)$ at grid points (x_m, y_n) on graph paper. Such drawings resemble direction fields for ODEs, but there is a distinction: The vectors of the vector field have varying magnitude, displayed by giving them different lengths, and each has a direction and so is drawn as an arrow. The elements of a direction field should all be drawn with the same length, and they have no direction. Figure 3.3 shows the vector field $\mathbf{v}(x, y) = x\mathbf{i} - y\mathbf{j}$ that corresponds to the uncoupled system

$$x' = x$$
$$y' = -y.$$

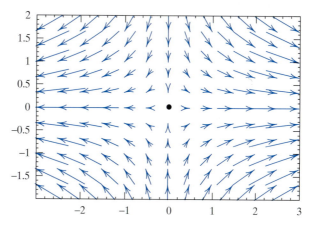

FIGURE 3.3   The vector field $x\mathbf{i} - y\mathbf{j}$.

The orbits are not shown, but you should verify that they consist of the rays of the x-axis, directed away from the origin; the rays of the y-axis, directed toward the origin; and components of the hyperbolas $xy = C$, directed downward in the first and second quadrants, and directed upward in the third and fourth quadrants.

We have seen that an autonomous system of two ODEs can be displayed as a vector field in the plane. It is also useful to interpret a vector field

$$\mathbf{v}(x, y) = f(x, y)\mathbf{i} + g(x, y)\mathbf{j}$$

as a system of ODEs,

$$x' = f(x, y)$$
$$y' = g(x, y).$$

If we think of $\mathbf{v}(x, y)$ as the velocity vector for a particle located at (x, y), then the particle will follow an orbit of this system. In the context of vector fields, it is customary to refer to orbits as integral curves. Thus, an **integral curve** of a vector field $\mathbf{v}(x, y)$ is a curve with parametric equations $(x, y) = (\phi(t), \psi(t))$ such that the tangent vector $\phi'(t)\mathbf{i} + \psi'(t)\mathbf{j}$ is identical to the field vector $\mathbf{v}(x, y)$ at each point of the curve; that is, for all t,

$$\mathbf{v}(\phi(t), \psi(t)) = \phi'(t)\mathbf{i} + \psi'(t)\mathbf{j}.$$

A point (x_1, y_1) where $\mathbf{v}(x_1, y_1) = \mathbf{0}$ is called a **stationary point**. The solution of the system $x' = f(x, y)$, $y' = g(x, y)$ that satisfies the initial condition $(x(0), y(0)) = (x_1, y_1)$ is $x(t) \equiv x_1$, $y(t) \equiv y_1$, since $f(x_1, y_1) = g(x_1, y_1) = 0$. Thus the "integral curve" through (x_1, y_1) isn't a curve at all; it's just the stationary point by itself.

Displaying a solution of a system

In Chapter 4, we will learn how to solve systems such as the one in the following example. The example shows how to check the validity of a solution that was given.

Example 3.2.3   Show that

$$x(t) = e^t \cos(10t)$$
$$y(t) = e^t \sin(10t)$$

satisfies the system

$$x' = x - 10y$$
$$y' = 10x + y. \tag{3.11}$$

Solution. By the product rule for differentiation,

$$x'(t) = e^t \cos(10t) - 10e^t \sin(10t) = x(t) - 10y(t),$$

and

$$y'(t) = e^t \sin(10t) + 10e^t \cos(10t) = y(t) + 10x(t). \qquad \square$$

There are several ways to display graphs of solutions of systems. Our first plot, Figure 3.4, shows the components $x(t)$ and $y(t)$ of the solution of (3.11) found in Example 3.2.3, plotted together.

We see that the solution functions are oscillating, with increasing amplitude. The orbit, displayed with the vector field representation of the system (3.11), is shown in Figure 3.5. The orbit is an outward spiral, which is certainly consistent with the graphs shown in Figure 3.4.

Using a three-dimensional plot, we can display the graph of (x, y) as a function of t. See Figure 3.6, in which the t-axis, for the independent variable, is vertical, and the x, y-plane is the "axis" for the dependent variables.

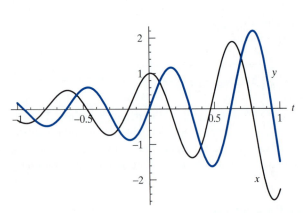

FIGURE 3.4 Component graphs of a solution $(x(t), y(t))$ of the system (3.11).

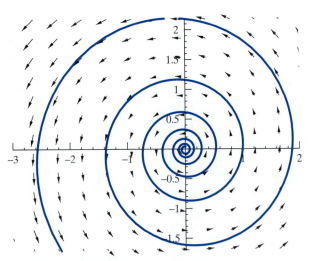

FIGURE 3.5 Orbit of a solution $(x(t), y(t))$ of the system (3.11). The vector field representation of the system of ODEs is also displayed.

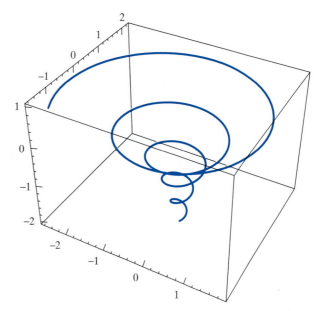

FIGURE 3.6   Graph of a solution $(x(t), y(t))$ of the system (3.11).

EXERCISES

In Exercises 1–8, show that the given pair of functions is a solution of the system of differential equations, and sketch the orbit determined by the solution. If a second-order ODE is given, replace it with a system of two first-order equations and proceed.

1.
$$\begin{cases} x' = 2y^2/t \\ y' = y^2/x \end{cases}$$
$x = t^2, \; y = t$

2. $y'' = y$; draw the orbits corresponding to three solutions:
$y = -e^{-t}, \; y = e^t, \; \text{and} \; y = e^t - e^{-t}.$

3.
$$\begin{cases} x' = y \\ y' = x \end{cases}$$
$x = e^{-t}, \; y = -e^{-t}$

4.
$$\begin{cases} x' = y \\ y' = x \end{cases}$$
$x = \cosh(t), \; y = \sinh(t).$
   ◆ **Hint:** Remember the identity $\cosh^2(t) - \sinh^2(t) = 1$.

5. $y'' = y^{-3}$; solution, $y = \sqrt{t^2 + 1}$

6.
$$\begin{cases} x' = xy \\ y' = x^2 \end{cases}$$
$x = \sec(t), \; y = \tan(t)$

7. $y'' = \frac{1}{2}t \, y' - y$; solution, $y = t^2 - 2$

8.
$$\begin{cases} x' = x - 2y \\ y' = 2x + y \end{cases}$$
$x = e^t \cos(2t), \; y = e^t \sin(2t)$

In Exercises 9–12, find stationary points for the given vector fields. Draw some of the integral curves (the vector fields are drawn to assist you).

9. $\mathbf{v}(x, y) = [4y\mathbf{i} - x\mathbf{j}]/16$

10. $\mathbf{v}(x, y) = [(y^2 - y)\mathbf{i} + x\mathbf{j}]/4$

11. $\mathbf{v}(x, y) = [(x^2 + 1)y\mathbf{i} - 2x\mathbf{j}]/16$

12. $\mathbf{v}(x, y) = [x\mathbf{i} + 2y\mathbf{j}]/8$

13. Use a CAS to plot the graphs of the component functions, the two-dimensional orbits, and the three-dimensional graphs of the solutions of the following systems.

a.
$$\begin{cases} x' = 5y \\ y' = -5x \end{cases}$$
with initial conditions $x(0) = 0$, $y(0) = 1$;
$0 \le t \le 2\pi$

b.
$$\begin{cases} x' = 2x - 6y \\ y' = x - 2y \end{cases}$$
with initial conditions $x(0) = 1$, $y(0) = 0$;
$0 \le t \le 4\pi$

c.
$$\begin{cases} x' = 2y + \sin(t) \\ y' = -2x - \sin(2t) \end{cases}$$
with initial conditions $x(0) = 0$, $y(0) = 0$;
$-2\pi \le t \le 4\pi$

d.
$$\begin{cases} x' = x + y \\ y' = t - x - y \end{cases}$$
with initial conditions $x(0) = 0$, $y(0) = 0$;
$-2 \le t \le 4$

14. Draw graphs of the component functions of the solution

$$(x, y) = (\sinh(t), \cosh(t))$$

of

$$x' = y; \; x(0) = 0$$
$$y' = x; \; y(0) = 1,$$

and plot its orbit.

3.3   A User's Guide to IVP Solvers

An **IVP solver** is a numerical method for computing solutions of IVPs.

We already know of one: Euler's method. Although it was presented in the context of a single ODE, it is easily modified to approximate solutions of IVPs involving systems of ODEs. Recall that Euler's method works by replacing the ODE by an easily solved difference equation. For a system,

$$x' = f(t, x, y); \; x(t_0) = A$$
$$y' = g(t, x, y); \; y(t_0) = B,$$

we use a system of difference equations

$$\Delta x_j = hf(t_i, x_i, y_i); \; x_0 = A$$
$$\Delta y_j = hg(t_i, x_i, y_i); \; y_0 = B,$$

where h is the time step, and t_i is the time after i steps have been taken; that is, $t_i = t_0 + i \cdot h$.

Using this system, we obtain sequences $\{x_i\}$ and $\{y_i\}$. If $x = \phi(t)$, $y = \psi(t)$ denotes the solution of the IVP, then x_i is an approximation of $\phi(t_i)$ and y_i approximates $\psi(t_i)$.

Example 3.3.1   Plot the orbit of the system

$$x' = 2x + 2y$$
$$y' = -x + 4y$$

with initial conditions $x(0) = 1$, $y(0) = 0$, for t ranging from -1 to 1.

Solution. It is possible to solve this system analytically, or to use an IVP solver incorporated in a CAS to produce a numerical solution. We will use a spreadsheet, with a moderately large step size, $h = 0.1$. Our spreadsheet, shown in Figure 3.7, has five columns, representing t, x, y, Δx, and Δy—the latter two are labeled Dx and Dy. The calculations are done in two blocks, the first for $0 \le t \le 1$, and the second for $0 \ge t \ge -1$.

The formulas shown in Figure 3.7 were entered once in each block and then copied. Thus, the entry in cell D2, +0.1*(2*B2+2*C2) was typed, copied onto the clipboard, and then pasted into the cells D3 through D12, and so on. This formula says to multiply the entry in cell B2 (the value of x_0) by 2 and add the result to $2\times$ the entry in cell C2 (the value of y_0) and then multiply this sum by 0.1. The result is $0.1(2x_0 + 2y_0)$, which is equal to Δx_1. The entry in cell B3 adds the entries in cells B2 and D2 to form $x_1 = x_0 + \Delta x_1$.

The spreadsheet in Figure 3.7 shows the formulas instead of numerical values. The numerical values and the orbit of the computed solution are shown in Figure 3.8. The graph was made by selecting xy graph in the graphics type menu, with x values from column B and series values from column C.   ❏

To assess the precision of the approximation just computed, see Figure 3.9.

The strength of Euler's method is its simplicity. As we have seen, the accumulated error varies proportionately with the step size. Since the amount of effort is inversely proportional to the time step, we can expect to use 10 times as much computer time to improve accuracy by 1 decimal place. It is worthwhile to seek a more efficient method.

Corel Quattro Pro - D:\BOOK\SPREADSHEETS\Euler.wb3

File Edit View Insert Format Data Graphics Tools Window Help

	A	B	C	D	E	F	G
1	t	x	y	Dx	Dy		
2	0	1	0	+0.1*(2*B2+2*C2)	+0.1*(-B2+4*C2)		
3	+0.1+A2	+B2+D2	+C2+E2	+0.1*(2*B3+2*C3)	+0.1*(-B3+4*C3)		
4	+0.1+A3	+B3+D3	+C3+E3	+0.1*(2*B4+2*C4)	+0.1*(-B4+4*C4)		
5	+0.1+A4	+B4+D4	+C4+E4	+0.1*(2*B5+2*C5)	+0.1*(-B5+4*C5)		
6	+0.1+A5	+B5+D5	+C5+E5	+0.1*(2*B6+2*C6)	+0.1*(-B6+4*C6)		
7	+0.1+A6	+B6+D6	+C6+E6	+0.1*(2*B7+2*C7)	+0.1*(-B7+4*C7)		
8	+0.1+A7	+B7+D7	+C7+E7	+0.1*(2*B8+2*C8)	+0.1*(-B8+4*C8)		
9	+0.1+A8	+B8+D8	+C8+E8	+0.1*(2*B9+2*C9)	+0.1*(-B9+4*C9)		
10	+0.1+A9	+B9+D9	+C9+E9	+0.1*(2*B10+2*C10)	+0.1*(-B10+4*C10)		
11	+0.1+A10	+B10+D10	+C10+E10	+0.1*(2*B11+2*C11)	+0.1*(-B11+4*C11)		
12	+0.1+A11	+B11+D11	+C11+E11	+0.1*(2*B12+2*C12)	+0.1*(-B12+4*C12)		
13							
14							
15	t	x	y	Dx	Dy		
16	0	1	0	-0.1*(2*B16+2*C16)	-0.1*(-B16+4*C16)		
17	-0.1+A16	+B16+D16	+C16+E16	-0.1*(2*B17+2*C17)	-0.1*(-B17+4*C17)		
18	-0.1+A17	+B17+D17	+C17+E17	-0.1*(2*B18+2*C18)	-0.1*(-B18+4*C18)		
19	-0.1+A18	+B18+D18	+C18+E18	-0.1*(2*B19+2*C19)	-0.1*(-B19+4*C19)		
20	-0.1+A19	+B19+D19	+C19+E19	-0.1*(2*B20+2*C20)	-0.1*(-B20+4*C20)		
21	-0.1+A20	+B20+D20	+C20+E20	-0.1*(2*B21+2*C21)	-0.1*(-B21+4*C21)		
22	-0.1+A21	+B21+D21	+C21+E21	-0.1*(2*B22+2*C22)	-0.1*(-B22+4*C22)		
23	-0.1+A22	+B22+D22	+C22+E22	-0.1*(2*B23+2*C23)	-0.1*(-B23+4*C23)		
24	-0.1+A23	+B23+D23	+C23+E23	-0.1*(2*B24+2*C24)	-0.1*(-B24+4*C24)		
25	-0.1+A24	+B24+D24	+C24+E24	-0.1*(2*B25+2*C25)	-0.1*(-B25+4*C25)		
26	-0.1+A25	+B25+D25	+C25+E25	-0.1*(2*B26+2*C26)	-0.1*(-B26+4*C26)		
27							
28							
29							

FIGURE 3.7 Spreadsheet for approximating the solution of $x' = 2x + 2y$, $y' = -x + 4y$, with initial conditions $(x(0), y(0)) = (1, 0)$, by Euler's method with step size $h = 0.1$.

A **classical IVP solver** is an IVP solver in which the user specifies a time step that will not be altered when the calculation is in progress. The **order** is a number n such that the accumulated error varies proportionately with $|h|^n$, where h is the time step. Euler's method, for example, is a classical method of order 1. Classical solvers can be constructed with arbitrarily high order. Software for two types of classical methods, Runge-Kutta methods and predictor methods, is packaged with CAS systems. The Runge-Kutta methods (the Euler method is one of these) operate by setting

$$\Delta y_m = h \times (\text{a weighted average of values of } f(t, y)), \quad \text{with } t_m \le t \le t_{m+1}.$$

For example, the "improved Euler method" is typical. Given (t_m, y_m), set

$$\bar{y}_{m+1} = y_m + hf(t_m, y_m) \quad \text{and} \quad t_{m+1} = t_m + h.$$

Then set

$$\Delta y_m = \frac{1}{2}(f(t_m, y_m) + f(t_{m+1}, \bar{y}_{m+1})).$$

This method is second order. It requires two function evaluations per step, but the payoff is that the accumulated error is proportional to h^2. Thus it will take about $\sqrt{10}$ times as much computer time to improve accuracy by one decimal place.

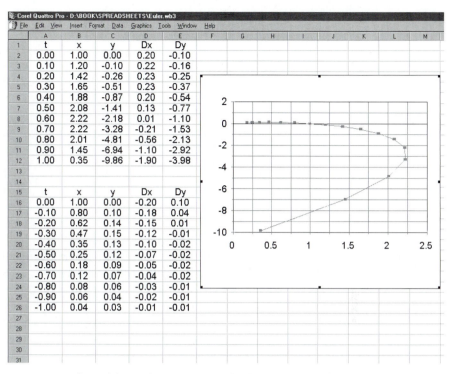

FIGURE 3.8   Spreadsheet calculation and orbit plot for the IVP, $x' = 2x + 2y$, $y' = -x + 4y$; $(x(0), y(0)) = (1, 0)$,. The formulas in the spreadsheet are shown in Figure 3.7.

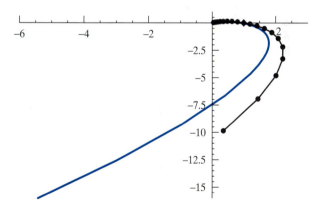

FIGURE 3.9   Orbit of $x' = 2x + 2y$, $y' = -x + 4y$, with initial condition $x(0) = 1$, $y(0) = 0$. The upper curve was drawn from the analytic solution, and the lower one was drawn from the approximation obtained by Euler's method in Example 3.3.1.

The most commonly used classical methods are of order 4. To obtain a fourth-order Runge-Kutta method, it turns out that four function evaluations are required per step. For the "RK4" method, we define

$$k_1 = f(t_m, y_m)$$
$$k_2 = f(t_m + 0.5h, y_m + 0.5k_1 h)$$
$$k_3 = f(t_m + 0.5h, y_m + 0.5k_2 h)$$
$$k_4 = f(t_m + h, y_m + k_3 h),$$

and then

$$\Delta y_m = h \left(\frac{1}{6} k_1 + \frac{1}{3} k_2 + \frac{1}{3} k_3 + \frac{1}{6} k_4 \right).$$

To obtain an additional decimal place of accuracy with the RK4 method, we expect to do $\sqrt[4]{10} \approx 1.8$ times as much work.

The predictor methods work a little differently. The Adams-Bashforth method (also a fourth-order method) calculates Δy_m as follows: Let $p_0 = f(t_m, y_m)$, $p_1 = f(t_{m-1}, y_{m-1})$, $p_2 = f(t_{m-2}, y_{m-2})$, $p_3 = f(t_{m-3}, y_{m-3})$. Then

$$h \times (\text{a weighted average of } p_0 \ldots p_3).$$

To be precise,

$$\Delta y_m = h \left(\frac{55}{24} p_0 - \frac{59}{24} p_1 + \frac{37}{24} p_2 - \frac{9}{24} p_3 \right).$$

The advantage of this method is that it recycles previously computed data and re-quires only one new function evaluation per step. This strength is also a weakness, because the error in the computed value of y_m is recycled and can cause accuracy to degenerate. To control error propagation, the Adams-Bashforth method can be coupled with a correction stage. The resulting method is known as the Adams-Bashforth-Moulton method, and it works as follows. Numerical analysts refer to the sequence of calculations by the acronym PECE.

Predict Use the Adams-Bashforth method to find a tentative value for y_{m+1}:

$$\bar{y}_{m+1} = y_m + h \left(\frac{55}{24} p_0 - \frac{59}{24} p_1 + \frac{37}{24} p_2 - \frac{9}{24} p_3 \right).$$

Evaluate Set $p_{-1} = f(t_{m+1}, \bar{y}_{m+1})$.

Correct Recalculate y_{m+1}, as

$$y_{m+1} = y_m + h \left(\frac{9}{24} p_{-1} + \frac{19}{24} p_0 - \frac{5}{24} p_1 + \frac{1}{24} p_2 \right).$$

Evaluate Calculate $f(t_{m+1}, y_{m+1})$ to prepare for the next step.

The improved Euler method that was just mentioned is a predictor-corrector method as well as a Runge-Kutta method, since $\bar{y}_{m+1}$ is predicted by the Euler method and used in a correction phase to get an improved approximation of the solution at $t = t_{m+1}$.

The classical methods are not used for high-precision work, because it is hard to assess the accuracy of a solution without recomputing the whole thing with smaller time steps and making comparisons. Methods that change the step size as the computation progresses are called **dynamic methods**. The Runge-Kutta-Fehlberg (RKF45) method, supplied with most CAS systems, is widely used. With this method, each value of y_{m+1} is calculated twice in each step, first using the RK4 method, followed by a Runge-Kutta method of order 5. When the two computed values differ by more than a specified tolerance, both are rejected, the time step is reduced, and the process is repeated. There is a provision to stop when a minimum time step is reached: This avoids having the program running forever if, for example, the solution has a vertical asymptote.

Other dynamic methods are designed for either increased precision or to address problems encountered with certain types of higher-order equations.

It is not practical to use a spreadsheet with dynamic IVP solvers. Fortunately, the programs supplied with CAS systems are excellent and easy to use.

It may be convenient to use an IVP solver independently of a CAS, and several such programs are available. Programs that implement classical solvers—the Euler, fourth order Runge-Kutta, and Adams-Bashforth-Moulton methods—as well as the Runge-Kutta-Fehlberg method, which incorporates dynamic step size allocation, can be obtained by following links on the text web site, http://www.prenhall.com/conrad. Typically, these programs have graphical output: either the graph of the solution of an IVP is drawn, or in the case of a system, the orbit of the solution in the phase plane may be drawn. It is up to the user to specify parameters, including the step size, the number of steps to be taken, the initial conditions, and the scale and location of the coordinate axes on the computer screen.

EXERCISES

1. The solution of the IVP

$$x' = -y \quad x(0) = 1$$
$$y' = x \quad y(0) = 0$$

is $(x, y) = (\cos(t), \sin(t))$. Use the RKF45 algorithm with this IVP to estimate $\sin(1)$. Compare the answer you get to the value from your calculator (be sure it is set to radians!).

2. The solution of the IVP

$$x' = -4y \quad x(0) = 1$$
$$y' = x \quad y(0) = 0$$

is $(x, y) = \left(\cos(2t), \frac{1}{2}\sin(2t)\right)$. Thus, the identity $x^2 + 4y^2 = 1$ is satisfied, and the orbit is an ellipse. Use an IVP solver to draw this ellipse. If you use a crude method like Euler's, do you get a closed curve? Repeat the experiment with a classical fourth-order method, and again with a dynamic method.

3. Consider the following system, which depends on a parameter ϵ:

$$x' = -y + \epsilon x(x^2 + y^2 - 1) \quad x(0) = A$$
$$y' = x + \epsilon y(x^2 + y^2 - 1) \quad y(0) = 0.$$

When $\epsilon = 0$ this system is the same as that of Exercise 1, and if $A = 1$ as well, the solution of the IVP is the same as in Exercise 1. This problem investigates what happens when $\epsilon \neq 0$ and $A \neq 1$. Let $r(x, y)$ denote the distance of the point (x, y) to the origin. Differentiation of $r^2 = x^2 + y^2$ with respect to t yields

$$r r' = x x' + y y'.$$

Furthermore, if $\theta = \arg(x + i y)$ is the polar angle,

$$\theta' = \frac{x y' - y x'}{x^2 + y^2}.$$

Show that if $(x(t), y(t))$ is the solution of the IVP, then

$$r' = \epsilon r(r^2 - 1)$$

and draw a phase diagram for this IVP. Furthermore, show that $\theta' = 1$. Thus orbits rotate counterclockwise about the origin, spiraling outward if r is increasing and inward if r is decreasing.

a. Put $\epsilon = -.25$ and $A = .05$. Use the RKF45 algorithm to draw the orbit of the solution for $0 \leq 30$. Explain how the orbit shown is consistent with the aforementioned conclusions.

b. Repeat the experiment with the following data.

  i. $(\epsilon, A) = (-.25, 4)$

  ii. $(\epsilon, A) = (.25, .95)$

  iii. $(\epsilon, A) = (.25, 1.05)$

4. The solution $\phi(t)$ of the IVP

$$y' = t^2 + y^2; \quad y(0) = 0 \qquad (3.12)$$

has two vertical asymptotes, $t = c$ and $t = -c$. Estimate the value of c.

5. Produce an accurate plot of the solution $\phi(t)$ of the IVP (3.12), for $-c < t < c$ ranging from the negative asymptote $t = -c$ to the positive asymptote $t = c$. Change the time step to 0.05, and use Euler's method to generate an approximate solution. Reduce the time step by a factor of 2 repeatedly. Describe precisely the effect that reducing the time step has on the error. Repeat the experiment with a second-order method and the classical RK4 method.

6. Draw the phase diagram for the ODE $y' = 4y - y^3$, and sketch the graphs of four solutions with different asymptotic behavior. Use an IVP solver to graph the four solutions with initial conditions that you should select so that the graphs resemble your sketches.

7. Exercise 6 suggests a way to find real roots of a polynomial $f(y)$ by using an IVP solver. The roots will be horizontal asymptotes of solutions of the differential equation $y' = f(y)$, so by plotting numerous solutions, they can be located.

◆ **Hint:** If you try to solve $y' = f(y)$, where $f(y)$ is a high degree polynomial, your computer will probably complain of overflow problems. In this case, try replacing $f(y)$ with $f(y)/10^N$. Here you would choose N to be a positive integer large enough to eliminate the overflow, but small enough so that the solution you are plotting "levels off" soon.

a. Try the method on a polynomial with known roots: $f(y) = (2y - 1)(2y + 1)^2(2y + 3)^3(2y + 5)$. Can you locate all of the roots (including the multiple ones)?

b. The polynomial

$$f(y) = y^7 - 49y^6 + 882y^5 - 7350y^4 + 29400y^3$$
$$- 52920y^2 + 35280y - 5040$$

has seven real roots in the interval $[0, 20]$. Approximate them.

8. The differential equation

$$y' + .05y = 2\sin(2t) + \sin(3t)$$

has a periodic solution. Use an IVP solver to plot the graph of this solution.

9. Use the computer to determine and graph several solutions of the following differential equations, and note any periodic solutions.

a. $y' + 0.1y = \sin(2t)$

b. $y' - y = \sin(20t)$

c. $y' + 10y = \sin(100t)$

10. Let $S(t) = +1$ if the integer part of t is even (i.e., $t \in [0, 1), [2, 3),$ etc.) and $S(t) = -1$ otherwise. This function is sometimes called a square wave. Graph the periodic solution of

$$y' + y = S(t).$$

*3.4 Autonomous Systems

A **phase portrait** for an autonomous system of ODEs,

$$\left. \begin{array}{l} x' = f(x, y) \\ y' = g(x, y) \end{array} \right\} \qquad (3.13)$$

is a drawing of the phase plane, showing a representative collection of orbits of the system.

Example 3.4.1 Draw a phase portrait of the system

$$x' = 2x$$
$$y' = y.$$

Solution. The system is uncoupled, and each equation in the system is a homogeneous linear ODE. The general solution of the first equation is $y = Ce^{2t}$, and the general solution of the second is $y = De^t$, where C and D are constants. An orbit in the phase plane is therefore described by parametric equations

$$\left. \begin{array}{l} x = Ce^{2t} \\ y = De^t \end{array} \right\}. \tag{3.14}$$

To eliminate t from these equations, note that $y^2 = D^2 e^{2t}$. Therefore, if $D \neq 0$,

$$\frac{x}{y^2} = \frac{C}{D^2} = \text{constant.}$$

Therefore, $x = Ay^2$, where A is a constant. The orbit follows a parabola with a horizontal axis of symmetry, opening to the right if $A > 0$, and to the left if $A < 0$. If $A = 0$, the orbit follows the y-axis. Finally, if $D = 0$, then $y = 0$: The orbit follows the x-axis.

Each parabola or axis actually consists of three orbits. The origin is a stationary point, and this an orbit by itself. The other two orbits follow the parabola or axis, leading away from the origin. Figure 3.10 is a phase portrait of the system (3.14). This figure includes the vector field, but normally the vector field is omitted. ❑

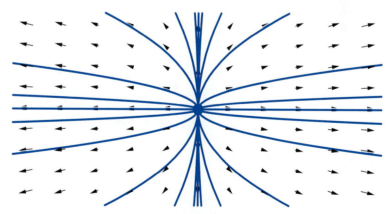

FIGURE 3.10 Vector field and phase portrait for the system of ODEs in Example 3.14.

Example 3.4.1 is unusual because we were able to determine the general solution of the system. Often, we must approach the problem of drawing a phase portrait by relying on qualitative analysis of the system and an IVP solver. The qualitative analysis should be done first, as it will aid us in choosing initial points for the IVP solver to use. Start by identifying the *nullclines*.

A set of points where $f(x, y) = 0$ is called an *x*-**nullcline** of the system (3.13). If $(x, y) = (\phi(t), \psi(t))$ is a solution whose orbit crosses an *x*-nullcline at a point $(x_1, y_1) = (\phi(t_1), \psi(t_1))$, then $\phi'(t_1) = f(x_1, y_1) = 0$. If $\psi'(t_1) = 0$ as well, then (x_1, y_1) is an stationary point and the solution in question is constant. If $\psi'(t_1) = g(x_1, y_1) \neq 0$, the orbit must have a vertical direction at (x_1, y_1), which will be upward if $g(x_1, y_1) > 0$ and downward if $g(x_1, y_1) < 0$. The *x*-nullcline should be marked with vertical arrows according to the sign of $g(x, y)$. Similarly, a *y*-nullcline is a set of points where $g(x, y) = 0$, and nonconstant orbits have horizontal tangents at points where they cross a *y*-nullcline. We mark the *y*-nullcline with horizontal arrows, pointed right where $f(x, y) > 0$ and pointed left where $f(x, y) < 0$.

Stationary points are located at the intersections of the *x*- and *y*-nullclines. They can be found by solving the pair of equations $f(x, y) = 0$, $g(x, y) = 0$.

Example 3.4.2   Find the nullclines and stationary points of the system

$$x' = x(x - 4y)$$
$$y' = y(y^2 - 2x - 9).$$

Solution. The *x*-nullclines are given by the equation

$$x(x - 4y) = 0.$$

Since this equation is satisfied if either $x = 0$ or if $x = 4y$, the *x*-nullclines are the *y*-axis and the line $x = 4y$.

The *y*-nullclines are determined by the equation

$$y(y^2 - 2x - 9) = 0,$$

and hence they are the *x*-axis and the parabola $x = (y^2 - 9)/2$.

The stationary points lie at the intersection of *x*-nullclines with *y*-nullclines. The axes meet at $(0, 0)$, and the *y*-nullcline $x = (y^2 - 9)/2$ meets the *y*-axis at $(0, \pm 3)$. Finally, the *x*-nullcline $x = 4y$ meets the *y*-nullcline $x = (y^2 - 9)/2$, where $y^2 - 8y - 9 = 0$. The solutions are $y = -1$ and $y = 9$, so $(-4, -1)$ and $(36, 9)$ are stationary points. Figure 3.11 shows the nullclines and stationary points. ❑

Integrals

An **integral** of a system of autonomous ODEs is a function $F(x, y)$ such that for every solution $x = \phi(t)$, $y = \psi(t)$ of the system, $F(\phi(t), \psi(t))$ is a constant function of t on the interval where $(\phi(t), \psi(t))$ is defined. This definition can be compared to the definition of integral of a single ODE, found on page 40. For example, it is known that all solutions of the system

$$x' = -y$$
$$y' = x$$

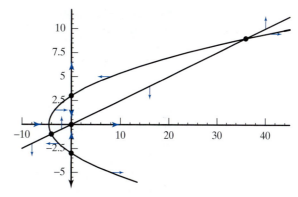

FIGURE 3.11   Stationary points and nullclines for the system of ODEs in Example 3.4.2. The x-nullcline is marked with vertical arrows indicating the direction of orbital crossing; the y-nullcline is marked with horizontal arrows. Stationary points are located where the two nullclines intersect.

have the form $x = R\cos(t - C)$, $y = R\sin(t - C)$, where R and C are arbitrary constants. Hence the trigonometric identity $\cos^2(\theta) + \sin^2(\theta) = 1$ implies that $x^2 + y^2 = R^2$. In other words, $F(x, y) = x^2 + y^2$ is an integral of this system.

Finding an integral of a system by using the general solution of the system is not practical, and a fact about parametric equations leads to a better approach. If $x = \phi(t)$, $y = \psi(t)$ are parametric equations for a curve, then the slope of the tangent to the curve at a point $(x_0, y_0) = (\phi(t_0), \psi(t_0))$ is equal to

$$\frac{dy}{dx} = \frac{dy/dt}{dx/dt} = \frac{\psi'(t_0)}{\phi'(t_0)}.$$

If $x = \phi(t)$, $y = \psi(t)$ are solutions of the system, then $\phi'(t) = f(\phi(t), \psi(t))$ and $\psi'(t) = g(\phi(t), \psi(t))$. Thus, the slope of the orbit passing through a point (x_0, y_0) is

$$\frac{dy}{dx} = \frac{\psi'(t_0)}{\phi'(t_0)} = \frac{g(x_1, y_1)}{f(x_1, y_1)}.$$

The orbit therefore satisfies the *first-order* ODE,

$$\frac{dy}{dx} = \frac{g(x, y)}{f(x, y)}.$$

This ODE is often written in differential form, as $g(x, y)\,dx - f(x, y)\,dy = 0$.

Finding an integral

Let $F(x, y)$ be an integral for the ODE

$$g(x, y)\, dx - f(x, y)\, dy = 0. \tag{3.15}$$

Then $F(x, y)$ is also an integral of the system

$$x' = f(x, y)$$
$$y' = g(x, y).$$

Example 3.4.3   Draw a phase portrait of the system

$$x' = -p^2 y$$
$$y' = q^2 x,$$

where p and q are nonzero constants.

Solution. We will find an integral of the system by finding an integral for the corresponding ODE

$$q^2 x\, dx + p^2 y\, dy = 0,$$

which is easily done since it is exact (and separable as well). The solution is

$$q^2 x^2 + p^2 y^2 = C. \tag{3.16}$$

We can assume $C \geq 0$, since otherwise equation (3.16) represents the empty set. When $C = 0$ we have the origin, which is a stationary point. If $C > 0$, put $a^2 = C/q^2$ and $b^2 = C/p^2$. Dividing equation (3.16) by C, we have

$$\frac{x^2}{a^2} + \frac{y^2}{b^2} = 1,$$

which is the equation of an ellipse, with principal axes of lengths $2a$ and $2b$ on the two coordinate axes. Since $x' = -p^2 y$, x is decreasing (directed to the left) on the upper half-plane $y > 0$, and increasing (directed to the right) on the lower half-plane. Hence the orbit is directed counterclockwise.

What do these orbits have in common? Each is an ellipse, for which the ratio of length of the vertical axis to the length of the horizontal axis is equal to

$$\frac{b}{a} = \frac{\sqrt{C}/q}{\sqrt{C}/p} = \frac{p}{q}.$$

Thus the orbits are the origin, and a family of similar ellipses. The phase portrait is shown in Figure 3.12. ❑

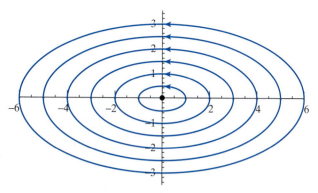

FIGURE 3.12   Orbits of $x' = -p^2 y$, $y' = q^2 x$.

The second-order ODE, $y'' + y^2 = 0$, can be replaced by the system

$$y' = v$$
$$v' = -y^2. \tag{3.17}$$

Example 3.4.4   Find an integral of the system (3.17) and draw its phase portrait.

Solution.   An integral $F(y, v)$ of the ODE

$$y^2 \, dy + v \, dv = 0$$

will also be an integral of the system (3.17). This ODE is exact, and we find

$$F(y, v) = \frac{1}{3} y^3 + \frac{1}{2} v^2.$$

The orbits lie on level curves of $F(y, v)$. Before drawing the phase portrait, we note that the y-nullcline is the v-axis, and the v-nullcline is the y-axis. There is one stationary point, the origin. The phase portrait shown in Figure 3.13 was drawn as a contour plot of the level curves of $F(y, v)$. The vector field

$$\mathbf{v}(y, v) = v \, \mathbf{i} - y^2 \mathbf{j}$$

was superimposed on the contour plot.   ❑

Can orbits intersect?

We will see that for autonomous systems, the answer is no. Is there an orbit passing through each point of the plane? Here the answer is yes. The proofs are based on the existence and uniqueness theorems for systems of ODEs. These theorems are modifications of the single-equation existence and uniqueness theorems stated in Section 2.4. As in the case of the single equation, there is a continuity requirement for f and g for the existence statement, and a Lipschitz condition is also needed for uniqueness.

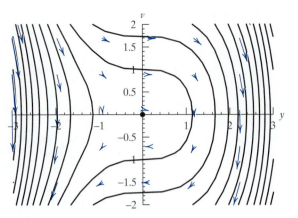

FIGURE 3.13   Phase portrait of $y'' = -y^2$: See Example 3.4.4.

Consider an IVP (where the ODEs are not necessarily autonomous),

$$\left. \begin{array}{l} x' = f(t, x, y); \quad x(t_0) = x_0 \\ y' = g(t, x, y); \quad y(t_0) = y_0 \end{array} \right\}. \tag{3.18}$$

The functions f and g depend on three variables and are required to be continuous in a *box* in three-dimensional space, $\mathbf{R}^3$, rather than a rectangle in the t, y-plane. We will denote by $\mathcal{B}$ the box

$$\mathcal{B} = \{(t, x, y) : |t - t_0| < k, |x - x_0| < l, |y - y_0| < m\},$$

centered at the point $(t_0, x_0, y_0) \in \mathbf{R}^3$. Here k, l, and m are positive numbers that specify the size of the box.

To formulate the Lipschitz condition that we need for a system, it is helpful to consider (3.18) as a single IVP where the dependent variable is a vector. Let $\mathbf{v}$ denote the vector whose components are the dependent variables x and y, and let $\mathbf{F}(t, \mathbf{v})$ be the vector function whose components are $f(t, x, y)$ and $g(t, x, y)$. In vector terms, the IVP (3.18) is stated as follows:

$$\frac{d\mathbf{v}}{dt} = \mathbf{F}(t, \mathbf{v}); \quad \mathbf{v}(t_0) = \mathbf{v}_0,$$

where $\mathbf{v}_0$ is the initial vector, with components (x_0, y_0).

We will use absolute value signs to denote distance in the plane: thus if $\mathbf{v}_1 = (x_1, y_1)$ and $\mathbf{v}_2 = (x_2, y_2)$,

$$|\mathbf{v}_1 - \mathbf{v}_2| = \sqrt{(x_1 - x_2)^2 + (y_1 - y_2)^2}.$$

With these definitions in mind, the function $\mathbf{F}(t, \mathbf{v})$ is said to satisfy a Lipschitz condition with respect to $\mathbf{v}$ in the box $\mathcal{B}$ if there is a constant K such that for all $(t, \mathbf{v}_1), (t, \mathbf{v}_2) \in \mathcal{B}$,

$$|\mathbf{F}(t, \mathbf{v}_1) - \mathbf{F}(t, \mathbf{v}_2)| \le K \, |\mathbf{v}_1 - \mathbf{v}_2|.$$

Our reason for using vector notation here is economy. The Lipschitz condition can be expressed without it, but it looks more complicated that way. See Exercise 18 at the end of this section.

Theorem 3.1

Existence Suppose that $f(t, x, y)$ and $g(t, x, y)$ are each continuous at every point in the box $\mathcal{B}$. Then there is a number $h > 0$ and there are functions $\phi(t)$ and $\psi(t)$, defined on the interval $t_0 - h < t < t_0 + h$ such that $x = \phi(t)$, $y = \psi(t)$ is a solution of the IVP (3.18).

Uniqueness If the vector function $\mathbf{F}(t, \mathbf{v}) = (f(t, x, y), g(t, x, y))$ satisfies a Lipschitz condition with respect to $\mathbf{v}$ in $\mathcal{B}$, and $x = \bar{\phi}(t)$, $y = \bar{\psi}(t)$ is another solution of the IVP (3.18), then there is a positive number h_1 such that for all t with $t_0 - h_1 < t < t_0 + h_1$, $\phi(t) = \bar{\phi}(t)$ and $\psi(t) = \bar{\psi}(t)$.

Continuous Dependence Assume that $\mathbf{F}(t, \mathbf{v})$ satisfies a Lipschitz condition with respect to $\mathbf{v}$ in $\mathcal{B}$. If $\mathbf{u} = (u_1, u_2)$ is a vector such that $(t_0, \mathbf{u}) \in \mathcal{B}$, let $\phi(t, \mathbf{u})$ denote the solution of

$$\frac{d\mathbf{v}}{dt} = \mathbf{F}(t, \mathbf{v}); \; \mathbf{v}(t_0) = \mathbf{u}.$$

Then there is a box $\mathcal{B}^* \subseteq \mathcal{B}$ such that $\phi(t, \mathbf{u})$ is continuous as a function of the three variables (t, u_1, u_2) with domain the box $\mathcal{B}^*$.

The proof of this theorem involves the same ideas as those of the single-equation theorems 2.3, 2.5, and 2.6, and is omitted.

Now let's adapt the uniqueness theorem to the question of intersecting orbits. If a system is not autonomous, it is certainly possible for orbits to intersect. If $(x, y) = (\phi_1(t), \psi_1(t))$ and $(x, y) = (\phi_2(t), \psi_2(t))$ are solutions of a system,

$$x' = f(t, x, y)$$
$$y' = g(t, x, y),$$

their orbits could intersect at a point

$$(x^*, y^*) = (\phi_1(t_1), \psi_1(t_1)) = (\phi_2(t_2), \psi_2(t_2)),$$

and as long as $t_1 \neq t_2$ this would be consistent with the uniqueness theorem. In the case of *autonomous* systems, we can show that the orbits do not intersect. To apply the uniqueness theorem, the following lemma is needed.

Lemma 3.4.1

Let $(x, y) = (\phi(t), \psi(t))$ be a solution of the autonomous system

$$x' = f(x, y)$$
$$y' = g(x, y).$$

Then for any t_0, $(x, y) = (\phi(t + t_0), \psi(t + t_0))$ is also a solution.

Proof Put $\bar{\phi}(t) = \phi(t + t_0)$ and $\bar{\psi}(t) = \psi(t + t_0)$. By the chain rule, $\frac{d\bar{\phi}}{dt} = \phi'(t + t_0)$ and $\frac{d\bar{\psi}}{dt} = \psi'(t + t_0)$. Since for any u,

$$\phi'(u) = f(\phi(u), \psi(u)),$$

it follows that

$$\frac{d\bar{\phi}}{dt} = \phi'(t + t_0) = f(\phi(t + t_0), \psi(t + t_0))$$

$$= f(\bar{\phi}(t), \bar{\psi}(t)).$$

Similarly, $\frac{d\bar{\psi}}{dt} = g(\bar{\phi}(t), \bar{\psi}(t))$, and therefore $(x, y) = (\bar{\phi}(t), \bar{\psi}(t))$ is a solution. ∎

The proof of this lemma was rather simple, and you should consider why it does not work for systems that are not autonomous.

Theorem 3.2
___

Suppose that the functions $f(x, y)$ and $g(x, y)$ are continuous and satisfy a Lipschitz condition in a rectangle $\mathcal{D} \subset \mathbf{R}^2$. Then for any point $(x^*, y^*) \in \mathcal{D}$, there is exactly one orbit of the system

$$\left. \begin{array}{l} x' = f(x, y) \\ y' = g(x, y) \end{array} \right\} \tag{3.19}$$

that contains (x^*, y^*).

Proof The existence and uniqueness theorems imply that there is exactly one solution, which we will denote $(x, y) = (\phi_1(t), \psi_1(t))$, of the system (3.19) with initial values $(x(0), y(0)) = (x^*, y^*)$. However, for any $t_0 \neq 0$, there is also a solution, which we will denote $(x, y) = (\phi_2(t), \psi_2(t))$, with initial values $(x(t_0), y(t_0)) = (x^*, y^*)$. Both solutions will describe an orbit that passes through (x^*, y^*).

We have to show that the two solutions describe the same orbit. Put $\bar{\phi}_2(t) = \phi_2(t + t_0)$, $\bar{\psi}_2(t) = \psi_2(t + t_0)$. By Lemma 3.4.1, $(x, y) = (\bar{\phi}_2(t), \bar{\psi}_2(t))$ is a solution of the system (3.19), and it describes the same orbit that (ϕ_2, ψ_2) does. Since $\bar{\phi}_2(0) = \phi_2(t_0) = x_0$, and $\bar{\psi}_2(0) = y_0$, it follows that $(\bar{\phi}_2, \bar{\psi}_2)$ satisfies the same initial conditions as (ϕ_1, ψ_1). It follows from the uniqueness theorem that $\phi_1(t) = \phi_2(t + t_0)$ and $\psi_1(t) = \psi_2(t + t_0)$ for all t such that all four functions are defined. ∎

Closed orbits

Theorem 3.2 does not preclude an orbit from intersecting *itself*. This would happen if the system (3.13) had a solution $(x, y) = (\phi(t), \psi(t))$ such that for some pair of numbers $t_1 \neq t_2$,

$$(\phi(t_1), \psi(t_1)) = (\phi(t_2), \psi(t_2)). \tag{3.20}$$

For example, if an orbit is simply an stationary point, the functions ϕ and ψ would be constant; it therefore intersects itself infinitely often.

Theorem 3.3
___

Let $(x, y) = (\phi(t), \psi(t))$ be a solution of the system (3.13) such that equation (3.20) holds for distinct numbers $t_1 < t_2$. Then ϕ and ψ are T-periodic, where $T = t_2 - t_1$, and the orbit described by $(\phi(t), \psi(t))$ is a closed curve.

Proof By Lemma 3.4.1, $(x, y) = (\phi(t + T), \psi(t + T))$ is also a solution of the system (3.13), and since it satisfies the same initial conditions as $(\phi(t), \psi(t))$ at $t = t_1$, we can use the uniqueness theorem to conclude that $(\phi(t+T), \psi(t+T)) \equiv (\phi(t), \psi(t))$. Therefore, the functions ϕ and ψ are T-periodic. A curve that is described by periodic parametric equations must be closed, so the corresponding orbit is closed. ∎

Van der Pol's equation

To illustrate Theorem 3.3, let's consider the second-order ODE,

$$x'' + c(x^2 - 1) x' + x = 0,$$

which is known as *van der Pol's equation*. The number c is a constant that we can vary to produce a variety of examples. Van der Pol's equation was developed in the 1920s as a model for the behavior of a triode vacuum tube. The dependent variable x represents the current, as a function of time, in an electrical circuit that includes a triode tube. (A modern circuit would probably have a semiconductor device in place of the triode.) We can replace van der Pol's equation with a system

$$\left. \begin{aligned} x' &= y \\ y' &= -x + c(1 - x^2)\, y \end{aligned} \right\}. \tag{3.21}$$

The case $c = 0$. If $c = 0$, the system (3.21) is identical to the system that we examined in Example 3.4.3, where both parameters p and q have the value 1. We found there that all orbits, except the stationary point at the origin, were closed ellipses (in our case, since $p = q$, the orbits are actually circles). Therefore, Theorem 3.3 tells us that all of the solutions except the stationary solution $x \equiv 0$, $y \equiv 0$ are periodic. In fact, you can easily verify that the orbit passing through the point $(0, 1)$ on the y-axis is traced by the solution

$$x = a \sin(t)$$
$$y = a \cos(t)$$

and is thus 2π-periodic.

The case $c > 0$. When $c = 0$, we are dismissing the important middle term of van der Pol's equation. Thus, let's use a positive value for c and see what happens.

Example 3.4.5 Draw the x- and y-nullclines of the van der Pol system (3.21), with $c = \frac{1}{2}$.

Solution. The x-nullcline is given by the equation $y = 0$ and is thus the x-axis. This means that all orbits cross the x-axis in the vertical direction. Further inspection of the system (3.21) shows that when $y = 0$ we have $y' = -x$; thus the orbits crossing the positive x-axis are directed downward, and the orbits crossing the negative part of the x-axis are directed upward.

The y-nullcline has the equation

$$-x + c(1 - x^2)\, y = 0.$$

We will substitute $c = \frac{1}{2}$ and solve for y to get

$$y = \frac{2x}{1 - x^2}.$$

The orbits will be directed to the right as they cross the y-nullcline in the upper half-plane, and in the lower half plane they are directed to the left. Figure 3.14 shows the nullclines. ❑

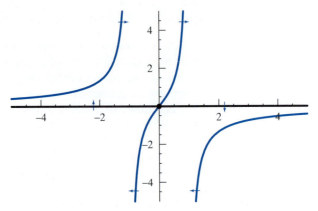

FIGURE 3.14 Nullclines of the van der Pol system (3.21), where $c = 1/2$.

Figure 3.14 indicates that orbits of the van der Pol system travel around the origin in a clockwise direction, but they may not be closed orbits—it is possible that they are spirals. Figure 3.15 displays segments of two orbits. I have chosen initial points $(x_0, y_0) = (1, 0)$ and $(x_0, y_0) = (3, 0)$ for these orbits, and followed each of them until it returned to the positive x-axis. You can see that neither orbit is closed, since neither returns to its starting point.

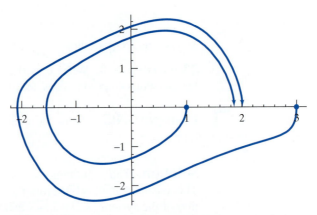

FIGURE 3.15 Segments of two orbits of the van der Pol system (3.21), with $c = 1/2$.

Let

$$x = \phi(t, u)$$
$$y = \psi(t, u)$$

denote the solution of the van der Pol system (3.21) with $c = \frac{1}{2}$ and initial point $(\phi(0), \psi(0)) = (u, 0)$. By the *continuous dependence* statement in Theorem 3.1, the functions ϕ and ψ are continuous functions of (t, u). Let $g(u)$ be the function, defined for all $u \in [1, 3]$, as follows: First let $s(u)$ be the least positive number such that $\psi(s(u), u) = 0$ and $\phi(s(u), u) > 0$. Thus, the orbit starting at $(u, 0)$ on the positive x-axis will return to the positive x-axis for the first time at $t = s(u)$. Now let

$$g(u) = \phi(s(u), u)$$

be the x-coordinate of the solution with initial point at $(u, 0)$, when it first returns to the positive x-axis.

It follows from the continuity of ϕ and ψ (and the implicit function theorem) that g is continuous. Now refer to Figure 3.15. You will see that $g(1) \approx 1.86$ and $g(3) \approx 2.02$. Thus the function $h(u) = g(u) - u$ is continuous, with $h(1) \approx -0.86 < 0$, and $h(3) \approx 0.98 > 0$. By the intermediate value theorem, there is a number $u^* \in (1, 3)$ with $h(u^*) = 0$; that is, $g(u^*) = u^*$. The solution $(x, y) = (\phi(t, u^*), \psi(t, u^*))$ thus satisfies

$$(\phi(0, u^*), \psi(0, u^*)) = (u^*, 0) = (\phi(s(u^*), u^*), \psi(s(u^*), u^*)).$$

Now it follows from Theorem 3.3 that $\phi(t, u^*)$ and $\psi(0, u^*))$ are periodic functions of t, and the orbit that they describe is a closed curve. This orbit is shown in Figure 3.16.

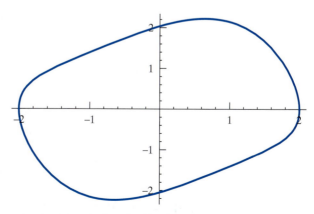

FIGURE 3.16   A closed orbit of the van der Pol system (3.21), with $c = 1/2$.

EXERCISES

In Exercises 1–7, sketch the nullclines and find the stationary points of the given system of ODEs. Then use an IVP solver to draw a few orbits.

1. $\begin{cases} x' = x^2 + y^2 \\ y' = x^2 - y^2 \end{cases}$
2. $\begin{cases} x' = x(x + y) \\ y' = y(2x - y) \end{cases}$

3. $\begin{cases} x' = -x \\ y' = 4y - y^2, \end{cases}$
4. $\begin{cases} x' = x(x - y^2) \\ y' = y(x - 4) \end{cases}$

5. $\begin{cases} x' = x^2 \\ y' = y \end{cases}$
6. $\begin{cases} x' = y + 1 \\ y' = y \end{cases}$

7. $\begin{cases} x' = x + y - 2 \\ y' = x - y \end{cases}$

Find integrals for the systems of ODEs in Exercises 8– 13.

8. $\begin{cases} x' = x \\ y' = y + x^2 \end{cases}$
9. $\begin{cases} x' = 2x \\ y' = x + y \end{cases}$

10. $\begin{cases} x' = x(1 - y) \\ y' = y(x - 1) \end{cases}$
11. $\begin{cases} x' = 3x \\ y' = 5y \end{cases}$

12. $\begin{cases} x' = 3y \\ y' = 5x \end{cases}$
13. $\begin{cases} x' = x^2 + y^2 \\ y' = -2xy \end{cases}$

14. Suppose that each orbit of an autonomous system of ODEs is an ellipse of the form $x^2 + 4y^2 = C$, where $C \geq 0$. Show that the origin is a stationary point.

15. Show that the graph of any solution of the nonautonomous first-order ODE,

$$\frac{dy}{dx} = f(x, y),$$

passing through the initial point (x_0, y_0), is the same as the orbit of the *autonomous system*

$$\frac{dx}{dt} = 1$$

$$\frac{dy}{dt} = f(x, y)$$

passing through the same point.

16. Exercise 15 shows that a nonautonomous first-order ODE can be treated as a system of two autonomous ODEs. Find a way to treat a nonautonomous system of two ODEs as a system of three autonomous ODEs.

17. Prove the existence and uniqueness theorem for second-order ODEs: *Suppose that $f(t, x, y)$ is continuous in a box B in $\mathbf{R}^3$ and satisfies a Lipschitz condition there with respect to the variables x, y (specify this Lipschitz condition). Then, given $(t_0, x_0, y_0) \in B$, there is a unique solution of the IVP*

$$\frac{d^2x}{dt^2} = f\left(t, x, \frac{dx}{dt}\right) \quad x(t_0) = x_0 \quad \frac{dx}{dt}\bigg|_{t=t_0} = y_0.$$

Base your proof on Theorem 3.1.

18. Write out the Lipschitz condition appearing on page 118 without using vector notation.

19. Consider the system

$$ax' + by' = f(t, x, y)$$
$$cx' + dy' = g(t, x, y)$$

with initial conditions $x(t_0) = x_0$, $y(t_0) = y_0$. Show that if the functions $f(t, x, y)$, $g(t, x, y)$ are continuous and satisfy a Lipschitz condition with respect to (x, y) in a box centered at (t_0, x_0, y_0) and if $ad - bc \neq 0$, then this IVP has a unique solution.

20. Here is a nice application of Green's theorem from advanced calculus. (People who haven't had advanced calculus are excused). Let $\mathbf{v}(x, y) = \begin{bmatrix} f(x, y) \\ g(x, y) \end{bmatrix}$ be a vector field defined on a rectangular domain $\mathcal{R}$, and suppose that for all $(x, y) \in \mathcal{R}$, div $\mathbf{v}(x, y) > 0$. Show that the system

$$x' = f(x, y)$$
$$y' = g(x, y)$$

has no closed orbits in $\mathcal{R}$. This fact, first proved in 1900 by the Swedish mathematician Ivar O. Bendixson, is called **Bendixson's criterion.**

*3.5 Populations of Interacting Species

Suppose that two species, A and B, occupy the same environment. It may happen that A is a predator species and B is its prey, or that A and B compete for the same food supply. In either case, the population of one of the species cannot be successfully modeled without considering the influence of the other. We will consider simple models involving systems of two ODEs.

Denote the populations of the two species by x and y. The derivatives $x'(t)$ and $y'(t)$ represent the absolute growth rates of these populations, and the *relative growth rates* are $x'(t)/x(t)$ and $y'(t)/y(t)$, respectively. Our model posits that each relative growth rate is a function of x and y. It can thus be expressed as a system of two ODEs,

$$\left. \begin{aligned} x' &= x\, f(x, y) \\ y' &= y\, g(x, y) \end{aligned} \right\}. \tag{3.22}$$

The logistic equation

$$x' = kx(1 - x/C)$$

is a one-species model similar to the system (3.22). We used it in Section 2.5 to study the dynamics of the population x of a single species in an environment with a carrying capacity of C individuals. In this model, the relative growth rate of x is a linear function of x, and in the two-species models that we will consider, the relative growth rate for each species will be a linear function of x and y.

The Lotka-Volterra equations

The Lotka-Volterra equations,

$$\left. \begin{aligned} x' &= x(a - by) \\ y' &= cy(x - d) \end{aligned} \right\}, \tag{3.23}$$

were the first differential equations model for the populations of more than one species. It was developed in 1926 by the Italian mathematician Vito Volterra as a model to explain the fluctuating populations of predator and prey fish in the Adriatic Sea. The same system was proposed in 1920 by an American biophysicist, Alfred Lotka, to determine the rate of a hypothetical chemical reaction (see Exercise 2 at the end of this section). Lotka's intention was to show that the concentrations of chemicals involved in a reaction could vary periodically.[1] In Volterra's model, x is the prey population, and y represents the population of a predator whose sole source of food is this prey. The relative growth rate of the prey species is set equal to $a - by$, indicating that in the absence of the predator, the number of prey will increase exponentially, but the relative rate of increase decreases linearly with the predator population. Of course, even in the absence of the predator, the prey population would be limited by the availability of resources, but the predator prevents the prey population from approaching a saturation level. The predator keeps the prey from starvation. The second equation of the system (3.23) indicates that the predator *will* starve without the prey: When $x = 0$ it reduces to $y' = -cdy$; thus $y = y_0 e^{-cdt}$ (where y_0 is the initial population). The parameter d represents the minimum prey population necessary to support predators.

---

[1] Oscillating chemical reactions are no longer hypothetical. The first observations of an oscillating reaction were made in 1951 by B. P. Belousov, a chemist in the former Soviet Union.

The system (3.23) has two stationary points. These are the solutions of

$$x(a - by) = 0$$
$$cy(x - d) = 0,$$

$(x, y) = (0, 0)$ and $(x, y) = (d, b/a)$. The stationary point at the origin is not unexpected: If both species are extinct, they will stay extinct. The other stationary point represents a situation in which the appetite of the predator is exactly matched by the reproductive rate of the prey. Our experience with the logistic equation suggests that the stationary point at the origin is probably unstable, and the stationary point at $(x, y) = (d, b/a)$ is stable. While this is true, it is also an oversimplification.

We can apply the procedure introduced in Section 3.4 to find an integral for the system (3.23), by solving the ODE

$$cy(x - d)\, dx - x(a - by)\, dy = 0.$$

This ODE is separable, and upon dividing through by $-xy$ we can integrate to obtain

$$\int c\left(\frac{d}{x} - 1\right) dx + \int \left(\frac{a}{y} - b\right) dy = \text{constant}$$

and it follows that

$$F(x, y) = cd \ln(x) + a \ln(y) - cx - by$$

is an integral.

To keep the notation simple, we will use the notation F_x, F_y, and so on, for the partial derivatives $\frac{\partial F}{\partial x}$, $\frac{\partial F}{\partial y}$, and so on. The critical points of $F(x, y)$ are found by setting F_x and F_y equal to 0 and solving for x and y:

$$F_x(x, y) = c(d/x - 1) = 0$$
$$F_y(x, y) = a/y - b = 0.$$

The stationary point $(d, a/b)$ is the only critical point, and we will apply the second derivative test to it. Since $F_{xx}(d, a/b) = -c/d$, $F_{yy}(d, a/b) = -b^2/a$, and $F_{xy} \equiv 0$,

$$F_{xx}(d, a/b) F_{yy}(d, a/b) - [F_{xy}(d, a/b)]^2 = \left(-\frac{c}{d}\right)\left(-\frac{b^2}{a}\right) > 0,$$

and we conclude that $F(x, y)$ has an absolute maximum at $(x, y) = (d, a/b)$.

It follows that the level curves of $F(x, y)$ in the first quadrant—which are the orbits of the Lotka-Volterra equations—are closed curves (see Figures 3.17 and 3.18) surrounding the stationary point at $(d, a/b)$. Orbits do not converge to this stationary point as $t \to \infty$, but they don't diverge away from it either. Since the orbits are closed curves, we conclude that the solutions of the Lotka-Volterra equations are periodic functions of time, and hence that the populations of predator and of prey will oscillate.

The Lotka-Volterra system is only a starting point in the study of populations of predators and their prey. Exercise 3 on page 664 points out a serious difficulty with the model.

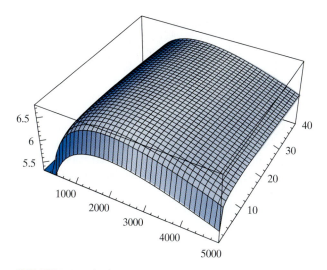

FIGURE 3.17   An integral for the Lotka-Volterra equations. The orbits are level curves. The parameters are $a = 0.1$, $b = 0.01$, $c = 0.0005$, and $d = 2000$. Thus, the stationary point is located at $(x, y) = (2000, 10)$.

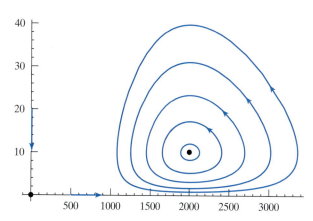

FIGURE 3.18   Orbits of the Lotka-Volterra equations.

Competing species

We will now consider an ecosystem in which two species A and B, with populations x and y, compete for resources. The organisms may be plants, competing for nutrients and sunlight, grazing animals competing for forage, or predators in competition for prey. The model is thus applicable to a single level of the food chain.

If the species B were removed from the environment, the population of the species A would grow according to the logistic equation, which we will write in the form

$$\frac{dx}{dt} = ax(C - x),$$

where C is its carrying capacity, and $a = k/C$ is the ratio of the initial relative growth rate to the carrying capacity. The species B will consume some of the resources required by A. This is taken into account by subtracting from the carrying capacity an amount to reflect the rate at which the resources needed by species A are consumed by species B. Thus the relative growth rate for the species A is

$$a(C - ry - x),$$

where r is the rate at which species B consumes the resources.

The relative growth rate of B is, by the same reasoning, proportional to $L - sx - y$, where L denotes the carrying capacity for B and s is the rate at which A consumes the resources that B needs. Putting these together, we have the system

$$\frac{1}{x}\frac{dx}{dt} = a(C - x - ry)$$

$$\frac{1}{y}\frac{dy}{dt} = d(L - sx - y).$$

We have thus derived the **competing species equations,**

$$\left. \begin{array}{l} x' = x(k - ax - by) \\ y' = y(l - cx - dy) \end{array} \right\}, \tag{3.24}$$

where $k = aC$, $b = ar$, $l = dL$, and $c = ds$.

The system (3.24) has four stationary points. Three of these occur when one or both of the species are extinct. For example, if A is extinct ($x = 0$), then the species B can have the stationary populations $y = 0$ (also extinct) or L (its carrying capacity). Furthermore, in this case the orbits would just be those of the phase diagram of the logistic equation.

The fourth stationary point, (x_1, y_1), is located at the intersection of the lines

$$ax + by = k, \quad \text{and} \tag{3.25}$$
$$cx + dy = l. \tag{3.26}$$

The line (3.25) is an x-nullcline, and the line (3.26) is a y-nullcline. If the lines do not intersect in the first quadrant, this stationary point is not of interest, because neither population can be negative.

Figure 3.19 shows four possible configurations of the nullclines, depending on the relative size of the parameters. The intercepts of the line (3.25) are at $(C, 0)$ and

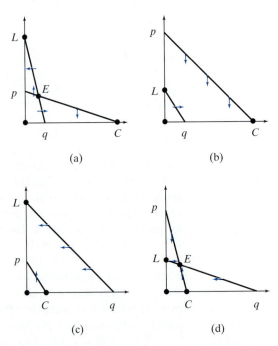

FIGURE 3.19   Four possible configurations for the nullclines of the system (3.24). The equilibria are marked with heavy dots; the x-nullcline is marked with vertical arrows, and the y-nullcline is marked with horizontal arrows.

$p = (0, k/b)$, and since $(C, 0)$ is a stationary point, it is marked with a heavy dot. The intercepts of the line (3.26) are $q = (l/c, 0)$ and the stationary point $(0, L)$. In configurations (a) and (d) the stationary point $E = (x_1, y_1)$ is also present. The arrows on the nullclines indicate the direction orbits must take when they cross. Thus, the x-nullcline has vertical arrows. The part of the x-nullcline that lies above the y-nullcline is in a region where $y' < 0$, so the arrows point down. When the x-nullcline is below the y-nullcline, the arrows point up, for $y' > 0$. The directions of the arrows marking the y-nullclines are explained in the same way.

It is not feasible to find an integral for the system (3.24), and to understand it we will need to use a different approach. A closed polygon $\mathcal{T}$ in the phase plane is called a **trap** if every orbit that crosses an edge or vertex of $\mathcal{T}$ is directed toward the interior. Thus, orbits can enter $\mathcal{T}$, but once they do so, they are trapped, since they cannot cross any edge from the inside outward. Identifying traps is an important technique for studying autonomous systems of ODEs. For the system (3.24) we will use traps whose edges are nullclines.

Configurations (b) and (d) are the easiest to analyze, and we will start with (b). The quadrilateral $LqCP$ is a trap, because all orbits are directed downward as they cross the x-nullcline (the line pC), to the right as they cross the y-nullcline (the line Lq), and no orbits cross the axes. Furthermore, all orbits starting outside this quadrilateral must either enter it, or converge to one of the stationary points C or L.

Inside the trap, x' is positive (it doesn't become negative until the other side of the x-nullcline is reached), and, since the quadrilateral lies above the y-nullcline, Lq, y' is negative. With orbits trapped inside and directed downward and to the right, it can be shown that they all converge to the stationary point $(C, 0)$. The conclusion is that in configuration (b), species A is dominant, and B will become extinct. The same reasoning shows that exactly the reverse is true in configuration (c): Species B is dominant, and A dies out. The phase portrait corresponding to configuration (b) is shown in Figure 3.20.

In configuration (a) the quadrilateral $OCEL$ (O stands for the origin) is a trap. Within this quadrilateral, the triangles pEL and qCE are also traps. In triangle pEL, orbits are directed upward and to the left; and in qCE they are directed downward and to the right. Thus orbits starting near E will converge to L or C if they enter the interior of one of these triangles. In particular, consider the situation when the initial population is inside the quadrilateral $OqEp$. Its orbit must either converge to E, or enter one of the aforementioned triangles, and hence converge to C or L. It can be shown that there is one orbit in $OqEp$ that does converge to E; it is called the **separatrix**. Orbits that start above the separatrix will cross the edge Ep, entering the triangle where all orbits converge to L; and orbits starting below the separatrix will enter the triangle qCE and converge to E. This situation is called **competitive exclusion**: The species with the initial population advantage dominates the other, which subsequently dies out. The phase portrait in this case is shown in Figure 3.21.

Now let us consider a happier situation: configuration (d). The quadrilateral $OqEp$ is a trap, and within it, the triangles LEp and CqE are also traps. The distinction between this and configuration (a) is that in this case, the orbits inside the triangles are directed toward E instead of away from it. Since every orbit either must converge directly to E or enter one of these triangles, we conclude that all

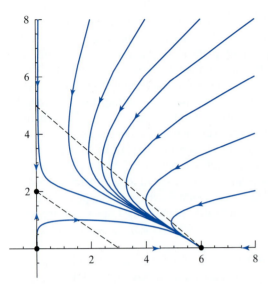

FIGURE 3.20   In this phase portrait, the parameters correspond to the configuration shown in Figure 3.19(b). The species whose population is represented by x dominates the species represented by y regardless of the initial conditions. The nullclines are represented by dashed lines, and the parameters are $(k, a, b, l, c, d) = (30, 5, 6, 6, 2, 3)$.

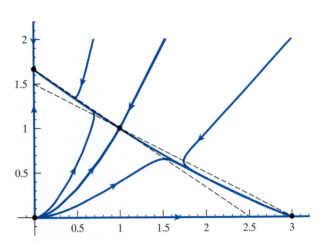

FIGURE 3.21   Competitive exclusion: the phase portrait of the system (3.24), when the nullclines (shown as dashed lines) are configured as in Figure 3.19(a). The parameter values are $k = 3$, $a = 1$, $b = 2$, $l = 5$, $c = 2$, and $d = 3$.

orbits converge to E. Here we have a stable equilibrium in which the species share the resources. The phase portrait is displayed as Figure 3.22.

It is interesting to speculate on which of these cases is applicable in familiar situations, such as crabgrass in the lawn, benign versus pathogenic bacteria in the body, introduction of new species of fish in a pond, and so on. Conclusions reached by using this model are not to be trusted without further study, because the model is based on many simplifying assumptions.

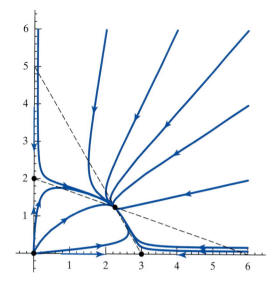

FIGURE 3.22   When the population growth parameters correspond to the configuration shown in Figure 3.19(d), the two species will coexist with populations represented by the stationary point in the first quadrant. The initial conditions have no effect on the outcome, as long as both species are present. Nullclines are represented by dashed lines, and the parameters are
$(k, a, b, c, l, c, d) = (15, 5, 3, 12, 2, 6)$.

EXERCISES

1. A population of insect pests is to be controlled by introducing predator insects. The objective is to keep the maximum number of pests present at any time as small as possible. Assuming that the Lotka-Volterra model is valid in this situation,

   a. If the growth parameters a, b, c, d of the Lotka-Volterra equations on page 125 are known, what is the optimum number of predators to release?

   b. Evaluate this strategy: Apply a pesticide to reduce the number of pests as much as possible; when the pesticide has degraded, introduce predators to keep the pests under control.

   c. If the number of pests in the environment is initially small, is it preferable to introduce the predators immediately or to wait for a while?

2. Given a chemical reaction $Q + R \rightarrow S$, the *law of mass action* states that the rate at which the product S is produced

is proportional to the product of the concentrations of Q and R. Thus, if x, y, and z denote the concentrations of Q, R, and S, respectively, then

$$\frac{dz}{dt} = kxy,$$

where k is constant.

Lotka[2] considered a hypothetical sequence of three reactions

$$P + Q \rightarrow 2Q, \quad Q + R \rightarrow 2R, \quad R \rightarrow S.$$

Suppose that the concentration of P is held fixed by replenishing that reactant at the same rate that it is consumed. Let x and y denote the concentrations of Q and R, and derive the Lotka-Volterra equations as a model of this chemical system.

3. Two species of fish inhabit a pond; neither is the prey of the other. The pond's carrying capacity is 1000 fish of species

[2] A. J. Lotka, "Undamped oscillations derived from the law of mass action," *J. Amer. Chem. Soc.* **42** (1920), pp. 1595–1599.

A, or 1500 of species B. Also, if there are 800 A-fish and 800 B-fish, the population will be stationary. Set up a system of ODEs to model the fish populations, and draw a sketch of the phase plane, identifying traps. How will the populations stabilize, assuming both populations are initially positive?

4. Repeat Exercise 3 under the assumption that the the carrying capacity of the pond for each individual species is unchanged, but the combination of A- and B-fish in the pond for a stationary population is

   a. 800 A-fish and 250 B-fish

   b. 600 fish of each species

5. **Symbiosis.** If the relationship between two species is mutually beneficial, the species are said to be in symbiosis. Lichens are familiar examples. While each lichen appears to be a single organism, in fact it consists of two species, an alga and a fungus, in symbiosis. Draw a the nullclines for the two-species model,

$$x' = ax(K - x + By)$$
$$y' = dy(L + Cx - y),$$

   and identify any traps. You will need to consider the cases where $BC \geq 1$ and $BC < 1$ separately.

6. This problem is for readers who have had an advanced calculus course. Let $(x, y) = (\phi(t), \psi(t))$ be a solution of the Lotka-Volterra equations, and let T denote its period. The average populations are

$$\bar{x} = \frac{1}{T} \int_0^T \phi(t)\, dt \quad \text{and} \quad \bar{y} = \frac{1}{T} \int_0^T \psi(t)\, dt.$$

   Prove that $\bar{x} = d$ and $\bar{y} = a/b$.

   ◆ **Hint:** One proof is based on Green's theorem.

7. This problem does not require advanced calculus, but readers who have not had this course will have to take the result of Exercise 6 on faith.

   Volterra's predator-prey model[3] was based on data from fish catches in the Adriatic Sea. Volterra observed that the ratio of predators to prey varied periodically, and this is reflected by his mathematical model. Let R denote the catch rate: that is, the fraction of the population (of both species) that is caught. Taking fishing into account, the Lotka-Volterra equations become

$$x' = x(a - by) - Rx$$
$$y' = cy(x - d) - Ry.$$

   Assuming that $a > R$, show that as the catch rate increases, the average population of prey increases, and the average predator population decreases. You may use the result of Exercise 6. Although the Lotka-Volterra equations oversimplify, this effect is substantiated by Volterra's data. For the same reason, it is unwise to apply a pesticide to "help" an insect predator control a pest species: The result will be an increase in the average number of pests.

3.6 Glossary

$\mathbf{R}^3$ Three-dimensional space.

Autonomous system   A system of ODEs in which the independent variable does not appear on the right side.

Bendixson's criterion   If a vector field, defined of a rectangle $\mathcal{R}$ in the plane, has nonvanishing divergence, then there are no closed integral curves in $\mathcal{R}$.

Classical IVP solver   An IVP solver that has a fixed time step.

Competing species equations   A system of ODEs that models the dynamics of the populations of two species that compete for the same resources. See page 128.

Competitive exclusion   A situation in which a species with an initial population advantage is able to overwhelm its competitor.

Dynamic IVP solver   An IVP solver that automatically adjusts its time step to maintain precision.

IVP solver   A numerical method for computing solutions of initial value problems.

Integral   (of a system of ODEs) A function $F(x, y)$ that is constant on all orbits of the system.

Integral curve   (of a vector field $\mathbf{v}(x, y) = f(x, y)\mathbf{i} + g(x, y)\mathbf{j}$) A curve given parametrically by $x = p(t)$, $y = q(t)$, where for all t, the tangent vector $p'(t)\mathbf{i} + q'(t)\mathbf{j}$ is equal

---

[3] V. Volterra, "Variazione fluttuazioni del numero d'individui in specie animali conviventi," *Mem. Acad. Lincei* **2** (1926), pp. 31–113. English translation: "Variations and fluctuations in the population of animals living together," in *Animal Ecology*, by R. N. Chapman (New York: McGraw-Hill, 1931), pp. 409–448.

to the vector at $(p(t), q(t))$, that is,

$$\mathbf{v}(p(t), q(t)) = p'(t)\mathbf{i} + q'(t)\mathbf{j}.$$

Lipschitz condition See page 118.

Lotka-Volterra equations A system of ODEs that is a model for the population dynamics of a predator and its prey. See page 125.

Nullcline (for a system $x' = f(x, y)$, $y' = g(x, y)$) The set of points where $f(x, y) = 0$ (the x-nullcline) of $g(x, y) = 0$ (the y-nullcline).

Order (of an IVP solver) A number n such that the accumulated error varies proportionately with the nth power of the time step.

Orbit The graph in the phase plane of parametric equations given by a solution of a system of two ODEs. It is customary to mark orbits to indicate the direction of increasing t.

Phase plane The plane in which the coordinates correspond to the values of the dependent variables of a system of two ODEs.

Phase portrait A drawing of the phase plane, showing a representative collection of orbits of a given system of two ODEs.

Replace an ODE The system

$$\left. \begin{array}{l} \dfrac{dy}{dt} = v \\[2mm] \dfrac{dv}{dt} = f(t, y, v) \end{array} \right\} \qquad (3.27)$$

is said to replace the second-order ODE

$$\frac{d^2 y}{dt^2} = f\left(t, y, \frac{dy}{dt}\right), \qquad (3.28)$$

since $(y, v) = (\phi(t), \psi(t))$ is a solution of (3.27) if and only if $y = \phi(t)$ satisfies (3.28) and $\psi = \phi'$.

Separatrix An orbit of an autonomous system of two ODEs that separates orbits with different limiting behavior.

Stationary point (of a vector field $\mathbf{v} = f(x, y)\mathbf{i} + g(x, y)\mathbf{j}$) A point (x_1, y_1) such that $\mathbf{v}(x_1, y_1) = \mathbf{0}$. An integral curve containing a stationary point consists of the stationary point alone.

(of a system of ODEs, $x' = f(x, y)$, $y' = g(x, y)$) A point (x_1, y_1) such that $f(x_1, y_1) = g(x_1, y_1) = 0$

System of ODEs Two or more ODEs to be treated simultaneously. To be well posed, the system must have one equation for each dependent variable, and only one independent variable.

Trap A closed polygon in the phase plane such that all orbits that cross its edges or vertices are pointed inward.

Uncoupled system A system of two ODEs in which one of the dependent variables does not appear in one of the equations.

Vector field A function, defined on a subset of the xy plane, that assigns to each point (x, y) in its domain a vector $\mathbf{v}(x, y) = f(x, y)\mathbf{i} + g(x, y)\mathbf{j}$ in the plane.

3.7 Chapter Review

1. Solve the initial value problem

$$x' = x + e^t; \; x(0) = 0$$
$$y' = xy^2; \quad y(0) = 1.$$

2. Find the general solution of

$$x' = -x$$
$$y' = y.$$

Also find an integral for this system and draw its phase portrait.

3. Repeat Exercise 2 for the system

$$x' = x$$
$$y' = 2y.$$

4. Suppose that $F'(y) = f(y)$. Find an integral for the system that replaces the ODE $y'' + f(y) = 0$. Apply the result to the following ODEs, in which α denotes a constant parameter:

   a. $y'' = -\alpha^2 \sin(y)$ (the pendulum equation)

   b. $y'' = -\alpha^2 y$ (the linearized pendulum)

   c. $y'' = -\alpha^2 y^{-2}$ (a falling body)

5. Find systems of first-order ODEs to replace the given second-order ODEs.

   a. $y'' + y = 0$

   b. $y'' = y' - t^2 \sin(y)$

6. Given that $y = e^t \cos(t)$ is a solution of the ODE

$$y'' = 2y' - 2y,$$

sketch the corresponding orbit of the system that replaces this ODE.

7. Find an integral for the system

$$x' = -x(x + 4y)$$
$$y' = 2y(x + y).$$

8. Find the stationary points of the following vector fields:

a. $\mathbf{v} = (x + y - 2)\mathbf{i} + (x - 3y + 2)\mathbf{j}$

b. $\mathbf{v} = (y - x)\mathbf{i} + (y - x^3)\mathbf{j}$

c. $\mathbf{v} = x(x + y + 4)\mathbf{i} + y(x + 5y)\mathbf{j}$

9. Write out the systems of ODEs corresponding to the vector fields in Exercise 8 and sketch their nullclines.

10. Draw the nullclines of the system

$$\left.\begin{array}{l} x' = (x + y)(y^2 - 1) \\ y' = (x - y)(x^2 - 1) \end{array}\right\}$$

and identify a trap.

11. Use an IVP solver to draw several integral curves for each of the vector fields in Exercise 8. Choose initial points near (but not at!) the stationary points, and let the time variable range from -5 to 5.

12. Use an IVP solver to plot the graph of the solution of the damped pendulum equation, $y'' = -.05y' - \sin(y)$, with initial conditions $y(0) = 0.25$, $y'(0) = 0$. Repeat with the "linearized version" of the equation, $y'' = -.05y' - y$, with the same initial conditions.

PART II

Linear Differential Equations

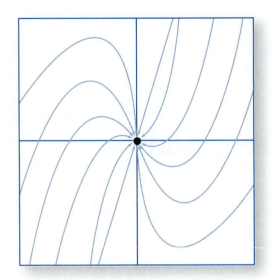

CHAPTER 4

Systems of Linear Differential Equations

4.1 The Initial Value Problem

A system of linear, first-order ODEs,

$$\left. \begin{array}{l} x' = p_1(t)\, x + q_1(t)\, y + r_1(t) \\ y' = p_2(t)\, x + q_2(t)\, y + r_2(t) \end{array} \right\}, \tag{4.1}$$

expresses the rate of change of each dependent variable as a function of the independent variable and both dependent variables. Thus, although each equation in the system (4.1) is first order, we cannot directly solve for x without knowing y, or for y without having first determined x. The problem is that the two equations are *coupled*. This kind of coupling appears in many applications, such as mechanical systems involving several weights and springs (the details are given in Section 5.7), parallel electrical circuits (which we will study in Section 6.2), and in economic models involving supplies of various commodities.

This chapter is an introduction to the general properties of linear systems. If the coefficients $p_i(t)$ and $q_i(t)$ are constants, we will learn how to decouple the two equations of the system (4.1) and thus to solve it. Systems with variable coefficients rarely decouple.

Systems in matrix form

We will use matrix notation for systems. A **matrix** is a rectangular array of numbers or functions. If there are m rows and n columns, the matrix is called an $m \times n$ matrix. For example, a 2×2 matrix would be a square array

$$\begin{bmatrix} a_{11} & a_{12} \\ a_{21} & a_{22} \end{bmatrix}.$$

This matrix uses the standard notation: the number or function that is located in the horizontal row i and vertical column j is a_{ij}. These a_{ij}'s are called the **entries** of the matrix.

A matrix with just one column is called a **column vector**. Boldface letters $\mathbf{v}$, $\mathbf{a}$, and so on will be used for column vectors, and capital letters A, B, and so on will denote more general matrices. **Scalars** (that is, constants and real-valued functions) will be denoted with lowercase italic letters x, a, t, and so on.

It is possible to add matrices and to multiply them, provided that the dimensions of the matrices are compatible. Any two matrices of the same dimensions can be added; each entry of $C = A + B$ is the sum of the corresponding entries of A and B. Thus $c_{ij} = a_{ij} + b_{ij}$.

The **matrix product** $P = AB$ of an $m \times n$ matrix A and a $p \times q$ matrix B can be formed if and only if $n = p$; that is, the number of *columns* of A is equal to the number of *rows* of B. When this compatibility condition holds, P will be a matrix of dimension $m \times q$, with as many rows as A has, and as many columns as B. For example, if A is a 2×2 matrix and B is a two-dimensional column vector (a 2×1 matrix), AB would be a two-dimensional column vector. The formula for the product is

$$\begin{bmatrix} a_{11} & a_{12} \\ a_{21} & a_{22} \end{bmatrix} \begin{bmatrix} b_1 \\ b_2 \end{bmatrix} = \begin{bmatrix} a_{11}b_1 + a_{12}b_2 \\ a_{21}b_1 + a_{22}b_2 \end{bmatrix}.$$

More generally, if A is an $m \times k$ matrix and B is a $k \times n$ matrix, the entries of the product $C = AB$ are computed as follows:

$$c_{ij} = a_{i1}b_{1j} + a_{i2}b_{2j} + \cdots + a_{ik}b_{kj}.$$

Thus, each of the k consecutive entries of row i of A is multiplied by the corresponding entry in column j of B, and then these products are added together. Matrix multiplication satisfies the *associative law:* Given three matrices A, B, and C, of dimensions such that the products $P = AB$ and $Q = BC$ are defined, it always holds that $PC = AQ$; in other words,

$$(AB)C = A(BC).$$

Matrix multiplication also satisfies the *distributive laws:* Assume that A and A' are matrices of the same dimensions and that B is a matrix of dimensions such that AB is defined. Let B' be a matrix of the same dimensions as B. Then

$$(A + A')B = AB + A'B \quad \text{and} \quad A(B + B') = AB + AB'.$$

Matrix algebra shares these properties with the arithmetic of real numbers. However, other familiar rules do not hold for matrix multiplication. The commutative law does not hold in general. If A is a $p \times q$ matrix and B is $q \times p$, then unless $p = q$, AB and BA are of different sizes ($p \times p$ and $q \times q$, respectively); thus $AB \neq BA$. Even if A and B are square matrices, so that $p = q$, it often happens that $AB \neq BA$.

To convert the system (4.1) to matrix form, let the column vector

$$\mathbf{v} = \begin{bmatrix} x \\ y \end{bmatrix}$$

represent the dependent variables, and the matrix

$$A(t) = \begin{bmatrix} p_1(t) & q_1(t) \\ p_2(t) & q_2(t) \end{bmatrix}$$

be the **coefficient matrix**. The vector function

$$\mathbf{r}(t) = \begin{bmatrix} r_1(t) \\ r_2(t) \end{bmatrix}$$

is the **source vector**. To see that the matrix ODE

$$\mathbf{v}' = A(t)\,\mathbf{v} + \mathbf{r}(t)$$

is equivalent to the system (4.1), we have only to multiply the matrix A and the column vector $\mathbf{v}$ to obtain the column vector

$$A(t)\,\mathbf{v} = \begin{bmatrix} p_1(t)\,x + q_1(t)\,y \\ p_2(t)\,x + q_2(t)\,y \end{bmatrix},$$

add the column vector $\mathbf{r}(t)$,

$$\begin{bmatrix} p_1(t)\,x + q_1(t)\,y \\ p_2(t)\,x + q_2(t)\,y \end{bmatrix} + \begin{bmatrix} r_1(t) \\ r_2(t) \end{bmatrix} = \begin{bmatrix} p_1(t)\,x + q_1(t)\,y + r_1(t) \\ p_2(t)\,x + q_2(t)\,y + r_2(t) \end{bmatrix}$$

and write out the matrix equation as

$$\begin{bmatrix} x' \\ y' \end{bmatrix} = \begin{bmatrix} p_1(t)\,x + q_1(t)\,y + r_1(t) \\ p_2(t)\,x + q_2(t)\,y + r_2(t) \end{bmatrix}.$$

Example 4.1.1   Put the system

$$x' = y$$
$$y' = 4x$$

in matrix form.

Solution. We need to find a coefficient matrix

$$A = \begin{bmatrix} a_{11} & a_{12} \\ a_{21} & a_{22} \end{bmatrix}$$

such that

$$A \begin{bmatrix} x \\ y \end{bmatrix} = \begin{bmatrix} a_{11}\,x + a_{12}\,y \\ a_{21}\,x + a_{22}\,y \end{bmatrix} = \begin{bmatrix} y \\ 4x \end{bmatrix}.$$

It follows that $a_{21} = 1$, $a_{12} = 4$, and the other entries are 0. Hence

$$A = \begin{bmatrix} 0 & 1 \\ 4 & 0 \end{bmatrix},$$

and the matrix form of the system is $\mathbf{v}' = A\,\mathbf{v}$.

It is not difficult to check that $x = e^{2t}$, $y = 2 e^{2t}$ is a solution of the system that we considered in Example 4.1.1. We can put this solution in vector notation as

$$\mathbf{v} = \begin{bmatrix} e^{2t} \\ 2 e^{2t} \end{bmatrix}.$$

It is possible to multiply any matrix by a scalar constant or function: Just multiply each entry by the scalar. Thus we can write the preceding vector solution as

$$\mathbf{v} = e^{2t} \begin{bmatrix} 1 \\ 2 \end{bmatrix}.$$

Example 4.1.2 Show that

$$\mathbf{w} = e^{-2t} \begin{bmatrix} 1 \\ -2 \end{bmatrix}$$

is also a solution of the system in Example 4.1.1.

Solution. Let

$$\mathbf{c} = \begin{bmatrix} 1 \\ -2 \end{bmatrix},$$

so that $\mathbf{w} = e^{-2t} \mathbf{c}$. We need to show that

$$\mathbf{w}' = A \cdot (e^{-2t} \mathbf{c}). \tag{4.2}$$

By matrix multiplication,

$$A\mathbf{c} = \begin{bmatrix} 0 & 1 \\ 4 & 0 \end{bmatrix} \begin{bmatrix} 1 \\ -2 \end{bmatrix} = \begin{bmatrix} (0)(1) + (1)(-2) \\ (4)(1) + (0)(-2) \end{bmatrix} = \begin{bmatrix} -2 \\ 4 \end{bmatrix} = -2\mathbf{c}.$$

Thus

$$\mathbf{w}' = \frac{d}{dt}(e^{-2t} \mathbf{c}) = e^{-2t}(-2) \mathbf{c} = e^{-2t} A \mathbf{c}.$$

We can now conclude that equation (4.2) holds, since $e^{-2t} A \mathbf{c} = A \cdot (e^{-2t} \mathbf{c})$. ❑

With vector notation, the existence and uniqueness theorem (Theorem 3.1) for systems looks the same as the existence and uniqueness theorems for first-order ODEs that we encountered in Chapter 2. The following theorem is for linear systems only; it differs from Theorem 3.1 in that the domain of the solution of an IVP involving a *linear* system can be specified.

Theorem 4.1

Suppose that the entries of the coefficient matrix $A(t)$ and the components of the source vector $\mathbf{f}(t)$ are continuous on an interval (a, b) containing the initial point t_0. Then the IVP

$$\mathbf{v}' = A(t)\, \mathbf{v} + \mathbf{f}(t); \ \ \mathbf{v}(t_0) = \mathbf{v}_0$$

has a solution $\mathbf{v}(t)$ that is defined for all $t \in (a, b)$. If $\mathbf{u}(t)$ is another solution of the same IVP, then $\mathbf{u}(t) = \mathbf{v}(t)$ for all $t \in (a, b)$.

Homogeneous systems

A system of ODEs

$$\mathbf{v}' = A(t)\,\mathbf{v}$$

with a zero source vector is said to be **homogeneous**. The following corollary of Theorem 4.1 is a useful property of homogeneous systems.

Corollary 4.1.1

Let $A(t)$ be an $n \times n$ matrix whose entries are continuous on the interval (a, b), and let $\mathbf{v}(t)$ be a solution of the homogeneous system $\mathbf{v}' = A(t)\,\mathbf{v}$. If, for some $t_0 \in (a, b)$, $\mathbf{v}(t_0) = \mathbf{0}$, then $\mathbf{v}(t) \equiv \mathbf{0}$ on (a, b).

Proof Notice that $\mathbf{v}(t)$ and $\mathbf{z}(t) \equiv \mathbf{0}$ are solutions of the IVP

$$\mathbf{v}' = A(t)\,\mathbf{v}; \ \ \mathbf{v}(t_0) = \mathbf{0}.$$

By Theorem 4.1, it follows that $\mathbf{v}(t) = \mathbf{z}(t)$ for all $t \in (a, b)$. ∎

If $\mathbf{v}_1(t)$ and $\mathbf{v}_2(t)$ are vector functions and c_1 and c_2 are scalar constants, the vector function

$$\mathbf{v}_3(t) = c_1\,\mathbf{v}_1(t) + c_2\,\mathbf{v}_2(t) \tag{4.3}$$

is called a **linear combination** of $\mathbf{v}_1$ and $\mathbf{v}_2$ (it is possible to form linear combinations of three or more vector functions in the same way).

Now suppose that $\mathbf{v}_1(t)$ and $\mathbf{v}_2(t)$ in (4.3) are solutions of a homogeneous linear system of ODEs, $\mathbf{v}' = A\mathbf{v}$. Then

$$\begin{aligned}\mathbf{v}_3' &= c_1\,\mathbf{v}_1' + c_2\,\mathbf{v}_2' \\ &= c_1\,A\,\mathbf{v}_1 + c_2\,A\,\mathbf{v}_2.\end{aligned}$$

Therefore, by the distributive law for matrix multiplication,

$$\begin{aligned}\mathbf{v}_3' &= A\,(c_1\,\mathbf{v}_1 + c_2\,\mathbf{v}_2) \\ &= A(t)\,\mathbf{v}_3.\end{aligned}$$

The result of this calculation can be summarized as the following theorem.

Theorem 4.2

Let $\mathbf{v}_1(t), \mathbf{v}_2(t), \ldots, \mathbf{v}_n(t)$ be solutions of a linear homogeneous system, $\mathbf{v}' = A(t)\,\mathbf{v}$. Then every linear combination

$$C_1\mathbf{v}_1 + C_2\mathbf{v}_2 + \cdots + C_n\mathbf{v}_n$$

of these solutions is also a solution.

If we have just one solution $\mathbf{v}_1(t)$, then Theorem 4.2 tells us that the family of functions $c_1\mathbf{v}_1(t)$ is an infinite family of solutions [unless $v_1(t) \equiv 0$, for then the family would not be infinite]. If there are two solutions available, $c_1\mathbf{v}_1(t) + c_2\mathbf{v}_2(t)$ is a larger family of solutions, unless $\mathbf{v}_2(t)$ happens to be a constant multiple of $\mathbf{v}_1(t)$.

In Example 4.1.2 we found that

$$\mathbf{v}_1 = e^{2t}\begin{bmatrix} 1 \\ 2 \end{bmatrix} \quad \text{and} \quad \mathbf{v}_2 = e^{-2t}\begin{bmatrix} 1 \\ -2 \end{bmatrix}$$

are solutions of the system

$$\mathbf{v}' = \begin{bmatrix} 0 & 1 \\ 4 & 0 \end{bmatrix}\mathbf{v}.$$

These solutions can be assembled into a family

$$\mathbf{v} = c_1\,e^{2t}\begin{bmatrix} 1 \\ 2 \end{bmatrix} + c_2\,e^{-2t}\begin{bmatrix} 1 \\ -2 \end{bmatrix}.$$

The functions $\mathbf{v}_1(t)$ and $\mathbf{v}_2(t)$ are said to be **linearly independent** if neither function is equal to a constant scalar multiple of the other. More generally, we will say that a set of vector functions $S = \{\mathbf{v}_1(t), \mathbf{v}_2(t), \ldots, \mathbf{v}_m(t)\}$ is linearly independent if none of the $\mathbf{v}_i(t)$ can be expressed as a linear combination of the other $(m-1)$ vector functions in S.

A similar definition applies to vectors in the plane (or n-dimensional space $\mathbf{R}^n$). We say that a set of constant vectors, $S = \{\mathbf{a}_1, \mathbf{a}_2, \ldots, \mathbf{a}_m\}$ is linearly independent if it is impossible to express any of the vectors in S as a linear combination of the other vectors in S.

Theorem 4.3

Let $A(t)$ be an $n \times n$ matrix whose entries are continuous functions on an interval (a, b), and let t_0 be any point in (a, b). Then solutions $\mathbf{v}_1(t), \mathbf{v}_2(t) \ldots, \mathbf{v}_n(t)$ of the linear homogeneous system $\mathbf{v}' = A(t)\mathbf{v}$ are linearly independent if and only if, as vectors in $\mathbf{R}^n$,

$$\mathbf{v}_1(t_0), \mathbf{v}_2(t_0), \ldots, \mathbf{v}_n(t_0)$$

are linearly independent.

Proof   For simplicity, let $n = 2$. If $\mathbf{v}_1(t_0)$ and $\mathbf{v}_2(t_0)$ are not linearly independent, then we can assume that there is a number c such that $\mathbf{v}_2(t_0) = c\mathbf{v}_1(t_0)$. Furthermore,

$$\mathbf{v}_3(t) = c\,\mathbf{v}_1(t) - \mathbf{v}_2(t)$$

is a linear combination of solutions of $\mathbf{v}' = A(t)\mathbf{v}$ and is therefore also a solution. Since $\mathbf{v}_3(t_0) = \mathbf{0}$, it follows from Corollary 4.1.1 that $\mathbf{v}_3(t) = \mathbf{0}$ for all t, and hence $\mathbf{v}_2(t) = c\mathbf{v}_1(t)$ for all $t \in (a, b)$.

Thus, if $\mathbf{v}_1$ and $\mathbf{v}_2$ *are* linearly independent as vector functions on (a, b), the initial vectors $\mathbf{v}_1(t_0)$ and $\mathbf{v}_2(t_0)$ must also be linearly independent.

To complete the proof we have to show that if $\mathbf{v}_1(t_0)$ and $\mathbf{v}_2(t_0)$ are linearly independent, then the vector functions $\mathbf{v}_1$ and $\mathbf{v}_2$ are linearly independent. If there were a scalar c such that

$$\mathbf{v}_2(t) = c \, \mathbf{v}_1(t)$$

or all $t \in (a, b)$, then

$$\mathbf{v}_2(t_0) = c \, \mathbf{v}_2(t_0).$$

This contradicts our assumption that $\mathbf{v}_1(t_0)$ and $\mathbf{v}_2(t_0)$ are linearly independent. ∎

Theorem 4.3 implies that the solution vectors $\mathbf{v}_1(t), \mathbf{v}_2(t), \ldots, \mathbf{v}_n(t)$ are either linearly independent for all t in the domain (a, b), or they are linearly dependent for all $t \in (a, b)$. They cannot be linearly independent at some points of the domain (a, b) and linearly dependent at others.

Example 4.1.3   Show that the solutions

$$\mathbf{v}_1 = e^{2t} \begin{bmatrix} 1 \\ 2 \end{bmatrix} \quad \text{and} \quad \mathbf{v}_2 = e^{-2t} \begin{bmatrix} 1 \\ -2 \end{bmatrix}$$

of the system

$$\mathbf{v}' = \begin{bmatrix} 0 & 1 \\ 4 & 0 \end{bmatrix} \mathbf{v}$$

are linearly independent.

Solution. Since $\mathbf{v}_2(0) = \begin{bmatrix} 1 \\ -2 \end{bmatrix}$ is not a scalar multiple of $\mathbf{v}_2(0) = \begin{bmatrix} 1 \\ 2 \end{bmatrix}$, the two solutions are independent. ❑

We are now almost ready for a full description of the general solution of a homogeneous linear system of ODEs. Just one more fact from linear algebra is needed: *If $S = \{\mathbf{a}_1, \mathbf{a}_2, \ldots, \mathbf{a}_n\}$ is a set of linearly independent vectors in $\mathbf{R}^n$, then every vector in $\mathbf{R}^n$ can be expressed as a linear combination of the vectors in S.*

Theorem 4.4

Let $A(t)$ be an $n \times n$ matrix of functions, all continuous on an interval (a, b), and let $t_0 \in (a, b)$. Given a set S of n linearly independent vectors in $\mathbf{R}^n$, say, $S = \{\mathbf{a}_1, \mathbf{a}_2, \ldots, \mathbf{a}_n\}$, let $\mathbf{v}_1(t), \mathbf{v}_2(t), \ldots, \mathbf{v}_n(t)$ be the solutions of

$$\mathbf{v}_i' = A(t) \, \mathbf{v}_i \tag{4.4}$$

with initial conditions $\mathbf{v}_i(t_0) = \mathbf{a}_i$.

Then if $\mathbf{v} = \mathbf{w}(t)$ is any solution of the system (4.4) that is defined on the interval (a, b), there are unique constants $c_1, c_2, \ldots, c_n$ such that

$$\mathbf{w} = c_1 \mathbf{v}_1 + c_2 \mathbf{v}_2 + \cdots + c_n \mathbf{v}_n. \tag{4.5}$$

Proof To keep the notation as simple as possible, we take $n = 2$. The vector $\mathbf{w}(t_0)$ can be expressed as a linear combination

$$\mathbf{w}(t_0) = c_1\,\mathbf{a}_1 + c_2\,\mathbf{a}_2$$
$$= c_1\,\mathbf{v}_1(t_0) + c_2\,\mathbf{v}_2(t_0).$$

Hence $\mathbf{z}(t) = \mathbf{w}(t) - (c_1\,\mathbf{v}_1(t) + c_2\,\mathbf{v}_2(t))$ a solution of the IVP

$$\mathbf{z}' = A(t)\,\mathbf{z}; \; \mathbf{z}(t_0) = \mathbf{0}.$$

By Corollary 4.1.1, $\mathbf{z}(t) \equiv 0$ on (a, b). This shows that there are constants c_i such that equation (4.5) holds. We must show that c_1 and c_2 are the only pair of constants such that (4.5) holds. Assume that another pair of constants c_i' exists, with $\mathbf{w}(t) = c_1'\,\mathbf{v}_1(t) + c_2'\,\mathbf{v}_2(t)$, and that $c_2' \neq c_2$. Then, for $t = t_0$,

$$c_1\,\mathbf{a}_1 + c_2\,\mathbf{a}_2 = c_1'\,\mathbf{a}_1 + c_2'\,\mathbf{a}_2$$

and hence

$$\mathbf{a}_2 = -\left(\frac{c_1 - c_1'}{c_2 - c_2'}\right)\mathbf{a}_1.$$

Since the $\mathbf{a}_i$ are linearly independent, this is impossible. ∎

The **general solution** of a system of ODEs is a family of solutions that encompasses all solutions of the system.

Corollary 4.1.2

Let $A(t)$ be an $n \times n$ matrix of functions that are continuous on the interval (a, b). Let $t_0 \in (a, b)$, and let $\mathbf{v}_1(t), \mathbf{v}_2(t), \ldots, \mathbf{v}_n(t)$ be solutions of $\mathbf{v}' = A(t)\mathbf{v}$ such that the set of vectors

$$S = \{\mathbf{v}_1(t_0), \mathbf{v}_2(t_0), \ldots, \mathbf{v}_n(t_0)\}$$

is linearly independent. Then the general solution of $\mathbf{v}' = A(t)\mathbf{v}$ is

$$\mathbf{v} = c_1\,\mathbf{v}_1 + c_2\,\mathbf{v}_2 + \cdots + c_n\mathbf{v}_n.$$

Example 4.1.4   Solve the IVP

$$\begin{array}{ll} x' = y; & x(0) = 1 \\ y' = 4x; & y(0) = 0. \end{array}$$

Solution. In matrix form, this system is

$$\begin{bmatrix} x' \\ y' \end{bmatrix} = \begin{bmatrix} 0 & 1 \\ 4 & 0 \end{bmatrix}\begin{bmatrix} x \\ y \end{bmatrix}.$$

In Example 4.1.3 we identified linearly independent solutions

$$\mathbf{v}_1(t) = e^{2t} \begin{bmatrix} 1 \\ 2 \end{bmatrix} \quad \text{and} \quad \mathbf{v}_2(t) = e^{-2t} \begin{bmatrix} 1 \\ -2 \end{bmatrix}.$$

Therefore, the general solution is $\mathbf{v} = c_1\mathbf{v}_1 + c_2\mathbf{v}_2$, and we need to determine values for the coefficients so that

$$\mathbf{v}(0) = \begin{bmatrix} 1 \\ 0 \end{bmatrix}.$$

Thus we need to solve

$$c_1 \begin{bmatrix} 1 \\ 2 \end{bmatrix} + c_2 \begin{bmatrix} 1 \\ -2 \end{bmatrix} = \begin{bmatrix} 1 \\ 0 \end{bmatrix},$$

which is equivalent to the pair of equations

$$c_1 + c_2 = 1$$
$$2c_1 - 2c_2 = 0.$$

Thus $c_1 = c_2 = \frac{1}{2}$. The solution is $x = \frac{1}{2}e^{2t} + \frac{1}{2}e^{-2t} = \cosh(2t)$ and $y = e^{2t} - e^{-2t} = 2\sinh(2t)$. ☐

EXERCISES

1. Show that

$$\mathbf{v}_1(t) = e^{3t} \begin{bmatrix} 1 \\ 1 \end{bmatrix} \quad \text{and} \quad \mathbf{v}_2(t) = e^{-t} \begin{bmatrix} 1 \\ -1 \end{bmatrix}$$

are solutions of the system

$$\begin{bmatrix} x' \\ y' \end{bmatrix} = \begin{bmatrix} 1 & 2 \\ 2 & 1 \end{bmatrix} \begin{bmatrix} x \\ y \end{bmatrix}.$$

Find the general solution.

2. Show that

$$\begin{bmatrix} x(t) \\ y(t) \end{bmatrix} = \begin{bmatrix} 3t - 5 \\ -6t + 4 \end{bmatrix}$$

is a solution of the system

$$\begin{bmatrix} x' \\ y' \end{bmatrix} = \begin{bmatrix} 1 & 2 \\ 2 & 1 \end{bmatrix} \begin{bmatrix} x \\ y \end{bmatrix} + \begin{bmatrix} 9t \\ 0 \end{bmatrix}.$$

In Exercises 3–7, solve the IVP.

3. $\begin{bmatrix} x' \\ y' \end{bmatrix} = \begin{bmatrix} 2 & 0 \\ 0 & 3 \end{bmatrix} \begin{bmatrix} x \\ y \end{bmatrix}$; $\begin{bmatrix} x(0) \\ y(0) \end{bmatrix} = \begin{bmatrix} 1 \\ 2 \end{bmatrix}$

4. $\begin{bmatrix} x' \\ y' \end{bmatrix} = \begin{bmatrix} -1 & 0 \\ 0 & 1 \end{bmatrix} \begin{bmatrix} x \\ y \end{bmatrix}$; $\begin{bmatrix} x(0) \\ y(0) \end{bmatrix} = \begin{bmatrix} 2 \\ 0 \end{bmatrix}$

5. $\dfrac{d}{dt} \begin{bmatrix} x \\ y \end{bmatrix} = \begin{bmatrix} 1 & 2 \\ 2 & 1 \end{bmatrix} \begin{bmatrix} x \\ y \end{bmatrix}$; $\begin{bmatrix} x(0) \\ y(0) \end{bmatrix} = \begin{bmatrix} 1 \\ 0 \end{bmatrix}$

(See Exercise 1.)

6. $\dfrac{d}{dt} \begin{bmatrix} x \\ y \end{bmatrix} = \begin{bmatrix} 1 & 2 \\ 2 & 1 \end{bmatrix} \begin{bmatrix} x \\ y \end{bmatrix}$; $\begin{bmatrix} x(0) \\ y(0) \end{bmatrix} = \begin{bmatrix} 0 \\ 5 \end{bmatrix}$

(See Exercise 1.)

7. $\begin{bmatrix} x' \\ y' \end{bmatrix} = \begin{bmatrix} 0 & 0 \\ -1 & 1 \end{bmatrix} \begin{bmatrix} x \\ y \end{bmatrix}$; $\begin{bmatrix} x(0) \\ y(0) \end{bmatrix} = \begin{bmatrix} 1 \\ 1 \end{bmatrix}$

In Exercises 8–14, write the IVP in matrix form. For each homogeneous system, use Corollary 4.1.2 to verify that the given family of solutions is actually the general solution. Finally, use the general solution to solve the IVP.

8. $x' = 4x + 3y;\ x(0) = 2$
$y' = 3x - 4y;\ y(0) = 4$

General solution: $\begin{bmatrix} x \\ y \end{bmatrix} = c_1 e^{5t} \begin{bmatrix} 3 \\ 1 \end{bmatrix} + c_2 e^{-5t} \begin{bmatrix} 1 \\ -3 \end{bmatrix}$

9. $\begin{cases} x' = 2x + 3y; \ x(0) = -1 \\ y' = -x - 2y; \ y(0) = 1 \end{cases}$

General solution: $\begin{bmatrix} x \\ y \end{bmatrix} = c_1 e^t \begin{bmatrix} 3 \\ -1 \end{bmatrix} + c_2 e^{-t} \begin{bmatrix} 1 \\ -1 \end{bmatrix}$

10. $\begin{cases} x' = x + 2y + 2e^t; \ x(0) = 0 \\ y' = -x - y; \quad\quad\ y(0) = 0 \end{cases}$

General solution:

$$\begin{bmatrix} x \\ y \end{bmatrix} = e^t \begin{bmatrix} 2 \\ -1 \end{bmatrix} + c_1 \begin{bmatrix} 2\cos(t) \\ -\cos(t) - \sin(t) \end{bmatrix} + c_2 \begin{bmatrix} 2\sin(t) \\ -\sin(t) + \cos(t) \end{bmatrix}$$

11. $\begin{cases} x' = 2x - y; \quad\quad x(0) = 1 \\ y' = 4x - 2y + 2; \ y(0) = -1 \end{cases}$

General solution:

$$\begin{bmatrix} x \\ y \end{bmatrix} = \begin{bmatrix} -t^2 \\ 2t - 2t^2 \end{bmatrix} + c_1 \begin{bmatrix} 1 \\ 2 \end{bmatrix} + c_2 \begin{bmatrix} 2t + 1 \\ 4t \end{bmatrix}$$

12. $\begin{cases} x' = 4x + 3y; \ x(0) = 7 \\ y' = -2x - y; \ y(0) = -2 \end{cases}$

General solution: $\begin{bmatrix} x \\ y \end{bmatrix} = c_1 e^t \begin{bmatrix} 1 \\ -1 \end{bmatrix} + c_2 e^{2t} \begin{bmatrix} 3 \\ -2 \end{bmatrix}$

13. $\begin{cases} x' = y + \cosh t; \ x(0) = 2 \\ y' = x + \sinh t; \ y(0) = 2 \end{cases}$

General solution: $\begin{bmatrix} x \\ y \end{bmatrix} = (c_1 + t) \begin{bmatrix} \cosh t \\ \sinh t \end{bmatrix} + c_2 \begin{bmatrix} \sinh t \\ \cosh t \end{bmatrix}$

14. $\begin{cases} x' = 2x + 2y + e^t; \ x(0) = 0 \\ y' = 3x + y; \quad\quad\ y(0) = -2 \end{cases}$

General solution:

$$\begin{bmatrix} x \\ y \end{bmatrix} = c_1 e^{-t} \begin{bmatrix} 2 \\ -3 \end{bmatrix} + c_2 e^{4t} \begin{bmatrix} 1 \\ 1 \end{bmatrix} + e^t \begin{bmatrix} -5 \\ 3 \end{bmatrix}$$

15. A linear second-order ODE is an ODE that can be put in the form

$$y'' + p(t)y' + q(t)y = r(t). \tag{4.6}$$

The functions $p(t)$ and $q(t)$ are the *coefficients* of the ODE, and $r(t)$ is the *source*. Show that a linear second-order ODE can be replaced with a linear system of first-order ODEs, and write the system corresponding to equation (4.6) in matrix form.

16. State and prove, as a corollary to Theorem 4.1, an existence and uniqueness theorem for second-order linear ODEs.

4.2 Systems with Constant Coefficients

The general solution of the linear homogeneous ODE,

$$\frac{dy}{dt} = a(t)\, y,$$

is found by substituting $y = C\, e^{f(t)}$. This yields

$$C\, e^{f(t)} f'(t) = a(t)(C\, e^{f(t)}).$$

After dividing through by $Ce^{f(t)}$, we obtain $f'(t) = a(t)$. It follows that

$$y = C\, e^{\int a(t)\, dt}$$

is a family of solutions. By Proposition 1.2.1, it is the general solution.

This idea can be used to solve a system of linear homogeneous ODEs if the coefficient matrix of the system is a constant matrix. Thus, let us consider a homogeneous system

$$\mathbf{v}' = A\mathbf{v}, \tag{4.7}$$

where A is an $n \times n$ constant matrix. We will substitute $\mathbf{v} = e^{st}\mathbf{b}$, where $\mathbf{b}$ is a constant vector and s is a constant that will be determined. Then $\mathbf{v}' = se^{st}\mathbf{b}$, and $A\mathbf{v} = e^{st}A\mathbf{b}$, so with this substitution, (4.7) becomes

$$se^{st}\mathbf{b} = e^{st}A\mathbf{b}.$$

We may divide by the nonzero factor e^{st} to obtain

$$s\mathbf{b} = A\mathbf{b}. \tag{4.8}$$

Equation (4.8) will hold only if the vector $A\mathbf{b}$ is a scalar multiple of $\mathbf{b}$ (and s is the scalar). A vector $\mathbf{b} \neq \mathbf{0}$ with this property is called a **characteristic vector** of the matrix A, and s is a **characteristic root**[1] of that matrix. Since there is an association between s and $\mathbf{b}$, we say that $\mathbf{b}$ **belongs to** s, and s belongs to $\mathbf{b}$. We can state what has been found as the following theorem.

Theorem 4.5

If $\mathbf{b}$ is a characteristic vector of the matrix A, and s is a characteristic root belonging to $\mathbf{b}$, then

$$\mathbf{v}(t) = e^{st}\mathbf{b}$$

is a solution of the system (4.7).

Let

$$I = \begin{bmatrix} 1 & 0 \\ 0 & 1 \end{bmatrix}$$

be the 2×2 **identity matrix**. The matrix I has the property that $I\,\mathbf{b} = \mathbf{b}$ for any vector $\mathbf{b}$.

For any scalar r, the matrix

$$rI = \begin{bmatrix} r & 0 \\ 0 & r \end{bmatrix}$$

is called a **scalar matrix**. If $A = rI$ is a scalar matrix and $\mathbf{b} \neq \mathbf{0}$, then $A\mathbf{b} = s\mathbf{b}$ if and only if $s = r$; thus r is the only characteristic root. Every nonzero vector is a characteristic vector belonging to r. The family of solutions of

$$\mathbf{v}' = rI\mathbf{v}$$

therefore has the form $\mathbf{v} = \mathbf{c}e^{rt}$; here the constant vector $\mathbf{c}$ is arbitrary.

If A isn't a scalar matrix, we will need to find a way to calculate characteristic roots and vectors of 2×2 matrices in order to deploy Theorem 4.5. Let

$$A = \begin{bmatrix} a & b \\ c & d \end{bmatrix}$$

---

[1]Linear algebra texts use the terminology *eigenvector* and *eigenvalue* (*eigen* is German for *characteristic*).

be a matrix, assumed not to be scalar. If $\mathbf{b} \neq \mathbf{0}$ is a characteristic vector and s is its characteristic root, then

$$A\mathbf{b} = sI\,\mathbf{b},$$

which can be rewritten as

$$(A - sI)\mathbf{b} = \mathbf{0}. \tag{4.9}$$

A matrix C with the property that there is a nonzero vector $\mathbf{d}$ such that $C\mathbf{d} = \mathbf{0}$ is said to be **singular**. We can interpret (4.9) as saying that the matrix $A - sI$ is singular. To find the values of s for which the matrix $A - sI$ is singular, we will use the determinant.

Let

$$B = \begin{bmatrix} a & c \\ b & d \end{bmatrix}$$

be a 2×2 matrix whose entries can be constants or functions. The **determinant** of B is defined to be a scalar that is calculated by the formula $\det(B) = a\,d - b\,c$.

Example 4.2.1   Find the determinant of

$$\begin{bmatrix} 9 & 7 \\ 3 & 5 \end{bmatrix}.$$

Solution.

$$\det \begin{bmatrix} 9 & 7 \\ 3 & 5 \end{bmatrix} = 9 \cdot 5 - 3 \cdot 7 = 24 \qquad \Box$$

The determinant can be defined for square matrices of any size, but in this text we will not consider determinants of matrices larger than 2×2. The following proposition will be familiar to all who have studied linear algebra, where it is shown to be valid for $n \times n$ matrices.

Proposition 4.2.1

Let $C = \begin{bmatrix} m & n \\ p & q \end{bmatrix}$ be a 2×2 matrix. Then C is singular if and only if $\det(C) = 0$.

Proof   Assume that C is singular. Then there is a vector $\mathbf{b} = \begin{bmatrix} h \\ k \end{bmatrix}$ (where h and k are not both 0), such that $C\mathbf{b} = \mathbf{0}$. By matrix multiplication,

$$C\mathbf{b} = \begin{bmatrix} mh + nk \\ ph + qk \end{bmatrix} = \begin{bmatrix} 0 \\ 0 \end{bmatrix}.$$

Assuming for definiteness that $h \neq 0$,

$$m = -\frac{k}{h}n \quad \text{and} \quad p = -\frac{k}{h}q.$$

It follows that

$$\det(C) = mq - pn = -\frac{k}{h}nq + \frac{k}{h}qn = 0.$$

Conversely, assume that $\det(C) = 0$. Since $mq = np$, you can easily show that

$$C\begin{bmatrix} n \\ -m \end{bmatrix} = C\begin{bmatrix} q \\ -p \end{bmatrix} = \mathbf{0}.$$

If C has any nonzero entries, then at least one of the vectors $\begin{bmatrix} n \\ -m \end{bmatrix}$ and $\begin{bmatrix} q \\ -p \end{bmatrix}$ is nonzero, and hence C is singular.

If all of the entries of C are equal to 0, then $C\mathbf{b} = \mathbf{0}$ for every vector $\mathbf{b}$, and hence C is singular. ∎

By equation (4.9) and Proposition 4.2.1, s is a characteristic root of A if and only if

$$\det(A - sI) = 0.$$

Since

$$A - sI = \begin{bmatrix} a-s & b \\ c & d-s \end{bmatrix},$$

we have the following formula:

$$\det(A - sI) = (a-s)(d-s) - bc$$
$$= s^2 - (a+d)s + (ad - bc).$$

The sum of the diagonal entries of a square matrix M is called the **trace** of M, and it is denoted $\mathrm{tr}(M)$. Thus, $\mathrm{tr}(A) = a + d$. Since $\det(A) = ad - bc$, we see that

$$\det(A - sI) = s^2 - \mathrm{tr}(A)\, s + \det(A).$$

Thus, the characteristic roots of A are the roots of the quadratic equation

$$s^2 - \mathrm{tr}(A)\, s + \det(A) = 0. \tag{4.10}$$

Equation (4.10) is called the **characteristic equation** of A.

Let A be a 2×2 matrix that is not of scalar type. If r_1 is a characteristic root of a matrix A, a characteristic vector $\mathbf{b} = \begin{bmatrix} h \\ k \end{bmatrix} \neq \mathbf{0}$ belonging to r_1 satisfies the matrix equation,

$$A\,\mathbf{b} = r_1\,\mathbf{b}.$$

This matrix equation can be replaced by two linear equations in which the unknowns are the components h and k of $\mathbf{b}$. If we were to graph the two equations, we would find that they represent the same line. Thus, any vector $\mathbf{b}$ whose components satisfy one of the equations will automatically satisfy the other, and there is an infinite family of solutions.

□ Example 4.2.2   Find the characteristic roots and vectors of

$$A = \begin{bmatrix} 1 & 2 \\ 3 & 2 \end{bmatrix}. \tag{4.11}$$

Solution. Since $\text{tr}(A) = 1 + 2 = 3$ and $\det(A) = (1)(2) - (3)(2) = -4$, the characteristic equation of A is

$$s^2 - 3s - 4 = 0$$

and it follows that the characteristic roots are $s = -1$ and $s = 4$.

The characteristic vectors belonging to $s = -1$ satisfy the equation $A\mathbf{b} = -\mathbf{b}$. Set

$$\mathbf{b} = \begin{bmatrix} h \\ k \end{bmatrix}.$$

Then

$$A\mathbf{b} = \begin{bmatrix} h + 2k \\ 3h + 2k \end{bmatrix} = \begin{bmatrix} -h \\ -k \end{bmatrix}.$$

Hence $h + 2k = -h$ and $3h + 2k = -k$. Both of these equations reduce to $k = -h$, so any nonzero vector that is a scalar multiple of

$$\mathbf{b}_1 = \begin{bmatrix} 1 \\ -1 \end{bmatrix}$$

is a characteristic vector belonging to -1.

To find a characteristic vector belonging to $s = 4$, we have to solve

$$A\mathbf{b} = \begin{bmatrix} 1 & 2 \\ 3 & 2 \end{bmatrix} \begin{bmatrix} h \\ k \end{bmatrix} = \begin{bmatrix} 4h \\ 4k \end{bmatrix}.$$

This is equivalent to solving the equations

$$h + 2k = 4h$$
$$3h + 2k = 4k.$$

These equations both reduce to $3h = 2k$, so it follows that any nonzero scalar multiple of

$$\mathbf{b}_2 = \begin{bmatrix} 2 \\ 3 \end{bmatrix}$$

is a characteristic vector belonging to 4.   ❑

To determine the general solution of a system of ODEs, it is helpful to know that solutions corresponding to different characteristic roots are linearly independent. The following theorem from linear algebra can be applied for this purpose.

Theorem 4.6

Let $s_1, s_2, \ldots, s_m$ be distinct characteristic roots of a matrix A, and let $\mathbf{b}_1, \mathbf{b}_2, \ldots, \mathbf{b}_m$ be characteristic vectors belonging to them. Then $\{\mathbf{b}_1, \mathbf{b}_2, \ldots, \mathbf{b}_m\}$ is linearly independent.

Proof for $m = 2$.   Suppose that $\mathbf{b}_2 = c\mathbf{b}_1$. Then $A\mathbf{b}_2 = cA\mathbf{b}_1 = cs_1\mathbf{b}_1 = s_1\mathbf{b}_2$. However, $\mathbf{b}_2$ is a characteristic vector belonging to s_2. Hence $A\mathbf{b}_2 = s_2\mathbf{b}_2$ It follows that $s_1 = s_2$, a contradiction. ∎

Corollary 4.2.2

Let A be a 2×2 matrix that has two distinct characteristic roots s_1 and s_2, and let $\mathbf{b}_1$ and $\mathbf{b}_2$ be characteristic vectors belonging to them. Then the general solution of $\mathbf{v}' = A\mathbf{v}$ is

$$\mathbf{v}(t) = c_1 e^{s_1 t}\mathbf{b}_1 + c_2 e^{s_2 t}\mathbf{b}_2. \tag{4.12}$$

Proof   Let $\mathbf{v}_1(t) = e^{s_1 t}\mathbf{b}_1$ and $\mathbf{v}_2(t) = e^{s_2 t}\mathbf{b}_2$. Then by Theorem 4.6, $\mathbf{v}_1(0) = \mathbf{b}_1$ and $\mathbf{v}_2(0) = \mathbf{b}_2$ are linearly independent. By Corollary 4.1.2 on page 143, it follows that the general solution can be expressed as in (4.12). ∎

Example 4.2.3   Find the general solution of the system

$$x' = 4x - 4y$$
$$y' = 3x - 3y.$$

Solution. Let

$$\mathbf{v} = \begin{bmatrix} x \\ y \end{bmatrix} \quad \text{and} \quad A = \begin{bmatrix} 4 & -4 \\ 3 & -3 \end{bmatrix}$$

so that our system is equivalent to $\mathbf{v}' = A\mathbf{v}$. Since $\operatorname{tr}(A) = 1$ and $\det(A) = 0$, the characteristic equation is $s^2 - s = 0$ and the characteristic roots are 0 and 1. Let

$$\mathbf{b}_1 = \begin{bmatrix} h \\ k \end{bmatrix}$$

be a characteristic vector belonging to 0. Then

$$4h - 4k = 0$$
$$3h - 3k = 0.$$

This reduces to $h = k$, so we can take

$$\mathbf{b}_1 = \begin{bmatrix} 1 \\ 1 \end{bmatrix}.$$

Now suppose that

$$\mathbf{b}_2 = \begin{bmatrix} h \\ k \end{bmatrix}$$

is a characteristic vector belonging to 1. Then

$$4h - 4k = h$$
$$3h - 3k = k,$$

so that $3h = 4k$. Hence we can set

$$\mathbf{b}_2 = \begin{bmatrix} 4 \\ 3 \end{bmatrix}.$$

The general solution to the matrix equation is

$$\mathbf{v} = c_1 e^{0t} \begin{bmatrix} 1 \\ 1 \end{bmatrix} + c_2 e^{1t} \begin{bmatrix} 4 \\ 3 \end{bmatrix}.$$

Thus $x = c_1 + 4c_2 e^t$ and $y = c_1 + 3c_2 e^t$. ❑

Example 4.2.4 Solve the IVP

$$\mathbf{v}' = \begin{bmatrix} 2 & 4 \\ 3 & -2 \end{bmatrix} \mathbf{v}; \ \mathbf{v}(0) = \begin{bmatrix} 1 \\ 0 \end{bmatrix}.$$

Solution. The trace of the coefficient matrix is 0, the determinant is -16, and the characteristic equation is $s^2 - 16 = 0$. The characteristic roots are ± 4. To find a characteristic vector belonging to 4, we solve

$$\begin{bmatrix} 2 & 4 \\ 3 & -2 \end{bmatrix} \begin{bmatrix} h \\ k \end{bmatrix} = \begin{bmatrix} 4h \\ 4k \end{bmatrix}$$

or

$$2h + 4k = 4h$$
$$3h - 2k = 4k.$$

Both of these equations reduce to $2k = h$, so

$$\mathbf{b}_1 = \begin{bmatrix} 2 \\ 1 \end{bmatrix}$$

is a characteristic vector belonging to 4. Turning to -4,

$$\mathbf{b}_2 = \begin{bmatrix} h \\ k \end{bmatrix}$$

is a characteristic vector if

$$2h + 4k = -4h$$
$$3h - 2k = -4k.$$

These equations reduce to $2k = -3h$ so we will put $h = 2$ and $k = -3$.

The general solution is

$$\mathbf{v}(t) = c_1 e^{4t} \begin{bmatrix} 2 \\ 1 \end{bmatrix} + c_2 e^{-4t} \begin{bmatrix} 2 \\ -3 \end{bmatrix}.$$

Thus

$$\mathbf{v}(0) = \begin{bmatrix} 2c_1 + 2c_2 \\ c_1 - 3c_2 \end{bmatrix}.$$

Since the initial condition specifies that $\mathbf{v}(0) = \begin{bmatrix} 1 \\ 0 \end{bmatrix}$, we can find c_1 and c_2 by solving

$$2c_1 + 2c_2 = 1$$
$$c_1 - 3c_2 = 0.$$

The solutions are $c_1 = 3/8$, $c_2 = 1/8$. We can now assemble the solution of the IVP:

$$\mathbf{v} = \begin{bmatrix} \frac{3}{4}e^{4t} + \frac{1}{4}e^{-4t} \\ \frac{3}{8}e^{4t} - \frac{3}{8}e^{-4t} \end{bmatrix}. \qquad \Box$$

Double characteristic roots

If the characteristic equation of a matrix A has a double root r, and $\mathbf{b}_1$ is a characteristic vector, then the system $\mathbf{v}' = A\mathbf{v}$ has the family of solutions $\mathbf{v} = c\, e^{rt}\, \mathbf{b}_1$. If there is a second, independent characteristic vector $\mathbf{b}_2$ belonging to r, then every vector in the plane can be expressed as a linear combination of $\mathbf{b}_1$ and $\mathbf{b}_2$. Since

$$A(c_1\mathbf{b}_1 + c_2\mathbf{b}_2) = c_1 A\mathbf{b}_1 + c_2 A\mathbf{b}_2$$
$$= r(c_1\mathbf{b}_1 + c_2\mathbf{b}_2),$$

every vector in the plane is a characteristic vector belonging to r; that is, $A\mathbf{v} = rI\mathbf{v}$ for all $\mathbf{v}$. Hence $A = rI$ is a scalar matrix, and the general solution of $\mathbf{v}' = A\mathbf{v}$ is $\mathbf{v} = e^{rt}\mathbf{c}$, where $\mathbf{c}$ is an arbitrary constant vector.

Not all 2×2 matrices with double characteristic roots are scalar matrices, and here is an example of such a matrix. The characteristic polynomial of

$$A = \begin{bmatrix} r & 0 \\ 1 & r \end{bmatrix}$$

is $s^2 - 2rs + r^2 = (s - r)^2$, so r is a double characteristic root. A is not a scalar matrix, so it cannot have two independent characteristic vectors. To see how to find the general solution of $\mathbf{v}' = A\mathbf{v}$, let us put

$$\mathbf{v} = \begin{bmatrix} x \\ y \end{bmatrix}$$

and write the matrix equation as the system

$$x' = rx$$
$$y' = x + ry.$$

This system is uncoupled. The solution of the first equation is $x = c\,e^{rt}$, and we substitute this into the second equation to obtain

$$y' = ry + c\,e^{rt}.$$

We will solve this inhomogeneous first-order equation by the method of variation of constants, starting with the homogeneous solution $y_h(t) = e^{rt}$. Set $y = w\,e^{rt}$, where w is a new dependent variable, to obtain

$$w'\,e^{rt} + rw\,e^{rt} = rw\,e^{rt} + c\,e^{rt},$$

which can be simplified as $w' = c$. Thus $w = ct + b$ and we have $y = (ct + b)\,e^{rt}$. In vector form,

$$\mathbf{v} = e^{rt}\left(c\begin{bmatrix}1\\t\end{bmatrix} + d\begin{bmatrix}0\\1\end{bmatrix}\right).$$

This solution is a linear combination of two solutions,

$$\mathbf{v}_1(t) = e^{rt}\begin{bmatrix}1\\t\end{bmatrix} = e^{rt}(\mathbf{i} + t\mathbf{j})$$

and

$$\mathbf{v}_2(t) = e^{rt}\begin{bmatrix}0\\1\end{bmatrix} = e^{rt}\mathbf{j},$$

where $\mathbf{i}$ and $\mathbf{j}$ are the standard basis vectors,

$$\mathbf{i} = \begin{bmatrix}1\\0\end{bmatrix} \quad \text{and} \quad \mathbf{j} = \begin{bmatrix}0\\1\end{bmatrix}.$$

The solution $\mathbf{v}_2$ derives from the characteristic root r and its characteristic vector, $\mathbf{j}$.

   Let us now turn to an arbitrary matrix A that has a double characteristic root r. We will assume that A is not a scalar matrix, since scalar matrices are handled differently.

   Following the preceding example, we will try to derive a solution of the form $\mathbf{v}(t) = e^{rt}(t\mathbf{b} + \mathbf{c})$. By the product rule for differentiation,

$$\mathbf{v}' = e^{rt}[r(t\mathbf{b} + \mathbf{c}) + \mathbf{b}].$$

Since $\mathbf{v}' = A\mathbf{v}$ and

$$A\mathbf{v} = e^{rt}(tA\mathbf{b} + A\mathbf{c}),$$

it follows that $\mathbf{v}$ is a solution if and only if for all t,

$$r(t\mathbf{b} + \mathbf{c}) + \mathbf{b} = t A\mathbf{b} + A\mathbf{c}. \tag{4.13}$$

Setting $t = 0$ in (4.13), we have

$$\mathbf{b} = (A - rI)\mathbf{c}. \tag{4.14}$$

On the other hand, if we differentiate (4.13) with respect to t, we have

$$A\mathbf{b} = r\mathbf{b}. \tag{4.15}$$

Equations (4.14) and (4.15) must both be satisfied. The following theorem from linear algebra tells us that since r is a double characteristic root of A, equation (4.15) will be satisfied automatically if (4.14) holds.

Theorem 4.7   **Cayley-Hamilton Theorem**

Let B be a 2×2 matrix, and let

$$s^2 - \operatorname{tr}(B)\, s + \det(B) = 0$$

be its characteristic equation. Let B^2 be the matrix formed by multiplying the matrix by itself. Then

$$B^2 - \operatorname{tr}(B)\, B + \det(B)I = 0I$$

Although stated here in terms of 2×2 matrices, the Cayley-Hamilton theorem holds for square matrices of any size. It can be summarized to say that "every square matrix satisfies its own characteristic equation."

Proof of the Cayley-Hamilton theorem   If $B = \begin{bmatrix} a & b \\ c & d \end{bmatrix}$, then $\operatorname{tr}(B) = a + d$ and $\det(B) = ad - bc$. By matrix multiplication,

$$B^2 = \begin{bmatrix} a^2 + bc & ab + bd \\ ac + cd & bc + d^2 \end{bmatrix}$$

and

$$\operatorname{tr}(B)\, B = \begin{bmatrix} a^2 + ad & b(a + d) \\ c(a + d) & ad + d^2 \end{bmatrix}.$$

It follows that

$$B^2 - \operatorname{tr}(B)\, B = \begin{bmatrix} bc - ad & 0 \\ 0 & bc - ad \end{bmatrix} = -\det(B)\, I. \qquad \blacksquare$$

Since r is a double characteristic root of A, its characteristic equation is $(s - r)^2 = 0$. By the Cayley-Hamilton theorem, $(A - rI)^2 = 0I$. Since $A \neq rI$, we can choose a vector $\mathbf{c} \neq \mathbf{0}$ that is *not* a characteristic vector: $A\mathbf{c} \neq r\mathbf{c}$. Put $\mathbf{b} = (A - rI)\mathbf{c}$; then (4.14) holds. Furthermore,

$$(A - rI)\mathbf{b} = (A - rI)^2\mathbf{c} = 0I\mathbf{c} = \mathbf{0}.$$

It follows that $A\mathbf{b} = r\mathbf{b}$: Thus (4.15) also holds. Hence $\mathbf{v}(t) = e^{s_0 t}(t\mathbf{b} + \mathbf{c})$ is a solution of the system $\mathbf{v}' = A\mathbf{v}$.

Matrices with Double Characteristic Roots

Let $A \neq rI$ be a 2×2 matrix that has a double characteristic root $s = r$. Choose a vector $\mathbf{c}$ such that $A\mathbf{c} \neq r\mathbf{c}$ and define $\mathbf{b} = A\mathbf{c} - r\mathbf{c}$. Then $\mathbf{v}_1(t) = e^{rt}\mathbf{b}$ and $\mathbf{v}_2(t) = e^{rt}(t\mathbf{b} + \mathbf{c})$ are linearly independent solutions of the system

$$\mathbf{v}' = A\mathbf{v}.$$

Example 4.2.5   Find the general solution of the system $\mathbf{v}' = A\mathbf{v}$, where $A = \begin{bmatrix} 1 & -2 \\ 2 & -3 \end{bmatrix}$.

Solution. The characteristic equation of A is $s^2 + 2s + 1 = (s + 1)^2 = 0$, which has the double root -1. Since $A \neq -I$, we start by finding a vector $\mathbf{c}$ that is not characteristic. One of the standard basis vectors will do: They cannot both be characteristic (why?). Put $\mathbf{c} = \begin{bmatrix} 1 \\ 0 \end{bmatrix}$; it is easy to see that $\mathbf{c}$ is not a characteristic vector, because $A\mathbf{c} \neq -\mathbf{c}$. Set

$$\mathbf{b} = A\mathbf{c} - (-1)\mathbf{c} = \begin{bmatrix} 1 \\ 2 \end{bmatrix} + \begin{bmatrix} 1 \\ 0 \end{bmatrix} = \begin{bmatrix} 2 \\ 2 \end{bmatrix}.$$

Then $\mathbf{v}_1(t) = e^{-t}\mathbf{b}$ and $\mathbf{v}_2(t) = e^{-t}(t\mathbf{b} + \mathbf{c})$ are independent solutions. The general solution is

$$\mathbf{v} = c_1 \begin{bmatrix} 2e^{-t} \\ 2e^{-t} \end{bmatrix} + c_2 \begin{bmatrix} (1 + 2t)e^{-t} \\ 2te^{-t} \end{bmatrix}.$$

EXERCISES

In Exercises 1–5, find all characteristic roots and vectors of the given matrix.

1. $\begin{bmatrix} 1 & 0 \\ 0 & -4 \end{bmatrix}$ **2.** $\begin{bmatrix} 2 & 3 \\ 1 & -2 \end{bmatrix}$ **3.** $\begin{bmatrix} 1 & 1 \\ 0 & -4 \end{bmatrix}$

4. $\begin{bmatrix} 3 & 3 \\ -3 & -3 \end{bmatrix}$ **5.** $\begin{bmatrix} 2 & 3 \\ -3 & 4 \end{bmatrix}$

6. Show that the matrix $\begin{bmatrix} 1 & -2 \\ 2 & 1 \end{bmatrix}$ has no real characteristic roots.

7. Let $s = r$ be a characteristic root of $A = \begin{bmatrix} a & b \\ c & d \end{bmatrix}$.

   a. Show that $\begin{bmatrix} b \\ r-a \end{bmatrix}$ is a characteristic vector belonging to r unless $b = 0$ and $r = a$.

   b. Show that if $b = 0$, then $s = a$ is a characteristic root, and $\begin{bmatrix} a-d \\ c \end{bmatrix}$ is a characteristic vector belonging to it, unless $a = d$ and $c = 0$ as well.

   c. How do you find the characteristic vectors if $a = d$ and $b = c = 0$?

In Exercises 8–13, find the general solution for each system.

8. $\mathbf{v}' = \begin{bmatrix} 3 & 4 \\ -2 & -3 \end{bmatrix}\mathbf{v}$ 9. $\mathbf{v}' = \begin{bmatrix} 3 & 2 \\ 1 & 2 \end{bmatrix}\mathbf{v}$

10. $\begin{cases} x' = -3x - y \\ y' = x - y \end{cases}$ 11. $\mathbf{v}' = \begin{bmatrix} 0 & 1 \\ 0 & 0 \end{bmatrix}\mathbf{v}$

12. $\begin{cases} x' = x - y \\ y' = x - y \end{cases}$ 13. $\begin{cases} x' = x + 2y \\ y' = 2x + y \end{cases}$

In Exercises 14–18, solve the IVP.

14. $\mathbf{v}' = \begin{bmatrix} 1 & 2 \\ -1 & 4 \end{bmatrix}\mathbf{v}; \ \mathbf{v}(0) = \begin{bmatrix} 1 \\ 0 \end{bmatrix}$

15. $\mathbf{v}' = \begin{bmatrix} -1 & 1 \\ -1 & -3 \end{bmatrix}\mathbf{v}; \ \mathbf{v}(0) = \begin{bmatrix} 0 \\ 1 \end{bmatrix}$

16. $\begin{cases} x' = x + y; \ x(0) = 4 \\ y' = x + 2y; \ y(0) = 2 \end{cases}$

17. $\begin{cases} x' = 5x - 2y \ x(0) = -1 \\ y' = 2x + y \ \ y(0) = -2 \end{cases}$

18. $\begin{cases} x' = x + y; \ x(0) = 3 \\ y' = x + y; \ y(0) = 1 \end{cases}$

19. Show that the characteristic equation of the system that replaces the second-order linear ODE

$$y'' + py' + qy = 0$$

is $s^2 + ps + q = 0$.

20. Let A be a 2×2 matrix with distinct characteristic roots s_1 and s_2, and let $\mathbf{c}$ be a vector that is *not* characteristic. Show that $(A - s_2 I)\mathbf{c}$ is a characteristic vector belonging to s_1, and that $(A - s_1 I)\mathbf{c}$ is a characteristic vector belonging to s_2.

4.3 Systems with Oscillating Solutions

We have already observed that $x = \cos(t)$, $y = \sin(t)$ is a solution of the linear system

$$\left. \begin{array}{l} x' = -y \\ y' = x. \end{array} \right\} \tag{4.16}$$

This system can be converted to vector form as $\mathbf{v}' = A\mathbf{v}$, where

$$A = \begin{bmatrix} 0 & -1 \\ 1 & 0 \end{bmatrix}.$$

Since $\operatorname{tr}(A) = 0$ and $\det(A) = 1$, the characteristic equation of A, $s^2 + 1 = 0$, has no real roots. It does have two *imaginary* roots, $\pm i$, where $i = \sqrt{-1}$. The system (4.16) is typical of the case where there are no real characteristic roots: All such systems have solutions that oscillate.

Euler's formula

To understand systems that have no real characteristic roots, we need to work with complex numbers. The following review is for the convenience of readers who lack experience with the complex number system.

The complex number system is an extension of the real number system formed by adjoining "imaginary numbers" to serve as square roots of negative real numbers. Thus, a complex number z can be expressed as

$$z = x + iy,$$

where x and y are real numbers. The operations of arithmetic, $+, -, \times, \div$, are all defined for complex numbers, and they satisfy the same commutative, associative, distributive, and existence of inverse laws that the real numbers do. In addition, complex arithmetic has one operation that real arithmetic does not have: conjugation. The conjugate of $z = x + iy$ is $\bar{z} = x - iy$. (It is standard practice to use $\bar{z}$ to denote the conjugate of z.) For example, if $z = 3 + 4i$, then $\bar{z} = 3 - 4i$. Here is how to do complex arithmetic:

Addition: Use vector addition:

$$(x_1 + iy_1) + (x_2 + iy_2) = (x_1 + x_2) + i(y_1 + y_2)$$

Thus, $(3 + 4i) + (5 - i) - 6 = 2 + 3i$.

Multiplication: Use binomial multiplication, *and* $i^2 = -1$. Thus

$$(x_1 + iy_1)(x_2 + iy_2) = x_1x_2 + i(x_1y_2 + x_2y_1) + y_1y_2i^2$$
$$= (x_1x_2 - y_1y_2) + i(x_1y_2 + x_2y_1).$$

Thus, $(12 - i) \times (1 + i) = (12 - i^2) + 12i - i = 13 + 11i$. Notice that $z\bar{z} = x^2 + y^2$ is real and positive (unless $z = 0$). If $z = 3 + 4i$, then $z\bar{z} = 3^2 + 4^2 = 25$.

Division: Multiply the numerator and the denominator by the conjugate of the denominator. The resulting quotient will be the same, but with a real denominator.

$$\frac{x_1 + iy_1}{x_2 + iy_2} = \left(\frac{x_1 + iy_1}{x_2 + iy_2}\right)\left(\frac{x_2 - iy_2}{x_2 - iy_2}\right)$$

$$= \left(\frac{x_1x_2 + y_1y_2}{x_2^2 + y_2^2}\right) + i\left(\frac{-x_1y_2 + x_2y_1}{x_2^2 + y_2^2}\right)$$

Thus,

$$\frac{2 + i}{3 + 4i} = \frac{(2 + i)(3 - 4i)}{25} = \frac{10 - 5i}{25} = 0.4 - 0.2i.$$

Just as it is customary to visualize real numbers as points on a line, complex numbers are visualized as points in a plane, called the **complex plane**; see Figure 4.1. The horizontal axis of the complex plane is identified with the real line and is called the **real axis**. The vertical axis is the **imaginary axis**.

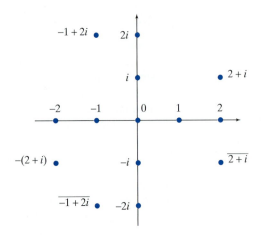

FIGURE 4.1   The complex plane.

The distance from $z = x + iy$ to 0 in the complex plane is, by the usual distance formula from analytic geometry, $\sqrt{x^2 + y^2}$. This distance is the **absolute value** of z and is denoted $|z|$. Notice that $|z| = \sqrt{z\,\bar{z}}$. All of the usual properties of absolute value for real numbers are still true. For example, the triangle inequality

$$|z_1 + z_2| \le |z_1| + |z_2|$$

can be interpreted as saying that one side of a triangle must be shorter than the sum of the lengths of the other two sides (see Figure 4.2).

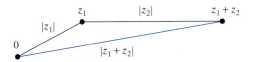

FIGURE 4.2   The triangle inequality:
$|z_1 + z_2| \le |z_1| + |z_2|$.

The components of a complex number $z = x + iy$ are the **real** and **imaginary** **parts** of s, $\mathrm{Re}(z) = x$ and $\mathrm{Im}(z) = y$, respectively. By convention, the imaginary part of a complex number is a real number. For example, $\mathrm{Im}(3 + 4i) = 4$, not $4i$.

Polar representation of complex numbers. The *polar representation* of a complex number is useful for multiplying and dividing complex numbers, and for raising them to powers. It is not useful for complex addition. The polar representation of $z = x + iy$ is simply a set of polar coordinates (r, θ) for the point (x, y), in which $r \ge 0$. Thus, $r = \sqrt{x^2 + y^2} = |z|$. The angle θ is called the **argument** of z and is denoted $\arg(z)$. The representation $z = x + iy$ can be recovered from the polar representation by the formulas $x = r\cos(\theta)$ and $y = r\sin(\theta)$; hence

$$z = r(\cos(\theta) + i\sin(\theta)).$$

To compute $\arg(z)$, use the inverse cosine function:

$$\arg(z) = \begin{cases} \cos^{-1}\left(\frac{\text{Re}(z)}{|z|}\right) & \text{if } \text{Im}(z) \geq 0 \\[2mm] -\cos^{-1}\left(\frac{\text{Re}(z)}{|z|}\right) & \text{if } \text{Im}(z) < 0. \end{cases}$$

The range of the argument as defined here is $-\pi < \arg(z) \leq \pi$, but we will follow the custom of considering $\arg(z)$ as a function with values taken mod 2π. For example, if I say $\arg(1 - \sqrt{3}i) = -\frac{\pi}{3}$ and you say $\arg(1 - \sqrt{3}i) = \frac{5\pi}{3}$, then we agree. Figure 4.3 illustrates the polar representation (r, θ) of $-1 + 2i$, where $r = |-1 + 2i| = \sqrt{5}$ and $\theta = \arg(-1 + 2i) = \cos^{-1}(-1/\sqrt{5})$, approximately $117°$.

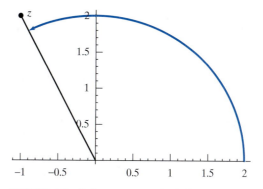

FIGURE 4.3   Polar representation of a complex number. The length of the heavy line to the point representing the complex number $z = -1 + 2i$ is $r = |z|$, and the radian measure of the directed arc shown is $\theta = \arg(z)$.

The following formula was published in 1748 by Leonhard Euler:

$$e^{i\theta} = \cos(\theta) + i\sin(\theta).$$

We will take this formula as the definition of the exponential of an imaginary number. It can be justified by substituting $x = i\theta$ in the Maclaurin series for e^x:

$$e^{i\theta} = \sum_{n=0}^{\infty} \frac{(i\theta)^n}{n!}.$$

In this series, the even terms are real and the odd terms are imaginary, because $(i\theta)^{2m} = (-1)^m \theta^{2m}$ and $(i\theta)^{2m+1} = (-1)^m i\theta^{2m+1}$. Therefore,

$$e^{i\theta} = \sum_{m=0}^{\infty} (-1)^m \frac{\theta^{2m}}{(2m)!} + i\sum_{m=0}^{\infty} (-1)^m \frac{\theta^{2m+1}}{(2m+1)!}. \tag{4.17}$$

The first sum on the right side of equation (4.17) is the Maclaurin series for the cosine function, and the second is i times the Maclaurin series for the sine.

Euler's formula enables us to extend our definition of the exponential to all complex numbers, by using the identity $e^{u+v} = e^u e^v$, with $u = \lambda$ and $v = i\omega$. Thus

$$e^{\lambda+i\omega} = e^\lambda e^{i\omega}$$
$$= e^\lambda (\cos(\omega) + i\sin(\omega)).$$

To make use of the complex exponential, we need to be able to differentiate it. If $f(t) = p(t) + iq(t)$ is a complex-valued function, where $p(t)$ and $q(t)$ are real-valued functions of a real variable t, we define its derivative as

$$f'(t) = p'(t) + i\,q'(t).$$

The following proposition shows that the derivative of e^{at} is equal to $a\,e^{at}$, even if a is a complex number.

Proposition 4.3.1

Let $a = \lambda + i\omega$ be a complex constant. Then

$$\frac{d}{dt}e^{at} = a\,e^{at}.$$

Proof

$$\frac{d}{dt}e^{at} = \frac{d}{dt}\{e^{\lambda t}[\cos(\omega t) + i\sin(\omega t)]\}$$
$$= \frac{d}{dt}[e^{\lambda t}\cos(\omega t)] + i\frac{d}{dt}[e^{\lambda t}\sin(\omega t)]$$
$$= [\lambda e^{\lambda t}\cos(\omega t) - e^{\lambda t}\omega\sin(\omega t)] + i[\lambda e^{\lambda t}\sin(\omega t) + \omega e^{\lambda t}\cos(\omega t)]$$
$$= e^{\lambda t}[\lambda\cos(\omega t) + i\omega\cos(\omega t) - \omega\sin(\omega t) + i\lambda\sin(\omega t)]$$
$$= e^{\lambda t}[(\lambda + i\omega)\cos(\omega t) + (i\omega + \lambda)i\sin(\omega t)]$$
$$= a\,e^{\lambda t}[\cos(\omega t) + i\sin(\omega t)] = a\,e^{at} \qquad \blacksquare$$

Let A be a square matrix with real entries. The characteristic polynomial of A may have some complex roots, and these are considered to be complex characteristic roots. For example, the characteristic polynomial of the matrix

$$A = \begin{bmatrix} 1 & -1 \\ 1 & 1 \end{bmatrix}$$

is $s^2 - 2s + 2$. The characteristic roots of A can be calculated by using the quadratic formula:

$$s_1, s_2 = \frac{2 \pm \sqrt{(-2)^2 - 8}}{2} = 1 \pm i.$$

Since complex arithmetic follows the same rules as real arithmetic, we can work with matrices and vectors that have complex entries just as we do when the entries are real. Thus, to find a characteristic vector $\mathbf{b}_1 = \begin{bmatrix} h \\ k \end{bmatrix}$ belonging to $s_1 = 1 + i$, we solve $(A - s_1 I)\mathbf{b}_1 = \mathbf{0}$:

$$\left(\begin{bmatrix} 1 & -1 \\ 1 & 1 \end{bmatrix} - \begin{bmatrix} 1+i & 0 \\ 0 & 1+i \end{bmatrix} \right) \begin{bmatrix} h \\ k \end{bmatrix} = \begin{bmatrix} 0 \\ 0 \end{bmatrix}.$$

After simplifying, this matrix equation reduces to two complex equations:

$$-i\,h - k = 0$$
$$h - i\,k = 0.$$

Observe that the second of these equations can be obtained from the first by multiplying through by i. Thus for any $h \neq 0$, $k = -i\,h$ will satisfy both equations. Let's put $h = 1$; then

$$\mathbf{b}_1 = \begin{bmatrix} 1 \\ -i \end{bmatrix}$$

is a characteristic vector belonging to $1 + i$.

Suppose that s is a complex characteristic root of a square matrix A, and $\mathbf{b}$ is a characteristic vector belonging to s. Put $\mathbf{v}(t) = e^{st}\mathbf{v}$. By Proposition 4.3.1, $\mathbf{v}'(t) = se^{st}\mathbf{b} = s\mathbf{v}$. Since $A\mathbf{v}(t) = e^{st}A\mathbf{b} = e^{st}s\mathbf{b} = s\mathbf{v}$, it follows that $\mathbf{v}(t)$ is a solution of $\mathbf{v}' = A\mathbf{v}$. Returning to our matrix $A = \begin{bmatrix} 1 & -1 \\ 1 & 1 \end{bmatrix}$, we see that

$$\mathbf{v}(t) = e^{(1+i)t} \begin{bmatrix} 1 \\ -i \end{bmatrix} = \begin{bmatrix} e^t(\cos(t) + i\sin(t)) \\ e^t(\sin(t) - i\cos(t)) \end{bmatrix}$$

is a solution of $\mathbf{v}' = A\mathbf{v}$.

This solution is complex since $\mathbf{b}$ is a complex vector, but we require a real solution, since $\mathbf{v}' = A\mathbf{v}$ is a real system of ODEs. By the following theorem, we can exchange a complex solution of a system of linear, homogeneous ODEs with real coefficients for *two* real solutions.

Theorem 4.8

Let $\mathbf{z}(t)$ be a complex solution of a homogeneous linear ODE or system of ODEs with real coefficients. Then $\mathbf{v}_1(t) = \text{Re}[\mathbf{z}(t)]$ and $\mathbf{v}_2(t) = \text{Im}[\mathbf{z}(t)]$ are solutions as well.

Proof For simplicity, let's assume the system takes the form $\mathbf{v}' = A\mathbf{v}$, where A is a real matrix (not necessarily constant). It is given that $\mathbf{z}(t) = \mathbf{v}_1(t) + i\mathbf{v}_2(t)$ is a complex-valued solution. Expand $\mathbf{z}' = A\mathbf{z}$ as

$$\mathbf{v}_1' + i\mathbf{v}_2' = A\mathbf{v}_1 + iA\mathbf{v}_2. \tag{4.18}$$

Equating real and imaginary parts of both sides of (4.18), we see that $\mathbf{v}_1' = A\mathbf{v}_1$, and $\mathbf{v}_2' = A\mathbf{v}_2$. It follows that $\mathbf{v}_1(t)$ and $\mathbf{v}_2(t)$ are solutions. ∎

By Theorem 4.8, the complex solution $\mathbf{v}(t) = \begin{bmatrix} e^t(\cos(t) + i\sin(t)) \\ e^t(\sin(t) - i\cos(t)) \end{bmatrix}$ of $\mathbf{v}' = \begin{bmatrix} 1 & -1 \\ 1 & 1 \end{bmatrix} \mathbf{v}$ yields two real solutions,

$$\mathbf{v}_1(t) = e^t \begin{bmatrix} \cos(t) \\ \sin(t) \end{bmatrix} \quad \text{and} \quad \mathbf{v}_2(t) = e^t \begin{bmatrix} \sin(t) \\ -\cos(t) \end{bmatrix}.$$

Since $\mathbf{v}_1(0) = \begin{bmatrix} 1 \\ 0 \end{bmatrix}$ and $\mathbf{v}_2(0) = \begin{bmatrix} 0 \\ -1 \end{bmatrix}$ are linearly independent, the solutions $\mathbf{v}_1(t)$ and $\mathbf{v}_2(t)$ are linearly independent.

Example 4.3.1   Find the general solution of the system $\mathbf{v}' = A\mathbf{v}$, where

$$A = \begin{bmatrix} -5 & -4 \\ 5 & -1 \end{bmatrix}.$$

Solution. Since $\text{tr}(A) = -6$ and $\det(A) = 25$, the characteristic equation of A is $s^2 + 6s + 25 = 0$. By the quadratic formula, the characteristic roots are

$$s = \frac{-6 \pm \sqrt{6^2 - 4 \times 25}}{2} = \frac{-6 \pm \sqrt{-64}}{2} = -3 \pm 4i.$$

To find a complex characteristic vector

$$\mathbf{b} = \begin{bmatrix} h \\ k \end{bmatrix}$$

(h and k will be complex numbers) belonging to $s = -3 + 4i$, we need to solve

$$\begin{bmatrix} -5 & -4 \\ 5 & -1 \end{bmatrix} \begin{bmatrix} h \\ k \end{bmatrix} = (-3 + 4i) \begin{bmatrix} h \\ k \end{bmatrix},$$

or

$$-5h - 4k = (-3 + 4i)h$$
$$5h - k = (-3 + 4i)k.$$

Adding $5h$ to both sides of the first equation and dividing through by 2 yields $-2k = (1 + 2i)h$. The second equation can be rewritten as $5h = -(2 - 4i)k$, and when we multiply both sides by $2 + 4i$ it becomes $5(2 + 4i)h = -20k$, obviously equivalent to the first equation. Thus we will put

$$\mathbf{b} = \begin{bmatrix} -2 \\ 1 + 2i \end{bmatrix}.$$

By Theorem 4.8, the real and the imaginary parts of

$$e^{(-3+4i)t}\mathbf{b} = e^{-3t} \begin{bmatrix} -2(\cos(4t) + i\sin(4t)) \\ \cos(4t) + i\sin(4t) + 2i(\cos(4t) + i\sin(4t)) \end{bmatrix}$$

$$= e^{-3t} \begin{bmatrix} -2\cos(4t) & - & 2i\sin(4t)) \\ \cos(4t) - 2\sin(4t) & + & i(2\cos(4t) + \sin(4t)) \end{bmatrix}$$

are solutions. These solutions are $v_1(t) = e^{-3t} \begin{bmatrix} -2\cos(4t) \\ \cos(4t) - 2\sin(4t) \end{bmatrix}$ and

$v_2(t) = e^{-3t} \begin{bmatrix} -2\sin(4t) \\ 2\cos(4t) + \sin(4t) \end{bmatrix}$. Since

$$v_1(0), v_2(0) = \begin{bmatrix} -2 \\ 1 \end{bmatrix}, \begin{bmatrix} 0 \\ 2 \end{bmatrix}$$

are linearly independent, Corollary 4.1.2 on page 143 tells us that the general solution is

$$v(t) = c_1 v_1(t) + c_2 v_2(t).$$ ☐

If we exchange a complex solution of a linear system for two real solutions, how can we be sure that the real solutions are linearly independent? The following theorem (proof in Exercise 11 at the end of this section) guarantees that if the solution is of the form $v(t) = e^{at}b$, where a is a complex number with nonzero imaginary part, then the real and imaginary parts of $v(t)$ will be linearly independent.

Theorem 4.9

Let $s = \lambda + i\omega$, where $\omega \neq 0$, be a complex characteristic root of a real matrix A, and $b = h + ik$ be a characteristic vector belonging to s. Then h and k are linearly independent.

Corollary 4.3.2

Let A be a 2×2 real matrix that has a characteristic root $\lambda + i\omega$ with $\omega \neq 0$, and let $h + ik$ be a characteristic vector belonging to that characteristic root. Then the general solution of

$$v' = Av$$

is

$$v(t) = c_1 e^{\lambda t} (\cos(\omega t)h - \sin(\omega t)k) + c_2 e^{\lambda t} (\sin(\omega t)h + \cos(\omega t)k),   \quad (4.19)$$

where c_1 and c_2 are constants.

EXERCISES

1. Find the reciprocal of $2 + i$.

2. Locate each of the complex numbers on the complex plane, calculate $\arg(z)$, $|z|$, and find $\lambda + \omega i$ such that $z = e^{\lambda + \omega i}$.

    a. $z = 2$ b. $z = -2$ c. $z = 1 - i$

    d. $z = -1 + i\sqrt{3}$ e. $z = -2i$

3. Find the sixth roots of 1 (there are six of them), and locate them on the complex plane.

    ◆ **Hint:** $1 = e^{2n\pi i}$.

4. Show that $e^{\bar{s}} = \overline{e^s}$.

5. Show that $|e^{\lambda + i\omega}| = e^{\lambda}$ (λ and ω are real).

6. Find all solutions of

$$e^{2s} + 2e^s + 2 = 0.$$

In Exercises 7–10, find the general solution.

7. $\begin{cases} x' = -\frac{1}{2}x - \frac{1}{2}y \\ y' = \frac{1}{2}x - \frac{1}{2}y \end{cases}$

8. $\begin{cases} x' = 2x + 5y \\ y' = -x \end{cases}$

9. $\begin{cases} x' = 4x - 8y \\ y' = 4x - 4y \end{cases}$

10. $\begin{cases} x' = -4x - 41y \\ y' = \frac{1}{2}x - 3y \end{cases}$

11. Let $s = \lambda + i\omega$, where $\omega \neq 0$, be a complex characteristic root of a real matrix A, and $\mathbf{b} = \mathbf{h} + i\mathbf{k}$ be a characteristic vector belonging to s.

     a. Show that $\bar{\mathbf{b}} = \mathbf{h} - i\mathbf{k}$ is a characteristic vector corresponding to $\bar{s} = \lambda - \omega i$.

     b. Using Theorem 4.6 on page 150, show that $\mathbf{b}$ and $\bar{\mathbf{b}}$ are linearly independent.

     c. Prove Theorem 4.9.

     d. Prove Corollary 4.3.2.

12. Let A denote a 2×2 matrix such that $\operatorname{tr} A = 0$ and $\det A < 0$. Show that the orbits of the system $\mathbf{v}' = A\mathbf{v}$ are hyperbolas, their asymptotes, and the origin.

   ◆ **Hint:** Imitate the case where $\det A > 0$.

4.4   General Solution of a Linear System

The method for solving a system of linear ODEs,

$$\mathbf{v}' = A(t)\mathbf{v} + \mathbf{f}(t), \tag{4.20}$$

where $A(t)$ is an $n \times n$ matrix, is basically the same as the method we used in Section 1.3 to solve a single first-order linear ODE. We start by finding the general solution

$$\mathbf{v}_h(t) = c_1\mathbf{v}_1(t) + c_2\mathbf{v}_2(t) + \cdots + c_n\mathbf{v}_n(t)$$

of the **associated homogeneous system,**

$$\mathbf{v}' = A(t)\mathbf{v}. \tag{4.21}$$

The method of variation of constants can be applied to determine a particular solution $\mathbf{v}_p(t)$ of an inhomogeneous system (4.20). The following theorem, which is a generalization of Theorem 1.2, assembles the general solution of the system.

Theorem 4.10

Suppose that all entries of the coefficient matrix $A(t)$ and source vector $\mathbf{f}(t)$ in (4.20) are continuous on an interval (a, b). Let $\mathbf{v}_p(t)$ be a particular solution of the system (4.20), and let $\mathbf{v}_1(t), \mathbf{v}_2(t), \ldots, \mathbf{v}_n(t)$ be linearly independent solutions of the associated homogeneous system, (4.21), all defined on (a, b). Then the general solution of the system (4.20) on (a, b) is

$$\mathbf{v}(t) = \mathbf{v}_p(t) + c_1\mathbf{v}_1(t) + c_2\mathbf{v}_2(t) + \cdots + c_n\mathbf{v}_n(t).$$

Proof Let $\mathbf{v}(t)$ be an arbitrary solution of the inhomogeneous system (4.20), and put $\mathbf{y}(t) = \mathbf{v}(t) - \mathbf{v}_p(t)$. Then

$$\begin{aligned} \mathbf{y}' &= \mathbf{v}' - \mathbf{v}'_p \\ &= [A(t)\mathbf{v} + \mathbf{f}(t)] - [A(t)\mathbf{v}_p + \mathbf{f}(t)] \\ &= A(t)\mathbf{v} - A(t)\mathbf{v}_p = A\mathbf{y}. \end{aligned}$$

It follows that $\mathbf{y}(t)$ is a solution of (4.21) and thus is a linear combination of the solutions $\mathbf{v}_i(t)$. Therefore, there are constants c_i such that

$$\mathbf{y}(t) = c_1\mathbf{v}_1(t) + c_2\mathbf{v}_2(t) + \cdots + c_n\mathbf{v}_n(t).$$

Conversely, suppose that $\mathbf{v}(t) = \mathbf{v}_p(t) + \mathbf{v}_h(t)$, where $\mathbf{v}_h(t) = c_1\mathbf{v}_1(t) + c_2\mathbf{v}_2(t) + \cdots + c_n\mathbf{v}_n(t)$. Then

$$\frac{d}{dt}(\mathbf{v}_p + \mathbf{v}_h(t)) = \mathbf{v}'_p + \mathbf{v}'_h(t)$$
$$= A(t)\mathbf{v}_p + \mathbf{f}(t) + A(t)\mathbf{v}_h$$
$$= A(t)(\mathbf{v}_p + \mathbf{v}_h) + \mathbf{f}(t),$$

which shows that $\mathbf{v}(t) = \mathbf{v}_p(t) + \mathbf{v}_h(t)$ is a solution of (4.20). ∎

Matrix solutions of homogeneous linear systems

Let $\mathcal{X}(t)$ be an $n \times n$ matrix whose entries are functions of t. The matrix formed by differentiating each entry of $\mathcal{X}$ will be denoted $\mathcal{X}'$. If

$$\mathcal{X}' = A(t)\mathcal{X}, \tag{4.22}$$

then $\mathcal{X}(t)$ is called a **matrix solution** of equation (4.21).

Proposition 4.4.1

Let the columns of the matrix $\mathcal{X}(t)$ be denoted $\mathbf{v}_1(t), \mathbf{v}_2(t), \ldots, \mathbf{v}_n(t)$. Then $\mathcal{X}(t)$ is a matrix solution of the homogeneous system (4.21) if and only if the $\mathbf{v}_i(t)$ are solutions of (4.21) as vector functions.

Proof The columns of $\mathcal{X}'$ are $\mathbf{v}'_1, \mathbf{v}'_2, \ldots, \mathbf{v}'_n$. Furthermore, the columns of $A(t)\mathcal{X}$ are $A(t)\mathbf{v}_i$, for $i = 1, 2, \ldots, n$. It follows each column of a matrix solution of (4.21) is itself a (vector) solution, and conversely that if each $\mathbf{v}_i(t)$ is a solution, then $\mathcal{X}(t)$ is a matrix solution. ∎

Example 4.4.1   Find a matrix solution of

$$\mathbf{v}' = \begin{bmatrix} 0 & 1 \\ 1 & 0 \end{bmatrix} \mathbf{v}. \tag{4.23}$$

Solution. The characteristic equation of

$$A = \begin{bmatrix} 0 & 1 \\ 1 & 0 \end{bmatrix}$$

is $s^2 - 1 = 0$; hence the characteristic roots are ± 1. We can use characteristic vectors

$$\mathbf{e}_1 = \begin{bmatrix} 1 \\ 1 \end{bmatrix} \quad \text{and} \quad \mathbf{e}_2 = \begin{bmatrix} 1 \\ -1 \end{bmatrix}$$

belonging to 1 and -1, respectively, to obtain the solutions $\mathbf{v}_1(t) = e^t \mathbf{e}_1$ and $\mathbf{v}_2(t) = e^{-t} \mathbf{e}_2$. Since

$$\mathbf{v}_1(t) = \begin{bmatrix} e^t \\ e^t \end{bmatrix} \quad \text{and} \quad \mathbf{v}_2(t) = \begin{bmatrix} e^{-t} \\ -e^{-t} \end{bmatrix},$$

$$\mathcal{X}(t) = \begin{bmatrix} e^t & e^{-t} \\ e^t & -e^{-t} \end{bmatrix}$$

is a matrix solution. ❑

Proposition 4.4.2

Suppose that $\mathcal{X}(t)$ is a matrix solution of (4.21), and let $\mathbf{c} \in \mathbf{R}^n$ be a constant vector. Then the vector function

$$\mathbf{v}(t) = \mathcal{X}(t)\mathbf{c}$$

is a solution of the system (4.21).

Proof Let $\mathbf{v}_1, \mathbf{v}_2, \ldots, \mathbf{v}_n$ denote the columns of $\mathcal{X}$. By the definition of matrix multiplication,

$$\mathcal{X}\mathbf{c} = c_1\mathbf{v}_1 + c_2\mathbf{v}_2 + \cdots + c_n\mathbf{v}_n$$

is a linear combination of solutions of the system (4.21) and thus itself a solution of the system. ∎

Suppose that the entries of the coefficient matrix $A(t)$ are continuous functions of t on some interval (a, b).

> **DEFINITION** A matrix solution $\mathcal{X}(t)$ of equation (4.21) is called a **fundamental matrix solution** if there is a point $t_0 \in (a, b)$ such that the columns of $\mathcal{X}(t_0)$ are linearly independent.

An $n \times n$ matrix P is said to be **nonsingular** if it is not singular—that is, $P\mathbf{d} = \mathbf{0}$ if and only if $\mathbf{d} = \mathbf{0}$, where $\mathbf{d}$ is an arbitrary n-dimensional column vector. It is a fundamental result from linear algebra that P is nonsingular if and only if, as vectors, the columns of P are linearly independent. The proof is Exercise 13 at the end of this section. Thus we can say that a matrix solution of the system (4.21) is a fundamental matrix solution if and only if $\mathcal{X}(t_0)$ is nonsingular. By Theorem 4.3 on page 141, it follows that $\mathcal{X}(t)$ is a fundamental matrix solution if and only if $\mathcal{X}(t)$ is nonsingular for all $t \in (a, b)$. This is a special property of matrix solutions of homogeneous linear systems of ODEs: It is possible for matrix functions in general to be nonsingular in some places, and not in others.

Example 4.4.2 Show that the matrix solution of the system (4.23) given in Example 4.4.1 is a fundamental matrix solution.

Solution. The columns of

$$X(0) = \begin{bmatrix} 1 & 1 \\ 1 & -1 \end{bmatrix}$$

are linearly independent.   ☐

Theorem 4.11

Suppose that the entries of the coefficient matrix $A(t)$ are continuous on the interval (a, b). Then the system $\mathbf{v}' = A(t)\mathbf{v}$ has a fundamental matrix solution $X(t)$, defined for $t \in (a, b)$.

Proof To keep the notation simple, we will take $n = 2$. Let

$$\mathbf{e}_1 = \begin{bmatrix} 1 \\ 0 \end{bmatrix} \quad \text{and} \quad \mathbf{e}_2 = \begin{bmatrix} 0 \\ 1 \end{bmatrix}$$

denote the standard basis vectors of $\mathbf{R}^2$. Choose a point $t_0 \in (a, b)$. By the existence and uniqueness theorem (Theorem 4.1) (stated on page 139), the initial value problems

$$\mathbf{v}' = A(t)\mathbf{v}; \ \mathbf{v}(t_0) = \mathbf{e}_i \quad (i = 1, 2)$$

have unique solutions, defined on (a, b). We will denote these solutions by $\mathbf{v}_1(t)$ and $\mathbf{v}_2(t)$, and let $X(t)$ be the matrix whose columns are $\mathbf{v}_1(t)$ and $\mathbf{v}_2(t)$. By Proposition 4.4.1, $X(t)$ is a matrix solution of the system (4.21), and since $X(t_0)$ is the identity matrix, it is nonsingular when $t = t_0$.   ■

The general solution of a homogeneous system can be expressed in terms of a fundamental matrix solution.

Theorem 4.12

Let $A(t)$ be a matrix whose entries are continuous on an interval (a, b), and suppose that $X(t)$ is a fundamental matrix solution of $\mathbf{v}' = A(t)\mathbf{v}$, defined on (a, b). If $\mathbf{v}(t)$ is a solution of the system, also defined on (a, b), then there is a constant vector $\mathbf{c}$ such that $\mathbf{v}(t) = X(t)\mathbf{c}$.

Proof Again, we take $n = 2$ Let $\mathbf{v}_1(t)$ and $\mathbf{v}_2(t)$ denote the columns of $X(t)$, and $t_0 \in (a, b)$. Since the $\mathbf{v}_i$ are independent, Corollary 4.1.2 on page 143 tells us that the general solution of $\mathbf{v}' = A(t)\mathbf{v}$ is

$$\mathbf{v} = c_1\mathbf{v}_1 + c_2\mathbf{v}_2.$$

Let $\mathbf{c} = \begin{bmatrix} c_1 \\ c_2 \end{bmatrix}$; then $X\mathbf{c} = \mathbf{v}$.   ■

Example 4.4.3   Find the general solution of the system that was considered in Example 4.4.1.

Solution. In Example 4.4.2, it was seen that

$$X = \begin{bmatrix} e^t & e^{-t} \\ e^t & -e^{-t} \end{bmatrix}$$

is a fundamental matrix solution. The general solution is

$$\mathbf{v} = X\mathbf{c}$$

$$= \begin{bmatrix} e^t & e^{-t} \\ e^t & -e^{-t} \end{bmatrix} \begin{bmatrix} c_1 \\ c_2 \end{bmatrix}$$

$$= \begin{bmatrix} c_1 e^t + c_2 e^{-t} \\ c_1 e^t - c_2 e^{-t} \end{bmatrix}$$

$$= c_1 e^t \begin{bmatrix} 1 \\ 1 \end{bmatrix} + c_2 e^{-t} \begin{bmatrix} 1 \\ -1 \end{bmatrix}. \qquad \square$$

Variation of constants

The method of variation of constants is used to find a particular solution of an inhomogeneous system. Recall from Section 1.3, where the method was introduced for the scalar case, that the general solution of the associated homogeneous equation was a required input. For systems, it is convenient to start with a fundamental matrix solution of the associated homogeneous system.

The key to adapting the method for matrix equations is the process of division by a matrix. It turns out that this is possible if and only if the matrix in question is nonsingular.

Let P be a square matrix. A square matrix Q is called the **inverse matrix** of P if $PQ = QP = I$, where I is the identity matrix. We will use the customary notation P^{-1} for the inverse matrix of P. The following theorem is from linear algebra. Although it is true for square matrices of any size, our proof only works for 2×2 matrices.

Theorem 4.13

Let

$$P = \begin{bmatrix} a & c \\ b & d \end{bmatrix}$$

be a nonsingular matrix. Then P has an inverse matrix.

Proof By Proposition 4.2.1 on page 147, $\det(P) = ad - bc$ is equal to 0 if and only if P is singular.

Thus, if P is nonsingular, $\det(P) \neq 0$. Define a matrix Q as

$$Q = \begin{bmatrix} d/\det(P) & -c/\det(P) \\ -b/\det(P) & a/\det(P) \end{bmatrix}.$$

You can verify that $PQ = I$ and $QP = I$ by matrix multiplication. ■

We will now see how to find a particular solution of an inhomogeneous system of ODEs. Let $\mathcal{X}(t)$ be a fundamental matrix solution of a homogeneous system $\mathbf{v}' = A(t)\mathbf{v}$. To find a particular solution of an inhomogeneous system

$$\mathbf{v}' = A(t)\mathbf{v} + \mathbf{f}(t), \tag{4.24}$$

set

$$\mathbf{v}_p(t) = \mathcal{X}(t)\mathbf{w}(t),$$

where $\mathbf{w}(t)$ is a vector function that will be determined. By the product rule for differentiation,

$$\begin{aligned}
\mathbf{v}'_p &= [d\mathcal{X}(t)/dt]\mathbf{w}(t) + \mathcal{X}(t)\mathbf{w}'(t) \\
&= A(t)\mathcal{X}(t)\mathbf{w}(t) + \mathcal{X}(t)\mathbf{w}'(t) \\
&= A(t)\mathbf{v}_p(t) + \mathcal{X}(t)\mathbf{w}'(t).
\end{aligned}$$

It follows that $\mathbf{v}_p(t)$ is a particular solution of the system (4.24) if and only if

$$\mathcal{X}(t)\mathbf{w}'(t) = \mathbf{f}(t). \tag{4.25}$$

Since $\mathcal{X}(t)$ is nonsingular, it has an inverse $\mathcal{X}^{-1}(t)$. Multiplying (4.25) by this, we have

$$\mathbf{w}'(t) = \mathcal{X}^{-1}(t)\mathbf{f}(t).$$

If $t_0 \in (a, b)$, then

$$\mathbf{w}(t) = \int_{t_0}^{t} \mathcal{X}^{-1}(s)\mathbf{f}(s)\,ds.$$

Now that $\mathbf{w}(t)$ has been determined we can find $\mathbf{v}_p(t)$ by multiplying by $\mathcal{X}(t)$. This can be summarized as follows:

Variation of Constants Formula

Let $\mathcal{X}(t)$ be a fundamental matrix solution of $\mathbf{v}' = A(t)\mathbf{v}$, where $A(t)$ is a matrix whose entries are all continuous on an interval (a, b). If $\mathbf{f}(t)$ is a vector function, also continuous on (a, b), and $t_0 \in (a, b)$, then

$$\mathbf{v}_p(t) = \mathcal{X}(t) \int_{t_0}^{t} \mathcal{X}^{-1}(s)\mathbf{f}(s)\,ds$$

is a particular solution of the inhomogeneous equation

$$\mathbf{v}' = A(t)\mathbf{v} + \mathbf{f}(t).$$

It is not necessary to memorize this formula; you just need to remember to substitute $\mathbf{v} = \mathcal{X}(t)\mathbf{w}$ in (4.24).

⬜ **Example 4.4.4** Find a particular solution of

$$x_1' = x_2 - e^{-t}$$
$$x_2' = x_1 + e^{-t}.$$

Solution. In Example 4.4.1, it was shown that $X(t) = \begin{bmatrix} e^t & e^{-t} \\ e^t & -e^{-t} \end{bmatrix}$ is a funda-

mental matrix solution of the associated homogeneous system (4.23). Set $\mathbf{v}_p(t) = X(t)\mathbf{w}(t)$. Thus

$$\begin{bmatrix} x_1 \\ x_2 \end{bmatrix} = \begin{bmatrix} e^t & e^{-t} \\ e^t & -e^{-t} \end{bmatrix} \begin{bmatrix} w_1 \\ w_2 \end{bmatrix} = \begin{bmatrix} e^t w_1 + e^{-t} w_2 \\ e^t w_1 - e^{-t} w_2 \end{bmatrix}.$$

Differentiating and using the product rule, we have

$$\begin{bmatrix} x_1' \\ x_2' \end{bmatrix} = \begin{bmatrix} e^t w_1' + e^{-t} w_2' \\ e^t w_1' - e^{-t} w_2' \end{bmatrix} + \begin{bmatrix} e^t w_1 - e^{-t} w_2 \\ e^t w_1 + e^{-t} w_2 \end{bmatrix}$$

and the right side of our system is

$$\begin{bmatrix} x_2 - e^{-t} \\ x_1 + e^{-t} \end{bmatrix} = \begin{bmatrix} e^t w_1 - e^{-t} w_2 \\ e^t w_1 + e^{-t} w_2 \end{bmatrix} + \begin{bmatrix} -e^{-t} \\ e^t \end{bmatrix}.$$

After canceling, we have

$$e^t w_1' + e^{-t} w_2' = -e^{-t}$$
$$e^t w_1' - e^{-t} w_2' = e^{-t}.$$

The solution of this equation is $w_1' = 0$, $w_2' = -1$, or

$$\mathbf{w}'(t) = \begin{bmatrix} 0 \\ -1 \end{bmatrix}.$$

Thus $\mathbf{w}(t) = \begin{bmatrix} 0 \\ -t \end{bmatrix}$, and we have the particular solution

$$\mathbf{v}_p(t) = \begin{bmatrix} e^t & e^{-t} \\ e^t & -e^{-t} \end{bmatrix} \begin{bmatrix} 0 \\ -t \end{bmatrix} = \begin{bmatrix} -te^{-t} \\ te^{-t} \end{bmatrix}.$$

The general solution is $\mathbf{v}_p(t) + \mathbf{v}_h(t)$, where $\mathbf{v}_h(t) = X(t)\mathbf{c}$ denotes the general solution of the associated homogeneous equation. ⬜

EXERCISES

In Exercises 1–4, find a fundamental matrix solution of the system.

1. $\begin{cases} x' = \frac{3}{2}x + \frac{1}{2}y \\ y' = \frac{1}{2}x + \frac{3}{2}y \end{cases}$

2. $\begin{cases} x' = -3x + y \\ y' = -5x - y \end{cases}$

3. $\begin{cases} x' = 2x - 4y \\ y' = 5x - 2y \end{cases}$

4. $\begin{cases} x' = -x - y \\ y' = x - 3y \end{cases}$

In Exercises 5–8, use the method of variation of constants to find a particular solution of the system of ODEs, and then write down the general solution. Fundamental matrix solutions of the associated were found in Exercises 1–4.

5. $\begin{cases} x' = \frac{3}{2}x + \frac{1}{2}y + b_1(t) \\ y' = \frac{1}{2}x + \frac{3}{2}y + b_2(t), \end{cases}$ where

a. $\begin{bmatrix} b_1 \\ b_2 \end{bmatrix} = \begin{bmatrix} e^t \\ e^t \end{bmatrix}$ b. $\begin{bmatrix} b_1 \\ b_2 \end{bmatrix} = \begin{bmatrix} e^t \\ -e^t \end{bmatrix}$

c. $\begin{bmatrix} b_1 \\ b_2 \end{bmatrix} = \begin{bmatrix} te^t \\ 0 \end{bmatrix}$

6. $\begin{cases} x' + 3x - y = -3e^{-2t} \\ y' + 5x + y = e^{-2t} \end{cases}$

7. $\begin{bmatrix} x' \\ y' \end{bmatrix} = \begin{bmatrix} 2 & -4 \\ 5 & -2 \end{bmatrix} \begin{bmatrix} x \\ y \end{bmatrix} + \begin{bmatrix} -32\cos(4t) \\ 0 \end{bmatrix}$

8. $\begin{bmatrix} x' \\ y' \end{bmatrix} + \begin{bmatrix} 1 & 1 \\ -1 & 3 \end{bmatrix} \begin{bmatrix} x \\ y \end{bmatrix} = \begin{bmatrix} b_1(t) \\ b_2(t) \end{bmatrix}$, where

a. $\begin{bmatrix} b_1(t) \\ b_2(t) \end{bmatrix} = e^{-2t}\begin{bmatrix} 1 \\ 3 \end{bmatrix}$ b. $\begin{bmatrix} b_1(t) \\ b_2(t) \end{bmatrix} = \begin{bmatrix} 2e^{-2t} \\ 0 \end{bmatrix}$

9. Show that
$$X = \begin{bmatrix} 2t^2 & t^3 \\ -t^2 & -t^3 \end{bmatrix}$$
is a fundamental matrix solution of
$$tx' = x - 2y$$
$$ty' = x + 4y.$$

Using variation of constants, find a particular solution of
$$tx' = x - 2y + b_1(t)$$
$$ty' = x + 4y + b_2(t),$$
where

a. $\begin{bmatrix} b_1(t) \\ b_2(t) \end{bmatrix} = \begin{bmatrix} t \\ 2t \end{bmatrix}$ b. $\begin{bmatrix} b_1(t) \\ b_2(t) \end{bmatrix} = \begin{bmatrix} t^3 \\ -t^3 \end{bmatrix}$

◆ **Warning:** The method of variation of constants is based on the assumption that the system has the form $\mathbf{v}' = A(t)\mathbf{v} + \mathbf{b}(t)$; the system in this problem must be written in that form before the method can be used.

10. Find the general solution of each of the following inhomogeneous systems.

a. $\begin{cases} x' = x + y + t^{-1} \\ y' = -x - y \end{cases}$ b. $\begin{cases} x' = x + y + t^{-1} \\ y' = -(x + y + t^{-1}) \end{cases}$

c. $\begin{cases} x' = 4x + 5y + 5\sin(3t) \\ y' = -5x - 4y + 3\cos(3t) - 4\sin(3t) \end{cases}$

d. $\begin{cases} x' = \sec^3(t) - y \\ y' = x - \sec^3(t) \end{cases}$

11. Let C be a constant matrix, and suppose that $X(t)$ is a matrix solution of $\mathbf{v}' = A(t)\mathbf{v}$. Show that XC is also a matrix solution.

12. Let $X_1(t)$ and $X_2(t)$ be fundamental matrix solutions of the linear system of differential equations $\mathbf{v}' = A(t)\mathbf{v}$, defined on an interval $\mathcal{I}$. Show that $X_1^{-1} \cdot X_2$ is a constant matrix.

13. Let Q be an $n \times n$ matrix. Show that Q is nonsingular if and only if the columns of q are linearly independent.

4.5   The Exponential of a Matrix

The exponential function $f(t) = e^{at}$ is used to represent the solution of the IVP, $y' = ay;\ y(0) = 1$. To generalize this observation to systems of linear IVPs, let us recall that e^{at} can be expanded as a power series as follows:

$$e^{at} = 1 + at + \frac{a^2}{2!}t^2 + \frac{a^3}{3!}t^3 + \cdots + \frac{a^n}{n!}t^n + \cdots.$$

It is possible to consider a series in which the constant a is replaced by a constant matrix A. By convention, $A^0 = I$ is the identity matrix, $A^1 = A$, $A^2 = A \times A$, and so on. Thus, for an $m \times m$ matrix A, let us define

$$e^{At} = I + At + \frac{A^2}{2!}t^2 + \cdots + \frac{A^n}{n!}t^n + \cdots. \tag{4.26}$$

With any power series, the first thing to consider is convergence. Let $p_{ij,k}$ denote the i,jth entry of A^k. Since $A^0 = I$, and $A^1 = A$,

$$p_{ij,0} = \begin{cases} 1 & \text{if } i = j \\ 0 & \text{if } i \neq j \end{cases},$$

$p_{ij,1} = a_{ij}$, and so on. Equation (4.26) defines e^{At} to be the $m \times m$ matrix whose i,jth entry is

$$p_{ij} = \sum_{k=0}^{\infty} \frac{1}{k!} p_{ij,k}\, t^k, \tag{4.27}$$

provided that this series is convergent for each i, j.

Theorem 4.14

For any $m \times m$ matrix A, the series (4.27) converges for all values of t.

Proof We will show that

$$\sum_{k=0}^{\infty} \left| \frac{1}{k!} p_{ij,k}\, t^k \right| \tag{4.28}$$

is convergent. Choose a number B such that for each entry a_{ij} of A, $|a_{ij}| \leq B$. We will estimate the magnitude of $p_{ij,k}$, where i and j are arbitrary, by showing that

$$|p_{ij,k}| \leq m^{k-1} B^k \quad \text{for } k \geq 1. \tag{4.29}$$

The proof that (4.29) holds for all $k \geq 1$ is by mathematical induction. If $k = 1$, the estimate holds because $|p_{ij,k}| = |a_{ij}| \leq B$. Let $k \geq 2$, and assume that (4.29) holds for all powers less than k. Then

$$|p_{ij,k-1}| \leq m^{k-2} B^{k-1}.$$

Since $A^k = A \cdot A^{k-1}$,

$$p_{ij,k} = a_{i1} p_{1j,k-1} + a_{i2} p_{2j,k-1} + \cdots + a_{im} p_{mj,k-1}.$$

Starting with the triangle inequality, we have

$$|p_{ij,k}| \leq |a_{i1}||p_{1j,k-1}| + |a_{i2}||p_{2j,k-1}| + \cdots + |a_{im}||p_{mj,k-1}|$$
$$\leq \underbrace{B \cdot m^{k-2} B^{k-1} + B \cdot m^{k-2} B^{k-1} + \cdots + B \cdot m^{k-2} B^{k-1}}_{m \text{ terms}}$$
$$= m^{k-1} B^k.$$

Thus, our assumption that (4.29) holds for all powers $< k$ implies that it holds for the kth power as well. By the principle of mathematical induction, (4.29) is valid for all $k \geq 1$. Then

$$\frac{1}{k!}|p_{ij,k}t^k| \leq \frac{m^{k-1}B^k}{k!}|t^k|,$$

which implies that the series (4.28) is dominated by the series

$$\sum_{k=0}^{\infty} \frac{m^{k-1}B^k}{k!}|t|^k.$$

The latter series converges since it is the Taylor series expansion of $f(t) = \frac{1}{m}e^{mB|t|}$. By the comparison test, the series (4.28) also converges. This means that the series (4.27) defining p_{ij} converges absolutely. Since every absolutely convergent series is convergent, the proof is complete. ∎

As a simple example, suppose that Z denotes the zero matrix (all entries equal 0). Then the only nonzero term in the series

$$I + Zt + \frac{1}{2}Z^2t^2 + \cdots$$

defining e^{Zt} is I. It follows that $e^{Zt} = I$. The next example is also fairly simple, because the series has only a finite number of nonzero terms.

Example 4.5.1   Let

$$A = \begin{bmatrix} 0 & 1 & 0 \\ 0 & 0 & 1 \\ 0 & 0 & 0 \end{bmatrix}.$$

Calculate e^{At}.

Solution. By matrix multiplication,

$$A^2 = \begin{bmatrix} 0 & 0 & 1 \\ 0 & 0 & 0 \\ 0 & 0 & 0 \end{bmatrix},$$

and $A^k = 0$ for $k \geq 3$. Hence

$$e^{At} = I + At + \frac{1}{2}(A)^2t^2$$

$$= \begin{bmatrix} 1 & t & \frac{1}{2}t^2 \\ 0 & 1 & t \\ 0 & 0 & 1 \end{bmatrix}.$$ □

⬜ **Example 4.5.2**   Calculate e^{At}, where

$$A = \begin{bmatrix} 1 & 0 \\ 0 & 2 \end{bmatrix}.$$

Solution. Experimenting, we find that $A^2 = \begin{bmatrix} 1 & 0 \\ 0 & 4 \end{bmatrix}$, $A^3 = \begin{bmatrix} 1 & 0 \\ 0 & 8 \end{bmatrix}$, and $A^3 = \begin{bmatrix} 1 & 0 \\ 0 & 16 \end{bmatrix}$. This leads us to guess that

$$A^n = \begin{bmatrix} 1 & 0 \\ 0 & 2^n \end{bmatrix}.$$

To verify that this is true, make the following observation: If

$$B = \begin{bmatrix} a & 0 \\ 0 & b \end{bmatrix} \quad \text{and} \quad C = \begin{bmatrix} c & 0 \\ 0 & d \end{bmatrix}$$

are diagonal matrices, then

$$BC = \begin{bmatrix} ac & 0 \\ 0 & bd \end{bmatrix}$$

is also a diagonal matrix whose diagonal entries are the products of the respective diagonal entries of B and C. It follows that

$$A^n = \begin{bmatrix} 1^n & 0 \\ 0 & 2^n \end{bmatrix}.$$

Thus

$$e^{At} = \sum_{k=0}^{\infty} \frac{1}{k!} A^k t^k$$

$$= \begin{bmatrix} \sum_{k=0}^{\infty} \frac{t^k}{k!} & 0 \\ 0 & \sum_{k=0}^{\infty} \frac{(2t)^k}{k!} \end{bmatrix}$$

$$= \begin{bmatrix} e^t & 0 \\ 0 & e^{2t} \end{bmatrix}. \qquad ⬜$$

The matrix exponential shares many properties with the ordinary exponential function, but there are limitations. The proof that $e^{r+s} = e^r e^s$ does not emphasize the commutative law, $rs = sr$, but it uses that fact. Multiplication of square matrices is not always commutative, so the matrix version of that identity is as follows.

Theorem 4.15
_____

Let A and B be $n \times n$ matrices that commute: $AB = BA$. Then

$$e^{A+B} = e^A e^B.$$

The proof of Theorem 4.15 is left to you (see Exercise 3).

The following result is the reason that the matrix exponential is included in this text: It provides a simple way to express the fundamental matrix solution of any linear system of first order differential equations with constant coefficients.

Theorem 4.16

Let A be a square matrix whose entries are constants. Then

$$de^{At}/dt = A \cdot e^{At}.$$

The proof of Theorem 4.16 uses a lemma that states a version of the chain rule applying to matrix polynomials. If $f(x) = c_0 + c_1 x + c_2 x^2 + \cdots c_k x^k$ is a polynomial, we will let $f(At)$ denote the matrix

$$f(At) = c_0 I + c_1 A t + c_2 A^2 t^2 + \cdots + c_k A^k t^k.$$

Lemma 4.5.1

Let $f(x)$ be a polynomial, and let $f'(x)$ denote the derivative of f with respect to x. Then

$$d[f(At)]/dt = A \cdot f'(At).$$

Proof Let us consider a special case, where $f(x) = c x^k$ is a monomial. Then $f'(x) = c k x^{k-1}$, so

$$A \cdot f'(At) = A \cdot (c k A^{k-1} t^{k-1}) = c k A^k t^{k-1}.$$

On the other hand,

$$d[f(At)]/dt = \frac{d}{dt}(c A^k t^k) = c A^k (k t^{k-1}).$$

Since these are equal, the lemma is true in this special case.

To complete the proof, we apply this special case to each term of the polynomial $f(x)$. Thus

$$\frac{d}{dt}(c_0 I + c_1 A t + \cdots + c_k A^k t^k) = A c_1 I + A c_2 2At + \cdots A c_k k A^{k-1} t^{k-1}$$

$$= A \cdot f'(At). \qquad \blacksquare$$

Proof of Theorem 4.16.   Let $F_m(At)$ denote the partial sum

$$F_m(At) = \sum_{k=0}^{m} \frac{1}{k!}(At)^k.$$

By Lemma 4.5.1, $dF_m(At)/dt = A \cdot F'_m(At)$. Since

$$F'_m(At) = \sum_{k=0}^{m} \frac{1}{k!}k(At)^{k-1}$$

$$= F_{m-1}(At),$$

it follows that $\lim_{m\to\infty} F'_m(At) = \lim_{m\to\infty} F_{m-1}(At) = e^{At}.$

It is also true that $\lim_{m \to \infty} F_m(At) = e^{At}$. Hence

$$\lim_{m \to \infty} d[F_m(At)]/dt = A \cdot e^{At},$$

and it might appear that the proof is complete. However, it still should be checked that

$$\lim_{m \to \infty} \left(\frac{d}{dt}[F_m(At)] \right) = \frac{d}{dt}[\lim_{m \to \infty} F_m(t)];$$

that is, that the limit of the derivatives is the derivative of the limit. The proof of this fact is omitted.   ∎

Theorem 4.16 implies that $\mathcal{X}(t) = e^{At}$ is a matrix solution of the system $\mathbf{x}' = A\mathbf{x}$. Since $\mathcal{X}(0) = e^{A \cdot 0} = I$ is nonsingular, $\mathcal{X}(t)$ is a fundamental matrix solution. We have thus established the following theorem.

Theorem 4.17

Let A be an $n \times n$ constant matrix. Then $e^{At} = \sum_{k=0}^{\infty} \frac{1}{k!} A^k t^k$ is a fundamental matrix solution of the system

$$\mathbf{x}' = A\mathbf{x},$$

and the general solution is

$$\mathbf{x}(t) = e^{At} \mathbf{c},$$

where the entries of the column vector $\mathbf{c}$ are the constants.

Calculation

Let A be a square matrix. In some cases, it is convenient to calculate e^{At} by direct substitution into the power series, as we did in Example 4.5.1. The usual approach is to start by finding a fundamental matrix solution $\mathcal{X}(t)$ of

$$\mathbf{v}' = A\mathbf{v}. \tag{4.30}$$

Let $\mathcal{Y}(t) = \mathcal{X}(t)[\mathcal{X}(0)]^{-1}$, which will also be a fundamental matrix solution (recall that $\mathcal{X}(t)C$ is a fundamental matrix solution for any nonsingular constant matrix C). Since e^{At} and $\mathcal{Y}(t)$ are each fundamental matrix solutions of (4.30), and both satisfy the initial condition

$$\mathcal{Y}(0) = e^{A \cdot 0} = I,$$

$\mathcal{Y}(t) = e^{At}$ for all t by the uniqueness theorem.

For example, we find in Example 4.3.1 on page 162 that

$$\mathbf{v}_1(t) = e^{-3t} \begin{bmatrix} -2\cos(4t) \\ \cos(4t) - 2\sin(4t) \end{bmatrix} \quad \text{and} \quad \mathbf{v}_2(t) = e^{-3t} \begin{bmatrix} -2\sin(4t) \\ 2\cos(4t) + \sin(4t) \end{bmatrix}$$

are solutions of $\mathbf{v}' = A\mathbf{v}$, where

$$A = \begin{bmatrix} -5 & -4 \\ 5 & -1 \end{bmatrix}.$$

Hence

$$\mathcal{X}(t) = e^{-3t} \begin{bmatrix} -2\cos(4t) & -2\sin(4t) \\ (\cos(4t) - 2\sin(4t)) & (2\cos(4t) + \sin(4t)) \end{bmatrix}$$

is a fundamental matrix solution. Substitute $t = 0$ to find

$$\mathcal{X}(0) = \begin{bmatrix} -2 & 0 \\ 1 & 2 \end{bmatrix}.$$

Using the formula developed in the proof of Theorem 4.13 on page 168, we find that

$$\mathcal{X}(0)^{-1} = \begin{bmatrix} \frac{1}{2} & 0 \\ -\frac{1}{4} & \frac{1}{2} \end{bmatrix}$$

Therefore,

$$e^{At} = \mathcal{X}(t)[\mathcal{X}(0)]^{-1}$$

$$= e^{-3t} \begin{bmatrix} -2\cos(4t) & -2\sin(4t) \\ (\cos(4t) - 2\sin(4t)) & (2\cos(4t) + \sin(4t)) \end{bmatrix} \begin{bmatrix} \frac{1}{2} & 0 \\ -\frac{1}{4} & \frac{1}{2} \end{bmatrix}$$

$$= e^{-3t} \begin{bmatrix} \left(\cos(4t) - \frac{1}{2}\sin(4t)\right) & -\sin(4t) \\ \frac{5}{4}\sin(4t) & \left(\cos(4t) + \frac{1}{2}\sin(4t)\right) \end{bmatrix}.$$

Seen in this way, calculation of the matrix exponential is an application of linear algebra. For some matrices, the process is complicated by multiple characteristic roots, complex roots, and so on. Fortunately, the task of computing the matrix exponential can be assigned to the computer. The prominent CAS packages (*Maple*, *Mathematica*, and MATLAB) all support the matrix exponential.

Inhomogeneous equations

When applying the variation of constants formula to a system

$$\mathbf{v}' = A\mathbf{v} + \mathbf{f}(t), \tag{4.31}$$

where A is an $n \times n$ constant matrix and $\mathbf{v}(t)$ and $\mathbf{f}(t)$ are n-dimensional vector functions, we can use the matrix exponential e^{At} as the fundamental matrix solution of the associated homogeneous system, $\mathbf{v}' = A\mathbf{v}$.

Thus, set $\mathbf{v}(t) = e^{At}\mathbf{z}(t)$. By the product rule for differentiation,

$$\mathbf{v}'(t) = A\,e^{At}\mathbf{z}(t) + e^{At}\mathbf{z}'(t)$$
$$= A\mathbf{v}(t) + e^{At}\mathbf{z}'(t).$$

Substituting this expression for $\mathbf{v}'(t)$ in (4.31),

$$A\mathbf{v}(t) + e^{At}\mathbf{z}'(t) = A\mathbf{v}(t) + \mathbf{f}(t)$$

and hence $e^{At}\mathbf{z}'(t) = \mathbf{f}(t)$. The inverse matrix of e^{At} is simply e^{-At}, so

$$\mathbf{z}'(t) = e^{-At}f(t).$$

Any antiderivative of $e^{-At}f(t)$ can be used as $\mathbf{z}(t)$; we will take

$$\mathbf{z}(t) = \int_0^t e^{-As}f(s)\,ds.$$

Then $\mathbf{v}(t) = e^{At}\mathbf{z}(t)$, or

$$\mathbf{v}(t) = e^{At}\int_0^t e^{-As}f(s)\,ds.$$

Since e^{At} does not depend on the variable s of integration,

$$\mathbf{v}(t) = \int_0^t e^{At}e^{-As}f(s)\,ds.$$

Noting that At and As are commuting matrices, we can combine exponentials to obtain a result:

Variation of Constants Formula

Systems with Constant Coefficients

The general solution of

$$\mathbf{v}' = A\mathbf{v} + \mathbf{f}(t),$$

where $\mathbf{v}$ and $\mathbf{f}$ are $n \times 1$ matrices and $\mathbf{A}$ is an $n \times n$ constant matrix, is

$$\mathbf{v}(t) = \int_0^t e^{A(t-s)}\mathbf{f}(s)\,ds + e^{At}\mathbf{c}.$$

Example 4.4.4 on page 170 was to illustrate the variation of constants method for finding a particular solution of an inhomogeneous system. We will return to the system of that example, to see how the variation of constants formula works.

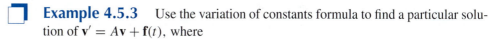

Example 4.5.3 Use the variation of constants formula to find a particular solution of $\mathbf{v}' = A\mathbf{v} + \mathbf{f}(t)$, where

$$A = \begin{bmatrix} 0 & 1 \\ 1 & 0 \end{bmatrix} \quad \text{and} \quad \mathbf{f} = e^{-t}\begin{bmatrix} -1 \\ 1 \end{bmatrix}.$$

Solution. The first step is to calculate e^{At}. Since $\text{tr}(A) = 0$ and $\det(A) = -1$, the characteristic equation of A is $s^2 - 1 = 0$. Thus, the characteristic roots are ± 1. To find a characteristic vector belonging to 1, we solve $(A - I)\begin{bmatrix} h \\ k \end{bmatrix} = \mathbf{0}$:

$$-h + k = 0$$
$$h - k = 0.$$

One solution is $\begin{bmatrix} h \\ k \end{bmatrix} = \begin{bmatrix} 1 \\ 1 \end{bmatrix}$. In the same way, we see that $\begin{bmatrix} 1 \\ -1 \end{bmatrix}$ is a characteristic vector belonging to -1. Thus the vector functions $e^t\begin{bmatrix} 1 \\ 1 \end{bmatrix}$ and $e^{-t}\begin{bmatrix} 1 \\ -1 \end{bmatrix}$ are solutions of the homogeneous system $\mathbf{v}' = A\mathbf{v}$. We can put them together to form a matrix solution

$$\mathcal{X}(t) = \begin{bmatrix} e^t & e^{-t} \\ e^t & -e^{-t} \end{bmatrix}.$$

Then

$$e^{At} = \mathcal{X}(t)(\mathcal{X}(0))^{-1}$$

$$= \begin{bmatrix} e^t & e^{-t} \\ e^t & -e^{-t} \end{bmatrix}\begin{bmatrix} 1 & 1 \\ 1 & -1 \end{bmatrix}^{-1}$$

$$= \begin{bmatrix} e^t & e^{-t} \\ e^t & -e^{-t} \end{bmatrix}\begin{bmatrix} \frac{1}{2} & \frac{1}{2} \\ \frac{1}{2} & -\frac{1}{2} \end{bmatrix}$$

$$= \begin{bmatrix} \cosh(t) & \sinh(t) \\ \sinh(t) & \cosh(t) \end{bmatrix}.$$

We can now apply the variation of constants formula. A particular solution of $\mathbf{v}' = A\mathbf{v} + \mathbf{f}(t)$ will be given by

$$\mathbf{v}(t) = \int_0^t \begin{bmatrix} \cosh(t - s) & \sinh(t - s) \\ \sinh(t - s) & \cosh(t - s) \end{bmatrix}\begin{bmatrix} -1 \\ 1 \end{bmatrix} e^{-s}\, ds$$

$$= \int_0^t \begin{bmatrix} -e^{s-t-s} \\ e^{s-t-s} \end{bmatrix} ds$$

$$= s\begin{bmatrix} -e^{-t} \\ e^{-t} \end{bmatrix}\Bigg|_{s=0}^t$$

$$= te^{-t}\begin{bmatrix} -1 \\ 1 \end{bmatrix}.$$

EXERCISES

1. Show that if $A = \begin{bmatrix} 0 & 1 & 0 & 0 \\ 0 & 0 & 1 & 0 \\ 0 & 0 & 0 & 1 \\ 0 & 0 & 0 & 0 \end{bmatrix}$, then $A^4 = 0$. Hence

calculate e^{At}.

2. Show that $\exp \begin{bmatrix} t & 1 \\ 0 & t \end{bmatrix} = \begin{bmatrix} e^t & e^t \\ 0 & e^t \end{bmatrix}$.

3. Prove Theorem 4.15.

◆ **Hint:** Review the proof that $e^{s+t} = e^s e^t$, based on the

formula $e^t = \sum_{k=0}^{\infty} \frac{1}{k!} t^k$. Where is the commutative law used?

4. Let $A(t)$ be a square matrix whose entries are functions of t, and let $B(t) = dA/dt$. Show that if $A(t)$ and $B(t)$ commute, then

$$e^{A(t)} = \sum_{k=0}^{\infty} \frac{1}{k!} [A(t)]^k$$

is a fundamental matrix solution of the system $\mathbf{x}' = B(t)\mathbf{x}$.

5. Show that if $A(t) = \begin{bmatrix} t & 1 \\ 0 & 2t \end{bmatrix}$, then A and dA/dt do not commute.

6. Let $A = \begin{bmatrix} 0 & 1 & -1 \\ 0 & 0 & 2 \\ 0 & 0 & 0 \end{bmatrix}$, and $B = \begin{bmatrix} 1 & 1 & 0 \\ 0 & 1 & 2 \\ 0 & 0 & 1 \end{bmatrix}$. Show

that $e^A = B$.

7. Compute all powers of

$$A = \begin{bmatrix} 0 & a & b \\ 0 & 0 & c \\ 0 & 0 & 0 \end{bmatrix},$$

and hence compute

a. e^{At} b. $e^{(A+\lambda I)t}$, where λ is a scalar constant

Use the result of part (b) to solve the following IVP:

$$\left. \begin{array}{l} x' = -2x + 3y + z; \; x(0) = 3 \\ y' = \quad\quad -2y - 4z; \; y(0) = 0 \\ z' = \quad\quad\quad\quad -2z; \; z(0) = 1 \end{array} \right\}.$$

8. Let $K = \begin{bmatrix} 0 & -1 \\ 1 & 0 \end{bmatrix}$. Determine all powers of K and hence compute

a. e^{Kt} b. $e^{(K+\lambda I)t}$, where λ is a scalar constant

Use the result of part (b) to solve the following IVP:

$$\left. \begin{array}{l} x' = -x - y; \; x(0) = 1 \\ y' = \quad x - y; \; y(0) = 0 \end{array} \right\}.$$

In Exercises 9–16, calculate the matrix e^{At}.

9. $A = \begin{bmatrix} 1 & 1 \\ -2 & -1 \end{bmatrix}$ **10.** $A = \begin{bmatrix} 0 & 4 \\ 2 & 2 \end{bmatrix}$

11. $A = \begin{bmatrix} -1 & 2 \\ -3 & 3 \end{bmatrix}$ **12.** $A = \begin{bmatrix} -3 & 4 \\ -4 & -1 \end{bmatrix}$

13. $A = \begin{bmatrix} -3 & -2 \\ 2 & 1 \end{bmatrix}$ **14.** $A = \begin{bmatrix} 2 & -5 \\ 4 & -2 \end{bmatrix}$

15. $A = \begin{bmatrix} 2 & 1 & 0 \\ -1 & 0 & 0 \\ 0 & 0 & -1 \end{bmatrix}$ **16.** $A = \begin{bmatrix} 0 & 1 & 2 \\ 0 & 0 & 1 \\ 0 & 0 & 0 \end{bmatrix}$

The following two exercises are intended to be challenging.

17. Let $\mathcal{P}$ be the vector space of all polynomials, and let $D : \mathcal{P} \longrightarrow \mathcal{P}$ denote the differentiation operator. Show that if $f(x) \in \mathcal{P}$, then

$$(e^{tD} f)(x) = f(x + t).$$

18. Let $A(t)$ be an $n \times n$ matrix whose entries are functions of t. Show that if the matrices $A(t)$ and $A(u)$ commute for all t and u, then $A(t)$ and $A'(t) = dA/dt$ commute. Hence find a fundamental matrix solution for $\mathbf{x}' = A'(t)\mathbf{x}$.

4.6 Glossary

e^{At} The matrix exponential.

I The identity matrix.

Im(z)   The imaginary part of $z = x + iy$: $\text{Im}(x + iy) = y$.

Re(z)   The real part of $z = x + iy$: $\text{Re}(x + iy) = x$.

Associated homogeneous system   [of a system of ODEs, $\mathbf{v}' = A(t)\mathbf{v} + \mathbf{b}(t)$] The system obtained by removing the source term: $\mathbf{v}' = A(t)\mathbf{v}$.

Absolute value   The distance of a complex number $z = x + iy$ to 0. The formula is $|z| = \sqrt{x^2 + y^2}$.

Argument   (of a complex number z) The angle made by a line connecting z to the origin with the positive real axis in the complex plane.

$$\arg(z) = \begin{cases} \cos^{-1}\left(\frac{\text{Re}(z)}{|z|}\right) & \text{if } \text{Im}(z) \geq 0 \\ -\cos^{-1}\left(\frac{\text{Re}(z)}{|z|}\right) & \text{if } \text{Im}(z) < 0 \end{cases}$$

Belongs to   Corresponding characteristic roots and characteristic vectors of a matrix are said to belong to each other.

Cayley-Hamilton theorem   Let $f(s) = 0$ be the characteristic equation of a matrix A. Then $f(A)$ is equal to the zero matrix.

Characteristic equation   A polynomial equation used to determine the characteristic roots of a matrix A. For an $n \times n$ matrix A, this is

$$\det(A - sI) = 0.$$

For a 2×2 matrix

$$A = \begin{bmatrix} a & c \\ b & d \end{bmatrix},$$

the characteristic equation is $s^2 - ps + q = 0$, where $p = \text{trace}(A) = a + d$, and $q = \det(A)$.

Characteristic root   (of a matrix A) A scalar s such that for some characteristic vector $\mathbf{b}$, $A\mathbf{b} = s\mathbf{b}$.

Characteristic vector   (of a matrix A). A nonzero vector $\mathbf{b}$ such that $A\mathbf{b} = s\mathbf{b}$ for some scalar s.

Coefficient matrix   The matrix $A(t)$ in the system

$$\mathbf{v}' = A(t)\mathbf{v} + \mathbf{b}(t).$$

Column vector   A matrix that has only one column. We use boldface letters for column vectors.

Complex plane   The ordinary xy-plane, in which the point (x, y) is made to correspond the complex number $z = x + iy$.

Determinant   A scalar-valued function of square matrices. For a 2×2 matrix

$$A = \begin{bmatrix} a & c \\ b & d \end{bmatrix},$$

$\det(A) = ad - bc$. Determinants are defined for $n \times n$ matrices in linear algebra courses.

Entry   A scalar that appears at a specified location in a matrix.

Exponential   (of a matrix A) A matrix function, denoted e^{At}, that is the sum of the power series

$$e^{At} = I + At + \frac{1}{2}A^2t^2 + \cdots + \frac{1}{n!}A^nt^n + \cdots.$$

If A is a constant matrix, e^{At} is the fundamental matrix solution of $\mathbf{v}' = A\mathbf{v}$ having initial value $\mathcal{X}(0) = I$.

Fundamental matrix solution   A matrix solution $\mathcal{X}(t)$ that is a nonsingular square matrix.

Homogeneous system   A system of linear ODEs that is satisfied by the vector function $\mathbf{v}(t) \equiv \mathbf{0}$. A homogeneous system always has a zero source vector.

Imaginary axis   The vertical axis of the complex plane, which corresponds to complex numbers $z = 0 + iy$.

Imaginary part   (of a complex number $z = x + iy$) $\text{Im}(z) = y$.

Identity matrix   The matrix

$$I = \begin{bmatrix} 1 & 0 & 0 & \cdots & 0 \\ 0 & 1 & 0 & \cdots & 0 \\ 0 & 0 & 1 & \cdots & 0 \\ 0 & 0 & 0 & \ddots & 0 \\ 0 & 0 & 0 & \cdots & 1 \end{bmatrix}.$$

I has the property that for all $\mathbf{b} \in \mathbf{R}^n$, $I\mathbf{b} = \mathbf{b}$.

Inverse matrix   (of a square matrix A) A matrix B such that $AB = BA = I$.

Linear combination   A sum of constant multiples of vector functions. For example, the set of linear combinations of $\mathbf{v}_1(t)$ and $\mathbf{v}_2(t)$ consists of all vector functions

$$c_1\mathbf{v}_1(t) + c_2\mathbf{v}_2(t).$$

Linearly independent   A set of vectors for which the only linear combination that is equal to the zero vector is the one with all coefficients equal to 0. For example, two nonzero vectors $\mathbf{v}_1$ and $\mathbf{v}_2$ are linearly independent if $\mathbf{v}_2$ is not equal to a constant scalar multiple of $\mathbf{v}_1$.

Matrix   A rectangular array of scalars.

Matrix product   The result of multiplying an $m \times n$ matrix A with entries a_{ij} and a $p \times q$ matrix B with entries b_{kl} is defined if and only if $n = p$; in this case it is an $m \times q$ matrix C whose entries are

$$c_{rs} = a_{r1}b_{1s} + a_{r2}b_{2s} + \cdots + a_{rn}b_{ns}.$$

Matrix solution   [of a homogeneous system of ODEs, $\mathbf{v}' = A(t)\mathbf{v}$] A matrix function $\mathcal{X}(t)$ such that $d\mathcal{X}(t)/dt = A(t)\mathcal{X}(t)$.

Nonsingular   A matrix that is not singular: A is nonsingular if and only if $A\mathbf{b} \neq \mathbf{0}$ for all $\mathbf{b} \neq \mathbf{0}$.

Real axis   The horizontal axis of the complex plane, which corresponds to numbers $z = x + 0i$.

Real part   (of a complex number $z = x + iy$) $\mathrm{Re}(z) = x$.

Scalar   A real or complex number, or a real or complex-valued expression. We use lowercase italic letters for scalars.

Singular   A matrix A such that $A\mathbf{b} = \mathbf{0}$ for some nonzero vector $\mathbf{b}$.

Source vector   The vector function $\mathbf{b}(t)$ in the system

$$\mathbf{v}' = A(t)\mathbf{v} + \mathbf{b}(t).$$

Trace   The sum of the diagonal entries of a square matrix.

Variation of constants   A method for finding a particular solution of an inhomogeneous system. Given a fundamental matrix solution $\mathcal{X}$ of $\mathbf{v}' = A\mathbf{v}$, substitute $\mathbf{v} = \mathcal{X}\mathbf{w}$ in

$$\mathbf{v}' = A\mathbf{v} + \mathbf{f}(t),$$

where $\mathbf{w}$ is to be determined. The resulting equation is readily simplified and integrated to find $\mathbf{w}$. If A is a constant matrix, then we can take $\mathbf{X}(t) = e^{At}$. Then the following formula for the solution results:

$$\mathbf{v}(t) = \int_0^t e^{A(t-s)} f(s)\, ds + e^{At}\mathbf{c}.$$

4.7 Chapter Review

1. Solve the initial value problem

$$x' = 2x + y; \ x(0) = 1$$
$$y' = x + 2y; \ y(0) = 0.$$

2. Find a fundamental matrix solution for each differential equation.

   a. $\begin{cases} x' = x - 2y \\ y' = 5x - y \end{cases}$ b. $\begin{cases} x' = x - y \\ y' = x + y \end{cases}$

   c. $\begin{bmatrix} x \\ y \end{bmatrix}' = \begin{bmatrix} 1 & 2 \\ 0 & 1 \end{bmatrix} \begin{bmatrix} x \\ y \end{bmatrix}$

   d. $\mathbf{x}' = A\mathbf{x}$, where

   $$A = \begin{bmatrix} 1 & 2 & 3 \\ 0 & 2 & 3 \\ 0 & 0 & 3 \end{bmatrix}$$

3. Find the general solution of

$$x' = x + 5y + \tan 2t$$
$$y' = -x - y,$$

given that $\mathcal{X}(t) = \begin{bmatrix} 2\sin 2t - \cos 2t & \sin 2t + 2\cos 2t \\ \cos 2t & -\sin 2t \end{bmatrix}$ is a fundamental matrix solution of the associated homogeneous equation.

4. Suppose A is an $n \times n$ matrix with independent characteristic vectors $\mathbf{b}_1, \ldots, \mathbf{b}_n$. Denote the corresponding characteristic roots by $r_1, \ldots, r_n$. Let $E(t)$ denote the diagonal matrix whose diagonal entries are $e^{r_1 t}, \ldots, e^{r_n t}$, and let B denote the matrix whose columns are the characteristic vectors $\mathbf{b}_1, \ldots, \mathbf{b}_n$. Show that $B \cdot E(t)$ is a fundamental matrix solution of the system $\mathbf{x}' = A\mathbf{x}$. Find an expression for e^{At}.

5. Use the result of Exercise 4 to calculate e^{At}, where

   a. $A = \begin{bmatrix} -2 & 2 & 0 & 4 \\ 3 & 22 & -36 & -28 \\ 1 & 9 & -16 & -19 \\ 1 & 8 & -11 & -12 \end{bmatrix}$.

   The characteristic roots of A are 9.3471, -11.0194, -2.3942, and -3.9330, and the corresponding characteristic vectors are

   $$\begin{bmatrix} .2908 \\ 1 \\ .1229 \\ .3251 \end{bmatrix}, \begin{bmatrix} -.3458 \\ 1 \\ .6709 \\ -.0113 \end{bmatrix}, \begin{bmatrix} 1 \\ -.1744 \\ -.0260 \\ -.0113 \end{bmatrix},$$

   and $\begin{bmatrix} 1 \\ -.9075 \\ -.5474 \\ -.0296 \end{bmatrix}$.

   b. $A = \begin{bmatrix} -10 & 17 & 14 & -9 & 19 \\ 15 & 17 & -6 & 16 & -43 \\ -15 & 19 & 9 & -15 & 41 \\ 1 & -2 & -2 & 1 & -1 \\ -7 & 9 & 4 & -7 & 20 \end{bmatrix}$.

   Here are three characteristic vectors:

   $$\begin{bmatrix} 1 \\ .01 + .98i \\ 1.11 + .004i \\ -.12 - .003i \\ .53 - .001i \end{bmatrix}, \begin{bmatrix} 1 \\ -.002 + .01i \\ .4167 + .08i \\ -.14 - .01i \\ .24 - .01i \end{bmatrix},$$

   and $\begin{bmatrix} 1 \\ .014 \\ .142 \\ -.85 \\ .02 \end{bmatrix}$.

The corresponding characteristic roots are $16.8 + 16.7i$, $1.58 + 1.14i$, and 0.21.

c. $A = \begin{bmatrix} -4 & 0 & -4 & 4 & -3 \\ 2 & -5 & 10 & -5 & 4 \\ -2 & 4 & -10 & 5 & -4 \\ 0 & 1 & 0 & -2 & 1 \\ 0 & 1 & -4 & 2 & -2 \end{bmatrix}$.

If you are not using a computer, skip the matrix multiplications and part (c).

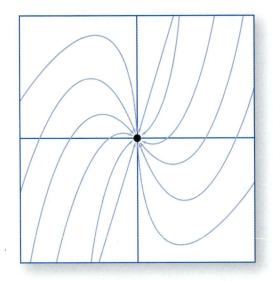

CHAPTER 5

Second-Order Equations

5.1 The Initial Value Problem

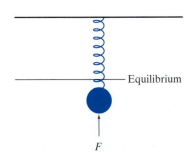

—— Equilibrium

F

Figure 5.1 A mechanical system. The arrow indicates the direction of the restoring force, which is upward since the spring has been stretched beyond the equilibrium position.

A linear second-order ODE is an equation of the form

$$a_2(t)\frac{d^2y}{dt^2} + a_1(t)\frac{dy}{dt} + a_0(t)y = f(t). \tag{5.1}$$

The terminology that we used for linear first-order ODEs and linear systems of ODEs will be adapted for second-order equations. Thus, the functions $a_i(t)$ are the **coefficient functions**, and $f(t)$ is the **source term**.

To visualize equation (5.1) in physical terms, consider a weight that is attached to a spring, as in Figure 5.1. We are interested in predicting the motion of the weight, given its initial position and velocity.

Newton's second law of motion states that the mass m of an object, multiplied by its acceleration a, is equal to the sum of the forces (often called the net force) applied to the object; that is, $F = ma$. We can identify four forces that might be considered:

- A force due to stretching or compressing the spring
- The force of gravity
- A frictional force
- An external force that might, for example, be due to a magnetic field

We will use the MKSA system of units: force in newtons, distance in meters, time in seconds, and mass in kilograms. If y denotes the position of the weight on a vertical axis, then the velocity of the weight is the first derivative of y with respect to time, and acceleration is given by the second derivative, $a = \frac{d^2y}{dt^2}$. Thus, Newton's second law says that

$$m\frac{d^2y}{dt^2} = F_{\text{spring}} + F_{\text{gravity}} + F_{\text{friction}} + F_{\text{external}}. \tag{5.2}$$

Although m is usually constant, we can conceive of situations where it is not. Perhaps the weight is a leaking pail of water.

Let's consider the spring force, F_{spring}, first. If a spring is stretched or compressed, it exerts a force, called the **restoring force**, directed to restore the spring to its natural length. Let x denote the extension of the spring beyond its natural length. The spring resists stretching, so that $F_{spring} < 0$ when $x > 0$, and $F_{spring} > 0$ when $x < 0$, because spring also resists compression. Mathematical models for mechanical systems are often based on the assumption that the spring force is a function $F_{spring}(x)$ of its extension.[1] The model that we will consider now is the **linear model**, which assumes that the restoring force is negatively proportional to the extension x of the spring:

$$F_{spring}(x) = -k\,x. \tag{5.3}$$

In (5.3), the constant k is called the **stiffness** of the spring. Its units are nt/m. A spring with high stiffness requires more force for the same extension than does a spring with low stiffness. For example, the spring in a scale designed to weigh swordfish will be stiffer than the one in a scale for weighing brook trout.

The linear model has been named Hooke's law, to honor the seventeenth-century English physicist Robert Hooke.

The next force to consider is that of gravity, $F_{gravity}$, which produces a constant acceleration of $g \approx 9.8$ m/sec^2. Thus, $F_{gravity} = mg$ newtons. The force of gravity will balance the spring force when $mg = k\,x$, and $x = \frac{mg}{k}$ is called the **equilibrium position** of the weight—we are now assuming that the mass is constant. In the absence of an external force, the weight will remain motionless if it is initially stopped in the equilibrium position.

Let y denote the displacement of the weight from this equilibrium position. Thus

$$y = x - \frac{mg}{k}.$$

Now suppose that the spring is extended by x meters beyond its natural length. Then

$$F_{spring} + F_{gravity} = mg - k\,x$$
$$= -k\left(x - \frac{mg}{k}\right) = -k\,y.$$

Thus, by using y rather than x to measure displacements, we can combine the spring and gravitational forces.

In the absence of friction, equation (5.2) reduces to the linear second order ODE, $my'' = -ky + F_{external}$, or, in the format of (5.1),

$$my'' + ky = f(t),$$

where $f(t)$ represents the external force.

---

[1] There are models that include the effect of aging on the spring, and for these the restoring force is a function of x and t (time). More complicated models incorporate the entire history of the spring in order to include the effect of wear.

The force due to friction is called the **damping force**. It resists motion; that is, its direction is opposite to the velocity. We will use a linear model and assume that the magnitude of the damping force is proportional to the velocity. Thus

$$F_{\text{friction}} = -b\,y'.$$

The constant b is the called the **damping**, and it is measured in kg/sec. With the linear model of friction included, equation (5.2) takes the form of a second-order linear ODE like (5.1):

$$my'' + by' + ky = f(t). \tag{5.4}$$

Although the parameters m, b, and k are usually taken to be constants, it is easy to conceive of situations in which they would be functions of t.

Phase plane analysis

By dividing the ODE (5.1) through by $a_2(t)$, we can obtain a simpler equation,

$$y'' + p(t)\,y' + q(t)\,y = g(t), \tag{5.5}$$

where $p(t) = a_1(t)/a_2(t)$, $q(t) = a_0(t)/a_2(t)$, and $g(t) = f(t)/a_2(t)$.

The ODE (5.5) is in turn equivalent to the system

$$\left.\begin{aligned} y' &= v \\ v' &= g(t) - p(t)v - q(t)y \end{aligned}\right\}, \tag{5.6}$$

in which the second equation is the same as (5.5), except that y' is replaced with v. Recall from Chapter 3 that $(y, v) = (\phi(t), \phi'(t))$ is a solution of (5.6) if and only if $y = \phi(t)$ is a solution of (5.5).

An initial value problem involving the system (5.6) would assign an initial condition of the form

$$\begin{bmatrix} y(t_0) \\ v(t_0) \end{bmatrix} = \begin{bmatrix} a \\ b \end{bmatrix}. \tag{5.7}$$

Since $v = y'$, this can be translated as a pair of initial conditions for equation (5.5):

$$\left.\begin{aligned} y(t_0) &= a \\ y'(t_0) &= b \end{aligned}\right\}. \tag{5.8}$$

The existence and uniqueness theorem for linear systems of ODEs, stated on page 139 as Theorem 4.1, is all that we need to prove the existence and uniqueness theorem for second-order linear ODEs.

Theorem 5.1 Existence and Uniqueness

Let (c, d) be an open interval such that

- $t_0 \in (c, d)$.
- Each function $g(t)$, $p(t)$, and $q(t)$ is continuous on (c, d).

Then there is a unique function $y = \phi(t)$, defined for all $t \in (c, d)$, that satisfies the ODE (5.5), and the initial conditions (5.8).

Proof Let

$$\begin{bmatrix} y \\ v \end{bmatrix} = \begin{bmatrix} \phi(t) \\ \psi(t) \end{bmatrix}$$

be a solution of the system (5.6), satisfying the initial condition (5.7). The existence of this solution on the interval (c, d) is guaranteed by Theorem 4.1. By the first equation of the system (5.6), $\psi(t) = \phi'(t)$. Hence $y = \phi(t)$ satisfies the initial conditions (5.8), as well as the ODE (5.5). This establishes existence.

Suppose that $\bar{\phi}(t)$ is another solution of the same IVP, (5.5), (5.8) on the interval (c, d). We would like to prove that $\phi(t) = \bar{\phi}(t)$ for all t in (c, d). Put $\bar{\psi}(t) = \bar{\phi}'(t)$, so that

$$\begin{bmatrix} y \\ v \end{bmatrix} = \begin{bmatrix} \bar{\phi}(t) \\ \bar{\psi}(t) \end{bmatrix}$$

is a solution of the system IVP (5.6), (5.7), and hence by Theorem 4.1,

$$\begin{bmatrix} \bar{\phi}(t) \\ \bar{\psi}(t) \end{bmatrix} = \begin{bmatrix} \phi(t) \\ \psi(t) \end{bmatrix}$$

for all $t \in (c, d)$. It follows that $\bar{\phi}(t) = \phi(t)$ on (c, d).   ■

A number t_1 where $g(t)$, $p(t)$, and $q(t)$ are all continuous is called an **ordinary point** of equation (5.5). A point where one or more of these is discontinuous is a **singular point**.

We will call the interval (c, d) the **domain of the IVP** (5.5), (5.8) if

- $t_0 \in (c, d)$,
- all points in (c, d) are ordinary points, and
- (c, d) is the largest interval with these properties.

This means that finite endpoints of the domain of an IVP must be singular points. For example, if equation (5.5) has no singular points, then its domain is the real line, $(-\infty, \infty)$. On the other hand, if t_0 is a singular point, the IVP does not have a domain, and the existence and uniqueness theorem gives us no information about it.

Example 5.1.1

a. Find the singular points of

$$\sin(t)y'' + (t^2 - t)y' - \frac{1}{2t - 1}y = 0. \tag{5.9}$$

b. Find the domain of initial value problems taking initial values at $t_0 = 1$.
c. Show that the IVP that couples (5.9) with the initial condition $y(0) = a$, $y'(0) = b$ has no solution if $a \neq 0$.

Solution. Divide (5.9) by the coefficient of y'' to obtain

$$y'' + \frac{t^2 - t}{\sin(t)}y' - \frac{1}{(2t - 1)\sin(t)}y = 0.$$

a. The singular points are the points where either

$$p(t) = \frac{t^2 - t}{\sin(t)} \quad \text{or} \quad q(t) = -\frac{1}{(2t - 1)\sin(t)}$$

are discontinuous. The function $p(t)$ has discontinuities at $t = n\pi$, where n is an integer, but the discontinuity at $t = 0$ is removable: It is not hard to verify that $\lim_{t \to 0} p(t) = -1$. On the other hand, $g(t)$ is discontinuous at all integer multiples of π, and at $t = 1/2$ as well. Thus, the singular points for the system (5.1) are $t = n\pi$, where $\sin(t) = 0$, and $t = 1/2$, where $2t - 1 = 0$, are singular.

b. The domain is $(1/2, \pi)$, since this is the interval whose endpoints are the singular points nearest to the initial point 1. Every initial value problem with initial values taken at the initial point $t_0 = 1$ has a unique solution that is defined on $(1/2, \pi)$.

c. If $\phi(t)$ satisfies equation (5.9), then for $t = 0$,

$$\sin(0) \cdot \phi''(0) + 0 \cdot \phi'(0) + \phi(0) = 0,$$

and it follows that $\phi(0) = 0$. Therefore, the initial condition $y(0) = a$ cannot be satisfied if $a \neq 0$. ❑

A linear ODE with zero source term is said to be **homogeneous**. A typical homogeneous second-order equation has the form

$$\frac{d^2 y}{dt^2} + p(t)\frac{dy}{dt} + q(t)y = 0. \tag{5.10}$$

Just as it was important to examine homogeneous systems before turning to inhomogeneous ones, we will find it useful to focus on the special properties of homogeneous ODEs. The following corollary of the existence and uniqueness theorem is essentially the same as Corollary 4.1.1 on page 140. This time, the proof is up to you. The term **homogeneous initial conditions** that is used in the statement of the corollary refers to initial conditions that would be satisfied by $y \equiv 0$. In the case of a second-order ODE with an ordinary point t_0, this would mean

$$y(t_0) = y'(t_0) = 0. \tag{5.11}$$

Corollary 5.1.1

Let $y = \phi(t)$ be a solution of the initial value problem involving a linear homogeneous ODE with homogeneous initial conditions at an ordinary point t_0. If (c, d) is the domain of this initial value problem, then $\phi(t) = 0$ for all $c < t < d$.

Example 5.1.2 The initial value problem

$$t^2 y'' - 3ty' + 4y = 0; \quad y(0) = y'(0) = 0$$

has the nonzero solution $y = t^2$. How can this be reconciled with Corollary 5.1.1?

Solution. Divide through by the coefficient of y'' to obtain

$$y'' - \frac{3}{t}y' + \frac{4}{t^2}y = 0.$$

The initial point $t = 0$ is a singular point, since

$$p(t) = -\frac{3}{t} \quad \text{and} \quad q(t) = \frac{4}{t^2}$$

have discontinuities there. Therefore, Corollary 5.1.1 does not apply.

If the homogeneous initial conditions had been taken at $t_0 = 1$ rather than $t_0 = 0$, then the initial value problem would have $(0, \infty)$ as its domain, and we would conclude that the solution would be identically zero on the interval $(0, \infty)$. ❑

EXERCISES

1. Which of the following are linear differential equations?

    a. $(\sin(t)y')' + 3y = 0$ b. $y'' = y/t$

    c. $y'' = t/y$ d. $y''' = (t^4 + 19t^2 - \frac{1}{\pi})y'$

2. Find the singular points of the following ODEs.

    a. $(1 - t^2)\,y'' - 2t\,y' + 6\,y = 0$

    b. $y'' - t\,y = 0$

    c. $\cos(t)\,y'' - t\,y' + (t^2 - 1)\,y = 1$

    d. $y'' - \tan(t)\,y = 0$

3. If $y = \phi(t)$ is a solution of $t\,y'' + y' + t\,y = 0$ and $\phi''(0)$ is defined, show that

    a. $\phi'(0) = 0$, and

    b. if $\phi(0) > 0$, then $\phi(0)$ is a relative maximum.

    ◆ **Hint:** Differentiate the ODE.

4. Find the domain of each IVP.

    a. $(1 - t^2)\,y'' - 2t\,y' + 6\,y = 0$; $y(0) = 1$, $y'(0) = 0$

    b. $y'' - t\,y = 0$; $y(0) = 1$, $y'(0) = 0$

    c. $\cos(t)\,y'' - t\,y' + (t^2 - 1)\,y = 1$; $y(\pi) = 1$, $y'(\pi) = 0$

    d. $y'' - \tan(t)\,y = 0$; $y(1) = 1$, $y'(1) = 0$

    e. $(1 - t)(2 - t)\,y'' + \sec(t)\,y' + \dfrac{y}{3t + 1} = 0$; $y(1.8) = 0$, $y'(1.8) = 0$

5. Prove Corollary 5.1.1.

6. Show that if $n \geq 2$, then $y = t^n$ is not a solution of any second-order linear homogeneous ODE that has an ordinary point at $t = 0$.

7. Suppose that $y = \phi_1(t)$ and $y = \phi_2(t)$ are solutions of a second-order linear homogeneous ODE that has an ordinary point at $t = 0$, and that $\phi_1(0) = \phi_2(0) = 0$. Show that

$$\phi_2'(0)\phi_1(t) - \phi_1'(0)\phi_2(t) \equiv 0.$$

5.2   Homogeneous Equations

A linear ODE

$$a_2(t)\,y'' + a_1(t)\,y' + a_0(t)\,y = f(t)$$

can be written simply as $\mathcal{L}(y) = f(t)$, where

$$\mathcal{L}(y) = a_2(t)\,y'' + a_1(t)\,y' + a_0(t)\,y.$$

$\mathcal{L}$ is an **operator** that acts on functions that have second derivatives.

An operator is a function whose domain and range consist of real-valued functions, instead of numbers. You are familiar with derivative operators, $\frac{d}{dt}$, $\frac{d^2}{dt^2}$, and so on, and with integral operators, such as $f \mapsto \int_a^t f(s)\,ds$. Each of these operators accepts a function that meets certain requirements (such as differentiability) and produces as output another function.

The important feature of the operator $\mathcal{L}$ is that it is **linear**; that is, the following two easily verified properties hold:

- For any sufficiently differentiable functions $\phi_1(t)$ and $\phi_2(t)$,

$$\mathcal{L}(\phi_1 + \phi_2)(t) = \mathcal{L}(\phi_1)(t) + \mathcal{L}(\phi_2)(t).$$

- For any sufficiently differentiable function $\phi(t)$ and any constant C,

$$\mathcal{L}(C\phi)(t) = C\mathcal{L}(\phi)(t).$$

Since $\mathcal{L}$ is formed by adding multiples of derivative operators $\frac{d^k}{dt^k}$, it is called a **linear differential operator**. We will use the acronym **LDO** for linear differential operator.

A point t_0 is said to be an **ordinary point** of $\mathcal{L}$ if and only if each quotient $a_1(t)/a_2(t)$ and $a_0(t)/a_2(t)$ is continuous on an open interval containing t_0. If t_0 is not an ordinary point, it is called a **singular point** of $\mathcal{L}$.

Theorem 5.2

Let $y_1(t)$ and $y_2(t)$ be solutions of a homogeneous linear ODE $\mathcal{L}(y) = 0$. Then for any constants C_1 and C_2,

$$y = C_1 y_1(t) + C_2 y_2(t) \tag{5.12}$$

is also a solution.

The solution (5.12) is called a **linear combination** of the functions $y_1(t)$ and $y_2(t)$. More generally, a linear combination of m functions, $y_1(t), y_2(t), \ldots, y_m(t)$ is an expression of the form

$$C_1 y_1 + C_2 y_2 + \cdots + C_m y_m.$$

We could state Theorem 5.2 compactly as "Any linear combination of solutions of a homogeneous linear ODE is also a solution." When the constants C_1 and C_2 are assigned specific values, the expression (5.12) represents a particular solution of $\mathcal{L}(y) = 0$. We will often leave the values of the constants unspecified in order to present a family of solutions.

Proof of Theorem 5.2   Since $\mathcal{L}$ is a linear operator, it follows that

$$
\begin{aligned}
\mathcal{L}(C_1 y_1 + C_2 y_2) &= \mathcal{L}(C_1 y_1) + \mathcal{L}(C_2 y_2) \\
&= C_1 \mathcal{L}(y_1) + C_2 \mathcal{L}(y_2) \\
&= 0
\end{aligned}
$$

since it is given that $\mathcal{L}(y_1) = 0$ and $\mathcal{L}(y_2) = 0$.   ■

Example 5.2.1   Given that $y = 1$ and $y = t^2$ are solutions of

$$ty'' - y' = 0,$$

find a family of solutions.

Solution. The ODE is linear and homogeneous. Therefore, by Theorem 5.2, any linear combination of 1 and t^2 is a solution. Thus, for any constants C_1 and C_2, $y = C_1 t^2 + C_2$ is a solution.

Given a homogeneous linear ODE $\mathcal{L}(y) = 0$, it is our objective to find a set of solutions $\{y_1(t), \ldots, y_k(t)\}$ such that the general solution is of the form

$$y = C_1 y_1(t) + \cdots + C_k y_k(t). \tag{5.13}$$

We would like our set of solutions to be minimal, and we will achieve this by requiring the set of solutions $\{y_1(t), \ldots, y_k(t)\}$ to be **linearly independent**. By definition, $\{y_1(t), \ldots, y_k(t)\}$ is linearly independent if

$$C_1 y_1(t) + C_2 y_2(t) + \cdots + C_k y_k(t) \equiv 0 \tag{5.14}$$

if and only if $C_1 = C_2 = \cdots = C_k = 0$.

If a set of solutions is not linearly independent, then there must be a linear combination of the form (5.14) in which at least one coefficient is nonzero—say $C_k \neq 0$. Then

$$y_k = -\frac{1}{C_k}(C_1 y_1 + \cdots + C_{k-1} y_{k-1}). \tag{5.15}$$

In this case, if a function ϕ is equal to a linear combination of $\{y_1(t), \ldots, y_k(t)\}$, then ϕ can equally well be expressed as a linear combination involving only $y_1(t)$ through $y_{k-1}(t)$, as follows. If

$$\phi(t) = A_1 y_1(t) + A_2 y_2(t) + \cdots + A_k y_k(t),$$

then we could use (5.15) to replace $y_k(t)$ and thus to obtain

$$\phi(t) = \left(A_1 - \frac{C_1}{C_k} A_k\right) y_1(t) + \left(A_2 - \frac{C_2}{C_k} A_k\right) y_2(t) + \cdots$$
$$+ \left(A_{k-1} - \frac{C_{k-1}}{C_k} A_k\right) y_{k-1}(t).$$

Example 5.2.2   Show that $\{1, t, t^2\}$ is linearly independent.

Solution. Consider a linear combination

$$f(t) = C_1 + C_2 t + C_3 t^2 \equiv 0.$$

Then $f''(t) \equiv 0$. Since $f''(t) = 2C_3$, $C_3 = 0$. Therefore, $f'(t) = C_2$, but this implies $C_2 = 0$, since $f'(t) \equiv 0$, too. It follows that $f(t) \equiv C_1$, so $C_1 = 0$ as well. We have proved that only the zero combination of $\{1, t, t^2\}$ is equal to the 0-polynomial, so the set is linearly independent.

A set S of solutions of $\mathcal{L}(y) = 0$, defined on a common interval (c, d), is said to be a **fundamental set** of solutions if

  i. Every solution of $\mathcal{L}(y) = 0$ that is defined on (c, d) can be expressed as a linear combination of the functions in S, and

 ii. S is linearly independent.

This definition only applies to *homogeneous* linear equations. That's because if the differential equation is not homogeneous, not all linear combinations of solutions will be solutions (see Exercise 10 at the end of this section).

If we verify that S is a fundamental set of solutions, then requirement (i) assures that every solution can be expressed in terms of the solutions in S. Requirement (ii) is included so that S will be a minimal set of solutions that satisfies property (i).

Theorem 5.3

Let t_0 be an ordinary point for the second order homogeneous linear ODE

$$a_2(t)y'' + a_1(t)y' + a_0(t)y = 0. \tag{5.16}$$

Let $\phi_1(t)$ be the solution of (5.16) with initial values

$$\phi_1(t_0) = 1, \quad \phi_1'(t_0) = 0$$

and let $\phi_2(t)$ be the solution with

$$\phi_2(t_0) = 0, \quad \phi_2'(t_0) = 1.$$

Then $\{\phi_1, \phi_2\}$ is a fundamental set of solutions on the interval (c, d) that is the domain of these IVPs.

Proof

  i. Let $\psi(t)$ be any solution of (5.16) that is defined on (c, d), and let $a = \psi(t_0)$, $b = \psi'(t_0)$. The linear combination

$$\phi(t) = a\phi_1(t) + b\phi_2(t)$$

satisfies the same initial conditions as $\psi(t)$; thus by the uniqueness statement of Theorem 5.1, $\psi(t) = \phi(t)$ for all $t \in (c, d)$. Hence ψ can be expressed as a linear combination of ϕ_1 and ϕ_2.

 ii. $S = \{\phi_1, \phi_2\}$ is linearly independent: Suppose that

$$C_1\phi_1(t) + C_2\phi_2(t) = 0 \tag{5.17}$$

for all $t \in (c, d)$. We will show that $C_1 = C_2 = 0$. For $t = t_0$, (5.17) reduces to $C_1 = 0$, since $\phi_1(t_0) = 1$ and $\phi_2(t_0) = 0$. Differentiation of (5.17) yields $C_1\phi_1'(t) + C_2\phi_2'(t) = 0$ for all $t \in (c, d)$. Since $\phi_1'(t_0) = 0$ and $\phi_2'(t_0) = 1$, we can substitute $t = t_0$ to show $C_2 = 0$. ∎

Theorem 5.3 identifies a fundamental set of solutions, $\{\phi_1(t), \phi_2(t)\}$, of a second-order linear homogeneous ODE. If we are given another set of solutions, how can we tell if it is a fundamental set? The following theorem provides a way to answer this question.

Theorem 5.4

Let $\mathcal{S} = \{\phi_1(t), \ldots, \phi_n(t)\}$ and $\mathcal{T} = \{\psi_1(t), \ldots, \psi_m(t)\}$ be linearly independent sets of functions. If every function in $\mathcal{T}$ can be expressed as a linear combination of the functions in $\mathcal{S}$, then $m \leq n$.

The proof is omitted, since it can be found in practically any linear algebra text (look in the chapter about the dimension of a vector space).

Corollary 5.2.1

Let $\mathcal{T}$ be a linearly independent set of solutions of a second-order linear homogeneous ODE $\mathcal{L}(y) = 0$, all defined at an ordinary point t_0. Then $\mathcal{T}$ is a fundamental set of solutions if and only if $\mathcal{T}$ consists of exactly two solutions.

Proof Let $\mathcal{S} = \{\phi_1, \phi_2\}$ be the fundamental set of solutions described in Theorem 5.3. Since every solution in $\mathcal{T}$ can be expressed as a linear combination of ϕ_1 and ϕ_2, Theorem 5.4 tells us that $\mathcal{T}$ contains no more than two solutions.

Since $\mathcal{T}$ is also a fundamental set of solutions, every solution of the ODE can also be expressed as a linear combination of the functions in $\mathcal{T}$. This means that the two functions in $\mathcal{S}$ can be expressed as linear combinations of the functions in $\mathcal{T}$. Applying Theorem 5.4 again, we conclude that $\mathcal{T}$ contains at least two solutions. ∎

Example 5.2.3 Verify that $\mathcal{S} = \{\cosh(2t), \sinh(2t)\}$ is a fundamental set of solutions of $y'' - 4y = 0$.

Solution. The differentiation formulas

$$\frac{d^2}{dt^2} \cosh(at) = a^2 \cosh(at) \quad \text{and} \quad \frac{d^2}{dt^2} \sinh(at) = a^2 \sinh(at)$$

establish that $y = \cosh(2t)$ and $y = \sinh(2t)$ are solutions of the ODE. The ODE in question is second order, and the set S consists of two solutions. We therefore only need to verify that the solutions are linearly independent. Since neither solution is identically 0, we only have to check that $\cosh(2t) \neq C \sinh(2t)$. But for $t = 0$ we have $\cosh(0) = 1$ and $\sinh(0) = 0$. Since $1 = C \cdot 0$ is impossible, the solutions are linearly independent. ☐

The **general solution** of a second-order differential equation on an interval (c, d) is a family of solutions that contains all solutions of the equation that are defined on (c, d). Since the linear span of a fundamental set of solutions is the entire family of solutions, the general solution of a homogeneous linear equation is

$$y = C_1 \phi_1(t) + C_2 \phi_2(t),$$

where $\{\phi_1(t), \phi_2(t)\}$ is a given fundamental set of solutions. Thus, Example 5.2.3 shows that the general solution of $y'' - 4y = 0$ can be expressed as

$$y = C_1 \cosh(2t) + C_2 \sinh(2t).$$

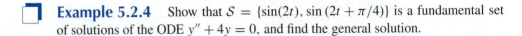

Example 5.2.4   Show that $\mathcal{S} = \{\sin(2t), \sin(2t + \pi/4)\}$ is a fundamental set of solutions of the ODE $y'' + 4y = 0$, and find the general solution.

Solution. The first step (omitted) is to make sure that $\phi_1(t) = \sin(2t)$ and $\phi_2(t) = \sin(2t + \pi/4)$ are solutions of the ODE. By Corollary 5.2.1 we can show that $\mathcal{S}$ is a fundamental set of solutions by showing that it is linearly independent. Neither solution is identically 0, and it is easy to see that $\sin(2t + \pi/4)$ is not a constant multiple of $\sin(2t)$. Hence the solutions are linearly independent.   ❑

The Wronskian

The functions $\phi_1(t), \phi_2(t)$ are solutions of the ODE

$$y'' + p(t)y' + q(t)y = 0 \qquad\qquad (5.18)$$

if and only if

$$\mathbf{v}_1(t) = \begin{bmatrix} \phi_1(t) \\ \phi_1'(t) \end{bmatrix} \quad \text{and} \quad \mathbf{v}_2(t) = \begin{bmatrix} \phi_2(t) \\ \phi_2'(t) \end{bmatrix}$$

are solutions of the system

$$\begin{aligned} y' &= u \\ u' &= -q(t)y - p(t)u. \end{aligned} \qquad\qquad (5.19)$$

The solutions $\mathbf{v}_1(t)$ and $\mathbf{v}_2(t)$ are the columns of a matrix solution

$$\mathcal{X}(t) = \begin{bmatrix} \phi_1(t) & \phi_2(t) \\ \phi_1'(t) & \phi_2'(t) \end{bmatrix}$$

of (5.19).

Proposition 5.2.2
_____

$\{\phi_1, \phi_2\}$ is a fundamental set of solutions of (5.18) if and only if

$$\mathcal{X}(t) = \begin{bmatrix} \phi_1(t) & \phi_2(t) \\ \phi_1'(t) & \phi_2'(t) \end{bmatrix}$$

is a fundamental matrix solution of (5.19).

Proof Recall that $\mathcal{X}$ is a fundamental matrix solution if and only if at each ordinary point t_0, the matrix $\mathcal{X}(t_0)$ is nonsingular.

Assume that $\mathcal{X}(t_0)$ is nonsingular. We want to show that ϕ_1 and ϕ_2 are linearly independent. If either function is identically 0, then the corresponding column of $\mathcal{X}(t)$ would be zeros, and hence $\mathcal{X}$ would be singular. If $\phi_2(t) \equiv C\phi_1(t)$, where C is constant, then $\mathbf{v}_2(t) = C\mathbf{v}_1(t)$. Since $\mathbf{v}_1$ and $\mathbf{v}_2$ are the columns of $\mathcal{X}$, it follows that $\mathcal{X}(t_0) \begin{bmatrix} C \\ -1 \end{bmatrix} = \mathbf{0}$, and again, $\mathcal{X}(t_0)$ is singular. It follows that ϕ_1 and ϕ_2 are linearly independent.

Conversely, assume that $\{\phi_1, \phi_2\}$ is a linearly independent set of solutions. Now we want to prove that $\mathcal{X}(t_0)$ is nonsingular. Suppose that to the contrary $\mathcal{X}$ is singular. Then there is a nonzero vector $\mathbf{p} = \begin{bmatrix} p_1 \\ p_2 \end{bmatrix}$ such that

$$\mathcal{X}(t_0)\mathbf{p} = \mathbf{0}. \tag{5.20}$$

Let $\psi(t) = p_1\phi_1(t) + p_2\phi_2(t)$. Then (5.20) implies that $\psi(t_0) = 0$ and $\psi'(t_0) = 0$. It follows from Corollary 5.1.1 that $\psi(t) \equiv 0$. Since the constants p_1 and p_2 are not both equal to zero, it follows that ϕ_1 and ϕ_2 are linearly dependent, a contradiction. ∎

The **Wronskian** of a pair of solutions, $\phi_1(t)$, $\phi_2(t)$, of a homogeneous second-order linear ODE is defined to be the determinant of the corresponding matrix solution, and is denoted $W[\phi_1, \phi_2](t)$. Thus,

$$W[\phi_1, \phi_2](t) = \det \begin{bmatrix} \phi_1(t) & \phi_2(t) \\ \phi_1'(t) & \phi_2'(t) \end{bmatrix}.$$

The following theorem describes the remarkable property of the Wronskian.

Theorem 5.5

Let $\phi_1(t)$, $\phi_2(t)$ be solutions of the homogeneous linear ODE (5.18), defined on an interval (c, d) containing no singular points.

  i. If $\{\phi_1(t), \phi_2(t)\}$ is linearly independent, then

$$W[\phi_1, \phi_2](t) \neq 0$$

  for all $t \in (c, d)$.
  ii. If $\{\phi_1(t), \phi_2(t)\}$ is linearly dependent, then

$$W[\phi_1, \phi_2](t) \equiv 0$$

  on (c, d).

Proof

  i. Assume that $\{\phi_1(t), \phi_2(t)\}$ is linearly independent. By Proposition 5.2.2, the corresponding matrix solution $\mathcal{X}$ of (5.19) is a fundamental matrix solution. Thus, $\mathcal{X}(t)$ is nonsingular at every ordinary point. By Proposition 4.2.1 on page 147, it follows that $W[\phi_1, \phi_2](t) = \det \mathcal{X}(t) \neq 0$ for every ordinary point t.

ii. If $\{\phi_1(t), \phi_2(t)\}$ is linearly dependent, then the matrix solution $\mathcal{X}(t)$ of (5.19) is singular for all t. Again referring to Proposition 4.2.1, $W[\phi_1, \phi_2](t) = 0$ for all t. ∎

Example 5.2.5 Show that $\{t, t^2\}$ is a fundamental set of solutions of the linear homogeneous ODE,

$$t^2 y'' - 2ty' + 2y = 0,$$

and that $W[t, t^2](t) = t^2$. Resolve the paradox that $W[t, t^2](0) = 0$, apparently contradicting Theorem 5.5.

Solution. It is easily verified that $y = t$ and $y = t^2$ are solutions. Since t^2 is not a constant multiple of t, they are linearly independent.

$$W[t, t^2](t) = \det \begin{bmatrix} t & t^2 \\ 1 & 2t \end{bmatrix} = t^2$$

Divide the ODE by the coefficient t^2 of y'' to obtain

$$y'' - \frac{2}{t} y' + \frac{2}{t^2} y = 0.$$

Thus $t = 0$ is a singular point, due to the discontinuity of both coefficient functions. Theorem 5.5 only guarantees that $W[t, t^2](t) \neq 0$ at ordinary points. ❑

Proposition 5.2.3

Let $\phi_1(t)$ and $\phi_2(t)$ be solutions of the differential equation

$$y'' + p(t)y' + q(t)y = 0,$$

and let $W(t) = W[\phi_1, \phi_2](t)$. Then

$$W' + p(t)W = 0.$$

The proof is Exercise 25 at the end of this section. With Proposition 5.2.3 it is possible to determine the Wronskian of two solutions of a homogeneous linear differential equation even if no formulas for the solutions are available, as is the case in the following example.

Example 5.2.6 Find $W(t) = W[\phi_1, \phi_2](t)$, where the ϕ_i are solutions of

$$e^t y'' - ty' + \ln(1 + t^2)y = 0$$

with initial values $(\phi_1(0), \phi_1'(0)) = (1, 0)$, and $(\phi_2(0), \phi_2'(0)) = (0, 1)$.

Solution. Divide the ODE by the coefficient e^t of y'' to obtain

$$y'' - te^{-t}\, y' + \ln(1 + t^2)e^{-t}\, y = 0.$$

By Proposition 5.2.3, $W(t)$ satisfies the differential equation

$$W' - te^{-t} W = 0 \tag{5.21}$$

and the initial value is $W(0) = \phi_1(0)\phi_2'(0) - \phi_2(0)\phi_1'(0) = 1$. Since equation (5.21) is a homogeneous linear first-order equation, its solution can be found by the method of Section 1.3. We obtain $W(t) = \exp(-(t+1)e^{-t})$. ❑

EXERCISES

1. Show that every polynomial of degree ≤ 3 can be expressed as a linear combination of

$$\{1, (t-1), (t-1)^2, (t-1)^3\}.$$

2. Which polynomials can be expressed as linear combinations of

$$\{(t-1), (t-1)^2, (t-1)^3\}?$$

◆ **Hint:** Find a property common to all polynomials $f(t) = C_1(t-1) + C_2(t-1)^2 + C_3(t-1)^3$.

3. Express $\cos(t - \frac{\pi}{4})$ as a linear combination of $\cos(t)$ and $\sin(t)$.

4. Is $\tan(t - \frac{\pi}{4})$ a linear combination of $\tan(t)$ and $\cot(t)$? Explain your answer.

5. Express $f(t) = t$ as a linear combination of

$$S = \{t^2, (t+1)^2, (t+2)^2\}.$$

6. a. Verify that $S = \{e^t, e^{-t}\}$ is a set of solutions of the differential equation $y'' - y = 0$.

   b. Find the general solution.

7. Find a fundamental set of solutions of $y'' = 0$.

8. Find all solutions of the form $y = t^n$ of the differential equation

$$t^2 y'' - ty' - 3y = 0,$$

   and then find the general solution.

9. Find all solutions of the form $y = e^{nt}$ of the differential equation

$$4y'' + 3y' - y = 0,$$

   and then find the general solution.

10. Suppose that $y_1(t)$ and $y_2(t)$ be solutions of an inhomogeneous linear ODE, $\mathcal{L}(y) = f(t)$, where $f(t) \neq 0$. Show that a linear combination $y = C_1 y_1(t) + C_2 y_2(t)$ of these solutions will also be a solution if and only if

$$C_1 + C_2 = 1.$$

11. Find a particular solution of the differential equation

$$ty'' - y' + ty = 5t^4 - 3t^2 + 18$$

    that is of the form $y = C_1 t^3 + C_2 t$.

12. Let $\phi_1(t)$, $\phi_2(t)$, and $\phi_3(t)$ be solutions of a second-order, homogeneous linear ODE, all defined on an interval (c, d) containing no singular points. Show that $\{\phi_1(t), \phi_2(t), \phi_3(t)\}$ is linearly dependent.

In Exercises 13–19, show that the given set S is a fundamental set of solutions, and find the general solution.

13. $y'' + 16y = 0$; $S = \{\sin 4t, \sin 4(t - \pi/3)\}$

14. $y'' + 4y' + 3y = 0$; $S = \{e^{-3t}, e^{-t}\}$

15. $y'' = 0$; $S = \{t + 1, t - 1\}$

16. $t^2 y'' + ty' - y = 0$ on $(0, \infty)$; $S = \left\{t + \dfrac{1}{t}, 2t + \dfrac{1}{t}\right\}$

17. $(ty')' = 0$ on $(0, \infty)$; $S = \{1, \ln(t)\}$

18. $ty'' - y' = 0$ on $(0, \infty)$; $S = \{1, t^2/2\}$

19. $t^2 y'' - ty' + y = 0$ on $(-\infty, 0)$; $S = \{t, t \ln(-t)\}$

20. For each ODE, specify a set of initial conditions $y(t_0) = a$, $y'(t_0) = b$, that are *not satisfied* by any solution.
    ◆ **Hint:** Start by finding t_0.
    a. $t^2 y'' + (t+1)y' - y = 0$
    b. $(1 - t^2)y'' - 2t\, y' + (1 + t^2)y = 0$
    c. $\cos(t)y'' - \sin(t)y' + y = 0$
    d. $(t - 1)y'' + 3\, y' + y = t$

21. Let $\phi_1(t) = t^3$ and $\phi_2(t) = |t|^3$. Show that $\{\phi_1, \phi_2\}$ is linearly independent, although $W[\phi_1, \phi_2](t) \equiv 0$. Does this contradict Theorem 5.5?

22. Let $\phi_1(t)$ and $\phi_2(t)$ be solutions of the linear homogeneous ODE,

$$y'' + p(t)y' + q(t)y = 0,$$

such that $\phi_1 \not\equiv 0$. Show that if t_0 is an ordinary point and $\phi_1'(t_0) = \phi_2'(t_0) = 0$, then there is a constant K such that $\phi_2 \equiv K\phi_1$.

23. Let $\{\phi_1(t), \phi_2(t)\}$ be a fundamental set of solutions of the linear homogeneous ODE,

$$y'' + p(t)y' + q(t)y = 0$$

on an interval (c, d) containing no singular points. Show that

a. ϕ_1 and ϕ_2 cannot both vanish at the same point $t_0 \in (c, d)$.

b. ϕ_1 and ϕ_2 cannot both have a relative extremum at the same point $t_0 \in (c, d)$.

c. if ϕ_1 and ϕ_2 both have an inflection point at $t_0 \in (c, d)$, then $p(t_0) = q(t_0) = 0$.

24. Prove that the solution $\phi(t)$ of the IVP

$$(t^2 + 1)y'' + (t^2 - 1)y = 0; \ y(0) = 1, \ y'(0) = 0$$

is an even function; that is, $\phi(-t) = \phi(t)$ for all t.

25. Prove Proposition 5.2.3.

26. Show that the Wronskian of any pair of solutions of

$$y'' + q(t)y = 0$$

is constant.

27. Determine the Wronskian of two solutions $\phi_1(t)$ and $\phi_2(t)$ of

$$(1 - t^2)y'' - 2ty' + 6y = 0,$$

given that $(\phi_1(0), \phi_1'(0)) = (1, 2)$ and $(\phi_2(0), \phi_2'(0)) = (-3, 1)$.

5.3   Homogeneous Equations with Constant Coefficients

Our approach to a homogeneous second-order linear ODE with constant coefficients,

$$y'' + p\,y' + q\,y = 0, \tag{5.22}$$

will be the same as the method used to solve linear first-order ODEs and systems with constant coefficients. We substitute $y = C\,e^{st}$ (where s is a constant to be determined), $y' = se^{st}$, and $y'' = s^2 e^{st}$ in equation (5.22) to obtain

$$s^2 e^{st} + ps\,e^{st} + q\,e^{st} = 0.$$

Dividing through by e^{st} results in the quadratic equation

$$s^2 + ps + q = 0, \tag{5.23}$$

called the **characteristic equation** of the ODE (5.22). In Exercise 17 you will be asked to verify that it is the same as the characteristic equation of the corresponding system,

$$\left. \begin{array}{l} y' = v \\ v' = -qy - pv \end{array} \right\}. \tag{5.24}$$

Each root $s = r_1$ of (5.23) leads to a solution $y = e^{r_1 t}$ of (5.22). If the characteristic equation has two distinct roots $s = r_1$ and $s = r_2$, then we can easily verify that $\mathcal{S} = \{e^{r_1 t}, e^{r_2 t}\}$ is linearly independent. Therefore, $\mathcal{S}$ is a fundamental set of solutions.

Example 5.3.1   Find a fundamental set of solutions of the differential equation

$$y'' + 4y' + 3y = 0 \qquad (5.25)$$

and thus find the general solution.

Solution. The characteristic equation is

$$s^2 + 4s + 3 = 0.$$

Since the roots of this quadratic equation are $r_1 = -1$ and $r_2 = -3$, $y = e^{-t}$ and $y = e^{-3t}$ are solutions. Since e^{-3t} is not a constant multiple of e^{-t}, the solutions are linearly independent. Hence $S = \{e^{-t}, e^{-3t}\}$ is a fundamental set of solutions of equation (5.25). The general solution is

$$y = C_1 e^{-t} + C_2 e^{-3t}. \qquad \square$$

In the next example, we will consider an equation with imaginary characteristic roots.

Example 5.3.2   Find the general solution of

$$y'' + k^2 y = 0. \qquad (5.26)$$

Solution. The characteristic equation is $s^2 + k^2 = 0$, so the characteristic roots are $s = \pm ki$. We have thus found two solutions, $e^{\pm kit}$. These solutions are complex valued, and we need real-valued solutions. By Euler's formula,

$$e^{ikt} = \cos(kt) + i \sin(kt).$$

Set $\phi_1(t) = \cos(kt)$ and $\phi_2(t) = \sin(kt)$; thus $\phi_1(t) = \mathrm{Re}[e^{ikt}]$ and $\phi_2(t) = \mathrm{Im}[e^{ikt}]$. Since $\phi_1(t) + i\phi_2(t)$ is a solution of equation (5.26),

$$(\phi_1 + i\phi_2)'' + k^2(\phi_1 + i\phi_2) = 0.$$

After rearrangement,

$$(\phi_1'' + k^2\phi_1) + i \cdot (\phi_2'' + k^2\phi_2) = 0 + i \cdot 0. \qquad (5.27)$$

The real part of equation (5.27) states that ϕ_1 is a solution of (5.26), and the imaginary part states that ϕ_2 is a solution. Therefore,

$$y = C_1 \cos(kt) + C_2 \sin(kt) \qquad (5.28)$$

is a family of solutions of equation (5.26). Since $\cos(kt)$ is not equal to a constant multiple of $\sin(kt)$, the solutions are independent, and (5.28) is the general solution.
$\square$

The solution of Example 5.3.2 was an application of Theorem 4.8 on page 161. The following proposition is a special case of Theorem 4.8, and the proof is left to you.

Proposition 5.3.1

Let $y = \phi(t)$ be a complex-valued solution of the homogeneous linear ODE

$$y'' + p(t)y' + q(t)y = 0,$$

where the coefficient functions $p(t)$ and $q(t)$ are assumed to be real valued. Then each of the functions $\phi_1(t) = \text{Re}[\phi(t)]$ and $\phi_2(t) = \text{Im}[\phi(t)]$ is a real-valued solution.

Proposition 5.3.1 is stated so that it applies to equations with variable coefficients, and we will use it in that context in Chapter 7.

Example 5.3.3   Find a fundamental set of solutions of

$$y'' + 4y' + 13y = 0 \tag{5.29}$$

and hence find the general solution.

Solution. The characteristic equation is

$$s^2 + 4s + 13 = 0.$$

The characteristic roots are $\frac{1}{2}(-4 \pm \sqrt{16 - 52}) = -2 \pm 3i$. The solution of equation (5.29) corresponding to $r_1 = -2 + 3i$ is

$$\phi(t) = e^{(-2+3i)t}$$
$$= e^{-2t}\cos(3t) + ie^{-2t}\sin(3t).$$

Now, by Proposition 5.3.1,

$$\phi_1(t) = \text{Re}[\phi(t)] = e^{-2t}\cos(3t) \quad \text{and} \quad \phi_2(t) = \text{Im}[\phi(t)] = e^{-2t}\sin(3t)$$

are solutions of equation (5.29). Since these solutions are linearly independent, they form a fundamental set. The general solution is

$$y = C_1 e^{-2t}\cos(3t) + C_2 e^{-2t}\cos(3t)$$
$$= e^{-2t}(C_1\cos(3t) + C_2\cos(3t)). \qquad \square$$

Double roots

Suppose that the characteristic equation of

$$y'' + py' + qy = 0 \tag{5.30}$$

has a double root r_1. Since $\phi_1(t) = e^{r_1 t}$ cannot by itself form a fundamental set of solutions, we need to find a second solution that is not a constant multiple of ϕ_1. One approach would be to find the fundamental matrix solution of the equivalent system, using the method summarized in the box on page 155. The following proposition, which will be useful for other purposes as well, provides an alternative approach.

Proposition 5.3.2

Let

$$\mathcal{L}(y) = a_2 y'' + a_1 y' + a_0 y$$

be a second-order LDO with constant coefficients, let $v(t)$ be a function that has a second derivative on an interval (c, d), and let k be a real or complex constant. Finally, let $f(s) = a_0 s^2 + a_1 s + a_0$ be the characteristic polynomial of $\mathcal{L}$. Then the following formula is valid:

$$\mathcal{L}(v(t) e^{kt}) = (f(k) v(t) + f'(k) v'(t) + \frac{1}{2} f''(k) v''(t)) e^{kt}. \tag{5.31}$$

In formula (5.31), $f'(k)$ and $f''(k)$ represent derivatives of $f(s)$, evaluated for $s = k$. Thus $f'(k) = 2a_2 k + a_1$ and $f''(k) = 2a_2$. The derivatives of $v(t)$ are with respect to t. The proposition can be generalized readily to apply to higher-order LDOs with constant coefficients.

Proof of Proposition 5.3.2. Let $y = v(t) e^{kt}$. By the product rule for differentiation,

$$y' = (kv + v')e^{kt}, \quad \text{and}$$
$$y'' = (k^2 v + 2kv' + v'')e^{kt}.$$

Thus

$$\mathcal{L}(y) = [a_2(k^2 v + 2kv' + v'') + a_1(kv + v') + a_0 v]e^{kt}.$$

Rearranging, we have

$$\mathcal{L}(y) = [(a_2 k^2 + a_1 k + a_0)v + (2a_2 k + a_1)v + a_2 v]e^{kt}$$
$$= (f(k) v(t) + f'(k) v'(t) + \frac{1}{2} f''(k) v''(t))e^{kt}. \quad \blacksquare$$

Proposition 5.3.3

If equation (5.30) has a double characteristic root r_1, then $\{e^{r_1 t}, t\, e^{r_1 t}\}$ is a fundamental set of solutions.

Proof Let $\mathcal{L}$ be the linear operator forming the left side of (5.30), and let $f(s)$ be the characteristic polynomial. Since $s = r_1$ is a double root of $f(s) = 0$, the graph of f as a function of s has a minimum at $s = r_1$. It follows that $f(r_1) = 0$ *and* $f'(r_1) = 0$ as well, while $f''(r_1) = 2$. Thus, for any differentiable function $v(t)$, Proposition 5.3.2 tells us that

$$\mathcal{L}(v(t)e^{r_1 t}) = (0 \cdot v(t) + 0 \cdot v'(t) + \frac{1}{2} f''(r_1)v''(t))e^{r_1 t}$$
$$= v''(t)e^{r_1 t}.$$

Thus, $\mathcal{L}(v(t)e^{r_1 t}) = 0$ if and only if $v''(t) \equiv 0$.; that is, if and only if $v = C_1 t + C_2$, where C_1 and C_2 are constants. The general solution of (5.30) is therefore

$$y = (C_1 t + C_2)e^{r_1 t},$$

and the fundamental set of solutions is as stated. $\blacksquare$

⬜ **Example 5.3.4** Find the general solution of

$$y'' + 6y' + 9y = 0.$$

Solution. The characteristic equation has the double root $s = -3$. By Proposition 5.3.3, $\{e^{-3t}, te^{-3t}\}$ is a fundamental set of solutions. The general solution is $y = C_1 e^{-3t} + C_2 t e^{-2t}$. ⬜

The following box summarizes the method for finding solutions of homogeneous linear second-order equations with constant coefficients.

Solutions of Second-Order Equations

The characteristic equation of

$$a_2 y'' + a_1 y' + a_0 y = 0 \tag{5.32}$$

is the quadratic equation

$$a_2 s^2 + a_1 s + a_0 = 0.$$

The roots of the characteristic equation are called the characteristic roots. The general solution of equation (5.32) is

  i. $y = C_1 e^{r_1 t} + C_2 e^{r_2 t}$, if the characteristic roots, r_1 and r_2, are real and distinct

  ii. $y = e^{pt}(C_1 \cos(qt) + C_2 \sin(qt))$, if the characteristic roots are a conjugate complex pair $p \pm iq$

  iii. $y = e^{r_1 t}(C_1 + C_2 t)$, if there is a double characteristic root r_1

In each of the preceding expressions, C_1 and C_2 represent constants whose values may be specified arbitrarily.

EXERCISES

In Exercises 1–16, find the characteristic roots and the general solution of the ODE.

1. $y'' - 4y = 0$

2. $y'' - 4y' = 0$

3. $y'' = 0$

4. $y'' - 25y = 0$

5. $y'' + 25y = 0$

6. $y'' + 4y' - y = 0$

7. $y'' + 4y' + 4y = 0$

8. $y'' + 5y' + 4y = 0$

9. $y'' + 2y' + 4y = 0$

10. $y'' + 8y' + 25y = 0$

11. $y'' - 14y' + 49y = 0$

12. $y'' + 3y' - 4y = 0$

13. $2y'' + y' - y = 0$

14. $5y'' + 3y' - 2y = 0$

15. $3y'' + 2y' + \frac{1}{4}y = 0$

16. $y'' + y' + y = 0$

17. Show that the characteristic equation of the ODE $y'' + p\,y' + q\,y = 0$ is the same as that of the coefficient matrix of the equivalent system,

$$\left. \begin{array}{r} y' = v \\ v' = -q\,y - p\,v \end{array} \right\} . \tag{5.33}$$

18. Suppose that in the ODE presented in Exercise 17, $p = -2r_1$ and $q = r_1^2$, so that there is a double characteristic root. Find the general solution of the system (5.33), using the method described on page 155, and hence provide another proof for Proposition 5.3.3.

19. Show that the following sets of functions are linearly independent:

a. $\{e^{r_1 t}, e^{r_2 t}\}$, where $r_1 \neq r_2$

b. $\{e^{pt} \cos(q\,t), e^{pt} \sin(q\,t)\}$, where $q \neq 0$

c. $\{e^{r_1 t}, t e^{r_1 t}\}$

20. Suppose that $\phi(t) = C_1 e^{r_1 t} + C_2 e^{r_2 t}$, where $r_1 \neq r_2$. Show that if for some t_0, $\phi(t_0) = \phi'(t_0) = 0$, then $C_1 = C_2 = 0$.

21. Repeat Exercise 20 for $\phi(t) = e^{pt}[C_1 \cos(qt) + C_2 \sin(qt)]$, where $q \neq 0$.

22. Repeat Exercise 20 for $\phi(t) = e^{r_1 t}(C_1 t + C_2)$.

23. Prove Proposition 5.3.1.

5.4 Free Vibrations

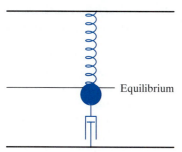

Figure 5.2 A damped mechanical system with one weight.

Figure 5.2 is a schematic drawing of a mechanical system with one weight of mass m and one spring of stiffness k. The displacement of the weight below its equilibrium position is x, and its velocity is y. The symbol that appears below the weight is called a **dashpot** The dashpot resembles a shock absorber, except that unlike some shock absorbers, there is no built-in spring. A dashpot represents a source of friction that obeys the linear model.

We saw in Section 5.1 that the following ODE applies to a system with a weight of mass m constrained to move on an axis, subject to the forces of a linear spring of stiffness k, a dashpot with damping b, and an external force $f(t)$:

$$my'' + by' + ky = f(t).$$

Here, y represents the displacement of the weight from its equilibrium position. If there is no external force ($f(t) \equiv 0$), the motion of the system is said to be **free**. In this case the ODE is linear, homogeneous, and with constant coefficients.

In the following example, we will examine free vibrations of an undamped system.

 Example 5.4.1   A weight whose mass is 1 kilogram is attached to a linear spring suspended vertically from a support. The spring is stretched 0.2 meter as a result of attaching this weight. The weight is held stationary 0.3 meter below this equilibrium position and released. Describe the motion of the system.

Solution. We can determine the stiffness by observing that the weight exerts a force of $mg = 9.8$ newtons on the spring and stretches it 0.2 meter. Thus $k = 9.8/0.2 = 49$ newtons per meter.

Let $y(t)$ denote the displacement of the weight below the equilibrium position t seconds after release. Then the velocity of the weight is $v(t) = y'(t)$ and the acceleration is $a(t) = y''(t)$.

The net force on the weight is $F = -49y$. By Newton's second law of motion, $F = m\,a(t)$, where $m = 1$ is the mass (in kilograms). Hence $y(t)$ is a solution of the ODE

$$y'' = -49y.$$

The initial data, given in the statement of the problem, are $y(0) = 0.3$, and $y'(0) = v(0) = 0$ (because the weight was stationary before release). Therefore, $y(t)$ is the solution of the IVP

$$y'' + 49y = 0; \quad y(0) = 0.3, \quad y'(0) = 0.$$

The characteristic roots are $\pm 7i$. Therefore, the general solution of the ODE is

$$y(t) = C_1 \cos(7t) + C_2 \sin(7t),$$

where C_1 and C_2 are constants. To determine their values, we use the initial conditions:

$$y(0) = C_1 \cos(0) + C_2 \sin(0) = C_1 = 0.3$$
$$y'(0) = -7C_1 \sin(0) + 7C_2 \cos(0) = 7C_2 = 0.$$

The displacement function is therefore $y(t) = 0.3 \cos(7t)$. ◻

The vector $\begin{bmatrix} x(t) \\ v(t) \end{bmatrix} = 0.3 \begin{bmatrix} \cos(7t) \\ -7\sin(7t) \end{bmatrix}$, whose components are the position and velocity of the vibrating weight in Example 5.4.1 is called the **phase vector** of the mechanical system. The components of this phase vector are simple harmonic oscillations [that is, functions of the form $a \sin(kt)$ or $a \cos(kt)$]. For this reason, the weight is said to vibrate in **simple harmonic motion**. The frequency of vibration, 7, is revealed by the characteristic roots $\pm 7i$ of the ODE. The model is physically unrealistic, because, like many models that ignore friction, its solution implies perpetual motion. In the next example, we will correct this by introducing some damping to the system. As we will see, the system behaves very differently, depending on the amount of damping there is.

Example 5.4.2 Assume that the mechanical system in Example 5.4.1 is damped. If the initial conditions are the same as in Example 5.4.1, determine the subsequent motion of the weight, where the damping is

- a. $b = 2$ kilograms per second
- b. $b = 14$ kilograms per second
- c. $b = 50$ kilograms per second

Solution. Let $y(t)$ denote the distance of the weight from its equilibrium position. Then $y(t)$ satisfies the ODE

$$y'' + by' + 49y = 0, \tag{5.34}$$

with initial conditions $y(0) = 0.3$ and $y'(0) = 0$.

a. The characteristic equation is $s^2 + 2s + 49 = 0$. Using the quadratic formula, we find the pair of complex characteristic roots,

$$s = -1 \pm \sqrt{48}i.$$

Therefore, the general solution is

$$y = e^{-t}\left[C_1 \cos(\sqrt{48}t) + C_2 \sin(\sqrt{48}t)\right].$$

Thus, $y(0) = C_1$. Differentiating, we have

$$y' = e^{-t}\left[(-C_1 + \sqrt{48}C_2)\cos(\sqrt{48}t) + (-\sqrt{48}C_1 - C_2)\sin(\sqrt{48}t)\right]$$

and hence $y'(0) = -C_1 + \sqrt{48}C_2$. Thus, the initial conditions are equivalent to two equations

$$\begin{aligned} C_1 &= .3 \\ -C_1 + \sqrt{48}C_2 &= 0. \end{aligned}$$

Hence, $C_1 = .3$, $C_2 = .3/\sqrt{48} \approx .04$, and

$$y \approx e^{-t}[.3\cos(6.9t) + .04\sin(6.9t)],$$

where $\sqrt{48}$ has been approximated as 6.9.

b. The characteristic equation, $s^2 + 14s + 49 = 0$, has a double characteristic root $r = -7$. Therefore, the general solution is $y = (C_1 t + C_2)e^{-7t}$, with $y' = (-7C_1 t + C_1 - 7C_2)e^{-7t}$. Thus, $y(0) = C_2$, and $y'(0) = C_1 - 7C_2$. The initial data imply that the following equations hold:

$$\begin{aligned} C_2 &= .3 \\ C_1 - 7C_2 &= 0. \end{aligned}$$

Therefore, $C_1 = 2.1$ and $C_2 = .3$ The equation of motion is

$$y = (2.1t + .3)e^{-7t}.$$

c. The characteristic equation, $s^2 + 50s + 49 = 0$, has two real roots, $s = -1$ and $s = -49$. This leads to the general solution $y = C_1 e^{-t} + C_2 e^{-49t}$. Since $y' = -C_1 e^{-t} - 49C_2 e^{-49t}$, we have $y(0) = C_1 + C_2$ and $y'(0) = -C_1 - 49C_2$. The initial conditions translate into the equations

$$\begin{aligned} C_1 + C_2 &= .3 \\ -C_1 - 49C_2 &= 0, \end{aligned}$$

so $C_1 = 49/160$ and $C_2 = -1/160$. The equation of motion is

$$y = \frac{49}{160}e^{-t} - \frac{1}{160}e^{-49t}. \qquad \square$$

Figures 5.3 through 5.5 show the graphs of $y(t)$ in each case of Example 5.4.2. The motion in case (a) is said to be an *underdamped vibration*. The *damping* in this case is insufficient to prevent oscillation. Underdamped motion is desirable in some cases (for example, a pendulum), and undesirable in others (for example, a swinging door, and your car's suspension).

The graph shown in Figure 5.3 shows a function that is a product $\phi(t) = e^{-\lambda t}p(t)$ of a decaying exponential function $e^{-\lambda t}$ and an oscillating periodic function $p(t)$. The graph of $\phi(t)$ oscillates between $-e^{-\lambda t}$ and $+e^{-\lambda t}$, and the graphs of these functions are said to form an **envelope** for the graph of $\phi(t)$. The envelope is shown as a pair of dashed curves in Figure 5.3.

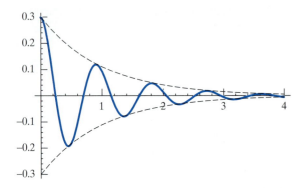

FIGURE 5.3   Motion of an underdamped system. The envelope, indicated as the dashed curves, is the graph of exponential factors.

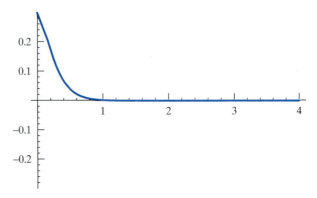

FIGURE 5.4   Motion of a critically damped system. The initial conditions are the same as for the system depicted in Figure 5.3.

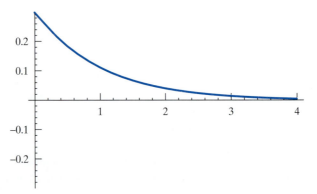

FIGURE 5.5   Motion of an overdamped system. The initial conditions are the same as for the system depicted in Figure 5.3.

The system of Example 5.4.2, case (b), is *critically damped*. The damping constant in a critically damped system has the minimum value that is sufficient to prevent oscillation. The displacement of a critically damped system generally has the form $\phi(t) = e^{-\lambda t}(C_1 t + C_2)$, where C_2 represents the initial displacement, and the initial velocity is $C_1 - \lambda C_2$. If C_1 and C_2 are of opposite sign, the graph of $\phi(t)$ will cross the t-axis at a point $t_1 > 0$. There can be no more than one zero for $\phi(t)$, since the linear factor can have only one zero, and the decaying exponential factor guarantees that the displacement will converge to 0 as $t \to \infty$.

The system of Example 5.4.2(c) is *overdamped*, which means that the damping is greater than the minimum sufficient to prevent oscillation. The displacement of an overdamped system generally has the form

$$\phi(t) = C_1 e^{-\lambda t} + C_2 e^{-\mu t}.$$

As in the case of the critically damped system, the displacement of a critically damped system can have at most one zero. See Exercise 8 at the end of this section.

The overdamped system approaches the equilibrium position more slowly than a critically damped system does. In fact, it can be shown that the critically damped system approaches equilibrium as fast as is possible without oscillation (see Exercise 13 on page 245).

The characteristic roots of the differential equation (5.34) determine the nature of the corresponding mechanical system. If there are two distinct negative roots, the system is overdamped; if there is a double characteristic root, the system is critically damped; and if there is a conjugate pair of complex roots, then the system is underdamped. The discriminant of the characteristic equation, which is $D = b^2 - 4mk$, can be used to distinguish between these cases quickly. If $D > 0$, the system is overdamped, if $D = 0$ it is critically damped; and if $D < 0$ the system is underdamped.

All damped mechanical systems eventually cease to have noticeable motion— in effect they stop after a while. Our mathematical model, if it is any good, should reflect this physical fact.

Let us say that an LDO

$$\mathcal{L}(y) = a_2(t)y'' + a_1(t)y' + a_0(t)y$$

is *stable* if, for every solution $y(t)$ of the homogeneous equation $\mathcal{L}(y) = 0$,

$$\lim_{t \to \infty} y(t) = 0.$$

If the LDO $(\mathcal{L}(y) = my'' + by' + ky$ is stable, then the free motion of a mechanical system with mass m, damping b, and stiffness k eventually ceases.

Proposition 5.4.1

The constant coefficient LDO $(\mathcal{L}(y) = my'' + by' + ky$ is stable if m, b, and k are positive.

Proof The characteristic roots of $\mathcal{L}(y) = 0$ are the roots of the quadratic equation $mr^2 + br + k = 0$. If r_1 is a real characteristic root, then $br_1 = -(mr_1^2 + k)$. The product of b and r_1 is therefore negative, and hence r_1 is negative.

If the characteristic roots form a conjugate complex pair $p \pm iq$, then, by the quadratic formula, $p = -\frac{b}{2m}$, which is negative.

Therefore, the solution $y(t)$ can be written either as

$$y(t) = C_1 e^{r_1 t} + C_2 e^{r_2 t},$$

where r_1 and r_2 are negative; or as

$$y(t) = e^{r_1 t}(C_1 t + C_2),$$

where again $r_1 < 0$; or as

$$e^{pt}[C_1 \cos(qt) + C_2 \sin(qt)]$$

with $p < 0$. In each case, every term has a decaying exponential factor. It follows that $\lim_{t \to \infty} y(t) = 0$. ∎

The natural unit of time

The coefficients of the ODE that models a damped mechanical system will change if we measure time, distance, or mass in different units. It is possible to use this fact to advantage by choosing a new time unit that will simplify the ODE, eliminating the mass and stiffness as parameters to be considered. Let us assume that the mechanical parameters of equation (5.34) are denominated in the meter-kilogram-second system. Define a new time unit of a seconds, where we will determine the value of a later. If we let T denote the elapsed time in the new units, $T = a^{-1}t$. For example, if $a = 60$, the T would be the time, denominated in minutes.

Conversion to the new time units is given by the formulas

$$\frac{dy}{dt} = \frac{dy}{dT}\frac{dT}{dt} = \frac{dy}{dT} \cdot a^{-1}$$

for the velocity, and

$$\frac{d^2y}{dt^2} = \frac{d}{dt}\left(\frac{dy}{dt}\right) = \frac{d}{dt}\left(\frac{dy}{dT}a^{-1}\right)$$

$$= \frac{d^2y}{dT^2} \cdot a^{-2}$$

for the acceleration. With the aid of these formulas, we can write equation (5.34) in the form

$$ma^{-2}\frac{d^2y}{dT^2} + ba^{-1}\frac{dy}{dT} + ky = 0.$$

Thus, for $a = \sqrt{\frac{m}{k}}$, we have

$$k\frac{d^2y}{dT^2} + b\sqrt{\frac{k}{m}}\frac{dy}{dT} + ky = 0,$$

or, after dividing, through by k

$$\frac{d^2y}{dT^2} + \frac{b}{\sqrt{km}}\frac{dy}{dT} + y = 0. \tag{5.35}$$

The number $B = \frac{b}{\sqrt{km}}$ will be called the **damping ratio** of the mass spring system. Its value is independent of the units in which the mass, damping, and stiffness parameters are denominated. The damping ratio measures the relative strength of the forces in the system. You can show that the system is

$$\begin{array}{ll} \text{underdamped} & \text{if } B < 2; \\ \text{critically damped} & \text{if } B = 2, \text{ and} \\ \text{overdamped} & \text{if } B > 2. \end{array}$$

The units of time in which T in equation (5.35) is denominated are the *natural units* for the mechanical system.

EXERCISES

1. A mechanical system has mass 200 kilograms, damping 300 kilograms per second, and spring constant

   a. $k = 0$ b. $k = 100$

   c. $k = 112.5$ d. $k = 125$

   newtons per meter. The mass is initially at the equilibrium position, with an initial velocity of 20 meters per second. In each case, determine the displacement function and the maximum displacement.

2. A mechanical system has $m = 1$ kilogram, and $k = 2500$ newtons per meter. At time zero, the mass is moved 0.1 meter beyond the equilibrium position and released. Find the displacement function $x(t)$, taking the following values for the damping (the units are kilograms per second).

   a. $b = 0$ b. $b = 60$

   c. $b = 100$ d. $b = 125$

   Sketch the graphs of all four displacement functions.

3. Classify the mechanical systems with the parameters listed as underdamped, critically damped, or overdamped. In each case, m denotes the mass, in kilograms; b denotes the damping constant, in kilograms per second; and k denotes the stiffness, in newtons per meter.

   a. $m = 1, b = 2, k = 2$

   b. $m = 1, b = 2, k = 0.5$ c. $m = 1, b = 2, k = 1$

4. Prove the following statement:

   $a_2 y'' + a_1 y' + a_0 y = 0$ *is stable if and only if the coefficients a_0, a_1, and a_2 are all of the same sign.*

5. Suppose that $a_2 y'' + a_1 y' + a_0 y = 0$ is stable, and let $\psi(t)$ be a particular solution of the inhomogeneous equation

   $$a_2 y'' + a_1 y' + a_0 y = f(t). \qquad (5.36)$$

   Show that if $\bar{\psi}(t)$ is any other solution of equation (5.36), then $\bar{\psi}(t)$ is asymptotic to $\psi(t)$; in other words, show that $\lim_{t \to \infty}(\bar{\psi}(t) - \psi(t)) = 0$.

6. A mechanical system has a mass of 100 kilograms, and a spring constant of 8.1×10^5 newtons per meter. If the system is critically damped, what would the damping constant be?

7. A damped mechanical system has parameters $m = 5$ kilograms, $b = 10$ kilograms per meter, and $k = 10$ newtons per meter. At time $t = 0$, the weight is at the equilibrium position with an initial velocity of 100 meters per second. Determine the motion of the weight as a function of time.

8. Show that the displacement function of an overdamped system, $\phi(t) = C_1 e^{-\lambda t} + C_2 e^{-\mu t}$, can have at most one

positive zero, $t_1 > 0$. Under what conditions on the coefficients C_1 and C_2 will there be a positive zero?

9. Show that if the displacement function $\phi(t)$ of a critically damped or overdamped mechanical system has a zero t_1— that is, $\phi(t_1) = 0$—then there are numbers t_2 and t_3 such that $t_1 < t_2 < t_3$ and $\phi'(t_2) = \phi''(t_3) = 0$.

   ◆ *Hint:* To see why this happens, sketch a rough graph of a nonoscillating displacement function that has a zero.

10. A damped mechanical system has mass 1 kilogram, damping constant 2 kilograms per second, and stiffness 1 newton per meter (the system is critically damped). Initially, the weight is drawn 0.2 meters past the equilibrium position, and then pushed toward the equilibrium position with a velocity of 1.2 meters per second.

    a. When will the weight reach the equilibrium position?

    b. Will the weight cross the equilibrium position again? If so, when?

    c. Determine the maximum displacement of the weight, and the time that the maximum displacement occurs.

11. The mechanical system in Exercise 10 is replaced by a system in which the mass, damping, and stiffness are 1 kilogram, 5 kilograms per second, and 6 newtons per meter, respectively. The initial conditions are the same. Answer the questions in Exercise 10.

12. The mechanical system in Exercise 10 is replaced by a system in which the mass, damping, and stiffness are 1 kilogram, 2 kilograms per second, and 2 newtons per meter, respectively. The initial conditions are the same. Answer the questions in Exercise 10.

13. Explain why the units of damping are kilograms per second.

14. A mechanical system has mass m, damping b, and stiffness k. Let $y(t)$ denote the displacement function. The total energy $E(t)$ of the system is the sum of the kinetic energy,

    $$K(t) = \frac{1}{2}m \left(\frac{dy}{dt}\right)^2,$$

    and the potential energy,

    $$U(t) = \frac{1}{2}ky^2.$$

    Show that the total energy of the system decreases at a rate proportional to the kinetic energy, and find the constant of proportionality. In other words, show that

    $$\frac{dE}{dt} = -c K(t),$$

    and determine the value of c.

15. For an overdamped mechanical system, it is sometimes preferable to choose a time unit so that $T = \frac{b}{m}t$. Write a differential equation for the motion of a mechanical system in terms of this time scale.

16. Devise a unit of time for an underdamped mass spring system so that the frequency of oscillation denominated in this unit is 1. Convert equation (5.34) to this time scale. How do the parameters of this new equation reflect the behavior of the system?

5.5 Inhomogeneous Equations

If a weight of mass m is subject to the forces of a linear spring with stiffness k, a dashpot with damping b, and an external force $f(t)$, then we have seen that the displacement y satisfies the linear ODE with constant coefficients,

$$my'' + by' + ky = g(t), \tag{5.37}$$

or

$$\mathcal{L}(y) = g(t),$$

where $\mathcal{L}(y) = my'' + by' + ky$. We can ascribe the motion of the weight to two causes: the external force, and the natural motion the system would have if unforced. Specifically, let $y_p(t)$ denote a particular solution of (5.37), and let $y_h(t)$ be a solution of $\mathcal{L}(y) = 0$. Then

$$\mathcal{L}(y_p + y_h) = \mathcal{L}(y_p) + \mathcal{L}(y_h).$$

Since $\mathcal{L}(y_p) = g(t)$ and $\mathcal{L}(y_h) = 0$, we see that

$$y = y_p(t) + y_h(t)$$

is also a solution of (5.37). We have a way to determine $y_h(t)$, since we know how to find the general solution of a homogeneous linear ODE with constant coefficients. We will develop some methods for determining a particular solution of an inhomogeneous linear ODE.

Here is an example to illustrate the first method, called **undetermined coefficients**. To find a particular solution of

$$y'' + 4y = e^t, \tag{5.38}$$

substitute $y = Ae^t$, where A is a constant. Since $y'' = Ae^t$, we find

$$Ae^t + 4Ae^t = e^t.$$

Cancelling e^t and simplifying shows $5A = 1$, so $A = \frac{1}{5}$. The particular solution is $y = \frac{1}{5}e^t$. In this calculation, we substituted an **undetermined coefficient expression** $y = Ae^t$ in the ODE and then solved for the undetermined coefficient, A.

The method of undetermined coefficients works only in limited circumstances. The ODE must be linear with constant coefficients, and the source term must be a polynomial, an exponential function, or a sine or cosine.[2]

---

[2]It also works for products of such functions.

The first step in using the method is to select the undetermined coefficient expression. Given an inhomogeneous equation with constant coefficients

$$a_2 y'' + a_1 y' + a_0 y = C e^{kt}, \qquad (5.39)$$

let $\mathcal{L}(y) = a_2 y'' + a_1 y' + a_0 y$ denote the LDO that forms the left side, and let $f(s) = a_2 s^2 + a_1 s + a_0$ be the characteristic polynomial. By Proposition 5.3.2 on page 201,

$$\mathcal{L}(v(t)e^{kt}) = (f(k)v(t) + f'(k)v'(t) + \frac{1}{2}f''(k)v''(t))e^{kt}. \qquad (5.40)$$

Thus, if $f(k) \neq 0$—that is, if k isn't a characteristic root of $\mathcal{L}(y) = 0$—then we can put $v = C/f(k)$ (a constant). By (5.40), $\mathcal{L}(ve^{kt}) = f(k)ve^{kt} = Ce^{kt}$ because $v' = v'' = 0$.

If k is a characteristic root of $\mathcal{L}(y) = 0$ so that $f(k) = 0$, then when v is constant, (5.40) simply tells us that $\mathcal{L}(ve^{kt}) = 0$. In this case, if $f'(k) \neq 0$—that is, k is not a double root—we can put $v = \frac{C}{f'(k)}t$. Again, by (5.40), $\mathcal{L}(v(t)e^{kt}) = f'(k)v'(t)e^{kt} = Ce^{kt}$.

If k is a double root, then $f(k) = f'(k) = 0$, so we let $v(t) = \frac{C}{f''(k)}t^2$. Then $v'' = \frac{2C}{f''(k)}$, and again $\mathcal{L}(v(t)e^{kt}) = \frac{1}{2}f''(k)v''(t)e^{kt} = Ce^{kt}$.

It isn't practical to memorize the preceding formulas. Instead, we just calculate $\mathcal{L}(Ae^{kt})$, where A is an "undetermined coefficient." The result will be of the form Be^{kt}. Provided that $B \neq 0$, we can put $y_p = \frac{C}{B}e^{kt}$. If the $B = 0$, then k was a characteristic root. In this case $\mathcal{L}(Ate^{kt}) = Be^{kt}$. Again, if $B \neq 0$ we can take $y_p = \frac{C}{B}t\,e^{kt}$.

If k is a double characteristic root, then $\mathcal{L}(At^2e^{kt}) = Be^{kt}$, and (5.40) tells us that $B \neq 0$. Thus, put $y_p = \frac{C}{B}t^2\,e^{kt}$.

Example 5.5.1   Find a particular solution of

$$5y'' + 7y' + 2y = 14e^{-t},$$

and then find the general solution.

Solution. Let $\mathcal{L}(y) = 5y'' + 7y' + 2y$. Then

$$\mathcal{L}(A\,e^{-t}) = A(5(-1)^2 + 7(-1) + 2)e^{-t} = 0,$$

indicating that -1 is a characteristic root of $\mathcal{L} = 0$. Therefore, we compute $\mathcal{L}(At\,e^{-t})$. Since $(A\,t\,e^{-t})' = A(1-t)e^{-t}$ and $(A\,t\,e^{-t})'' = A(t-2)e^{-t}$,

$$\mathcal{L}(A\,t\,e^{-t}) = A(5(t-2) + 7(1-t) + 2t)e^{-t} = -3\,A\,e^{-t}.$$

Set $-3\,A\,e^{-t} = 14e^{-t}$, to obtain $A = -14/3$, and

$$y_p = -\frac{14}{3}te^{-t}$$

is a particular solution.

If y_g is any other solution of $\mathcal{L}(y) = 14e^{-t}$, then

$$\mathcal{L}(y_g - y_p) = \mathcal{L}(y_1) - \mathcal{L}(y_p) = 0.$$

Therefore, $y_g - y_p$ is a solution of the homogeneous equation $\mathcal{L}(y) = 0$. The characteristic roots are the roots of

$$5s^2 + 7s + 2 = (s + 1)(5s + 2) = 0,$$

or -1 and $-\frac{2}{5} = -0.4$. Thus, the general solution of $\mathcal{L}(y) = 0$ is

$$y_g - y_p = C_1 e^{-t} + C_2 e^{-0.4t}.$$

The general solution of $\mathcal{L}(y) = 14e^{-t}$ is therefore

$$y_g = \left(C_1 - \frac{14}{3}t \right) e^{-t} + C_2 e^{-0.4t}. \qquad \square$$

Example 5.5.1 uses a general principle, known as the **principle of superposition**, that we will state as a theorem.

Theorem 5.6
_____

Let $\mathcal{L}$ be an LDO of order n and let $y_p(t)$ denote a particular solution of the inhomogeneous linear ODE

$$\mathcal{L}(y) = f(t), \tag{5.41}$$

and let $y_h(t)$ denote the family of functions that is the general solution of the homogeneous equation,

$$\mathcal{L}(y) = 0. \tag{5.42}$$

Then the general solution $y_g(t)$ of (5.41) can be expressed as the sum of y_p and y_h:

$$y_g(t) = y_p(t) + y_h(t). \tag{5.43}$$

Proof We need to show that every function in the family (5.43) is a solution of (5.41). Since $\mathcal{L}$ is linear, $\mathcal{L}(y_p + y_h) = \mathcal{L}(y_p) + \mathcal{L}(y_h)$. Since it is given that $\mathcal{L}(y_p) = f$ and $\mathcal{L}(y_h) = 0$, it follows that $\mathcal{L}(y_g) = f$, as required.

We also need to show that there are no solutions of (5.41) that do not belong to the family (5.43). Suppose that $\psi(t)$ is a solution of (5.41), and put $\phi(t) = \psi(t) - y_p(t)$. Then

$$\mathcal{L}(\psi) = \mathcal{L}(\phi) - \mathcal{L}(y_p) = f - f = 0.$$

In other words, ϕ satisfies (5.42) and thus belongs to the family y_h. Since $\psi = y_p + \phi$, it follows that ψ belongs to the family y_g. ∎

Example 5.5.2   Find a particular solution of

$$4y'' + 4y' + y = -12\,e^{-t/2} \tag{5.44}$$

and then find the general solution.

Solution. Notice that $-\frac{1}{2}$ is a double characteristic root of $\mathcal{L}(y) = 0$, where $\mathcal{L}(y) = 4y'' + 4y' + y$. Therefore, the general solution of the homogeneous equation $\mathcal{L}(y) = 0$ is

$$y_h = (C_1 + C_2 t)e^{-t/2}.$$

We won't bother to calculate $\mathcal{L}(Ae^{-t/2})$ or $\mathcal{L}(Ate^{-t/2})$, since we know that $-\frac{1}{2}$ is a double characteristic root. If $y_p = A\,t^2 e^{-t/2}$, then

$$y_p' = A\left(2t - \frac{1}{2}t^2\right)e^{-t/2}$$

$$y_p'' = A\left(2 - 2t + \frac{1}{4}t^2\right)e^{-t/2}.$$

Thus $\mathcal{L}(y_p) = 8\,Ae^{t/2} = -12\,e^{t/2}$ if $A = -3/2$, and

$$y_p = -\frac{3}{2}t^2 e^{-t/2}$$

is a particular solution. The general solution is the sum of $Y - p$ and y_h,

$$y_g = \left(C_1 + C_2 t - \frac{3}{2}t^2\right)e^{-t/2}.$$ ❑

The method of undetermined coefficients is also applicable if the source term is a polynomial. To see why, suppose that $v(t)$ is a polynomial and put $k = 0$ in (5.40). Then we have

$$\mathcal{L}(v(t)) = f(0)v(t) + f'(0)v'(t) + \frac{1}{2}f''(0)v''(t),$$

which will be a polynomial. Thus, let the undetermined coefficient expression be a polynomial with undetermined coefficients, as in the following example.

Example 5.5.3   Find a particular solution of

$$y'' + 4y = 16t^2, \tag{5.45}$$

and then find the general solution.

Solution. The source term is a quadratic polynomial, so we will try to find a solution that is also a quadratic polynomial.

Put $\mathcal{L}(y) = y'' + 4y$. Set $y_p(t) = A_2 t^2 + A_1 t + A_0$. Then $y_p'' = 2A_2$, and therefore

$$\mathcal{L}(y_p) = 2A_2 + 4(A_2 t^2 + A_1 t + A_0) = 16t^2 + 0t + 0. \tag{5.46}$$

Now we equate the coefficients of the powers of t in equation (5.46) and find that the A_i must satisfy the system of equations

$$4A_2 = 16$$
$$4A_1 = 0$$
$$2A_2 + 4A_0 = 0.$$

Thus $A_2 = 4$, $A_1 = 0$, and $A_0 = -2$. It follows that

$$y_p(t) = 4t^2 - 2$$

is a particular solution of equation (5.46).

Since the characteristic roots of $\mathcal{L}(y) = 0$ are $\pm 2i$, the general solution of this homogeneous ODE is

$$y_h = C_1 \cos(2t) + C_2 \sin(2t).$$

The general solution of (5.45) is the sum of y_p and y_h,

$$y = 4t^2 - 2 + C_1 \cos(2t) + C_2 \sin(2t). \qquad \blacksquare$$

Notice that in Example 5.5.3 it would incorrect to use an undetermined coefficient expression $y_p(t) = At^2$, even though the source is of this form.

The method of undetermined coefficients can be applied when the source term is a constant multiple of $\cos(\omega t)$ or $\sin(\omega t)$. To find a particular solution of

$$\mathcal{L}(y) = A_1 \cos(\omega t) + A_2 \sin(\omega t),$$

where $\mathcal{L}$ is an LDO with constant coefficients and A_1 and A_2 are given constants, set

$$y_p = B_1 \cos(\omega t) + B_2 \sin(\omega t), \tag{5.47}$$

where B_1 and B_2 are undetermined coefficients. To understand what will happen, remember that $\cos(\omega t) = \text{Re}(e^{i\omega t})$ and $\sin(\omega t) = \text{Im}(e^{i\omega t})$.

If $i\omega$ is not a characteristic root of $\mathcal{L}(y) = 0$, then by (5.40), with $v = 1$ and $k = i\omega$,

$$\mathcal{L}(e^{i\omega t}) = f(i\omega)e^{i\omega t}.$$

Equating real and imaginary parts of this equation,

$$\mathcal{L}(\cos(\omega t)) = \text{Re}(f(i\omega)e^{i\omega t}).$$

Note that the identity $\text{Im}(p + iq) = -i\,\text{Re}(p + iq)$ holds for any complex number $p + iq$. Thus,

$$\mathcal{L}(\sin(\omega t)) = \text{Im}(f(i\omega)e^{i\omega t}) = \text{Re}(-if(i\omega)e^{i\omega t}).$$

Thus

$$\begin{aligned}
\mathcal{L}(y_p) &= B_1\mathcal{L}(\cos(\omega t)) + B_2\mathcal{L}(\sin(\omega t)) \\
&= B_1\,\text{Re}(f(i\omega)e^{i\omega t}) + B_2\,\text{Re}(-if(i\omega)e^{i\omega t}) \\
&= \text{Re}((B_1 - iB_2)f(i\omega)e^{i\omega t}).
\end{aligned}$$

Since $A_1\cos(\omega t) + A_2\sin(\omega t) = \text{Re}((A_1 - iA_2)e^{i\omega t})$, the equation

$$A_1 - iA_2 = \frac{B_1 - iB_2}{f(i\omega)}$$

determines B_1 and B_2 simultaneously, provided that $f(i\omega) \neq 0$. In practice, we don't use this equation to determine B_1 and B_2; it just tells us that if $f(i\omega) \neq 0$, it is possible to use the undetermined coefficient expression (5.47).

Example 5.5.4   Find the general solution of

$$3y'' + 7y' - 6y = 2\sin(4t). \tag{5.48}$$

Solution. Let $\mathcal{L}(y) = 3y'' + 7y' - 6y$. The characteristic equation of $\mathcal{L}(y) = 0$,

$$3s^2 + 7s - 6 = 0$$

factors as $(3s - 2)(s + 3) = 0$. Thus the characteristic roots are $\frac{2}{3}$ and -3, and therefore the general solution of $\mathcal{L}(y) = 0$ is

$$y_h = C_1 e^{2t/3} + C_2 e^{3t}.$$

Now we will find a particular solutions y_p of $\mathcal{L}(y) = -2\sin(4t)$. The sum of y_h and y_p is the general solution of equation (5.48).

Note that

$$\begin{aligned}
\mathcal{L}(\cos(4t)) &= 3(-16\cos(4t)) + 7(-4\sin(4t)) - 6\cos(4t) \\
&= -54\cos(4t) - 28\sin(4t).
\end{aligned}$$

Similarly,

$$\mathcal{L}(\sin(4t)) = 28\cos(4t) - 54\sin(4t).$$

Thus, we should take $y_p = B_1\cos(4t) + B_2\sin(4t)$, even though the source only involves the sine. Then

$$\begin{aligned}
\mathcal{L}(y_p) &= B_1\mathcal{L}(\cos(4t)) + B_2\mathcal{L}(\sin(4t)) \\
&= (-54B_1 + 28B_2)\cos(4t) + (-28B_1 - 54B_2)\sin(4t).
\end{aligned}$$

Since $\mathcal{L}(y_p) = -2\sin(4t)$, B_1 and B_2 must satisfy

$$-54B_1 + 28B_2 = 0 \quad \text{(so that } \mathcal{L}(y_2) \text{ has no cosine term)}$$
$$-28B_1 - 54B_2 = -2$$

Solving for B_1 and B_2 yields $B_1 = 14/925$ and $B_2 = 27/925$. The general solution of equation (5.48) is

$$y = \frac{14}{925}\cos(4t) + \frac{27}{925}\sin(4t) + C_1 e^{2t/3} + C_2 e^{-3t}. \qquad \square$$

In a case where $i\omega$ is a characteristic root of $\mathcal{L}(y) = 0$, (5.40) tells us that the appropriate undetermined coefficient expression for a particular solution of $\mathcal{L}(y) = Ce^{i\omega t}$ is $y_p = Bte^{i\omega t}$. Therefore, to solve $\mathcal{L}(y) = C\cos(\omega t)$ or $\mathcal{L}(y) = C\sin(\omega t))$, we should use an undetermined coefficient expression of the form

$$y_p = t(B_1\cos(\omega t) + B_2\sin(\omega t)). \qquad (5.49)$$

Example 5.5.5   Solve the IVP

$$y'' + \omega^2 y = \cos(\omega t); \quad y(0) = y'(0) = 0. \qquad (5.50)$$

Solution. The characteristic roots are $\pm\omega i$, and this leads us to the homogeneous solution,

$$y_h = C_1\cos(\omega t) + C_2\sin(\omega t).$$

Let y_p be given by (5.49). Then

$$y'_p = B_1\cos(\omega t) + B_2\sin(\omega t) + t q(B_2\cos(\omega t) - B_1\sin(\omega t))$$
$$y''_p = 2\omega(B_2\cos(\omega t) - B_1\sin(\omega t)) - t\omega^2(B_1\cos(\omega t) + B_2\sin(\omega t)).$$

Thus $\mathcal{L}(y_p) = 2\omega(B_2\cos(\omega t) - B_1\sin(\omega t)) = \cos(\omega t)$. It follows that

$$2\omega B_2 = 1$$
$$-2\omega B_1 = 0.$$

The particular solution is $y_p = (t\sin(\omega t))/(2\omega)$, and the general solution is

$$y = \frac{t}{2\omega}\sin(\omega t) + C_1\cos(\omega t) + C_2\sin(\omega t).$$

The initial conditions are $y(0) = C_1 = 0$ and $y'(0) = \omega C_2 = 0$. Thus $C_1 = C_2 = 0$, and the solution of the IVP is

$$y = \frac{t}{2\omega}\sin(\omega t). \qquad \square$$

Figure 5.6 is a graph of the solution of Example 5.5.5, and of the source function, when $\omega = \pi$. Notice that the solution is unbounded, although the source is bounded. This phenomenon is known as *resonance*. It can cause bridges to buckle, and it is behind the working of many electronic devices that amplify signals.

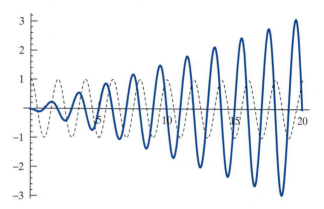

FIGURE 5.6   The solution of the initial value problem in Example 5.5.5 (solid curve) and the source term (dashed curve). This unbounded solution is called a resonant solution.

It sometimes happens that the source term $g(t)$ in a linear ODE $\mathcal{L}(y) = g(t)$ is equal to a sum of the form $A_1 e^{at} + A_2 e^{bt}$. To use the method of undetermined coefficients in such a case, find particular solutions y_1 and y_2 of $\mathcal{L}(y) = e^{at}$ and $\mathcal{L}(y) = e^{bt}$, respectively. Then $y_p = y_1 + y_2$.

Example 5.5.6   Find the general solution of

$$y'' - y' - 2y = \cosh(t).$$

Solution. Let $\mathcal{L}(y) = y'' - y' - 2y$. The characteristic roots of $\mathcal{L}(y) = 0$ are -1 and 2. Thus the general solution of $\mathcal{L}(y) = 0$ is $y_h = C_1 e^{-t} + C_2 e^{2t}$.

The source function $\cosh(t)$ can be expressed as a linear combination of of exponential functions:

$$\cosh(t) = \frac{1}{2}e^t + \frac{1}{2}e^{-t}. \tag{5.51}$$

Let us start by finding a particular solution $y_1(t)$ of $\mathcal{L}(y) = e^t$. Since 1 is not a characteristic root, put $y_1 = Ae^t$. Then

$$\mathcal{L}(y_1) = A(e^t - e^t - 2e^t) = A(-2e^t) = e^t$$

so $A = -\frac{1}{2}$ and $y_1 = -\frac{1}{2}e^t$.

A particular solution $y_2(t)$ of $\mathcal{L}(y) = e^{-t}$ will have the form $y_2 = Bte^{-t}$:

$$\mathcal{L}(y_2) = B((te^{-t} - 2e^{-t}) - (-te^{-t} + e^{-t}) - 2te^{-t} = B(-3e^{-t}) = e^{-t}.$$

Thus $B = -\frac{1}{3}$ and $y_2 = -\frac{1}{3}te^{-t}$. By referring to (5.51), we can assemble the general solution:

$$y = \frac{1}{2}y_1 + \frac{1}{2}y_2 + y_h$$

$$= -\frac{1}{4}e^t + \left(C_1 - \frac{1}{6}t\right)e^{-t} + C_2e^{2t}. \qquad \square$$

EXERCISES

1. Identify the equations to which the method of undetermined coefficients is applicable.

   a. $y'' + 3y' + y = t^3$ b. $y'' + yy' = t^2$

   c. $y'' + 9y = \sin(3t)$

   d. $-18y'' + 36y' - 5y = \tan(9t)$

   e. $y'' + 4y = 0$ f. $y'' - y = t\ln(t + 2)$

   g. $y'' - y = t^{-1}$ h. $y'' + ty' - 12y = t - 12$

In Exercises 2–9, find a particular solution of the ODE by using the method of undetermined coefficients.

2. $y'' + y = t^2$

3. $y'' + 3y' + 2y = e^{2t}$

4. $y'' + 3y' + 2y = e^{2t} - e^{-t}$

5. $y'' - y = e^{-t}$

6. $75y'' - y' = 4$

7. $y'' - y = \sin(2t) + 2\cos(2t)$

8. $4y'' + y = \cos\left(\dfrac{t}{2}\right)$

9. $y'' + 2y' + y = e^{-t}$

In Exercises 10–13, find the general solution.

10. $y'' + y = t^2$

11. $y'' - 4y' + 2y = e^{2t}$

12. $y'' + 9y = 54\cos\left(3t - \frac{\pi}{6}\right)$

13. $y'' - 2y' + y = 1 + e^t$

Solve the initial value problems in Exercises 14–22.

14. $y'' + y = t^2$; $y(0) = 1$, $y'(0) = 0$

15. $y'' - 4y' + 2y = e^{2t}$; homogeneous initial conditions at $t = 0$

16. $y'' + 9y = \cos(3t)$; homogeneous initial conditions at $t = 0$

17. $y'' + 9y = \cos(3t)$; homogeneous initial conditions at $t = \frac{\pi}{12}$

18. $y'' + 4y' + 4y = e^{-2t}$; $y(0) = 0$, $y'(0) = 1$

19. $y'' - y = \sinh t$; homogeneous initial conditions at $t = \ln(2)$

20. $y'' + 7y' + 12y = 1 - t^2$; $y(0) = 1$, $y'(0) = -1$

21. $y'' + 2y' + 2y = 10\sin(t)$; $y(0) = 0$, $y'(0) = -6$

22. $y'' - 2y' + y = 1 + e^t$; homogeneous initial conditions at $t = 0$

23. a. Let the characteristic polynomial for the differential equation $\mathcal{L}(y) = 0$ be $f(s) = a_2s^2 + a_1s + a_0$. Show that

$$\mathcal{L}(t^p e^{kt}) = (t^p f(k) + pt^{p-1}f'(k)$$
$$+ \frac{p(p-1)}{2}f''(k))e^{kt}.$$

   b. Suppose that $s = k$ is a double characteristic root; that is, $f(k) = f'(k) = 0$. Give a formula for the general solution of

$$\mathcal{L}(y) = e^{kt}(c_n t^n + c_{n-1}t^{n-1} + \cdots + c_0).$$

24. Solve the initial value problem

$$y'' + 9y = 36\cos(\omega t); \quad y(0) = y'(0) = 0$$

(your answer will depend on the parameter ω).

   a. Show that the solution is *bounded* if and only if $\omega \neq 3$.

   b. Use a computer to graph the solutions for $\omega = 1$, $\omega = 2.5$, $\omega = 2.9$, and $\omega = 3$.

   c. Use l'Hôpital's rule to derive a formula for the solution when $\omega = 3$ from the formula for the solution when $\omega \neq 3$.

25. This problem requires a computer equipped with an IVP solver. You may find that some numerical methods don't give the correct results. An adaptive solver, such as the Runge-Kutta-Fehlberg method, is the best choice.

   a. Plot the solution of

$$y'' + y = \sin(t)$$

with homogeneous initial conditions at $t = 0$, over the interval $0 \leq t \leq M$, where M should be as large as practical (at least 100). Explain the pattern that you see.

b. Perturb the equation in part (a) to see what happens: Plot the solution of

$$y'' + 1.2y = \sin(t)$$

with the same initial conditions, and over the same interval. For an explanation of what you will see, read the material on *beats,* starting on page 223.

c. Try a nonconstant perturbation of the equation in

part (a): Plot the solution of

$$y'' + \left[1 + \frac{1}{2}\sin(\pi t)\right] y = \sin(t).$$

Does the picture change?

d. Try a perturbation that is "in step:" plot the solution of

$$y'' + \left[1 + \frac{1}{2}\sin(t)\right] y = \sin(t).$$

5.6   Forced Vibrations

Forced vibrations occur when a mechanical system is subjected to an external force that is a periodic function of time. An automobile driving on a bumpy road is an example: The springs react to forces resulting from the motion over the bumps. These forces depend on the speed of the automobile and the location of the bumps and are functions of time that are periodic if the bumps are located at regular intervals. Automobiles are equipped with numerous springs, shock absorbers, and so on and are thus considerably more complicated than the simple systems that we will be considering. As we will see in Chapter 6, analogous phenomena occur in electrical circuits.

We will start with a frictionless system, consisting of a weight of mass m and a spring with stiffness k, modeled by the ODE

$$my'' + ky = f(t), \tag{5.52}$$

where $f(t)$ represents an external force. Take $f(t) = A\sin(\omega t)$, where ω is the frequency of the forcing function, in radians per second. In cycles per second, or **hertz**, the frequency is $\frac{\omega}{2\pi}$. The parameter A of the forcing function is the amplitude (in meters).

The characteristic roots of $my'' + ky = 0$ are $\pm i\beta$, where $\beta^2 = \frac{k}{m}$. The parameter β is called the **resonant frequency** of the mechanical system. The resonant frequency is the frequency of free vibrations of the system, as we saw in Section 5.4.

The nonresonant case When $k \neq \omega^2$,

$$my'' + ky = A\sin(\omega t). \tag{5.53}$$

We can find a particular solution by substituting $y_p = B_1\cos(\omega t) + B_2\sin(\omega t)$. Then

$$m\,y_p'' + k\,y_p = (-m\omega^2 + k)[B_1\cos(\omega t) + B_2\sin(\omega t)] = A\sin(\omega t).$$

Therefore, $B_1 = 0$ and $B_2 = \frac{A}{-m\omega^2+k}$. Since $k = m\beta^2$,

$$y_p = \frac{A}{m(\beta^2 - \omega^2)}\sin(\omega t).$$

The general solution of equation (5.53) is the sum of y_p and the general solution

$$y_h = C_1 \cos(\beta t) + C_2 \sin(\beta t)$$

of the associated homogeneous equation:

$$y(t) = \frac{A}{m(\beta^2 - \omega^2)} \sin(\omega t) + C_1 \cos(\beta t) + C_2 \sin(\beta t).$$

We will refer particular solution where $C_1 = C_2 = 0$ as the ω-**periodic** solution of equation (5.53). Notice that the period in this case is $2\pi/\omega$. If either of the parameters C_i is nonzero, the solution is usually not periodic. See Exercise 6.

The amplitude of the ω-periodic solution is $\frac{A}{m|\beta^2 - \omega^2|}$ meters. If the forcing frequency ω is varied so that it approaches the resonant frequency β, this amplitude increases without bound.

Example 5.6.1   A mechanical system has parameters $k = 8000$ newtons per meter and $m = 5$ kilograms. Determine its resonant frequency.

If an external force of $400 \sin(\omega t)$ newtons is applied to the system, find the amplitude of the resulting ω-periodic motion, and sketch its graph as a function of ω.

Solution. Let $\mathcal{L}(y) = 5y'' + 8000y$. The characteristic roots of $\mathcal{L}(y) = 0$ are $\pm\beta i$, where β is the resonant frequency. Thus,

$$\beta = \sqrt{\frac{8000}{5}} = 40 \text{ radians per second} \approx 6.4 \text{ hertz.}$$

The displacement y of the mechanical system satisfies the ODE

$$\mathcal{L}(y) = 400 \sin(\omega t).$$

The ω-periodic solution is of the form $y_p = B_1 \cos(\omega t) + B_2 \sin(\omega t)$. Since

$$\mathcal{L}(y_p) = (8000 - 5\omega^2)(B_1 \cos(\omega t) + B_2 \sin(\omega t)) = 400 \sin(\omega t),$$

$B_1 = 0$ and

$$B_2 = \frac{400}{-5\omega^2 + 8000}.$$

Thus $y_p = B_2 \sin(\omega t)$. The amplitude of y_p is $B(\omega) = |B_2|$. The graph of $B(\omega)$, shown in Figure 5.7, is called the **frequency-response curve** for the system.   ❑

The resonant case   If the frequency ω of the external force matches a natural frequency of free vibrations β of a mechanical system, resonance occurs. To study resonance, we will find a solution of

$$my'' + m\beta^2 y = A \sin(\beta t) \tag{5.54}$$

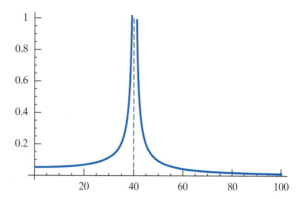

FIGURE 5.7   Frequency-response curve for an
undamped mechanical system.

by using the method of undetermined coefficients. Let $\mathcal{L}(y) = m(y'' + \beta^2 y)$. Since βi is a characteristic root of $\mathcal{L}(y) = 0$, there is a particular solution of the form

$$y_p = t(B_1 \cos(\beta t) + B_2 \sin(\beta t)).$$

Then

$$\mathcal{L}(y_p) = m[t(-\beta^2(B_1 \cos(\beta t) + B_2 \sin(\beta t)))$$
$$+ 2\beta(-B_1 \sin(\beta t) + B_2 \cos(\beta t)) + \beta^2 t(B_1 \cos(\beta t) + B_2 \sin(\beta t))]$$
$$= 2m\beta[B_2 \cos(\beta t) - B_1 \sin(\beta t)] = A \sin(\beta t).$$

Therefore, $B_2 = 0$, and

$$B_1 = -\frac{A}{2m\beta}.$$

Thus, $y = -\frac{A}{2m\beta} t \cos(\beta t)$ is a particular solution of

$$my'' + ky = A \sin(\beta t),$$

and the general solution is

$$y = \left(B_1 - \frac{A}{2m\beta} t \right) \cos(\beta t) + B_2 \sin(\beta t).$$

If the amplitude A of the forcing term is nonzero, equation (5.54) *has no* periodic solutions because it sustains oscillations of ever increasing magnitude. In the absence of all damping forces, a mechanical system would eventually be destroyed by resonant motion. Resonance is seen in everyday experience. The motion of a backyard swing is modeled approximately by the ODE

$$L\theta'' + g\theta = h(t),$$

where θ is the angular displacement of the swing, L denotes the length of the swing (that is, the distance from the center of mass of the child on the swing to the point where the swing is suspended), and g is the gravitational acceleration. The forcing function $f(t)$ represents the action of someone who is pushing the swing. For a full discussion of pendulums (a swing is essentially a forced pendulum), see page 413.

It is natural to push a backyard swing with a frequency equal to the swing's resonant frequency. The resulting motion is bounded for two reasons: first, friction is present; and second, the motion of the swing is not accurately modelled by linear ODE. The approximate pendulum equation given above is only valid for small amplitudes of swing.

Suspension bridges must be designed to avoid destructive resonance resulting from wind stress. In 1940, the Tacoma Narrows Bridge, which crossed Puget Sound in the state of Washington, collapsed in a 70-kilometer-per-hour wind. The periodic force was a result of the same phenomenon that causes a flag to flap periodically in a steady wind.[3]

Example 5.6.2   The mechanical system discussed in Example 5.6.1 is subjected to the external force $400 \sin(40t)$. Assume that the system is initially at rest in the equilibrium position. Sketch the graph of the displacement function.

Solution. To find a particular solution of $\mathcal{L}(y) = 400 \sin(40t)$, where $\mathcal{L}$ is the operator defined in Example 5.6.1, set $y_p = t[B_1 \cos(40t) + B_2 \sin(40t)]$. The factor of t is necessary because $40i$ is a characteristic root of $\mathcal{L}(y) = 0$. Then

$$\mathcal{L}(y_p) = 5[1600t\,(B_1 \cos(40t) + B_2 \sin(40t)) + 80(-B_1 \sin(40t) + B_2 \cos(40t))$$
$$+ 8000(B_1 \cos(40t) + B_2 \sin(40t))]$$
$$= 400(-B_1 \sin(40t) + B_2 \cos(40t)) = 400 \sin(40t).$$

Thus $B_1 = -1$, and $B_2 = 0$, so that $y_p = -t \cos(40t)$. The general solution of $\mathcal{L}(y) = 0$ is $y_h = C_1 \cos(40t) + C_2 \sin(40t)$. Thus, the general solution of

$$5y'' + 8000x = 400 \sin(40t)$$

is

$$y = (C_1 - t) \cos(40t) + C_2 \sin(40t).$$

The initial conditions are $y(0) = 0$ and $y'(0) = 0$. Since $y(0) = C_1$, we have $C_1 = 0$. Therefore,

$$y(t) = -t \cos(40t) + C_2 \sin(40t), \quad \text{and}$$
$$y'(t) = (40C_2 - 1) \cos(40t) + 40t \sin(40t).$$

It follows that $y'(0) = 40C_2 - 1 = 0$, so $C_2 = \frac{1}{40}$. The graph of the solution,

$$y(t) = -t \cos 40t + \frac{1}{40} \sin(40t),$$

is shown in Figure 5.8. ❑

---

[3]The linear model that we are studying is only valid for small amplitude vibrations of the bridge. For a recent discussion of the Tacoma Narrows incident, see "Large torsional oscillations in suspension bridges visited again: vertical forcing creates torsional response," by P. J. McKenna and Cillian O'Tuama, *American Mathematical Monthly* 108 (October 2001) pp. 738–745. This article includes instructions for downloading software that you can use to perform your own experiments.

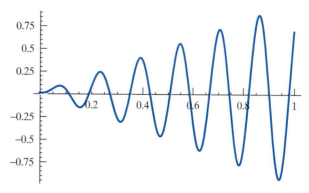

FIGURE 5.8   Resonant motion of an undamped mechanical system.

Beats

Consider an initial value problem

$$my'' + ky = \sin(\omega t); \quad y(0) = 0, \ y'(0) = 0. \qquad (5.55)$$

We will assume that this is the nonresonant case: $k \neq m\omega^2$. It is of interest to see what happens when ω is relatively close to the resonant frequency. If you have an IVP solver, you may wish to experiment with this initial value problem before reading the remainder of this section. Put $m = 1$, and $k = 100$, and plot the solutions for several different values of ω: 8, 9, 9.5, 10 (the resonant frequency, for comparison purposes), 10.5, 11, and 12 are suggested.

The general solution of the differential equation $my'' + ky = \sin(\omega t)$ is

$$y = \frac{1}{m(\beta^2 - \omega^2)} \sin(\omega t) + C_1 \cos(\beta t) + B_2 \sin(\beta t),$$

where $\beta = \sqrt{k/m}$. The initial conditions give

$$y(0) = C_1 = 0$$

and

$$y'(0) = \frac{\omega}{m(\beta^2 - \omega^2)} + \beta C_2 = 0$$

so that

$$V_2 = \frac{\omega}{\beta} \cdot \frac{1}{m(\beta^2 - \omega^2)}.$$

The solution is therefore

$$y = \psi(t) = \frac{1}{m(\beta^2 - \omega^2)} \left[\sin(\omega t) - \frac{\omega}{\beta} \sin(\beta t) \right]. \qquad (5.56)$$

To simplify this expression, let $\theta = \frac{1}{2}(\omega + \beta)$ and $\delta = \frac{1}{2}(\omega - \beta)$. Since we are assuming that β and ω are approximately the same, $\theta \approx \omega$, and δ, which is called the **beat frequency**, is small.

Since $\omega = \theta + \delta$ and $\beta = \theta - \delta$, the addition formula for the sine shows that

$$\sin(\omega t) = \sin(\theta t)\cos(\delta t) + \sin(\delta t)\cos(\theta t)$$
$$\sin(\beta t) = \sin(\theta t)\cos(\delta t) - \sin(\delta t)\cos(\theta t).$$

Substituting these expressions into equation (5.56) and simplifying, we find that

$$\psi(t) = \frac{1}{m(\beta^2 - \omega^2)}\left[\left(1 - \frac{\omega}{\beta}\right)\sin(\theta t)\cos(\delta t) + \left(1 + \frac{\omega}{\beta}\right)\sin(\delta t)\cos(\theta t)\right].$$

If the ratio $\frac{\omega}{\beta}$ is nearly equal to 1, the term $\left(1 - \frac{\omega}{\beta}\right)\sin(\theta t)\cos(\delta t)$ is negligible, and $1 + \frac{\omega}{\beta} \approx 2$. Therefore,

$$\psi(t) \approx \frac{2}{m(\beta^2 - \omega^2)}\sin(\delta t)\cos(\theta t).$$

Notice that $\delta \to 0$ as $\omega \to \beta$, and the period of $\cos(\delta t)$ (which is equal to $2\pi/\delta$) becomes longer. At the same time, the period of $\sin(\theta t)$ doesn't change much, since $\theta \approx \omega$. The graph of $\psi(t)$ thus appears to oscillate within an envelope defined by the graphs of

$$y = \pm\frac{2}{m(\beta^2 - \omega^2)}\sin(\delta t),$$

as shown in Figure 5.9. In the figure, we have taken $m = 1$, $\beta = 10.25$, and $\omega = 9.75$, so that $\delta = 0.25$, $\theta = 10$ and $2/[m(\beta^2 - \omega^2)] = 0.2$.

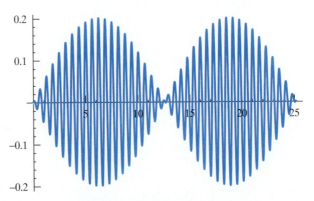

FIGURE 5.9   Beats: the graph of $x = 0.2\sin(0.25t)\cos(10t)$.

The predominant feature in Figure 5.9 is the low-frequency oscillation, called **beats.** This phenomenon occurs in functions that are linear combinations of oscillations with frequencies that are relatively close, so that they can alternately reinforce and cancel each other. In the case of forced vibrations, the two frequencies are the forcing frequency and the frequency of free vibrations.

When tuning a musical instrument, the frequency of the beats measures the difference between the vibration frequency of a tuning fork and that of the instrument's string. Another application of beats is to FM radio. A radio receiver has an oscillating circuit whose resonant frequency is equal to the carrier frequency of the transmitter. The signal produced by the transmitter is the sum of the carrier frequency and the frequency of the sound which the radio receiver is to produce. The sound frequency may range from 40 to 20,000 hertz, which is small in comparison to the carrier frequency, which is typically about 10^8 hertz. The beats in the receiver's oscillating circuit are amplified to produce the sound that we hear.

Damped systems

Let us consider a mechanical system with one weight of mass m, damping constant b, a spring of stiffness k, and an external force $f(t)$. The governing ODE is

$$my''(t) + by'(t) + ky(t) = f(t). \tag{5.57}$$

We will take $f(t) = Ae^{i\omega t}$, because it is simpler to determine the amplitude and the phase of the solution when working with the complex exponential function than it is when working with sines and cosines. To apply our results when $f(t) = A\sin(\omega t)$ in equation (5.57), we would work with the complex ODE

$$mz''(t) + bz'(t) + kz(t) = Ae^{i\omega t}. \tag{5.58}$$

When we have found a solution $z(t)$ of (5.58), the corresponding solution of equation (5.57) would be $y(t) = \text{Im}[z(t)]$.

Example 5.6.3   A mechanical system has mass of 5 kilograms, damping of 2 kilograms per second, and spring constant of 5.2 newtons per meter. An external force of $f(t) = 10\sin(t)$ newtons is applied. Assuming that the mass is initially at rest in its equilibrium position, determine the motion of the system.

Solution. The motion is determined by the IVP

$$5y'' + 2y' + 5.2y = 10\sin(t); \ \ y(0) = y'(0) = 0. \tag{5.59}$$

Let $\mathcal{L}(y) = 5y'' + 2y' + 5.2y$. If z_p is a particular solution of

$$5z'' + 2z' + 5.2z = 10e^{it}, \tag{5.60}$$

then $\text{Im}(z_p)$ will be a particular solution of $\mathcal{L}(y) = 10\sin(t)$. We can put $z_p = Ce^{it}$; then

$$\mathcal{L}(z_p) = 5(-Ce^{it}) + 2(iCe^{it}) + 5.2Ce^{it} = 10e^{it}.$$

After canceling the factor e^{it}, we find that

$$C = \frac{10}{.2 + 2i}.$$

Express the denominator of C in polar form as $re^{i\theta}$, where $\theta = \arg(.2 + 2i)$, and

$$r - |.2 + 2i| = \sqrt{.2^2 + 2^2} \approx 2.01.$$

Thus

$$C = \frac{10}{r}e^{-i\theta} \approx 4.975e^{-1.471i}.$$

It follows that the complex particular solution is

$$z_p(t) = 4.975e^{i(t-1.471)}.$$

Taking the imaginary part, we have a particular solution for equation (5.59):

$$y = \psi(t) = 4.975\sin(t - 1.471).$$

The characteristic roots are

$$r_1, r_2 = \frac{-2 \pm \sqrt{4 - 104}}{10} = -0.2 \pm i.$$

Hence the general solution of the associated homogeneous equation is

$$y = e^{-0.2t}[C_1 \cos(t) + C_2 \sin(t)].$$

The general solution of equation (5.59) is therefore

$$y = 4.975\sin(t - 1.471) + e^{-0.2t}[C_1 \cos(t) + C_2 \sin(t)].$$

Thus, $y(0) = 0 = -4.975\sin(1.471) + C_1$ and $y'(0) = 0 = 4.975\cos(1.471) - 0.2C_1 + C_2$. It follows that $C_1 = 4.975\sin(1.471) = 4.950$ and $C_2 = 0.2C_1 - 4.975\cos(1.471) = 0.494$. The displacement function is thus

$$y = 4.975\sin(t - 1.471) + e^{-0.2t}[4.950\cos(t) + 0.494\sin(t)]. \qquad \square$$

The motion of the weight in Example 5.6.3 has two components: a **periodic solution** of equation (5.59), and a **transient**. The transient is a solution of the homogeneous equation that decays to zero as $t \to \infty$.

The characteristic roots of equation (5.58) cannot be pure imaginary unless there is no damping ($b = 0$). Since resonance can only occur when $i\omega$ is a characteristic root, damped systems do not actually resonate. However, if the damping is relatively small, the response of the mechanical system will be sensitive to the frequency of the forcing function, and vibrations with magnitudes that are potentially destructive can occur.

The solution $z = \psi(t) = Ce^{i\omega t}$ of (5.58) found by the method of undetermined coefficients is periodic. By Proposition 5.4.1 on page 207, all solutions $z = \phi(t)$ of the associated homogeneous equation have the property that $\lim_{t \to \infty} \phi(t) = 0$. This can be restated to say that every solution $z = \bar{\psi}(t)$ of (5.58) has the property

$$\lim_{t \to \infty} (\bar{\psi}(t) - \psi(t)) = 0$$

independently of the initial conditions. In other words, every solution of (5.58) is the sum of the same periodic solution $\psi(t)$ and a transient that is determined by the initial conditions. Since the transient decays exponentially, its effect is noticeable only for a short time: See Exercise 20.

Let $\mathcal{L}(z) = mz'' + bz' + kz$ be the linear operator on the left side of (5.58). Following the solution of Example 5.6.3,

$$\mathcal{L}(Ce^{i\omega t}) = (-m\omega^2 + ib\omega + k)e^{i\omega t} = Ae^{i\omega t}.$$

We can cancel the factor $e^{i\omega t}$ and solve for the undetermined coefficient C to obtain

$$C = \frac{A}{(k - m\omega^2) + b\omega i}. \tag{5.61}$$

The polar form for the complex number in the denominator of formula (5.61) is

$$(k - m\omega^2) + b\omega i = r(\omega)e^{i\theta(\omega)},$$

where

$$r(\omega) = |(k - m\omega^2) + b\omega i|$$

$$= \sqrt{(k - m\omega^2)^2 + b^2\omega^2}$$

and

$$\theta(\omega) = \arg[(k - m\omega^2) + b\omega i]$$

$$= \cos^{-1}\left(\frac{k - m\omega^2}{r(\omega)}\right).$$

Question   The range of the inverse cosine is $[0, \pi]$. How do we know that $\theta(\omega)$ must fall within this range?

At frequency ω, $r(\omega)$ is called the **mechanical resistance**, and $\theta(\omega)$ is the **phase lag**, of the mechanical system.

The complex parameter C in equation (5.61) is

$$C = \frac{A}{r(\omega)}e^{-i\theta(\omega)},$$

and so the periodic solution of the complex equation is

$$z_p(t) = \frac{A}{r(\omega)}e^{i(\omega t - \theta(\omega))}. \tag{5.62}$$

To recover the solution $y_p(t)$ of $\mathcal{L}(y) = A\sin(\omega t)$, where the source term is real, take the imaginary part of $z_p(t)$:

$$y_p(t) = \frac{A}{r(\omega)}\sin(\omega t - \theta(\omega)).$$

The mechanical resistance is the ratio of the amplitude of the forcing function to the amplitude of the periodic motion of the system. By equation (5.62), $\theta(\omega)$ is the amount by which the phase of the displacement function lags behind the phase of the forcing function.

The graph of $|C(\omega)| = \frac{A}{r(\omega)}$ is the *frequency-response* curve for the damped mechanical system. The frequency ω at which $|C(\omega)|$ is maximum is called the **natural frequency** of the system. The natural frequency depends on the parameters of the system, m, b, and k.

Example 5.6.4   A mechanical system has parameters $m = 5$ kilograms and $k = 8000$ newtons per meter. The damping is $b = 20$ kilograms per second, and the forcing function is $f(t) = 400\sin(\omega t)$ newtons. Draw the frequency-response curve, and determine the natural frequency.

Solution. The ODE $\mathcal{L}(z) = 400e^{i\omega t}$, where

$$\mathcal{L}(z) = 5z'' + 20z' + 8000z,$$

will serve as a model for the system. The displacement is $y = \mathrm{Im}(z(t))$, and the amplitude of the displacement is $|z(t)|$.

Put $z_p(t) = Ce^{i\omega t}$. Then

$$\mathcal{L}(z_p) = (-5\omega^2 + 20i\omega + 8000)Ce^{i\omega t} = 400e^{i\omega t}.$$

Therefore,

$$C = \frac{400}{(8000 - 5\omega^2) + 20i\omega}$$

$$= \frac{400}{r(\omega)e^{i\theta(\omega)}},$$

where $\theta(\omega) = \arg((8000 - 5\omega^2) + 20i\omega)$ and the mechanical resistance is

$$r(\omega) = |(8000 - 5\omega^2) + 20i\omega|$$

$$= \sqrt{(8000 - 5\omega^2)^2 + 400\omega^2}$$

$$= \sqrt{25\omega^4 - 79600\omega^2 + 64000000}.$$

By completing the square, we find that

$$r(\omega) = 5\sqrt{(\omega^2 - 1592)^2 + 25536}.$$

The minimum value or $r(\omega)$ occurs when $\omega = \sqrt{1592} \approx 39.9$ radians per second. Thus is the natural frequency of the system. At this frequency, the amplitude of oscillation is

$$\frac{400}{r(\sqrt{1592})} \approx 0.5006 \text{ meters.}$$

The frequency-response curve, shown in Figure 5.10, is the graph of

$$|C(\omega)| = \frac{400}{r(\omega)}.$$

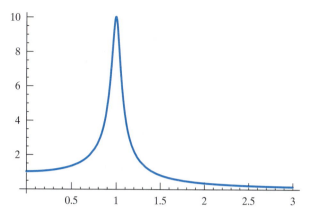

FIGURE 5.10 Frequency-response curve for an underdamped mechanical system.

EXERCISES

1. Draw the frequency-response curve for an undamped mechanical system in which $m = 100$ kilograms and $k = 8.1 \times 10^5$ newtons per meter.

2. Draw the frequency-response curve for an undamped mechanical system in which $m = 1$ kilogram and $k = 196$ newtons per meter.

3. In an undamped mechanical system, does the resonant frequency increase or decrease

   a. when the mass is increased?

   b. when the spring is replaced with a stiffer one?

4. An undamped mechanical system has $m = 4$ kilograms and $k = 9$ kilograms. A force of $f(t) = \sin(\pi t/2)$ is applied.

   a. Assuming that the mass is initially at rest in its equilibrium position, determine the motion.

   b. Show that the motion is not periodic.

   c. Find initial conditions such that the motion will be periodic.

5. The parameters of an undamped mechanical system are $m = 0.25$ kilograms and $k = 400$ newtons per meter.

   a. Describe the motion of the system if, initially, the mass is displaced 0.3 meters from the equilibrium position and given an initial velocity of 10 meters per second toward the equilibrium position.

   b. Suppose that the system is given an external force of $50 \sin(20t)$ newtons and that the system is initially at rest in the equilibrium position. Describe the motion of the system.

   c. Is the motion of the forced system considered in part (b) periodic? If so, what is the period?

6. Let $y(t) = A \cos(\omega t + \phi) + B \cos(\alpha t + \theta)$.

   a. Show that y is a periodic function of t if either $A = 0$ or $B = 0$, and find the period in each case.

   b. Show that if A and B are both nonzero, then $y(t)$ is still periodic if $\frac{\omega}{\alpha}$ is a rational number. What will the period be in this case?

   c. Assume that A and B are both nonzero and that $\frac{\omega}{\alpha}$ is irrational. Show that y is not a periodic function.[4]

7. Solve the initial value problem

   $$y'' + 1.002001y = \sin(0.999t); \quad y(0) = 1, \ y'(0) = 0.$$

   Express your answer as the sum of a beat term and other terms. To what extent are the other terms negligible?

8. Solve the initial value problem

   $$y'' + 104448400y = \sin(9780t); \quad y(0) = 0, \ y'(0) = 0.$$

   Express your answer as the sum of a beat term and other terms. To what extent are the other terms negligible?

9. A mechanical system has a mass of 4 kilograms and a stiffness of 9 newtons per meter. The forcing function is a simple harmonic oscillation with period 4. Will beats be observed? Taking the forcing function to be of the form

---

[4]In this case, $y(t)$ is an *almost periodic* function. For a brief discussion of almost periodic functions, see the appendix of the book *Ordinary Differential Equations,* by Jack Hale (second edition published by Krieger (Malabar, Florida) in 1980).

$\sin(kt)$ and homogeneous initial conditions, determine the beat amplitude and frequency.

10. A tuning fork vibrates at 440 hertz. When the A key above middle C on our piano is struck, and we listen carefully to the tuning fork (which was *not* struck), we hear an oscillating sound with a frequency of 1 hertz. Explain how this can be used to tune the A string (a correctly tuned A string will vibrate at 440 hertz).

11. **Beats experiment**   Set up your IVP solver to plot solutions of $y'' + y = \sin(kt)$ with homogeneous initial conditions, over the interval $[0, 30\pi]$, where k is allowed to assume various values.

   a. Plot solutions for the following values of k: 1, 1.05, 1.1, 0.95, 0.9, and 0.8. You should observe both resonance and beats.

   b. Plot y' as a function of t for each of the above values of t.

   c. Plot the orbit of each solution in the phase plane.

   d. This is a two-person game. Your opponent generates a random number A between 0.5 and 1.5, by using a random number generator (if the generator produces random numbers in the interval $[0, 1]$, that's fine: Just add 0.5 to the random number that was generated). You select a number k (devise your own strategy for choosing k) and tell your opponent to display a graph of the solution of $y'' + A^2 y = \sin(kx)$, with homogeneous initial conditions, on the interval $[0, 30\pi]$. Based on the beats in the graph that you see, you can guess two possible values of A. Refine your guesses by trying new values of k. Your score in this game is $-\log_{10} |A - \hat{A}|$, where $\hat{A}$ is your best estimate of A after you have seen **five** graphs. Now trade places with your opponent and start over. The person with the highest score wins.

12. A mechanical system has mass 1 kilogram; damping 2 kilograms per second; and spring constant 257 newtons per meter. A force of $f(t) = 20\sin(32t)$ newtons is applied. Determine the periodic motion of the mass.

13. The mechanical system in Exercise 12 is given a force of $20\sin(t)$. Determine the periodic motion.

14. The parameters of a mechanical system are $(m, b, k) = (1, 2, 1)$. The forcing function is $\sin(\omega t)$ Find the mechanical resistance as a function of ω.

15. The parameters of a mechanical system are $(m, b, k) = (1, \sqrt{2}, 1)$. The forcing function is $\sin(\omega t)$ Draw the frequency-response curve.

16. The parameters of a mechanical system are $(m, b, k) = (1, 1, 1)$. The forcing function is $\sin(\omega t)$ Find the frequency ω for which the amplitude of the periodic motion is at a maximum.

17. Find the mechanical resistance and phase lag, for mechanical systems with the following parameters. The forcing function is $f(t) = \sin(\omega t)$ newtons:

   a. Mass, 10 kilograms. Damping, 1 kilogram per second. Stiffness, 16,000 newtons per meter. Forcing frequency, $\omega = 40$ radians per second.

   b. Mass, 10 kilograms. Damping, 80 kilograms per second. Stiffness, 16,000 newtons per meter. Forcing frequency, $\omega = 40$ radians per second.

   c. Mass, 10 kilograms. Damping, 800 kilograms per second. Stiffness, 16,000 newtons per meter. Forcing frequency, $\omega = 1$ radian per second.

   d. Mass, 0.1 kilogram. Damping, 40 kilograms per second. Stiffness, 16,000 newtons per meter. Forcing frequency, $\omega = 280$ radians per second.

18. Draw frequency-response curves for the mechanical systems described in Exercise 17, and note the point on the curve corresponding to the prescribed frequency.

19. This exercise concerns a mechanical system in which the parameters are $(m, k) = (1, 1)$. Let b denote the damping.

   a. Find the frequency $\omega(b)$ for which the mechanical resistance is at a minimum.

   b. Show that, in particular, if the system is *overdamped* $(b > 2)$, the resistance is an increasing function of the frequency.

   c. What happens if the system is critically damped?

   d. Is it possible for the resistance of an underdamped system to be an increasing function of the frequency?

   e. Find a number b_0 such that the natural frequency is 0 when $b \geq b_0$, and the natural frequencies of all systems with $b < b_0$ are nonzero.

   f. Show that when $b < b_0$, the natural frequency decreases if the damping is increased.

20. Show that the ODE

$$my''(t) + by'(t) + ky(t) = C\cos(\alpha t),$$

where the parameters m, b, k, C, and α are positive, has a periodic solution $y = \psi(t)$. Also show that $\psi(t)$ is the *only* solution that is periodic, and that if $\bar{\psi}(t)$ is other solution,

$$\lim_{t \to \infty} (\bar{\psi}(t) - \psi(t)) = 0.$$

21. **Nonlinear springs**   A spring S has restoring force

$$F_S(y) = -(ky + ly^3 + \cdots)$$

when extended by y units. This spring is said to be symmetric, because $F_S(-y) = -F_S(y)$. We will always assume that $k > 0$. If $l > 0$, the spring is said to be *hard*, while if $l < 0$ the spring is *soft*.

A pendulum can be treated as a nonlinear spring. Let L be the length of the pendulum arm, let g be the gravitational acceleration, and let θ denote the angular displacement of the pendulum arm from the vertical position. The ODE

$$L\theta'' + g\sin\theta = 0 \qquad (5.63)$$

will be derived on page 000. Thus, a pendulum is analogous to a mechanical system where θ is the displacement, L is the mass, and the restoring force due to the spring is

$$F_S = -g\sin\theta = -g\left(\theta - \frac{1}{6}\theta^3 + \cdots\right).$$

The coefficients are $k = g$ and $l = -\frac{1}{6}g$, and since $l < 0$, the pendulum acts as a soft spring.

For a nonlinear spring, the *stiffness* is defined to be the ratio $G_S(y) = -\frac{1}{y}F_S(y)$ of the magnitude of the restoring force divided by the extension. Linear springs have constant stiffness, since $F_S(y) = -ky$, and $G_S(y) \equiv k$, but the stiffness of a nonlinear spring depends on its extension.

a. Show that S is a soft spring if and only if $G_S''(0) < 0$, and S is a hard spring if and only if $G_S''(0) > 0$.

b. For an undamped linear mechanical system, the period of oscillation is independent of the amplitude. If an undamped mechanical system involves a symmetric nonlinear spring, how is the period related to the amplitude? Consider hard and soft springs separately. You need not prove your answer: This question asks you to speculate.

c. Using an IVP solver, plot solutions of the following initial value problems over the interval $[0, 10]$, comparing linear and hard springs. Choose a fourth-order method with a step of 0.05. Note the period of each solution. Is your speculation in part (b) confirmed?

   i. $y'' = -y - ly^3$; $y(0) = 0$, $y'(0) = 1$, for $l = 0$, 0.1, and 0.2.

ii. Use the same differential equation as in part (a), but with initial conditions $y(0) = 0$, $y'(0) = 2$.

iii. Use the same differential equation as in part (a), but with initial conditions $y(0) = 0$, $y'(0) = 3$.

d. Repeat the computations in question 3 for soft springs, using $l = 0$, -0.1, and -0.2.

22. Nonlinear resonance   We have seen that nonlinear springs do not have a natural frequency of vibration (the frequency depends on the amplitude of oscillation). Can resonance occur in forced vibrations of a nonlinear spring? Specifically, can we choose a frequency ω such that the ODE

$$y'' + ky + ly^3 = \sin(\omega t) \qquad (5.64)$$

with homogeneous initial conditions has an unbounded solution?

a. Set $k = 1$ and $l = 0.1$ in equation (5.64). Plot the graphs of the solution (with homogeneous initial conditions), using several different values of ω. Do you observe resonance? Beats?

b. Estimate, to two decimal places, the forcing frequency that gives the greatest amplitude solution.

c. Change the initial conditions to $y(0) = 0$, $y'(0) = 2$. Does the forcing frequency that gives the greatest amplitude solution remain the same? Give a plausible explanation of the results that you get.

d. Repeat part (a), using $l = -0.1$ (a soft spring). You will get unbounded solutions, but is there a better explanation than resonance?

23. If p and q are positive numbers, then $\arg(p + iq)$ can be represented in terms of inverse trigonometric functions in three ways: $\arctan(q/p)$, $\arcsin(q/\sqrt{p^2 + q^2})$, or $\arccos(p/\sqrt{p^2 + q^2})$. Which inverse trigonometric function is best to represent the phase lag, $\theta(\omega) = \arg[(k - m\omega^2) + ib\omega]$?

*5.7 Modes of Vibration

To construct a model for a mechanical system involving two or more weights, or in which the weights can move in more than one dimension, a system of ODEs is needed. A mechanical system is said to have n **degrees of freedom** if it is possible to specify the positions of all of the weights in the system with the values of n variables $x_1(t), \ldots, x_n(t)$. The systems that we have studied up to this point all have one degree of freedom because in each case there was one weight in motion, constrained to move on an axis. A system with one weight that can move in three dimensions would have three degrees of freedom, one for each coordinate in $\mathbf{R}^3$. A system with two weights constrained to move on an axis would have two degrees of freedom; if the two weights could move without constraint, there would be six degrees of freedom.

A system of ODEs that serves as a model for a mechanical system having n degrees of freedom requires n dependent variables x_i representing the positions of the weights. These are the components of a vector $\mathbf{x}$ called the **configuration vector**. The **velocity vector v** is defined as the derivative of $\mathbf{x}$. Thus, there are an additional n variables $v_i = x_i'$ representing velocity. The **phase vector** combines all $2n$ components as

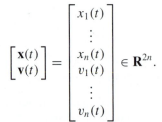

$$\begin{bmatrix} \mathbf{x}(t) \\ \mathbf{v}(t) \end{bmatrix} = \begin{bmatrix} x_1(t) \\ \vdots \\ x_n(t) \\ v_1(t) \\ \vdots \\ v_n(t) \end{bmatrix} \in \mathbf{R}^{2n}.$$

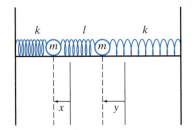

Figure 5.11. A mass spring system. In the positions shown, the displacements x and y are negative.

Figure 5.11 shows a mechanical system in which weights with identical masses of m kilograms are held between two walls by linear springs. The weights are constrained to move on a horizontal axis. The outer two springs are identical, with stiffness of k newtons per meter, while the stiffness of the middle spring is l newtons per meter.

In a system with more than one weight that can move, we must be careful in defining the concept of displacement. The system is in equilibrium when each weight is motionless, and the net force on each weight is zero. The **equilibrium position** of a particular weight in the system is the position of that weight when the system is in equilibrium.

Let $x_1(t)$ and $x_2(t)$ denote the displacements, measured in meters, of weights 1 and 2 from their respective equilibrium positions. The velocities will be $v_1(t)$ and $v_2(t)$, respectively, in meters per second. In each case, the positive direction is to the right.

The weights are subject only to the forces due to the springs. The extension of the left spring is x_1 meters, and it exerts a force of $-kx_1$ newtons on weight 1. The middle spring is extended $x_2 - x_1$ meters. It exerts equal, but oppositely directed, forces on each of the weighs. Thus the force on weight 1 is $l(x_2 - x_1)$ newtons, while the force on weight 2 is $-l(x_2 - x_1)$ newtons. The right spring is compressed x_2 meters. It therefore exerts a force of $-kx_2$ newtons on weight 2.

Example 5.7.1   Find a system of ODEs that will serve as a mathematical model for the mechanical system in Figure 5.11, assuming that $m = 3$, $k = 3$, and $l = 12$.

Solution. The net force on weight 1 is $-kx_1 + l(x_2 - x_1)$. Thus, by Newton's second law,

$$mx_1'' = -kx_1 + l(x_2 - x_1). \tag{5.65}$$

Similarly,

$$mx_2'' = -kx_2 - l(x_2 - x_1). \tag{5.66}$$

We can combine (5.65) and (5.66) to obtain a homogeneous system of second-order ODEs. With $m = 3$, $k = 3$, and $l = 12$, this system is

$$3x_1'' = -15x_1 + 12x_2$$
$$3x_2'' = 12x_1 - 15x_2.$$

We can divide both equations by 3, and express the resulting system in matrix form as

$$\mathbf{x}'' = B\mathbf{x}, \tag{5.67}$$

where $\mathbf{x} = \begin{bmatrix} x_1 \\ x_2 \end{bmatrix}$ and $B = \begin{bmatrix} -5 & 4 \\ 4 & -5 \end{bmatrix}$. ❑

The system (5.67) consists of second order ODEs, each of which can be replaced by two first-order equations. If $v_1 = x_1'$ and $v_2 = x_2'$ denote the velocities of the weights, then the following system is equivalent to (5.67):

$$x_1' = v_1$$
$$x_2' = v_2$$
$$v_1' = -5x_1 + 4x_2$$
$$v_2' = 4x_1 - 5x_2.$$

Thus, the initial conditions should take the form

$$\begin{bmatrix} x_1(t_0) \\ x_2(t_0) \\ v_1(t_0) \\ v_2(t_0) \end{bmatrix} = \begin{bmatrix} p_1 \\ p_2 \\ q_1 \\ q_2 \end{bmatrix}.$$

These can be expressed more briefly as $\begin{bmatrix} \mathbf{x}(t_0) \\ \mathbf{v}(t_0) \end{bmatrix} = \begin{bmatrix} \mathbf{p} \\ \mathbf{q} \end{bmatrix}$, where $\mathbf{x}(t)$ is the configuration vector, with initial configuration $\mathbf{p}$; and $\mathbf{v}(t)$ is the velocity vector, with initial velocity $\mathbf{q}$. In other words, the initial conditions for (5.67) are given by specifying an initial phase vector: $\mathbf{p}$ and $\mathbf{q}$ represent the position and velocity of weights 1 and 2 at an initial time t_0. By the existence and uniqueness theorem, if the initial position and velocity of each weight is known, then the subsequent motion of the system is completely predictable.

The linear model for a frictionless mechanical system with n weights that are connected together and constrained to move horizontally on an axis takes the form

$$M\mathbf{x}'' = B\mathbf{x}. \tag{5.68}$$

Here, $\mathbf{x}$ is the vector of displacements, and

$$M = \begin{bmatrix} m_1 & 0 & 0 & \cdots & 0 \\ 0 & m_2 & 0 & \cdots & 0 \\ 0 & 0 & m_3 & \cdots & 0 \\ \vdots & \vdots & \vdots & \ddots & \vdots \\ 0 & 0 & 0 & \cdots & m_n \end{bmatrix}$$

is the diagonal matrix that specifies the masses of the weights in the system.

The matrix B is called the **coupling matrix**. The jth column of B is the force vector that results when $x_j = 1$, and $x_k = 0$ for $k \neq j$.

Example 5.7.2   Find the mass matrix and the coupling matrix for a mechanical system with two weights, with masses m_1 and m_2, suspended horizontally between three springs with stiffness k_0, k_1, k_2, as shown in Figure 5.12.

Solution. As each weight is constrained to move on a common axis, the system has two degrees of freedom. The mass matrix is

$$M = \begin{bmatrix} m_1 & 0 \\ 0 & m_2 \end{bmatrix}.$$

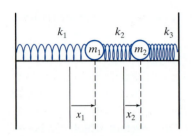

Figure 5.12 The mechanical system of Example 5.7.2. In the positions shown, both displacements are positive.

The first column of the coupling matrix is the force vector that results when the displacement vector is $\begin{bmatrix} x_1 \\ x_2 \end{bmatrix} = \begin{bmatrix} 1 \\ 0 \end{bmatrix}$. In this case, the left spring is extended 1 unit, the middle spring is extended -1 unit, and the right spring is extended 0 units. The force on the left object is $-(k_0 + k_1)$ and the force on the right object is k_1. Therefore, column 1 of the coupling matrix is $\begin{bmatrix} -(k_0 + k_1) \\ k_1 \end{bmatrix}$.

If $\begin{bmatrix} x_1 \\ x_2 \end{bmatrix} = \begin{bmatrix} 0 \\ 1 \end{bmatrix}$, then the three springs have extensions $0, 1, -1$, respectively, and the forces on the objects are k_1 and $-(k_1 + k_2)$, respectively. This tells us the second column of the coupling matrix. Hence,

$$B = \begin{bmatrix} -(k_0 + k_1) & k_1 \\ k_1 & -(k_1 + k_2) \end{bmatrix}.$$

The matrix B found in Example 5.7.2 is **symmetric**—that is, $b_{ij} = b_{ji}$ for $1 \leq i, j \leq n$. This is true for any frictionless mechanical system in which weights are coupled with springs in a one-dimensional array and constrained so that they all move on the same axis. To see why, let's number the weights so that weight 1 is leftmost, followed by weight 2, and so on, consecutively until weight n. There are $n + 1$ springs, with spring 0 connecting weight 1 to a fixed support, spring 2 connecting weights 1 and 2, and so on until we reach spring n, which connects weight n and the other fixed support. The mass of weight i is m_i, and spring j has stiffness k_j.

Notice that b_{ij} is the force applied to weight i resulting from the configuration where $x_j = 1$ and $x_m = 0$ for $m \neq j$. Thus $b_{ij} = b_{ji} = 0$ unless either $j = i$ or weights i and j are directly coupled by a spring—that is, either $j = i - 1$ or $j = i + 1$. Suppose that $j = i + 1$. Then in our configuration, spring $j - 1$ has an extension of $+1$, spring j has extension -1, and the other springs all have extensions of 0. The force applied to weight $j - 1$ is directed to the right and has magnitude k_{j-1}. The force on weight j is $-(k_{j-1} + k_j)$, and provided that $j + 1 \leq n$, the force on weight $j + 1$ is k_{j+1}, since it is to the right of spring j, which is compressed. We can summarize this by saying $b_{(j-1)j} = k_{j-1}$, $b_{jj} = -(k_{j-1} + k_j)$, $b_{(j+1)j} = k_j$, and $b_{mj} = 0$ for $m \notin \{j - 1, j, j + 1\}$. In other words, if $i = j - 1$ then $b_{ij} = k_i$. On the other hand, $b_{ji} = b_{(i+1)i} = k_i$ as well, and this shows that B is symmetric.

Consider a system of n second-order ODEs that has the form (5.68), where $\mathbf{x} \in \mathbf{R}^n$ and M, B are $n \times n$ matrices. Let's determine all solutions of the form

$$\mathbf{x}(t) = e^{st}\mathbf{c}. \tag{5.69}$$

Since $\mathbf{x}''(t) = s^2 e^{st}\mathbf{c}$, we have

$$s^2 e^{st} M\mathbf{c} = e^{st} B\mathbf{c}.$$

After dividing through by e^{st}, we find that (5.69) is a solution of (5.68) if and only if $B\mathbf{c} = s^2 M\mathbf{c}$, or $M^{-1}B\mathbf{c} = s^2\mathbf{c}$. This means that $r = s^2$ is a characteristic root of $M^{-1}B$ and $\mathbf{c}$ is a characteristic vector belonging to r. A theorem about symmetric matrices from linear algebra will now assist us. It uses a term, **positive definite**, that will be explained in advance. Let A be a symmetric $n \times n$ matrix. Given any vector $\mathbf{v} \in \mathbf{R}^n$, let $\mathbf{v}^*$ be the transpose of $\mathbf{v}$. Thus, we consider $\mathbf{v}$ to be a column— an $n \times 1$ matrix—and $\mathbf{v}^*$ is a row, or $1 \times n$ matrix. The product $\mathbf{v}^* A\mathbf{v}$ is a 1×1 matrix, which is just a real number. We say that A is positive definite if

$$\mathbf{v}^* A\mathbf{v} > 0$$

for all $\mathbf{v} \neq \mathbf{0}$. For example, a diagonal matrix with positive diagonal entries, such as the mass matrix M, is positive definite.

Theorem 5.7

Let M and B be symmetric $n \times n$ matrices, where M is also positive definite. Then there are real numbers $r_1, r_2, \ldots, r_n$, not necessarily distinct, and corresponding vectors $\mathbf{c}_1, \mathbf{c}_2, \ldots, \mathbf{c}_n$ such that

- $B\mathbf{c}_i = r_i M\mathbf{c}_i$ for $i = 1, 2, \ldots, n$.
- $\{\mathbf{c}_1, \mathbf{c}_2, \ldots, \mathbf{c}_n\}$ is linearly independent.
- $\mathbf{c}_i^* M\mathbf{c}_j = 0$ for $i \neq j$.

In the special case where M is the identity matrix, Theorem 5.7 tells us that all characteristic roots of a symmetric matrix B are real and simple, and the characteristic vectors are orthogonal. The proof, which can be found in linear algebra texts, is omitted. The vectors $\mathbf{c}_i$ are called the **principal axes** of B with respect to M. The principal axes are the characteristic vectors of $M^{-1}B$, and the r_j are the characteristic roots that belong to them.

The phase vector corresponding to a solution of (5.68) of the form (5.69) is

$$\begin{bmatrix} \mathbf{x}(t) \\ \mathbf{v}(t) \end{bmatrix} = e^{st} \begin{bmatrix} \mathbf{c} \\ s\mathbf{c} \end{bmatrix}.$$

Thus, $\mathbf{c}$ is the configuration of the system at $t = 0$, and $s\mathbf{c}$ is the initial velocity vector. There are two independent solutions of (5.68) corresponding to each nonzero characteristic root r_j of $M^{-1}B$. These have configurations $\mathbf{x}_j^{\pm}(t) = e^{\pm\sqrt{r_j}t}\mathbf{c}_j$, and the initial phase vectors $\begin{bmatrix} \mathbf{c}_j \\ \pm\sqrt{r_j}\mathbf{c}_j \end{bmatrix}$ are linearly independent. If $r_j = s_j^2 > 0$, the solution $\mathbf{x}_j^+(t) = e^{s_j t}\mathbf{c}_j$ is unbounded as $t \to \infty$ while $\mathbf{x}_j^-(t)$ is unbounded as $t \to -\infty$. If $r_j = -\omega_j^2$ is negative, the solutions are

$$\mathbf{x}_j^\pm(t) = e^{\pm i\omega_j t}\mathbf{c}_j$$
$$= (\cos(\omega_j t) \pm i\sin(\omega_j t))\mathbf{c}_j.$$

The real and imaginary parts of $\mathbf{x}_j^\pm(t)$ are themselves solutions, with initial phase vectors

$$\begin{bmatrix} \cos(0)\mathbf{c}_j \\ \omega_j \sin(0)\mathbf{c}_j \end{bmatrix} = \begin{bmatrix} \mathbf{c}_j \\ 0 \end{bmatrix} \quad \text{and} \quad \begin{bmatrix} \sin(0)\mathbf{c}_j \\ \omega_j \cos(0)\mathbf{c}_j \end{bmatrix} = \begin{bmatrix} 0 \\ \omega_j \mathbf{c}_j \end{bmatrix}$$

that are linearly independent. These solutions are bounded as $t \to \pm\infty$. As the motion of a mechanical system involving springs, no friction, and no source of energy is expected to be bounded, the hypothesis that all characteristic roots of $M^{-1}B$ are negative is plausible. In fact, this is true, but the proof requires a knowledge of advanced linear algebra and is omitted.

Example 5.7.3 Find the characteristic roots and principal axes of the matrix B in Example 5.7.1, and use them to explain the motion of the mechanical system.

Solution. Since $\text{tr}(B) = -10$ and $\det(B) = 9$, the characteristic equation of B is

$$s^2 + 10s + 9 = 0,$$

and hence the characteristic roots are $r_1 = -9$ and $r_2 = -1$. The corresponding principal axes are the characteristic vectors. To calculate them, we have to solve $(B - r_j I)\mathbf{c}_j = 0$. Thus, the principal axis corresponding to r_1 is the solution of

$$\begin{bmatrix} 4 & 4 \\ 4 & 4 \end{bmatrix} \begin{bmatrix} h \\ k \end{bmatrix} = \begin{bmatrix} 0 \\ 0 \end{bmatrix}.$$

Thus we can take $\mathbf{c}_1 = \begin{bmatrix} 1 \\ -1 \end{bmatrix}$. You can easily verify that the principal axis for r_2 is $\mathbf{c}_2 = \begin{bmatrix} 1 \\ 1 \end{bmatrix}$. The solutions corresponding to $r_1, \mathbf{c}_1$ are linear combinations of

$$\begin{bmatrix} x_1 \\ x_2 \end{bmatrix} = \begin{bmatrix} \cos(3t) \\ -\cos(3t) \end{bmatrix} \quad \text{and} \quad \begin{bmatrix} x_1 \\ x_2 \end{bmatrix} = \begin{bmatrix} \sin(3t) \\ -\sin(3t) \end{bmatrix}.$$

These have the common property of being simple harmonic oscillations with angular frequency 3 radians per second, where the displacement of object 1 is exactly opposite the displacement of object 2. Similarly, the solutions corresponding to $r_2, \mathbf{c}_2$ are linear combinations of

$$\begin{bmatrix} x_1 \\ x_2 \end{bmatrix} = \begin{bmatrix} \cos(t) \\ \cos(t) \end{bmatrix} \quad \text{and} \quad \begin{bmatrix} x_1 \\ x_2 \end{bmatrix} = \begin{bmatrix} \sin(t) \\ \sin(t) \end{bmatrix},$$

which are simple harmonic oscillations with angular frequency 1 radian per second. In this case the objects vibrate in phase with each other: Their displacements are always equal. ❑

DEFINITION   Suppose that the motion of a mechanical system with n degrees of freedom is modeled by a system (5.68), $M\mathbf{x}'' = B\mathbf{x}$, where $\mathbf{x} \in \mathbf{R}^n$ is the configuration vector. We will say that ω is a **fundamental frequency** of the system if $-\omega^2$ is a characteristic root of $M^{-1}B$. The corresponding principal axis of B with respect to M is called the **fundamental mode of vibration** for that frequency.

A mechanical system with n degrees of freedom has n fundamental frequencies, $\omega_1, \omega_2, \ldots, \omega_n$, which are not necessarily distinct. There are two independent solutions of (5.68) corresponding to each frequency ω_j, $\mathbf{x}(t) = \cos(\omega_j t)\mathbf{c}_j$ and $\mathbf{x}(t) = \sin(\omega_j t)\mathbf{c}_j$. Each of these solutions is a simple harmonic oscillation with frequency ω_j along the principal axis $\mathbf{c}_j$. The trigonometric identity $\sin(\omega_j t) \equiv \cos(\omega_j t - \pi/2)$ tells us that these two solutions are closely related—one lags behind the other but they follow the same orbit.

If a mechanical system has two or more modes of vibration, its motion will normally incorporate all of the modes at once, and will appear to be chaotic. With special initial conditions, the system will exhibit motion in only one mode; then the system will be in simple harmonic motion. This means that the weights vibrate with a common frequency, and each displacement function has the form

$$x_j = a_j \cos(k\,t - \theta_j).$$

The constant a_j represents the amplitude of vibration about the equilibrium position of the weight, and θ_j is the phase of vibration. These will not be the same for each weight: Only the frequency is common to all of the weights.

Example 5.7.4   Figure 5.13 shows a mass spring system with three weights and four identical springs. As indicated in the figure, the mass of the weights on the left and right are 2 kilograms, while the middle weight has a mass of 1 kilogram. The stiffness of each spring is 1 newton per meter. Find its fundamental frequencies and modes of vibration.

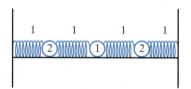

Figure 5.13 The mechanical system for Example 5.7.4.

Solution.   There are three degrees of freedom. Let's number the springs 0, 1, 2, and 3, going from left to right. The weights will be numbered 1, 2, 3.

$$\text{The mass matrix is } M = \begin{bmatrix} 2 & 0 & 0 \\ 0 & 1 & 0 \\ 0 & 0 & 2 \end{bmatrix}.$$

To determine the coupling matrix, consider first the situation where the displacements are $(x_1, x_2, x_3) = (1, 0, 0)$. The weight 1 is subject to forces due to the extension of spring 0, and the compression of spring 1. The total force is -2. Weight 2 is only subject to the force from spring 1, which is compressed. Thus the force on weight 2 is $+1$. The force on weight 3 is 0, since springs 2 and 3 are not extended or compressed. It follows that the first column of B is $\begin{bmatrix} -2 \\ 1 \\ 0 \end{bmatrix}$. Now consider what happens when $(x_1, x_2, x_3) = (0, 1, 0)$. This means springs 0 and 3 are unaffected, while spring 1 is extended 1 unit and spring 2 is compressed 1 unit. The forces on weights 1 and 3 will be $+1$, while the force on object 2 is -2. Thus, the second column of B is $\begin{bmatrix} 1 \\ -2 \\ 1 \end{bmatrix}$. It will be left to you to show that column 3 of B

is $\begin{bmatrix} 0 \\ 1 \\ -2 \end{bmatrix}$. It follows that the motion of the weights in this system is governed by

$$\begin{bmatrix} 2 & 0 & 0 \\ 0 & 1 & 0 \\ 0 & 0 & 2 \end{bmatrix} \begin{bmatrix} x_1'' \\ x_2'' \\ x_3'' \end{bmatrix} = \begin{bmatrix} -2 & 1 & 0 \\ 1 & -2 & 1 \\ 0 & 1 & -2 \end{bmatrix} \begin{bmatrix} x_1 \\ x_2 \\ x_3 \end{bmatrix}.$$

To solve this system, we will need to determine the characteristic roots and vectors of

$$M^{-1}B = \begin{bmatrix} -1 & \frac{1}{2} & 0 \\ 1 & -2 & 1 \\ 0 & \frac{1}{2} & -1 \end{bmatrix}.$$

While it is certainly possible to do this with pencil and paper—if you are good at linear algebra, you can do it—we will relegate the task to a CAS. We will find three characteristic roots, $r_1 = -1$ and $r_2, r_3 = -\frac{3}{2} \pm \frac{\sqrt{5}}{2}$, and characteristic vectors (which are the principal axes of B with respect to M),

$$\mathbf{c}_1 = \begin{bmatrix} 1 \\ 0 \\ -1 \end{bmatrix}, \quad \text{and} \quad \mathbf{c}_2, \mathbf{c}_3 = \begin{bmatrix} 1 \\ -1 \pm \sqrt{5} \\ 1 \end{bmatrix}.$$

In the first mode of vibration, weight 2 remains fixed, while weights 1 and 3 vibrate, in opposite phase, with a frequency of 1 radian per second. Both vibrations have the same amplitude. In the second mode of vibration, all three weights vibrate in phase. The vibrations of weights 1 and 3 have the same amplitude. The ratio of the amplitude of vibration of the middle weight to that of the outer weights is given by the second coordinate of the principal axis describing the mode of vibration. Thus, in the second mode the frequency of vibrations is $\sqrt{r_2} \approx 0.618$, and weight 2 vibrates with amplitude $(\sqrt{5} - 1) \approx 1.236$ times the amplitude of weights 1 and 3, and the same frequency. In the third mode, weight 2 vibrates in opposite phase to weights 1 and 3. The frequency is $\sqrt{r_3} \approx 1.618$ and the middle weight vibrates with an amplitude of $(\sqrt{5} + 1) \approx 3.236$ times the amplitude of the vibrations of the outer two weights. ❑

EXERCISES

1. An undamped mass spring system in which the mass and stiffness are $m = 3$ kilograms and $k = 147$ newtons per meter, respectively. Determine the fundamental frequency and its mode of vibration.

2. Describe the mode of vibration with frequency 3 radians per second for the mechanical system in Example 5.7.3.

3. Show that the vibrations of the mechanical system in Example 5.7.3 are always 2π-periodic.

4. Show that, except for the fundamental modes of vibration, none of the vibrations of the mechanical system in Example 5.7.4 are periodic.

A CAS is recommended for computing charactteristic roots and vectors in the following exercises.

5. Consider the mechanical system of Example 5.7.4, modified by replacing the middle weight with a 4 kilogram weight.

   a. Determine the fundamental frequencies.

   b. Show that if the initial displacements of the left and right masses are equal, but in opposite directions, the middle weight is initially in its equilibrium position, and all three weights are initially motionless, then the system will be in simple harmonic motion. What will the frequency be?

   c. In the other two fundamental modes of vibration, the initial displacements of the left and right masses are equal, and in the *same* direction. What what is the initial displacement of the middle weight in each case?

6. In the mechanical system of Figure 5.11, suppose that the stiffness of each of the left and middle springs is equal to 1, and that the stiffness of the right spring is equal to 2. The mass of the left weight is equal to 2, and the mass of the right weight is equal to 1. What sets of initial conditions result in simple harmonic motion? Describe the simple harmonic motion.

7. Two objects of equal mass are connected by a spring. This system is suspended from the ceiling by a second spring; see Figure 5.14. Take the spring constants to be 1 for the lower spring, and 2 for the upper one. Find the fundamental frequencies and modes of vibration for this mechanical system.

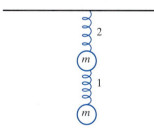

FIGURE 5.14   The mechanical system for Exercise 7.

8. Figure 5.15 shows a mechanical system in which the masses and springs are as labeled. Find the fundamental frequencies and modes of vibration.

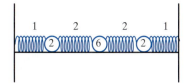

FIGURE 5.15   The mechanical system for Exercise 8.

9. A weight is suspended between two linear springs as in Figure 5.16. It is constrained to move in the plane, so there are two degrees of freedom. Derive a system of two second-order ODEs that governs the motion of the object. Is it a linear system?

FIGURE 5.16   A system with two degrees of freedom.

10. Two weights of mass 1 are connected by a spring with stiffness 2. This system is suspended from the ceiling by a second spring whose stiffness is 1 as in Figure 5.17. The vibrations are damped by a dashpot with damping constant $b = 0.1$ placed between the objects as shown. The vibrations between the upper object and the ceiling remain undamped.

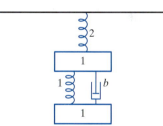

FIGURE 5.17   A semidamped mechanical system. The damping is $b = 0.1$.

Derive a system of two second-order ODEs in which the dependent variables are the displacements x_1 and x_2 of the two weights. Using y_1 and y_2 to denote the velocities of the weights, convert your system into a system of four first-order ODEs. Use a CAS to find the characteristic roots and vectors of the coefficient matrix, and give a verbal description of the motion of this mechanical system.

5.8 **Variation of Constants**

Consider an inhomogeneous linear ODE,

$$a_2(t)y'' + a_1(t)y' + a_0(t)y = b(t). \tag{5.70}$$

By Theorem 5.6 on page 212, we can find the general solution of (5.70) by determining a particular solution $y_p(x)$, and adding to that the general solution $y_h(t)$ of the associated homogeneous equation $\mathcal{L}(y) = 0$, where

$$\mathcal{L}(y) = a_2(t)y'' + a_1(t)y' + a_0(t)y$$

is the operator that forms the left side of (5.70). We will now focus on the method of variation of constants, which can be used to find a particular solution of (5.70) even in cases where the method of undetermined coefficients does not apply, such as equations with variable coefficients, and equations where the source is not a product of an exponential function and a polynomial.

We have encountered this method in solving inhomogeneous first-order equations and systems, and by converting (5.70) into a system, we will be able to take advantage of results that we previously obtained (see page 169). It is necessary to have available the general solution

$$y = C_1\phi_1(t) + C_2\phi_2(t) \tag{5.71}$$

of the associated homogeneous equation $\mathcal{L}(y) = 0$ in order to proceed, just as it was in the case of systems of differential equations.

As usual, divide (5.70) by $a_2(t)$ to obtain

$$y'' + p(x)y' + q(x)y = f(x), \tag{5.72}$$

where $p(t) = a_1(t)/a_2(t)$, $q(t) = a_0(t)/a_2(t)$, and $f(t) = b(t)/a_2(t)$. Equation (5.72) is equivalent to the system

$$\left. \begin{array}{l} y' = v \\ v' = f(t) - q(t)\,y - p(t)\,v \end{array} \right\}, \tag{5.73}$$

or $\mathbf{u}' = A\mathbf{u} + \mathbf{b}$, where

$$\mathbf{u} = \begin{bmatrix} y \\ v \end{bmatrix}, \quad A = \begin{bmatrix} 0 & 1 \\ -q(t) & -p(t) \end{bmatrix}, \quad \text{and} \quad \mathbf{b} = \begin{bmatrix} 0 \\ f(t) \end{bmatrix}.$$

The solutions of the system $\mathbf{u}' = A\mathbf{u}$ corresponding to $\phi_1(t)$ and $\phi_2(t)$ are

$$\mathbf{u}_1 = \begin{bmatrix} \phi_1(t) \\ \phi_1'(t) \end{bmatrix} \quad \text{and} \quad \mathbf{u}_2 = \begin{bmatrix} \phi_2(t) \\ \phi_2'(t) \end{bmatrix},$$

and these combine to give the fundamental matrix solution

$$\mathcal{X}(t) = \begin{bmatrix} \phi_1(t) & \phi_2(t) \\ \phi_1'(t) & \phi_2'(t) \end{bmatrix}.$$

We will attempt to find a solution of the inhomogeneous system (5.73) of the form $\mathbf{u}(t) = \mathcal{X}(t)\mathbf{w}(t)$, where

$$\mathbf{w} = \begin{bmatrix} w_1(t) \\ w_2(t) \end{bmatrix}$$

is a vector function whose components $w_i(t)$ are to be determined. When this expression is substituted into (5.73), we obtain by using the product rule to differentiate $\mathcal{X}(t)\mathbf{w}(t)$,

$$\mathcal{X}'\mathbf{w} + \mathcal{X}\mathbf{w}' = A\mathcal{X}\mathbf{w} + \mathbf{b}.$$

Since $\mathcal{X}' = A\mathcal{X}$, the two terms involving $\mathbf{w}$ can be canceled to obtain

$$\mathcal{X}\mathbf{w}' = \mathbf{b}.$$

Using matrix multiplication, we find that $w_1'(t)$ and $w_2'(t)$ must satisfy the linear equations

$$\left. \begin{array}{r} w_1'\phi_1 + w_2'\phi_2 = 0 \\ w_1'\phi_1' + w_2'\phi_2' = f(t). \end{array} \right\} \tag{5.74}$$

Solving (5.74) for the w_i' and integrating leads to a particular solution of the system (5.73), $\mathbf{u} = \mathcal{X}\mathbf{w}$, or

$$\begin{bmatrix} y \\ v \end{bmatrix} = \mathcal{X}(t) \begin{bmatrix} w_1(t) \\ w_2(t) \end{bmatrix} = \begin{bmatrix} w_1(t)\phi_1(t) + w_2(t)\phi_2(t) \\ w_1(t)\phi_1'(t) + w_2(t)\phi_2'(t) \end{bmatrix}.$$

The first row is a particular solution $y = \psi(t)$ of the original ODE (5.70), where

$$\psi(t) = w_1(t)\phi_1(t) + w_2(t)\phi_2(t).$$

Since $\psi(t)$ is formed from the general solution (5.71) of $\mathcal{L}(y) = 0$ by replacing the constants C_1 and C_2 by variables w_1 and w_2, the method is called "variation of constants."

Example 5.8.1   Find the general solution of

$$y'' + 9y = \sec(3t).$$

Solution. The corresponding system is

$$\mathbf{u}' = A\mathbf{u} + \mathbf{b}, \tag{5.75}$$

where

$$A = \begin{bmatrix} 0 & 1 \\ -9 & 0 \end{bmatrix} \quad \text{and} \quad \mathbf{b} = \begin{bmatrix} 0 \\ \sec(3t) \end{bmatrix}.$$

The general solution of the associated homogeneous equation, $y'' + 9y = 0$, is

$$y = C_1 \cos(3t) + C_2 \sin(3t).$$

This leads us to the fundamental matrix solution

$$\mathcal{X}(t) = \begin{bmatrix} \cos(3t) & \sin(3t) \\ -3\sin(3t) & 3\cos(3t) \end{bmatrix}$$

of $\mathbf{u}' = A\mathbf{u}$. We seek a solution of (5.75) of the form $\mathbf{u} = \mathcal{X}\mathbf{w}$. As we have seen, the unknown vector $\mathbf{w}$ satisfies

$$\mathcal{X}\mathbf{w}' = \mathbf{b},$$

or

$$w_1' \cos(3t) + w_2' \sin(3t) = 0$$
$$-3w_1' \sin(3t) + 3w_2' \cos(3t) = \sec(3t).$$

To find w_1', multiply the first equation by $3\cos(3t)$, the second by $\sin(3t)$, and subtract. This gives $3w_1'[\cos^2(3t) + \sin^2(3t)] = -\tan(3t)$, and thus $w_1' = -\frac{1}{3}\tan(3t)$. A similar calculation shows that $w_2' = \frac{1}{3}$. Integrating these expressions yields $w_1 = \frac{1}{9} \ln|\cos(3t)|$ and $w_2 = \frac{1}{3}t$. The particular solution that we have found is

$$\psi(t) = w_1\phi_1 + w_2\phi_2$$
$$= \frac{1}{9} \ln|\cos(3t)| \cos(3t) + \frac{1}{3}t \sin(3t).$$

The general solution is

$$y = \psi(t) + C_1 \cos(3t) + C_2 \sin(3t)$$
$$= \left[\frac{1}{9} \ln|\cos(3t)| + C_1\right] \cos(3t) + \left[\frac{t}{3} + C_2\right] \sin(3t). \qquad \square$$

We will now derive the **variation of constants formula** for a particular solution of equation (5.72). We do so by solving the system (5.74) of linear equations for w_1' and w_2', obtaining

$$w_1'(t) = -\frac{\phi_2(t)f(t)}{\phi_1(t)\phi_2'(t) - \phi_2(t)\phi_1'(t)}$$

$$w_2'(t) = \frac{\phi_1(t)f(t)}{\phi_1(t)\phi_2'(t) - \phi_2(t)\phi_1'(t)}.$$

The expression in both denominators is $\phi_1(t)\phi_2'(t) - \phi_2(t)\phi_1'(t)$, which we recognize as the Wronskian, $W[\phi_1, \phi_2](t)$. Since the Wronskian is nonzero at all ordinary points of (5.72), we are assured that the functions $w_j(t)$ are defined at all ordinary points.

The particular solution $\psi(t)$ is given by the formula $\psi = \phi_1 w_1 + \phi_2 w_2$. Furthermore, we can integrate our formulas for $w'_1(t)$ and $w'_2(t)$ to obtain

$$w'_1(t) = \int_{t_0}^{t} \frac{\phi_2(s) f(s)}{W[\phi_1, \phi_2](s)} \, ds \quad \text{and} \quad w'_2(t) = \int_{t_0}^{t} \frac{\phi_1(s) f(s)}{W[\phi_1, \phi_2](s)} \, ds,$$

where t_0 denotes a specified ordinary point of the ODE (5.72). Therefore,

$$\psi(t) = -\phi_1(t) \int_{t_0}^{t} \frac{\phi_2(s) f(s)}{W[\phi_1, \phi_2](s)} \, ds + \phi_2(t) \int_{t_0}^{t} \frac{\phi_1(s) f(s)}{W[\phi_1, \phi_2](s)} \, ds,$$

and these integrals can be combined and rearranged to yield the following **variation of constants formula**:

$$\psi(t) = \int_{t_0}^{t} \frac{[\phi_1(s)\phi_2(t) - \phi_1(t)\phi_2(s)]}{W[\phi_1, \phi_2](s)} f(s) \, ds.$$

The integral that appears in the formula is a definite integral with respect to s. The variable t appears both as a limit of the integral and in the integrand. The result of the integration is a function of t alone, as we would expect.

Define a function of two variables,

$$K(t, s) = \frac{[\phi_1(s)\phi_2(t) - \phi_1(t)\phi_2(s)]}{W[\phi_1, \phi_2](s)}.$$

In terms of this function, the variation of constants formula can be written

$$\psi(t) = \int_{t_0}^{t} K(t, s) f(s) \, ds. \tag{5.76}$$

Example 5.8.2   Find a formula for a particular solution of

$$y'' + 3y' + 2y = f(t).$$

Solution. Let $\mathcal{L}(y) = y'' + 3y' + 2y$. The characteristic roots of the associated homogeneous equation $\mathcal{L}(y) = 0$ are -1 and -2, so $\{e^{-t}, e^{-2t}\}$ is a fundamental set of solutions. We need to find functions w_1 and w_2 such that $y = w_1 e^{-t} + w_2 e^{-2t}$ is a particular solution of $\mathcal{L}(y) = f(t)$. The fundamental matrix solution of the system corresponding to $\mathcal{L}(y) = 0$,

$$y' = v$$
$$v' = -2y - 3v,$$

is $\mathcal{X} = \begin{bmatrix} e^{-t} & e^{-2t} \\ -e^{-t} & -2e^{-2t} \end{bmatrix}$. The unknown functions w_1 and w_2 are determined by the system

$$\mathcal{X} \begin{bmatrix} w'_1 \\ w'_2 \end{bmatrix} = \begin{bmatrix} 0 \\ f(t) \end{bmatrix},$$

or

$$e^{-t}w_1' + e^{-2t}w_2' = 0$$
$$-e^{-t}w_1' - 2e^{-2t}w_2' = f(t).$$

This yields

$$w_1' = -\frac{e^{-2t}f(t)}{(-e^{-3t})} = e^t f(t), \quad \text{and} \quad w_2' = \frac{e^{-t}f(t)}{(-e^{-3t})} = -e^{2t}f(t).$$

We can integrate each of these over an interval $[0, t]$ to obtain

$$w_1(t) = \int_0^t e^s f(s)\, ds \quad \text{and} \quad w_2(t) = -\int_0^t e^{-2s} f(s)\, ds.$$

The particular solution is

$$y = w_1 e^{-t} + w_2 e^{-2t}$$
$$= \int_0^t (e^{-(t-s)} - e^{-2(t-s)}) f(s)\, ds. \qquad \square$$

For any function $f(t)$ that is continuous on some open interval containing t_0, define $\mathcal{M}(b)(t) = \int_{t_0}^t K(t, s) f(s)\, ds$. We can view $\mathcal{M}$ as an inverse operator to

$$\mathcal{L}(y) = a_2(t)y'' + a_1(t)y' + a_0(t)y,$$

since $\mathcal{L} \circ \mathcal{M}(f(t)) = f(t)$ [this is just another way of saying that $y = \mathcal{M}(b)$ is a solution of $\mathcal{L}(y) = b(t)$].

Example 5.8.3   Find an inverse operator for $\mathcal{L}(y) = y'' + 4y$.

Solution. Let $\phi_1(t) = \cos(2t)$ and $\phi_2(t) = \sin(2t)$. Then $\{\phi_1, \phi_2\}$ is a fundamental set of solutions of $\mathcal{L}(y) = 0$, and the Wronskian is

$$W[\cos 2t, \sin 2t](t) = (\cos(2t))(\sin(2t))' - (\sin(2t))(\cos(2t))' = 2.$$

Thus,

$$K(t, s) = \frac{1}{2}[\cos(2s)\sin(2t) - \cos(2t)\sin(2s)]$$

$$= \frac{1}{2}\sin 2(t - s).$$

Hence $\mathcal{M}(b)(t) = \int_0^t \frac{1}{2}\sin 2(t - s)b(s)\, ds$ is an inverse operator to $\mathcal{L}$. $\qquad \square$

EXERCISES

In Exercises 1–11, use the method of variation of constants to find a particular solution of the given differential equation. If the equation has variable coefficients, the general solution of the associated homogeneous equation is given.

1. $y'' + 3y' + 2y = e^{2t}$ **2.** $y'' + y = t^2$

3. $y'' + 7y' + 12y = \cos 2t$ **4.** $y'' + 9y = \sec^3 3t$

5. $t^2 y'' + 4ty' + 2y = \ln t.$ $[y_h = C_1 t^{-1} + C_2 t^{-2}]$

6. $(t^2 + t)y'' + (4t + 2)y' + 2y = t^3.$ $[y_h = C_1 t^{-1} + C_2 (t+1)^{-1}]$

7. $y'' + 4y = \sin 2t$ **8.** $y'' - 6y' + 9y = te^{3t}$

9. $y'' - 6y' + 9y = \dfrac{1}{t} e^{3t}$ **10.** $y'' + y = \tan t$

11. $4y'' + 4y' + y = e^{-t/2} \ln t$

12. Use the Leibniz rule to find ψ'', where

$$\psi(t) = \int_{t_0}^{t} \frac{[\phi_1(s)\phi_2(t) - \phi_1(t)\phi_2(s)]}{a_2(s)W[\phi_1, \phi_2](s)} b(s)\, ds.$$

Use the result of your calculation to verify the variation of constants formula for second-order equations.

13. This exercise demonstrates that a critically damped mechanical system approaches equilibrium faster than an overdamped system. Consider two mechanical systems, with identical masses and stiffness. Use the natural time scale (which will be the same for both systems). The first system will be overdamped, so that it is governed by the differential equation

$$\frac{d^2 x}{dt^2} + B\frac{dx}{dt} + x = 0,$$

with $B > 2$. The second is critically damped and is modeled by the differential equation

$$\frac{d^2 y}{dt^2} + 2\frac{dy}{dt} + y = 0.$$

Each mass is initially at rest, with a displacement $x(0) = 1$. Show that $x(t) > y(t)$ for all $t > 0$.

◆ **Hint:** Let $z(t) = x(t) - y(t)$. Show that $z(t)$ is the solution of

$$z'' + 2z' + z = f(t),$$

with homogeneous initial conditions, where $f(t) = (2 - B)\frac{dx}{dt} > 0$ for all $t > 0$. Use the variation of constants formula to show that $z(t) > 0$ for all $t > 0$.

14. Find initial conditions so that the overdamped system in Exercise 13 approaches equilibrium faster than the critically damped system with the same initial conditions.

15. Find the general solution of

$$t^2 y'' + 3ty' + y = \frac{1}{t \ln|t|}$$

given that $\phi_1(t) = t^{-1}$ and $\phi_2(t) = t^{-1} \ln|t|$ are linearly independent solutions of the associated homogeneous equation.

16. The general solution of

$$t^2 y'' + ty' + (t^2 - \frac{1}{4})y = 0$$

is $y_h(t) = t^{-\frac{1}{2}}(C_1 \cos t + C_2 \sin t)$. Find the general solution of

$$t^2 y'' + ty' + (t^2 - \frac{1}{4})y = t^{\frac{5}{2}}.$$

17. Use variation of constants to find an integral formula for solutions of

$$y'' + 2y' + y = b(t)$$

with homogeneous initial conditions at $t = 0$.

18. Notice that $y = t$ and $y = t \ln t$ satisfy $t^2 y'' - ty' + y = 0$. Use this information to derive an inverse operator for

$$\mathcal{L}(y) = t^2 y'' - ty' + y.$$

19. Find the general solution of $y'' + 4ty' + (4t^2 + 2)y = (t^2 + \frac{1}{2})e^{-t^2}$. The general solution of the associated homogeneous equation is $y_h = e^{-t^2}(C_1 + C_2 t)$.

*5.9 **Reduction of Order**

You can verify that $y = t$ is a particular solution of the ODE

$$y'' - t\, y + y = 0. \tag{5.77}$$

We will use a procedure known as **reduction of order** to find a second solution $y = \phi_2(t)$, such that $\{t, \phi_2(t)\}$ is linearly independent.

In its most general form, the reduction of order procedure is designed to find the general solution of a second-order linear ODE,

$$a_2(t)y'' + a_1(t)y' + a_0(t)y = b(t). \tag{5.78}$$

It can be applied if a nontrivial solution $\phi_1(t)$ of the associated homogeneous equation

$$a_2(t)y'' + a_1(t)y' + a_0(t)y = 0 \tag{5.79}$$

is known.

The procedure can be described very briefly: Substitute $y = v \cdot \phi_1(t)$ in equation (5.78), where v is a new dependent variable. When simplified, equation (5.78) becomes a first-order linear ODE with dependent variable $w = v'$.

Example 5.9.1   Find the general solution of equation (5.77).

Solution. We have noticed that $y = t$ is a particular solution, so put $y = vt$. Then $y' = v't + v$ and $y'' = v''t + 2v'$. Substituting these converts (5.77) into the following equation:

$$(v''t + 2v') - t(v't + v) + tv = 0,$$

and this can be put in simplified form as

$$tv'' + (2 - t^2)v' = 0. \tag{5.80}$$

Since equation (5.80) has v'' and v' terms, but no v term, we can put $w = v'$, and $w' = v''$ and convert (5.80) to the first-order equation

$$tw' + (2 - t^2)w = 0.$$

The general solution is

$$w = C \exp\left(-\int \frac{2 - t^2}{t}\, dt\right) = \frac{C}{t^2} e^{t^2/2}.$$

To find v, we must find an antiderivative of w. There is no elementary function whose derivative is w, so we leave v in integral form:

$$v = C \int \frac{e^{t^2/2}}{t^2}\, dt. \tag{5.81}$$

We have found a family of functions v with the property that every solution of equation (5.77) is of the form

$$y = Ctv + Dt,$$

where v belongs to this family. It is disquieting to notice that none of the functions in (5.81) are defined at $t = 0$, even though 0 is an ordinary point. We can resolve this problem with the help of power series. Since

$$e^u = 1 + u + \frac{1}{2!}u^2 + \cdots + \frac{1}{n!}u^n + \cdots ,$$

$$\frac{e^{t^2/2}}{t^2} = \frac{1}{t^2} + \frac{1}{2} + \frac{1}{2! \cdot 2^2}t^2 + \cdots + \frac{1}{n! \cdot 2^n}t^{2n-2} + \cdots ,$$

and

$$\int \frac{e^{t^2/2}}{t^2} dt = -\frac{1}{t} + \frac{1}{2}t + \frac{1}{2! \cdot 2^2 \cdot 3}t^3 + \cdots + \frac{1}{n! \cdot 2^n(2n-1)}t^{2n-1} + \cdots + D,$$

where D is an integration constant. Finally,

$$y = C\left(-1 + \frac{t^2}{2} + \frac{t^4}{2! \cdot 2^2 \cdot 3} + \cdots + \frac{t^{2n}}{n! \cdot 2^n(2n-1)} + \cdots \right) + Dt$$

is a presentation of the general solution in a form that makes it evident that it is defined at $t = 0$.   ❑

Example 5.9.1 makes it clear that solutions of simple linear ODEs can force us to step outside the realm of elementary functions. We will frequently have to leave answers "in integral form."

To see why the reduction of order procedure works in general, recall that we are substituting $y = v\phi_1(t)$ in equation (5.78). By the product rule, $y' = v'\phi_1(t) + vy_1'(t)$ and

$$y'' = v''\phi_1(t) + 2v'\phi_1'(t) + v\phi_1''(t).$$

Equation (5.78) takes the form

$$a_2(t)[v''\phi_1(t) + 2v'\phi_1'(t) + v\phi_1''(t)]$$
$$+ a_1(t)[v'\phi_1(t) + vy_1'(t)] + a_0(t)v\phi_1(t) = b(t).$$

We can rearrange this equation as follows:

$$a_2(t)\phi_1(t)v'' + [2a_2(t)y_1'(t) + a_1(t)\phi_1(t)]v'$$
$$+ [a_2(t)\phi_1''(t) + a_1(t)\phi_1'(t) + a_0(t)\phi_1(t)]v = b(t).$$

Since $\phi_1(t)$ is a solution of the homogeneous equation (5.79), the coefficient of v is equal to zero for all t. Hence the new equation does not involve v, but just v' and v'', and we can therefore put $w = v'$, $w' = v''$ as we did in Example 5.9.1. Then w is a solution of the linear first-order equation

$$q_1(t)w' + q_0(t)w = b(t), \tag{5.82}$$

where $q_1(t) = a_2(t)\phi_1(t)$, and $q_0(t) = 2a_2(t)\phi_1'(t) + a_1(t)\phi_1(t)$. The general solution of equation (5.82) is a family of functions $w(t) = w_p(t) + C_1 w_1(t)$, where $w_p(t)$ is a particular solution of equation (5.82) and w_1 is a solution of the associated homogeneous equation. Let $v_p(t) = \int w_p(t)\,dt$ and $v_1(t) = \int w_1(t)\,dt$. Since $v(t)$ is an antiderivative of $w(t)$,

$$v(t) = v_p(t) + C_1 v_1(t) + C_2,$$

where C_2 is a constant of integration. The final step is to multiply by $\phi_1(t)$ and obtain the following family of solutions of equation (5.78):

$$y(t) = v_p(t)\phi_1(t) + C_1 v_1(t)\phi_1(t) + C_2\phi_1(t). \tag{5.83}$$

We conclude this section with two further examples of the reduction of order procedure.

Example 5.9.2   Given that $\phi_1(t) = t^{-2}$ is a solution of

$$ty'' + (3 - 2t)y' - 4y = 0, \tag{5.84}$$

use the method of reduction of order to find the general solution.

Solution. Set $y = v\,t^{-2}$. Then $y' = -2v\,t^{-3} + v't^{-2}$, and $y'' = 6v\,t^{-4} - 4v't^{-3} + v''t^{-2}$. Substituting these expressions into equation (5.84), we have

$$t(6vt^{-4} - 4v't^{-3} + v''t^{-2}) + (3 - 2t)(-2vt^{-3} + v't^{-2}) - 4vt^{-2} = 0.$$

After canceling out the terms involving v and simplifying, we obtain

$$t^{-1}v'' - (t^{-2} + 2t^{-1})v' = 0.$$

Now set $w = v'$ and multiply through by t; the result is the linear first-order equation

$$w' - (t^{-1} + 2)w = 0. \tag{5.85}$$

The general solution of equation (5.85) is

$$w = C_1 \exp\left(\int (t^{-1} + 2)\,dt\right)$$

$$= C_1 e^{\ln|t| + 2t}$$

$$= C_1 t e^{2t}.$$

Therefore,

$$v = C_1 \int t e^{2t}\,dt$$

$$= C_1\left(\frac{1}{2}t - \frac{1}{4}\right)e^{2t} + C_2.$$

Then

$$y(t) = vt^{-2}$$

$$= C_1\left(\frac{1}{2}t^{-1} - \frac{1}{4}t^{-2}\right)e^{2t} + C_2 t^{-2}. \qquad\square$$

⬜ **Example 5.9.3**   Find the general solution of

$$t^2 y'' - t y' + y = t \ln(t), \tag{5.86}$$

given that $y = t$ is a solution of the associated homogeneous equation

$$t^2 y'' - t y' + y = 0.$$

Solution. Substitute $y = vt$, $y' = v + tv'$, and $y'' = 2v' + tv''$ in equation (5.86). After simplifying, the resulting equation is

$$t^3 v'' + t^2 v' = t \ln(t).$$

Set $w = v'$ and divide through by t^3; then

$$w' + t^{-1} w = t^{-2} \ln(t). \tag{5.87}$$

The general solution of the associated homogeneous equation $w' + t^{-1} w = 0$ of equation (5.87) is $w = C e^{-\ln(t)} = t^{-1}$. Using variation of constants, we substitute $w = u t^{-1}$ and $w' = u t^{-1} - u t^{-2}$ in equation (5.87) and simplify to get $u' t^{-1} = t^{-2} \ln(t)$, and hence

$$u = \int \frac{1}{t} \ln t \, dt = \frac{1}{2} [\ln(t)]^2 + C_1,$$

so

$$w(t) = u t^{-1} = \frac{1}{2t} [\ln(t)]^2 + C_1 \frac{1}{t}.$$

Then

$$v(t) = \int w(t) \, dt$$

$$= \frac{1}{6} [\ln(t)]^3 + C_1 \ln(t) + C_2,$$

and

$$y(t) = t v(t)$$

$$= \frac{1}{6} t [\ln(t)]^3 + C_1 t \ln(t) + C_2 t. \qquad \square$$

EXERCISES

Use reduction of order to find the general solution of each ODE.

1. $t y'' + (4 + t) y' + 3y = 0$; $\phi_1(t) = t^{-3}$

2. $t y'' + (4 + t) y' + 3y = -e^{-t}$; $\phi_1(t) = t^{-3}$

3. $t y'' + (4 + t) y' + 3y = t + 1$; $\phi_1(t) = t^{-3}$

4. $y'' - 2y' + y = \frac{1}{t} e^t$

5. $y'' + y = \tan(t)$

6. $y'' + 16y = \sec^3(4t)$

7. $y'' + 6y' + 8y = t^2 e^{-2t}$

8. $t^2 y'' - t y' + y = \dfrac{1}{t}$; $\phi_1(t) = t$

9. $t y'' - [t \tan(t) + 1] y' + [\tan(t) - t \sec^2(t)] y = 0$; $\phi_1 = \sec(t)$

10. $y'' + (2t - 1) y' + (t^2 - t + 1) y = 0$; $\phi_1(t) = e^{-t^2/2}$

11. $y'' + (2t - 1) y' + (t^2 - t + 1) y = e^{-t^2/2}$; $\phi_1(t) = e^{-t^2/2}$

5.10 Glossary

$\mathcal{D}$ The differentiation operator: $\mathcal{D} y = y'$.

Associated homogeneous equation   The homogeneous ODE $\mathcal{L}(y) = 0$ formed from an inhomogeneous equation

$$\mathcal{L}(y) = b(t)$$

by replacing the source $b(t)$ with 0.

Beats   The low-frequency envelope of the graph of a particular solution of $m x'' + k x = \sin(\omega t)$ that is observed when ω is close to the resonant frequency.

Characteristic equation   (for a second-order homogeneous linear ODE with constant coefficients, $a_2 y'' + a_1 y' + a_0 y = 0$) The quadratic equation obtained by substituting $y = e^{st}$, and simplifying:

$$a_2 s^2 + a_1 s + a_0 = 0.$$

The characteristic equation of a homogeneous linear ODE of order n is a polynomial equation of degree n that is obtained in the same way.

Characteristic root   A root of the characteristic equation.

Coefficient function   The functions $a_i(t)$ appearing in the linear ODE,

$$a_2(t) y'' + a_1(t) y' + a_0(t) y = b(t).$$

Damping   The magnitude of the damping force divided by the magnitude of the velocity.

Damping force   A force in a mechanical system that is due to friction. It is assumed to be proportional to velocity.

Damping ratio   (of a mechanical system with one degree of freedom) The ratio $\dfrac{b}{\sqrt{km}}$, where b is the damping, k is the stiffness, and m is the mass of the weight. This ratio is dimensionless, and remains the same regardless of the units in which the parameters of the system are denominated.

Degrees of freedom   (of a mechanical system) The number of variables that must be specified in order to describe the position of each weight in the system.

Domain of an IVP   The largest interval in the domain of the independent vector, containing the initial point and no singular points.

Equilibrium position   The position of a weight in a mechanical system when the net force on each of the weights in the system is 0.

Frequency-response curve   A graph of the amplitude of the periodic motion resulting from a periodic force applied to a mechanical system, as a function of the frequency.

Fundamental frequency   A frequency at which a mechanical system can vibrate in simple harmonic motion when no external force is applied. k is a fundamental frequency of the system if and only if ki is a characteristic root of the coefficient matrix of the corresponding system of ODEs.

Fundamental set of solutions   A set $\mathcal{S} = \{\phi_1(t), \dots, \phi_n(t)\}$ of solutions of a homogeneous linear ODE $\mathcal{L}(y) = 0$ with the property that every solution ψ of $\mathcal{L}(y) = 0$ can be expressed uniquely as a linear combination

$$\psi = C_1 \phi_1 + \cdots + C_n \phi_n.$$

If the order of the ODE is n, then every fundamental set of solutions will be a linearly independent set consisting of n solutions.

General solution   A family of solutions of a differential equation, encompassing all of the solutions of the differential equation that are defined on a given interval.

Homogeneous equation   A linear differential equation for which $y \equiv 0$ is a solution. For example, a second-order homogeneous equation would have the form

$$a_2(t) y'' + a_1(t) y' + a_0(t) y = 0.$$

Homogeneous initial conditions   Initial conditions that would be satisfied by $y \equiv 0$. For example, homogeneous equations for a second-order ODE at an ordinary point t_0 would take the form $y(t_0) = y'(t_0) = 0$.

LDO   Acronym for linear differential operator.

Linear combination   Given a set of functions $\mathcal{S} = \{\phi_1(t), \dots, \phi_n(t)\}$, any function of the form

$$y(t) = C_1 \phi_1(t) + \cdots + C_n \phi_n(t)$$

is said to be a linear combination of $\mathcal{S}$.

Linear differential operator   An operator of the form

$$\mathcal{L}(y) = a_n(t) y^{(n)} + \cdots + a_1(t) y' + a_0(t) y.$$

Linear model   A mathematical model that postulates a linear relationship. For example, Hooke's law says that the restoring force F_S of a spring is negatively proportional to its extension x past its natural length: $F_S = -k\,x$. We have also used a linear model for the damping force. In this context, the damping force F_F is negatively proportional to the velocity v; that is, $F_F = -bv$.

Linear operator   An operator $\mathcal{L}$ such that for any functions f and g for which $\mathcal{L}(f)$ and $\mathcal{L}(g)$ exist, and for any constants C and D,

$$\mathcal{L}(Cf + Dg) = C\mathcal{L}(f) + D\mathcal{L}(g).$$

Linearly independent   A set of functions,

$$S = \{\phi_1(t), \phi_2(t), \ldots, \phi_n(t)\},$$

such that no function in S is identically equal to 0 and no function in S can be expressed as a linear combination that the only linear combination of the other functions in S. When $n = 2$ this means that ϕ_2 is not equal to a constant multiple of ϕ_1. For $n > 2$ a convenient test for linear independence is to show that that the only linear combination of the functions in S that is identically 0 is

$$0 \cdot \phi_1(t) + 0 \cdot \phi_2(t) + \cdots + 0 \cdot \phi_n(t).$$

Mechanical system   A collection of one or more devices, such as springs and weights, attached to one another and subject to various forces.

Mode of vibration   A simple harmonic motion of a mechanical system.

Natural unit of time   (of a mechanical system with one degree of freedom) $\sqrt{\frac{m}{k}}$ time units, where m denotes the mass of the weight, and k is the stiffness of the spring. One natural unit represents the same length of time, regardless of the system of units used.

Newton's second law of motion   The net force F applied to an object is equal to the product of its mass m and its acceleration a: $F = ma$.

Operator   A function that converts functions to other functions. One example is the differentiation operator, $\mathcal{D}(f)(t) = f'(t)$.

Ordinary point

(of a linear ODE $a_2(t)y'' + a_1(t)y' + a_0(t)y = b(t)$

or of an LDO $\mathcal{L}(y) = a_2(t)y'' + a_1(t)y' + a_0(t)y)$

A point t_0 such that each of the functions $p(t) = a_1(t)/a_2(t)$, $q(t) = a_0(t)/a_2(t)$, and (only in the case of the ODE) $f(t) = b(t)/a_2(t)$ is continuous at in an open interval containing t_0.

Periodic solution   A solution $x = \psi(t)$ of the ODE

$$m\,x'' + b\,x' + k\,x = A\,f(t),$$

where $f(t) = \sin(\omega t)$, $\cos(\omega t)$, or $e^{i\,\omega t}$ that is periodic, with period $2\pi/\omega$.

Phase vector   The vector whose components are the positions and velocities of the weights in a mechanical system.

Reduction of order   Use of a known solution of a homogeneous linear ODE to determine the general solution.

Resonant solution   A solution of

$$y'' + k\,y = f(t)$$

that is unbounded, where $f(t)$ is a bounded periodic function.

Restoring force   The force exerted by a spring to resist extension or compression.

Simple harmonic motion   A mechanical system is in simple harmonic motion if the displacement of each weight is $x_i = a_i \sin(k\,t - \theta_i)$, where a_i is the amplitude of the vibration, θ_i is the phase of vibration, and k is the frequency (different weights may have different amplitudes and phases, but all have the same frequency).

Singular point   (of a linear ODE or a LDO) A point that is not an ordinary point.

Source   The term $b(t)$ in the linear differential equation

$$a_2(t)y'' + a_1(t)y' + a_0(t)y = b(t).$$

Stiffness   The magnitude of the restoring force of a spring divided by the extension of the spring beyond its natural length.

Transient   A solution of a homogeneous linear ODE that decays to 0 as $t \to \infty$.

Variation of constants formula   A formula that gives the a particular solution of the ODE

$$a_2(t)y'' + a_1(t)y' + a_0(t)y = b(t),$$

as follows:

$$y(t) = \int_{t_0}^{t} \frac{\phi_1(s)\phi_2(t) - \phi_1(t)\phi_2(s)}{a_2(s)W[\phi_1, \phi_2](s)} b(s)\, dt,$$

where $\{\phi_1, \phi_2\}$ is any fundamental set of solutions of the associated homogeneous equation, and $W[\phi_1, \phi_2]$ is their Wronskian.

Wronskian   (of functions $\phi_1(t)$ and $\phi_2(t)$)

$$W[\phi_1, \phi_2](t) = \det \begin{bmatrix} \phi_1(t) & \phi_2(t) \\ \phi_1'(t) & \phi_2'(t) \end{bmatrix}$$
$$= \phi_1(t)\phi_2'(t) - \phi_2(t)\phi_1'(t).$$

5.11 Chapter Review

In Exercises 1–7, find the general solution.

1. $2y'' - 7y' + 5y = 0$ **2.** $2y'' + 2y' + y = 0$

3. $y'' - 6y' + 9y = 0$ **4.** $y'' + y = t^3$

5. $y'' + y = \sin(t)$ **6.** $y'' - 4y' + 4y = te^{-2t}$

7. $y'' - 4y' + 4y = e^{2t}$

In Exercises 8–12, solve the IVP.

8. $y'' + .2y' + .01y = 0$; $y(0) = 0$, $y'(0) = 2$

9. $y'' + y = te^{-2t}$; homogeneous initial conditions at $x = 0$

10. $y'' + 3y' + 2y = \sinh t$; homogeneous initial conditions at $x = 0$

11. $y'' + y = 0$; $y(0) = 1$, $y'(0) = -2$

12. $y'' + 7y' + 12y = e^{-t}$; $y(0) = 0$, $y'(0) = 1$

13. Find a fundamental set of solutions for each ODE.

    a. $y'' + 3y' + 2y = 0$ b. $y'' + y' + 2y = 0$

    c. $\mathcal{D}^2(y \cos t) = 0$

14. Find particular solutions of the following ODEs by using the variation of constants method. In each case, S denotes a fundamental set of solutions of the associated homogeneous equation.

    a. $y'' = te^t$; $S = \{t, t - 2\}$

    b. $t^2 y'' + 2ty' - 6y = t^2$; $S = \{t^2, t^{-3}\}$

    c. $y'' + 10y' + 25y = e^{-5t} \ln(t)$ for $t > 0$; $S = \{e^{-5t}, te^{-5t}\}$

    d. $y'' - 4y = \operatorname{sech}(2t)$; $S = \{\cosh(2t), \sinh(2t)\}$

15. Find an inverse operator for $\mathcal{L}(y) = y'' + 2y' + y$.

16. Use reduction of order to find the general solution on the designated interval.

    a. $y'' + 6y' + 9y = 0$ on **R**; $\phi_1 = e^{-3t}$

    b. $t^2 y'' + 3ty' + y = 0$ on $(0, \infty)$; $\phi_1 = t^{-1}$

    c. $t^2 y'' + 3ty' + y = \ln(t)$ on $(0, \infty)$; $\phi_1 = t^{-1}$

    d. $ty'' + 2(t - 1)y' - 4y = 0$ on $(0, \infty)$; $\phi_1 = e^{-2t}$

17. Derive a system of differential equations to model the mechanical system depicted in Figure 5.18. Determine the fundamental frequencies and modes of vibration for the system.

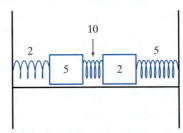

FIGURE 5.18   A mechanical system: See Exercise 17. The masses are given in kilograms, and the stiffness of each spring is specified in newtons per meter.

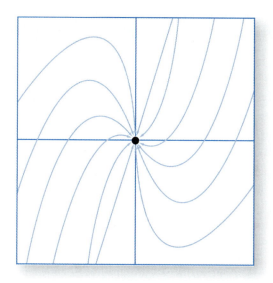

CHAPTER 6

The Laplace Transform

6.1 Introduction

This chapter is about an approach to solving differential equations that involves the Laplace transform. The Laplace transform is an *integral transform,* and before we start, I will explain the strategy behind the use of integral transforms. We are given a problem, such as a differential equation, that involves an unknown function. We will assume that the problem has a unique solution, $y(t)$. Thus, if the problem is a differential equation, it should come with initial conditions.

An **integral transform** is an operator of the form

$$I(f) = \int_a^b K(s,t)\, f(t)\, dt. \tag{6.1}$$

The function $K(s,t)$—called the kernel of the transform—and the limits of integration, a and b (either or both of these may be infinite), are specified and determine the particular integral transform. For the Laplace transform, $a = 0$, $b = \infty$, and the kernel is $K(s,t) = e^{-st}$. When evaluating the integral in (6.1), the variable s is treated as a constant. The transform $I(f)$ of f is then a function of s.

The reason to use integral transforms is that it may be easier to find the transform $Y(s) = I(y)$ of $y(t)$ than it is to derive a formula for $y(t)$ itself. Thus, we take an indirect approach: Solve the *transformed problem* to determine $Y(s)$. Since the function $Y(s)$ has no direct relation to the original problem, we have to undo the transform to find the actual solution $y(t)$.

Here is an important property shared by all integral transforms.

Proposition 6.1.1

Let $f(t)$ and $g(t)$ be functions that have transforms $I(f)$ and $I(g)$, as defined in (6.1), and let C be a constant. Then

$$I(C \cdot f) = C \cdot I(f), \text{ and}$$
$$I(f + g) = I(f) + I(g).$$

We can restate Proposition 6.1.1 as saying that an integral transform is a linear operator.

Proof of Proposition 6.1.1.   Let $F(s) = I(f)$ and $G(s) = I(g)$. Then

$$I(f + g) = \int_a^b K(s, t)[f(t) + g(t)]\, dt$$

$$= \int_a^b K(s, t)\, f(t)\, dt + \int_a^b K(s, t)\, g(t)\, dt$$

$$= F(s) + G(s)$$

and

$$I(C \cdot f) = \int_a^b K(s, t)\, C\, f(t)\, dt = C \cdot F(s). \qquad \blacksquare$$

The Laplace transform has a strange history. It is named for the French mathematician P. S. Laplace, who was not its inventor. It appeared in a book by Laplace published in 1812, but it can be found in a paper by L. Euler that was written in 1737. Neither Euler nor Laplace were responsible for the widespread recognition of the Laplace transform as a useful way to solve certain ODE problems. It was popularized in the late 1800s by an English physicist, Oliver Heaviside, who developed a method to solve linear ODEs with constant coefficients that he called "the operational calculus." Heaviside was interested in modeling the behavior of electrical circuits, and wanted to be able to handle discontinuous source terms that might result from switches being turned on and off. His operational calculus did this very well, but it was not mathematically rigorous. It became accepted as a legitimate technique when it was realized that the operational calculus method was equivalent to the Laplace transform method for solving ODEs.

Let $f(t)$ be a function that is defined for all $t \geq 0$. The Laplace transform

$$L(f) = \int_0^\infty f(t)e^{-st}\, dt \tag{6.2}$$

provides a bridge between two "domains," the **time domain** and the **transform domain**. The time domain is the positive real axis, with the variable t representing time. An IVP that serves as a model for a physical process would be stated in the time domain. The domain consisting of all numbers s where the integral defining the Laplace transform is finite is the transform domain. Even in applied problems, formulas in the transform domain have no physical significance.

IVPs involving linear ODEs with constant coefficients in the time domain become algebraic equations in the transform domain. An ODE model for a complicated mechanical system or electrical circuit can be replaced by a problem that can be solved *without calculus* in the transform domain. When the solution is found there, an inverse transform is used to compute the solution in the time domain.

The Laplace transform is an improper integral, so by definition,

$$L(f) = \lim_{B \to \infty} \int_0^B f(t)e^{-st}\, dt.$$

The function $f(t)$ has a Laplace transform if there is a number s such that this limit exists. The set of all such s forms the domain of definition for $F(s) = L(f)$.

Example 6.1.1   Calculate the Laplace transform of the constant function $f(t) \equiv 1$.

Solution.

$$L(1) = \int_0^\infty e^{-st}\, dt$$

$$= \lim_{B \to \infty} \int_0^B e^{-st}\, dt$$

$$= \lim_{B \to \infty} -\frac{1}{s}e^{-st}\Big|_0^B$$

$$= -\frac{1}{s}\lim_{B \to \infty}(e^{-sB} - 1)$$

If $s > 0$, this limit is $\frac{1}{s}$, since $e^{-sB} = \frac{1}{e^{sB}} \to 0$ as $B \to \infty$. If $s \leq 0$, the limit is infinite. Therefore, $L(1) = \frac{1}{s}$ for $s > 0$.   ❑

Example 6.1.2   Find $L(e^{kt})$.

Solution.

$$L(e^{kt}) = \int_0^\infty e^{-st}e^{kt}\, dt$$

This integral diverges if $s \leq k$; for $s > k$,

$$L(e^{kt}) = \int_0^\infty e^{-(s-k)t}\, dt$$

$$= \lim_{B \to \infty} \int_0^B e^{-(s-k)t}\, dt$$

$$= \lim_{B \to \infty} -\frac{e^{-(s-k)t}}{s - k}\Big|_0^B$$

$$= \lim_{B \to \infty} \frac{1}{s - k}(1 - e^{-(s-k)B})$$

$$= \frac{1}{s - k}.$$ ❑

The following example shows that it is possible for a discontinuous function to have a Laplace transform.

Example 6.1.3   Find $L(y)$, where

$$y(t) = \begin{cases} 1 & \text{for } 0 \le t \le 1 \\ 0 & \text{for } t > 1. \end{cases}$$

Solution.

$$L(y) = \int_0^\infty y(t)e^{-st}\, dt$$

$$= \int_0^1 1 \cdot e^{-st}\, dt + \int_1^\infty 0 \cdot e^{-st}\, dt$$

$$= \frac{1}{s}(1 - e^{-s})$$

□

Not all functions have Laplace transforms, since the defining integral,

$$\int_0^\infty f(t)e^{-st}\, dt,$$

might not have a finite value for any value of s.

Example 6.1.4   Which of the following functions have Laplace transforms?

a. $f(t) = \dfrac{1}{t}$ b. $f(t) = \dfrac{1}{\sqrt{t}}$ c. $f(t) = t^t$

Solution.

a. $f(t) = \frac{1}{t}$ does not have a Laplace transform because for all s and $B > 0$, $\int_0^B \frac{e^{-st}\, dt}{t}$ is a divergent improper integral, due to the singularity at $t = 0$.

b.   Let

$$I_1 = \int_0^1 \frac{1}{\sqrt{t}}e^{-st}\, dt \quad \text{and} \quad I_2 = \int_1^\infty \frac{1}{\sqrt{t}}e^{-st}\, dt.$$

Then $I = \int_0^\infty \frac{1}{\sqrt{t}}e^{-st}\, dt$ converges if and only if both I_1 and I_2 converge, because $I = I_1 + I_2$. Let us assume the $s > 0$. Then $\frac{1}{\sqrt{t}}e^{-st}$ for all $t > 0$. The comparison test for improper integrals thus guarantees that I_1 is convergent, because $\int_0^1 \frac{1}{\sqrt{t}}\, dt$ converges. If $t \ge 1$, then the inequality $\frac{1}{\sqrt{t}}e^{-st} \le e^{-st}$ holds. The comparison test can be used again to prove that I_2 converges, because $\int_1^\infty e^{-st}\, dt$ is convergent. We have thus shown that $\int_0^\infty \frac{1}{\sqrt{t}}e^{-st}\, dt$ is convergent for all $s > 0$: Thus the Laplace transform of $\frac{1}{\sqrt{t}}$ is defined for $s > 0$.

c. Since $t^t = e^{t \ln t}$,

$$\int_0^\infty f(t)e^{-st}\, dt = \int_0^\infty e^{t(\ln(t)-s)}\, dt.$$

If s_1 is a fixed value of s, then

$$\lim_{t\to\infty} \ln(t) - s_1 = \infty.$$

It follows that $e^{-s_1 t} f(t) \to \infty$. Since $\int_0^\infty f(t)e^{-st}\, dt$ must diverge if $\lim_{t\to\infty} f(t)e^{-st} \neq 0$, it follows that $\int_0^\infty f(t)e^{-st}\, dt$ diverges for each s, and that f does not have a Laplace transform. ❑

The following are useful criteria that—taken together—imply a function has a Laplace transform.

Locally integrable A function $f(t)$ is locally integrable if for every $B > 0$, the definite integral

$$\int_0^B |f(t)|\, dt$$

has a finite value. Examples would include functions that are continuous on $[0, \infty)$, as well as many that are not, such as $f(t) = t^{-1/2}$.

Piecewise continuous A function $f(t)$ is piecewise continuous on $[0, \infty)$ if either

i. There is a sequence

$$t_0 = 0 < t_1 < t_2 < \cdots < t_n < \cdots$$

with $\lim_{n\to\infty} t_n = +\infty$, such that f is continuous on each interval (t_{n-1}, t_n) and the following one-sided limits exist and are finite:

$$\lim_{t\to t_n^+} f(t) \text{ for } n \geq 0, \quad \text{and} \quad \lim_{t\to t_n^-} f(t) \text{ for } n \geq 1.$$

ii. There are numbers

$$t_0 = 0 < t_1 < t_2 < \cdots < t_m$$

such that for $n = 1, 2, \ldots m$, f is continuous on each interval (t_{n-1}, t_n), as well as (t_m, ∞), and the following one-sided limits exist and are finite:

$$\lim_{t\to t_n^+} f(t) \text{ for } 0 \leq n \leq m, \text{ and } \lim_{t\to t_n^-} f(t) \text{ for } 1 \leq n \leq m.$$

See Example 6.1.3 for a function that is piecewise continuous but not continuous. We will work with many piecewise continuous functions in this chapter, and as you can show, any function that is piecewise continuous on $[0, \infty)$ is locally integrable (see Exercise 8 at the end of this section).

Exponential order A function $f(t)$ is of exponential order if there is a number K such that

$$\lim_{t\to\infty} e^{-Kt} f(t) = 0.$$

Proposition 6.1.2

If f is a function that is locally integrable and of exponential order, then there is a number s_1 such that the Laplace transform $F(s) = L(f)$ is defined for all $s > s_1$. Furthermore, $\lim_{s\to\infty} F(s) = 0$.

Proof If f is locally integrable, then for $s > 0$

$$\int_0^B e^{-st}|f(t)|\,dt < \int_0^B |f(t)|\,dt < \infty$$

for all $B > 0$. Hence $\int_0^B e^{-st} f(t)\,dt$ is finite as well. Assuming that f is also of exponential order, there is a number K such that

$$\lim_{t\to\infty} e^{-Kt} f(t) = 0. \tag{6.3}$$

Let $s_1 = K$ if $K > 0$; otherwise $s_1 = 0$. By (6.3) there is a number M such that $e^{-s_1 t}|f(t)| < 1$ for all $t > M$. Then $|f(t)| < e^{s_1 t}$ and hence $e^{-st}|f(t)| < e^{-(s-s_1)t}$ when $t > M$. For $s > s_1$ and $B > M$ we have

$$\int_0^B e^{-st}|f(t)|\,dt \le \int_0^M |f(t)|\,dt + \int_M^B e^{-(s-s_1)t}\,dt. \tag{6.4}$$

As $B \to \infty$ the first integral on the right side of (6.4) is not affected, and the second converges to $\frac{e^{-Ms}}{s-s_1}$. Hence

$$\int_0^B e^{-st}|f(t)|\,dt$$

is a bounded, increasing function of B. It follows that $\int_0^\infty e^{-st} f(t)\,dt$ converges. The proof that $L(f) = F(s) \to 0$ as $s \to \infty$ is given in outline form in Exercise 9 at the end of this section. ∎

In order to apply the Laplace transform to differential equations, it is necessary to have a formula for the Laplace transform of the derivative of a function.

Proposition 6.1.3

Let $f(t)$ be a function such that both f and f' have Laplace transforms. Then $L(f') = sL(f) - f(0)$.

Proof Because f' has a Laplace transform, there is a number s_1 such that

$$L(f') = \int_0^\infty f'(t)e^{-st}\,dt = \lim_{B\to\infty} \int_0^B f'(t)e^{-st}\,dt,$$

provided that $s > s_1$. For any value of s we can integrate by parts to obtain

$$\int_0^B f'(t)e^{-st}\, dt = \int_0^B e^{-st}\, df(t)$$

$$= e^{-st} f(t)|_{t=0}^B - \int_0^B f(t)\, d(e^{-st})$$

$$= e^{-sB} f(B) - f(0) + s \int_0^B f(t)e^{-st}\, dt.$$

Since $f(t)$ has a Laplace transform, there is a number s_2 such that the integral $\int_0^\infty e^{-st} f(t)\, dt$ converges and $\lim_{B\to\infty} e^{-sB} f(B) = 0$ for $s > s_2$ It follows that for $s > \max\{s_1, s_2\}$,

$$L(f') = s \int_0^\infty e^{-st} f(t)\, dt - f(0). \qquad \blacksquare$$

To find the Laplace transform of higher derivatives, it is only necessary to use Proposition 6.1.3 repeatedly. For example, if f'' also has a Laplace transform, then

$$L(f'') = sL(f') - f'(0)$$
$$= s[sL(f)) - f(0)] - f'(0)$$
$$= s^2 L(f) - sf(0) - f'(0).$$

Similarly, you can verify that, provided that f''' has a Laplace transform too,

$$L(f''') = s^3 F(s) - s^2 f(0) - sf'(0) - f''(0),$$

and, in general,

$$L(f^{(n)}) = s^n F(s) - s^{n-1} f(0) - \cdots - sf^{(n-2)}(0) - f^{(n-1)}(0)$$

if all the derivatives mentioned have Laplace transforms.

Proposition 6.1.3 has the following corollary, which will be useful in our study of electrical circuits.

Corollary 6.1.4

Let $f(t)$ be a function that has a Laplace transform, and let

$$g(t) = \int_0^t f(\tau)\, d\tau.$$

If g also has a Laplace transform, then

$$L(g) = \frac{1}{s} L(f).$$

The proof of Corollary 6.1.4 is Exercise 12 at the end of this section.

Now consider an IVP,

$$a_2 y'' + a_1 y' + a_0 y = f(t); \quad y(0) = b_0, \, y'(0) = b_1, \tag{6.5}$$

in which the a_i and b_i are constants, and $f(t)$ is a function that has a Laplace transform. By the existence and uniqueness theorem, this IVP has a unique solution $y(t)$. Let us assume that y, y', and y'' all have Laplace transforms and denote the Laplace transform of y by $Y(s)$. Let us take the Laplace transform of each side of the ODE part of (6.5):

$$L(a_2 y'' + a_1 y' + a_0 y) = L(f).$$

Since L is a linear operator and the coefficients of the ODE are constant,

$$a_2 L(y'') + a_1 L(y') + a_0 L(y) = L(f). \tag{6.6}$$

Let $F(s) = L(f)$. We can replace $L(y)$ in (6.6) with $Y(s)$, and, by Proposition 6.1.3, $L(y') = sY(s) - y(0)$ and $L(y'') = s^2 Y(s) - sy(0) - y'(0)$. Referring to the initial conditions, we know that $y(0) = b_0$ and $y'(0) = b_1$. Hence equation (6.6) can be rewritten as

$$a_2(s^2 Y(s) - s b_0 - b_1) + a_1(s Y(s) - b_0) + a_0 Y(s) = F(s). \tag{6.7}$$

This equation is not a differential equation and can be solved for $Y(s)$ by using simple algebra. We call equation (6.7) the Laplace transform of the IVP (6.5).

To find the solution $y(t)$ of the original IVP, it is necessary to compute the **inverse Laplace transform** of $Y(s)$. This is usually done by referring to a table of Laplace transforms such as the one on page 275. The following simple example shows how the procedure works. (With an IVP this simple, most people would not use the Laplace transform!)

Example 6.1.5 Use the Laplace transform method to solve the IVP

$$y' - k y = 0; \quad y(0) = 1,$$

where k is constant.

Solution. Let $y(t)$ denote the solution of the IVP, and $Y(s) = L(y)$. Then $L(y') = sY(s) - y(0)$. The initial condition says $y(0) = 1$, so $L(y') = sY - 1$. The transformed equation is therefore $sY - 1 - kY = 0$, or

$$(s - k)Y = 1.$$

It follows that $Y = \frac{1}{s-k}$. Now let's refer to Example 6.1.2, in which we found that $L(e^{kt}) = \frac{1}{s-k}$. We therefore conclude that the solution of the IVP is $y = e^{kt}$. ❑

How do we know that there isn't some function, other than e^{kt}, whose Laplace transform is $\frac{1}{s-k}$? If there were such a function, which we'll call $f(t)$, then the solution of the IVP in Example 6.1.5 could be either e^{kt} or $f(t)$. To avoid this ambiguity, we need the following uniqueness theorem for inverse Laplace transforms.

Theorem 6.1

Let $f_1(t)$ and $f_2(t)$ be continuous on $[0, \infty)$, and suppose that both functions have Laplace transforms, $F_1(s)$ and $F_2(s)$. If $F_1(s) = F_2(s)$ for all s, where both F_1 and F_2 are defined, then $f_1(t) = f_2(t)$ for all $t \geq 0$.

The proof of Theorem 6.1 is beyond the scope of this text.

EXERCISES

1. Find the Laplace transform of $f(t) = t - 2$.

2. Let $f(t) = \begin{cases} 0 & \text{if } 0 \leq t \leq 2 \\ t - 2 & \text{otherwise.} \end{cases}$ Calculate $L(f)$.

3. Calculate $L(|t - 4|)$.

4. Let $f(t) = \begin{cases} 0 & \text{if } t < a \\ 1 & \text{if } t \geq a. \end{cases}$ Calculate $L(f)$.

5. Find the Laplace transform of $f(t) = \begin{cases} 0 & \text{if } t < 1 \\ 1 & \text{if } 1 \leq t < 2 \\ 2 & \text{if } t \geq 2. \end{cases}$

6. Let $f(t) = \lfloor t \rfloor$ be the greatest integer function. Show that

$$L(f) = \sum_{n=0}^{\infty} \frac{n}{s}(e^{-ns} - e^{-(n+1)s}).$$

7. Which of the following functions have Laplace transforms?

   a. $f(t) = \dfrac{1}{t + 1}$

   b. $f(t) = \dfrac{\sin(\pi t)}{t - \lfloor t \rfloor}$, where $\lfloor t \rfloor$ is the greatest integer $\leq t$

   c. $f(t) = e^{3t}$ d. $f(t) = 2^t$

   e. $f(t) = e^{t^2}$ f. $f(t) = e^{-t^2}$

   g. $f(t) = t^3 \sinh t$ h. $f(t) = \lfloor t \rfloor!$

8. Show that a function that is piecewise continuous on $[0, \infty)$ is locally integrable.

   ◆ **Hint:** Start by proving that if f is continuous on (a, b) and that the limits

   $$\lim_{t \to a^+} f(t) \quad \text{and} \quad \lim_{t \to b^-} f(t)$$

   are each finite, then $\int_a^b f(t)\,dt$ exists and is finite.

9. Let $F(s)$ denote the Laplace transform of a locally integrable function $f(t)$ that is of exponential order. This exercise outlines the proof that $\lim_{s \to \infty} F(s) = 0$. Since $f(t)$ has a Laplace transform, there is a number s_1 such that

   $$F(s) = \int_0^{\infty} f(t)e^{-st}\,dt$$

   converges for all $s > s_1$. We will assume that $s_1 \geq 0$.

   a. Assuming that $f(t) \geq 0$ for all $t \geq 0$, show that for $s > s_1$, and $u > 0$,

   $$F(s) \leq \int_0^u f(t)\,dt + e^{-(s-s_1)u} \int_u^{\infty} e^{-s_1 t} f(t)\,dt.$$

   b. Let $G(u) = \int_0^u f(t)\,dt$ Show that $G(u)$ is continuous, and, for every $u > 0$, and $s > s_1$,

   $$F(s) \leq G(u) + e^{-(s-s_1)u} F(s_1).$$

   c. Given $\epsilon > 0$, show that $u > 0$ can be found such that $G(u) < \frac{\epsilon}{2}$.

   d. Show that there is a number M such that for all $s > M$,

   $$e^{-(s-s_1)u} F(s_1) < \frac{\epsilon}{2},$$

   and complete the proof for the case when $f(t) \geq 0$ for all $t \geq 0$.

   e. If $f(t)$ doesn't satisfy the assumption that $f(t) \geq 0$ for all $t \geq 0$, show that there are functions f_+ and f_-, both nonnegative, locally integrable and of exponential order, such that for all t, we have $f(t) = f_+(t) - f_-(t)$. Then appeal to the linearity of the Laplace transform to complete the proof.

10. It is known that $\cos(kt)$ is the solution of the IVP,

    $$y'' + k^2 y = 0; \quad y(0) = 1, \; y'(0) = 0.$$

    Transform this IVP to an algebraic equation in the transform domain, and solve that equation. You will thus obtain a formula for $L(\cos(kt))$. Refer to the table of Laplace transforms on page 275 to see if your answer is correct.

11. Use the method of Exercise 10 to find $L(\sin(kt))$.

12. Prove Corollary 6.1.4.

    ◆ **Hint:** $g'(t) = f(t)$.

13. Solve the differential-integral IVP,

$$y' = -4 \int_0^t y(\tau)\, d\tau; \ \ y(0) = 1.$$

You may find Corollary 6.1.4 and the result of Exercise 10 to be of use.

14. Solve the integral equation $y = 1 + \int_0^t y(u)\, du$.

15. Derive the following formula for the Laplace transform of

e^{At}, where A is an $n \times n$ constant matrix:

$$L(e^{At}) = (sI - A)^{-1}.$$

◆ **Hint:** e^{At} is the solution of the IVP,

$$\frac{d}{dt}\mathcal{X} - A\mathcal{X} = 0; \ \ \mathcal{X}(0) = I,$$

where the dependent variable, $\mathcal{X}$, is an $n \times n$ matrix function of t.

6.2 Electrical Circuits

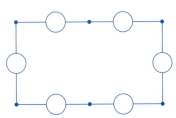

Figure 6.1 Arrangement of components in a series circuit. The circles represent the components, and the heavy dots are nodes.

A **series circuit** is a collection of electrical components connected by conducting wires so as to form a simple loop, as in Figure 6.1. The components that we will consider are the ones that can be modeled by linear differential equations: *capacitors, resistors, choke coils,* and various *sources of electromotive force,* such as batteries or generators.

Capacitors, resistors, and choke coils

A **capacitor** is a device for storing electrical charge. It can be thought of as a reservoir that holds electrons. It actually consists of two plates, separated by a thin electrical insulator. One plate has an excess of electrons, and the other has a deficit. The charge imbalance is due to a difference of **electrical potential** between the two plates of the capacitor. Electrical potential at a point in a circuit is roughly analogous to water pressure in a reservoir. In the circuit shown in Figure 6.2 it is maintained by the source of electromotive force, which acts as an electron pump. In a capacitor, the number of excess electrons on the negative plate is proportional to the electrical potential difference between the plates.

Figure 6.2 A circuit containing a capacitor **C** and a battery **E**. **A** and **B** are nodes.

Let x denote the charge held by the capacitor, and let $V(\mathbf{A})$ and $V(\mathbf{B})$ denote the electrical potential at the two plates **A** and **B** of the capacitor. **Faraday's law** expresses the proportionality between the potential difference and the charge held:

$$x = C[V(\mathbf{B}) - V(\mathbf{A})].$$

The constant C is called the **capacitance** of the capacitor. Faraday's law is named for the English physicist Michael Faraday.

Electrical charge is measured in **coulombs,** with each coulomb representing the charge of 6.24146×10^{18} electrons, and electrical potential is measured in **volts**. The unit of capacitance is the **farad.** If the capacitance of a capacitor is 1 farad, then a potential difference of 1 volt will cause the capacitor to hold a charge of 1 coulomb. If, in Figure 6.2, the potential difference due to the battery is V volts, then Faraday's law would say that $x = CV$ coulombs.

Let us now turn to another component found in almost all electrical circuits: the resistor. Just as the potential drop between the nodes of a capacitor is related to the charge, the potential drop across the nodes of a resistor is related to the **current**. An electrical current is a flow electrons in an electrical conductor. The rate of flow is measured in **amperes**; 1 ampere is a rate of 1 coulomb per second.

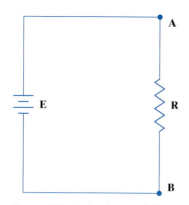

Figure 6.3 A circuit containing a resistor **R** and a battery **E**. **A** and **B** are nodes.

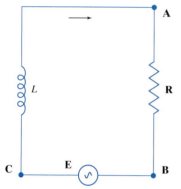

Figure 6.4 An *RL* series circuit. The points labeled **A**, **B**, and **C** are nodes, and the arrow indicates that a current flowing clockwise is to be considered positive. The inductor is labeled *L* and the resistor is labeled **R**. The source of electromotive force is a generator (labeled **E**).

Figure 6.3 is a circuit diagram for a series circuit containing a resistor and a battery. **Ohm's law**, published in 1827 by the German physicist G. S. Ohm, states that the potential difference $V(\mathbf{B}) - V(\mathbf{A})$ between the ends of a resistor is proportional to the current. The constant of proportionality, denoted R, is the **resistance**. Thus,

$$V(\mathbf{B}) - V(\mathbf{A}) = Ry,$$

where y is the current, in amperes. Resistance is measured in **ohms.** If a 1-ampere current flows in a 1 ohm resistor, there will be a potential difference of 1 volt.

The third type of circuit component that we will use is the **choke coil**. The potential drop across the nodes of a choke coil is related to the rate of change of current. In 1831 the American physicist J. Henry showed that a coil of wire carrying a variable current $y(t)$ produces a potential difference proportional to the $\frac{dy}{dt}$. The constant of proportionality, denoted L, is the **inductance**. Figure 6.4 is a diagram of an RL circuit; that is, a circuit that contains a choke coil (the element labelled L). **Henry's law** states that the potential difference between two terminals **C** and **B** of an inductor is proportional to the rate of change of the current. Thus,

$$V(\mathbf{C}) - V(\mathbf{B}) = L\frac{dy}{dt},$$

where y is the current carried by the choke coil. Inductance is measured in **henries**. A choke coil with $L = 1$ henry will produce a potential difference of 1 volt if $dy/dt = 1$ ampere per second.

ODE models of electrical circuits

An *RLC* circuit is an electrical circuit that involves just capacitors, resistors, inductors, and a source of electromotive force. Each component of the circuit is governed by an equation relating the potential difference across its nodes to its charge, current, or derivative of current. These equations can be combined to obtain an ODE model for the circuit by using **Kirchhoff's laws**, which are stated as follows.

   I. The net sum of currents entering any node of a circuit is 0.
   II. In each simple loop of a circuit, the sum of the potential differences across the elements of the loop is equal to 0.

These laws are named for a German physicist, G. R. Kirchhoff, who published them in 1845.

The first law is only used in parallel circuits, which have nodes where three or more conductors are joined. In a series circuit, which forms a simple loop, only the second law is needed.

Let us consider a series circuit containing one capacitor of capacitance C, one resistor with resistance R, one choke coil of inductance L, and an electromotive force $E(t)$. We are going to derive a differential-integral equation in which the dependent variable is y, the current.

In any series circuit, the rate at which charge flows into the capacitor is equal to the current, since by definition, the current is the rate of flow of charge. Thus, if x denotes the charge held by the capacitor,

$$y = \frac{dx}{dt}.$$

We can also express this relation as

$$x = x_0 + \int_0^t y(\tau)\,d\tau, \tag{6.8}$$

where x_0 is the charge held by the capacitor when $t = 0$.

If we refer to the *RLC* circuit shown in Figure 6.5 and apply Faraday's law, we see that

$$x = C[V(\mathbf{P}) - V(\mathbf{B})].$$

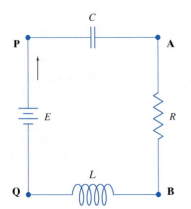

By Ohm's law,

$$V(\mathbf{B}) - V(\mathbf{A}) = R\,y,$$

and by Henry's law,

$$V(\mathbf{A}) - V(\mathbf{Q}) = L\,y'.$$

Figure 6.5 A circuit containing a resistor R, a choke coil L, a capacitor C, and a battery E. **A**, **B**, **Q**, and **P** are nodes.

Finally, $V(\mathbf{Q}) - V(\mathbf{P}) = -E$. By Kirchhoff's second law,

$$(V(\mathbf{P}) - V(\mathbf{B})) + (V(\mathbf{B}) - V(\mathbf{A})) + (V(\mathbf{A}) - V(\mathbf{Q})) + (V(\mathbf{Q}) - V(\mathbf{P})) = 0.$$

The laws of Faraday, Ohm, and Henry transform this equation to

$$\frac{1}{C}x + R\,y + L\,y' - E = 0.$$

We can combine this result with equation (6.8) to obtain the following mathematical model for the current in the circuit:

$$Ly' + Ry + \frac{1}{C}\left(x_0 + \int_0^t y(\tau)\,d\tau\right) = E(t). \tag{6.9}$$

Example 6.2.1   A circuit consists of a 0.001-farad capacitor and a choke coil with inductance 0.1 henry. The circuit is superconducting; that is, the resistance is 0. There is no electromotive force. If the capacitor has an initial charge of 10^{-4} and the initial current is 0, find the current as a function of time.

Solution. If we put the parameters $L = 0.1$, $R = 0$, $C^{-1} = 1000$, $x_0 = 10^{-4}$, and $E(t) \equiv 0$ in equation (6.9), we have

$$0.1y' + 1000\left(10^{-4} + \int_0^t y(\tau)\,d\tau\right) = 0.$$

By linearity of the Laplace transform,

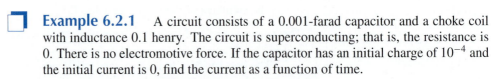

$$0.1L(y') + 1000\left(10^{-4}\frac{1}{s} + L\left[\int_0^t y(\tau)\,d\tau\right]\right) = 0,$$

with initial condition $y(0) = 0$. Assume that the solution $y(t)$ of this IVP, as well as its derivative and antiderivatives have Laplace transforms. Using Proposition 6.1.3 and Corollary 6.1.4, we have

$$0.1sY - y(0) + \frac{1}{10s} + 1000\frac{1}{s}Y = 0,$$

where we are given that $y(0) = 0$. This equation can be solved for Y to obtain

$$Y = -\frac{1}{s^2 + 10^4}.$$

In Section 6.3 we will show that $L[\sin(kt)] = \frac{k}{s^2+k^2}$. Since the denominator in our expression for $Y(s)$ is $s^2 + 10^4 = s^2 + 100^2$, we can put $k = 100$. Thus

$$Y = -\frac{1}{100}L[\sin(100t)].$$

Thus, the current of this circuit is $y = -\frac{1}{100}\sin(100t)$.   ❑

Example 6.2.2   In an *RLC* series circuit, $R = 5$ ohms, $L = 0.1$ henry, and $C = 0.01$ farad. The circuit is connected to a 12-volt battery, and at time 0, the capacitor holds no charge, and the current is 0. Determine the current as a function of time.

Solution. Putting the given parameters in equation (6.9), we obtain the IVP,

$$0.1y' + 5y + 100 \int_0^t y(\tau)\,d\tau = 12; \quad y(0) = 0.$$

Now take the Laplace transform, with $Y = L(y)$, $sY = L(y')$,

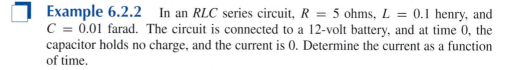

$$\frac{1}{s}Y = L\left(\int_0^t y(\tau)\,d\tau\right),$$

and $12/s = L(12)$. The result is

$$0.1sY + 5Y + \frac{100}{s}Y = \frac{12}{s},$$

which can be solved for Y as

$$Y = \frac{120}{s^2 + 50s + 1000}.$$

The method for finding the inverse Laplace transform of this expression is presented in Section 6.3. The result is

$$y = \frac{120}{\sqrt{375}}e^{-25t}\sin(\sqrt{375}t).$$

This current is oscillating (it behaves as an underdamped mechanical system does) and transient. Very soon, it will be negligible. When $y = y' = 0$, we can ignore the resistor and the coil: Only the battery and the capacitor count, and we have $x = CE = 0.12$ coulomb.   ❑

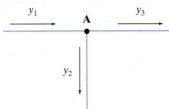

Figure 6.6 A node in a parallel circuit.

Figure 6.6 shows a node in a parallel circuit. The arrows indicate the positive direction of flow for each current, so when a given current has a negative value, it flows in the direction opposite to the arrow. For example, if $y_1 < 0$, the current flows to the left from the node. Kirchhoff's first law tells us that the currents satisfy the equation

$$y_2 = y_1 - y_3. \tag{6.10}$$

Now consider a parallel circuit involving node **A** as shown in Figure 6.7. The potential differences between the nodes of the circuit are calculated as follows. By *Ohm's law*,

$$V(\mathbf{X}) - V(\mathbf{Y}) = R_1 y_1,$$
$$V(\mathbf{A}) - V(\mathbf{B}) = R_2 y_2, \quad \text{and}$$
$$V(\mathbf{H}) - V(\mathbf{K}) = R_3 y_3.$$

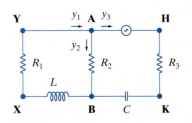

Figure 6.7. A parallel circuit.

By *Henry's law*,

$$V(\mathbf{B}) - V(\mathbf{X}) = L y_1',$$

and, finally, by *Faraday's law*,

$$V(\mathbf{K}) - V(\mathbf{B}) = \frac{1}{C} x,$$

where x denotes the charge held by the capacitor.

Since no component intervenes between the nodes **A** and **Y**, $V(\mathbf{A}) = V(\mathbf{Y})$. Kirchhoff's second law, applied to the left loop of the circuit, yields

$$(V(\mathbf{Y}) - V(\mathbf{X})) + (V(\mathbf{B}) - V(\mathbf{A})) + (V(\mathbf{X}) - V(\mathbf{B})) = 0,$$

or

$$R_1 y_1 + R_2 y_2 + L y_1' = 0. \tag{6.11}$$

Turning to the right loop, we have

$$(V(\mathbf{H}) - V(\mathbf{K})) + (V(\mathbf{K}) - V(\mathbf{B})) + (V(\mathbf{B}) - V(\mathbf{A})) + (V(\mathbf{A}) - V(\mathbf{H})),$$

or

$$R_3 y_3 + \frac{1}{C} x - R_2 y_2 - E(t) = 0. \tag{6.12}$$

We can use the relation $x' = y_3$ to eliminate y_3 from equations equations (6.10) and (6.12). Equation (6.10) can then be used to eliminate y_2 from equations (6.11) and (6.12). The resulting system, with dependent variables y_1 and x, is

$$R_1 y_1 + L y_1' + R_2(y_1 - x') = 0$$
$$R_3 x' + \frac{1}{C} x - R_2(y_1 - x') = E(t).$$

These equations can be simplified as

$$\left. \begin{array}{l} L y_1' - R_2 x' + (R_1 + R_2) y_1 = 0 \\[2mm] (R_2 + R_3) x' - R_2 y_1 + \frac{1}{C} x = E(t). \end{array} \right\} \tag{6.13}$$

Example 6.2.3   In the circuit shown in Figure 6.7, the resistances, measured in ohms, are $R_1 = 3 \times 10^{-3}$, $R_2 = 2 \times 10^{-3}$, and $R_3 = 200$. The inductance is $L = .05$ henries, and the capacitance is $C = .05$ farads. These parameters are subject to an error of not more than 1%. The electromotive force is $E(t) = 100 \sin(10t)$, and, initially, when $t = 0$, the capacitor is discharged and current is not flowing. Write down an initial value problem that will serve as a mathematical model for the circuit, and solve it by using the Laplace transform.

Solution. The coefficient of x' in the second equation of the system (6.13) is $R_2 + R_3$. We will consider R_2 to be negligible here, since it is less than the 2 ohm error that is possible in R_3. R_2 may not be negligible as the coefficient of y_1. The system (6.13), with the given electrical parameters, is

$$\left. \begin{aligned} 0.05 y_1' - 0.002 x' + 0.005 y_1 &= 0 \\ 200 x' - 0.002 y_1 + 20 x &= 100 \sin(10t). \end{aligned} \right\} \tag{6.14}$$

The initial conditions are $y_1(0) = 0$ and $x(0) = 0$.

Let $X = L(x)$ and $Y = L(y_1)$ be the Laplace transforms of the variables in (6.14); then $sX = L(x')$ and $sY = L(y_1')$. Since $L(100 \sin(10t)) = 1000/(s^2 + 100)$, the transformed system is

$$-0.002 s X + (0.05s + 0.005) Y = 0$$

$$(200s + 20) X - 0.002 Y = \frac{1000}{s^2 + 100}.$$

The solution of this system is

$$X = -\frac{1000(0.05s + 0.005)}{(s^2 + 100)[(0.002)^2 s - (0.05s + 0.005)(200s + 20)]}$$

$$\approx \frac{5}{(s^2 + 100)(s + .1)},$$

where we have treated the term $(0.002)^2 s$ in the denominator as negligible. Similarly,

$$Y \approx \frac{.2s}{(s^2 + 100)(s + 0.1)^2}.$$

The final step is to take inverse Laplace transforms. The procedure for this task are presented in the next two sections. To obtain a quick answer now, we will use a CAS to find the inverse transform. The results are as follows (I have rounded coefficients to one significant digit):

$$x = 0.05 \, e^{-0.1t} - 0.05 \, \cos(10t) + 0.0005 \, \sin(10t)$$

$$y_1 = 0.002 \, e^{-0.1t} - 0.0002 \, t \, e^{-0.1t} - 0.002 \, \cos(10t) + 0.00004 \, \sin(10t).$$

Thus, the charge and current oscillate with a frequency of 10 radians per second, after the decay of some transients. ❑

EXERCISES

1. An *RLC* series circuit has $R = 2$ ohms and $L = 0.1$ henry. At time zero, it is connected to a battery with a constant electromotive force of 1 volt. The capacitor has no initial charge. Find the current $y(t)$ if the capacitor has capacitance

   a. 0.15625 farad   b. 0.1 farad   c. 0.08 farad

2. In the *RLC* series circuit of Exercise 1, find in each case, $\lim_{t\to\infty} x(t)$, where $x(t)$ is the charge held by the capacitor, in coulombs.

3. An *RLC* series circuit produces an oscillating current with a frequency of $10/\pi$ cycles per second (20 radians per second) when connected to a battery. Given that the resistance is 0.2 ohm and the inductance is 0.1 henry, calculate the capacitance.

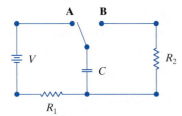

FIGURE 6.8   See Exercise 4.
$V = 6$ volts, $C = 10^{-4}$ farad,
$R_1 = 10^4$ ohms, and $R_2 = 1$ ohm.

4. The circuit diagram in Figure 6.8 represents a device that can produce a remarkably large current for a short period of time. It is a simplified version of an electronic flash used by photographers. The switch is held at node **A** for 2 seconds, and then moved to node **B**. Calculate the current that flows through the resistor on the right side after the switch is moved to node **B**, assuming that, before the process begins, the capacitor holds no charge.

   ✔ *Answer:* $y = -0.8e^{-10000(t-2)}$ amperes

5. An *RLC* series circuit has the following parameters. Derive an ODE that will serve as a model for the circuit. Solve the equation, always assuming that at $t = 0$, no current is flowing, and the capacitor is not charged. You may omit the inverse Laplace transform step, but you should use an IVP solver to plot the graph of the solution if you do.

   a. $R = 7.5$ ohms, $L = 0.1$ henry, $C = 10^{-2}$ farad, and EMF (electromotive force) $= 1000$ volts. Graph the current in the circuit as a function of time, and determine its peak value. Describe the behavior of the system after a long time (5 seconds) has elapsed.

   b. $R = 7.5$ ohms, $L = 0.1$ henry, $C = 10^{-2}$ farad, and EMF $= 1000 \sin(32t)$ volts. Plot graphs of the charge $x(t)$ and the current $y(t)$ as functions of time over the

   intervals $0 \le t \le 0.5$ second, and $4.5 \le t \le 5$ seconds. Does this system settle into a periodic solution? If you change the initial conditions, does that alter the periodic solution that is observed after time has elapsed? Compare the periods of the current and capacitor charge with that of the EMF.

   c. $R = 0.75$ ohms, $L = 0.1$ henry, $C = 10^{-2}$ farad, and EMF $= 1000 \sin(32t)$ volts. How does the reduced resistance affect the periodic solution? If the amplitude is magnified, what is the magnification factor?

6. a. Produce graphs of the current and capacitor charge for the solution of Example 6.2.3 covering the time interval $0 \le t \le 5$. Does it appear that either or both of these are periodic? (You may have to play with the vertical scale of the graphs to see clearly what's going on.)

   b. Some of the coefficients in the system (6.14) are rather small in magnitude. Starting with the coefficient of y_1 in the first equation, eliminate the coefficients in increasing order of magnitude until you observe a change in one of the solution functions $y_1(t)$ or $x(t)$.

   c. Suppose that the capacitor has an initial charge of 0.01 coulomb. How does this affect the solution?

7. Derive systems of two first-order equations modeling the circuits diagrammed below. In each case, the dependent variables should be the current y carried by the choke coil, and the charge x held by the capacitor.

   a.

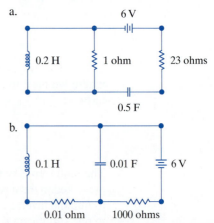

   b.

8. Solve each of the systems in Exercise 7, assuming homogeneous initial conditions. You are permitted to use a CAS to take the inverse Laplace transforms if you wish, or to use an IVP solver to graph the solutions. Graph the current and charge functions on a time scale of $0 \le t \le 5$. Adjust the vertical dimension of each graph to display it clearly. What distinguishes the two circuits from one another?

6.3 Properties of the Laplace Transform

In this section, we will develop a table of Laplace transforms that will greatly simplify calculations. It will almost never be necessary to calculate a Laplace transform by evaluating the integral as we did in Section 6.1. We will start with the Laplace transform of t^n.

Proposition 6.3.1

Let n be a nonnegative integer.

$$L(t^n) = \frac{n!}{s^{n+1}}.$$

Proof Substitute $u = st$ in the integral

$$\int_0^\infty t^n e^{-st}\, dt = L(t^n).$$

Since $t = u/s$, $dt = du/s$, and u runs from 0 to ∞ as t runs from 0 to ∞,

$$L(t^n) = \int_0^\infty \left(\frac{u}{s}\right)^n e^{-u} \left(\frac{du}{s}\right)$$

$$= \frac{1}{s^{n+1}} \int_0^\infty u^n e^{-u}\, du.$$

To complete the calculation, we just have to show that

$$\int_0^\infty u^n e^{-u}\, du = n!. \tag{6.15}$$

This is done by mathematical induction. For $n = 0$, we have $0! = 1$ by convention, and $\int_0^\infty e^{-u}\, du = e^0 = 1$. This starts the induction. To advance it we use the formula

$$\int_0^\infty u^{n+1} e^{-u}\, du = (n+1) \int_0^\infty u^n e^{-u}\, du, \tag{6.16}$$

which holds for any real $n > -1$. The proof of this formula is Exercise 25 at the end of this section. If equation (6.15) holds for an integer $n \geq 0$, then by formula (6.16),

$$\int_0^\infty u^{n+1} e^{-u}\, du = (n+1) \int_0^\infty u^n e^{-u}\, du = (n+1)n! = (n+1)!.$$

If follows that (6.15) holds for $n + 1$. By the principle of mathematical induction, it holds for all integers $n \geq 0$. ■

We can now compute the Laplace transform of any polynomial.

Example 6.3.1   Find the Laplace transform of

$$f(t) = t^4 - \frac{1}{2}t^3 + 40t^2 - 5t + 1.$$

Solution. Using Proposition 6.3.1, $L(t^4) = \frac{24}{s^5}$, $L(t^3) = \frac{6}{s^4}$, $L(t^2) = \frac{2}{s^3}$, $L(t) = \frac{1}{s^2}$, and $L(1) = \frac{1}{s}$. Since the Laplace transform is a linear operator,

$$L(f) = \frac{24}{s^5} + \frac{1}{2}\frac{6}{s^4} + 40\frac{2}{s^3} - 5\frac{1}{s^2} + \frac{1}{s}.$$

In our calculation of $L(t^n)$, it is not necessary for n be an integer. Euler was the first to notice that formula (6.15) can be used to define factorials of numbers that are not integers. To avoid confusion, Euler defined a function, known as the **gamma function** by

$$\Gamma(z) = \int_0^\infty u^{z-1} e^{-u}\, du$$

for $z > 0$. Thus, for any integer $n \geq 0$, $n! = \Gamma(n+1)$. Formula (6.16) can be translated to an identity for the gamma function, $\Gamma(z+1) = z\Gamma(z)$ (this is Exercise 26).

In terms of the gamma function, Proposition 6.3.1 can be rephrased as an identity: For all real $n > -1$,

$$L(t^n) = \frac{\Gamma(n+1)}{s^{n+1}}.$$

For example, it is known that $\Gamma\left(\frac{1}{2}\right) = \sqrt{\pi}$; hence we can calculate the Laplace transform of $1/\sqrt{t}$ as follows:

$$L\left(t^{-1/2}\right) = \frac{\Gamma(1/2)}{s^{1/2}} = \sqrt{\frac{\pi}{s}}.$$

In Example 6.1.2 on page 255, we found that $L(e^{kt}) = 1/(s-k)$. The following proposition gives a generalization of this fact, which is the special case $f(t) = 1$, and $F(s) = 1/s$.

Proposition 6.3.2
_____

Let $F(s)$ denote the Laplace transform of a function $f(t)$. Then $L[e^{kt} f(t)] = F(s-k)$.

Proof

$$L[e^{kt} f(t)] = \int_0^\infty e^{-st} e^{kt} f(t)\, dt$$

$$= \int_0^\infty e^{-(s-k)t} f(t)\, dt$$

$$= F(s-k)$$

Example 6.3.2   Find the Laplace transform of $g(t) = t^2 e^{3t}$.

Solution. Let $F(s) = L(t^2) = 2/s^3$. Then

$$L(g) = F(s - 3) = \frac{2}{(s - 3)^3}.$$

If $f(t)$ is a complex-valued function—that is, $f(t) = g(t) + ih(t)$, where g and h are real valued—then $f(t)$ will have a Laplace transform if $g(t)$ and $h(t)$ do. Since the Laplace transform is a linear operator,

$$L(f) = L(g) + iL(h).$$

This observation allows us to treat the sine and cosine functions as the imaginary and real parts of the complex exponential function.

Example 6.3.3   Find the Laplace transforms of

  a. $\sin(bt)$ b. $\cos(bt)$

Solution. Put $f(t) = e^{ibt}$. Then, if $F(s)$ denotes the Laplace transform of $f(t)$,

$$
\begin{aligned}
F(s) &= \frac{1}{s - ib} \\
&= \frac{s + ib}{s^2 + b^2} \\
&= \frac{s}{s^2 + b^2} + i\frac{b}{s^2 + b^2}.
\end{aligned}
$$

Therefore,

$$L[\cos(bt)] = \operatorname{Re}[F(s)] = \frac{s}{s^2 + b^2},$$

and

$$L[\sin(bt)] = \operatorname{Im}[F(s)] = \frac{b}{s^2 + b^2}.$$

The following proposition is directed toward calculating inverse Laplace transforms.

Proposition 6.3.3

  i. $L[e^{kt}\cos(bt)] = \dfrac{s - k}{(s - k)^2 + b^2}$ ii. $L[e^{kt}\sin(bt)] = \dfrac{b}{(s - k)^2 + b^2}$

  iii. $L[e^{kt}\cosh(bt)] = \dfrac{s - k}{(s - k)^2 - b^2}$ iv. $L[e^{kt}\sinh(bt)] = \dfrac{b}{(s - k)^2 - b^2}$

Proof We will prove formulas (i) and (iii). The other two formulas can be derived in the same way. Let

$$F(s) = L[\cos(bt)] = \frac{s}{s^2 + b^2}$$

(see Example 6.3.3). By Proposition 6.3.2,

$$L[e^{kt} \cos(bt)] = F(s - k) = \frac{s - k}{(s - k)^2 + b^2}.$$

Now let $G(s) = L[\cosh(bt)]$. By the linearity property,

$$G(s) = \frac{1}{2}[L(e^{bt}) + L(e^{-bt})]$$

$$= \frac{1}{2}\left[\frac{1}{s - b} + \frac{1}{s + b}\right]$$

$$= \frac{s}{s^2 - b^2}.$$

Again, by Proposition 6.3.2

$$L[e^{kt} \cosh(bt)] = G(s - k) = \frac{s - k}{(s - k)^2 - b^2}. \qquad \blacksquare$$

Proposition 6.3.3 can be used to calculate the inverse Laplace transform of certain *rational functions* $F(s) = \frac{p(s)}{q(s)}$, where $p(s)$ and $q(s)$ are polynomials. The denominator $q(s)$ must be quadratic, because the denominators of the expressions in Proposition 6.3.3 are quadratic, while the numerator, $p(s)$, must be of degree 0 or 1. By completing the square, we can put $q(s)$ in the form $(s - k)^2 \pm b^2$. If the $\pm$ sign is $+$, we will refer to formulas (i) and (ii); if it is $-$, formulas (iii) and (iv) will be used.

Assuming that $q(s) = (s - k)^2 + b^2$ to be specific, let

$$F(s) = \frac{as + c}{(s - k)^2 + b^2}.$$

In Example 6.3.4, it is shown how to find coefficients A and C such that $A(s - k) + Cb = as + c$. It then follows that

$$L^{-1}(F) = e^{kt}[A \cos(bt) + C \sin(bt)],$$

where $L^{-1}(F)$ denotes the inverse Laplace transform of the function $F(s)$.

Example 6.3.4   Find the inverse Laplace transform of

$$F(s) = \frac{3s - 7}{s^2 + 8s + 13}.$$

Solution. Complete the square for $q(s) = s^2 + 8s + 13$: $q(s) = (s + 4)^2 - 3$. It follows that $k = -4$ and $b = \sqrt{3}$. Now we turn to finding coefficients A and C such that

$$F(s) = A \frac{s + 4}{(s + 4)^2 - 3} + C \frac{\sqrt{3}}{(s + 4)^2 - 3}.$$

The coefficients must satisfy the equation

$$A(s + 4) + C\sqrt{3} = 3s - 7.$$

Equating coefficients of s, $A = 3$. Thus $C\sqrt{3} = -7 - 4A$, and $C = -19/\sqrt{3}$. We conclude that

$$L^{-1}(F) = e^{-4t}\left(3\cosh(\sqrt{3}t) - \frac{19}{\sqrt{3}}\sinh(\sqrt{3}t)\right). \qquad \square$$

Example 6.3.5   Use the Laplace transform to solve the IVP

$$y'' + 6y' + 18y = 0; \quad y(0) = 2, \quad y'(0) = -3.$$

Solution. Let $y(t)$ denote the solution of the IVP, and put $Y(s) = L(y)$. Then $L(y') = sY - y(0)$ and $L(y'') = s^2 Y - sy(0) - y'(0)$. The initial conditions are $y(0) = 2$ and $y'(0) = -3$; hence $L(y') = sY - 2$ and $L(y'') = s^2 Y - 2s + 3$. The Laplace transform of the ODE is

$$(s^2 Y - 2s + 3) + 6(sY - 2) + 18Y = 0.$$

Solving for Y,

$$Y = \frac{2s - 3 + 12}{s^2 + 6s + 18}.$$

To complete the solution, we must take the *inverse Laplace transform.* By completing the square in the denominator in our expression for Y we find

$$Y = \frac{2s + 9}{(s + 3)^2 + 9} = A \frac{s + 3}{(s + 3)^2 + 9} + B \frac{3}{(s + 3)^2 + 9}.$$

By Proposition 6.3.3, (i) and (ii), it follows that the inverse Laplace transform of Y is

$$y = e^{-3t}(A\cos(3t) + B\sin(3t)),$$

where the constants A and B are determined by

$$A(s + 3) + B(3) = 2s + 9.$$

Equate the coefficients of s to see that $A = 2$. Then $2s + 6 + 3B = 2s + 9$ so $B = 1$. It follows that the solution of the IVP is

$$y = e^{-3t}(2\cos(3t) + \sin(3t)). \qquad \square$$

The formula for the Laplace transform of the higher derivatives of a function makes it possible to solve differential equations of order three or more, as in the following example.

Example 6.3.6 Use the Laplace transform to solve of the initial value problem

$$y''' - 3y'' + 3y' - y = t^2 e^t; \quad y(0) = 0, \quad y'(0) = 0, \quad y''(0) = 0.$$

Solution. Let $Y(s) = L(y)$. Then because of the homogeneous initial conditions, $L(y') = sY$, $L(y'') = s^2Y$, and $L(y''') = s^3Y$. It follows that

$$L(y''' - 3y'' + 3y' - y) = s^3Y - 3s^2Y + 3sY - Y$$
$$= (s - 1)^3 Y.$$

The Laplace transform of the differential equation is

$$(s - 1)^3 Y(s) = L[t^2 e^t]$$
$$= \frac{2}{(s - 1)^3}.$$

It follows that $Y(s) = 2/(s - 1)^6$. Since $L(t^5 e^t) = 5!/(s - 1)^6$,

$$y = \frac{2}{5!} t^5 e^t = \frac{1}{60} t^5 e^t.$$ ❑

In the following example, we will use the Laplace transform to solve a system of linear equations with constant coefficients.

Example 6.3.7 Solve the initial value problem

$$x' = x - y; \quad x(0) = 0$$
$$y' = 2x - y; \quad y(0) = -1.$$

Solution. Let $X(s) = L(x)$ and $Y(s) = L(y)$. Using the initial conditions and Proposition 6.1.3, we see that

$$L(x') = sX \text{ and } L(y') = sY + 1.$$

Thus X and Y satisfy the *algebraic* system of equations

$$sX = X - Y$$
$$sY + 1 = 2X - Y.$$

Let us simplify the equations:

$$\left. \begin{array}{r} (s - 1)X + \quad\quad Y = \quad 0 \\ -2X + (s + 1)Y = -1. \end{array} \right\} \tag{6.17}$$

The system (6.17) can be solved by elimination. Multiply the first equation by $s + 1$ and subtract the second equation to obtain $(s^2 + 1)X = 1$. Thus we find that $X = \frac{1}{s^2+1}$. To determine Y, note that the first equation tells us that $Y = -(s-1)X = -\frac{s-1}{s^2+1}$.

Taking the inverse Laplace transform of X is easy: By Example 6.3.3, $L[\sin(t)] = \frac{1}{s^2+1}$. It follows that $x(t) = \sin(t)$. To find the inverse Laplace transform of Y, we break it into two fractions,

$$Y = \frac{1}{s^2 + 1} - \frac{s}{s^2 + 1},$$

and refer to Example 6.3.3 again. We find that $y(t) = \sin(t) - \cos(t)$. ❑

The following table is a summary of the facts about the Laplace transform that we have derived so far. In the table, $f(t)$ denotes an arbitrary function that has a Laplace transform; $F(s)$ denotes the Laplace transform of $f(t)$.

Table 6.1 Table of Laplace transforms.

Function	Transform	Function	Transform
1	$\dfrac{1}{s}$	$tf(t)$	$-F'(s)$
t	$\dfrac{1}{s^2}$	$e^{kt} f(t)$	$F(s - k)$
t^2	$\dfrac{2}{s^3}$	$f'(t)$	$sF(s) - f(0)$
t^n	$\dfrac{n!}{s^{n+1}} = \dfrac{\Gamma(n+1)}{s^{n+1}}$	$f''(t)$	$s^2 F(s) - sf(0) - f'(0)$
e^{kt}	$\dfrac{1}{s - k}$	$f^{(n)}(t)$	$s^n F(s) - s^{n-1} f(0) - \cdots - f^{(n-1)}(0)$
$f(kt)$	$\dfrac{1}{k}F\left(\dfrac{s}{k}\right)$	$\displaystyle\int_0^t f(u)\, du$	$\dfrac{1}{s}F(s)$
$\cos(bt)$	$\dfrac{s}{s^2 + b^2}$	$e^{kt} \cos(bt)$	$\dfrac{s - k}{(s - k)^2 + b^2}$
$\sin(bt)$	$\dfrac{b}{s^2 + b^2}$	$e^{kt} \sin(bt)$	$\dfrac{b}{(s - k)^2 + b^2}$
$\cosh(bt)$	$\dfrac{s}{s^2 - b^2}$	$e^{kt} \cosh(bt)$	$\dfrac{s - k}{(s - k)^2 - b^2}$
$\sinh(bt)$	$\dfrac{b}{s^2 - b^2}$	$e^{kt} \sinh(bt)$	$\dfrac{b}{(s - k)^2 - b^2}$

A larger table appears at the end of this chapter, along with references to tables that professionals would use in their work.

EXERCISES

In Exercises 1–24, calculate the Laplace transform.

1. $L(t^3 + 5t^2 + 2t - 1)$ **2.** $L[(t+2)^3]$

3. $L(te^t)$ **4.** $L(t - e^{-t})$

5. $L[\sinh(t)]$ **6.** $L[\cos(5t)]$

7. $L[e^t \sin(2t)]$ **8.** $L[t\cos(5t)]$

9. $L[te^t \sin(2t)]$ **10.** $L(t^2 e^t)$

11. $f(t) = e^{-3t}(t^2 - 3t + 5)$

12. $f(t) = (t - 5)\sin(2t)$ **13.** $f(t) = (te^t)^3$

14. $f(t) = t^2 \sin(9t)$ **15.** $f(t) = t^{100} e^{7t}$

16. $f(t) = \sin(3t)\cos(8t)$.

◆ **Hint:** $\sin(u)\cos(v) = \frac{1}{2}(\sin(u+v) + \sin(u-v))$

17. $f(t) = \sin(t)\sinh(t)$ **18.** $f(t) = t\sinh(\sqrt{2}t)$

19. $f(t) = \int_0^t ue^{3u}\cos(5u)\,du$ **20.** $f(t) = \sin^2(t)$

21. $f(t) = [t + \sin(t)]^2$ **22.** $f(t) = [\sin(t) + \cos(t)]^2$

23. $f(t) = \dfrac{d}{dt}\{(t^2 + 3t - 6)e^{3t}\}$ **24.** $L[(3 + e^t)^3]$

25. Prove that formula (6.16) holds for all $n > -1$ by integrating by parts.

26. Prove that for all $z > 0$, $\Gamma(z + 1) = z\Gamma(z)$.

27. Find $L(\sqrt{t})$.

28. Calculate $L([te^{-t}]^{1/2})$.   **29.** Calculate $L(\sin t/\sqrt{t})$.

30. The Maclaurin series for the exponential function is

$$e^t = \sum_{n=0}^{\infty} \frac{t^n}{n!}.$$

Take the Laplace transform of each term of the series, and show that the result is a geometric series whose sum is $1/(s-1)$.

Use the Laplace transform method to solve each of the initial value problems in Exercises 31–42.

31. $y'' + 4y' + 4y = 0$; $y(0) = 0$, $y'(0) = 1$

32. $y'' + 2y' + y = te^{-t}$; $y(0) = 0$, $y'(0) = 1$

33. $y'' + 2y' + 2y = 0$; $y(0) = 1$, $y'(0) = -1$

34. $y''' - 3y'' + 3y' - y = e^t$; homogeneous initial conditions at $t = 0$

35. $y'' + 8y' + 25y = 0$; $y(0) = 1$, $y'(0) = 3$

36. $y'' + 4y' + 3y = 2e^{-t}$; $y(0) = 0$, $y'(0) = 1$

37. $y''' + y'' - 2y = 0$; $y(0) = 1$, $y'(0) = -4$, $y''(0) = 6$

38. $\begin{cases} x' = y; \quad x(0) = 1 \\ y' = -9x; \ y(0) = 0 \end{cases}$ **39.** $\begin{cases} x' = y; \ x(0) = 1 \\ y' = x; \ y(0) = 1 \end{cases}$

40. $\begin{cases} x' = x - y; \ x(0) = 1 \\ y' = x + y; \ y(0) = 0 \end{cases}$ **41.** $\begin{cases} x' = 3x - 2y; \ x(0) = 3 \\ y' = 2x - y; \quad y(0) = 0 \end{cases}$

42. $y'' + 5y' + 4y = 0$; $y(0) = 1$, $y'(0) = -4$

43. Use the Laplace transform to solve the initial value problem

$$y' + \int_0^t y(u)\,du = 1; \quad y(0) = 1.$$

44. Let $f(t)$ be a function that has a Laplace transform, and put $F(s) = L(f)$. If k denotes a positive constant, prove that

$$L[f(kt)] = \frac{1}{k} F\left(\frac{s}{k}\right).$$

Use this result to derive formulas for the Laplace transforms of

a. e^{kt}, given that $L(e^t) = \dfrac{1}{s-1}$

b. $\cos(kt)$, given that $L[\cos(t)] = \dfrac{s}{s^2+1}$

c. $\sin(kt)$, given that $L[\sin(t)] = \dfrac{1}{s^2+1}$

6.4 Inverse Transforms of Rational Functions

To compute the inverse Laplace transform of a rational expression

$$X(s) = \frac{p(s)}{q(s)},$$

we would like to use a table of Laplace transforms. Since it is not feasible to put all such expressions in a table, it is necessary to split $X(s)$ into a sum of partial fractions and look up the inverse transforms of the simpler fractions. The partial fractions algorithm is covered in calculus texts, and we will review it now. Before starting, let us recall that to expand $X(s)$ in partial fractions, it is necessary for $X(s)$ to be a **proper rational function**. This means that the degree of the numerator polynomial $p(s)$ must be strictly less than the degree of the denominator, $q(s)$. Since a rational expression $X(s)$ is proper if and only if $\lim_{s\to\infty} X(s) = 0$, Proposition 6.1.2 on page 258 assures us that if $X(s)$ has an inverse Laplace transform, then $X(s)$ is proper.

The first step of the partial fractions algorithm is to factor the denominator $q(s)$. We will assume that all factors are powers of linear binomials $(s - k)$ or powers of irreducible quadratics $(s^2 + \alpha s + \beta)$.

The factors of $q(s)$ determine the form of the partial fraction decomposition of $F(s)$. Corresponding to a factor $(s - k)^m$, there will be the following partial fractions:

$$\frac{C_1}{s - k} + \frac{C_2}{(s - k)^2} + \cdots + \frac{C_m}{(s - k)^m}, \tag{6.18}$$

where the coefficients $C_1, \ldots, C_m$ are undetermined.

An irreducible quadratic factor $s^2 + \alpha s + \beta$ should be written as a completed square $(s - k)^2 + b^2$. Corresponding to it there are two partial fractions, with undetermined coefficients A and C:

$$\frac{A(s - k)}{(s - k)^2 + b^2} + \frac{Cb}{(s - k)^2 + b^2}. \tag{6.19}$$

For example, if $q(s)$ has a factor

$$s^2 + s + 2 = (s + 1/2)^2 + (\sqrt{7/4})^2,$$

where $k = -1/2$ and $b = \sqrt{7/4}$, there would be the partial fractions

$$\frac{A(s + 1/2)}{(s + 1/2)^2 + (\sqrt{7/4})^2} + \frac{C\sqrt{7/4}}{(s + 1/2)^2 + (\sqrt{7/4})^2}.$$

Discussion of factors $[(s - k)^2 + b^2]^m$ ($m > 1$) that are powers of irreducible quadratics will be postponed until Section 6.7 (see page 307).

To calculate the undetermined coefficients in the partial fractions expression, multiply the entire partial fraction expression by $q(s)$, thus canceling all of the denominators, and equate the result with the numerator $p(s)$ of $F(s)$.

Once the values of the undetermined coefficients are calculated, it is a simple matter to take the inverse Laplace transform. The inverse transform of the expression (6.18) is

$$C_1 e^{kt} + C_2 t e^{kt} + \cdots + \frac{C_m}{(m - 1)!} t^{m-1} e^{kt},$$

and the inverse transform of the expression (6.19) is

$$A e^{kt} \cos(bt) + C e^{kt} \sin(bt).$$

Example 6.4.1   Find the inverse Laplace transform of

$$X(s) = \frac{s^2 + 2s + 3}{(s+1)(s+2)(s+3)}.$$

Solution. The partial fractions expansion, with undetermined coefficients, is

$$X(s) = \frac{A}{s+1} + \frac{B}{s+2} + \frac{C}{s+3}.$$

Multiplying through by $q(s) = (s+1)(s+2)(s+3)$, we have

$$s^2 + 2s + 3 = A(s+2)(s+3) + B(s+1)(s+3) + C(s+1)(s+2). \quad (6.20)$$

Setting $s = -1$ in equation (6.20), we have $2 = A \cdot 1 \cdot 2 + B \cdot 0 + C \cdot 0$; hence $A = 1$. Putting $s = -2$ in equation (6.20) leads to $B = -3$; with $s = -3$, we find that $C = 3$. Thus

$$X(s) = \frac{1}{s+1} - \frac{3}{s+2} + \frac{3}{s+3},$$

and hence $L^{-1}(F) = e^{-t} - 3e^{-2t} + 3e^{-3t}$.   ❑

Example 6.4.2   Solve the IVP

$$y'' + 4y' + 3y = -6e^{-3t}; \quad y(0) = 2, \quad y'(0) = -1. \quad (6.21)$$

Solution. Let $y(t)$ denote the solution of the IVP, and $Y(s) = L(y)$ Then

$$L(y') = sY - y(0) = sY - 2,$$
$$L(y'') = s^2 Y - sy(0) - y'(0) = s^2 Y - 2s + 1, \text{ and}$$
$$L(-6e^{-3t}) = -\frac{6}{s+3}.$$

Hence the transformed ODE is

$$s^2 Y - 2s + 1 + 4(sY - 2) + 3Y = -\frac{6}{s+3},$$

which we solve for Y to get

$$Y = \frac{2s + 7}{s^2 + 4s + 3} - \frac{6}{(s^2 + 4s + 3)(s+3)}.$$

Since $s^2 + 4s + 3 = (s+3)(s+1)$,

$$Y = \frac{2s + 7}{(s+3)(s+1)} - \frac{6}{(s+3)^2(s+1)},$$

It is not necessary to "simplify" the expression by combining the fractions over a common denominator, but we note that the least common denominator is $q(s) = (s+3)^2(s+1)$, because that determines the format of the partial fractions expression,

$$Y = \frac{A_1}{s+3} + \frac{A_2}{(s+3)^2} + \frac{B}{s+1}.$$

After multiplying through by $q(s)$ we have

$$(2s+7)(s+3) - 6$$
$$= A_1(s+3)(s+1) + A_2(s+1) + B(s+3)^2. \tag{6.22}$$

Setting $s = -3$ yields $-6 = -2A_2$, so $A_2 = 3$. If $s = -1$, we have $4 = 4B$, and $B = 1$. After substituting these values into equation (6.22) and simplifying the result, we have

$$2s^2 + 13s + 15 = A_1(s^2 + 4s + 3) + 3(s+1) + (s^2 + 6s + 9),$$

or $s^2 + 4s + 3 = A_1(s^2 + 4s + 3)$. It follows that $A_1 = 1$. Thus,

$$Y = \frac{1}{s+3} + \frac{3}{(s+3)^2} + \frac{1}{s+1}.$$

The solution of the initial value problem is the inverse Laplace transform of Y:

$$y(t) = (3t+1)e^{-3t} + e^{-t}. \qquad \blacksquare$$

Example 6.4.3   Solve the initial value problem

$$y'' - 4y' + 5y = 24e^t; \quad y(0) = y'(0) = 0.$$

Solution. Let $Y(s)$ denote the solution of the initial value problem. Then

$$L(y') = sL(y) - y(0) = sY(s),$$

and similarly $L(y'') = s^2Y(s)$. The Laplace transform of the source term is

$$L(24e^t) = \frac{24}{s-1}.$$

The transformed problem is therefore

$$s^2Y - 4sY + 5Y = \frac{24}{s-1}.$$

Solving for Y, we have

$$Y = \frac{24}{(s^2 - 4s + 5)(s-1)} = \frac{24}{[(s-2)^2 + 1](s-1)}. \tag{6.23}$$

The partial fractions decomposition of Y is

$$\frac{24}{[(s-2)^2 + 1](s-1)} = \frac{A(s-2)+B}{(s-2)^2+1} + \frac{C}{s-1}.$$

Multiply through by $[(s-2)^2 + 1](s-1)$ to get

$$24 = [A(s-2)+B](s-1) + C[(s-2)^2 + 1]. \tag{6.24}$$

Set $s = 1$ in equation (6.24) to determine the coefficient C:

$$24 = C \cdot 2$$

so $C = 12$. By using $s = 2$ we obtain $24 = B + C$; thus $B = 12$. Now, with these values of B and C, equation (6.24) reduces to

$$24 = (A+12)s^2 + (-3A - 36)s + (2A + 48)$$
$$= (A+12)(s^2 - 3s + 2) + 24.$$

It follows that $A = -12$, and therefore the solution is

$$y(t) = 12e^{2t}[\sin(t) - \cos(t)] + 12e^t. \qquad \square$$

In the following example, we will investigate the periodic and transient currents in an *RLC* series circuit.

Example 6.4.4 An *RLC* series circuit has $R = 16$ ohms, $L = 0.1$ henry, and $C = 0.001$ farad. The EMF is $E(t) = 100 \sin(60t)$ volts. The initial current is 0, and the initial charge of the capacitor is 0.1 coulomb. Find the current as a function of time, and show that it is the sum of a periodic current and a transient.

Solution. Let y denote the current carried by the circuit and x be the capacitor charge. Substitute the parameters in the differential-integral equation (6.9) to obtain

$$0.1y' + 16y + 1000\left(x_0 + \int_0^t y(\tau)\,d\tau\right) = 100\sin 60t. \tag{6.25}$$

The initial charge is $x_0 = 0.1$, and the initial current is $y(0) = 0$. The transform of equation (6.25) is therefore

$$0.1(sY) + 16Y + 1000\left(\frac{0.1}{s} + \frac{1}{s}Y\right) = 100\frac{60}{s^2 + 60^2}.$$

This can be solved to obtain

$$Y = -\frac{100}{0.1s^2 + 16s + 1000} + \frac{6000s}{(s^2 + 60^2)(0.1s^2 + 16s + 1000)}$$

and, after simplifying and completing the square,

$$Y = -\frac{1000}{(s+80)^2 + 60^2} + \frac{60000s}{(s^2 + 60^2)((s+80)^2 + 60^2)}.$$

To find the inverse Laplace transform, we have to apply a partial fractions decomposition to the second term (the first term is already in good shape). The format of the decomposition is

$$\frac{60000s}{(s^2 + 60^2)((s + 80)^2 + 60^2)} = \frac{As + 60B}{s^2 + 60^2} + \frac{C(s + 80) + 60D}{(s + 80)^2 + 60^2}.$$

Multiply this out to obtain

$$60000s = (As + 60B)((s + 80)^2 + 60^2) + (C(s + 80) + 60D)(s^2 + 60^2)$$

or

$$60000s = (A + C)s^3 + (160A + 60B + 80C + 60D)s^2$$
$$+ (10000A + 9600B + 3600C)s$$
$$+ (600000B + 288000C + 216000D).$$

Equating the coefficients of like powers of s,

$$0s^3 = (A + C)s^3$$
$$0s^2 = (160A + 60B + 80C + 60D)s^2$$
$$60000s = (10000A + 9600B + 3600C)s$$
$$0 = (600000B + 288000C + 216000D).$$

The solution of this system is

$$A = \frac{150}{52}, \quad B = \frac{225}{52}, \quad C = -\frac{150}{52}, \quad \text{and} \quad D = -\frac{425}{52}.$$

The inverse transform of the first term of Y is

$$y_1 = \frac{1000}{60}e^{-80t}\sin(60t),$$

which is a transient. The inverse transform of the second term of Y is $y_2 + y_3$, where

$$y_2 = A\cos(60t) + B\sin(60t) = \frac{1}{52}(150\cos(60t) + 225\sin(60t))$$

is periodic, and

$$y_3 = e^{-80t}(C\cos(60t) + D\sin(60t)) = -\frac{e^{-80t}}{52}(150\cos(60t) + 425\sin(60t))$$

is transient.   ❑

Electrical impedance If an electromotive force $E(t) = A \sin(\omega t)$ is applied to an RLC series circuit, the current will be the sum of a periodic current and a transient. To see why, take the Laplace transform of the integral-differential equation (6.9):

$$L(sY - y_0) + R(sY) + \frac{1}{C}\left(\frac{x_0}{s} + \frac{1}{s}Y\right) = \frac{A\omega}{s^2 + \omega^2}.$$

Solving for Y gives the following result:

$$Y = \frac{Ls y_0 - x_0/C}{Ls^2 + Rs + 1/C} + \frac{As\omega}{(s^2 + \omega^2)(Ls^2 + Rs + 1/C)}.$$

In partial fractions, Y will have the form

$$Y = \frac{Ps + Q\omega}{s^2 + \omega^2} + \frac{Ms + N}{Ls^2 + Rs + 1/C}.$$

The inverse transform of the first term is

$$P \cos(\omega t) + Q \sin(\omega t),$$

which is periodic. The inverse transform of the second term is a solution of the homogeneous linear ODE

$$Ly'' + Ry' + y/C = 0.$$

Since L, R, and C are positive, Proposition 5.4.1 on page 207 tells us that this ODE is stable, which means that all solutions are transient.

The amplitude of the periodic current is proportional to the amplitude of the EMF, but the phase is shifted. Electrical engineers have found the simplest model for this phenomenon involves the complex exponential function as the EMF. Thus, we will determine the periodic solution of the differential-integral equation

$$Ly' + Ry + \frac{1}{C}\int_0^t y(\tau)\, d\tau = e^{i\omega t}. \tag{6.26}$$

The initial conditions can only influence the transient (why?), so we will just assume that the initial current and charge are 0.

The Laplace transform of (6.26) is

$$LsY + RY + \frac{1}{Cs}Y = \frac{1}{s - i\omega}.$$

Now solve for Y:

$$Y = \frac{s}{(Ls^2 + Rs + 1/C)(s - i\omega)} = \frac{A}{s - i\omega} + \frac{Bs + C}{Ls^2 + Rs + 1/C}.$$

We will not bother to compute B and C, since they are coefficients of transient terms. Multiplying through, we have

$$s = A(Ls^2 + Rs + 1/C) + (Bs + C)(s - i\omega).$$

To calculate A, let $s = i\omega$. Then

$$i\omega = A(-L\omega^2 + iR\omega + 1/C),$$

and

$$A = \frac{i\omega}{-L\omega^2 + iR\omega + 1/C}.$$

The **electrical impedance** of the circuit is the complex ratio of the EMF to the periodic current. We will denote the impedance $P(\omega)$. Since $y = Ae^{i\omega t}$ is the periodic solution of equation (6.26),

$$P(\omega) = \frac{1}{A} = R + i\left(L\omega - \frac{1}{C\omega}\right). \tag{6.27}$$

Notice that the resistance R is the real part of the impedance. The imaginary part is known as the **reactance.** If the EMF is measured in volts and the current is measured in amperes, the units of impedance, resistance, and reactance are ohms.

Example 6.4.5   Find the impedance $P(\omega)$ of an RLC series circuit with parameters $R = 2$ ohms, $L = .1$ henry, and $C = .01$ farad. For what value of ω is $|P(\omega)|$ at a minimum?

Solution. By formula (6.27),

$$P(\omega) = \frac{0.02\omega + i(0.001\omega^2 - 1)}{0.01\omega}$$

$$= 2 + i\left(0.1\omega - \frac{100}{\omega}\right).$$

Since

$$|P(\omega)|^2 = 2^2 + \left(0.1\omega - \frac{100}{\omega}\right)^2,$$

$|P(\omega)|$ is at a minimum when the reactance $0.1\omega - \frac{100}{\omega} = 0$; that is, for $\omega = \sqrt{1000}$ radians per second. At this frequency, the impedance is equal to the resistance: 2 ohms. ❑

The significance of finding the frequency ω that minimizes $|P(\omega)|$, as we did in Example 6.4.5, is that for that frequency, the circuit acts as a dc (direct current) circuit: The current is given by dividing the EMF by the resistance.

The electrical-mechanical analogy

We have been using a model for *RLC* series circuits in which the dependent variable is the current. If the charge x held by the capacitor is used as the dependent variable instead, then

$$x = x_o + \int_0^t y(\tau)\,d\tau,$$

$x' = y$, and $x'' = y'$. Thus, the integral-differential equation (6.9) is replaced by an ODE,

$$Lx'' + Rx' + \frac{1}{C}x = E(t). \tag{6.28}$$

This is the same kind of ODE that we have used as a mathematical model for a damped mechanical system. Table 6.2 displays the correspondence between electrical and mechanical parameters that results because both physical systems are governed by the same ODE.

Table 6.2 Corresponding parameters for mechanical systems and electrical circuits.

Mechanical System	Electrical Circuit
mass	inductance
damping	resistance
stiffness	$\frac{1}{\text{capacitance}}$
external force	electromotive force
displacement	electrical charge
velocity	current

An analog computer is an electrical circuit that simulates some other physical process. The simplest analog computer is an *RLC* series circuit, which can be devised to model any physical process governed by a second-order linear differential equation with constant coefficients. For example, Table 6.2 tells us that if the resistance, inductance, and capacitance are set correctly, current measurements in the analog computer can be interpreted as velocities in a mechanical system.

The design principle for an analog computer is that the ODE for the *RLC* series circuit must be the same as the ODE that models the mechanical system. We are free to choose convenient units, and often the units used for the circuit will be its natural units, and the units for the mechanical system will also be natural ones.

In Exercise 27, you are asked to show that the ODE

$$Lx'' + Rx' + \frac{1}{C}x = Ae^{i\omega t},$$

in which A denotes the amplitude of the EMF, measured in volts, can be rescaled as

$$\frac{d^2 X}{dT^2} + B\frac{dX}{dT} + X = e^{i\Omega t}$$

by changing units so that

$$T = \frac{1}{\sqrt{LC}}t, \quad B = \sqrt{\frac{C}{L}}R, \quad X = \frac{1}{CA}x, \quad \text{and} \quad \Omega = \omega\sqrt{LC}.$$

Example 6.4.6   A mechanical system has the following parameters: $m = 0.5$ kilograms, $b = 2$ kilograms per second, and $k = 2$ newtons per meter. Describe an *RLC* series circuit that serves as an analog computer for the mechanical system but that reacts twice as fast as the mechanical system.

Solution. The mechanical system is governed by the ODE

$$0.5\frac{d^2y}{dt^2} + 2\frac{dy}{dt} + 2y = 0, \tag{6.29}$$

where y is the displacement of the weight from its equilibrium position. By changing the unit of time, we can rescale the equation as

$$\frac{d^2y}{dT^2} + B\frac{dy}{dT} + y = 0, \tag{6.30}$$

where $T = at$, and a is to be determined. By the chain rule,

$$\frac{dy}{dt} = a\frac{dy}{dT} \quad \text{and} \quad \frac{d^2y}{dt^2} = a^2\frac{d^2y}{dT^2}.$$

Let's divide (6.29) by $m = 0.5$ and convert to the new time scale. The following equation results:

$$a^2\frac{d^2y}{dT^2} + 4a\frac{dy}{dT} + 4y = 0.$$

By dividing through by a^2, we obtain

$$\frac{d^2y}{dT^2} + \frac{4}{a}\frac{dy}{dT} + \frac{4}{a^2}y = 0,$$

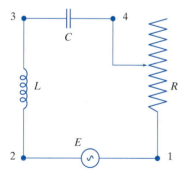

Figure 6.9. See Example 6.4.7.

and it follows that with $a = 2$ the rescaled version of (6.29) will match (6.30). The damping ratio is $B = 4/a = 2$.

We will look for an *RLC* series circuit whose rescaled equation is the same, and with time $\tau = 2t = T$, because the circuit is to react twice as fast as the mass spring system. Since the conversion from seconds to the natural time scale for the circuit is $\tau = \sqrt{LC}T$, It follows that $\sqrt{LC} = 1$ and $R = 2\sqrt{L/C}$. There are many solutions—for example $L = C = 1$ and $R = 2$.

Example 6.4.7   Figure 6.9 shows an *RLC* series circuit in which the resistance is variable, but the inductance and capacitance are fixed as 0.1 henry and .01 farad, respectively. Find the resistance setting that would be appropriate for this circuit to serve as an analog computer for the mechanical system of Example 6.4.6.

Solution. As noted in the solution of Example 6.4.6, $R = 2\sqrt{L/C} = 2\sqrt{10}$ ohms. The conversion from the natural time scale to seconds for the circuit is $\tau = \sqrt{LC}T = \sqrt{0.001}T$. The measure of time for the mass spring system will be given by the variable t; it was shown in the solution of Example 6.4.6 that $T = 2t$. Hence $\tau = 2\sqrt{.001}t \approx 0.063t$. The time scale for the circuit is thus slower than the mechanical system by a factor of approximately 0.063. ❑

EXERCISES

In Exercises 1–8, find the inverse Laplace transform of the given expression.

1. $\dfrac{6}{s^3 + s^2}$

2. $\dfrac{2s + 9}{s^2 + 3s + 2}$

3. $\dfrac{s + 5}{s^2 + 6s + 10}$

4. $\dfrac{3s - 4}{s^2 + 4s + 4}$

5. $\dfrac{s^2 - s + 1}{s^4 - s^3}$

6. $\dfrac{s^3 - 4}{s^4 - 4s^2}$

7. $\dfrac{s - 2}{s^3 + 2s^2 + 2s}$

8. $\dfrac{3s^2 + 4s + 9}{(s^2 + 2s + 5)(s + 3)}$

In Exercises 9–25, use the Laplace transform to solve the initial value problem.

9. $y'' + 4y = 16t$; homogeneous initial conditions at $t = 0$

10. $2y'' + 7y' + 5y = \sin 3t$; homogeneous initial conditions at $t = 0$

11. $y'' + 4y = 3 \sin t$; $y(0) = 0$, $y'(0) = -1$

12. $y'' + y' + y = -2 \sin t$; $y(0) = 2$, $y'(0) = 0$

13. $y'' + y' + y + \displaystyle\int_0^t y(u)\,du = 0$; $y(0) = 1$, $y'(0) = 0$

14. $y' + 2y + 2\displaystyle\int_0^t y(u)\,du = e^{-t}$; $y(0) = 1$

15. $y' + 3y = e^{-t}[(2t + 1) - \sin t]$; $y(0) = 0$

16. $y' + 8\left[y + \displaystyle\int_0^t y(v)\,dv\right] = 1$; $y(0) = 1$

17. $y'' - \displaystyle\int_0^t y(v)\,dv = 0$; $y(0) = 0$, $y'(0) = 1$

18. $y'' - 3y' + 3y = \displaystyle\int_0^t (1 + y(v))\,dv$; homogeneous initial conditions at $t = 0$

19. $\begin{cases} x' = x + y; & x(0) = 1 \\ y' = -x - y - t; & y(0) = 0 \end{cases}$

20. $\begin{cases} x' = x + 2y; & x(0) = 1 \\ y' = -5x - y; & y(0) = -1 \end{cases}$

21. $\begin{cases} x' = -3x - 2y; & x(0) = 2 \\ y' = 2x + 2y; & y(0) = -\frac{5}{2} \end{cases}$

22. $\begin{cases} x' = -3x - y - t; & x(0) = 0 \\ y' = x - y + 1; & y(0) = 0 \end{cases}$

23. $y''' + 6y'' + 12y' + 8y = 96te^{-2t}$; homogeneous initial conditions at $t = 0$

24. $y'' + 6y' + 25y = 16e^{-3t}$; $y(0) = 1$, $y'(0) = 17$

25. $y^{(4)} - 16y = 0$; $y(0) = y'(0) = 0$, $y''(0) = 8$, $y'''(0) = 0$

26. Find the periodic solution of each of the following differential equations, and show that all solutions are the sum of the periodic solution and a transient. Initial conditions are omitted, because you are not supposed to compute the transient!

   a. $y'' + 2y' + 2y = \sin t$

   b. $y'' + .02y' + 1.0001y = \sin 2t$

   c. $y''' + 3y'' + 4y' + 2y = e^{it}$

   d. $y''' + 6y'' + 12y' + 8y = \sin \dfrac{t}{2}$

27. Find natural units of time and current for the following ODE model of an *RLC* circuit with an EMF of $A \sin \omega t$:

$$L\frac{d^2y}{dt^2} + R\frac{dy}{dt} + \frac{1}{C}y = \frac{d}{dt}[A \sin \omega t].$$

28. An *RLC* series circuit has an inductance of 0.01 henry, a resistance of 2 ohms, and a variable capacitor. The circuit will be used to model a mass spring system in which the parameters are $m = 2$ kilograms, $b = 1$ kilogram per second, and $k = 1000$ newtons per meter. What capacitance should be used? Compare the time scales on which the mass spring system and the circuit operate.

29. The **resonant frequency** of an *RLC* series circuit is the frequency ω of the EMF function $E(t) = e^{i\omega t}$ for which the absolute value of the electrical impedance is at a minimum.

   a. Show that the resonant frequency is independent of the resistance and is thus the same if the resistance is changed to make the circuit undamped, underdamped, or overdamped.

b. Give a formula for the resonant frequency in terms of the inductance and capacitance.

c. Show that if $P(\omega)$ denotes the electrical impedance of an *RLC* series circuit at frequency ω, then $\lim_{\omega \to 0} |P(\omega)|$ and $\lim_{\omega \to \infty} |P(\omega)|$ are both infinite.

30. An *RLC* series circuit has inductance of 0.01 henry and capacitance of 0.0025 farad. The resistance is variable. Calculate the resonant frequency.

31. The parameters of an *RLC* series circuit are $R = 8$ ohms, $L = .5$ henries, and $C = .025$ farads.

a. Is the circuit overdamped, critically damped, or underdamped?

b. Assume that the initial current is zero, and that no electromotive force is applied. If the capacitor has an initial charge of 1.25 coulomb, find the current $y(t)$.

c. Assume that an electromotive force $V(t) = 5-12 \sin 6t$ is applied. If the initial current is $y(0) = 0$ and the initial charge of the capacitor is $x(0) = 0$, find the charge $x(t)$.

d. Calculate the electrical impedance $P(\omega)$ of this circuit as a function of frequency.

e. For what frequency is $|P(\omega)|$ at a minimum?

32. An *RLC* series circuit has inductance of 0.1 henry and resistance of 2 ohms. The capacitance is variable. Sketch a graph of the resonant frequency as a function of the capacitance.

33. Suppose that $f(t)$ and $f'(t)$ have Laplace transforms, and let $F(s) = L(f(t))$. Show that

$$\lim_{s \to \infty} sF(s) = f(0).$$

34. Let $F(s)$ denote the Laplace transform of $f(t) = e^{-t^{-2}}$. Show that for all n, $f^{(n)}(0) = 0$, and hence that for any value of n,

$$\lim_{n \to \infty} s^n F(s) = 0.$$

35. Let A be an $n \times n$ matrix. In Exercise 15 on page 262, we saw that $L(e^{At}) = (sI - A)^{-1}$.

a. Suppose that

$$A = \begin{bmatrix} 3 & 4 \\ -2 & -3 \end{bmatrix}.$$

Show that

$$(sI - A)^{-1} = \frac{1}{s^2 - 1} \begin{bmatrix} s+3 & 4 \\ -2 & s-3 \end{bmatrix},$$

and hence that $e^{At} = \begin{bmatrix} 2e^t - e^{-t} & 2e^t - 2e^{-t} \\ -e^t + e^{-t} & -e^t + 2e^{-t} \end{bmatrix}.$

b. Calculate e^{At}, where

$$A = \begin{bmatrix} 1 & 1 \\ -1 & 3 \end{bmatrix}.$$

✔ **Answer:** $e^{2t} \begin{bmatrix} 1-t & t \\ -t & 1+t \end{bmatrix}.$

6.5   **The Unit Step Function**

The Laplace transform is especially useful in solving initial value problems where the source terms are discontinuous.

The simplest discontinuous function is the **unit step function,**

$$u(t) = \begin{cases} 0 & \text{if } t < 0, \text{ and} \\ 1 & \text{if } t \geq 0. \end{cases}$$

The unit step function is continuous everywhere except at $t = 0$, where the value jumps from 0 to 1. As can be seen in Figure 6.10, we can move the discontinuity to $t = a$ by using the function $u(t - a)$.

The magnitude of the discontinuity can be modified by using a scale factor. If we put $f(t) = ku(t - a)$, then the jump is from $f(t) = 0$ to $f(t) = k$ at $t = a$. Linear combinations of unit step functions can be used to represent more complicated discontinuous functions.

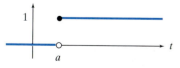

Figure 6.10 The step function $u(t - a)$, with jump at $t = a$.

⌐ **Example 6.5.1**   Express the function

$$f(t) = \begin{cases} 3 & \text{if } 0 \le t < 2 \\ 0 & \text{otherwise} \end{cases}$$

in terms of step functions.

Solution. The function $f(t)$ has discontinuities at $t = 0$ and at $t = 2$; hence we will try to find a linear combination of $u(t - 0)$ and $u(t - 2)$ that is equal to $f(t)$. If

$$f(t) = C_1 u(t) + C_2 u(t - 2),$$

then for $t < 0$ we would automatically have $f(t) = 0$. For $0 \le t < 2$, $f(t) = C_1$, and for $t \ge 2$, $f(t) = C_1 + C_2$. Hence $C_1 = 3$ and $C_1 + C_2 = 0$. It follows that $f(t) = 3[u(t) - u(t - 2)]$.   ❑

Step functions can be used to construct continuous functions.

Proposition 6.5.1

If $f(t)$ is continuous at $t = 0$, and $f(0) = 0$, then $g(t) = f(t - a)u(t - a)$ is continuous at $t = a$.

Proof An alternative way to present $g(t)$ would be

$$g(t) = \begin{cases} 0 & \text{if } t < a \\ f(t - a) & \text{if } t \ge a. \end{cases}$$

Since $g(t) = 0$ for $t < a$, $\lim_{t \to a^-} g(t) = 0$. For $t > a$, $g(t) = f(t - a)$, and thus

$$\lim_{t \to a^+} g(t) = \lim_{t \to a^+} f(t - a) = f(0) = 0,$$

because f is continuous at $t = 0$.   ■

For example, the function $c(t) = t\,u(t)$, which we will call the **corner function**, is continuous. The graph of $y = c(t)$ follows the negative t-axis to the origin, and the line $y = t$ for $t > 0$. The corner in the graph at the origin inspires the name. Notice that the corner function is an antiderivative of the unit step function.

The corner in the graph of $c(t - a) = (t - a)u(t - a)$ is located at $t = a$. This means that the slope jumps from 0 to 1 at $t = a$. Multiplying a corner function by a constant will alter the magnitude of the jump in slope. Thus, the slope of $k\,c(t - a)$ jumps from 0 to k at $t = a$. When corner functions, with corners at different locations, are added, the result is a graph with several corners joined by line segments.

Figure 6.11 shows the graph of a function that looks like a triangle and has three corners. At each corner of the triangle, we note the change of slope. At the corner $\left(\frac{3}{2}, 0\right)$, the slope changes from 0 to $\frac{3/2 - 0}{3 - 3/2} = 1$; at the corner $\left(3, \frac{3}{2}\right)$, the slope changes from 1 to $\frac{0 - 3/2}{5 - 3} = -\frac{3}{4}$; and at the corner $(5, 0)$, the slope changes from $-\frac{3}{4}$

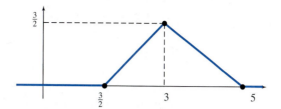

FIGURE 6.11   The triangle function.

to 0. We can present the triangle function as the sum of three corner functions. Each corner function is multiplied by the jump in the slope at the corner. Thus, the triangle function is

$$f(t) = c(t - 3/2) + \left(-\frac{3}{4} - 1\right) c(t - 3) + \left(0 - \left(-\frac{3}{4}\right)\right) u(t - 5)$$

$$= c(t - 3/2) - \frac{7}{4} c(t - 3) + \frac{3}{4} c(t - 5).$$

Laplace transform

The Laplace transform of $f(t) = u(t - a)$ is

$$F(s) = \int_0^\infty e^{-st} u(t - a) \, dt.$$

If $a \le 0$, then $u(t - a) = 1$ for $t \ge 0$. Hence

$$F(s) = \int_0^\infty e^{-st} (1) \, dt = L(1) = 1/s.$$

If $a > 0$, then

$$F(s) = \int_0^\infty e^{-st} u(t - a) \, dt$$

$$= \int_0^a e^{-st} (0) \, dt + \int_a^\infty e^{-st} (1) \, dt$$

$$= 0 - \frac{1}{s} e^{-st} \Big|_a^\infty = \frac{e^{-as}}{s}.$$

Example 6.5.2   Find the Laplace transform of

$$f(t) = 3(u(t) - u(t - 2)).$$

Solution. The Laplace transform of $f(t)$ is

$$F(s) = 3L[u(t)] - 3L[u(t - 2)]$$

$$= \frac{3}{s} - \frac{3}{s} e^{-2s}.$$

Step functions are used to "cut off" other functions. For example, the function

$$f(t) = \begin{cases} t & \text{if } t \geq 1 \\ 0 & \text{if } t < 1 \end{cases}$$

is obtained by cutting off the function $g(t) = t$; an expression for f using a step function is $f(t) = tu(t-1)$. The following proposition gives a formula for the Laplace transform of a function that has been cut off in this way.

Proposition 6.5.2

Let $f(t)$ be a function that has a Laplace transform. Then for any constant $a \geq 0$,

$$L[f(t-a)u(t-a)] = e^{-as}L(f).$$

Notice that the formula is *not* $L(f(t)u(t-a)) = \cdots$. The independent variable t is *translated* before cutting off the function.

Proof of Proposition 6.5.2.

$$L(f(t-a)u(t-a)) = \int_0^\infty e^{-st}f(t-a)u(t-a)\,dt$$

$$= \int_0^a 0\,dt + \int_a^\infty e^{-st}f(t-a)\,dt$$

Substituting $w = t - a$, $dw = dt$, we have

$$L(f(t-a)u(t-a)) = \int_0^\infty e^{-s(w+a)}f(w)\,dw$$

$$= e^{-as}\int_0^\infty e^{-sw}f(w)\,dw$$

$$= e^{-as}L(f).\qquad\blacksquare$$

Example 6.5.3   Find the Laplace transform of $c(t-a)$, where $a > 0$.

Solution. Since $c(t-a) = (t-a)u(t-a)$,

$$L(c(t-a)) = e^{-as}L(t) = \frac{e^{-as}}{s^2}. \qquad \square$$

Example 6.5.4   Find the Laplace transform of $tu(t-1)$.

Solution. Express the function in the form $f(t-1)u(t-1)$. If $f(t-1) = t$, what is f? To figure this out, put $\bar{t} = t - 1$ so that $t = \bar{t} + 1$ and $f(\bar{t}) = t = \bar{t} + 1$. Using the formula for the Laplace transform of a polynomial,

$$L(f) = L(t+1) = \frac{1}{s^2} + \frac{1}{s}.$$

Therefore, by Proposition 6.5.2,

$$L(f(t-1)u(t-1)) = \left(\frac{1}{s^2} + \frac{1}{s}\right)e^{-s}. \qquad \square$$

Proposition 6.5.2 can also be used to find inverse Laplace transforms. To find the inverse Laplace transform of a function $e^{-as}F(s)$, where $F(s)$ is a rational function, the first step is to find the inverse Laplace transform of $F(s)$. If $f(t) = L^{-1}(F)$, then $L^{-1}[e^{-as}F(s)] = f(t-a)u(t-a)$.

Example 6.5.5   Find the inverse Laplace transform of

$$G(s) = \frac{1 - e^{-\pi s}}{s(s^2 + 4)}.$$

Solution. The inverse Laplace transform of $F(s) = \frac{1}{s(s^2+4)}$ is found by computing the partial fractions expansion

$$F(s) = \frac{1}{4}\left(\frac{1}{s} - \frac{s}{s^2 + 4}\right).$$

Thus $L^{-1}(F(s)) = \frac{1}{4}[1 - \cos(2t)]$. Let $f(t)$ denote this function. Since $G(s) = F(s) - e^{-\pi s}F(s)$,

$$L^{-1}(G) = f(t) - f(t - \pi)u(t - \pi)$$
$$= \frac{1}{4}[1 - \cos(2t)] - \frac{1}{4}[1 - \cos(2(t - \pi))]u(t - \pi).$$

Since $\cos(2(t - \pi)) = \cos(2t)$,

$$L^{-1}(G) = \frac{1}{4}[1 - u(t - \pi)][1 - \cos(2t)].$$

The factor $1 - u(t - \pi) = 1$ for $0 \le t < \pi$ and vanishes for $t \ge \pi$. See Figure 6.12. ❑

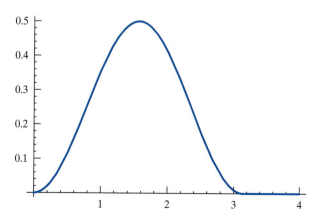

FIGURE 6.12   The inverse Laplace transform of $G(s) = \frac{1-e^{-\pi s}}{s(s^2+4)}$. See Example 6.5.5.

Example 6.5.6   Solve the initial value problem

$$y'' + 4y = g(t)$$

with homogeneous initial conditions at $t = 0$, where

$$g(t) = \begin{cases} 1 & \text{if } 0 \le t < \pi \\ 0 & \text{otherwise.} \end{cases}$$

Solution. In terms of the unit step function, $g(t) = 1 - u(t - \pi)$. The Laplace transform of the initial value problem is thus

$$L(y'') + 4L(y) = L(1) - L(u(t - \pi)).$$

Let $Y(s) = L(y)$, so that $s^2 Y(s) = L(y'')$. Thus

$$(s^2 + 4)Y = \frac{1}{s} - \frac{1}{s}e^{\pi s}.$$

It follows that $Y(s) = \frac{1 - e^{-\pi s}}{s(s^2+4)}$, and, by the result of Example 6.5.5,

$$y(t) = L^{-1}(Y(s)) = \frac{1}{4}[1 - u(t - \pi)][1 - \cos(2t)].$$ ❑

An RLC circuit

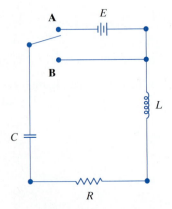

Figure 6.13 An *RLC* series circuit with a switch. $E = 600$ volts, $R = .3$ ohms, $L = .05$ henry, and $C = .1$ farad.

In the *RLC* circuit shown in Figure 6.13, the switch is held in position **A** for 0.2 seconds; then it is moved to position **B**. Suppose that we want to calculate the current carried by the circuit as a function of time, assuming that at $t = 0$, the current is 0, and the capacitor holds no charge.

We will represent the EMF of the circuit with a single discontinuous function:

$$E(t) = \begin{cases} 600 & \text{if } 0 \le t < 0.2 \\ 0 & \text{if } t \ge 0.2 \end{cases}$$

$$= 600[1 - u(t - 0.2)].$$

Let $y(t)$ denote the current. Then the differential-integral equation

$$.05y' + .3y + 10 \int_0^t y(u)\, du = E(t), \tag{6.31}$$

with a homogeneous initial condition at $t = 0$, serves as a mathematical model of the circuit.

Let $Y(s)$ denote the Laplace transform of $y(t)$. Since $y(0) = 0$, $L(y') = sY(s)$. Furthermore, $L\left(\int_0^t y(v)\, dv\right) = \frac{1}{s}Y(s)$, so the Laplace transform of equation (6.31) is

$$\left(.05s + .3 + \frac{10}{s}\right)Y = \frac{600}{s}(1 - e^{-.2s}).$$

Solving for Y, we have

$$Y = \frac{600(1 - e^{-.2s})}{.05s^2 + .3s + 10}.$$

To find the inverse Laplace transform of Y, we first calculate

$$L^{-1}\left(\frac{600}{.05s^2 + .3s + 10}\right) = L^{-1}\left(\frac{12000}{(s+3)^2 + 191}\right)$$

$$= \frac{12000}{\sqrt{191}}e^{-3t}\sin(\sqrt{191}t)$$

$$= Ce^{-3t}\sin(kt),$$

where $k = \sqrt{191}$ and $C = 12000/k \approx 868$. Then the inverse Laplace transform of $Y(s)$ is

$$y(t) = C[e^{-3t}\sin(kt) - e^{-3(t-0.2)}\sin(k(t - 0.2))u(t - 0.2)].$$

The current thus starts as a damped oscillating function. When the switch is thrown, another damped oscillating function is subtracted. The second function is identical to the first, but, because of the differing phase, cancellation does not occur. See Figure 6.14.

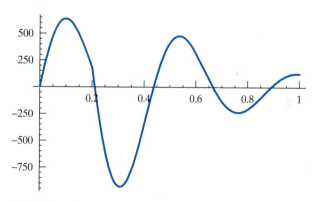

FIGURE 6.14   The current carried by the circuit shown in Figure 6.13.

EXERCISES

Use step functions and corner functions to produce expressions for the functions defined in Exercises 1–8.

1. $f(t) = \begin{cases} 0 & \text{if } t < 15, \text{ and} \\ 2 & \text{if } t \geq 15 \end{cases}$

2. $f(t) = \begin{cases} 0 & \text{if } t < 2, \text{ and} \\ 4 & \text{if } 2 \leq t < 5 \\ 0 & \text{if } 5 \leq t \end{cases}$

3. $f(t) = \begin{cases} 1 & \text{if } 0 < t < 2, \\ -1 & \text{if } 2 \leq t < 4, \text{ and} \\ 0 & \text{if } t \geq 4 \end{cases}$

4. $f(t) = \begin{cases} 0 & \text{if } t < 5, \text{ and} \\ 2t - 10 & \text{if } t \geq 5 \end{cases}$

5. $f(t) = |3t - 7|$

6. $f(t) = \begin{cases} e^{-2t} & \text{if } 0 \leq t < 1, \text{ and} \\ 0 & \text{if } t \geq 1 \end{cases}$

7. $f(t) = \begin{cases} t & \text{if } 0 < t < 1, \\ 2-t & \text{if } 1 \leq t < 3, \\ t-4 & \text{if } 3 \leq t < 4, \text{ and} \\ 0 & \text{if } t \geq 4 \end{cases}$

8. $f(t) = \begin{cases} \sin t & \text{if } 0 \leq t < \pi, \text{ and} \\ 0 & \text{if } t \geq \pi \end{cases}$

In Exercises 9–15, find the Laplace transform of the given function.

9. $f(t) = c(t-5)$ **10.** $f(t) = (t-1)^2 u(t-1)$

11. $f(t) = e^{3t}e^{-3}u(t-3)$ **12.** $f(t) = e^{-2t}(1-u(t-2))$

13. $f(t) = t^2 u(t-1)$

14. $f(t) = (t-\pi)u(t-\pi) - (t-2\pi)u(t-2\pi)$

15. $f(t) = (t^2 - 100)u(t-10)$

16. Find expressions for the functions whose graphs are shown in terms of step or corner functions, and calculate the Laplace transform of each.

a.

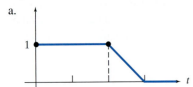

b.

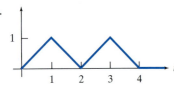

c.

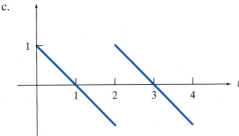

d.

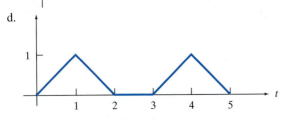

Find the inverse Laplace transforms of the expressions in Exercises 17–20.

17. $G(s) = \dfrac{1-e^{-s}}{s}$ **18.** $G(s) = \dfrac{1-e^{-s}}{s^2 + \pi^2}$

19. $G(s) = \dfrac{1+e^{-\pi s}}{s^2 + 1}$ **20.** $G(s) = \dfrac{2}{s^3}(1-e^{-2s})$

In Exercises 21–30, use the Laplace transform to solve the initial value problem. If the initial conditions are omitted, assume homogeneous initial conditions. Sketch a graph of the solution of each problem that is marked with an asterisk (*).

21. (*) $y'' + y = 1 - 2u(t-2\pi) + u(t-4\pi)$

22. $y'' - y = 4e^t(1 - u(t-1))$

23. (*) $y'' + 4y = 2(1 - u(t-\pi))\sin(2t)$

24. (*) $y'' + y = -c\left(t - \frac{\pi}{2}\right); \; y(0) = 1, \; y'(0) = 0$

25. $y'' + 2y' + y = u(t) - 2u(t-1) + u(t-2)$

26. (*) $y'' - 3y' + 2y = e^t - (e^t + 4e(t-2))u(t-1); \; y(0) = 0, \; y'(0) = -1$

27. (*) $y = u(t-1) + \int_0^t y(u)\,du$

28. $y' - y + 6\int_0^t y(u)\,du = u(t-3); \; y(0) = 1$

29. (*) $y' + \pi^2 \int_0^t y(u)\,du = 1 - u(t-1) + u(t-2); \; y(0) = 1$

30. (*) $y''' - y' = u(t-2)$

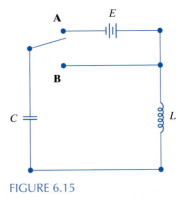

FIGURE 6.15

31. In the circuit shown in Figure 6.15, the switch is held in position **A** for π seconds, and then moved to position **B**. Find the current as a function of time, assuming that at $t = 0$, the current is 0, but the capacitor holds a charge of 0.6 coulombs.

6.6   The Delta Function

The **delta function** was defined in 1925 by the physicist P. A. M. Dirac, in connection with his work in quantum mechanics. The name is derived from Dirac's notation, $\delta(t)$.

The delta function is not really a function, although Dirac treated it as one. Let a denote a nonnegative constant. Then $\delta(t - a)$ denotes the "function" with the following properties:

$$\delta(t - a) = \begin{cases} 0 & \text{if } t \neq a, \\ \infty & \text{if } t = a; \end{cases} \tag{6.32}$$

and

$$\int_0^\infty \delta(t - a)\, dt = 1. \tag{6.33}$$

Because the delta function at $t = a$ has all of its effect at the time $t = a$, it is also called the *unit impulse* at $t = a$.

The defining properties of the delta function were difficult for mathematicians to assimilate. Since ∞ is not a number, the meaning of property (6.32) is not clear. Property (6.33) also seems wrong, since if a function is zero, except at one point, its integral ought to equal zero. Twenty years after Dirac defined the delta function, the French mathematician Laurent Schwartz defined a generalization of the concept of function, called a **distribution**. Delta functions are distributions, and Schwartz's theory placed them in the context of mathematical analysis.[1]

The theory of distributions is beyond the scope of this text. We will view the delta function as the "limit" of the **approximate delta function** f_ϵ. Let

$$f_\epsilon(t - a) = \begin{cases} \frac{1}{\epsilon} & \text{if } a \leq t < a + \epsilon \\ 0 & \text{otherwise.} \end{cases}$$

For every $\epsilon > 0$, and $a \geq 0$,

$$\int_0^\infty f_\epsilon(t - a)\, dt = 1,$$

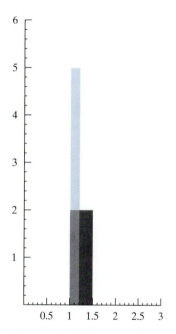

Figure 6.16. The approximate delta functions $f_{0.2}(t - 1)$ and $f_{0.5}(t - 1)$.

since this integral simply measures the area of a rectangle with width ϵ and height $1/\epsilon$ (see Figure 6.16). Further, $f_\epsilon(t - a) = 0$ except when $t \in [a, a + \epsilon)$. Thus, the approximate delta function assumes the characteristics of the delta function at $t = a$ as $\epsilon \to 0$.

In terms of step functions, $f_\epsilon(t - a) = \frac{1}{\epsilon}[u(t - a) - u(t - a - \epsilon)]$. Therefore, the Laplace transform of $f_\epsilon(t - a)$ is

$$L[f_\epsilon(t - a)] = \frac{1}{\epsilon}[L(u(t - a)) - L(u(t - a - \epsilon))]$$

$$= \frac{1}{\epsilon}\left[\frac{e^{-as}}{s} - \frac{e^{-(a+\epsilon)s}}{s}\right]$$

$$= e^{-as}\left[\frac{1 - e^{-\epsilon s}}{\epsilon s}\right].$$

---

[1] L. Schwartz, *Théorie des distributions*, Paris, Hermann, 1966. The first edition was published in 1950.

By l'Hôspital's rule, $\lim_{\epsilon \to 0} \frac{1 - e^{-\epsilon s}}{\epsilon s} = 1$. Therefore,

$$\lim_{\epsilon \to 0^+} L[f_\epsilon(t - a)] = e^{-as}.$$

This argument is intended to justify the formula

$$\boxed{L(\delta(t - a)) = e^{-as}.}$$

A proof can be given in the context of distribution theory.

The delta function can be multiplied by a continuous "ordinary" function, as follows:

Proposition 6.6.1

Suppose that $f(t)$ is continuous at $t = a$. Then

$$f(t)\delta(t - a) = f(a)\delta(t - a).$$

While this proposition is about distributions—making its proof inaccessible to us—consider the following intuitive argument: Since $\delta(t - a) = 0$ for $t \neq a$, the product $(f(t) - f(a))\delta(t - a) = 0$ for all t. The trouble with this "proof" is that it leaves unclear the reason that f needs to be continuous at $t = a$.

Example 6.6.1   Find the Laplace transform of

$$f(t) = t^2(\delta(t - 2) - \delta(t - 1) + \delta(t)).$$

Solution. By Proposition 6.6.1, $t^2\delta(t - 2) = 2^2\delta(t - 2)$, $t^2\delta(t - 1) = \delta(t - 1)$, and $t^2\delta(t) = 0$. Therefore,

$$L(f) = L(4\delta(t - 2)) - L(\delta(t - 1))$$
$$= 4e^{-2s} - e^{-s}.$$ ❑

The Laplace transform enables us to solve initial value problems involving linear differential equations with constant coefficients where the delta function appears in the source term.

Example 6.6.2   Solve the initial value problem

$$y'' + 4y' + 3y = -e^{-3t}\delta(t - 1); \quad y(0) = 0, \; y'(0) = 1.$$

Solution. Proposition 6.6.1 enables us to simplify the source term as $-e^{-3}\delta(t - 1)$. Put $Y = L(y)$ (assuming that y has a Laplace transform); then $L(y') = sY$ and $L(y'') = s^2Y - 1$. Furthermore, $L(-e^{-3}\delta(t - 1)) = -e^{-3}e^{-s}$. Therefore, the Laplace transform of the problem is

$$(s^2 + 4s + 3)Y = 1 - e^{-3}e^{-s}.$$

Solving for Y,

$$Y = \frac{1 - e^{-3-s}}{s^2 + 4s + 3}$$

$$= \frac{1 - e^{-3-s}}{2}\left[\frac{1}{s+1} - \frac{1}{s+3}\right].$$

The inverse Laplace transform of $\frac{1}{s+1} - \frac{1}{s+3}$ is $e^{-t} - e^{-3t}$. We can calculate $y(t)$ by using Proposition 6.5.2 on page 290:

$$y(t) = \frac{1}{2}[e^{-t} - e^{-3t}]u(t) - e^{-3}[e^{-(t-1)} - e^{-3(t-1)}]u(t-1). \qquad \Box$$

Physical applications of the delta function

In electrical circuits, the delta function can serve as a model for voltage spikes. In a spike event, the EMF applied to a circuit increases by thousands of volts and then decreases to a normal level, all in a time interval measured in billionths of a second, or *nanoseconds*. If the EMF is given by a function $E(t)$, then a spike occurring at time $t = a$ would change the EMF to $E_1(t) = E(t) + f(t - a)$, where f is a function whose value at $t - a$ is the spike voltage at time t. If the electrical potential is measured in volts, and time is in seconds, then $f(t - a)$ has a very large magnitude at $t = a$ and there is a number ϵ, whose order of magnitude is a few nanoseconds, such that $f(t - a) = 0$ for $|t - a| > \epsilon$. It would be nearly impossible to make the measurements necessary to graph f accurately. The best we can do is to approximate $f(t - a)$ by replacing it with a multiple $m\delta(t - a)$ of the delta function. The number m represents the magnitude of the impulse represented by the function f. Thus, we take

$$m = \int_{-\epsilon}^{\epsilon} f(s)\,ds \qquad (6.34)$$

so as to guarantee that

$$\int_{-\epsilon}^{\epsilon} f(s)\,ds = \int_{-\epsilon}^{\epsilon} m\delta(s)\,ds.$$

Equation (6.34) has the surprising consequence that the units in which the magnitude of an electrical potential spike is measured are volt-seconds, since this magnitude represents the area below a graph in a coordinate system where the vertical and horizontal axes are denominated in volts and seconds, respectively. If we use the delta function as a model for the potential spike, then the EMF would be

$$E_1(t) = E(t) + m\delta(t - a). \qquad (6.35)$$

Each term in (6.35) must have the same units: volts. The units of the parameter m are volt·seconds, and this is compensated for by the units of $\delta(t)$, which are inverse seconds.

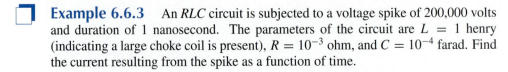

Example 6.6.3   An *RLC* circuit is subjected to a voltage spike of 200,000 volts and duration of 1 nanosecond. The parameters of the circuit are $L = 1$ henry (indicating a large choke coil is present), $R = 10^{-3}$ ohm, and $C = 10^{-4}$ farad. Find the current resulting from the spike as a function of time.

Solution. Let $y(t)$ denote the current, in amperes. Then y satisfies the differential-integral equation,

$$Ly' + Ry + \frac{1}{C}\left(x_0 + \int_0^t y(u)\,du\right) = E(t),$$

where x_0 denotes the initial charge of the capacitor, and $E(t)$ is the electrical potential. We will assume that $x_0 = 0$, since we are only interested in the consequences of the voltage spike—which could not have put an initial charge on the capacitor. We will approximate $E(t)$ as $m\delta(t)$, where m is the magnitude of the spike. Thus

$$m = \text{overvoltage} \times \text{duration}$$
$$= 200,000 \times 10^{-9}$$
$$= 2 \times 10^{-4} \text{ volt-seconds.}$$

Thus we have the differential-integral equation

$$y' + 10^{-3}y + 10^4 \int_0^t y(u)\,du = 10^{-4}\delta(t).$$

Assuming that $y(0) = 0$, the Laplace transform of this equation is

$$sY + 0.001Y + \frac{10000}{s}Y = 0.0001$$

so that

$$Y = \frac{0.0001s}{s^2 + 0.001s + 10000} \approx 0.0001\frac{s + 0.0005}{(s + 0.0005)^2 + 100^2}.$$

It follows that $y \approx 0.0001e^{-.0005t}\cos(100t)$.   ❑

The delta function is also used in mechanical problems. Consider a damped mechanical system with a weight of mass m, damping b, and a spring with stiffness k. If a force of $p\epsilon^{-1}$ newtons is applied from $t = 0$ until $t = \epsilon$, and then withdrawn, we can use the ODE

$$my'' + by' + ky = pf_\epsilon(t)$$

as a model for the displacement $y(t)$ of the weight. In the limiting case, where the force is considered to have infinite magnitude over a time interval of length zero, we have

$$my'' + by' + ky = p\delta(t). \tag{6.36}$$

Let us assume that the weight in this system is initially at rest in its equilibrium position. Then the Laplace transform of (6.36) is

$$(ms^2 + bs + k)Y = p, \tag{6.37}$$

where $Y = L(y)$. Compare this result with the same system when, instead of the impulsive force, the weight is given an initial velocity v_0, and its initial position is still the equilibrium position. In this case, $L(by') = bY$ and $L(my'') = ms^2Y - mv_0$. It follows that this unforced case, Y is determined by the equation

$$(ms^2 + bs + k)Y = mv_0. \tag{6.38}$$

Recall from physics that the momentum of an object of mass m and velocity $\mathbf{v}$ is defined to be the product $\mathbf{p} = m\mathbf{v}$. Thus, our comparison of (6.37) and (6.38) reveals that the following two initial situations lead to the same motion:

- The weight is in the equilibrium position and motionless; an impulsive force of magnitude p is applied.
- The weight is in the equilibrium position with initial momentum $mv_0 = p$; no force is applied.

For example, suppose that the weight of a simple mass-spring system is struck with a hammer. The **law of conservation of momentum** states that the sum of the momenta of the objects in a closed system is constant. Thus, if $\mathbf{p}_0$ denotes the momentum of the hammer before it strikes the weight, and $\mathbf{p}_w$ and $\mathbf{p}_h$ denote the momenta of the weight and hammer, respectively, afterward, then

$$\mathbf{p}_0 = \mathbf{p}_w + \mathbf{p}_h.$$

The kinetic energy of an object with mass m and velocity $\mathbf{v}$ is the scalar quantity given by the formula $k = \frac{1}{2}m\|\mathbf{v}\|^2$. The **law of conservation of energy** also applies to say that the total kinetic energy of the system after the hammer strikes is equal to the kinetic energy of the hammer before it struck. If k_0, k_h, and k_w denote the kinetic energy of the hammer before it struck, after it struck, and the kinetic energy of the weight after being struck, respectively, then we have

$$k_0 = k_h + k_w.$$

In Exercise 19 at the end of this section, you are asked to use the preceding two conservation laws show that

$$\mathbf{p}_h = \frac{m_h - m_w}{m_h + m_w}\mathbf{p}_0 \quad \text{and} \quad \mathbf{p}_w = \frac{2m_w}{m_h + m_w}\mathbf{p}_0,$$

where m_h and m_w denote the masses of the hammer and weight, respectively.

Example 6.6.4   The weight in a mechanical system has a mass of 0.1 kilogram, the damping is 0.02 kilogram per second, and the stiffness is 1000 newtons per meter. At $t = 0$ the weight, which is motionless and in its equilibrium position, is struck by a hammer of mass 0.5 kilogram, traveling at 30 meters per second. Find an ODE that that serves as a model for the displacement function of the weight, and solve it.

Solution. The momentum of the hammer before striking the weight is $\mathbf{p}_0 = 0.5 \times 30 = 15$ kilogram meter per second. The momentum imparted to the weight by the collision is

$$p_w = \frac{2(0.1)}{0.1 + 0.5}p_0 = 5 \text{ kilogram meter per second.}$$

Thus, the displacement y satisfies the ODE

$$0.1y'' + 0.02y' + 1000y = 5\delta(t)$$

with homogeneous initial conditions. Taking the Laplace transform, with $Y = L(y)$, we have

$$(0.1s^2 + 0.02s + 1000)Y = 5,$$

and hence

$$Y = \frac{50}{(s+0.1)^2 + 9999.99} \approx \frac{1}{2}\frac{100}{(s+0.1)^2 + 100^2} = \frac{1}{2}L(e^{-.01t}\sin(100t)).$$

It follows that the displacement of the weight is $y = \frac{1}{2}e^{-.01t}\sin(100t)$.   ❑

As a final example, we will consider a mechanical system that is subjected to repeated impulsive forces.

Example 6.6.5   A mechanical system consists of a weight of mass 1 kilogram connected to a spring of stiffness π^2 newtons per meter. There is no damping. At 1-second intervals, an impulsive force of 2 kilogram meters per second is applied. Assuming that the weight is initially at rest in its equilibrium position, determine the displacement function.

Solution. We can express the repeated impulses in the form

$$F = 2\delta(t) + 2\delta(t-1) + 2\delta(t-2) + \cdots .$$

Here, $\delta(t)$ delivers the first blow, $\delta(t-1)$ delivers the second, and so on. The sum of these delta functions is the external force on the mechanical system. Thus, the displacement function is determined by the ODE

$$y'' + \pi^2 y = 2\delta(t) + 2\delta(t-1) + 2\delta(t-2) + \cdots \qquad (6.39)$$

with homogeneous initial conditions. Let $Y = L(y)$. The Laplace transform of (6.39) is then

$$s^2 Y + \pi^2 Y = 2 + 2e^{-s} + 2e^{-2s} + \cdots .$$

Hence

$$Y = \frac{2}{\pi}\left(\frac{\pi}{s^2 + \pi^2} + \frac{\pi e^{-s}}{s^2 + \pi^2} + \frac{\pi e^{-2s}}{s^2 + \pi^2} + \cdots\right).$$

The inverse Laplace transform of $\frac{\pi}{s^2+\pi^2}$ is $\sin(\pi t)$, and if k is an integer,

$$
\begin{aligned}
L^{-1}\left(\frac{\pi e^{-ks}}{s^2 + \pi^2}\right) &= \sin(\pi(t-k))u(t-k) \\
&= (-1)^k \sin(\pi t)u(t-k) \\
&= \begin{cases} (-1)^k \sin(\pi t) & \text{if } k \le \lfloor t \rfloor \\ 0 & \text{if } k > \lfloor t \rfloor, \end{cases}
\end{aligned}
$$

where $\lfloor t \rfloor$ denotes the greatest integer $\le t$. Thus

$$
\begin{aligned}
L^{-1}(Y) &= \frac{2}{\pi}(\sin(\pi t) - \sin(\pi t)u(t-1) + \sin(\pi t)u(t-2) - \cdots \\
&\quad \pm \sin(\pi t)u(t - \lfloor t \rfloor)) \\
&= \frac{2}{\pi}\sin(\pi t)\underbrace{(1 - 1 + 1 - \cdots \pm 1)}_{\lfloor t \rfloor + 1 \text{ terms}} \\
&= \begin{cases} \frac{2}{\pi}\sin(\pi t) & \text{if } \lfloor t \rfloor \text{ is even} \\ 0 & \text{if } \lfloor t \rfloor \text{ is odd.} \end{cases}
\end{aligned}
$$

Thus, the first impulse sets the weight into simple harmonic motion, and the second impulse stops it. The third impulse starts the process again. The graph of the displacement function is shown in Figure 6.17. ❑

FIGURE 6.17 Displacement of a mechanical system that is repeatedly struck by a hammer. See Example 6.6.5.

Integrating the delta function

Since $g(t)\delta(t-a) = g(a)\delta(t-a)$ and $\int_0^\infty \delta(t-a)\,dt = 1$, it is plausible that, for $a \ge 0$,

$$
\int_0^\infty g(t)\delta(t-a)\,dt = g(a). \tag{6.40}
$$

Since the delta function is not a function, (6.40) cannot be proved by referring to theorems about integrals of functions. We will therefore accept equation (6.40) as a *definition,* provided that g is continuous on some interval (b, c) containing a. If $g(t)$ is discontinuous at $t = a$, then we will consider $\int_0^\infty g(t)\delta(t-a)\,dt$ to be undefined. The following theorem shows that (6.40) is consistent with the notion that $\delta(t-a) = \lim_{\epsilon \to 0^+} f_\epsilon(t-a)$.

Theorem 6.2

Suppose that $g(t)$ is continuous on an open interval (b, c) containing $t = a$, where $a \geq 0$. Then

$$\lim_{\epsilon \to 0^+} \int_0^\infty g(t) f_\epsilon(t - a)\, dt = g(a).$$

Proof Since $f_\epsilon(t - a) = 0$ for $t < a$ and $t > a + \epsilon$,

$$\int_0^\infty g(t) f_\epsilon(t - a)\, dt = \int_a^{a+\epsilon} g(t) f_\epsilon(t - a)\, dt.$$

Furthermore, $f_\epsilon(t - a) = 1/\epsilon$ for $a \leq t < a + \epsilon$, so

$$\int_a^{a+\epsilon} g(t) f_\epsilon(t - a)\, dt = \frac{1}{\epsilon} \int_a^{a+\epsilon} g(t)\, dt,$$

and this latter integral represents the average value of $g(t)$ on the interval $[a, a + \epsilon]$. If ϵ is sufficiently small, g will be continuous on this interval, and then, by the mean value theorem for integrals, there will be a number $\xi \in (a, a + \epsilon)$ such that

$$\frac{1}{\epsilon} \int_a^{a+\epsilon} g(t)\, dt = g(\xi).$$

As $\epsilon \to 0^+$, $\xi \to a$, and by the continuity of g, $g(\xi) \to g(a)$. ∎

EXERCISES

In each of Exercises 1–4, find the Laplace transform.

1. $\delta(t - 1) + \delta(t - 2) + 3\delta(t - 3)$ **2.** $e^{-t} - 5\delta(t - 1)$

3. $\sqrt{t^2 + 6t}\,\delta(t - 2)$ **4.** $(t - 1)\delta(t - 1)$

In Exercises 5–14, solve the initial value problem. Sketch graph of the solutions of Exercises 5–11. If the initial conditions are not stated, they are to be assumed to be homogeneous.

5. $y' = \delta(t - 1) + y;\ y(0) = 1$

6. $y'' + 4\pi^2 y = 2\delta\left(t - \frac{1}{2}\right) + \delta(t - 1);\ y(0) = 0,\ y'(0) = 1$

7. $y'' + 4\pi^2 y = -2\delta\left(t - \frac{1}{2}\right) + \delta(t - 1);\ y(0) = 0,\ y'(0) = 1$

8. $y'' + 2y' + y = 3\delta(t - 2);\ y(0) = 0,\ y'(0) = 1$

9. $y'' + 4y' + 29y = \delta(t - \pi);\ y(0) = 0,\ y'(0) = 1$

10. $y'' - 3y' + 2y = 1 - \delta(t - 2)$

11. $4y'' + 4y' + y = e^{-t/2}(1 + 4\delta(t - 1))$

12. $y''' + y'' + y' + y = \delta(t - \pi) + \delta(t - 2\pi);\ y(0) = 0,$ $y'(0) = 0,\ y''(0) = 1$

13. $y''' - 7y' + 6y = 1 - u(t - 1)$

14. $y^{(4)} + 2y'' + y = 2\delta(t - \pi) + \delta(t - 2\pi);\ y(0) = y'(0) = y''(0) = 0,\ y'''(0) = 1$

15. In the circuit of Example 6.6.3, the choke coil is replaced by one of inductance $L = 10^{-4}$ henry. Determine the current that results from the power spike as a function of time. Discuss the qualitative difference in the behavior of the circuit resulting from this change in inductance. What would happen if there were no choke coil at all in this circuit?

16. In the circuit shown, a voltage spike of 10^{-3} volt seconds occurs across the terminals labeled **A** and **B**. Determine the current in the resistor **R** as a function of time, assuming that before the spike occurs, no current is flowing, and the capacitor has no charge.

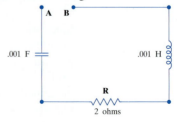

17. The weight in Example 6.6.4 is struck a second time by the hammer, $\pi/100$ seconds after the first impulse. Determine the displacement function for the weight.

18. A weight of 1 kilogram is suspended from a spring of stiffness 100 newtons per meter. A second, identical weight is attached to the first weight by an identical spring. Both weights are motionless in their equilibrium positions, when the bottom weight is propelled upward as the result of a collision with a hammer that also weighs 1 kilogram, and with velocity 50 meters per second. Determine the displacement functions of the two weights.

19. A hammer of mass m_h strikes a motionless weight of mass m_w. Let the initial momentum of the hammer be denoted p_0, and let the momenta of the hammer and weight post-collision be denoted p_h and p_w, respectively. Use the laws

of conservation of momentum and energy to show that

$$p_h = \frac{m_h - m_w}{m_h + m_w} p_0 \quad \text{and} \quad p_w = \frac{2m_w}{m_h + m_w} p_0.$$

◆ **Hint:** Show that in terms of momentum, the law of conservation of energy can be rephrased as

$$\frac{p_0^2}{m_h} = \frac{p_h^2}{m_h} + \frac{p_w^2}{m_w}$$

and proceed from there.

20. The Laplace transform of $\delta(t)$ is 1. How can this fact be reconciled with the statement in Proposition 6.1.2 on page 258 that if $F(s)$ is the Laplace transform of a function $f(t)$, then $\lim_{s \to \infty} F(s) = 0$?

6.7 Convolution

Suppose that there is given a linear ODE with constant coefficients,

$$a_2 y'' + a_1 y' + a_0 y = f(t), \tag{6.41}$$

where the coefficients a_i are known, but we want a formula for a solution that depends on the source function f. This type of problem is encountered with a mechanical system in which the mass and the spring and damping constants are fixed, but we need to compute the response to a variety of forcing functions.

The variation of constants formula would be one solution of this problem, but it is also possible to use the Laplace transform. Taking homogeneous initial conditions, the transform of (6.41) is

$$(a_2 s^2 + a_1 s + a_0)Y = F(s),$$

where $L(f) = F(s)$. Put

$$H(s) = \frac{1}{a_2 s^2 + a_1 s + a_0},$$

so that the Laplace transform of the solution of (6.41) is

$$Y(s) = H(s)F(s). \tag{6.42}$$

With (6.42) we have achieved half of our objective. The remaining task is to find the inverse Laplace transform of (6.42).

The function $H(s)$ is called the **transfer function** associated with the LDO

$$\mathcal{L}(y) = a_2 y'' + a_1 y' + a_0 y.$$

Let $\phi(t)$ denote the inverse Laplace transform of $H(s)$. The function $\phi(t)$ is called the **impulse response** for $\mathcal{L}$, because it turns out to be equal to the solution of the ODE,

$$\mathcal{L}(y) = \delta(t), \tag{6.43}$$

with homogeneous initial conditions at $t = 0$. Here is the proof: By (6.42), the Laplace transform of the solution of (6.43) is $Y(s) = H(s) \cdot L(\delta(t)) = H(s) \cdot 1$. Hence $y(t) = L^{-1}(H(s)) = \phi(t)$.

We will express the inverse Laplace transform of (6.42) in terms of the inverse transforms $\phi(t)$ and $f(t)$ of $H(s)$ and $F(s)$, respectively, by developing a new type of product of functions.

Let $f(t)$ and $g(t)$ be locally integrable functions defined for $t \geq 0$. The **convolution product** of f and g is a new function, denoted $f * g(t)$, given by the formula

$$f * g(t) = \int_0^t f(t-u)g(u)\, du.$$

Example 6.7.1 Let $f(t) = 1$ and $g(t) = t^2$. Calculate $f * g(t)$.

Solution. $1 * t^2 = \displaystyle\int_0^t 1 \cdot u^2\, du = \frac{1}{3}t^3$

Example 6.7.2 Calculate $(\sin t) * (\cos 2t)$.

Solution.

$$(\sin t) * (\cos 2t) = \int_0^t \sin(t-u)\cos(2u)\, du.$$

$$= \frac{1}{2}\int_0^t [\sin(t+u) + \sin(t-3u)]\, du,$$

where we have used the identity $\sin P \cos Q = \frac{1}{2}[\sin(P+Q) + \sin(P-Q)]$. Thus

$$(\sin t) * (\cos 2t) = \frac{1}{2}\left[-\cos(t+u) + \frac{1}{3}\cos(t-3u)\right]\Big|_0^t$$

$$= \frac{1}{3}(\cos t - \cos 2t).$$

Example 6.7.3 Calculate $e^{at} * e^{bt}$, where a and b are constants.

Solution. Assume that $a \neq b$. Then

$$e^{at} * e^{bt} = \int_0^t e^{a(t-u)}e^{bu}\, du$$

$$= e^{at}\int_0^t e^{(b-a)u}\, du$$

$$= e^{at} \cdot \frac{1}{b-a}(e^{(b-a)t} - 1)$$

$$= \frac{e^{bt} - e^{at}}{b-a}.$$

If $a = b$, then

$$e^{at} * e^{at} = e^{at} \int_0^t \underbrace{e^{(a-a)u}}_{1} \, du$$

$$= t e^{at}.$$ ❑

The convolution theorem

The following theorem relates the Laplace transform and the convolution product.

Theorem 6.3   Convolution Theorem

Let $f(t)$ and $g(t)$ be functions that have Laplace transforms. Then $f * g(t)$ has a Laplace transform as well, and

$$L(f * g) = L(f)L(g).$$

Proof   Let $F(s) = L(f)$ and $G(s) = L(g)$. By the definition of the Laplace transform,

$$F(s)G(s) = \int_0^\infty e^{-su} f(u) \, du \int_0^\infty e^{-sv} g(v) \, dv.$$

We can write this product as a double integral

$$F(s)G(s) = \int_0^\infty \int_0^\infty e^{-su} e^{-sv} f(u) g(v) \, du \, dv$$

$$= \int \int_{\mathcal{R}} e^{-s(u+v)} f(u) g(v) \, dA,$$

where $\mathcal{R}$ is the first quadrant.

The change of variables $r = v$, $t = u + v$ has unit Jacobian and takes the first quadrant to the region

$$\mathcal{R}' = \{(r, t) : 0 \le r \le t\}.$$

It follows that

$$F(s)G(s) = \int \int_{\mathcal{R}'} e^{-st} f(t - r) g(r) \, dA.$$

If we replace this integral over $\mathcal{R}'$ by an iterated integral, we have

$$F(s)G(s) = \int_0^\infty \int_0^t e^{-st} f(t - r) g(r) \, dr \, dt$$

$$= \int_0^\infty e^{-st} [f * g](t) \, dt$$

$$= L(f * g).$$ ■

Theorem 6.3 can be restated in terms of inverse Laplace transforms as follows: *"The inverse transform of a product $L(f)L(g)$ is a convolution product $f * g$."*

Recall that in (6.42), we determined that the Laplace transform of the solution $y(t)$ of $\mathcal{L}(y) = f(t)$ with homogeneous initial conditions at $t = 0$ was $Y(s) = H(s)F(s)$, where H is the transfer function associated with the operator $\mathcal{L}$ and $F = L(f)$. Applying the convolution theorem, we can take the inverse Laplace transform of this product to obtain $y(t) = \phi * f(t)$, or

$$y(t) = \int_0^t \phi(t - s)f(s)\,ds, \tag{6.44}$$

where $\phi = L^{-1}(H)$ is the impulse response of $\mathcal{L}$.

Example 6.7.4   Find a general formula for solutions of

$$y'' + 4y = f(t),$$

with homogeneous initial conditions at $t = 0$. Use the formula to determine the solution when $f(t) = \sin(2t)$.

Solution. The transfer function is $H(s) = 1/(s^2 + 4)$, so the impulse response is $\phi(t) = L^{-1}(H) = \frac{1}{2}\sin(2t)$. Hence the solution is $y(t) = \frac{1}{2}\sin(2t) * f(t)$, or

$$y(t) = \int_0^t \frac{1}{2}\sin(2(t - s))f(s)\,ds.$$

The same formula was obtained by variation of constants in Example 5.8.3 on page 244.

If $f(t) = \sin(2t)$, we have

$$y(t) = \frac{1}{2}\sin(2t) * \sin(2t)$$

$$= \frac{1}{2}\int_0^t \sin(2(t - s))\sin(2s)\,ds.$$

To evaluate this integral, we use the trigonometric identity

$$\sin(P)\sin(Q) = \frac{1}{2}[\cos(P - Q) - \cos(P + Q)].$$

Thus

$$y(t) = \frac{1}{4}\int_0^t [\cos(2(t - s) - 2s) - \cos(2(t - s) + 2s)]\,ds$$

$$= \frac{1}{4}\left[-\frac{1}{4}\sin(2t - 4s) - s\cos(2t)\right]\Big|_0^t$$

$$= \frac{1}{8}\sin(2t) - \frac{1}{4}t\cos(2t).$$

The convolution product of functions obeys the commutative, distributive, and associative laws, just as the ordinary product does. This is the content of the following proposition.

Proposition 6.7.1

Let $f(t)$, $g(t)$, and $h(t)$ denote functions that are locally integrable. Then

Commutative Law $f * g(t) = g * f(t)$

Distributive Law $f * (g + h)(t) = f * g(t) + f * h(t)$

Associative Law $f * (g * h)(t) = (f * g) * h(t)$

Furthermore, the delta function serves as the unit: For any function g that is continuous on $[0, \infty)$, $\delta * g = g$.

We will omit the proof of this proposition. However, the convolution theorem can be used to see that it is plausible. If each of the functions has a Laplace transform [denoted $F(s)$, $G(s)$, and $H(s)$, respectively], then

$$L(f * g) = F(s)G(s)$$
$$= G(s)F(s)$$
$$= L(g * f).$$

Since the Laplace transforms of $f * g$ and $g * f$ are equal, $f * g = g * f$. The other two laws can be justified in the same way.

To see that the delta function is the unit for the convolution product, we refer to equation (6.40) on page 301, which states that $\int_0^\infty \delta(t - s)g(s)\, ds = g(t)$. Since $\delta(t - s) = 0$ for $s > t$, it follows that

$$\delta * g(t) = \int_0^t \delta(t - s)g(s)\, ds = g(t). \qquad \blacksquare$$

If we want to calculate the convolution $\delta(t - a) * g(t)$, we may use the Laplace transform:

$$L(\delta(t - a) * g(t)) = L(\delta(t - a))L(g(t))$$
$$= e^{-as}G(s).$$

By Proposition 6.5.2 on page 290, it follows that

$$\delta(t - a) * g(t) = u(t - a)g(t - a). \tag{6.45}$$

It may be surprising that the unit for the convolution product is the delta function, rather than the constant function 1. However, we can refer to Example 6.7.1 to see that $1 * t^2 = t^3/3$, and see that 1 is certainly not the unit. In Exercise 2 you will be asked to show that the only function that satisfies $1 * f \equiv f$ is $f \equiv 0$.

The convolution theorem enables us to compute the inverse Laplace transform of expressions of the form

$$\frac{A(s - k) + Cb}{((s - k)^2 + b^2)^n}. \tag{6.46}$$

This is illustrated in the following example.

Example 6.7.5   Find the inverse Laplace transform of

$$Y = \frac{4}{s^5 + s^4 + 2s^3 + 2s^2 + s + 1}.$$

Solution. The denominator can be factored as $(s + 1)(s^2 + 1)^2$, and we take this as the cue to decompose Y as the following partial fractions expression:

$$Y = \frac{A}{s + 1} + \frac{Bs + C}{s^2 + 1} + \frac{Es + F}{(s^2 + 1)^2}.$$

Thus,

$$(s + 1)(s^2 + 1)^2 Y = 4$$
$$= A(s^2 + 1)^2 + (Bs + C)(s + 1)(s^2 + 1) + (Es + F)(s + 1)$$
$$= (A + B)s^4 + (B + C)s^3 + (2A + B + C + E)s^2$$
$$+ (B + C + E + F)s + A + C + F.$$

By equating the coefficients s^4, s^3, s^2, and s to 0, and equating the constant term $A + C + F$ to 1, we obtain the following system of equations:

$$\begin{aligned}
A + B &= 0 \\
B + C &= 0 \\
2A + B + C + E &= 0 \\
B + C + E + F &= 0 \\
A + C + F &= 4.
\end{aligned}$$

The solution of these equations is $A = 1$, $B = -1$, $C = 1$, $E = -2$, and $F = 2$. Thus

$$Y = \frac{1}{s + 1} + \frac{-s + 1}{s^2 + 1} + \frac{2}{s^2 + 1} \cdot \frac{-s + 1}{s^2 + 1}.$$

We can take the inverse Laplace transform of each term, using the convolution theorem for the last one:

$$y = e^{-t} - \cos(t) + \sin(t) + 2\sin(t) * (-\cos(t) + \sin(t)).$$

Finally, let's compute the convolution:

$$\sin(t) * (-\cos(t) + \sin(t)) = \int_0^t \sin(t - u)(-\cos(u) + \sin(u))\,du$$
$$= \frac{1}{2} \int_0^t (-\sin(t) - \sin(t - 2u) + \cos(t - 2u) - \cos(t))\,du$$
$$= -\frac{1}{2}[t(\cos(t) + \sin(t)) - \sin(t)].$$

Thus

$$y = e^{-t} - \cos(t) + 2\sin(t) - t(\cos(t) + \sin(t)). \qquad \square$$

EXERCISES

1. Assume that $f(t)$ has a Laplace transform. Use the convolution theorem to prove that

$$1 * f(t) = \int_0^t f(u)\,du.$$

2. Prove that if $1 * f(t) \equiv f(t)$, then $f(t) \equiv 0$.

◆ **Hint:** Use the result of Exercise 1.

In Exercises 3–6, calculate the convolution.

3. $t * \sin t$ **4.** $\sin 4t * e^{3t}$

5. $e^t * e^{-t}$ **6.** $e^t * e^{2t} * e^{3t}$

7. Calculate $\delta(t - 2\pi) * \sin(t)$.

8. Calculate $\delta(t - \ln(2)) * (e^t - 1)$.

9. Give a formula for the convolution product

$$\underbrace{1 * 1 * \cdots * 1}_{n}.$$

10. Calculate $L^{-1}\left(\dfrac{1}{(s^2 + 1)^2}\right)$.

11. Calculate $L^{-1}\left(\dfrac{s^2}{(s^2 + 1)^2}\right)$.

12. Use the result of Exercise 9 to develop a formula for $t^n * f(t)$.

13. Find a function $f(t)$ with the property that $f * f * f(t) = t^8$.

◆ **Hint:** Find $L(f)$ first.

14. Let $f(t) = (\sin 4t) * (\cos 3t)$. Calculate the Laplace transform of

a. $e^{2t} f(t)$ b. $f'(t)$

15. Show that $e^{at}[f * g(t)] = [e^{at} f(t)] * [e^{at} g(t)]$.

16. Show that $t[f * g(t)] = [tf(t)] * [g(t)] + [f(t)] * [tg(t)]$.

In Exercises 17–24, determine the transfer function and impulse response for the LDO $\mathcal{L}$. Write down an integral formula for the solution of the equation

$$\mathcal{L}(y) = g(t)$$

with homogeneous initial conditions at $t = 0$.

17. $\mathcal{L}(y) = y' + 4y$ **18.** $\mathcal{L}(y) = y'' + 4y' + 3y$

19. $\mathcal{L}(y) = 4y'' + y$ **20.** $\mathcal{L}(y) = 6y''' - 11y'' + 6y' - y$

21. $\mathcal{L}(y) = y^{(4)} + 2y'' + y$ **22.** $\mathcal{L}(y) = y^{(4)} - 2y'' + y$

23. $\mathcal{L}(y) = y' - \int_0^t e^{t-u} y(u)\,du$

24. $\mathcal{L}(y) = y'' - 9(te^{-2t}) * y$

25. Use the definition of the convolution product and equation (6.40) on page 301 to prove the formula (6.45) for $\delta(t - a) * g(t)$.

6.8   The Square Wave

A function $f(t)$ is **periodic** if there is a number $T > 0$ such that for all t, $f(t+T) = f(t)$. The number T is called a *period* of f. If T is a period of f, any integer multiple of T is also a period.

The trigonometric functions are familiar examples of periodic functions. Another example is the **square wave** $S(t)$, shown in Figure 6.18. There are several ways to write a formula for the square wave. For example, you can show that $S(t) = (-1)^{\lfloor t \rfloor}$, where $\lfloor t \rfloor$ denotes the greatest integer $\leq t$, and also that $S(t) = 2u(\sin(\pi t)) - 1$.

A third way of expressing $S(t)$ will be more convenient for us. For $t \geq 0$,

$$S(t) = 1 - 2u(t - 1) + 2u(t - 2) - 2u(t - 3) + 2u(t - 4) - 2u(t - 5) + \cdots .$$
$$(6.47)$$

While this looks like an infinite sum, for any given value of t, all terms after the one involving $u(t - \lfloor t \rfloor)$ are zero. The proof of (6.47) is as follows. The starting value of $S(t)$ is $S(0) = 1$. At each odd integer n, there is a jump discontinuity of -2; this is put in place by the term, $-2u(t - n)$. If n is even, the term $2u(t - n)$ causes a jump from -1 to 1 at $t = n$.

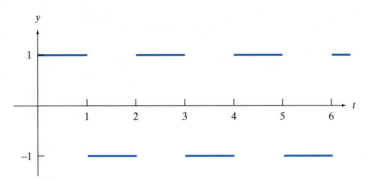

FIGURE 6.18   A square wave.

We can put (6.47) more compactly as

$$S(t) = 1 + 2\sum_{n=1}^{\infty}(-1)^n u(t-n), \text{ for } t \geq 0.$$

To compute the Laplace transform of $S(t)$, we work on this sum term by term:

$$L(S(t)) = L(1) + 2\sum_{n=1}^{\infty}(-1)^n L(u(t-n))$$

$$= \frac{1}{s} + \frac{2}{s}\sum_{n=1}^{\infty}(-1)^n e^{-ns}.$$

Example 6.8.1   Solve the IVP

$$y'' + k^2 y = S(t)$$

with homogeneous initial conditions at $t = 0$.

Solution. The transfer function for $\mathcal{L}(y) = y'' + k^2 y$ is $H(s) = \frac{1}{s^2+k^2}$. Denoting the Laplace transform of the solution of the IVP by Y,

$$Y = H(s)L(S(t))$$

$$= \frac{1}{s(s^2+k^2)}\left(1 + 2\sum_{n=1}^{\infty}(-1)^n e^{-ns}\right)$$

$$= \frac{1}{s(s^2+k^2)} + 2\sum_{n=1}^{\infty}(-1)^n \frac{e^{-ns}}{s(s^2+k^2)}.$$

To determine the inverse Laplace transform of Y, start with the partial fraction expansion,

$$\frac{1}{s(s^2+k^2)} = \frac{1}{k^2}\left(\frac{1}{s} - \frac{s}{s^2+k^2}\right),$$

and then take the inverse Laplace transform of each term of Y. Thus

$$L^{-1}\left(\frac{1}{s(s^2+k^2)}\right) = \frac{1}{k^2}(1-\cos(kt))$$

and

$$L^{-1}\left(\frac{e^{-ns}}{s(s^2+k^2)}\right) = \frac{1}{k^2}(1-\cos(k(t-n)))u(t-n).$$

The solution of the IVP is obtained by combining these expressions:

$$y = \frac{1}{k^2}\left(1-\cos(kt) + 2\sum_{n=1}^{\infty}(-1)^n(1-\cos(k(t-n)))u(t-n)\right)$$

$$= \frac{1}{k^2}\left(1 + 2\sum_{n=1}^{\infty}(-1)^n u(t-n) - \cos(kt) - 2\sum_{n=1}^{\infty}(-1)^n\cos(k(t-n))u(t-n)\right)$$

$$= \frac{1}{k^2}\left(S(t) - \cos(kt) - 2\sum_{n=1}^{\infty}(-1)^n\cos(k(t-n))u(t-n)\right).$$

Figure 6.19 displays a graphs of $y(t)$ for $k = 1$, π, 2π, and 3π. ❑

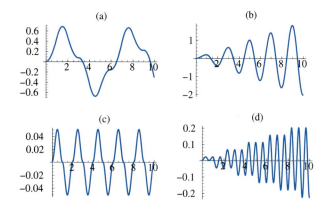

FIGURE 6.19   Solutions of $y'' + k^2 y = S(t)$ for (a) $k = 1$, (b) $k = \pi$, (c) $k = 2\pi$, and (d) $k = 3\pi$.

If we examine Figure 6.19 closely, we will see that all of the solutions are smooth functions, even though the formula for the solution includes the square wave, which is not continuous. To explain this, let

$$f(t) = (1-\cos(kt))/k^2.$$

It is easy to verify that $f(0) = 0$ and $f'(0) = 0$. Thus, for any number n, the function $f_n(t) = (-1)^n f(t-n)u(t-n)$ is a differentiable function on **R**. Since the solution of the IVP can be expressed as

$$y = f(t) + 2\sum_{n=1}^{\infty} f_n(t),$$

it too is differentiable.

When $k = 1$, the solution is bounded but not periodic. The reason is that f is $2\pi/k$-periodic; for $k = 1$ this means the period is 2π. Since $f(t-n) = 1-\cos(t-n)$ is out of phase with $f(t)$, the terms of the sum neither reinforce each other nor cancel each other.

When $k = \pi$, the solution is not bounded. In this case, f is 2-periodic, and

$$f(t - n) = 1 - \cos(\pi t - n\pi) = 1 - (-1)^n \cos(\pi t).$$

Therefore, the solution is

$$y = \frac{1}{\pi^2}\left(S(t) - \cos(\pi t) - 2\sum_{n=1}^{\infty}(-1)^n(-1)^n \cos(\pi t))u(t - n)\right)$$

$$= \frac{1}{\pi^2}\left(S(t) - \cos(\pi t)\left(1 + \sum_{n=1}^{\infty} u(t - n)\right)\right).$$

Since

$$\sum_{n=1}^{\infty} u(t - n) = u(t - 1) + u(t - 2) + \cdots u(t - \lfloor t \rfloor) = \lfloor t \rfloor$$

we can write the solution in this case as

$$y = \frac{1}{\pi^2}\left[S(t) - (1 + 2\lfloor t \rfloor)\cos(\pi t)\right].$$

This solution exhibits *resonance*. Although the source term is bounded, the solution is unbounded, due to the factor $(1 + 2\lfloor t \rfloor)$. We know that resonance occurs in solutions of the ODE

$$y'' + k^2 y = \sin(\pi t)$$

if the frequency of the impulse response of the LDO $y'' + k^2 y$ is equal to the forcing frequency; that is, if $k = \pi$. We can interpret this fact physically as a statement that a mechanical system will resonate if it is forced at its natural frequency of vibration. We have just found that if the source is a square wave, instead of a sine function, resonance will still occur when the forcing frequency matches the natural frequency.

When $k = 2\pi$, the graph appears to be that of a periodic function. This is explained as follows.

$$f_n(t) = \frac{(-1)^n}{4\pi^2}(1 - \cos(2\pi t - 2\pi n))u(t - n)$$

$$= \frac{(-1)^n}{4\pi^2}(1 - \cos(2\pi t))u(t - n)$$

$$= f(t)(-1)^n u(t - n)$$

Therefore,

$$y = f(t) + 2 \sum_{n=1}^{\infty} f_n(t)$$

$$= f(t) \left(1 + 2 \sum_{n=1}^{\infty} (-1)^n u(t - n) \right)$$

$$= f(t) S(t)$$

can be expressed as the product of two 2-periodic functions and is thus 2-periodic. Resonance occurs again when $k = 3\pi$. This can be seen in Figure 6.19, and it can be verified by showing that in this case, the solution is

$$y = \frac{1}{9\pi^2} \left(S(t) - (1 + 2\lfloor t \rfloor) \cos(3\pi t) \right).$$

The reason that the ODE $y'' + k^2 y = S(t)$ has resonant solutions for more than one value of k, while the ODE $y'' + k^2 y = \sin(\pi t)$ resonates only when $k = 1$, can best be explained in terms of trigonometric series. It can be shown that

$$S(t) = \frac{4}{\pi} \sin(\pi t) + \frac{4}{3\pi} \sin(3\pi t) + \frac{4}{5\pi} \sin(5\pi t) + \cdots$$

in the sense that the infinite series converges to $S(t)$ for all values of t except when t is an integer. When k is an odd multiple of π, the solution resonates with the term $\frac{4}{k} \sin(kt)$ of this series.

EXERCISES

1. Show that $S(t) = (-1)^{\lfloor t \rfloor}$.

2. Show that $S(t) = 2u(\sin(\pi t)) - 1$, except when t is an odd integer.

3. Let $\Lambda(t)$ be the sawtooth function whose graph appears in Figure 6.20.

  a. Express $\Lambda(t)$ as an infinite sum of corner functions. Your formula must be valid for all $t \geq 0$, but need not apply for negative t.

  b. Show that $\Lambda'(t) = -2S(t)$ except when t is an integer.

  c. Use the result of part (a) to calculate $L(\Lambda(t))$.

  d. Use the result of part (b) to calculate $L(\Lambda(t))$, and compare the result with your answer to part (c).

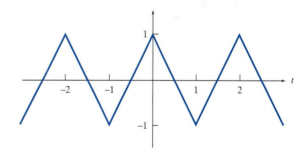

FIGURE 6.20   The sawtooth function.

4. Solve the IVP $y'' + k^2 y = \Lambda(t)$ with homogeneous initial conditions at $t = 0$.

5. Solve the IVP

$$y'' + y = \sum_{n=1}^{\infty} \delta(t - 2n\pi)$$

with homogeneous initial conditions at $t = 0$.

6. Show that the solution of $y'' + (3\pi)^2 y = S(t);\ y(0) = y'(0) = 0$ is

$$y = \frac{1}{9\pi^2}\left(S(t) - (1 + 2\lfloor t\rfloor)\cos(3\pi t)\right).$$

7. Show that all solutions of the ODE

$$y'' + (n\pi)^2 y = S(t)$$

are unbounded if n is an odd integer, and periodic if n is even.

8. Is the solution in Exercise 5 resonant? Draw its graph.

9. For which values of k is the solution of the IVP in Exercise 4 resonant? Find a formula for the resonant solutions.

The following two problems expand on a familiar topic: periodic solutions of linear ODEs

$$a_2 y'' + a_1 y' + a_0 y = b(t)$$

with positive constant coefficients a_i, where the source $b(t)$ is periodic.

10. This problem outlines the method for

$$y'' + 3y' + 2y = 2S(t). \tag{6.48}$$

a. Let $y(t)$ be the solution of (6.48) with homogeneous initial conditions. Show that

$$L(y) = \left(\frac{1}{s} - \frac{2}{s+1} + \frac{1}{s+2}\right)\left(2\sum_{k=0}^{\infty}(-1)^k e^{-ks} - 1\right).$$

b. Show that

$$y(t) = S(t) - 4p(t) + 2q(t) + l(t),$$

where

$$L(p) = \frac{1}{s+1}\sum_{k=0}^{\infty}(-1)^k e^{-ks},$$

$$L(q) = \frac{1}{s+2}\sum_{k=0}^{\infty}(-1)^k e^{-ks},$$

and $l(t)$ is transient.

c. Show that

$$p(t) = \sum_{k=0}^{\lfloor t\rfloor}(-1)^k e^{-(t-k)},$$

where $\lfloor t\rfloor$ denotes the greatest integer $\le t$.

d. Let $\theta(t) = t - \lfloor t\rfloor$ ($\theta(t)$ is 1-periodic). Use the formula for the sum of a finite geometric progression[2] to show that

$$p(t) = e^{-\theta(t)}\frac{1 + (-1)^{\lfloor t\rfloor}(-e)^{-\lfloor t+1\rfloor}}{1 + e^{-1}},$$

and conclude that $p(t) = \frac{e}{e+1}e^{-\theta(t)} + \text{transient}$.

e. Show that $q(t) = \frac{e^2}{e^2+1}e^{-2\theta(t)} + \text{transient}$.

f. Show that the periodic solution of equation (6.48) is

$$y_p(t) = S(t) - \frac{4e}{e+1}e^{-\theta(t)} + \frac{2e^2}{e^2+1}e^{-2\theta(t)}.$$

11. Find the periodic solution of each of the following differential equations (listed in order of difficulty).

a. $y'' + 3y' + 2y = \sum_{k=0}^{\infty}\delta(t-k)$

b. $y'' + 2y' + 2y = \sum_{k=0}^{\infty}(-1)^k\delta(t-k\pi)$

c. $y'' + 2y' + (1+\pi^2)y = S(t)$

12. Let $y(t)$ denote the centigrade temperature of the beer in a bottle. According to Newton's law of cooling,[3]

$$y'(t) = -k(y(t) - A(t)),$$

where $A(t)$ denotes the ambient temperature and k is the rate constant for cooling. For this beer bottle, $k = 0.4\ \text{hour}^{-1}$. The bottle is kept on the kitchen shelf during even-numbered hours, and in the refrigerator during odd-numbered hours. If the refrigerator temperature is $3°C$ and the kitchen temperature is $25°C$, find $y(t)$. Neglect transients; just the periodic solution is required. What are the minimum and maximum temperatures of the beer when the temperature oscillates periodically?

13. Let $f(t)$ be a continuous T-periodic function.

a. Show that

$$L(f) = \sum_{n=0}^{\infty}\int_{nT}^{(n+1)T} e^{-st} f(t)\,dt.$$

b. Show that

$$\int_{nT}^{(n+1)T} e^{-st} f(t)\,dt = e^{-snT}\int_0^T e^{-sw} f(w)\,dw.$$

c. Conclude that

$$L(f)(s) = \left(\int_0^T e^{-sw} f(w)\,dw\right)\sum_{n=0}^{\infty}(e^{-sT})^n.$$

---

[2] $1 - k + k^2 + \cdots + (-1)^N k^N = (1 + (-1)^N k^{N+1})/(1 + k)$

[3] See Section 1.3 (page 15).

d. Use the formula for the sum of a geometric series to simplify the preceding formula for $L(f)$.

14. Use the result of Exercise 13 to recalculate the Laplace transforms of $\sin(kt)$ and $\cos(kt)$.

15. Let $f(t) = c(\sin(t))$, where c is the corner function (f is called the half-wave rectified sine function). Sketch a graph of $f(t)$, and use the result of Exercise 13 to compute its Laplace transform.

16. Solve the IVP, $y'' + k^2 y = f(t)$, $y(0) = y'(0) = 0$, where f is the half-wave rectified sine function (see Exercise 15). For which k are there resonant solutions?

◆ **Hint:** Use the formula for $L(f)$ given by Exercise 13, part (c).

6.9 Glossary

$c(t)$ The corner function.

$\delta(t)$ The delta function.

$L(f)$ The Laplace transform of a function $f(t)$.

$L^{-1}(Y)$ The inverse Laplace transform fo Y.

$u(t)$ The unit step function.

Ampere   ampere A unit measuring current. 1 ampere is 1 coulomb per second.

Approximate delta function

$$f_\epsilon(t - a) = \begin{cases} \frac{1}{\epsilon} & \text{if } a \leq t < a + \epsilon \\ 0 & \text{otherwise.} \end{cases}$$

Capacitance   The ratio of the charge stored by a capacitor to the potential difference at its terminals.

Capacitor   A device for storing an electrical charge.

Choke coil   An electrical circuit component in which the rate of change of current is proportional to the potential difference between its terminals.

Convolution product   (of two functions $f(t)$ and $g(t)$) The function

$$f * g(t) = \int_0^t f(t - u)g(u)\,du.$$

Convolution theorem   If $L(f) = F(s)$ and $L(g) = G(s)$, then

$$L(f * g) = F(s)G(s).$$

Corner function

$$c(t) = \begin{cases} 0 & \text{if } t \leq 0 \\ t & \text{if } t > 0 \end{cases}$$

Coulomb   A unit of electrical charge, equal to the charge of 6.24146×10^{18} electrons.

Current   A flow of electrons in an electrical conductor.

Delta function   Not actually a function, $\delta(t - a)$ is defined by the property

$$\int_0^\infty \delta(t - a)g(t)\,dt = g(a)$$

for all functions g that are continuous near $t = a$.

Electrical impedance   (of a circuit) The complex ratio of the EMF $e^{i\omega t}$ to the resulting periodic current. It is given by the formula

$$P(\omega) = R + i\left(L\omega - \frac{1}{C\omega}\right),$$

where R is the resistance, L the inductance, and C the capacitance.

Electrical potential   The work required to move a unit of charge from one node to another is the difference in potential between those nodes.

Farad   A unit of capacitance. For each farad of capacitance, a 1-volt potential causes a capacitor to store 1 coulomb of charge.

Faraday's law   The charge held by a capacitor is proportional to the difference in potential across its plates: $x = CV$, where x is the charge, C is the capacitance, and V is the potential drop.

Gamma function

$$\Gamma(z) = \int_0^\infty u^{z-1} e^{-u}\,du$$

The identity $\Gamma(z + 1) = z\Gamma(z)$ holds for $z > 0$. Since $\Gamma(1) = 1$, we have $\Gamma(2) = 1$, $\Gamma(3) = 2$, $\Gamma(4) = 6$, and so on, with $\Gamma(n) = (n - 1)!$ for any positive integer n. Thus the gamma function generalizes the factorial, enabling us to take "factorials" of numbers that are not integers.

Henry   Unit of inductance. A current increase of 1 ampere per second will cause a 1-volt potential difference across the terminals of a 1-henry choke coil.

Henry's law   The potential drop across the terminals of an inductor, or choke coil, is proportional to the rate of change of the current:

$$V = L\frac{dy}{dt},$$

where V is the potential difference, L is the inductance, and y is the current.

Impulse response   [of an LDO, $\mathcal{L}(y) = a_2 y'' + a_1 y' + a_0 y$] The solution of the IVP,

$$\mathcal{L}(y) = 0, \ y(0) = 0, \ y'(0) = 1.$$

Alternatively, the solution of

$$\mathcal{L}(y) = \delta(t)$$

with homogeneous initial conditions.

Inductance   The ratio of the potential drop across an inductor, or choke coil, to the rate of change of current.

Integral transform   An operator that takes a function $f(t)$ to a function

$$\int_a^b K(s,t) f(t)\,dt,$$

where $K(s,t)$ is a specified function, called the kernel of the transform, and the limits of integration are also specified. The Laplace transform is an integral transform [the kernel is $K(s,t) = e^{-st}$].

Inverse Laplace transform   [of a function $Y(s)$] The function $f(t)$ with the property $L(f)(s) = Y(s)$.

Kirchhoff's laws

I.   The net sum of currents entering any node of a circuit is 0.

II.   In each simple loop of a circuit, the sum of the potential differences across the elements of the loop equals 0.

Laplace transform   [of a function $f(t)$]

$$L(f)(s) = \int_0^\infty f(t)e^{-st}\,dt$$

Ohm   A unit of resistance. A potential drop of 1 volt across a 1 ohm resistor causes a current of 1 ampere.

Ohm's law   The potential difference between the terminals of a resistor is proportional to the current it carries: $V = yR$, where V is the potential difference, y is the current, and R is the resistance.

Periodic function   A function f such that there is a fixed number T, called the period, with $f(t+T) = f(t)$ for all real t.

Proper rational function   A rational function $p(t)/q(t)$ in whcih the numerator polynomial $p(t)$ is of lesser degree that the denominator polynomial $q(t)$.

Reactance   The imaginary part of the impedance of an electrical circuit.

Resistance   A parameter associated with an electrical resistor: the potential difference across the resistor that produces a current of exactly one ampere. The resistance of an electrical circuit is the real part of its impedance.

Resonant frequency   The frequency at which the reactance vanishes.

Series circuit   An electrical circuit that is a simple loop. There are no nodes at which more than two conductors are joined.

Square wave   A periodic function

$$S(t) = \begin{cases} 1 & \text{if } \lfloor t \rfloor \text{ is even} \\ -1 & \text{if } \lfloor t \rfloor \text{ is odd.} \end{cases}$$

Time domain   The real line, with the coordinate representing time.

Transform domain   The subset of the real line on which the Laplace transform of a function is defined.

Transfer function   The Laplace transform of the impulse response.

Unit step function

$$u(t-a) = \begin{cases} 0 & \text{if } t < a \\ 1 & \text{if } t \geq a \end{cases}$$

Volt   A unit of electrical potential.

Table 6.3 Table of Laplace Transforms. More complete tables are available in the *Handbook of Mathematical Functions*, edited by M. Abramowitz and I. Stegun, Dover Publications, Inc., New York (1965); A. Erdélyi et al, Tables of Integral Transforms, McGraw-Hill, New York, (1954); and in *Laplace Transforms, Tables and Theorems*, by P. A. McCollum and B. F. Brown, Holt, Rinehart, and Winston (1965). The tables complement the capabilities of computer algebra systems (CAS), since they focus on piecewise continuous functions that are not analytic, while analytic functions are the forte of the CAS.

Function	Transform	Function	Transform
1	$\dfrac{1}{s}$	$e^{kt}f(t)$	$F(s-k)$
t	$\dfrac{1}{s^2}$	$f'(t)$	$sF(s)-f(0)$
t^2	$\dfrac{2}{s^3}$	$f''(t)$	$s^2F(s)-sf(0)-f'(0)$
t^n	$\dfrac{\Gamma(n+1)}{s^{n+1}}$	$f^{(n)}(t)$	$s^nF(s)-s^{n-1}f(0)-\cdots-f^{(n-1)}(0)$
e^{kt}	$\dfrac{1}{s-k}$	$\displaystyle\int_0^t f(u)\,du$	$\dfrac{1}{s}F(s)$
$f(kt)$	$\dfrac{1}{k}F\left(\dfrac{s}{k}\right)$	f is T-periodic	$\left[\displaystyle\int_0^T g(t)e^{-st}\,dt\right]\left[\displaystyle\sum_{n=0}^{\infty}e^{-nTs}\right]$
$tf(t)$	$-F'(s)$	$S\left(\dfrac{t}{k}\right)$	$\dfrac{1}{s}\left[2\displaystyle\sum_{n=0}^{\infty}(-1)^n e^{-nks}-1\right]$
$\cos(bt)$	$\dfrac{s}{s^2+b^2}$	$e^{kt}\cos(bt)$	$\dfrac{s-k}{(s-k)^2+b^2}$
$\sin(bt)$	$\dfrac{b}{s^2+b^2}$	$e^{kt}\sin(bt)$	$\dfrac{b}{(s-k)^2+b^2}$
$\cosh(bt)$	$\dfrac{s}{s^2-b^2}$	$e^{kt}\cosh(bt)$	$\dfrac{s-k}{(s-k)^2-b^2}$
$\sinh(bt)$	$\dfrac{b}{s^2-b^2}$	$e^{kt}\sinh(bt)$	$\dfrac{b}{(s-k)^2-b^2}$
$u(t-a)$	$\dfrac{1}{s}e^{-as}$	$te^{kt}\cos(bt)$	$\dfrac{1}{(s-k)^2+b^2}-\dfrac{2b^2}{[(s-k)^2+b^2]^2}$
$f(t-a)u(t-a)$	$F(s)e^{-as}$	$te^{kt}\sin(bt)$	$\dfrac{2b(s-k)}{[(s-k)^2+b^2]^2}$
$f*g(t)$	$F(s)G(s)$	$t^2e^{kt}\cos(bt)$	$\dfrac{2(s-k)}{[(s-k)^2+b^2]^2}-\dfrac{8b^2(s-k)}{[(s-k)^2+b^2]^3}$
$\delta(t-a)$	e^{-as}	$t^2e^{kt}\sin(bt)$	$\dfrac{6b}{[(s-k)^2+b^2]^2}-\dfrac{8b^3}{[(s-k)^2+b^2]^3}$

6.10 Chapter Review

1. Calculate $L(f)$, where $f(t) = \begin{cases} 2-t & \text{if } 0 \le t < 2 \\ 0 & \text{otherwise.} \end{cases}$

In Exercises 2–22, use the Laplace transform method to solve the initial value problem. When initial conditions are omitted, they are to be taken to be homogeneous at $t = 0$.

2. $y' + 3y = 5e^{-t}\cos t$

3. $y'' + 2y' - 3y = 0$; $y(0) = 1$, $y'(0) = -3$

4. $y'' + 5y' - 6y = 0$; $y(0) = 4$, $y'(0) = 39$

5. $y'' + 4y' + 4y = te^{2t+1}$

6. $y'' + 3y' + 2y = 0$; $y(0) = 1$, $y'(0) = 0$

7. $y'' + 4y' - 12y = 8e^{2t}$; $y(0) = 1$, $y'(0) = 5$

8. $y'' + 4y = e^{2t}\cos 2t$

9. $5y' + 2y - 3\int_0^t y(u)\,du + 8 = 0$

10. $y''' + y'' + y' + y = 0$; $y(0) = y'(0) = 0$, $y''(0) = 1$

11. $y^{(4)} + 4y''' + 6y'' + 4y' + y = t^2 e^{-t}$

12. $y = 1 + \int_0^t y(u)\,du$

13. $y'' + y = \delta(t) + 2\delta(t-\pi)$; $y(0) = 0$, $y'(0) = 1$

14. $y'' + 2y' + 2y = \delta(t) + 2\delta(t-\pi) + 3\delta(t-2\pi)$

15. $y'' + 2y' + (1+\pi^2)y = \delta(t) + \delta(t-1) + \delta(t-2)$

16. $y'' + \pi^2 y = u(t) - u(t-1)$

17. $y'' + 4\pi^2 y = S(t)$ ($S(t) = (-1)^{\lfloor t \rfloor}$ is the square wave)

18. $y'' + \pi^2 y = \sum_{n=0}^{\infty}(-1)^n \delta(t-n)$

19. $y'' + \pi^2 y = \sum_{n=0}^{\infty} \delta(t-n)$

20. $y'' + y = \{1 - u(t-\pi)\}\sin t$

21. $y^{(4)} - 16y = 16u(t-1)$; $y(0) = y'(0) = y'''(0) = 0$, $y''(0) = 4$

22. $y^{(4)} + 5y'' + 4y = \delta(t) - \delta(t-\pi) - \delta(t-2\pi)$

23. Find the transfer function and impulse response for the following differential operators:

a. $L(y) = y' + y$ b. $L(y) = y'' + y' + y$

c. $L(y) = 2y'' + 3y' + y$

d. $L(y) = y''' + y'' + y' + y$

24. Derive a system of differential equations to serve as a model for the *RLC* series circuit shown in Figure 6.21. Assume that at $t = 0$, the capacitor has no charge, and the current is 0. Determine the current $y(t)$ in the choke coil by using the Laplace transform method.

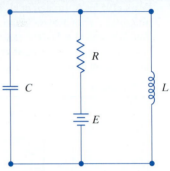

FIGURE 6.21   The electrical circuit for Exercise 24. The EMF is $E = 6$ volts, the resistance is $R = 100$ ohms, the capacitance is $C = 0.005$ farad, and the inductance is 0.01 henry.

25. Solve the initial value problem

$$4y' + \int_0^t e^{-(t-s)}y(s)\,ds = 0; \quad y(0) = 1.$$

26. Solve $y' - 2\int_0^t e^{t-u}y(u)\,du = e^t$ with a homogeneous initial condition at $t = 0$.

27. Give an example of

a. A function that is continuous on $[0, \infty)$ but does not have a Laplace transform

b. A discontinuous function that does have a Laplace transform

c. A function that has a Laplace transform but whose derivative does not

The remaining exercises are not review exercises; they are intended to be more challenging.

28. Calculate $L\left(\dfrac{1}{\lfloor t \rfloor + 1}\right)$.

29. In the *RLC* series circuit whose diagram is shown in Figure 6.22, the switch is moved back and forth between positions **A** and **B**. The amount of time spent in each position before reversing is $\frac{\pi}{10}$ seconds. Determine the stable periodic current.

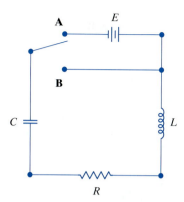

FIGURE 6.22   An *RLC* series circuit with a switch. $R = 1$ ohm, $L = .05$ henry, $C = 0.1$ farad, and $E = 6$ volts.

30. Let $f(t)$ be a function that is locally integrable and of exponential order, and let $Y(s) = L(f)$. Show that

$$\frac{dY}{ds} = L(-tf(t)).$$

31. Calculate the Laplace transform of the Bessel function $J_0(t)$. Given that both $J_0(t)$ and $J_0'(t)$ have Laplace transforms, all that you need to know to solve this problem is the result of the previous exercise, that $J_0(0) = 1$, and that $y = J_0(t)$ is a solution of Bessel's equation

$$ty'' + y' + ty = 0.$$

Use the result to derive a power series expansion of $J_0(t)$.

32. Calculate the following Laplace transforms.

    a. $L\left(\dfrac{\sin t}{t}\right)$ b. $L\left(\dfrac{\cos t - 1}{t^2}\right)$

CHAPTER 7

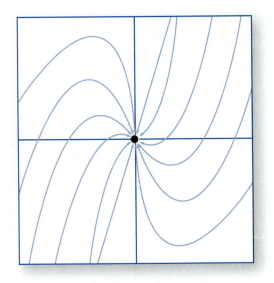

Linear Differential Equations with Variable Coefficients

7.1 Power Series

Our objective in this chapter is to develop solutions of linear ODEs with variable coefficients. Before starting, we should recognize that for most linear ODEs that have variable coefficients, analytic methods are not appropriate. It is preferable to rely on IVP solvers to calculate solutions numerically, and qualitative analysis to determine other features of solutions. Nevertheless, there are variable-coefficient linear ODEs that are important in applications and are susceptible to analytic methods. We will focus as much as possible on these equations.

Our tool will be power series. The idea is simple enough: To find a solution of

$$a_2(t)y'' + a_1(t)y' + a_0(t)y = 0,$$

substitute a power series $y = \sum_{n=0}^{\infty} A_n t^n$ with undetermined coefficients, and try to solve for the coefficients. To see how this works, let us consider a familiar ODE that has constant coefficients.

Example 7.1.1   Find the general solution of $y' - y = 0$.

Solution. Put $y = \sum_{n=0}^{\infty} A_n t^n$. Differentiating term by term, we find

$$y' = \sum_{n=0}^{\infty} n A_n t^{n-1}$$

$$= A_1 + 2A_2 t + 3A_3 t^2 + 4A_4 t^3 + \cdots .$$

Thus,

$$y' - y = (A_1 - A_0) + (2A_2 - A_1)t + (3A_3 - A_2)t^2 + \cdots$$
$$+ (nA_n - A_{n-1})t^{n-1} + \cdots .$$

Since $y' - y = 0$, each of the coefficients of this power series is equal to 0: $A_1 = A_0$, $A_2 = \frac{1}{2}A_1$, $A_3 = \frac{1}{3}A_2$, and, in general, $A_n = \frac{1}{n}A_{n-1}$. If we put $A_0 = C$, then $A_1 = C$, $A_2 = \frac{1}{2}C$, $A_3 = \frac{1}{3\cdot 2}C$, and, in general,

$$A_n = \frac{1}{n \cdot (n-1) \cdots 3 \cdot 2} C = \frac{1}{n!}C.$$

We have found that the solution can be represented by the power series

$$y = C \sum_{n=0}^{\infty} \frac{1}{n!} t^n.$$

We recognize this power series as $y = Ce^t$.   ❑

The solution of the ODE in this example was an elementary function whose power series is familiar to us. It often happens that a solution of an ODE with variable coefficients is not an elementary function, and in this case we have to glean what information we can from its power series expansion.

Review of power series

A **power series** is an infinite series of the form

$$\sum_{n=0}^{\infty} A_n (t - t_0)^n. \tag{7.1}$$

The set of real values of t for which the series converges is called the **convergence set** of the series. For any power series, one of the following statements is true of its convergence set:

  i. $S = \{t_0\}$.
  ii. S is an interval whose endpoints are $t_0 - R$ and $t_0 + R$, where R is some positive number.
  iii. $S = \mathbf{R}$.

In case (ii), the number R is called the **radius of convergence** of the power series. To complete the picture, the radius of convergence is said to be 0 in case (i), and ∞ in case (iii).

Two formulas for calculating the radius of convergence of the series (7.1) are as follows (each formula is based on the assumption that the limit in question exists):

$$R = \lim_{n \to \infty} \frac{1}{\sqrt[n]{|A_n|}} \tag{7.2}$$

$$= \lim_{n \to \infty} \left| \frac{A_n}{A_{n+1}} \right|. \tag{7.3}$$

Formula (7.2) is based on the **root test** for convergence of an infinite series: A series $\sum_{n=0}^{\infty} c_n$ is convergent if $\lim_{n \to \infty} \sqrt[n]{|c_n|} < 1$, and divergent if that limit is > 1. Applied to a power series, with $c_n = A_n(t - t_0)^n$, we see that the power series

$\sum_{n=0}^{\infty} A_n (t - t_0)^n$ is convergent for all t such that $\lim_{n \to \infty} \sqrt[n]{|A_n (t - t_0)^n|} < 1$; that is,

$$|t - t_0| < \frac{1}{\lim_{n \to \infty} \sqrt[n]{|A_n|}}.$$

The proof of formula (7.3) is based on the **ratio test**: Let

$$\rho = \lim_{n \to \infty} \left| \frac{c_{n+1}}{c_n} \right|$$

be the ratio of two consecutive terms of the power series $\sum_{n=0}^{\infty} c_n$. Then the series converges if $\rho < 1$, and diverges if $\rho > 1$. The test is indecisive if $\rho = 1$.

Differentiation of power series

A function $f(t)$ is said to be **analytic** at a point t_0 if there is a power series $\sum_{n=0}^{\infty} A_n (t - t_0)^n$ with radius of convergence $R > 0$ such that $f(t) = \sum_{n=0}^{\infty} A_n (t - t_0)^n$ for all t in an open interval I containing t_0. The derivative of f on I then has the power series given by differentiating term by term:

$$f'(t) = \sum_{n=1}^{\infty} n A_n (t - t_0)^{n-1}.$$

This differentiated power series also has radius of convergence equal to R, so the process can be continued. It follows that f is infinitely differentiable on I.

If $f(t)$ and its derivatives are known, then the coefficients A_n are determined by the formula

$$A_n = \frac{1}{n!} f^{(n)}(t_0). \tag{7.4}$$

This formula is derived by taking the nth derivative of the power series and substituting $t = t_0$. In particular, $A_0 = f(t_0)$, $A_1 = f'(t_0)$, and $A_2 = \frac{1}{2} f''(t_0)$. The following proposition is an immediate consequence of formula (7.4).

Proposition 7.1.1

Let $f(t) = \sum_{n=0}^{\infty} A_n (t - t_0)^n$ and $g(t) = \sum_{n=0}^{\infty} B_n (t - t_0)^n$ be functions that are analytic at t_0. If $f(t) = g(t)$ for all t in an open interval containing t_0, then the coefficients of the two power series are identical: $A_n = B_n$ for all $n \geq 0$.

Proof If $f(t)$ and $g(t)$ agree on an interval containing t_0, then their derivatives must also be identical on the interval. The coefficients of the power series, which are determined by formula (7.4), are therefore the same. ∎

What functions are analytic at a point t_0? We have seen that an analytic function must be infinitely differentiable. Given an infinitely differentiable function $f(t)$, Taylor's theorem states that for every $N \geq 1$,

$$f(t) = A_0 + A_1(t - t_0) + A_2(t - t_0)^2 + \cdots + A_N(t - t_0)^N + E_N(t),$$

where the coefficients A_n are given by formula (7.4). The size of the **remainder term** $E_N(t)$ can be estimated because

$$E_N(t) = \frac{1}{(N+1)!} f^{(N+1)}(\xi)(t - t_0)^{N+1}$$

for some number ξ between t_0 and t. The function $f(t)$ is analytic at t_0 if and only if $E_N(t) \to 0$ as $N \to \infty$ for all t near t_0. For example, the elementary transcendental functions e^t, $\ln t$, $\sin t$, and so on are analytic throughout their domains, as are the polynomials and rational functions. Some algebraic functions are not infinitely differentiable everywhere and are consequently not analytic at certain points. For example, $\sqrt[3]{t}$ is not analytic at $t = 0$. The simplest example of an infinitely differentiable function that fails to be analytic is

$$f(t) = \begin{cases} e^{-1/t^2} & \text{if } t \neq 0 \\ 0 & \text{if } t = 0, \end{cases} \tag{7.5}$$

which is not analytic at $t = 0$.

Sums and products of series

If $f(t) = \sum_{n=0}^{\infty} A_n(t - t_0)^n$ and $g(t) = \sum_{n=0}^{\infty} B_n(t - t_0)^n$ are analytic at $t = t_0$, then $f(t) + g(t)$ and $f(t) \cdot g(t)$ are also analytic there. Their sum is represented by the series

$$(f + g)(t) = \sum_{n=0}^{\infty} (A_n + B_n)(t - t_0)^n.$$

The product is more complicated: $(f \cdot g)(t) = \sum_{n=0}^{\infty} C_n(t - t_0)^n$, where

$$C_0 = A_0 B_0$$
$$C_1 = A_0 B_1 + A_1 B_0$$
$$C_2 = A_0 B_2 + A_1 B_1 + A_2 B_0$$
$$\vdots$$
$$C_n = \sum_{j+k=n} A_j B_k.$$

Unless one of the power series involved in a product is particularly simple, the coefficients of the product series may be difficult to express in closed form. However, if $g(t) = B(t - t_0)^m$ is a *monomial*, then the product series is just

$$f(t)g(t) = \sum_{n=0}^{\infty} B A_n(t - t_0)^{n+m}. \tag{7.6}$$

It will be convenient to perform an operation known as **index shifting** on such a series. The objective of index shifting is to obtain a power series expression in which the summation index n is the power of $(t - t_0)$. The series (7.6) can be written as

$$B A_0(t - t_0)^m + B A_1(t - t_0)^{m+1} + B A_2(t - t_0)^{m+2} + \cdots .$$

Notice that the coefficient of $(t - t_0)^n$ is BA_{n-m}; therefore,

$$f(t)g(t) = \sum_{n=m}^{\infty} BA_{n-m}(t - t_0)^n.$$

Index shifting can also be used to multiply series by binomials; for example,

$$(1 - 3t^2)\sum_{n=0}^{\infty} A_n t^n = 1 \cdot \sum_{n=0}^{\infty} A_n t^n - 3t^2 \cdot \left(\sum_{n=0}^{\infty} A_n t^n\right)$$

$$= \sum_{n=0}^{\infty} A_n t^n - \sum_{n=0}^{\infty} 3A_n t^{n+2}$$

$$= \sum_{n=0}^{\infty} A_n t^n - \sum_{n=2}^{\infty} 3A_{n-2} t^n$$

$$= A_0 + A_1 t + \sum_{n=2}^{\infty} (A_n - 3A_{n-2})t^n.$$

Question   Where did the terms $A_0 + A_1 t$ come from?

The following lemma will help to make index shifting a routine matter.

Lemma 7.1.2   Index Shifting

Suppose that $y(t) = \sum_{n=0}^{\infty} A_n t^n$, and k is a nonnegative integer. Then the following formulas hold:

$$t^k y(t) = \sum_{n=k}^{\infty} A_{n-k} t^n \tag{7.7}$$

$$t^k y'(t) = \sum_{n=k}^{\infty} (n - k + 1)A_{n-k+1} t^n \tag{7.8}$$

$$t^k y''(t) = \sum_{n=k}^{\infty} (n - k + 2)(n - k + 1)A_{n-k+2} t^n. \tag{7.9}$$

Proof   We will only prove formula (7.8), but the other two proofs are similar. Term-by-term differentiation of

$$y(t) = A_0 + \sum_{n=1}^{\infty} A_n t^n$$

shows that

$$t^k y'(t) = t^k \sum_{n=1}^{\infty} nA_n t^{n-1} = \sum_{n=1}^{\infty} nA_n t^{n+k-1}.$$

We handle the summation in a way analogous to integration by substitution. Put $m = n + k - 1$ We have to put the limits of summation in terms of m (from $m = k$ to ∞) and to replace n by $m - k + 1$ in two places. This gives

$$t^k y'(t) = \sum_{m=k}^{\infty} (m - k + 1)A_{m-k+1} t^m.$$

Now revert to n as the summation index.

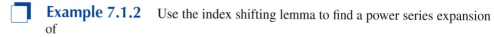

Example 7.1.2 Use the index shifting lemma to find a power series expansion of

$$(t^2 + t)y''(t) - 3ty'(t),$$

where

$$y(t) = \sum_{n=0}^{\infty} A_n t^n.$$

Solution. According to the lemma,

$$t^2 y''(t) = \sum_{n=2}^{\infty} n(n-1)A_n t^n$$

$$ty''(t) = 2A_2 t + \sum_{n=2}^{\infty} (n+1)n A_{n+1} t^n$$

$$-3ty'(t) = -3A_1 t + \sum_{n=2}^{\infty} -3n A_n t^n.$$

By adding these equations, we find that

$$(t^2 + t)y''(t) - 3ty'(t)$$

$$= (2A_2 - 3A_1)t + \sum_{n=2}^{\infty} [(n^2 - n)A_n + (n^2 + n)A_{n+1} - 3n A_n]t^n$$

$$= (2A_2 - 3A_1)t + \sum_{n=2}^{\infty} [(n^2 - 4n)A_n + (n^2 + n)A_{n+1}]t^n. \qquad \square$$

Approximation of functions

Although a power series may be convergent for all values of t, its value as a means of approximating the function it represents is limited. Consider the Taylor expansion

$$\sin(t) = \sum_{n=0}^{\infty} \frac{(-1)^n}{(2n+1)!} t^{2n+1},$$

whose radius of convergence is infinite. To approximate $\sin t$, a **truncation**

$$\mathrm{Tr}_N(t) = \sum_{n=1}^{N} \frac{(-1)^n}{(2n-1)!} t^{2n-1}$$

may be used. The precision of the approximation depends, of course, on N; but it depends more heavily on t. To achieve ten decimal place accuracy, the following are adequate approximations:

$$\mathrm{Tr}_1(t) = t \text{ for } |t| < 0.00066;$$

$$\mathrm{Tr}_2(t) = t - \frac{t^3}{6} \text{ for } |t| < 0.02;$$

$$\mathrm{Tr}_3(t) = t - \frac{t^3}{6} + \frac{t^5}{120} \text{ for } |t| < 0.114;$$

and

$$\text{Tr}_4(t) = t - \frac{t^3}{6} + \frac{t^5}{120} - \frac{t^7}{5040} \quad \text{for } |t| < 0.384.$$

Figure 7.1 displays the graph of $\sin(t)$, and the graphs of the four truncated series just listed. Notice that none of the series is a substitute for $\sin(t)$, but all approximate it pretty well near $t = 0$. Taking more terms will produce an approximation that is valid at a greater distance from the origin, but all truncations of the series are bound to break down eventually. In general, the *computational* purpose of a power series expanded about $t = t_0$ is to study the function it represents near $t = t_0$. In many cases it is sufficient to truncate the series after the first two or three nonzero terms.

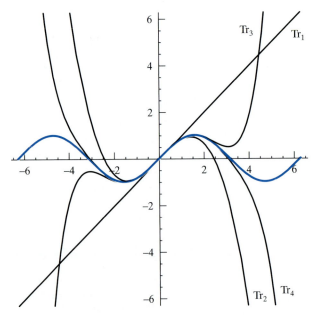

FIGURE 7.1 The graphs of $\sin(t)$, and the first four truncations of its Taylor series expansion at $t = 0$.

EXERCISES

In Exercises 1–6, multiply the power series.

1. $(1 + t) \sum_{n=0}^{\infty} A_n t^n$

2. $(1 + t)^2 \sum_{n=0}^{\infty} (-1)^n (n + 1) t^n$

3. $(2t - t^3) \sum_{n=0}^{\infty} A_n t^{2n}$ **4.** $(1 - t^2) \sum_{n=1}^{\infty} \frac{1}{n^2} t^n$

5. $(3t - 4t^2) \sum_{n=0}^{\infty} t^n$ **6.** $(t^2 - 1) \sum_{n=0}^{\infty} A_n (t - 1)^n$

7. Find the first four terms of the series representation of $e^t \sum_{n=0}^{\infty} A_n t^n$.

8. Let $\sum_{n=0}^{\infty} C_n (t - 1)^n$ denote the product series

$$\left(\sum_{n=0}^{\infty} (t - 1)^n \right) \left(\sum_{n=0}^{\infty} n(t - 1)^n \right).$$

Calculate C_0, C_1, C_2, C_3, and C_4.

In Exercises 9–14, use index shifting to find a power series expansion for the given expression. In these problems, $f(t) = \sum_{n=0}^{\infty} A_n t^n$.

9. $t^2 f''(t) + t f'(t) + f(t)$ **10.** $f'(t) - 12t f(t)$

11. $f''(t) - tf(t)$

12. $(1 - 3t^2)e^t$

13. $\frac{d^2}{dt^2}\cos t$. Recall that $\cos t = \sum_{n=0}^{\infty} \frac{(-1)^n}{(2n)!} t^{2n}$.

14. $\frac{d^2}{dt^2}\cos\sqrt{t}$

15. Let $\mathcal{L}(y) = y'' - 2ty' + y$. Give the series representation for $\mathcal{L}(y)$, if

a. $y = \sum_{n=0}^{\infty} 2^n t^n$

b. $y = e^t$

c. $y = \sum_{n=0}^{\infty}(t-1)^n$

d. $y = \sum_{n=0}^{\infty} A_n t^n$

e. $y = \sum_{n=0}^{\infty} A_n(t+1)^n$

16. Let $\mathcal{L}(y) = t^2 y'' + 4ty' + 2y$. Give the series representation

of

$$\mathcal{L}\left(\sum_{n=0}^{\infty} A_n t^n\right).$$

17. Prove formula (7.3) for the radius of convergence of a power series.

18. Show that the ratio test cannot be used to determine if a power series converges at the endpoints of its convergence set.

19. Let $f(t)$ denote the function defined in formula (7.5). Prove that f is infinitely differentiable, even at $t = 0$, and that for all $n \geq 0$, $f^{(n)}(0) = 0$. It follows that the Taylor coefficients A_n of $f(t)$ at $t_0 = 0$ are all 0. Obviously, $f(t) \neq 0$; hence f is not analytic at 0.

7.2 Solutions at an Ordinary Point

Consider an ODE

$$a_2(t)\, y'' + a_1(t)\, y' + a_0(t)\, y = 0,$$

where the coefficient functions are analytic at a point t_0. If $a_2(t_0) \neq 0$, it is easy to see that t_0 is an ordinary point.

Before considering a specific example, we need to introduce some notation. If $n = 2k + 1$ is an odd positive integer, $n!!$ will denote the product of all *odd* positive integers $\leq n$. Thus

$$(2k+1)!! = (2k+1)(2k-1)(2k-3)\cdots 1.$$

If $n = 2k$ is an even positive integer, $n!!$ is the product of all even positive integers $\leq n$; so that

$$n!! = n(n-2)(n-4)\cdots 2.$$

Finally, by convention, $0!! = 1$. Double factorial notation is used in reference works such as *Handbook of Mathematical Functions,* by M. Abramowitz and I. Stegun.[1] You should be aware that although there are two factorial signs, $n!!$ has about half as many factors as $n!$. For example, $10! = 3{,}628{,}800$ is much larger than $10!! = 3840$.

Example 7.2.1 Find a power series representation at $t_0 = 0$ for the general solution of

$$y'' + ty' + y = 0. \tag{7.10}$$

Solution. Let $\mathcal{L}(y) = y'' + ty' + y$. If $y = \sum_{n=0}^{\infty} A_n t^n$, then by the index shifting lemma,

$$ty' = \sum_{n=1}^{\infty} nA_n t^n$$

[1] New York: Dover Publications (1965).

and

$$y'' = \sum_{n=0}^{\infty} (n+2)(n+1)A_{n+2}t^n.$$

Therefore,

$$\mathcal{L}(y) = \sum_{n=0}^{\infty} (n+2)(n+1)A_{n+2}t^n + \sum_{n=1}^{\infty} nA_n t^n + \sum_{n=0}^{\infty} A_n t^n$$

$$= 2A_2 + A_0 + \sum_{n=1}^{\infty} [(n+2)(n+1)A_{n+2} + (n+1)A_n]t^n.$$

Since $\mathcal{L}(y) = 0$, we have $2A_2 + A_0 = 0$ and

$$(n+2)(n+1)A_{n+2} + (n+1)A_n = 0 \text{ for } n \geq 1.$$

Thus, $A_2 = -\frac{1}{2}A_0$, and for $n \geq 1$,

$$A_{n+2} = -\frac{1}{n+2}A_n. \tag{7.11}$$

We will obtain expressions for the coefficients in closed form. Set $A_0 = B$ and $A_1 = C$; these will be the parameters of the general solution. Thus, $A_2 = -\frac{1}{2}B$, and if n is even, we can apply equation (7.11) repeatedly to obtain $A_4 = -\frac{1}{4}A_2 = \left(-\frac{1}{4}\right)\left(-\frac{1}{2}\right)B$, and so on, with

$$A_{2k} = \left(-\frac{1}{2k}\right) \cdot \left(-\frac{1}{2k-2}\right) \cdots \left(-\frac{1}{2}\right) \cdot B$$

$$= \frac{(-1)^k}{(2k)!!}B.$$

Similarly, if $n = 2k + 1$ is odd,

$$A_{2k+1} = \left(-\frac{1}{2k+1}\right) \cdot \left(-\frac{1}{2k-1}\right) \cdots \left(-\frac{1}{3}\right) \cdot C$$

$$= \frac{(-1)^k}{(2k+1)!!}C.$$

We can obtain a fundamental set of solutions of equation (7.10) by choosing special values for the parameters. If we set $B = 1$ and $C = 0$, we have

$$y_1(t) = \sum_{m=0}^{\infty} \frac{(-1)^m}{(2m)!!}t^{2m}.$$

By the result of Exercise 8b at the end of this section, $(2m)!! = 2^m m!$. Hence

$$y_1(t) = \sum_{m=0}^{\infty} \frac{1}{m!}\left(-\frac{t^2}{2}\right)^m = e^{-t^2/2}.$$

The other solution, with $B = 0$ and $C = 1$, is

$$y_2(t) = \sum_{m=0}^{\infty} \frac{(-1)^m}{(2m+1)!!} t^{2m+1}.$$

This solution cannot be expressed in terms of elementary functions.

To verify that the general solution of equation (7.10) is

$$y = B y_1(t) + C y_2(t),$$

we need to show that $y_1(t)$ and $y_2(t)$ actually represent solutions, and that they are linearly independent.

These power series formally satisfy equation (7.10); therefore, if they have positive radii of convergence, they represent solutions of the ODE. To determine the radius of convergence of $y_2(t)$, refer to formula (7.3) on page 321:

$$R = \lim_{k \to \infty} \left| \frac{A_{2k-1}}{A_{2k+1}} \right| = \lim_{k \to \infty} (2k+1) = \infty.$$

It follows that $y_2(t)$ is a solution of equation (7.10). A similar calculation would show that $y_1(t)$ is also a solution.

If the two solutions were linearly dependent, then there would be a constant K such that $y_1(t) = K y_2(t)$ for all t. This is impossible, because for $t = 0$ this means that $1 = K \cdot 0$. Thus, the solutions are linearly independent.   ❑

The key step in Example 7.2.1 was the formula (7.11), which is called a **recursive formula** for the coefficients. If we are given A_0 and A_1, (7.11) can be used to generate the sequence of coefficients, one at a time. It is not always possible to obtain a recursive formula for the coefficients of a series solution of an ODE, but when one is available, it can often be used to obtain formulas for the coefficients, as we did in the example.

Our efforts to show that $y_1(t)$ and $y_2(t)$ are convergent power series were actually not necessary. The following theorem guarantees that if a linear ODE has analytic coefficients, then its solutions are analytic at ordinary points; in other words, the radius of convergence is positive.

Theorem 7.1   I. L. Fuchs, 1866

Let $y(t)$ be a solution of

$$a_2(t)y'' + a_1(t)y' + a_0(t)y = 0, \tag{7.12}$$

where the coefficients $a_i(t)$ are analytic at t_0, and $a_2(t_0) \neq 0$. Then $y(t)$ is analytic at t_0.

The proof of Theorem 7.1 will be omitted, because it is beyond the scope of this text.

Legendre's equation

Legendre's equation is

$$(1 - t^2)y'' - 2ty' + v(v + 1)y = 0, \tag{7.13}$$

where v is a parameter. By Theorem 7.1, every solution is analytic at $t = 0$.

The solutions of Legendre's equation are called *Legendre functions*. If the parameter v is a positive integer, one of the solutions is a polynomial that we will use in Section 11.3 to define certain harmonic functions on spherical domains. Legendre polynomials were first discussed in 1784 by the French mathematician, A. M. Legendre.

Example 7.2.2   Determine the Legendre functions for $v = 5$.

Solution. For $v = 5$, Legendre's equation is

$$\mathcal{L}(y) = (1 - t^2)y'' - 2ty' + 30y = 0.$$

If we set $y = \sum_{n=0}^{\infty} A_n t^n$, then by the index shifting lemma,

$$y'' = \sum_{n=0}^{\infty} (n + 2)(n + 1)A_{n+2}t^n,$$

$$t^2 y'' = \sum_{n=2}^{\infty} n(n - 1)A_n t^n, \text{ and}$$

$$ty' = \sum_{n=1}^{\infty} nA_n t^n.$$

It follows that

$$\mathcal{L}(y) = 2A_2 + 6A_3 t + \sum_{n=2}^{\infty} (n + 2)(n + 1)A_{n+2}t^n - \sum_{n=2}^{\infty} n(n - 1)A_n t^n$$

$$- 2A_1 t - 2\sum_{n=2}^{\infty} nA_n t^n + 30\left(A_0 + A_1 t + \sum_{n=2}^{\infty} A_n t^n\right)$$

$$= 2A_2 + 30A_0 + (6A_3 + 28A_0)t$$

$$+ \sum_{n=2}^{\infty} \{(n + 2)(n + 1)A_{n+2} + [-n(n - 1) - 2n + 30]A_n\}t^n$$

$$= 2A_2 + 30A_0 + (6A_3 + 28A_0)t$$

$$+ \sum_{n=2}^{\infty} [(n + 2)(n + 1)A_{n+2} - (n^2 + n - 30)A_n]t^n.$$

The solution of $\mathcal{L}(y) = 0$ is found by setting each coefficient of the power series equal to zero and solving for the undetermined coefficients A_n. Thus, the coefficients of t^0 are $2A_2 + 30A_0 = 0$; the coefficients of t^1 are $6A_3 + 28A_1 = 0$, and for $n > 1$,

$$(n + 2)(n + 1)A_{n+2} - (n^2 + n - 30)A_n = 0.$$

It follows that

$$A_2 = -15A_0 \quad \text{and} \quad A_3 = -\frac{14}{3}A_1.$$

Furthermore, we can factor $n^2 + n - 30$ as $(n+6)(n-5)$, and obtain the recursive formula,

$$A_{n+2} = \frac{(n+6)(n-5)}{(n+2)(n+1)}A_n.$$

Since A_{n+2} depends only on A_n, not A_{n-1}, it follows that if for some n, $A_n = 0$, then $A_{n+2k} = 0$ for all $k \geq 0$. For example, if $A_1 = 0$, then $A_n = 0$ for all odd integers n.

Let $y_1(t)$ be the solution with $y_1(0) = 1$ and $y_1'(0) = 0$. Then $A_0 = 1$, $A_1 = 0$, and

$$A_2 = -15 = \frac{(6)(-5)}{(2)(1)}, \quad A_4 = \frac{(8)(-3)(6)(-5)}{(4)(3)(2)(1)},$$

and so on, with

$$A_{2m} = \frac{(2m+4)(2m+2)\cdots(6)(2m-7)(2m-9)\cdots(-5)}{(2m)!}.$$

The odd coefficients A_{2m+1} of the series solution are all equal to 0. Thus

$$y_1(t) = 1 - 15t^2 + \sum_{m=2}^{\infty} \frac{(2m+4)(2m+2)\cdots(6)(2m-7)(2m-9)\cdots(-5)}{(2m)!}t^{2m}.$$

If $y_2(t)$ is the solution with $y_2(0) = 0$ and $y_2'(0) = 1$, then all of the even coefficients are zero. For the odd coefficients, we have

$$A_3 = -\frac{14}{3} = \frac{(7)(-4)}{(3)(2)}, \quad A_5 = \frac{(9)(-2)(7)(-4)}{(5)(4)(3)(2)}, \quad \text{and}$$

$$A_7 = \frac{(11)(0)(9)(-2)(7)(-4)}{7!} = 0.$$

It follows that $A_{7+2k} = 0$ for all $k \geq 0$. There are only three nonzero coefficients of this power series: $y_2(t)$ is the polynomial

$$y_2(t) = t - \frac{14}{3}t^3 + \frac{21}{5}t^5. \qquad \square$$

IVPs

To find the solution of an IVP

$$a_2(t)y'' + a_1(t)y' + a_0(t)y = 0; \quad y(t_0) = y_0, \ y'(t_0) = y_1,$$

where t_0 is an ordinary point, substitute $y = \sum_{n=0}^{\infty} A_n(t - t_0)^n$. Since $y(t_0) = A_0$ and $y'(t_0) = A_1$, the first two coefficients are determined by the initial conditions.

Example 7.2.3   Find a power series solution for the IVP

$$(t^2 - 4t)\, y'' + (4t - 8)\, y' + 2\, y = 0; \quad y(2) = 0, \quad y'(2) = 1.$$

Solution. Since the initial point is $t_0 = 2$, we will find a power series solution of the form

$$y = \sum_{n=0}^{\infty} A_n\, (t - 2)^n.$$

To simplify the calculation, put $\bar{t} = t - 2$. Then $t = \bar{t} + 2$. Furthermore,

$$y' = \frac{dy}{dt} = \frac{dy}{d\bar{t}},$$

so this substitution transforms the ODE to

$$(\bar{t}^2 - 4)\, y'' + 4\bar{t}\, y' + 2\, y = 0, \tag{7.14}$$

where $y = \sum_{n=0}^{\infty} A_n\, \bar{t}^n$. Then

$$\bar{t}^2\, y'' = \sum_{n=2}^{\infty} n\, (n - 1)\, A_n\, \bar{t}^n$$

$$4\bar{t}\, y' = \sum_{n=1}^{\infty} 4n\, A_n\, \bar{t}^n$$

$$= 4A_1\bar{t} + \sum_{n=2}^{\infty} 4n\, A_n\, \bar{t}^n,$$

and by index shifting,

$$-4\, y'' = \sum_{n=0}^{\infty} [-4\, (n + 2)\, (n + 1)\, A_{n+2}]\, \bar{t}^n$$

$$= -8A_2 - 24A_3\bar{t} + \sum_{n=2}^{\infty} [-4\, (n + 2)\, (n + 1)\, A_{n+2}]\, \bar{t}^n.$$

Substituting these expressions into equation (7.14) results in the equation

$$-8A_2 + 2A_0 + (-24A_3 + 6A_1)\bar{t}$$

$$+ \sum_{n=2}^{\infty} \{[n\, (n - 1) + 4n + 2]\, A_n - 4\, (n + 2)(n + 1)A_{n+2}\}\bar{t}^n = 0.$$

Since the constant term must equal 0, we have $-8A_2 + 2A_0 = 0$; hence $A_2 = \frac{1}{4}A_0$. Turning to the coefficient of $\bar{t}$, we have $-24A_3 + 6A_1 = 0$, so $A_3 = \frac{1}{4}A_1$. In general, by equating the coefficient of $\bar{t}^n$ to 0 we obtain a recursion formula:

$$A_{n+2} = \frac{n\, (n - 1) + 4n + 2}{4\, (n + 2)(n + 1)}\, A_n = \frac{1}{4}\, A_n.$$

The initial conditions imply that $A_0 = 0$ and $A_1 = 1$. Therefore, $a_2 = 0$, $A_3 = \frac{1}{4}$, and by the recursion formula $A_n = 0$ when n is even, and $A_{2m+1} = \frac{1}{4^m}$. Therefore, the solution is

$$y = \sum_{m=0}^{\infty} \frac{1}{4^m} \bar{t}^{2m+1} = \bar{t} \sum_{m=0}^{\infty} \left(\frac{\bar{t}^2}{4} \right)^m.$$

By recognizing the sum of a geometric series here, we can simplify this result:

$$y = \frac{\bar{t}}{1 - \frac{\bar{t}^2}{4}} = \frac{4(t-2)}{4 - (t-2)^2}.$$

Notice that this solution blows up at both singular points, $t = 0$ and $t = 4$, of the ODE. ◻

In the following example, we will consider an ODE in which one of the co-efficients is a not a polynomial. We will be unable to derive an expression for the coefficients of the power series representing the solution, or even a recursion formula.

Example 7.2.4   Let $y(t)$ be the solution of

$$(\cos t)y'' - t^2 y' + y = 0, \tag{7.15}$$

with initial conditions $y(0) = 0$, and $y'(0) = 1$. Find the coefficients of t^n in the Taylor series expansion of $y(t)$ at $t = 0$ for $0 \le n \le 5$.

Solution. Set $y = \sum_{n=0}^{\infty} A_n t^n$. The initial conditions imply that $A_0 = 0$ and $A_1 = 1$; hence,

$$y = t + A_2 t^2 + A_3 t^3 + \cdots.$$

We have $y' = 1 + 2A_2 t + 3A_3 t^2 + \cdots$, and hence

$$t^2 y' = t^2 + 2A_2 t^3 + 3A_3 t^4 + \cdots.$$

The power series that represents $(\cos t)y''$ is obtained by multiplying the power series

$$\cos t = \sum_{n=0}^{\infty} \frac{(-1)^n}{(2n)!} t^{2n} \quad \text{and} \quad y'' = \sum_{n=0}^{\infty} (n+2)(n+1) A_{n+2} t^n.$$

This product cannot be expressed simply, but it is straightforward to compute the terms up to any degree n. The terms up to degree 3 are

$$(\cos t)y'' = 2A_2 + 6A_3 t + (12A_4 - A_2)t^2 + (20A_5 - 3A_3)t^3 + \cdots.$$

Thus,

$$(\cos t)y'' - t^2 y' + y = 2A_2 + (1 + 6A_3)t + [(12A_4 - A_2) - 1 + A_2]t^2$$
$$+ [(20A_5 - 3A_3) - 2A_2 + A_3]t^3 + \cdots.$$

Each coefficient must equal 0, so we have

$$2A_2 = 0$$
$$1 + 6A_3 = 0$$
$$12A_4 - 1 = 0$$
$$20A_5 - 2A_3 - 2A_2 = 0.$$

By solving the first three equations, we find that $A_2 = 0$, $A_3 = -\frac{1}{6}$, and $A_4 = \frac{1}{12}$. These values may be substituted into the fourth equation; then

$$A_5 = \frac{1}{10}(A_2 + A_3) = -\frac{1}{60}.$$

The solution of the IVP is

$$y = t - \frac{1}{6}t^3 + \frac{1}{12}t^4 - \frac{1}{60}t^5 + \cdots. \tag{7.16}$$

Theorem 7.1 is especially useful in the context of Example 7.2.4. Since we do not know all of the terms of the power series that represents the solution, we have no direct way of testing it for convergence. It is only indirectly, through the theorem, that we know that the series (7.16) has a positive radius of convergence.

EXERCISES

1. Given that $f(t) = \frac{1}{t+1}$ satisfies the ODE

$$(t + 1)y' + y = 0,$$

find a power series expansion of $f(t)$ about $t_0 = 0$.

2. Use the ODE $y' - 2y = 0$ to derive a power series expansion for $f(t) = e^{2t}$.

3. Use the ODE $y' + 2ty = 0$ to derive a power series expansion for $f(t) = e^{-t^2}$.

4. Derive the power series

$$\sin t = \sum_{n=0}^{\infty} \frac{(-1)^n}{(2n+1)!}t^{2n+1} \quad \text{and} \quad \cos t = \sum_{n=0}^{\infty} \frac{(-1)^n}{(2n)!}t^{2n}$$

by using the fact that $\{\sin t, \cos t\}$ is a fundamental set of solutions of $y'' + y = 0$.

5. Find a linear ODE with constant coefficients satisfied by $y(t) = e^{2t} \cos 3t$, and use it to find a the first five terms of a power series expansion of $y(t)$.

6. Find power series representations of a fundamental set of solutions of $(1 - t^2)y'' + 2y = 0$.

7. Find the solution of the IVP

$$(1 + t^2)y'' + ty' - y = 0; \quad y(0) = 1, \quad y'(0) = 0$$

in the form of a power series.

8. The following problems are for practice with double factorial notation.

a. Show that for any positive integer n, $n! = n!!(n-1)!!$.

b. Show that if m is a nonnegative integer, then $(2m)!! = 2^m m!$.

9. Let $n!!! = \begin{cases} n(n-3)(n-6)\cdots 3 & \text{if } n = 3m, \\ n(n-3)(n-6)\cdots 1 & \text{if } n = 3m+1, \\ n(n-3)(n-6)\cdots 2 & \text{if } n = 3m+2. \end{cases}$

Use this "triple factorial" notation to present a fundamental set of solutions of Airy's equation

$$y'' - ty = 0,$$

expanded at $t = 0$.

10. It was shown in Example 7.2.1 that the ODE $y'' + ty' + y = 0$ has a fundamental set of solutions $\{y_1(t), y_2(t)\}$, where

$y_1(t) = e^{-t^2/2}$ and

$$y_2(t) = \sum_{m=0}^{\infty} \frac{(-1)^m}{(2m+1)!!} t^{2m+1}.$$

Use the method of reduction of order to find an integral expression for $y_2(t)$.

11. Show that if ν is a positive integer, then Legendre's equation [see equation (7.13) on page 330] has a solution that is a polynomial.

12. The Legendre polynomial $P_\nu(t)$ is the polynomial solution of equation (7.13), normalized so that $P_\nu(1) = 1$. Show that

$$P_5(t) = \frac{1}{8}(15t - 70t^3 + 63t^5).$$

13. Let $y_1(t)$ be the solution of Legendre's equation derived in Example 7.2.2 on page 330.

  a. Show that

$$y_1(t) = 1 - 15t^2 + 30t^4 - 10t^6$$

$$- \frac{15}{8} \sum_{m=4}^{\infty} \frac{(2m+4)!!(2m-7)!!}{(2m)!} t^{2m}.$$

  b. Show that the radius of convergence of the power series representing $y_1(t)$ is 1.

14. Solve the IVP

$$(t^2 - 4t)\,y'' + (4t - 8)\,y' + 2\,y = 0;$$

$$y(2) = 1, \ y'(2) = 0.$$

(See Example 7.2.3.)

In Exercises 15–25, find the solution of the IVP in the form of a power series. If it is feasible to do so, give the coefficients in closed form.

15. $y'' + 2ty' - y = 0; \ y(0) = 1, \ y'(0) = 0$

16. $(1 - t^2)y'' + 3ty' - 4y = 0; \ y(0) = 0, \ y'(0) = 1$

17. $(2t^2 - 4t + 3)y'' + 2(t - 1)\,y' - 2y = 0; \ y(1) = 0, \ y'(1) = 1$

18. $y'' - ty' - (t^2 + 1)y = 0; \ y(0) = 2, \ y'(0) = 1$

19. $(2t - t^2)y'' - 4(t - 1)y' - 2y = 0; \ y(1) = 0, \ y'(1) = 1$

20. $(1 + t^2)y'' - 2y = 0; \ y(0) = 1, \ y'(0) = 0$

21. $(t^2 - 1)y'' + 6ty' + 6y = 0; \ y(0) = 1, \ y'(0) = 0$

22. $y'' - (t + 2)\,y' + 3y = 0; \ y(-2) = y'(-2) = 1$

23. $(1 - t^2)y'' - 2ty' - \frac{1}{4}y = 0; \ y(0) = 1, \ y'(0) = 0$

24. $y'' - t\,y = 0; \ y(1) = 0, \ y'(1) = 1$

25. $(1 - t^2)y'' - 8ty' - 12y = 0; \ y(0) = 1, \ y'(0) = 0$

In Exercises 26–29, find the first four nonzero terms of power series representing a fundamental set of solutions of the given ODE, expanded at the given point t_0.

26. $y'' - (\sin t)y = 0; \ t_0 = 0$

27. $y'' + \sqrt{t}y' + y = 0; \ t_0 = 1$

28. $y'' + (\ln t)y = 0; \ t_0 = 1$

29. $(\cos t)y'' + e^t y = 0; \ t_0 = 0$

In each of Exercises 30–37, find a polynomial that satisfies the ODE.

30. $(1 - t^2)y'' + 12y = 0$

31. $y'' + ty' - 6y = 0$

32. $(1 + t^2)y'' + ty' - 25y = 0$

33. $(1 - t)^2 y'' + 3(1 - t)y' + 4y = 0$

34. $(1 + 2t + t^2)y'' - 6(1 + t)y' + 12y = 0$

35. $(t^2 - 1)y'' - 2ty' + 2y = 0$

36. $(t^2 - 1)y'' - 4ty' + 6y = 0$

37. $y'' - 2ty' + 2\nu y = 0$, for $\nu = 1, 3, 5$. (Hermite polynomials)

7.3   Cauchy-Euler Equations

A **Cauchy-Euler equation** is a linear ODE of the special form

$$a_2 t^2 y'' + a_1 t y + a_0 y = 0, \tag{7.17}$$

where a_0, a_1, and a_2 are constants. Every Cauchy-Euler equation has a singular point at 0, because the coefficient of y'' vanishes there, and no other singular points. IVPs with initial point $t_0 = 0$ are not well posed (the existence and uniqueness theorem doesn't apply). If $t_0 > 0$, we can expect solution of an IVP to be defined on $(0, \infty)$, and if $t_0 < 0$, the solution will be defined on $(-\infty, 0)$. To avoid confusion, we will make the blanket assumption that the variable t is positive. The case where t is assumed to be negative is Exercise 15 at the end of this section.

Cauchy-Euler equations can be solved by a method similar to the way that linear ODEs with constant coefficients are solved. Substitute $y = t^s$, where the exponent s will be determined later, to obtain

$$a_2 t^2 s(s-1) t^{s-2} + a_1 t s t^{s-1} + a_0 t^s = 0.$$

If we now divide through by t^s and simplify, we obtain a quadratic equation,

$$a_2 s^2 + (a_1 - a_2)s + a_0 = 0. \tag{7.18}$$

Thus, s is a root of equation (7.18) if and only if $y = t^s$ is a solution of equation (7.17). Equation (7.18) is called the **indicial equation** of the Cauchy-Euler equation (7.17), and its roots are the **exponents** of (7.17).

Example 7.3.1   Find the general solution of the Cauchy-Euler equation

$$2t^2 y'' + 9ty' + 5y = 0. \tag{7.19}$$

Solution. To find the indicial equation, substitute $y = t^s$, $t\, y' = s\, t^s$, and $t^2\, y'' = s(s-1)\, t^s$ in equation (7.19). The result is

$$2s(s-1)\, t^s + 9s\, t^s + 5t^s = 0$$

and we can obtain the indicial equation

$$2s^2 + 7s + 5 = 0$$

by dividing through by t^s. This quadratic can be factored as $(s+1)(2s+5) = 0$, so it is easily solved, and we find that the exponents are -1 and $-5/2$. Thus $y = t^{-1}$ and $y = t^{-5/2}$ are solutions of equation (7.19). They are linearly independent, because t^{-1} is not a constant multiple of $t^{-5/2}$. The general solution is

$$y = C_1 t^{-1} + C_2 t^{-5/2}. \qquad \square$$

Complex exponents. If $k = \sigma + i\tau$ and $t > 0$, then

$$t^k = t^\sigma t^{i\tau}$$
$$= t^\sigma e^{i\tau \ln t}$$
$$= t^\sigma [\cos(\tau \ln t) + i \sin(\tau \ln t)].$$

Thus, if k is a complex exponent of a Cauchy-Euler equation, then t^k is a complex-valued solution, defined on $(0, \infty)$. By Proposition 5.3.1 on page 200,

$$\mathrm{Re}(t^k) = t^\sigma \cos(\tau \ln(t)), \quad \text{and}$$
$$\mathrm{Im}(t^k) = t^\sigma \sin(\tau \ln(t))$$

are real-valued solutions.

Example 7.3.2   Find the general solution of

$$t^2 y'' + t y' + 36 y = 0$$

on $(0, \infty)$.

Solution. Substituting $y = t^s$, we have $s(s-1)t^s + st^s + 36t^s = 0$. Therefore, the indicial equation is $s^2 + 36 = 0$. The exponents are $s = \pm 6i$, and the corresponding solution is

$$y(t) = t^{6i} = e^{6i \, \ln(t)}$$
$$= \cos(6 \ln t) + i \sin(6 \ln t).$$

Taking real and imaginary parts, $\cos(6 \ln t)$ and $\sin(6 \ln t)$ are seen to be the real solutions, and

$$y(t) = A \cos(6 \ln t) + B \sin(6 \ln t)$$

is the general solution. ❑

Example 7.3.3   Find a fundamental set of solutions of

$$t^2 y'' + 5t y' + 5y = 0$$

on $(0, \infty)$.

Solution. The indicial equation (found by the same method as in the previous examples) is $s^2 + 4s + 5 = 0$, with roots $s = -2 \pm i$. This leads to a complex-valued solution

$$y_1(t) = t^{-2}[\cos(\ln t) + i \sin(\ln t)].$$

The real and imaginary parts of $y_1(t)$ are the real-valued solutions $y_2(t) = t^{-2} \cos(\ln t)$ and $y_3(t) = t^{-2} \sin(\ln t)$. It is left to you to verify that $\{y_2(t), y_3(t)\}$ is linearly independent, and thus a fundamental set of solutions. ❑

Double exponents. Let $\mathcal{L}(y) = t^2 y'' + pty' + qy$, and let $f(s) = 0$ be its indicial equation. Thus,

$$f(s) = s(s-1) + ps + q,$$

and for any s, we have $\mathcal{L}(y) = f(s)t^s$. If $f(s) = 0$ has a double root $s = k$, then we can factor f as $f(s) = (s - k)^2$. In other words,

$$\mathcal{L}(t^s) = (s - k)^2 t^s. \tag{7.20}$$

As usual, we have $\mathcal{L}(t^k) = 0$. We could find a second, independent solution by reduction of order, but there is a simpler way. Differentiate equation (7.20) with respect to s:

$$\frac{\partial}{\partial s} \mathcal{L}(t^s) = 2(s - k)t^s + (s - k)^2 t^s \ln t.$$

If we now set $s = k$, we obtain

$$\frac{\partial}{\partial s}\mathcal{L}(t^s)\bigg|_{s=k} = 0.$$

Since we can interchange the order of taking partial derivatives,

$$\frac{\partial}{\partial s}\mathcal{L}(y) = \frac{\partial}{\partial s}\left(t^2\frac{\partial^2 y}{\partial t^2} + pt\frac{\partial y}{\partial t} + qy\right)$$

$$= t^2\frac{\partial^3 y}{\partial s\partial t^2} + pt\frac{\partial^2 y}{\partial s\partial t} + q\frac{\partial y}{\partial s}$$

$$= \mathcal{L}\left(\frac{\partial y}{\partial s}\right).$$

Therefore,

$$\frac{\partial}{\partial s}\mathcal{L}(t^s)\bigg|_{s=k} = \mathcal{L}\left(\frac{\partial(t^s)}{\partial s}\bigg|_{s=k}\right),$$

and thus $y = \frac{\partial(t^s)}{\partial s}\big|_{s=k}$ is a solution of $\mathcal{L}(y) = 0$. Since $\frac{\partial}{\partial s}t^s = t^s\ln t$, our second solution is $y = t^k\ln(t)$.

Example 7.3.4   Find the general solution of

$$t^2y'' - ty' + y = 0.$$

Solution. Let $\mathcal{L}(y) = t^2y'' - ty' + y$. Then

$$\mathcal{L}(t^s) = (s^2 - 2s + 1)t^s,$$

so $s = 1$ is a double exponent. It follows that $y(t) = t^1$ and $y_2(t) = t\ln t$ form a fundamental set of solutions. The general solution is therefore

$$y = Bt + Ct\ln t. \qquad \square$$

Summary

A Cauchy-Euler equation is an ODE of the form

$$a_2t^2y'' + a_1ty' + a_0y = 0.$$

The indicial equation, $a_2s^2 + (a_1 - a_2)s + a_0 = 0$, is obtained by substituting $y = t^s$ in the ODE. If the indicial equation has distinct real roots $s = k$ and $s = l$, then $\{t^k, t^l\}$ is a fundamental set of solutions. If there is a conjugate pair of complex roots, $s = \sigma \pm i\tau$, with $\tau \neq 0$, then $\{t^\sigma\cos(\tau\ln(t)), t^\sigma\sin(\tau\ln(t))\}$ is a fundamental set of solutions. Finally, if $s = k$ is a double root, $\{t^k, t^k\ln(t)\}$ is a fundamental set of solutions.

The relation with constant coefficient equations

Let us substitute $t = e^x$, and $y(e^x) = Y(x)$ in the Cauchy-Euler equation

$$a_2\, t^2\, y'' + a_1\, t\, y' + a_0\, y = 0. \tag{7.21}$$

Then $\frac{dY}{dx} = \frac{dy}{dt}\frac{dt}{dx}$, and since $\frac{dt}{dx} = e^x = t$,

$$\frac{dY}{dx} = ty'.$$

Now differentiate again with respect to x, and remember to use the product rule:

$$\begin{aligned}
\frac{d^2Y}{dx^2} &= \frac{dt}{dx}y' + t\frac{dy'}{dt}\frac{dt}{dx} \\
&= ty' + t^2 y'' \\
&= \frac{dY}{dx} + t^2 y''.
\end{aligned}$$

It follows that $t^2 y'' = \frac{d^2Y}{dx^2} - \frac{dY}{dx}$. Thus, the substitution converts equation (7.21) into a constant coefficient equation

$$a_2\frac{d^2Y}{dx^2} + (a_1 - a_2)\frac{dY}{dx} + a_0\, Y = 0. \tag{7.22}$$

The characteristic equation of equation (7.22) is the same as the indicial equation of equation (7.21),

$$a_2 s^2 + (a_1 - a_2)s + a_0 = 0,$$

so if s is an exponent of (7.21) and t^s is the corresponding solution, then s is also a characteristic root of equation (7.22), corresponding to the solution $e^{sx} = t^s$.

EXERCISES

In Exercises 1–4, find the indicial equation.

1. $4t^2 y'' + y = 0$

2. $ty' + 12y = 0$

3. $ty'' - 9y' + \frac{15}{t}y = 0$

4. $y'' - 4t^{-2}y = 0$

In Exercises 5–10, find the general solution on $(0, \infty)$.

5. $t^2 y'' - 7ty' + 15y = 0$

6. $t^2 y'' - 12y = 0$

7. $4t^2 y'' - 8ty' + 9y = 0$

8. $t^2 y'' + ty' + 16y = 0$

9. $t^2 y'' - ty' + 2y = 0$

10. $y'' = 6t^{-2}y$

11. Show that the Cauchy-Euler equation $a_2\, t^2\, y'' + a_1\, t\, y' +$ $a_0\, y = 0$ is equivalent to a system of the form

$$t\, y' = v$$
$$t\, v' = b\, y + c\, v$$

and find the coefficients b and c in terms of the a_i.

12. Solve the IVP

$$4t^2 y'' - 7ty' + 6y = 0; \quad y(1) = 1,\ y'(1) = 2.$$

13. Solve the IVP

$$t^2 y'' + 5ty' + 4y = 0; \quad y(-1) = 1,\ y'(-1) = 0.$$

14. Let $a_2(t)$ and $a_0(t)$ be even functions, and let $a_1(t)$ be an odd function. Show that if $y = \phi(t)$ is a solution of the ODE

$$a_2(t)\, y'' + a_1(t)\, y' + a_0(t)\, y = 0,$$

then $y = \phi(-t)$ is also a solution.

15. Show that $y = \phi(t)$ is a solution of the Cauchy-Euler equation

$$a_2\, t^2\, y'' + a_1\, t\, y' + a_0\, y = 0 \qquad (7.23)$$

if and only if $y = \phi(-t)$ is a solution. Explain how this fact can be used to find the general solution of equation (7.23) on the interval $(-\infty, 0)$.

◆ **Hint:** Use the result of Exercise 14.

16. Show that an ODE of the form

$$a_2(t - t_0)^2 y'' + a_1(t - t_0)y' + a_0 y = 0,$$

where a_0, a_1 and a_2 are constants can be converted to a Cauchy-Euler equation by the change of variable $\bar{t} = t - t_0$.

Use this observation to find the general solutions of the following ODEs.

a. $(t + 5)^2 y'' - 2y = 0$

b. $t^2 y'' - t(2y'' - 4y') + y'' - 4y' + 2y = 0$

c. $y'' + \dfrac{1}{t - 2} y' - \dfrac{1}{(t - 2)^2} y = 0$

d. $(t + 3)^2 y'' + 5(t + 3)y' + 4y = 0$

e. $(t - 1)^2 y'' - 7(t - 1)y' + 25y = 0$

17. Solve the IVP

$$(t - 1)^2 y'' - (t - 1)y' + y = 0; \quad y(0) = 0, \ y'(0) = 2.$$

18. Solve the IVP

$$(t - 2)^2 y'' - (t - 2)y' + 10y = 0;$$
$$y(2 - e^\pi) = 0, \ y'(2 - e^\pi) = 1.$$

7.4 Regular Singular Points

The ODE

$$t^2 y'' + ty' + (t^2 - v^2)y = 0 \qquad (7.24)$$

resembles a Cauchy-Euler equation,

$$t^2 y'' + ty' - v^2 y = 0. \qquad (7.25)$$

To solve the Cauchy-Euler equation, we substitute $y = t^s$. Define a linear differential operator $\mathcal{L}_1$ by

$$\mathcal{L}_1(y) = t^2 y'' + ty' - v^2 y.$$

Then

$$\mathcal{L}_1(t^s) = t^2 s(s - 1)t^{s-2} + tst^{s-1} - v^2 t^s$$
$$= (s^2 - v^2)t^s$$
$$= f(s)t^s,$$

where $f(s) = s^2 - v^2$ is the **indicial polynomial** of $\mathcal{L}_1$. The exponents of equation (7.25) are $s = \pm v$, and hence the general solution is $y = At^v + Bt^{-v}$. These exponents will also be helpful in solving equation (7.24).

The operator associated with equation (7.24) is

$$\mathcal{L}(y) = t^2 y'' + ty' + (t^2 - v^2)y$$
$$= \mathcal{L}_1(y) + t^2 y.$$

If we set $y = t^s$, we have

$$\mathcal{L}(t^s) = f(s)t^s + t^{s+2}. \tag{7.26}$$

We will apply $\mathcal{L}$ to an expression

$$y(t, s) = \sum_{n=0}^{\infty} A_n t^{n+s},$$

where the coefficients except for A_0 are undetermined, but $A_0 = 1$. The special value for A_0 is justified because if $A_0 = 0$, then we should change s so that the first term of the series is nonzero. If $A_0 \neq 0$, we can multiply $y(t, s)$ by a constant to make $A_0 = 1$. Before proceeding farther, we take note of the warning label:

> **WARNING:** The following calculation is *formal*: In calculating the coefficients of the series, we are ignoring the issue of convergence, without which the solution would be meaningless.

By equation (7.26),

$$\mathcal{L}(y(t, s)) = \sum_{n=0}^{\infty} A_n \mathcal{L}(t^{n+s})$$

$$= \sum_{n=0}^{\infty} A_n(f(n + s)t^{n+s} + t^{n+s+2})$$

$$= \sum_{n=0}^{\infty} A_n f(n + s)t^{n+s} + \sum_{n=2}^{\infty} A_{n-2}t^{n+s},$$

where we have shifted indices in the second sum. The two sums can be recombined (remember our convention that $A_n = 0$ if n is negative), to obtain

$$\mathcal{L}(y(t, s)) = A_0 f(s)t^s + A_1 f(1 + s)t^{1+s} + \sum_{n=2}^{\infty}(A_n f(n + s) + A_{n-2})t^{n+s}.$$

Thus $\mathcal{L}(y(t, s)) = 0$ implies that $A_0 f(s) = 0$, $A_1 f(1 + s) = 0$, and

$$A_n f(n + s) = -A_{n-2} \ (n \geq 2). \tag{7.27}$$

Since $A_0 = 1$, it follows that $f(s) = 0$, so $s = \pm\nu$. We will put $s = \nu$, where ν is the positive exponent, in what follows. This choice ensures that $f(n + \nu) \neq 0$ for all positive integers n.

In particular, since $f(1 + \nu) \neq 0$, we conclude that $A_1 = 0$. Equation (7.27) now leads to the recursive formula

$$A_n = \frac{-1}{f(n + \nu)} A_{n-2} \quad \text{for } n \geq 2. \tag{7.28}$$

Since $A_1 = 0$, it follows that $A_n = 0$ if n is odd, and if $n = 2m$ is even,

$$A_{2m} = \frac{(-1)^m}{f(2 + \nu) \cdot f(4 + \nu) \cdots f(2m + \nu)}.$$

Since

$$f(2k + \nu) = (2k + \nu)^2 - \nu^2$$
$$= 4k(k + \nu),$$

the expression appearing as the denominator of A_{2m} is

$$f(2 + \nu)f(4 + \nu) \cdots f(2m + \nu) = [4(1 + \nu)][4(2)(2 + \nu)] \cdots [4m(m + \nu)]$$
$$= 4^m m!(1 + \nu)(2 + \nu) \cdots (m + \nu).$$

Therefore, we can express the solution $y(t, \nu)$ of equation (7.24) as

$$y(t, \nu) = \sum_{m=0}^{\infty} \frac{(-1)^m t^{2m+\nu}}{4^m m!(1 + \nu)(2 + \nu) \cdots (m + \nu)}, \qquad (7.29)$$

for $t > 0$.

It is necessary to validate this formal calculation (remember the warning label). Thus, we need to show that the power series $\sum_{m=0}^{\infty} A_{2m} t^{2m}$ has a positive radius of convergence. By formula (7.3) on page 321, the radius of convergence of $\sum_{m=0}^{\infty} A_{2m} x^m$ is

$$R = \lim_{n \to \infty} \left| \frac{A_{2m}}{A_{2m+2}} \right|.$$

Since

$$\left| \frac{A_{2m}}{A_{2m+2}} \right| = \frac{4^{m+1}(m + 1)!(1 + \nu)(2 + \nu) \cdots (m + \nu)(m + 1 + \nu)}{4^m m!(1 + \nu)(2 + \nu) \cdots (m + \nu)}$$
$$= 4(m + 1)(m + 1 + \nu)$$
$$\to \infty \text{ as } m \to \infty,$$

$R = \infty$. We can therefore be sure that $y(t, \nu)$ is a solution of equation (7.24).

To obtain a second solution we can start over and use the other root of $f(s) = 0$, $s = -\nu$. We can put $A_1 = 0$ to ensure that $A_1 f(1 - \nu) = 0$. The recursion formula derived from equation (7.27) with $s = -\nu$ is

$$A_{2m} = \frac{-1}{f(2m - \nu)} A_{2m-2}. \qquad (7.30)$$

Since $f(2m - \nu) = (2m - \nu)^2 - \nu^2 = 2m(2m - 2\nu)$, we will have a problem when ν is an integer: When we try to calculate $A_{2\nu}$, we have to divide by $2\nu(2\nu - 2\nu) = 0$. If ν is not an integer, then

$$y(t, -\nu) = \sum_{m=0}^{\infty} \frac{(-1)^m t^{2m-\nu}}{4^m m!(1 - \nu)(2 - \nu) \cdots (m - \nu)}$$

is a solution of equation (7.24), and $\{y(t, \nu), y(t, -\nu)\}$ is a fundamental set of solutions.

Equation (7.24) is important in mathematical physics. It is named for the German mathematician F. W. Bessel, who used it in a study of planetary motion. Solutions of Bessel's equation, called **Bessel functions**, are among the most frequently used special functions. They were actually discovered by Euler, long before Bessel was born, in a study of the vibrations of a drum.

Regular singular points

Our approach to solving Bessel's equation is based on a method developed in 1901 by another German mathematician, F. G. Frobenius.

The Frobenius method applies to linear ODEs of the form

$$p(t) \, t^2 \, y'' + q(t) \, t \, y' + r(t) \, y = 0. \tag{7.31}$$

For Bessel's equation, $p(t) = q(t) = 1$, and $r(t) = t^2 - v^2$.

An ODE that is equivalent to one in the form of equation (7.31), where the functions $q(t)/p(t)$ and $r(t)/p(t)$ are analytic at $t = 0$, is said to have a **regular singular point** at $t = 0$. More generally, an ODE has a regular singular point at $t = a$ if it can be put in the form

$$p(t) \, (t - a)^2 \, y'' + q(t) \, (t - a) \, y' + r(t) \, y = 0,$$

where $q(t)/p(t)$ and $r(t)/p(t)$ are analytic at $t = a$.

Example 7.4.1   Show that the **hypergeometric equation**

$$t(1 - t) \, y'' + [c - (a + b + 1)t] \, y' - a \, b \, y = 0,$$

has regular singular points at $t = 0$ and $t = 1$.

Solution. Multiply the equation by t to obtain

$$t^2 \, (1 - t) \, y'' + t \, [c - (a + b + 1) \, t]y' - a \, b \, t \, y = 0.$$

In this form, $p(t) = 1 - t$, $q(t) = c - (a + b + 1) \, t$, and $r(t) = -a \, b \, t$. All three functions are analytic, and $p(0) = 1$ implies that $q(t)/p(t)$ and $r(t)/p(t)$ are also analytic. It follows that 0 is a regular singular point.

If we multiply the hypergeometric equation by $t - 1$, we obtain

$$-t(t - 1)^2 y'' + [c - (a + b + 1) \, t](t - 1)y' - a \, b(t - 1) \, y = 0.$$

Now, $p(t) = -t$, and $q(t) = c - (a + b + 1) \, t$ and $r(t) = -a \, b$ as before. Since $p(1) \neq 0$, the quotients $q(t)/p(t)$ and $r(t)/p(t)$ represent analytic functions at $t = 1$, and hence the singular point at 1 is regular.   ❑

An ODE that has a singular point at $t = 0$, but cannot be put into the form of equation (7.31) has an **irregular singular point**. For example, $t = 0$ is an irregular singular point of the equation

$$t^2 \, y'' + (t + 1) \, y' + y = 0$$

because there is no function $q(t)$, analytic at $t = 0$, such that $tq(t)$ is equal to the coefficient $t + 1$ of y'. Irregular singular points are an advanced topic, beyond the scope of this text.

In equation (7.31), let $p(t) = \sum_{m=0}^{\infty} P_m t^m$, $q(t) = \sum_{m=0}^{\infty} Q_m t^m$, and $r(t) = \sum_{m=0}^{\infty} R_m t^m$. The polynomial

$$f_0(s) = P_0 \, s \, (s - 1) + Q_0 \, s + R_0$$

is called the **indicial polynomial** of the operator

$$\mathcal{L}(y) = p(t)\, t^2\, y'' + q(t)\, t\, y' + r(t)\, y$$

at the regular singular point $t = 0$. An analogous definition applies for nonzero singular points, and if an ODE has more than one regular singular point (we have seen that the hypergeometric equation does), then the indicial polynomials at the singular points may not be the same. The roots μ, ν of $f_0(s) = 0$ are called the **exponents** of equation (7.31). This is consistent with our usage of the term in the context of Cauchy-Euler equations and Bessel's equation. With Bessel's equation the exponent $\mu = -\nu$ could not be used to form a solution $y(t, \mu)$ if ν was an integer. This phenomenon is a frequent obstacle to solving equations with regular singular points, and inspires the following definition.

> **DEFINITION** Let μ and ν be the exponents of equation (7.31), chosen so that $\mu \le \nu$. If $\nu = \mu$, ν is said to be a **double exponent**. If $\nu - \mu$ is a positive integer, then μ is called a **blocked exponent**. Finally, an exponent that is not blocked is **unblocked**. The larger of the two exponents will always be unblocked, and both exponents are unblocked if the difference is not an integer.

Theorem 7.2

Let ν be an unblocked exponent of equation (7.31). Then there are coefficients A_n, $n = 0, 1, 2, \ldots$, with $A_0 = 1$, such that

$$y(t, \nu) = \sum_{n=0}^{\infty} A_n t^{n+\nu}$$

satisfies equation (7.31). The power series $\sum_{n=0}^{\infty} A_n t^n$ has a positive radius of convergence.

This theorem is due to L. Fuchs. If equation (7.31) has a double or blocked exponent, then Theorem 7.2 provides only one solution. The process for finding a second, linearly independent solution is will be covered in Section 7.6, but we have already seen a special case: the derivation of the second (logarithmic) solution of a Cauchy-Euler equation with a double exponent.

Before discussing the proof of Theorem 7.2, we will consider an example.

Example 7.4.2 Find a solution of the hypergeometric equation

$$t(1 - t)y'' - (2 + t)y' + 4y = 0.$$

Solution. We have $p(t) = 1 - t$, $q(t) = -(2 + t)$, and $r(t) = 4t$ (see Example 7.4.1); thus $P_0 = 1$, $Q_0 = -2$, and $R_0 = 0$. The indicial equation is

$$s(s - 1) - 2s = 0,$$

and the exponents are $\nu = 3$ and $\mu = 0$. We will apply the method of Frobenius to find a solution corresponding to the unblocked exponent 3. We have

$$y(t, 3) = \sum_{n=0}^{\infty} A_n t^{n+3},$$

in which $A_0 = 1$. Then by index shifting,

$$(2 + t)y' = 2\sum_{n=0}^{\infty} A_n(n + 3)t^{n+2} + \sum_{n=1}^{\infty} A_{n-1}(n + 2)t^{n+2}$$

$$= 6t^2 + \sum_{n=1}^{\infty} [2A_n(n + 3) + A_{n-1}(n + 2)]t^{n+2}.$$

Similarly,

$$t(1 - t)y'' = \sum_{n=0}^{\infty} A_n(n + 3)(n + 2)t^{n+2} - \sum_{n=1}^{\infty} A_{n-1}(n + 2)(n + 1)t^{n+2}$$

$$= 6t^2 + \sum_{n=1}^{\infty} [A_n(n + 3)(n + 2) - A_{n-1}(n + 2)(n + 1)]t^{n+2}.$$

It is not a coincidence that the two expressions have identical first terms. Let

$$\mathcal{L}(y) = t(1 - t)y'' - (2 + t)y' + 4y.$$

Then

$$\mathcal{L}(y(t, 3)) = f_0(3)t^2 + \text{higher-order terms},$$

where $f_0(s)$ is the indicial polynomial. Since $f_0(3) = 0$, the t^2 terms in $\mathcal{L}(y(t, 3))$ must cancel, and we have

$$\mathcal{L}(y(t, 3)) = \sum_{n=1}^{\infty} [A_n(n + 3)(n + 2) - A_{n-1}(n + 2)(n + 1)$$

$$- 2A_n(n + 3) - A_{n-1}(n + 2) + 4A_{n-1}]t^{n+2}$$

$$= \sum_{n=1}^{\infty} [A_n n(n + 3) - A_{n-1}n(n + 4)]t^{n+2}.$$

It follows that $\mathcal{L}(y(t, 3)) = 0$ if the coefficients satisfy the recursion formula

$$A_n = \frac{n + 4}{n + 3}A_{n-1}.$$

Since $A_0 = 1$, the recursion formula tells us that $A_1 = \frac{5}{4}, A_2 = \frac{6}{4}, \ldots, A_n = \frac{n+4}{4}$. The solution is

$$y(t, 3) = \frac{1}{4} \sum_{n=0}^{\infty} (n + 4)t^{n+3}.$$

It is not difficult to see that the radius of convergence of this power series is 1.   ◻

For more information about the solution of Example 7.4.2, see Exercises 9 and 10 at the end of this section.

Formal verification of Theorem 7.2 The proof of Theorem 7.2 involves two steps. In the first step, it is shown that there is a power series $\sum_{n=0}^{\infty} A_n t^n$ such that

$$y = t^{\nu} \sum_{n=0}^{\infty} A_n t^n \tag{7.32}$$

satisfies equation (7.31) in the *formal sense,* that is, without regard to convergence.

The second step in the proof of Theorem 7.2 is to show that the power series which was proved to exist has positive radius of convergence. We will omit this step, as it involves methods of mathematical analysis that are beyond the scope of this text. The book of Coddington and Levinson (a precise reference is given in the Suggestions for Further Reading at the end of this chapter) contains a proof of the second step.

By multiplying power series, we have

$$p(t)t^2 \frac{d^2}{dt^2} y(t, s) = \left(\sum_{m=0}^{\infty} P_m t^m \right) \left(\sum_{n=0}^{\infty} A_n (n+s)(n+s-1)t^{n+s} \right)$$

$$= \sum_{n=0}^{\infty} \sum_{k=0}^{n} P_{n-k} A_k (k+s)(k+s-1) t^{n+s}.$$

Similarly,

$$q(t)t \frac{d}{dt} y(t, s) = \sum_{n=0}^{\infty} \sum_{k=0}^{n} Q_{n-k} A_k (k+s) t^{n+s},$$

and

$$r(t)y(t, s) = \sum_{n=0}^{\infty} \sum_{k=0}^{n} R_{n-k} A_k t^{n+s}.$$

Hence

$$\mathcal{L}(y(t, s)) = \sum_{n=0}^{\infty} \sum_{k=0}^{n} [P_{n-k}(k+s)(k+s-1) + Q_{n-k}(k+s) + R_{n-k}] A_k t^{n+s}. \tag{7.33}$$

To simplify this expression, let

$$f_m(s) = P_m s(s-1) + Q_m s + R_m.$$

Notice that the indicial polynomial is $f_0(t)$. The expression in square brackets in the series on the right side of equation (7.33) is equal to $f_{n-k}(k+s)$, so that

$$\mathcal{L}(y(t, s)) = \sum_{n=0}^{\infty} \left(\sum_{k=0}^{n} f_{n-k}(k+s) A_k \right) t^{n+s}. \tag{7.34}$$

If $\mathcal{L}(y(t, s)) = 0$, then for each $n \geq 0$,

$$\sum_{k=0}^{n} f_{n-k}(k + s)A_k = 0. \tag{7.35}$$

For $n = 0$, equation (7.35) is just $A_0 f_0(s) = 0$. Since $A_0 = 1$, $f_0(s) = 0$; that is, s must be an exponent ν of equation (7.31). For $n = 1$, equation (7.35) is

$$f_0(1 + \nu)A_1 + f_1(\nu)A_0 = 0;$$

therefore, $A_1 = -\frac{f_1(\nu)A_0}{f_0(1+\nu)}$. By continuing in this way, we can calculate the coefficients $A_0, A_1, \ldots$, with the recursive formula

$$A_n = -\frac{1}{f_0(n + \nu)} \sum_{k=0}^{n-1} f_{n-k}(k + \nu)A_k. \tag{7.36}$$

It is here that we have to use the hypothesis that the exponent ν is unblocked. The coefficient A_n would be undefined if $f_0(n + \nu) = 0$. But, as we have noted, f_0 is just the indicial polynomial. Therefore, if $f_0(n + \nu) = 0$, then $\mu = n + \nu$ is an exponent for $\mathcal{L}(y) = 0$, and, since $\mu - \nu = n$ is a positive integer, ν would then be blocked. ∎

Singular points at $t = a$

If an ODE has a regular singular point $a \neq 0$, a change of variables $\tau = t - a$ can be used to move the singular point to the origin. The method of Frobenius can then be applied to find a series solution.

Example 7.4.3   The Legendre equation,

$$(1 - t^2)\, y'' - 2t\, y' + \nu(\nu + 1)\, y = 0,$$

has singular points at $t = \pm 1$. Show that the singular point at $t = 1$ is regular, and determine its exponents.

Solution. Let $\tau = t - 1$, and put $Y(\tau) = y(\tau + 1) = y(t)$. By the chain rule,

$$\frac{dy}{dt} = \frac{dY}{d\tau}\frac{d\tau}{dt} = \frac{dY}{d\tau}$$

and similarly $\frac{d^2 y}{dt^2} = \frac{d^2 Y}{d\tau^2}$. In terms of the variable τ, Legendre's equation is

$$(1 - (\tau + 1)^2)\, Y'' - 2(\tau + 1)\, Y' + \nu(\nu + 1)\, Y = 0,$$

which we simplify as

$$\tau(\tau + 2)\, Y'' + 2(\tau + 1)\, Y' - \nu(\nu + 1)\, Y = 0.$$

If we multiply through by τ, the resulting equation,

$$(\tau + 2)\,\tau^2 Y'' + 2(\tau + 1)\,\tau\, Y' - \nu(\nu + 1)\tau\, Y = 0,$$

is of the form (7.31), with $p(\tau) = \tau + 2$, $q(\tau) = -2(\tau + 1)$, and $r(\tau) = \nu(\nu + 1)\tau$. The indicial polynomial is therefore

$$f_0(s) = 2s(s - 1) + 2s = 2s^2.$$

Thus Legendre's equation has a double exponent 0 at the singular point $t = 1$.   ❑

EXERCISES

Which of the differential equations in Exercises 1–8 have regular singular points at 0? For those that do, find the indicial equation and calculate the exponents. Which exponents, if any, are blocked?

1. $y'' - \frac{1}{t}y = 0$ **2.** $t^2 y'' + y' - y = 0$

3. $(t^2 y')' + y = 0$ **4.** $(1 - \cos t)y'' + 2ty' + y = 0$

5. $\sin(t)y'' + y' + ty = 0$

6. $ty'' + y' + (t - 1)y = 0$

7. $t^3 y'' + \sin^2(t)y' + t(t - 1)y = 0$

8. $(t^n y')' + (t^{n-1}y)' = 0$

9. Show that the radius of convergence of the series $y(t, 3)$ which was found in the solution of Example 7.4.2 is 1.

10. Show that the solution $y(t, 3)$ of Example 7.4.2 is the rational function

$$y(t, 3) = \frac{t^3}{1 - t} + \frac{t^4}{4(1 - t)^2}.$$

11. Find the general solution of Bessel's equation

$$t^2 y'' + ty' + \left(t^2 - \frac{1}{4}\right)y = 0,$$

and express it in terms of elementary functions.

12. Laguerre polynomials are useful for approximating improper integrals of the form $\int_0^\infty f(t)\,dt$, using a method known as Gauss-Laguerre quadrature. For a positive integer N, the Laguerre polynomial $L_N(t)$ is defined by Laguerre's equation,

$$ty'' + (1 - t)y' + Ny = 0.$$

Show that Laguerre's equation has a solution that is a polynomial of degree N, and that all polynomial solutions are constant multiples of one another. The polynomial solution with constant term equal to 1 is $L_N(t)$.

13. Find $L_N(t)$ for $N = 1, 2$, and 3.

14. Find a solution of $y'' = \frac{1}{t}y$ (not $y \equiv 0$).

15. Find a solution of $t^2 y'' + t(t - 1)y' + y = 0$, by using the method of Frobenius. Find a second, independent solution by using the method of reduction of order.

16. Technically, one of the exponents of

$$4t^2 y'' - 4ty' + (3 - t^2)y = 0$$

is blocked. Show that nevertheless, the method of Frobenius works for both exponents and hence find the general solution.

17. Find the general solution of

$$2t^2 y'' + 3ty' - ty = 0.$$

18. Find the general solution of

$$ty'' + 2y' - ty = 0.$$

19. Find solutions of the hypergeometric equation

$$t(1 - t)y'' + (3 + t)y' - y = 0$$

at each of its singular points.

20. A system of ODEs of the form

$$\left.\begin{array}{l} t\,x' = a(t)\,x + b(t)\,y \\ t\,y' = c(t)\,x + d(t)\,y \end{array}\right\} \qquad (7.37)$$

has a regular singular point at $t = 0$ if the coefficient functions $a(t)$, $b(t)$, $c(t)$, and $d(t)$ are analytic at $t = 0$.

a. Given that the ODE

$$p(t) t^2 y'' + q(t) t y' + r(t) y = 0$$

has a regular singular point at $t = 0$, find an equivalent system of the form

$$t y' = v$$
$$t v' = c(t) y + d(t) v$$

and verify that it too has a regular singular point at $t = 0$.

b. Let

$$C(t) = \begin{bmatrix} a(t) & b(t) \\ c(t) & d(t) \end{bmatrix}$$

be the coefficient matrix of the system (7.37), which we can write as $t \, \mathbf{v}' = C(t) \mathbf{v}$. Show that

$$C(t) = \sum_{n=0}^{\infty} C_n t^n,$$

where each matrix C_n is a constant matrix.

c. Let μ and ν be the characteristic roots of C_0, and assume $\mu \le \nu$. Let $\mathbf{b}_0$ be a characteristic vector belonging

to ν. Ignoring convergence issues, show that the system (7.37) has a solution of the form

$$\mathbf{v} = \mathbf{b}_0 t^\nu + \mathbf{b}_1 t^{\nu+1} + \mathbf{b}_2 t^{\nu+2} + \cdots.$$

d. Under what conditions will there be a solution of the form

$$\mathbf{w} = \mathbf{c}_0 t^\mu + \mathbf{c}_1 t^{\mu+1} + \mathbf{c}_2 t^{\mu+2} + \cdots?$$

Explain your answer.

21. Show that $t = -1$ is a regular singular point of Legendre's equation with exponent 0 (see Example 7.4.3).

22. Find a series solution for Legendre's equation, expanded about the singular point $t = 1$, by using the method of Frobenius (see Example 7.4.3).

23. Show that when the parameter ν has an integer value, the series found in Exercise 22 is actually a polynomial. [The polynomial thus obtained is called the Legendre polynomial of order ν and denoted $P_\nu(t)$.]

24. Find all singular points of

$$t(1 - t^2)^2 y'' + (1 - t) y' + t y = 0$$

and calculate the exponents for each regular singular point.

7.5 Bessel Functions

In this section we will describe a fundamental set of solutions of Bessel's equation of order ν,

$$t^2 y'' + t y' + (t^2 - \nu^2) y = 0, \tag{7.38}$$

where the parameter ν can have any nonnegative real value.

Bessel functions of the first kind are among the most frequently encountered special functions. They are defined by the power series

$$J_\nu(t) = |t|^\nu \sum_{m=0}^{\infty} \frac{(-1)^m}{2^{2m+\nu} m! \Gamma(m + \nu + 1)} t^{2m}. \tag{7.39}$$

To understand this formula, we need to learn more about the gamma function,

$$\Gamma(z) = \int_0^\infty u^{z-1} e^{-u} \, du \text{ for } z > 0,$$

which was defined in Section 6.3. The graph of $y = \Gamma(z)$ is shown in Figure 7.2.

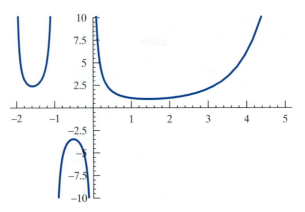

FIGURE 7.2   The Γ-function.

You will notice that the graph of the gamma function is carried to the left of the y-axis. This is unexpected, since the defining integral for the gamma function is divergent for $z \leq 0$. We found in Section 6.3 that if n is an integer ≥ 0, then $\Gamma(n + 1) = n!$, and that for all $z > 0$, $\Gamma(z + 1) = z\Gamma(z)$. This identity can be used to extend the definition[2] of $\Gamma(z)$ for $z < 0$. For example, with $z = -\frac{1}{2}$, we have

$$-\frac{1}{2}\Gamma\left(-\frac{1}{2}\right) = \Gamma\left(\frac{1}{2}\right)$$

and hence $\Gamma(-1/2) = -2\Gamma(1/2)$.

Let $L(z) = 1/\Gamma(z)$. The identity $\Gamma(z) = (z - 1)\Gamma(z - 1)$ can be written, in terms of L, as

$$L(z - 1) = (z - 1)L(z). \tag{7.40}$$

We can use formula (7.40) to extend the definition of $L(z)$ to cover all $z \in \mathbf{R}$, and the graph of L is shown in Figure 7.3.

Notice that the reciprocal gamma function appears to be continuous on $\mathbf{R}$. In fact, it is analytic, and can be represented by a power series with infinite radius of convergence. The figure also indicates that for integers $n \leq 0$, $L(n) = 0$. This important fact is left for you to prove.

To define Bessel functions of the second kind, we need to know more about the reciprocal gamma function. The necessary facts are stated in the following two lemmas, with proofs omitted. (The proofs are discussed in Exercises 4 and 5 at the end of this section.) The first lemma relates the reciprocal Γ-function to the sine.

---

[2]C. F. Gauss provided an alternative definition for the gamma function:

$$\Gamma(x) = \lim_{n \to \infty} \frac{n^z n!}{z(z + 1) \cdots (z + n)}.$$

This definition holds for all z except 0 and negative integers. For an amazing proof that the Euler and Gauss formulas for $\Gamma(z)$ are equal when $z > 0$, see *The Gamma Function*, by Emil Artin. Bibliographic information is provided in Suggestions for Further Reading (page 372).

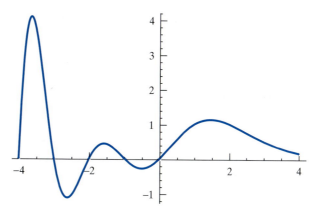

FIGURE 7.3   The reciprocal Γ-function.

Lemma 7.5.1

If ν is any real number, then $\pi L(\nu)L(1-\nu) = \sin(\pi\nu)$.

The formula for the derivative of the reciprocal gamma-function involves a new function $\phi(k)$, which is defined to be the sum of the reciprocals of the integers up to k. Thus,

$$\phi(k) = \sum_{m=1}^{k} \frac{1}{m}.$$

We will also need to refer to **Euler's constant** γ, which we will define[3] as

$$\gamma = -\int_0^\infty \ln(t)\,e^{-t}\,dt.$$

Euler's constant is an important number that has been calculated to great precision. To ten decimal places, $\gamma = 0.5772156649\ldots.$

Lemma 7.5.2

If k is an integer, then

$$L'(k+1) = \begin{cases} (\gamma - \phi(k))/k! & \text{if } k \geq 0 \\ -(-1)^k(|k|-1)! & \text{if } k < 0. \end{cases} \tag{7.41}$$

The proof of Lemma 7.5.2 is outlined in Exercise 5 at the end of this section.

We are now ready to tackle Bessel functions. We will start by finding the relationship between Bessel functions and the solution $y(t; \nu)$ of Bessel's equation defined by equation (7.29) on page 342.

---

[3]This definition is equivalent to the generally accepted one:

$$\gamma = \lim_{k\to\infty}[\phi(k) - \ln k];$$

however, we will not need to use this fact.

Proposition 7.5.3

$$J_\nu(t) = 2^{-\nu} L(\nu + 1) y(t; \nu).$$

It follows from this proposition that $J_\nu(t)$ is a solution of Bessel's equation, because it is a constant multiple of a solution.

Proof of Proposition 7.5.3.   The coefficients A_n of the series defining $y(t; \nu)$ were determined by setting $A_0 = 1$, and $A_1 = 0$. For $n \geq 2$, the recursion formula (7.28) on page 341 was used to express A_n in terms of A_{n-2}. We will show that the coefficient of t^n in the series (7.39) is $2^{-\nu} L(\nu + 1) A_n = \frac{1}{2^\nu \Gamma(\nu+1)} A_n$. This is obviously true if $n = 0$ or $n = 1$; to complete the proof we only need to notice that the same recursion formula holds for the coefficients of the series (7.39); that is,

$$\frac{(-1)^m}{2^{2m+\nu} m! \Gamma(m + \nu + 1)} = \frac{-1}{2m(2m + 2\nu)} \left(\frac{(-1)^{m-1}}{2^{2m-2+\nu}(m - 1)! \Gamma(m + \nu)} \right)$$

for $m \geq 1$.   ■

It was shown in Section 7.4 that if ν is not an integer, then

$$\{y(t; \nu), y(t; -\nu)\}$$

is a fundamental set of solutions of Bessel's equation.   Therefore, by Proposition 7.5.3,

$$\{J_\nu(t), J_{-\nu}(t)\}$$

is a also fundamental set of solutions if ν is not an integer. If ν is a positive integer, $y(t; -\nu)$ is undefined. One of the advantages of Bessel functions of the first kind is that $J_{-\nu}(t)$ is defined for all ν, although the following proposition implies that $\{J_\nu(t), J_{-\nu}(t)\}$ is linearly dependent if ν is an integer.

Lemma 7.5.4

The Wronskian of $J_\nu(t)$ and $J_{-\nu}(t)$ is given by the formula

$$W[J_\nu, J_{-\nu}](t) = -\frac{2}{\pi t} \sin(\pi \nu).$$

Proof By Proposition 5.2.3 on page 196, $W(t) = W[J_\nu, J_{-\nu}](t)$ satisfies the linear first-order differential equation

$$W' + \frac{1}{t} W = 0. \tag{7.42}$$

The general solution of equation (7.42) is $W(t) = C/t$, where C is constant. We will evaluate C by noting that $C \equiv t W(t)$. Thus,

$$C = t \det \begin{bmatrix} J_\nu(t) & J_{-\nu}(t) \\ J_\nu'(t) & J_{-\nu}'(t) \end{bmatrix}$$

$$= t \det \begin{bmatrix} \frac{t^\nu}{2^\nu \Gamma(\nu+1)} + \cdots & \frac{t^{-\nu}}{2^{-\nu} \Gamma(1-\nu)} + \cdots \\ \frac{\nu t^{\nu-1}}{2^\nu \Gamma(\nu+1)} + \cdots & \frac{-\nu t^{-(\nu+1)}}{2^{-\nu} \Gamma(1-\nu)} + \cdots \end{bmatrix}$$

$$= \frac{-\nu}{\Gamma(\nu + 1)\Gamma(1 - \nu)} - \frac{\nu}{\Gamma(1 - \nu)\Gamma(\nu + 1)} + \cdots.$$

The higher-order terms signified by $\cdots$ all involve t, so they must cancel out. Therefore,

$$C = -\frac{2v}{\Gamma(v+1)\Gamma(1-v)}.$$

Since $\Gamma(v+1) = v\Gamma(v)$,

$$C = -2L(v)L(1-v),$$

where L is the reciprocal gamma function.

By Lemma 7.5.1,

$$L(v)L(1-v) = \frac{1}{\pi}\sin(\pi v).$$

Hence, $W[J_v, J_{-v}](t) = -2\sin(\pi v)/(\pi t)$.   ∎

It follows from Lemma 7.5.4 that when v is an integer, the Wronskian is identically 0: J_v and J_{-v} are dependent.

Example 7.5.1   Show that $J_{-3}(t) = -J_3(t)$.

Solution. By definition,

$$J_{-3}(t) = \sum_{m=0}^{\infty} \frac{(-1)^m L(m-2)\, t^{2m-3}}{2^{2m-3} m!}$$

$$= \frac{L(-2)\, t^{-3}}{2^{-3} \cdot 0!} - \frac{t^{-1} L(-1)}{2^{-1} \cdot 1!} + \frac{L(0)\, t}{2 \cdot 2!} - \frac{L(1)\, t^3}{2^3 \cdot 3!} \cdots.$$

Since $L(n) = 0$ if n is an integer ≤ 0, the coefficients of t^{-3}, t^{-1}, and t vanish. Thus

$$J_{-3}(t) = \frac{-t^3}{2^3 \cdot 3! \cdot 0!} + \frac{t^5}{2^5 \cdot 4! \cdot 1!} + \frac{-t^7}{2^7 \cdot 5! \cdot 2!} \cdots$$

$$= -\sum_{n=0}^{\infty} \frac{(-1)^m t^{2m+3}}{2^{2m+3}(m+3)!\, m!}$$

$$= -J_3(t).$$ ❑

You can generalize the solution of Example 7.5.1 and prove the following lemma.

Lemma 7.5.5

If n is an integer, then $J_{-n}(t) = (-1)^n J_n(t)$.

If v is not an integer, the **Bessel function of the second kind** is defined by the formula

$$Y_v(t) = \frac{J_v(t)\cos(\pi v) - J_{-v}(t)}{\sin(\pi v)}. \tag{7.43}$$

Lemma 7.5.4 can be used to calculate the Wronskian of $J_v(t)$ and $Y_v(t)$ as follows:

$$W[J_v, Y_v](t) = W[J_v, J_v](t)\cot(\pi v) - W[J_v, J_{-v}](t)\csc(\pi v)$$

$$= 0 - \left(-\frac{2}{\pi t}\sin(\pi v)\right)\csc(\pi v)$$

$$= \frac{2}{\pi t}.$$

Notice that $W[J_v, Y_v](t)$ is independent of v.

Proposition 7.5.6

Let n be a nonnegative integer, and put

$$Y_n(t) = \lim_{v \to n} Y_v(t).$$

Then $Y_n(t)$ is a solution of Bessel's equation and $\{J_n(t), Y_n(t)\}$ is linearly independent.

Proof

$$\lim_{v \to n}[J_v(t)\cos(\pi v) - J_{-v}(t)] = J_n(t)\cos(\pi n) - J_{-n}(t)$$

Since $\cos(\pi n) = (-1)^n$, it follows from Lemma 7.5.5 that the limit is 0. Since $\lim_{v \to n}\sin(\pi v) = 0$, we can use l'Hôpital's rule:

$$\lim_{v \to n} Y_v(t) = \lim_{v \to n} \frac{\frac{\partial}{\partial v}[J_v(t)\cos(\pi v) - J_{-v}(t)]}{\frac{\partial}{\partial v}(\sin \pi v)}$$

$$= \frac{\left[\frac{\partial J_v(t)}{\partial v}\cos(\pi v) - \pi J_v(t)\sin(\pi v) - \frac{\partial J_{-v}(t)}{\partial v}\right]_{v=n}}{\pi \cos(\pi n)}$$

$$= \frac{1}{\pi}\frac{\left[\frac{\partial J_v(t)}{\partial v}(-1)^n - \pi J_v(t)\cdot 0 - \frac{\partial J_{-v}(t)}{\partial v}\right]_{v=n}}{(-1)^n}$$

$$= \frac{1}{\pi}\left[\frac{\partial J_v(t)}{\partial v} - (-1)^n\frac{\partial J_{-v}(t)}{\partial v}\right]_{v=n}. \tag{7.44}$$

It follows that the limit defining $Y_n(t)$ exists. Since $(-1)^n J_{-n}(t) = J_n(t)$, it might appear that the expression (7.44) vanishes. However, this relation between J_n and

J_{-n} does not imply any relation between the derivatives with respect to the parameter: Remember that Lemma 7.5.5 holds only if n is an integer.

Let $\mathcal{L}_\nu$ be the differential operator

$$\mathcal{L}_\nu(y) = t^2 y'' + t y' + (t^2 - \nu^2)y.$$

If ν is not an integer, then $\mathcal{L}_\nu(Y_\nu) = 0$. Since by definition,

$$Y_n(t) = \lim_{\nu \to n} Y_\nu(t),$$

$Y_\nu(t)$ is continuous, as a function of ν and t, at $\nu = n$. It can be shown that $Y_\nu'(t)$ and $Y_\nu''(t)$ are also continuous at $\nu = n$; we omit this proof. It follows that

$$\mathcal{L}_n(Y_n(t)) = \lim_{\nu \to n} \mathcal{L}_\nu(Y_\nu(t)) = 0,$$

so $Y_n(t)$ satisfies Bessel's equation.

Since $W[J_n, Y_n](t) = \lim_{\nu \to n} W[J_\nu, Y_\nu](t) = \frac{2}{\pi t}$, $\{J_n(t), Y_n(t)\}$ is linearly independent. ∎

The series representation of $Y_n(t)$ is rather complicated. It is obtained by carrying out the differentiations as indicated in (7.44), using the series (7.39) to represent J_ν and $J_{-\nu}$. As a result of differentiation of $t^{n+\nu}$, the series for y_ν involves a logarithmic term when ν is an integer. Lemma 7.5.2 is used to differentiate the reciprocal gamma function appearing in (7.39). The details of deriving the series for Y_ν when ν is an integer are omitted; we will be content with a statement of the result:

$$Y_n(t) = \frac{1}{\pi}\left\{ 2 J_n(t)\left[\ln\left(\frac{t}{2}\right) + \gamma\right] - \sum_{m=0}^{n-1} \frac{(n-m-1)!}{m!}\left(\frac{t}{2}\right)^{2m-n} \right.$$

$$\left. - \sum_{m=0}^{\infty} \frac{(-1)^m}{(m+n)!\,m!}[\phi(m) + \phi(m+n)]\left(\frac{t}{2}\right)^{2m+n} \right\}.$$

Modified Bessel functions

Solutions of the differential equation

$$t^2 y'' + t y' - (t^2 + \nu^2)y = 0 \tag{7.45}$$

are called **modified Bessel functions**. The standard fundamental set of solutions is $\{I_\nu(t), K_\nu(t)\}$, where

$$I_\nu(t) = \sum_{m=0}^{\infty} \frac{1}{m!\,\Gamma(m+\nu+1)}\left(\frac{t}{2}\right)^{2m+\nu},$$

and, if ν is not an integer,

$$K_\nu(t) = \frac{\pi}{2}\frac{I_\nu(t) - I_{-\nu}(t)}{\sin(\pi\nu)}.$$

If ν is an integer, the limiting value of $K_\nu(t)$ is used. I_ν and K_ν are called the modified Bessel functions of the first and second kinds, respectively. In some mathematical tables, they are called *hyperbolic Bessel functions*. The reason for this name is given in Exercise 28 at the end of this section.

Related differential equations

Let $y(t)$ denote either a solution of Bessel's equation or a solution of the modified Bessel equation (7.45). Set

$$w(x) = x^q\, y(a\, x^p), \qquad\qquad (7.46)$$

where a, p, and q are parameters. We will use w', w'' for $\frac{dw}{dx}$ and $\frac{d^2 w}{dx^2}$, respectively.

Proposition 7.5.7

If $y(t)$ is a solution of Bessel's equation, then $w(x)$ is a solution of the differential equation

$$x^2\, w'' + (1 - 2q)x\, w' + (a^2\, p^2\, x^{2p} - p^2\, v^2 + q^2)\, w = 0. \qquad (7.47)$$

If $y(t)$ is a solution of the modified Bessel equation, then $w(x)$ satisfies

$$x^2\, w'' + (1 - 2q)\, x\, w' - (a^2\, p^2\, x^{2p} + p^2\, v^2 - q^2)\, w = 0. \qquad (7.48)$$

Equations (7.47) and (7.48) have four parameters: a, p, q, and v, so there is a large family of differential equations whose solutions can be expressed in terms of Bessel functions. The proof of Proposition 7.5.7 is given in outline form in Exercise 27 at the end of this section.

Example 7.5.2   Express $\sin(x)$ in terms of Bessel functions.

Solution. The initial value problem

$$w'' + w = 0; \ w(0) = 0, \ w'(0) = 1$$

defines $\sin(x)$. If we multiply this differential equation by x^2, we will have a differential equation of the form (7.47):

$$x^2\, w'' + x^2 w = 0. \qquad\qquad (7.49)$$

Equate the coefficients of $x\, w'$ and of w in the left sides of equations (7.47) and (7.49),

$$1 - 2q = 0$$
$$a^2\, p^2\, x^{2p} - p^2\, v^2 + q^2 = x^2.$$

These equations yield $q = \frac{1}{2}$, $p = 1$, $a = 1$, and $v = \frac{1}{2}$. There is therefore a solution $y(t)$ of

$$t^2\, y'' + t\, y' + \left(t^2 - \frac{1}{4}\right) y = 0$$

such that $\sin(x) = \sqrt{x}\,y(x)$. Since $y(t) = AJ_{\frac{1}{2}}(t) + BJ_{-\frac{1}{2}}(t)$, it only remains to determine the constants A and B. According to equation (7.39), the series representation of $y(t)$ starts out as follows:

$$y(t) = B\frac{1}{2^{-1/2}\Gamma(\frac{1}{2})}t^{-\frac{1}{2}} + A\frac{1}{2^{1/2}\Gamma(\frac{3}{2})}t^{\frac{1}{2}} + \cdots .$$

Since $\Gamma\left(\frac{1}{2}\right) = \sqrt{\pi}$ and $\Gamma\left(\frac{3}{2}\right) = \frac{1}{2}\sqrt{\pi}$ (see Exercise 2 at the end of this section), this can be rewritten as

$$y(t) = \sqrt{\frac{2}{\pi}}\left[Bt^{-\frac{1}{2}} + At^{\frac{1}{2}}\right] + \cdots .$$

It follows that

$$\sin(x) = \sqrt{\frac{2}{\pi}}[B + Ax] + \cdots .$$

Setting $x = 0$, we see that $B = 0$; if we differentiate and set $x = 0$, we will see that $A = \sqrt{\frac{\pi}{2}}$. Therefore,

$$\sin(x) = \sqrt{\frac{\pi x}{2}}J_{\frac{1}{2}}(x). \qquad \square$$

Example 7.5.3   Find a fundamental set of solutions of

$$x^2 w'' - 3xw' + (x+4)w = 0. \tag{7.50}$$

Solution. The ODE can be placed in the form (7.47), where the parameters satisfy the equations

$$1 - 2q = -3$$
$$a^2 p^2 x^{2p} - p^2 v^2 + q^2 = x + 4.$$

The solutions are $q = 2$, $p = \frac{1}{2}$ (these two parameters should always be evaluated first: q by using the first equation; p by comparison of exponents of x in the second equation), $a = 2$ (compare coefficients of x in the second equation), and $v = 0$ (compare constants in the second equation). Therefore, if $y(t)$ is a solution of

$$t^2 y'' + ty' + t^2 y = 0, \tag{7.51}$$

then $w(x) = x^2 y(2\sqrt{x})$. Since $\{J_0(t), Y_0(t)\}$ is a fundamental set of solutions of equation (7.51), it follows that $\{x^2 J_0(2\sqrt{x}), x^2 Y_0(2\sqrt{x})\}$ is a fundamental set of solutions of equation (7.50). $\qquad \square$

Example 7.5.4   Find the general solution of

$$xw'' + 3w' - xw = 0 \tag{7.52}$$

on $(0, \infty)$.

Solution. After multiplying equation (7.52) by x, we find that it is in the form of equation (7.47), with

$$1 - 2q = 3$$

$$a^2 p^2 x^2 p - p^2 v^2 + q^2 = -x^2.$$

It is readily seen that $q = -1$, $p = 1$, and $v = 1$. However, a must be a solution of $a^2 = -1$. We should therefore use equation (7.48); we will have the same values for q, p, and v, but $a = 1$. We conclude that the general solution can be expressed in terms of *modified* Bessel functions:

$$w(x) = \frac{1}{x}[A\, I_1(x) + B\, K_1(x)].$$

□

EXERCISES

1. Integrate by substitution to show that

$$\Gamma\left(\frac{1}{2}\right) = 2\int_0^\infty e^{-v^2}\, dv.$$

2. Calculate $\Gamma(z)$ for $z = -\frac{1}{2}, \frac{1}{2}, \frac{3}{2}$, and $\frac{5}{2}$. You may use the result of problem 1 and the identity

$$\int_0^\infty e^{-v^2}\, dv = \frac{\sqrt{\pi}}{2}.$$

3. Let $L(n)$ denote the reciprocal gamma function. Show that $L(n) = 0$ if and only if n is an integer ≤ 0.

4. Let $g(v) = \frac{\pi}{\Gamma(v)\Gamma(1-v)}$. Without referring to Lemma 7.5.1,

  a. Show that $g(v+1) = -g(v)$.

  b. Show that g is 2-periodic.

  c. Show that $g(v) = \sin(\pi v)$ if v is an integer or half-integer.

5. Let $L(k)$ be the reciprocal gamma function, and

$$\psi(z) = \frac{d}{dz}\ln(\Gamma(z)).$$

Prove the following statements, and deduce that Lemma 7.5.2 holds:

  a. If z is not a negative integer, then

$$L'(z+1) = -L(z+1)\psi(z+1).$$

  b. $\psi(1) = -\gamma$

  c. $\psi(z+1) = \psi(z) + \frac{1}{z}$ unless z is an integer ≤ 0

  d. If z is an integer > 1, then $\psi(z+1) = \phi(z) - \gamma$.

  e. $\cos(\pi z) = L'(z)L'(1-z) - L(z)L'(1-z)$ for any real value of z

  f. If z is a positive integer, then $L'(1-z) = -(-1)^z(z-1)!$.

6. Write out the expression for $Y_3(x)$, using no summation signs. One use of each of the following notations is permitted: J_3, γ, ln, and an ellipsis $\cdots$.

7. Show that $\cos x = \sqrt{\frac{\pi x}{2}}J_{-1/2}(x)$.

8. Every CAS has Bessel functions. Use a CAS to plot graphs of $J_v(t)$ for $v = 0$, $\frac{1}{2}$, and 1, over the range $0 \leq t \leq 30$. Describe any similarities of the graphs. You should be able to explain the features of the graph of $J_{1/2}$ in detail.

9. Use a CAS to plot the graphs of $\sqrt{t}\, J_v(t)$ for $v = 0$, $\frac{1}{2}$, and 1, over the range $0 \leq t \leq 30$. Again, notice any similarities.

10. Repeat Exercises 8 and 9 for the Bessel functions of the second kind, $Y_v(t)$.

11. Show that $\sqrt{t}\, J_v(t)$ and $\sqrt{t}\, Y_v(t)$ satisfy the ODE

$$y'' + \left(1 + \frac{C}{t^2}\right)y = 0.$$

When t is large the solutions of this ODE will be approximately the same as solutions of $y'' + y = 0$ (sines and cosines).

12. The zeros of $J_v(t)$ are denoted $j_{v,n}$, with $j_{v,1}$ being the smallest, $j_{v,2}$ the second smallest, and so on. The zeros of Y_v are $y_{v,n}$. Using the graphs produced in Exercises 8–10, approximate $j_{v,n}$ and $y_{v,n}$ for $1 \leq n \leq 9$. Are the zeros evenly spaced, or nearly so? What is the approximate difference between consecutive pairs of zeros?

13. A Bessel identity

a. Give a power series expression for $J_\nu'(t)$.

b. Show that

$$J_\nu'(t) = J_{\nu-1}(t) - \frac{\nu}{t}J_\nu(t).$$

◆ **Hint:** $2m + \nu = 2(m + \nu) - \nu$.

In Exercises 14–19, find the general solution of the differential equation in terms of Bessel functions or modified Bessel functions.

14. $x^2 w'' - xw' + (x^2 - 1)w = 0$

15. $x^2 w'' - xw' + w = 0$. (Can the solution of *any* Cauchy-Euler equation be expressed in terms of Bessel functions? Why or why not?)

16. $(x - 1)^2 w'' + (x - 1)w' + (x^2 - 2x)w = 0$

17. $4x^2 w'' + 8xw' + (1 - x)w = 0$

18. $x^2 w'' - 2xw' + (x - 2)w = 0$

19. $w'' + x^2 w = 0$

20. The Airy function Ai (t) is defined to be the solution of the initial value problem

$$w'' - xw = 0; \quad w(0) = c_0, \; w'(0) = c_1,$$

where $c_0 = \frac{1}{3^{2/3}\Gamma(\frac{2}{3})} \approx 0.355$ and $c_1 = \frac{-1}{3^{1/3}\Gamma(\frac{1}{3})} \approx -0.259$.

a. Plot a graph of Ai (x) for $-5 \le x \le 10$. You may either use an ODE solver or a CAS (every CAS has Airy functions).

b. Explain why the solution oscillates for $x < 0$, but does not oscillate for $x > 0$.

◆ **Hint:** Compare Airy's equation with $y'' + c\,y = 0$, where c is constant.

c. Find an expression for Ai (x) in terms of a Bessel function *and* a modified Bessel function.

21. Express the hyperbolic sine and cosine functions in terms of modified Bessel functions.

22. Use a substitution of the form $w = y(ae^x)$ to express the general solutions of the following differential equations in terms of Bessel functions or modified Bessel functions.

a. $w'' + e^{2x}w = 0$ b. $4w'' - (e^{2x} + 1)w = 0$

23. **Riccati's equation.** The nonlinear first-order differential equation

$$y' + ay^2 = bt^n,$$

where a and b are constants, is known as Riccati's equation. Show that the substitution $y = \frac{v'}{av}$ converts Riccati's equation into a second-order linear equation with dependent variable v. Let $a = 2$, $b = -2$, and $n = 2$. Express the general solution of the Riccati equation with these parameters in terms of Bessel functions.

24. Find the general solution of the Riccati equation

$$y' + y^2 = 6t^{-2}.$$

25. Prove Lemma 7.5.5.

26. Show that $I_\nu(t)$ satisfies equation (7.45).

27. This exercise is an outline of the proof of Proposition 7.5.7. Fill in the details.

a. Starting with the chain rule, from which it follows that

$$\frac{dy}{dt} = \frac{d(x^{-q}w)}{dx}\left[\frac{dt}{dx}\right]^{-1},$$

show that $y' = \frac{1}{ap}x^{-p-q}(xw' - qw)$.

b. Show that

$$y'' = \frac{dy'}{dx}\left[\frac{dt}{dx}\right]^{-1}$$

$$= \frac{1}{(a\,p)^2}x^{-2p-q}[x^2 w'' + (1 - p - 2q)\,x\,w'$$

$$+ (q^2 + pq)\,w].$$

c. Substitute these expressions into the Bessel equation and the modified Bessel equation to derive equations (7.47) and (7.48), respectively.

28. Show that if n is an integer, and $x > 0$, then $I_n(t) = i^{-n}J_n(it)$ and $J_n(t) = i^{-n}I_n(it)$. [The relationship between $K_n(x)$ and the ordinary Bessel functions is more complicated.]

29. Show that if n is an integer, then $I_n(t) = I_{-n}(t)$, but otherwise I_n and I_{-n} are linearly independent.

◆ **Hint:** Use Exercise 28.

30. Show that $W[I_\nu, I_{-\nu}](t) = \frac{-2}{\pi t}\sin(\pi \nu)$.

31. Show that $W[I_\nu, K_\nu](t) = \frac{1}{t}$.

*7.6 Double and Blocked Exponents

Consider an ODE

$$\mathcal{L}(y) = p(t)\, t^2\, y'' + q(t)\, t\, y' + r(t)\, y = 0, \tag{7.53}$$

where $p(t) = \sum_{n=0}^{\infty} P_n\, t^n$, $q(t) = \sum_{n=0}^{\infty} Q_n\, t^n$, and $r(t) = \sum_{n=0}^{\infty} R_n\, t^n$ are analytic functions. The exponents μ, ν of equation (7.53) are the roots of the indicial equation:

$$P_0\, s\, (s-1) + Q_0\, s + P_0 = 0. \tag{7.54}$$

We will be interested in the cases where $\mu = \nu$ (a double exponent), and $\nu - \mu = N$, where N is a positive integer (μ is a blocked exponent). It is helpful to keep the examples of double and blocked exponents that we have already encountered in mind: Cauchy-Euler equations with double exponents, and Bessel equations with an integer parameter $\nu = n$.

The solutions of equation (7.53) that were defined in Section 7.4 had the form

$$y(t; s) = t^s + \sum_{n=1}^{\infty} A_n(s)\, t^{n+s},$$

but in the case of double exponents, and frequently for blocked exponents as well, it was not possible to find two independent solutions of this type. Since $y(t; s)$ is considered to be a function of two variables, the coefficients A_n, for $n \geq 1$, will turn out to be functions of s. Thus, if $s = \mu$ is a double or blocked exponent, we can use the technique of differentiating $y(t; s)$ with respect to s, and putting $s = \mu$ to find a second solution. This technique was employed in the case of the Cauchy-Euler equation with a double exponent, and it works for any regular singular point having a double exponent. It can be modified to find a solution belonging to a blocked exponent, too (recall that we differentiated with respect to a parameter to define Bessel functions of the second kind).

In the following discussion, we will always assume that $t > 0$, because $\frac{\partial t^s}{\partial s} = t^s \ln(t)$ is only defined for $t > 0$.

Let $f_n(s) = P_n\, s\, (s-1) + Q_n\, s + R_n$. Thus, the indicial equation is $f_0(s) = 0$. By equation (7.34) on page 346,

$$\mathcal{L}(y(t; s)) = \sum_{n=0}^{\infty} \left[\sum_{k=0}^{n} f_{n-k}\, (k+s)\, A_k(s) \right] t^{n+s}$$

$$= \sum_{n=0}^{\infty} \left[f_0(n+s)\, A_n(s) + \sum_{k=0}^{n-1} f_{n-k}\, (k+s)\, A_k(s) \right] t^{n+s}. \tag{7.55}$$

Starting with $A_0 = 1$, the objective is to choose $A_n(s)$ so that the coefficients of t^{n+s} on the right side of (7.55) are zero for $n > 0$, and consequently

$$\mathcal{L}(y(t; s)) = f_0(s)\, t^s.$$

Thus, put

$$A_n(s) = -\frac{1}{f_0(n+s)} \sum_{k=0}^{n-1} f_{n-k}(k+s) A_k(s)$$

for $n \geq 1$. For example, $A_1(s) = -\frac{f_1(s)}{f_0(1+s)}$, and

$$A_2(s) = -\frac{1}{f_0(2+s)}[f_2(s) + f_1(1+s)A_1(s)]$$
$$= -\frac{f_2(s)f_0(1+s) - f_1(1+s)f_1(s)}{f_0(1+s)f_0(2+s)},$$

and so on.

Exercise (do before reading further): Write an explicit formula for $A_3(s)$ in terms of f_0, f_1, f_2, and f_3.

Each $A_n(s)$ is a rational function of s. If $s = \mu$ is a zero of the denominator $q(s)$ of a rational function $f(s) = \frac{p(s)}{q(s)}$, and $p(\mu) \neq 0$, then μ is said to be a **pole** of $f(s)$. The order of μ as a zero of $q(s)$ is the order of the pole. Thus, if the pole μ is of order k, then $q(s) = (s - \mu)^k q_1(s)$, where $q_1(\mu) \neq 0$. The function

$$g(s) = (s - \mu)^k f(s) = \frac{p(s)}{q_1(s)} \quad \text{for } s \neq \mu$$

is not formally defined for $s = \mu$, but we can put $g(\mu) = p(\mu)/q_1(\mu)$ to obtain a function that is continuous at μ. By extension, poles of the coefficients $A_n(s)$ of the defining power series for $y(t; s)$ are also considered to be poles of $y(t; s)$.

Lemma 7.6.1

If a number μ is a pole of $y(t; s)$, then $\mu + N$ is an exponent of equation (7.53) for some integer $N > 0$.

Proof Suppose that μ is a pole of $y(t; s)$; then μ is also a pole of one of the coefficients $A_k(s)$. Since the denominator polynomial of $A_k(s)$ is

$$f_0(1+s) \cdot f_0(2+s) \cdots f_0(k+s),$$

$f_0(N + \mu) = 0$ for some integer N with $1 \leq N \leq k$, and hence $N + \mu$ is an exponent. ∎

By Lemma 7.6.1, an unblocked exponent of equation (7.53) cannot be a pole of $y(t; s)$.

Example 7.6.1   Find $y(t; s)$ for the hypergeometric equation,

$$t(1 - t)y'' + (1 - t)y' + y = 0, \tag{7.56}$$

and determine its poles.

Solution. If we multiply equation (7.56) by t, we will see that $P_0 = Q_0 = 1$, $P_1 = Q_1 = -1$, and $R_1 = 1$. All other coefficients are 0. The indicial polynomial is therefore $f_0(s) = s(s-1) + s = s^2$, and $f_1(s) = -s(s-1) - s + 1 = 1 - s^2$. It follows that 0 is a double exponent. We write

$$y(t; s) = \sum_{n=0}^{\infty} A_n(s) t^{n+s},$$

where $A_0(s) = 1$. We need to determine $A_n(s)$, for $n > 0$, so that $\mathcal{L}(y(t; s)) = f_0(s) t^{n+s}$, where $\mathcal{L}$ is the operator defined by the left side of (7.56). Index shifting shows that

$$\mathcal{L}(y(t; s)) = \sum_{n=0}^{\infty} [A_{n+1}(s)(n+s+1)(n+s) - A_n(s)(n+s)(n+s-1)$$

$$+ A_{n+1}(s)(n+s+1) - A_n(s)(n+s) + A_n(s)] t^{n+s}$$

$$= \sum_{n=0}^{\infty} [A_{n+1}(s)(n+s+1)^2 - A_n(s)((n+s)^2 - 1)] t^{n+s}.$$

Setting the coefficients equal to 0 shows that

$$A_{n+1}(s) = \frac{n+s-1}{n+s+1} A_n(s).$$

Starting with $n = 0$, we have

$$A_1 = \frac{s-1}{s+1} A_0 = \frac{s-1}{s+1},$$

$$A_2 = \frac{s}{s+2} A_1 = \frac{s(s-1)}{(s+2)(s+1)},$$

$$A_3 = \frac{s+1}{s+3} A_2 = \frac{(s+1)s(s-1)}{(s+3)(s+2)(s+1)} = \frac{s(s-1)}{(s+3)(s+2)}.$$

In each subsequent coefficient, cancellation occurs (check this), so that

$$A_n = \frac{s(s-1)}{(s+n)(s+n-1)}$$

for $n \geq 2$, and

$$y(t; s) = t^s + \frac{s-1}{s+1} t^{s+1} + \sum_{n=2}^{\infty} \frac{s(s-1)}{(s+n)(s+n-1)} t^{s+n}.$$

We can now set $s = 0$ and determine one solution of the ODE: $y(t; 0) = 1 - t$. The poles of $y(t; s)$ are the negative integers $-1, -2, \ldots$.   ☐

Double exponents

If equation (7.53) has a double exponent v, then v is unblocked, since there are no other exponents. It follows from Lemma 7.6.1 that $y(t; s)$ does not have a pole at $s = v$. Furthermore, since v is a double root, $f_0(s)$ can be factored as

$$f_0(s) = P_0 (s - v)^2.$$

Since $A_0 = 1$,

$$\mathcal{L}(y(t; s)) = P_0(s - v)^2 t^s. \tag{7.57}$$

By differentiating this identity with respect to s, we obtain

$$\mathcal{L}(z(t; s)) = P_0 [(s - v)^2 t^s \, \ln(t) + 2 (s - v) t^s], \tag{7.58}$$

where $z(t; s) = \frac{\partial}{\partial s} y(t; s)$. When we set $s = v$ in (7.57) and (7.58), the right side of each vanishes, so that $y(t; v)$ and $z(t; v)$ are solutions of (7.53).

Put $y_1(t) = y(t; v)$ and $y_2(t) = z(t; v)$. Let $B_n(s) = \frac{dA_n}{ds}$; then by the product rule,

$$z(t; s) = \frac{\partial}{\partial s} \left[\sum_{n=0}^{\infty} A_n t^{n+s} \right]$$

$$= \sum_{n=0}^{\infty} A_n t^{n+s} \, \ln(t) + \sum_{n=0}^{\infty} B_n t^{n+s}$$

$$= y(t; s) \ln(t) + \sum_{n=0}^{\infty} B_n t^{n+s}.$$

Since A_0 is constant, $B_0 = 0$. Therefore, substituting $s = v$ gives the second solution,

$$y_2(t) = y_1(t) \ln(t) + \sum_{n=1}^{\infty} B_n(v)t^{n+v}.$$

The result is summarized in the following theorem, attributed to Frobenius.

Theorem 7.3

If (7.53) has a double exponent v, then there is a fundamental set of solutions $\{y_1(t), y_2(t)\}$ of the form

$$y_1(t) = \sum_{n=0}^{\infty} A_n t^{n+v}$$

and

$$y_2(t) = y_1(t) \ln(t) + \sum_{n=1}^{\infty} B_n t^{n+v},$$

which is valid on an interval of the form $0 < t < R$ (R is positive and may be infinite).

The proof of the convergence assertion of Theorem 7.3 is beyond the scope of this text.

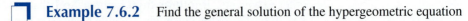

Example 7.6.2   Find the general solution of the hypergeometric equation

$$t(1 - t)y'' + (1 - t)y' + y = 0.$$

Solution. It was shown in Example 7.6.1 that $s = 0$ is a double exponent and that

$$y(t; s) = t^s + \frac{s - 1}{s + 1}t^{s+1} + \sum_{n=2}^{\infty} \frac{s(s - 1)}{(s + n)(s + n - 1)}t^{s+n}.$$

It follows that $\{y_1(t), y_2(t)\}$ is a fundamental set of solutions, where $y_1(t) = y(t; 0) = 1 - t$ and

$$y_2(t) = \frac{\partial}{\partial s}y(t; s)\bigg|_{s=0} = y_1(t)\ln(t) + \sum_{m=1}^{\infty} B_n(0)t^n.$$

By logarithmic differentiation,

$$B_n(s) = \frac{dA_n}{ds}$$

$$= A_n(s)\frac{d}{ds}\ln(A_n(s))$$

$$= A_n(s)\frac{d}{ds}[\ln(s) + \ln(s - 1) - \ln(s + n) - \ln(s + n - 1)]$$

$$= \frac{s(s - 1)}{(s + n)(s + n - 1)}\left[\frac{1}{s} + \frac{1}{s - 1} - \frac{1}{s + n} - \frac{1}{s + n - 1}\right]$$

$$= \frac{s - 1}{(s + n)(s + n - 1)}\left[1 + \frac{s}{s - 1} - \frac{s}{s + n} - \frac{s}{s + n - 1}\right]$$

for $n \geq 2$, and

$$B_1(s) = \frac{d}{ds}\frac{s - 1}{s + 1} = \frac{2}{(s + 1)^2}.$$

We now set $s = 0$ to obtain $B_1(0) = 2$, and for $n \geq 2$,

$$B_n(0) = \frac{-1}{n(n - 1)}.$$

Thus

$$y_2(t) = y_1(t)\ln(t) + 2t - \sum_{n=2}^{\infty} \frac{1}{n(n - 1)}t^n.$$

Let $S = \sum_{n=2}^{\infty} \frac{1}{n(n-1)}t^n$ denote the summation shown in the preceding formula. Taking the second derivative of S with respect to t, we have $S'' = \sum_{n=2}^{\infty} t^{n-2}$, which is a geometric series. Therefore, $S'' = \frac{1}{1-t}$. Taking two antiderivatives, and noting that $S(0) = S'(0) = 0$, we find that $S = (1-t)\ln(1-t) + t$. Since $y_1(t) = 1 - t$, the simplified form of $y_2(t)$ is

$$y_2(t) = (1-t)\ln(t) + 2t - (1-t)\ln(1-t) - t$$

$$= (1-t)\ln\left(\frac{t}{1-t}\right) + t.$$

This solution is valid for $0 < t < 1$, and on that interval, the general solution is

$$y = C_1(1-t) + C_2\left[(1-t)\ln\left(\frac{t}{1-t}\right) + t\right].$$ ❑

Blocked exponents

An exponent μ is **blocked** if there is another exponent $\nu = \mu + N$, where N is a positive integer. The solution $y_1(t) = y(t; \nu)$ is defined, but $y(t; \mu)$ may not be, because $s = \mu$ can be a pole of $y(t; s)$. If $y(t; s)$ has no pole at $s = \mu$, then $y_2(t) = y(t; \mu)$ is a solution such that $\{y_1, y_2\}$ is linearly independent.

Let us now assume that $y(t; s)$ indeed has a pole at $s = \mu$. This pole will make its appearance in the coefficient

$$A_N(s) = -\frac{1}{f_0(N+s)} \sum_{k=0}^{N-1} f_{N-k}(k+s) A_k(s)$$

because the denominator function is $f_0(N+s) = P_0(s-\mu)(N+s-\mu)$. This pole is of order 1, and furthermore, $A_n(s)$ may also have a pole of order 1 (inherited from A_N), when $n > N$. Put

$$u(t; s) = (s-\mu)\, y(t; s) = \sum_{n=0}^{\infty} C_n(s) t^{n+s}.$$

Since the pole of $y(t; s)$ is of order 1, $u(t; s)$ does not have a pole at $s = \mu$. Although we can calculate $u(t; \mu)$, the following lemma shows the result to be of limited value.

Lemma 7.6.2

Let $C = C_N(\mu)$. If $C = 0$, then $y(t; s)$ does not have a pole at $s = \mu$, and thus $\{y(t; \mu), y(t; \nu)\}$ is a fundamental set of solutions of (7.53). If $C \neq 0$, then

$$u(t; \mu) \equiv C\, y(t; \nu).$$

Thus, we can't combine $y(t; \nu)$ and $u(t; \mu)$ to obtain a fundamental set of solutions.

Proof of Lemma 7.6.2.   For $n < N$, the coefficients $A_n(s)$ do not have poles at $s = \mu$; hence

$$C_n(\mu) = (s-\mu)|_{s=\mu} A_n(\mu) = 0.$$

If $C = 0$, then $A_N(s)$ does not have a pole at $s = \mu$ either, and hence $y(t; s)$ doesn't have a pole at $s = \mu$. It follows that $y(t; \mu) = t^\mu + \cdots$ and $y(t; \nu) = t^\nu + \cdots$ are linearly independent solutions of (7.53).

Now let us assume that $C \neq 0$. When calculating $C_n(\mu)$ for $n > N$, we can use the recursion formula

$$C_n(\mu) = -\frac{1}{f_0(n + \mu)} \sum_{k=N}^{n-1} f_{n-k}(k + \mu) C_k(\mu),$$

where we have started the summation at $k = N$ because $C_k(\mu) = 0$ for $k < N$. Now let's shift some indices: For $n \geq N$, put $n = m + N$, and note that $n + \mu = m + \nu$, and

$$u(t; \mu) = \sum_{m=0}^{\infty} C_{m+N}(\mu) t^{m+\nu}.$$

The recursion formula for $C_{m+N}(\mu)$ can be recast as

$$C_{m+N}(\mu) = -\frac{1}{f_0(m + \nu)} \sum_{k=0}^{m-1} f_{m-k}(k + \nu) C_{k+N}(\mu).$$

Compare this with the recursion formula for $A_m(\nu)$:

$$A_m(\mu) = -\frac{1}{f_0(m + \nu)} \sum_{k=0}^{m-1} f_{m-k}(k + \nu) A_k(\mu).$$

These formulas are the same. However, the starting values are different: $A_0 = 1$ and $C_{0+N} = C$. Thus, $C_{m+N} = C A_m$. ∎

To get a solution belonging to the blocked exponent μ when $C \neq 0$, we can differentiate $u(t; s)$ with respect to s. Notice that

$$
\begin{aligned}
\mathcal{L}(u(t; s)) &= \mathcal{L}((s - \mu) y(t; s)) \\
&= (s - \mu) \mathcal{L}(y(t; s)) \\
&= (s - \mu) f_0(s) t^s \\
&= P_0 (s - \mu)^2 (s - \nu) t^s.
\end{aligned}
$$

Let $z(t; s) = \frac{\partial u}{\partial s}(t; s)$. Then

$$
\begin{aligned}
\mathcal{L}(z(t; s))) &= P_0 \frac{\partial}{\partial s}[(s - \mu)^2 (s - \nu) t^s] \\
&= P_0 (s - \mu)[2 (s - \nu) t^s + (s - \mu) t^s + (s - \mu)(s - \nu) t^s \ln(t)].
\end{aligned}
$$

It follows that $\mathcal{L}(z(t; \mu)) = 0$, and we put $y_2(t) = z(t; \mu)$. Then

$$y_2(t) = \sum_{n=0}^{\infty} C_n(\mu) t^{n+\mu} \ln(t) + \sum_{n=0}^{\infty} B_n(\mu) t^{n+\mu}, \tag{7.59}$$

where $B_n(s) = \frac{\partial C_n}{\partial s}$. Notice that although $C_n(\mu) = 0$ for $n < N$, it is not necessarily true that $B_n(\mu) = 0$ for these n. For example, $C_0(s) = s - \mu$, so $B_0(s) = 1$.

By Lemma 7.6.2, the first term of (7.59) is equal to $C\,y_1(t)\,\ln(t)$. Our conclusions are summarized in the following theorem, also due to Frobenius.

Theorem 7.4

Let v and μ denote the exponents of equation (7.53), and assume that $v = \mu + N$, where N is a positive integer. Then there are power series $\sum_{n=1}^{\infty} A_n\,t^n$ and $\sum_{n=1}^{\infty} B_n\,t^n$, each with a positive radius of convergence, and a constant C (possibly equal to 0), such that equation (7.53) has a fundamental set of solutions $\{y_1(t),\,y_2(t)\}$ defined for positive t up to the radius of convergence as follows:

$$y_1(t) = t^v + \sum_{n=1}^{\infty} A_n\,t^{n+v}$$

and

$$y_2(t) = C\,y_1(t)\,\ln(t) + t^\mu + \sum_{n=1}^{\infty} B_n\,t^{n+\mu}.$$

This completes the description of the method of Frobenius, as applied to second-order linear equations. The method is also applicable to higher-order equations with regular singular points.

Example 7.6.3   The equation

$$ty'' - y = 0 \tag{7.60}$$

has a regular singular point at $t = 0$, with one blocked exponent. Find series representations for a fundamental set of two solutions.

Solution. Let $\mathcal{L}(y) = ty'' - y$. We will look for solutions of $\mathcal{L}(y) = 0$ that have the form

$$y(t; s) = t^s + \sum_{n=1}^{\infty} A_n(s)t^{n+s} = t^s + A_1(s)t^{s+1} + \cdots.$$

Thus,

$$\mathcal{L}(y(t, s)) = s(s-1)t^{s-1} - t^s$$

$$+ \sum_{n=1}^{\infty} A_n(s)((n+s)(n+s-1)t^{n+s-1} - t^{n+s})$$

$$= s(s-1)t^{s-1} - t^s + A_1(s)[(s+1)st^s - t^{s+1}]$$

$$+ A_2(s)[(s+2)(s+1)t^{s+1} - t^{s+2}] + \cdots.$$

Let's rearrange this by grouping terms by power of t:

$$\mathcal{L}(y(t, s)) = s(s-1)t^{s-1} + \sum_{n=1}^{\infty}[(n+s+1)(n+s)A_{n+1}(s) - A_n(s)]t^{n+s}$$

$$= s(s-1)t^{s-1} + [(s+1)sA_1(s) - 1]t^s$$

$$+ [(s+2)(s+1)A_2(s) - A_1(s)]t^{s+1} + \cdots.$$

We will define the $A_n(s)$ for $n > 1$ by the recursion formula

$$A_{n+1}(s) = \frac{1}{(n+s+1)(n+s)} A_n(s),$$

starting with $A_0 = 1$. Thus $A_1 = \dfrac{1}{(s+1)s}$, $A_2 = \dfrac{1}{(s+2)(s+1)^2 s}$, $A_3 = \dfrac{1}{(s+3)(s+2)^2(s+1)^2 s}$, and so on.

With this choice of coefficients, $\mathcal{L}(y(t;s)) = s(s-1)t^{s-1}$. It follows that

$$y(t;1) = t\left(1 + \sum_{n=1}^{\infty} A_n(1)t^n\right)$$

$$= t + \frac{1}{2}t^2 + \frac{1}{3 \cdot 2^2}t^3 + \frac{1}{4 \cdot 3^2 \cdot 2^2}t^4 + \cdots$$

is a solution of $\mathcal{L}(y) = 0$. Denote this solution by $y_1(t)$. It is tempting to advance $y(t;0)$ as another solution, but we know better: Because 0 is a blocked exponent, $y(t;s)$ has a pole at $s = 0$, and $y(t;0)$ is undefined. To remove this pole, we consider the expression

$$u(t;s) = sy(t;s)$$

$$= st^s + \sum_{n=1}^{\infty} sA_n(s)t^{n+s}$$

$$= st^s + \frac{1}{s+1}t^{s+1} + \frac{1}{(s+2)(s+1)^2}t^{s+2}$$

$$+ \frac{1}{(s+3)(s+2)^2(s+1)^2}t^{s+3} + \cdots,$$

which does not have a pole at $s = 0$. If we set $s = 0$ in this expression, the result,

$$\sum_{n=1}^{\infty}(sA_n(s))|_{s=0}t^n = t + \frac{1}{2}t^2 + \frac{1}{3 \cdot 2^2}t^3 + \frac{1}{4 \cdot 3^2 \cdot 2^2}t^4 + \cdots,$$

is equal to $y_1(t)$ and is not a new solution. A new solution is obtained from $z(t;s) = \frac{\partial u}{\partial s}(t;s)$. Thus,

$$z(t;s) = t^s + st^s \ln(t) + \sum_{n=1}^{\infty} sA_n(s)t^{n+s}\ln(t) + \sum_{n=1}^{\infty} \frac{d}{ds}[sA_n(s)]t^{n+s}$$

$$= u(t;s)\ln(t) + t^s - \frac{1}{(s+1)^2}t^{s+1} - \frac{1}{(s+2)(s+1)^2}\left(\frac{1}{s+2} + \frac{2}{s+1}\right)t^{s+2}$$

$$- \frac{1}{(s+3)(s+2)^2(s+1)^2}\left(\frac{1}{s+3} + \frac{2}{s+2} + \frac{2}{s+1}\right)t^{s+3} - \cdots.$$

Since

$$L(z(t; s)) = \frac{\partial}{\partial s}[sL(y(t; s))]$$

$$= \frac{\partial}{\partial s}[s^2(s-1)t^s]$$

$$= [2s(s-1) + s^2 + s^2(s-1)\ln(t)]t^s,$$

it follows that $y_2(t) = z(t; 0)$ is a solution. Here are the first few terms of this solution [remember that $u(t; 0) = y_1(t)$]:

$$y_2(t) = y_1 \ln(t) + 1 - t - \frac{t^2}{2}\left(\frac{1}{2} + \frac{2}{1}\right) - \frac{t^3}{12}\left(\frac{1}{3} + \frac{2}{2} + \frac{2}{1}\right) + \cdots.$$

Since $y_2(t)$ isn't a constant multiple of $y_1(t)$, we have found a fundamental set of solutions: $\{y_1(t), y_2(t)\}$.   ❑

EXERCISES

1. Show that if equation (7.53) has a double exponent v, then the set of solutions $\{y_1(t), y_2(t)\}$ given by Theorem 7.3 is a fundamental set of solutions.

2. Suppose that $f_k(s) \equiv 0$ for $k \neq 0, p$, where p is a positive integer. Show that

$$A_n(s) = -\frac{f_p(n+s-p)}{f_0(n+s)}A_{n-p}(s).$$

3. Consider the ODE

$$t^2 y'' - (t + t^2)y' + y = 0.$$

Show that $f_0(s) = (s-1)^2$, $f_1(s) = -s$, and $f_n(s) = 0$ for $n > 1$.

4. (Continuation of Exercise 3) Find an expression for $A_n(s)$.

5. Suppose that for a certain ODE, $s = 1$ is a double exponent, and

$$y(t; s) = \sum_{n=0}^{\infty} \frac{1}{s(s+1)(s+2)\cdots(s+n-1)}t^{n+s}.$$

Find the general solution of that ODE. Be sure to ascertain that the series involved are convergent.

6. Find the ODE that is described in Exercise 5.

7. Show that the ODE $(4t^2 - t^4)y'' + y = 0$ has a regular singular point at 0 with a double exponent and find a fundamental set of solutions.

✔**Answer:** $y_1(t) = \sqrt{t}\left\{1 - \sum_{m=1}^{\infty}\frac{(4m-3)!!}{2^{2m+4}(m!)^2}t^{2m}\right\}$;

$y_2(t) = y_1(t)\ln(t)$

$\quad + \sqrt{t}\left\{1 - \sum_{m=1}^{\infty}\frac{(4m-3)!!}{2^{2m+4}(m!)^2}[2\phi(4m-1)\right.$

$\quad\quad \left. - \phi(2m-1) - \phi(m) - 1]t^{2m}\right\}$

8. This exercise concerns the ODE

$$t^2 y'' + ty' - \left(t + \frac{1}{4}\right)y = 0.$$

a. Show that the exponents are $\pm\frac{1}{2}$.

b. Derive the expression $y(t; s) = \sum_{n=0}^{\infty} A_n(s)t^{n+s}$, where $A_0 = 1$, $A_1 = (s + \frac{3}{2})^{-1}(s + \frac{1}{2})^{-1}$, and, for $n > 1$, the recursive formula for A_n is

$$A_n = \frac{A_{n-1}}{(s+n+\frac{1}{2})(s+n-\frac{1}{2})}.$$

c. Find the solution corresponding to the unblocked exponent.

✔**Answer:** $y_1(t) = \sqrt{t}\sum_{n=0}^{\infty}\frac{t^n}{(n+1)(n!)^2}$.

d. Show that $C_0 = s + \frac{1}{2}$, $C_1 = \frac{2}{2s+3}$ and, for $n > 1$,

$$C_n = \frac{C_{n-1}}{(s+n+\frac{1}{2})(s+n-\frac{1}{2})}.$$

e. Find the solution which corresponds to the blocked exponent.

✔ **Answer:** $y_2(t) = 2y_1(t)\ln(t)$

$+\frac{1}{\sqrt{t}}\left\{1 - t - \sum_{n=2}^{\infty}\left[\frac{1}{(n!)^2} + \frac{2\phi(n-1)}{n((n-1)!)^2}\right]t^n\right\}.$

Exercises 9–14 compare two ways of solving the ODE

$$t^2 y'' + (t^2 - t)\,y' + y = 0. \qquad (7.61)$$

9. Show that $s = 1$ is a double exponent of equation (7.61).

10. Use the method of Frobenius to find the solution $y_1(t) = y(t; 1)$. Show that $y_1(t) = t\,e^{-t}$.

11. It follows from Theorem 7.3 that

$$y_2(t) = t\,e^{-t}\ln t + \sum_{n=1}^{\infty} B_n t^{n+1}. \qquad (7.62)$$

Substitute the expression (7.62) into equation (7.61) and show that the coefficients B_n are determined by $B_0 = 0$ and the recursive formula

$$n\,B_n = -\left[B_{n-1} + \frac{(-1)^n}{n!}\right]. \qquad (7.63)$$

12. Let $K_n = (-1)^n n!\,B_n$. Show that (7.63) implies that $K_n = -\phi(n)$.

13. Use the method of reduction of order to show that

$$\bar{y}_2(t) = t\,e^{-t}\int_1^t \frac{e^\tau}{\tau}\,d\tau$$

is a solution of equation (7.61).

14. Show that

$$\int_1^t \frac{e^\tau}{\tau}\,d\tau = \ln t + e^t \sum_{n=1}^{\infty} B_n t^n, \qquad (7.64)$$

and hence that $\bar{y}_2(t) = y_2(t)$. One way of establishing the identity (7.64) is repeated integration by parts.

15. The ODE

$$t(1 - t)\,y'' - (2 + t)\,y' + 4\,y = 0$$

was shown in Example 7.4.2 on page 344 to have the exponents 0 and 3 at $t = 0$. Show that, although the exponent 0 is blocked, the coefficient C in the logarithmic solution vanishes. Find the solution at $t = 0$ corresponding to the exponent 0.

7.7 Glossary

$\phi(k)$ The function, with domain the positive integers, defined by

$$\phi(k) = 1 + \frac{1}{2} + \frac{1}{3} + \frac{1}{4} + \cdots + \frac{1}{k}.$$

γ See **Euler's constant**.

$N!!$ See **Double factorial**.

Analytic function   A function $f(t)$ is analytic at t_0 if there is a power series with radius of convergence $R > 0$ such that

$$f(t) = \sum_{n=0}^{\infty} A_n(t - t_0)^n \quad \text{for } |t - t_0| < R.$$

Bessel function   Solution of Bessel's equation,

$$t^2 y'' + t y' + (t^2 - v^2)y = 0.$$

Bessel functions of the first kind, denoted $J_v(t)$, are defined by a power series (see the list of power series on page 372). Bessel functions of the second kind are defined, in terms of Bessel functions of the first kind, by

$$Y_v(t) = J_v(t)\cot(\pi v) - J_{-v}(t)\csc(\pi v),$$

with a limiting value if v is an integer.

Blocked exponent   An exponent μ at a regular singular point such that $\mu + N$ is also an exponent for some positive integer N.

Cauchy-Euler equation   An ODE

$$P(t - t_0)^2 y'' + Q(t - t_0)y' + Ry = 0,$$

where P, Q, and R are constants.

Convergence set   The set of points where a power series converges. This set is an interval, centered at the point t_0 where the expansion is taken.

Double exponent   A double root of the indicial equation.

Euler's constant   The number

$$\gamma = -\int_0^{\infty} \ln(t)\,e^{-t}\,dt = 0.5772156649\ldots.$$

It is also true that $\gamma = \lim_{k\to\infty}(\phi(k) - \ln(k))$.

Exponents   The roots of the indicial equation.

Double factorial   (of a positive integer N) Let $m = 1$ if N is odd, and $m = 2$ if N is even. Then

$$N!! = N(N - 2)(N - 4)\cdots m.$$

Frobenius method   To find a solution of an ODE as a series expanded about a regular singular point t_0, substitute

$$y = |t - t_0|^\nu \left(1 + \sum_{n=1}^{\infty} A_n(t - t_0)^n\right), \qquad (7.65)$$

where ν is an unblocked exponent, and solve for the coefficients $A_1, A_2, \ldots$. If the exponents are distinct and both are unblocked, a fundamental set of solutions $\{y_1(t), y_2(t)\}$ can be obtained in this way. Otherwise, proceed as follows. If $s = \nu$ is a double exponent, then there is a second, independent solution of the form

$$y_2(t) = y_1(t) \ln|t - t_0| + |t - t_0|^\nu \sum_{n=0}^{\infty} B_n(t - t_0)^n.$$

The coefficients B_i can be determined by substitution of this expression in the ODE. If μ is an exponent blocked by the exponent $\nu = \mu + N$, a second, independent solution will take the form

$$y_2(t) = C\, y_1(t) \ln|t - t_0| + |t - t_0|^\mu \left(1 + \sum_{n=1}^{\infty} B_n(t - t_0)^n\right).$$

Fuchsian equation   A linear ODE, all of whose singular points (real, complex, or ∞) are regular.

Index shifting   An operation that can be performed on a power series, producing an equivalent power series that is summed over a different range of indices.

Indicial equation and polynomial   (of an ODE

$$p(t)(t - t_0)^2 y'' + q(t)(t - t_0)y' + r(t)y = 0$$

at a regular singular point t_0) The indicial polynomial is

$$f_0(s) = P_0\, s\, (s - 1) + Q_0\, s + R_0,$$

where $P_0 = p(t_0)$, $Q_0 = q(t_0)$, and $R_0 = r(t_0)$, and the indicial equation is $f_0(s) = 0$.

Irregular singular point   A singular point that is not a regular singular point.

Modified Bessel function   A solution of the modified Bessel equation,

$$t^2 y'' + t y' - (t^2 + \nu^2)y = 0.$$

Ordinary point   (of a linear ODE) A point t_0 is an ordinary point of the ODE

$$a_2(t)\, y'' + a_1(t)\, y' + a_0(t)y = 0$$

if $a_1(t)/a_2(t)$ and $a_0(t)/a_2(t)$ are analytic at t_0.

Pole   (of a function $f(s)$) A point s_0 such that $f(s_0)$ is undefined, but for some integer n, $(s - s_0)^n f(s)$ is analytic at s_0. For example, $f(s) = \frac{s}{s-2}$ has a pole at $s = 2$.

Power series   An infinite series whose terms are monomials in an expression $(t - t_0)$:

$$\sum_{n=0}^{\infty} A_n(t - t_0)^n.$$

Radius of convergence   Half the length of the convergence set of a power series.

Ratio test   Let $\rho = \lim_{n \to \infty} A_{n+1}/A_n$. The series $\sum_{n=0}^{\infty} A_n$ converges if $\rho < 1$, and diverges if $\rho > 1$. The test is inconclusive if $\rho = 1$.

Recursive formula   A formula used in defining a sequence of numbers or functions that specifies each term as a function of previous terms.

Regular singular point   A singular point t_0 of an ODE that can be put in the form

$$p(t)(t - t_0)^2 y'' + q(t)(t - t_0)y' + r(t)y = 0,$$

where $q(t)/p(t)$ and $r(t)/p(t)$ are analytic at t_0.

Remainder term   The function

$$E_N(t) = f(t) - \left(\sum_{m=0}^{N} \frac{1}{m!} f^{(m)}(t - t_0)^m\right),$$

where $f(t)$ is $N + 1$ times differentiable. According to Taylor's theorem,

$$E_N(t) = \frac{1}{(N+1)!} f^{(N+1)}(\xi)(t - t_0)^{N+1},$$

where ξ is a number between t_0 and t. If f is infinitely differentiable, $E_N(t)$ is the error that results from truncating the power series expansion at t_0 at N terms, and thus f is analytic at t_0 if and only if there is a number $\rho > 0$ such that

$$\lim_{N \to \infty} E_N(t) = 0$$

for all t such that $|t - t_0| < \rho$.

Root test   Let $\rho = \lim_{n \to \infty} \sqrt[n]{A_n}$. The series $\sum_{n=0}^{\infty} A_n$ converges if $\rho < 1$, and diverges if $\rho > 1$. The test is inconclusive if $\rho = 1$.

Truncation   Replacing an infinite series by the sum of a finite number of its terms.

Unblocked exponent   An exponent that is not blocked.

7.8 Suggestions for Further Readings

A good preparation for further study of linear ODEs with analytic coefficients is a course in complex analysis. Many complex analysis texts include proofs of Fuchs's theorem on solutions at an ordinary point, usually in the form of an existence and uniqueness theorem for *holomorphic ODEs*. The corresponding results for solutions at regular singular points are also proved in the context of complex analysis, but they are too specialized to find their way into introductory complex analysis texts. Proofs may be found in chapter IV of *Theory of Ordinary Differential Equations*, second edition, by E. A. Coddington and N. Levinson.[4] Coddington and Levinson also introduce the subject of irregular singular points.

Fuchsian equations are the focal point of the monograph, *Galois' Dream*, by Michio Kuga.[5] This unusual book has no prerequisites and is highly recommended as a complement to this chapter. You will find out that group theory and topology play an important role in the study of ODEs on the complex domain. The hypergeometric equation is an important Fuchsian equation and is the subject of a fascinating and accessible monograph, *Hypergeometric Equations, My Love*, by Masaaki Yoshida.[6]

The 39-page booklet entitled *The Gamma Function*, by Emil Artin,[7] is a classic of mathematical exposition and also highly recommended.

Bessel functions have been well explored in the nearly 150 years since they were first defined. For a summary of what is known, the reader may consult the *Handbook of Mathematical Functions*, edited by Milton Abramowitz and Irene Stegun.[8] *Introduction to Bessel Functions*, by Frank Bowman,[9] is a nicely written monograph that in a relatively short space covers many interesting properties and applications of Bessel functions.

List of Power Series

Unless noted to the contrary, the radius of convergence is infinite.

Algebraic functions.
Geometric series.

$$\sum_{n=0}^{\infty} t^n = \frac{1}{1-t} \quad \text{for } |t| < 1$$

Binomial series.

$$\sum_{n=0}^{\infty} \frac{\nu(\nu-1)(\nu-2)\cdots(\nu-n+1)}{n!} t^n = (1+t)^\nu \quad \text{for } |t| < 1$$

Elementary transcendental functions.

$$-\sum_{n=1}^{\infty} \frac{1}{n} t^n = \ln(1-t) \quad \text{for } |t| < 1$$

$$\sum_{n=0}^{\infty} \frac{k^n}{n!} t^n = e^{kt}$$

[4]Malabar, Fla.: Robert Krieger, 1984.

[5]English translation by Susan Addington and Motohico Mulase, Boston: Birkhäuser, 1993.

[6]Braunschweig/Wiesbaden: Verweg, 1997.

[7]Leipzig, (1931); English translation by Michael Butler: Holt, Rinehart, and Winston, 1964.

[8]New York: Dover Publications, 1965.

[9]New York: Dover Publications, 1958.

$$\sum_{n=0}^{\infty} \frac{(-k)^n}{(2n)!} t^{2n} = \cos(kt)$$

$$\sum_{n=0}^{\infty} \frac{(-k)^n}{(2n+1)!} t^{2n+1} = \sin(kt)$$

$$\sum_{n=0}^{\infty} \frac{(-1)^n}{2n+1} t^{2n+1} = \tan^{-1}(t)$$

Bessel functions.

$$\sum_{n=0}^{\infty} \frac{(-1)^n}{2^{2n+v} n! \Gamma(n+v+1)} t^{2n+v} = J_v(t) \quad \text{for } t > 0$$

$$\sum_{n=0}^{\infty} \frac{1}{2^{2n+v} n! \Gamma(n+v+1)} t^{2n+v} = I_v(t) \quad \text{for } t > 0$$

7.9 Chapter Review

Locate the singular points of the differential equations in Exercises 1–8. Include singular points at infinity, and classify each singular point as regular or irregular. Find the indicial equation at each regular singular point and determine the exponents. Are any exponents blocked? Can the solution be expressed in terms of Bessel functions or modified Bessel functions?

1. $y'' + y = 0$
2. $t^2 y'' + 13ty' + 5y = 0$
3. $(t^2 + 2t + 1)y'' + (t+1)y' + y = 0$
4. $\sin^2 ty'' + t(t - \pi)y' + 2y = 0$
5. $t^2 y'' + (t^6 - 2)y = 0$

6. $t^2 y'' + ty' - (t^2 + 1)y = 0$
7. $t(1 - t)y'' + (3 - t)y' + y = 0$
8. $t^4 y'' + 2t^3 y' + y = 0$
9. Find the general solution of $y'' + (1 - t)y' + y = 0$.
10. Find the general solution of $y'' + t^2 y = 0$
    a. as a linear combination of two power series, and
    b. in terms of Bessel functions
11. Find a fundamental set of solutions of

$$16t^2 y'' + 16(t^2 - t)y' + 15y = 0.$$

PART III

Nonlinear Differential Equations

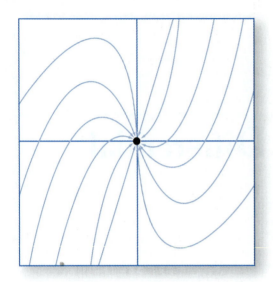

CHAPTER 8

Stability Theory

8.1 Phase Portraits of Linear Systems

Let A be a 2×2 constant matrix. We can use the characteristic roots and vectors of A to draw a phase portrait the system of ODEs,

$$\mathbf{v}' = A\mathbf{v}. \tag{8.1}$$

All linear systems have a stationary point at the origin, since $\mathbf{0}' = \mathbf{0} = A\mathbf{0}$. If A is singular, then there is a vector $\mathbf{b} \neq \mathbf{0}$, such that $A\mathbf{b} = \mathbf{0}$. In this case, all scalar multiples of $\mathbf{b}$ are stationary points, and we say that the system is **degenerate**. If A is nonsingular, then $\mathbf{0}$ is the only stationary point, and we say that (8.1) is **nondegenerate**.

Nodes

Let $A = rI$, where $r \neq 0$ is constant. Then the system $\mathbf{v}' = A\mathbf{v}$ can be written in the form

$$x' = r x$$
$$y' = r y$$

and the general solution is $x = c_1 e^{rt}$, $y = c_2 e^{rt}$. Since $y/x = c_2/c_1$ is constant, the orbits are half-lines directed toward the origin (if $r < 0$) or away from the origin (if $r > 0$). The phase portrait, shown as Figure 8.1, is called a **proper node**. We say that the node is **stable** if $r < 0$, because the solutions approach the origin with increasing time. If $r > 0$, the node is unstable.

Now let's consider the case of a matrix A that has unequal real characteristic roots. If

$$A = \begin{bmatrix} -2 & 0 \\ 0 & -1 \end{bmatrix},$$

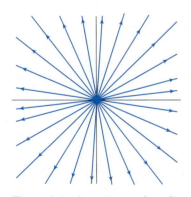

Figure 8.1. A proper node: the phase portrait of $\mathbf{v}' = I\mathbf{v}$.

the characteristic vectors are $\mathbf{i} = \begin{bmatrix} 1 \\ 0 \end{bmatrix}$ and $\mathbf{j} = \begin{bmatrix} 0 \\ 1 \end{bmatrix}$, with characteristic roots -2 and -1, respectively. The general solution of the system $\mathbf{v}' = A\mathbf{v}$ is

$$\mathbf{v} = c_1 e^{-2t}\mathbf{i} + c_2 e^{-t}\mathbf{j}.$$

Orbits where $c_2 = 0$ follow the x-axis toward the origin, and when $c_1 = 0$ instead, the orbit follows the y-axis toward the origin. If c_1 and c_2 are both nonzero, then the orbit is given by the parametric equations

$$x = c_1 e^{-2t}$$
$$y = c_2 e^{-t}.$$

Since $(e^{-t})^2 = e^{-2t}$, we can eliminate t from the parametric equations and obtain

$$\frac{x}{c_1} = \left(\frac{y}{c_2}\right)^2,$$

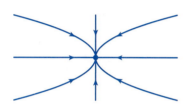

Figure 8.2. A stable improper node: phase portrait of $x' = -2x$, $y' = -y$.

or $x = a y^2$, where $a = c_1/c_2^2$. The orbits that do not follow either coordinate axis follow parabolas toward the origin. The phase portrait, shown in Figure 8.2, displays the orbits on the x and y-axes, and four other orbits. The reason that not all orbits are straight lines is that as t increases, $x = c_1 e^{-2t}$ decreases in magnitude faster than $y = c_2 e^{-t}$. This causes the orbits to become squashed against the y-axis, unless $c_2 = 0$. This phase portrait, a distorted version of the proper node, is called an **improper node**. It is stable, because all orbits converge to the origin.

In general, the phase portrait of a linear system of ODEs is called a **stable node** if every orbit is directed toward the origin as $t \to \infty$. If all orbits approach the origin as $t \to -\infty$, the phase portrait is an unstable node. Stable and unstable nodes can be recognized by calculating the characteristic roots.

Proposition 8.1.1

Let A be a 2×2 matrix with nonzero characteristic roots. The phase portrait of the system $\mathbf{v}' = A\mathbf{v}$ is a stable node if and only if both characteristic roots of A are negative, or are complex numbers with negative real parts, and it is an unstable node if and only if both characteristic roots are positive, or are complex numbers that have positive real parts.

The proof is left to you; see Exercise 29 at the end of this section.

Example 8.1.1   Draw a phase portrait of the system

$$x' = 2x - y$$
$$y' = -x + 2y.$$

Solution. The characteristic roots are 1 and 3, so the phase portrait is an unstable node. The vectors $\begin{bmatrix} 1 \\ 1 \end{bmatrix}$ and $\begin{bmatrix} 1 \\ -1 \end{bmatrix}$, respectively, are characteristic vectors belonging

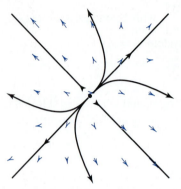

Figure 8.3 A node with two distinct, positive characteristic roots: See Example 8.1.1. The arrows in this figure show the vector field corresponding to the system of ODEs.

to these roots. It follows that the half-lines with slopes ± 1, directed away from the origin, are orbits of the system. The other orbits are not straight. As $t \to -\infty$, the second term of the general solution

$$\begin{bmatrix} x \\ y \end{bmatrix} = c_1 e^t \begin{bmatrix} 1 \\ 1 \end{bmatrix} + c_2 e^{3t} \begin{bmatrix} 1 \\ -1 \end{bmatrix}$$

is negligible; the orbit therefore recedes from the origin with slope 1. As t becomes large, the first term becomes negligible, so the slope approaches -1, the slope of the other characteristic direction. See Figure 8.3. ❑

If a 2×2 matrix A has a double characteristic root $r \neq 0$, and only one characteristic direction, then the phase portrait of

$$\begin{bmatrix} x' \\ y' \end{bmatrix} = A \begin{bmatrix} x \\ y \end{bmatrix} \tag{8.2}$$

is an improper node, stable if $r < 0$, and unstable if $r > 0$.

Example 8.1.2 Draw the phase portrait of

$$x' = r\,x$$
$$y' = k\,x + r\,y.$$

Solution. The system has a double characteristic root of r, and the characteristic direction is vertical. Assuming that $r > 0$, the phase portrait is an unstable node, with straight orbits emanating from the origin and following the y-axis.

The general solution of this system,

$$x = c\,e^{rt} \tag{8.3}$$
$$y = (kct + d)e^{rt}, \tag{8.4}$$

was derived on page 153 (for $k = 1$). We can eliminate t by solving (8.3) to get

$$t = \frac{1}{r}\ln(x/c) = \frac{1}{r}(\ln|x| - \ln|c|) \tag{8.5}$$

(the absolute value signs are necessary because x may be negative—we only know that x and c have the same sign). Now divide (8.4) by (8.3) to obtain $y/x = (kt + (d/c))$, and substitute (8.5) for t. This yields

$$\frac{y}{x} = \frac{k}{r}\ln|x| - \frac{k}{r}\ln|c|) + \frac{d}{c}.$$

Let $b = -\frac{k}{r}\ln|c|) + \frac{d}{c}$. Then

$$y = x\left(\frac{k}{r}\ln|x| + b\right). \tag{8.6}$$

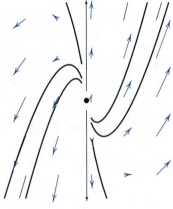

Figure 8.4. An unstable improper node corresponding to a double characteristic root: the vector field and phase portrait of $x' = x$, $y' = 2x + y$. See Example 8.1.2.

Each orbit follows the graph of (8.6) for some particular value of b. The phase portrait can be drawn by sketching graphs of equation (8.6) for various b. Figure 8.4 shows the phase portrait for $k = 2$ and $r = 1$. ❑

Saddles

If the characteristic roots r_1 and r_2 of a 2×2 matrix A have opposite signs (we'll assume $r_1 < 0$ and $r_2 > 0$), then exactly two orbits of $\mathbf{v}' = A\mathbf{v}$ will be pointed toward the origin, and two will be and pointed away from it. Let $\mathbf{e}_1$ and $\mathbf{e}_2$ be characteristic vectors belonging to r_1 and r_2, respectively.

The two orbits pointed toward the origin follow the half-lines terminating at the origin, with direction $\mathbf{e}_1$, and the two orbits pointed away from the origin follow the half-lines coming from the origin with direction $\mathbf{e}_2$. The lines through the origin in the $\mathbf{e}_1$ and $\mathbf{e}_2$ directions are called the **stable line** and the **unstable line**, respectively.

The remaining nonstationary orbits resemble hyperbolas, with the stable and unstable lines as asymptotes. As $t \to \infty$ the orbits approach the unstable line, and as $t \to -\infty$ they approach the stable line.

Example 8.1.3   Draw a phase portrait of the system

$$x' = -x$$
$$y' = x + y.$$

Solution. The characteristic roots of $A = \begin{bmatrix} -1 & 0 \\ 1 & 1 \end{bmatrix}$ are -1 and 1, corresponding to characteristic vectors

$$\mathbf{e}_1 = \begin{bmatrix} 2 \\ -1 \end{bmatrix} \quad \text{and} \quad \mathbf{e}_2 = \begin{bmatrix} 0 \\ 1 \end{bmatrix}.$$

These vectors determine the half-line orbits, which are drawn first. The remaining orbits are asymptotic to these. Figure 8.5 displays the phase portrait and the vector field.   ❑

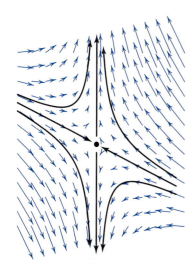

Figure 8.5. A saddle: See Example 8.1.3.

The phase portrait of a system having characteristic roots of opposite sign is called a **saddle**. To see why, let us find an integral for the system that we studied in Example 8.1.3. Referring to the box on page 116, an integral for this system can be found by determining an integral for the ODE

$$(x + y)\, dx + x\, dy = 0,$$

which happens to be in exact form. It is straightforward to obtain the integral $F(x, y) = \frac{1}{2}x^2 + xy$, and the orbits follow level curves $F(x, y) = C$. The origin is a critical point of F (that is, both partial derivatives of F vanish there), and by the second derivative test,

$$\frac{\partial^2 F}{\partial x^2} \frac{\partial^2 F}{\partial y^2} - \left(\frac{\partial^2 F}{\partial x \partial y} \right) = (1)(0) - (1)^2 < 0,$$

this critical point is a saddle point.

You can classify a linear system with constant coefficients as a saddle, degenerate, or a node with very little effort. The following lemma from linear algebra is the key. Although it is stated in terms of on 2×2 matrices, it holds for all square matrices.

Lemma 8.1.2

Let A be an 2×2 matrix, with characteristic roots r_1 and r_2. Then the trace of A is equal to the sum of the characteristic roots:

$$\text{tr}(A) = r_1 + r_2;$$

and the determinant is equal to the product of the characteristic roots:

$$\det(A) = r_1 \cdot r_2.$$

Proof The characteristic polynomial of A can be factored as $(s - r_1)(s - r_2)$. Multiply this out to obtain

$$s^2 - (r_1 + r_2)s + r_1 r_2 = s^2 - \text{tr}(A)\, s + \det(A).$$

It follows that $\text{tr}(A) = r_1 + r_2$ and $\det(A) = r_1 r_2$.   ∎

By the lemma, the characteristic roots of a 2×2 matrix A are of opposite sign if and only if $\det(A) < 0$, and 0 is a characteristic root if and only if $\det(A) = 0$. Thus, the phase portrait of the system $\mathbf{v}' = A\mathbf{v}$ is a saddle if and only if $\det(A) < 0$, and it is degenerate if and only if $\det(A) = 0$.

If $\det(A) > 0$, it is possible that A has no real characteristic roots at all. If we assume that A has real characteristic roots, then the characteristic roots must be of the same sign, and hence the phase portrait is a node. If the roots are positive, the node is unstable, and if they are negative, the node is stable. Since the trace of A is equal to the sum of the characteristic roots, it must be of the same sign as the roots, and thus serves as a stability indicator.

Classifying Phase Portraits of 2 x 2 Linear Systems

Real Characteristic Roots

Let A be a 2×2 matrix that has real characteristic roots (see Exercise 28). The phase portrait of

$$\mathbf{v}' = A\mathbf{v}$$

is

- Degenerate if $\det(A) = 0$,
- A saddle if $\det(A) < 0$,
- A stable node if $\text{tr}(A) < 0$ and $\det(A) > 0$, and
- An unstable node if $\text{tr}(A) > 0$ and $\det(A) > 0$.

Complex characteristic roots

Consider a system $\mathbf{v}' = A\mathbf{v}$, where A is a 2×2 matrix with real entries and complex characteristic roots. We will denote the roots, which are conjugate to each other, by $\lambda \pm i\omega$, and the characteristic vectors will be $\mathbf{h} \pm i\mathbf{k}$. The general solution is

$$\mathbf{v} = e^{\lambda t}[(c_1 \cos(\omega t) + c_2 \sin(t))\mathbf{h} + (-c_1 \sin(\omega t) + c_2 \cos(\omega t))\mathbf{k}],$$

where c_1 and c_2 are constants [see equation (4.19) on page 163]. If $\lambda > 0$, the exponential factor will cause all solutions to converge to 0 as $t \to -\infty$, and the phase portrait will be an unstable node. Similarly, if $\lambda < 0$, the phase portrait is a stable node. These are called **spiral nodes**, since the phase portraits turn out to be spirals, expanding outward in the unstable case and inward in the stable case.

If $\lambda = 0$, then $\mathbf{v}(t)$ is periodic (the period is $2\pi/\omega$), and hence the orbits are closed. A phase portrait in which all orbits are closed is called a **center**. These observations can be summarized as follows:

Classifying Phase Portraits of 2 x 2 Linear Systems

Complex Characteristic Roots

Let $\lambda \pm i\omega$ be the characteristic roots of a 2×2 real matrix A, and assume $\omega \neq 0$. Then the phase portrait of $\mathbf{v}' = A\mathbf{v}$ is

- a center if $\lambda = 0$;
- a stable node if $\lambda < 0$; and
- an unstable node if $\lambda > 0$.

Centers

The characteristic roots of a matrix A are pure imaginary numbers $\pm i\omega$ if and only if the characteristic equation of A is $s^2 + \omega^2 = 0$. By Lemma 8.1.2 on page 380, it follows that A has pure imaginary characteristic roots if and only if $\text{tr}(A) = 0$ and $\det(A) > 0$.

Let

$$A = \begin{bmatrix} a & b \\ c & -a \end{bmatrix} \tag{8.7}$$

be a matrix whose trace is 0. It follows from equation (3.15) on page 116 that an integral for the system

$$\left. \begin{array}{l} x' = ax + by \\ y' = cx - ay \end{array} \right\} \tag{8.8}$$

can be found by solving the differential equation

$$(cx - ay)\,dx - (ax + by)\,dy = 0. \tag{8.9}$$

By the exactness condition (see page 45), this equation is in exact form:

$$-a = \frac{\partial}{\partial y}(cx - ay) = \frac{\partial}{\partial x}[-(ax + by)].$$

Integrating equation (8.9) yields $F(x, y) = C$, where

$$F(x, y) = \frac{1}{2}c x^2 - a xy + \frac{1}{2}b y^2.$$

Recall from analytic geometry that an equation

$$px^2 + qxy + ry^2 = C$$

represents an ellipse if its discriminant $d = 4pr - q^2$ is positive, and a hyperbola if $d < 0$. The discriminant of $F(x, y) = C$ is equal to $bc - a^2 = -\det(A)$. Thus, if $\det A > 0$, the level curves of $F(x, y)$ are ellipses, and if $\det(A) < 0$, the level curves are hyperbolas. Since the orbits of (8.8) must follow the level curves, they are ellipses when $\det(A) > 0$. Since we are aware that when $\det(A) < 0$ the phase portrait is a saddle, it is not surprising that in this case the orbits follow hyperbolas.

To determine the direction of the orbits, let $(0, y_0)$ denote the point where a given orbit crosses the positive y-axis. At that point, $\frac{dx}{dt} = by_0$. Thus the orbit will be directed to the right at $(0, y_0)$ if $b > 0$; this indicates the clockwise direction. Similarly, if $b < 0$, the direction is counterclockwise.

Example 8.1.4 Draw a phase portrait of the system

$$x' = 2x + 8y$$
$$y' = -5x - 2y.$$

Solution. The matrix $\begin{bmatrix} 2 & 8 \\ -5 & -2 \end{bmatrix}$ has trace 0 and determinant 36. Therefore, the characteristic roots are $\pm 6i$. Each solution of the system is the real or imaginary part of a complex periodic function $e^{6it}\mathbf{f}$, where $\mathbf{f}$ denotes a complex characteristic vector of the matrix A belonging to the characteristic root $6i$. The period is $2\pi/6$, and the angular frequency is 6 radians per unit time.

The phase portrait is a center, and it can be drawn with the aid of the integral found by integrating the exact equation

$$(-5x - 2y)\, dx - (2x + 8y)\, dy = 0.$$

Thus, the orbits are the ellipses

$$\frac{5}{2}x^2 + 2xy + 4y^2 = c. \tag{8.10}$$

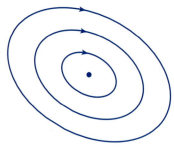

Figure 8.6. A center: See Example 8.1.4.

Since $b = 8 > 0$, the orientation is clockwise. See Figure 8.6. ❑

Optional: Drawing the elliptical orbits in Example 8.1.4

Let $R(t) = [x(t)]^2 + [y(t)]^2$. Then

$$\frac{dR}{dt} = 2x \cdot x' + 2y \cdot y'.$$

Substituting $x' = 2x + 8y$ and $y' = -5x - 2y$ (from the differential equations), we find that

$$\frac{dR}{dt} = 2x(2x + 8y) + 2y(-5x - 2y)$$
$$= 4x^2 + 6xy - 4y^2$$
$$= 2(2x - y)(x + 2y).$$

At the extreme points of the ellipse, $\frac{dR}{dt} = 0$, this will occur when $y = 2x$ and when $y = -\frac{1}{2}x$. These lines contain the axes of each orbit. You may wish to show that the longer axis is on the line $y = 2x$ and that the ratio of the lengths of the axes is 3.

Spiral nodes

Suppose that the characteristic roots of the system

$$\left. \begin{array}{l} x' = ax + by \\ y' = cx + dy \end{array} \right\} \tag{8.11}$$

are complex numbers $\lambda \pm i\omega$ with $\lambda, \omega \neq 0$. Let $\mathbf{h} + i\mathbf{k}$ be a characteristic vector belonging to $\lambda + i\omega$, so that one solution of the system (8.11) is (in vector form)

$$\mathbf{v} = \operatorname{Re}(e^{(\lambda + i\omega)t}(\mathbf{h} + i\mathbf{k}))$$
$$= e^{\lambda t}(\cos(\omega t)\mathbf{h} - \sin(\omega t)\mathbf{k}).$$

All orbits of the system (8.11) resemble this orbit and can in fact be obtained from it by a simple rotation and rescaling. Let

$$\mathbf{h} = \begin{bmatrix} p \\ q \end{bmatrix} \quad \text{and} \quad \mathbf{k} = \begin{bmatrix} m \\ n \end{bmatrix}.$$

Then $x = e^{\lambda t}(p \cos(\omega t) - m \sin(\omega t))$ and $y = e^{\lambda t}(q \cos(\omega t) - n \sin(\omega t))$ are oscillating functions that either converge to 0 (if $\lambda < 0$) as $t \to \infty$ or oscillate with exponentially increasing amplitude (if $\lambda > 0$). In the phase plane, the orbit will be a spiral that winds around the origin, directed inward if $\lambda < 0$ and outward if $\lambda > 0$. For this reason, nodes in which the characteristic roots are complex with nonzero real parts are called **spiral nodes.** Many texts use the term *focus* instead of *spiral node*.

Example 8.1.5   Draw a phase portrait for the system

$$\left. \begin{array}{l} x' = x - y \\ y' = x + y. \end{array} \right\} \tag{8.12}$$

Solution. The characteristic roots are $1 \pm i$. Let

$$\mathbf{h} + i\mathbf{k} = \begin{bmatrix} p \\ q \end{bmatrix} + i \begin{bmatrix} m \\ n \end{bmatrix}$$

be a characteristic vector belonging to $1 + i$. Then

$$\begin{bmatrix} 1 & -1 \\ 1 & 1 \end{bmatrix} \begin{bmatrix} p + im \\ q + in \end{bmatrix} = (1 + i) \begin{bmatrix} p + im \\ q + in \end{bmatrix}$$

or

$$p - k + i(m - n) = q - m + i(h + m)$$
$$p + k + i(m + n) = q - n + i(k - n).$$

The real part of the first equation implies $q = m$, and the imaginary part implies $p = -n$. If these equalities hold, then the second equation is also satisfied. We will take $q = m = 1$ and $p = -n = 0$; thus $\mathbf{h} = \mathbf{j}$ and $\mathbf{k} = \mathbf{i}$ are the standard basis vectors for the plane. The vector solution

$$\mathbf{v} = \mathrm{Re}(e^{(1+i)t}(\mathbf{h} + i\mathbf{k}) = e^t(\cos(t)\mathbf{j} - \sin(t)\mathbf{i})$$

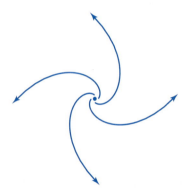

Figure 8.7. A spiral node: See Example 8.1.5.

can also be expressed by the equations $x = -e^t \sin(t)$, $y = e^t \cos(t)$.
    Put $\hat{x} = e^{-t}x = -\sin(t)$ and $\hat{y} = e^{-t}y = \cos(t)$. Then $(\hat{x}, \hat{y})$ follows the unit circle counterclockwise with frequency 1 radian per unit time. The orbits of the system (8.12) are outward spirals. The spiral $(x, y) = (-e^t \sin(t), e^t \cos(t))$ starts at $(0, 1)$ when $t = 0$ and crosses the positive y-axis when $t = 2n\pi$ at the point $y = e^{2n\pi}$. These crossing points form a geometric sequence on the y-axis, with ratio $e^{2\pi} \approx 535$. Unless a very large or very detailed drawing is made, it is unlikely that more than one crossing will be seen. The phase portrait in Figure 8.7 shows four orbits of the system. ❑

Example 8.1.6 Draw a phase portrait of the system

$$\left. \begin{aligned} x' &= x + 8y \\ y' &= -5x - 3y. \end{aligned} \right\} \tag{8.13}$$

Solution. The characteristic roots are $-1 \pm 6i$. From this information alone, we know that the orbits are spirals, directed inward toward the origin. Since x' is positive on the positive y-axis the orbit winds clockwise around the origin. The parameter $\omega = 6$ gives the angular velocity. Hence, each orbit crosses the positive x-axis once every $2\pi/6 = \pi/3$ units of time. The exponential factor in the solution is e^{-t}. It follows from this that the crossing points form a geometric sequence with ratio $e^{-\pi/3} \approx 0.35$.
    The procedure for finding a characteristic vector belonging to $-1 + 6i$ is the same as the one we used in Example 8.1.5; we find the characteristic vector

$$\mathbf{h} + i\mathbf{k} = \begin{bmatrix} 4 \\ -1 \end{bmatrix} + 3i \begin{bmatrix} 0 \\ 1 \end{bmatrix}.$$

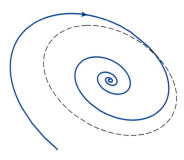

Figure 8.8. A spiral node. See Example 8.1.6. The dashed curve is the elliptical trajectory of $(\hat{x}, \hat{y})$.

This leads to the solution $\mathbf{v} = e^{-t}\, \text{Re}(e^{6it}(\mathbf{e} + i\mathbf{k})$, or

$$x = 4e^{-t}\cos(6t)$$
$$y = e^{-t}(-\cos(6t) - 3\sin(6t)).$$

Again, let $\hat{x} = e^t\, x = 4\cos(6t)$ and $\hat{y} = e^t\, y = -\cos(6t) - 3\sin(6t)$. Then $(\hat{x}, \hat{y})$ traces an ellipse in the phase plane (by determining the local maxima and minima of $R(t) = \hat{x}^2 + \hat{y}^2$ it is possible to determine the axes of this ellipse, which are oblique rather than horizontal and vertical). Figure 8.8 displays one orbit of the system (8.13), and the ellipse traced by $(\hat{x}, \hat{y})$.   ❑

Degenerate systems

If A is a 2×2 matrix that is singular, then the system $\mathbf{v}' = A\mathbf{v}$ is **degenerate**. All points of the phase plane are stationary if A is the zero matrix, and if $A \neq 0$ we still know that 0 is a characteristic root since there must be a nonzero vector $\mathbf{b}$ with

$$A\mathbf{b} = \mathbf{0} = 0 \cdot \mathbf{b};$$

thus $\mathbf{b}$ is a characteristic vector belonging to 0. The line through the origin with direction $\mathbf{b}$ consists of stationary points. Here are two examples of degenerate systems: in the first, there is a nonzero characteristic root; and in the second, 0 is a double root.

Example 8.1.7   Draw a phase portrait of the system

$$x' = -x$$
$$y' = 0.$$

Solution. This system is equivalent to the vector equation $\mathbf{v}' = A\mathbf{v}$, where

$$A = \begin{bmatrix} -1 & 0 \\ 0 & 0 \end{bmatrix}.$$

The characteristic roots are -1 and 0, and $\mathbf{i}$ and $\mathbf{j}$ are corresponding characteristic vectors. The stationary line is therefore the y-axis. The general solution is

$$\mathbf{v} = c_1 e^{-t}\mathbf{i} + c_2\mathbf{j} = \begin{bmatrix} c_1 e^{-t} \\ c_2 \end{bmatrix}$$

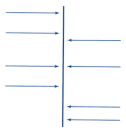

Figure 8.9. A degenerate system: See Example 8.1.7.

or $x = c_1 e^{-t}$, $y = c_2$. The characteristic vectors indicate the direction of the orbits, in this case horizontal, pointed toward the stationary line. See Figure 8.9.   ❑

Example 8.1.7 is typical of a degenerate system $\mathbf{v}' = A\mathbf{v}$, where A is a matrix whose characteristic roots are 0 and $r \neq 0$. Let $\mathbf{e}$ and $\mathbf{f}$ be characteristic vectors belonging to 0 and r, respectively. The stationary line has the direction of $\mathbf{e}$, and the nonstationary orbits will be half-lines parallel to $\mathbf{f}$, directed toward or away from the stationary line, depending on the sign of r. In our example, the nonstationary orbits were perpendicular to the stationary line, but in many cases the angle is oblique.

Figure 8.10. A degenerate system: 0 is a double characteristic root. See Example 8.1.8.

■ **Example 8.1.8**   Draw a phase portrait of the system

$$\left. \begin{array}{l} x' = 0 \\ y' = -x. \end{array} \right\} \tag{8.14}$$

Solution. The coefficient matrix $A = \begin{bmatrix} 0 & 0 \\ -1 & 0 \end{bmatrix}$ has 0 as a double characteristic root. As usual in this situation, we will start with a *non*characteristic vector—we can use $\mathbf{i} = \begin{bmatrix} 1 \\ 0 \end{bmatrix}$, since $A\mathbf{i} = -\mathbf{j}$, where $\mathbf{j} = \begin{bmatrix} 0 \\ 1 \end{bmatrix}$, which is not a multiple of $\mathbf{i}$. Furthermore, $\mathbf{e} = A\mathbf{i} - 0 \cdot \mathbf{i} = -\mathbf{j}$ *is* a characteristic vector. The stationary line is the y-axis, and since the characteristic vector is vertical, and the nonstationary orbits are straight lines. The lines to the left of the y-axis have the opposite direction to $\mathbf{e}$ (that is, they point upward), and those to the right point downward. See Figure 8.10. ❑

Notice that in this example, the nonstationary orbits are lines parallel to the stationary line, with opposite directions on either side of the stationary line. This is typical of degenerate systems of two linear equations when 0 is a double root. The only special property of the system in Example 8.1.8 is that the stationary line is vertical—it could have been any direction.

EXERCISES

In Exercises 1–12 select the phase portrait that corresponds to the system $\mathbf{v}' = A\mathbf{v}$, where A is the given matrix.

1. $A = \begin{bmatrix} 1 & -1 \\ 1 & -1 \end{bmatrix}$

2. $A = \begin{bmatrix} -3 & 1 \\ 1 & -1 \end{bmatrix}$

3. $A = \begin{bmatrix} 4 & -3 \\ 6 & -5 \end{bmatrix}$

4. $A = \begin{bmatrix} 2 & 0 \\ 0 & 1 \end{bmatrix}$

5. $A = \begin{bmatrix} 1 & 3 \\ 2 & 2 \end{bmatrix}$

6. $A = \begin{bmatrix} 2 & 1 \\ 1 & 2 \end{bmatrix}$

7. $A = \begin{bmatrix} 0 & -1 \\ -1 & 0 \end{bmatrix}$

8. $A = \begin{bmatrix} 1 & 0 \\ 0 & -1 \end{bmatrix}$

9. $A = \begin{bmatrix} 2 & 0 \\ 0 & 2 \end{bmatrix}$

10. $A = \begin{bmatrix} -4 & 3 \\ -3 & 2 \end{bmatrix}$

11. $A = \begin{bmatrix} -1 & 1 \\ 1 & -1 \end{bmatrix}$

12. $A = \begin{bmatrix} 3 & -2 \\ 4 & -3 \end{bmatrix}$

(a)

(b)

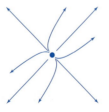

(c)

(d)

(e)

(f)

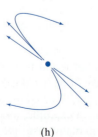

(g)

(h)

(i)

(j)

(k)

(l)

13. Draw a phase portrait for the system $x' = y' = x + y$.

14. Draw a phase portrait for the system

$$\begin{cases} x' = x + y \\ y' = -x - y. \end{cases}$$

15. Show that the each orbit of the system in Example 8.1.3 on page 379 is a branch of a hyperbola $x^2 + 2xy = c$, a half-line with slope $-\frac{1}{2}$, or a vertical half-line.

16. Draw a phase portrait of the system

$$\begin{cases} x' = x \\ y' = -3y. \end{cases}$$

Show that the orbits are horizontal or vertical half lines, and curves (not hyperbolas) satisfying the equation $x^3 y = $ constant.

In Exercises 17–26, draw the phase portrait of the given system, and classify it as degenerate, a saddle, a center, or a node (stable or unstable). Drawings of saddles must clearly indicate the stable and unstable lines.

17. $\begin{cases} x' = 2x + y \\ y' = 3x \end{cases}$ **18.** $\begin{cases} x' = 5x + 5y \\ y' = 3x + 7y \end{cases}$

19. $\begin{cases} x' = -3x + 5y \\ y' = -\frac{5}{2}x + 2y \end{cases}$ **20.** $\begin{cases} x' = x + y \\ y' = -5x - y \end{cases}$

21. $\begin{cases} x' = x - y \\ y' = 0 \end{cases}$ **22.** $\begin{cases} x' = 2x + y \\ y' = -x \end{cases}$

23. $\begin{cases} x' = 2x - 5y \\ y' = 17x - 2y \end{cases}$ **24.** $\begin{cases} x' = 3x - 10y \\ y' = 8x - 5y \end{cases}$

25. $\begin{cases} x' = 3x + y \\ y' = -5x - 3y \end{cases}$ **26.** $\begin{cases} x' = 8x + 16y \\ y' = 4x + 8y \end{cases}$

27. Show that if the phase portrait of Example 8.1.1 is rotated $45°$ clockwise, the nonvertical orbits will lie on the graphs of functions $y = kx^3$.

28. Let A be a 2×2 matrix. Show that

- if $\det(A) < \frac{1}{4}(\text{tr}(A))^2$, then A has two distinct characteristic real roots,
- if $\det(A) = \frac{1}{4}(\text{tr}(A))^2$, then A has a double characteristic root, and
- if $\det(A) > \frac{1}{4}(\text{tr}(A))^2$, then A has a no real characteristic roots.

29. Prove Proposition 8.1.1.

30. Construct, *if possible*, linear systems of two ODEs that fulfill the specifications. (Answers are not unique.)

a. A degenerate system; stationary line: $y = 2x$.

b. A nondegenerate system; the positive and negative half lines on the x-axis are orbits directed away from the origin; all other orbits converge to the origin.

c. A system in which all orbits except the stationary point at the origin approach ∞ as $t \to \infty$.

d. A system with exactly one nonstationary orbit that converges to the origin.

e. A system in which there are exactly two nonstationary orbits that converge to the origin.

f. A system in which there are exactly four nonstationary orbits that converge to the origin.

31. Consider a homogeneous linear system

$$\begin{aligned} x' &= ax + by \\ y' &= cx + dy. \end{aligned}$$

a. Show that if we substitute $y = vx$, where v is a new independent variable, the result is the following system:

$$\left. \begin{aligned} x' &= x(a + bv) \\ xv' + vx' &= x(c + dv). \end{aligned} \right\} \quad (8.15)$$

b. Multiply the first equation in the system (8.15) by v and subtract it from the second. After dividing through by x, you should obtain

$$v' = -bv^2 + (d - a)v + c. \quad (8.16)$$

c. Draw a phase diagram for equation (8.16). You will have to consider three cases, depending on the number of stationary points: 0, 1, or 2.

d. Show that the stationary points of equation (8.16) are the slopes of the characteristic vectors of the system (8.15).

e. Give an interpretation of the direction of the orbits in the phase diagram for equation (8.16), and explain how it is related to the phase portrait of the system (8.15).

f. What does the phase diagram indicate in the case where the quadratic equation $-bv^2 + (d - a)v + c = 0$ has no roots?

Match the systems in Exercises 32–37 with the phase portraits shown.

32. $\begin{cases} x' = x + 3y \\ y' = x - y \end{cases}$

33. $\begin{cases} x' = 4y \\ y' = -x \end{cases}$

34. $\begin{cases} x' = 2x - 4y \\ y' = 2x - 2y \end{cases}$

35. $\begin{cases} x' = -3x + 10y \\ y' = -2x + y \end{cases}$

36. $\begin{cases} x' = x + y \\ y' = -x + y \end{cases}$

37. $\begin{cases} x' = -4x - 4y \\ y' = 2x \end{cases}$

(a)   (b)   (c)   (d)   (e)   (f)

8.2 Stability of Linear Systems

The phase diagram of the ODE $y' = ay$ is ●←— if $a < 0$, and ←●→ if $a > 0$. In the former case, we would say that the stationary point $y = 0$ is **stable**, and in the latter, that it is **unstable**. In the stable case, each solution $y = \phi(t)$ has the property that $\phi(t) \to 0$ as $t \to \infty$, and every nonzero solution is unbounded as $t \to -\infty$. The reverse holds in the unstable case: Every nonzero solution $\phi(t)$ is unbounded as $t \to \infty$, but $\lim_{t \to -\infty} \phi(t) = 0$.

If A is a 2×2 matrix, the phase portrait is the key to understanding the stability of the origin as a stationary point of the system $\mathbf{v}' = A\mathbf{v}$. If the phase portrait is a node (spiral or otherwise), the situation is analogous to the one-dimensional case. When the characteristic roots of A are negative, or complex with negative real part—that is, they are located to the left of the imaginary axis in the complex plane—the node is stable. Then, with the exception of the solution $\mathbf{v} \equiv \mathbf{0}$, every solution converges to the origin as $t \to \infty$, and by following the orbits backward, we see that the solutions are unbounded as $t \to -\infty$. When the characteristic roots are both on the right side of the imaginary axis, the node is unstable, and the situation is reversed: Following the orbits backward shows that every solution converges to $\mathbf{0}$ as $t \to -\infty$; and all nonzero solutions are unbounded as $t \to \infty$.

Since there are systems whose phase portraits are not nodes, we will distinguish between two degrees of stability, depending on whether all solutions converge to the origin as $t \to \infty$ or merely remain near the origin for $t \geq 0$. A linear system $\mathbf{v}' = A\mathbf{v}$ is said to be

- **neutrally stable** if for every solution $\mathbf{v}(t)$ there is a number M such that the maximum distance from the origin to $\mathbf{v}(t)$ is less than M for all $t \geq 0$.
- **asymptotically stable** if $\lim_{t \to \infty} \mathbf{v}(t) = \mathbf{0}$ for every solution $\mathbf{v}(t)$

We have seen how to classify the phase portrait of any nondegenerate two-dimensional system as a node, a center, or a saddle. A two-dimensional system $\mathbf{v}' = A\mathbf{v}$ is asymptotically stable if and only if its phase portrait is a stable node. If the phase portrait is a saddle or an unstable node, then the system is *not* neutrally stable, because some orbits are unbounded as $t \to \infty$ in these cases. If the phase portrait is a center, then the orbits are ellipses. Recall from analytic geometry that the equation of any ellipse centered at the origin can be written in the form

$$\frac{x^2}{a^2} + \frac{y^2}{b^2} = 1,$$

provided that we rotate the coordinate axes (rotation preserves the distance of a point to the origin) to align them with the axes of the ellipse. The larger of the numbers a and b is the length of the major semiaxis of the ellipse, and it represents the maximum distance of a point on the ellipse to the origin. Thus, if $\mathbf{v}(t)$ is a solution of a two-dimensional system whose phase portrait is a center, then the maximum distance from $\mathbf{v}(t)$ to the origin will never be greater than the length of the major semiaxis of the orbit, and the system is therefore neutrally stable. The following proposition restates what we have just proved.

Proposition 8.2.1

Let A be a real 2×2 matrix with $\det(A) \neq 0$. The system $\mathbf{v}' = A\mathbf{v}$ is

- asymptotically stable if and only if $\det(A) > 0$ and $\operatorname{tr}(A) < 0$.
- neutrally stable if and only if $\det(A) > 0$, and $\operatorname{tr}(A) \leq 0$.

Proof Lemma 8.1.2 (page 380) states that the product of the characteristic roots of A is equal to $\det(A)$, and their sum is $\operatorname{tr}(A)$. Thus, if the two roots are real, they will both be negative if and only if $\det(A) > 0$ and $\operatorname{tr}(A) < 0$. If they are complex, they form a conjugate pair $\lambda \pm i\mu$, and their product, $\det(A) = \lambda^2 + \mu^2 > 0$. Furthermore, $\operatorname{tr}(A) = 2\lambda$ so that $\lambda \pm i\mu$ will both be located to the left of the imaginary axis if and only if $\operatorname{tr}(A) < 0$. In both the real and complex cases, $\det(A) > 0$ and $\operatorname{tr}(A) < 0$ if and only if the phase portrait is a stable node.

We have seen that a two-dimensional neutrally stable nondegenerate system has to be either a center or a stable node—this is equivalent to saying that the characteristic roots must be either to the left of the imaginary axis, or nonzero pure imaginary numbers. If the roots are pure imaginary numbers $\pm i\mu$, their sum must be equal to 0, and their product is $-i^2\mu^2 = \mu^2 > 0$. Thus $\operatorname{tr}(A) = 0$ and $\det(A) > 0$. Conversely, if $\operatorname{tr}(A) = 0$ and $\det(A) > 0$, then the characteristic equation for A is $s^2 + \det(A) = 0$, so the characteristic roots, $s = \pm\sqrt{-\det(A)}$, are pure imaginary. ∎

Proposition 8.2.1 covers nondegenerate systems, but before we can state a generalization that will apply to systems of n equations, we need to understand what happens with degenerate systems. Let A be a 2×2 matrix with $\det(A) = 0$. Then 0 is a characteristic root of A, and we can find a characteristic vector $\mathbf{b}$ belonging to 0. Every point on the line through the origin in the $\mathbf{b}$ direction belonging to 0 will be stationary, and we will call that line a *stationary line*. Since orbits of points on a stationary line do not approach the origin, the system is not asymptotically stable. We can still ask if it is neutrally stable, and there are four cases to consider:

Case I: $A = 0I$. Every point is stationary, and the system is neutrally stable.

Case II: $A \neq 0I$ but $\operatorname{tr}(A) = 0$. A has a double characteristic root, 0. The nonstationary orbits are lines parallel to the stationary line, as shown in Figure 8.10 (page 386). Since the nonstationary orbits are not bounded, the system is unstable.

Case III: tr(A) < 0 There are two characteristic roots, 0 and tr(A). The nonstationary orbits are half lines, parallel to the characteristic vector belonging to tr A, directed toward the stationary line, and terminating there. See Figure 8.9 on page 385. Thus, if $\mathbf{v}(t)$ is a solution, then for $t \geq 0$, $\mathbf{v}(t)$ will be located on a line segment with one endpoint at $\mathbf{v}(0)$ and the other endpoint on the stationary line, and $\mathbf{v}(t)$ will not be farther from the origin than the distance to the more remote of these two endpoints. It follows that the system is neutrally stable.

Case IV: tr(A) > 0 The phase portrait is as in case III, but the nonstationary orbits are directed away from the stationary line. The system is unstable.

If it were not for case II, we could say that a system is neutrally stable if and only if none of its characteristic roots are located to the right of the imaginary axis in the complex plane. While it is degenerate, situations analogous to case II are encountered in nondegenerate systems in higher dimensions. Consider the following example of an system of ODEs with *complex* coefficients:

Example 8.2.1   Let

$$A = \begin{bmatrix} i & 1 \\ 0 & i \end{bmatrix}.$$

Find the general solution of $\mathbf{v}' = A\mathbf{v}$ and determine whether of not the system is stable.

Solution. The characteristic equation of A is $s^2 - 2i\,s + i^2 = 0$, or $(s - i)^2 = 0$. Thus, i is a double characteristic root. It is easily checked that $\mathbf{b} = \begin{bmatrix} 0 \\ 1 \end{bmatrix}$ is not a characteristic vector, and we put $\mathbf{c} = (A - i\,I)\mathbf{b} = \begin{bmatrix} 1 \\ 0 \end{bmatrix}$, which is a characteristic vector. Thus the general solution is

$$\mathbf{v}(t) = k_1 e^{it}(t\mathbf{c} + \mathbf{b}) + k_2 e^{it}\mathbf{c}.$$

The solution corresponding to $(k_1, k_2) = (1, 0)$ is $\mathbf{v}_1(t) = e^{it}\begin{bmatrix} t \\ 1 \end{bmatrix}$. The distance from the origin to $\mathbf{v}_1(t)$ is equal to $|e^{it}|\sqrt{t^2 + 1}$, where $|e^{it}| = |\cos(t) + i\sin(t)| = 1$. This solution is not bounded and hence the system is unstable.   ❑

Example 8.2.1 shows that a nondegenerate system with all characteristic roots on the imaginary axis may be unstable. The key feature of the example was the double characteristic root, i. The number of characteristic roots (counted with multiplicity) of an $n \times n$ matrix is n. As complex characteristic roots of a real matrix must occur in conjugate pairs, this means that no real matrix smaller that 4×4 can have multiple complex characteristic roots: if, for example, i is a double characteristic root of a real matrix A, then $-i$ would also have to be a double characteristic root.

If $A = i\,I_2$, where I_2 is the 2×2 identity matrix, then $e^{At} = e^{it}I_2$ is a fundamental matrix solution of $\mathbf{v}' = A\mathbf{v}$. It follows that every solution has the form $\mathbf{v}(t) = e^{it}\mathbf{v}(0)$; and hence that $|\mathbf{v}(t)| = |e^{it}||\mathbf{v}(0)| = |\mathbf{v}(0)|$. Thus the system

$\mathbf{v}' = i\ I_2\mathbf{v}$, which, like the system in Example 8.2.1, has a double characteristic root i, is stable. The distinction between the two systems is that the one in Example 8.2.1 does not have two independent characteristic vectors, but $\mathbf{v}' = i\ I_2\mathbf{v}$ does.

Let us say that a characteristic root r of a matrix A is **semisimple** if the multiplicity of r as a solution of the characteristic equation of A is equal to the maximum number of linearly independent characteristic vectors belonging to r.

With this definition in mind we can state the general stability theorem for homogeneous theorem for homogeneous linear systems.

Theorem 8.1

Let A be an $n \times n$ matrix with real or complex entries. Then the system $\mathbf{v}' = A\mathbf{v}$ is stable if and only if all of the characteristic roots of A are located either on or to the left of the imaginary axis and, in addition, all characteristic roots that are on the imaginary axis are semisimple.

The system $\mathbf{v}' = A\mathbf{v}$ is asymptotically stable if and only if all characteristic roots of A are located strictly to the left of the imaginary axis.

To prove this theorem, we will show that if each characteristic root of A is located strictly to the left of the imaginary axis, then $\lim_{t \to \infty} e^{At}\mathbf{c} = \mathbf{0}$ for every $\mathbf{c} \in \mathbf{R}^n$. This will prove the second assertion of the theorem. To prove the first assertion, we will show that if, in addition to characteristic roots located in the left half-plane there are semisimple characteristic roots on the imaginary axis, then $e^{At}\mathbf{c}$ is bounded as $t \longrightarrow \infty$.

Since every solution $\mathbf{v}(t)$ of $\mathbf{v}' = A\mathbf{v}$ can be expressed as

$$\mathbf{v} = e^{At}\mathbf{c},$$

where $\mathbf{c} = \mathbf{v}(0)$, this is all that we need to do.

Recall (page 172) that e^{At} is equal to the sum of a convergent series:

$$e^{At} = I + tA + \frac{1}{2}t^2 A + \frac{1}{3!}t^3 A^3 + \cdots = \sum_{m=0}^{\infty} \frac{1}{m!}t^m A^m.$$

Let r be a characteristic root of A, and let $\mathbf{b}$ be a corresponding characteristic vector. Thus, $A\mathbf{b} = r\mathbf{b}$, and it isn't hard to see that for any $m \geq 0$,

$$A^m \mathbf{b} = r^m \mathbf{b}$$

as well. It follows that

$$e^{At}\mathbf{b} = \sum_{m=0}^{\infty} \frac{t^m}{m!} A^m \mathbf{b}$$

$$= \sum_{m=0}^{\infty} \frac{t^m}{m!} r^m \mathbf{b} = e^{rt}\mathbf{b}.$$

In other words, $\mathbf{b}$ is also a characteristic vector for e^{At} with characteristic root e^{rt}.

If all of the characteristic roots of A are semisimple (in this case, A is said to be **diagonalizable**), then any vector $\mathbf{c}$ can be expressed as a sum of characteristic vectors of A:

$$\mathbf{c} = \mathbf{b}_1 + \mathbf{b}_2 + \cdots \mathbf{b}_n.$$

It follows that

$$e^{At}\mathbf{c} = e^{r_1 t}\mathbf{b}_1 + e^{r_2 t}\mathbf{b}_2 + \cdots e^{r_n t}\mathbf{b}_n, \qquad (8.17)$$

where r_j is the characteristic root belonging to $\mathbf{b}_j$.

Let $r_j = \lambda_j + i\omega_j$. Then $e^{r_j t} = e^{\lambda_j t}(\cos(\omega_j t) + i\sin(\omega_j t))$ is bounded as $t \longrightarrow \infty$ when $\lambda_j \leq 0$. When $\lambda_j < 0$,

$$\lim_{t \to \infty} e^{r_j t} = 0.$$

Thus, as $t \longrightarrow \infty$, the right side of (8.17) is bounded if all of the characteristic roots lie on the imaginary axis or to its left, and it converges to 0 if all of the characteristic roots lie strictly to the left of the imaginary axis. This proves Theorem 8.1 in the case where the matrix A is diagonalizable.

Let r be a characteristic root of A. A **generalized characteristic vector** belonging to r is vector $\mathbf{b}$ such that

$$(A - rI)^p \mathbf{b} = \mathbf{0} \qquad (8.18)$$

for some integer $p \geq 1$. If (8.18) holds for $p = 1$, we have $A\mathbf{b} = r\mathbf{b}$; in other words, $\mathbf{b}$ is a characteristic vector belonging to r. Notice that if (8.18) holds for a value $p = p_0$, then it also holds for all $p > p_0$, since

$$(A - rI)^p \mathbf{b} = (A - rI)^{p - p_0} \underbrace{(A - rI)^{p_0}\mathbf{b}}_{= \mathbf{0}} = \mathbf{0}.$$

We can compute $e^{At}\mathbf{b}$ when $\mathbf{b}$ is a generalized characteristic vector by an indirect method, as follows. Notice that (8.18) implies that when it is expanded as a series, $e^{(A-rI)}\mathbf{b}$ has only a finite number of nonzero terms. Given that $(A - rI)^p \mathbf{b} = \mathbf{0}$,

$$e^{(A-rI)t}\mathbf{b} = \mathbf{b} + (A - rI)t\mathbf{b} + \frac{1}{2}(A - rI)^2 t^2 \mathbf{b} + \cdots$$
$$+ \frac{1}{(p-1)!}(A - rI)^{p-1} t^{p-1}\mathbf{b}.$$

Since the matrices $(A - rI)t$ and rIt commute, we have

$$e^{At} = e^{(A-rI)t}e^{rIt} = e^{rt}e^{(A-rI)t}.$$

Thus,

$$e^{At}\mathbf{b} = e^{rt}\left[I + (A - rI)t + \frac{1}{2}(A - rI)^2 t^2 + \cdots + \frac{1}{(p-1)!}(A - rI)^{p-1} t^{p-1} \right]\mathbf{b}.$$
$$(8.19)$$

Lemma 8.2.2

Let b be a generalized characteristic vector belonging to a characteristic root $r = \lambda + i\omega$ of a matrix A. Then

- If r is in the left half-plane, that is, $\lambda < 0$, then $\lim_{t\to\infty} e^{At}\mathbf{b} = \mathbf{0}$.
- If r is on the imaginary axis, that is, $\lambda = 0$, then $e^{At}\mathbf{b}$ is bounded as $t \longrightarrow \pm\infty$ if and only if $\mathbf{b}$ is actually a characteristic vector.
- If r is in the right half-plane, that is, $\lambda > 0$, then $e^{At}\mathbf{b}$ is unbounded as $t \longrightarrow \infty$.

Proof   For $\lambda < 0$, and $k \geq 1$, both numerator and denominator of

$$t^k e^{\lambda t} = \frac{t^k}{e^{-\lambda t}}$$

approach ∞ as $t \longrightarrow \infty$. By using l'Hôspital's rule k times, we can show that

$$\lim_{t\to\infty} \frac{t^k}{e^{-\lambda t}} = \lim_{t\to\infty} \frac{k!}{(-\lambda)^k e^{-\lambda t}} = 0.$$

Each term on the right side of (8.19) can be expressed as

$$e^{\lambda t} t^k (\cos(\omega t) + i \sin(\omega t))\frac{1}{(k!)}(A - rI)^k \mathbf{b}, \tag{8.20}$$

which converges to $\mathbf{0}$ as $t \to \infty$, provided that $\lambda < 0$. It follows that in this case, $\lim_{t\to\infty} e^{At}\mathbf{b} = \mathbf{0}$.

Due to the t^k factor, the expression (8.20) is bounded only if $k = 0$ when $\lambda = 0$. Thus, if $\mathbf{b}$ is not a characteristic vector, $e^{At}\mathbf{b}$ is not bounded.

Finally, if $\lambda > 0$, then $e^{\lambda t}$ is unbounded, and hence $e^{At}\mathbf{b}$ is unbounded.   ■

Proof of Theorem 8.1   We need the following theorem from linear algebra. Its proof is omitted.

Theorem 8.2

Let A be an $n \times n$ matrix. Then every vector in $\mathbf{R}^n$ can be expressed as a sum of at most n generalized characteristic vectors of A.

Let $\mathbf{v}(t)$ be a solution of $\mathbf{v}' = A\mathbf{v}$, and put $\mathbf{c} = \mathbf{v}(0)$. Then there are generalized characteristic vectors $\mathbf{b}_1, \mathbf{b}_2, \ldots, \mathbf{b}_m$ such that

$$\mathbf{c} = \mathbf{b}_1 + \mathbf{b}_2 + \cdots + \mathbf{b}_m.$$

Since $\mathbf{v}(t) = e^{At}\mathbf{c}$,

$$\mathbf{v}(t) = e^{At}\mathbf{b}_1 + e^{At}\mathbf{b}_2 + \cdots + e^{At}\mathbf{b}_m. \tag{8.21}$$

If all characteristic roots of A are located in the left half-plane, then by Lemma 8.2.2, each term on the right side of (8.21) converges to $\mathbf{0}$, and it follows that $\lim_{t\to\infty} \mathbf{v}(t) = \mathbf{0}$.

If all characteristic roots lie either on the imaginary axis or in the left half-plane, and the roots on the imaginary axis are semisimple, then we can still apply Lemma 8.2.2, because every generalized characteristic vector belonging to a semisimple characteristic root is actually a characteristic vector.   ■

EXERCISES

For Exercises 1–10, a matrix A is given. Determine if system $\mathbf{v}' = A\mathbf{v}$ is asymptotically stable. If it isn't asymptotically stable, is the system stable? If you conclude that the system is unstable, find an unbounded solution.

1. $A = \begin{bmatrix} 1 & 1 \\ -2 & -2 \end{bmatrix}$

2. $A = \begin{bmatrix} 2 & 1 \\ 3 & 0 \end{bmatrix}$

3. $A = \begin{bmatrix} 1 & 1 \\ -1 & 1 \end{bmatrix}$

4. $A = \begin{bmatrix} 1 & 1 \\ -1 & -1 \end{bmatrix}$

5. $A = \begin{bmatrix} 1 & 1 & 0 & 0 \\ -2 & -2 & 0 & 0 \\ 0 & 0 & 2 & 1 \\ 0 & 0 & 3 & 0 \end{bmatrix}$

6. $A = \begin{bmatrix} 0 & 1 \\ -1 & 0 \end{bmatrix}$

7. $A = \begin{bmatrix} 0 & 1 & 0 & 0 \\ -1 & 0 & 0 & 0 \\ 0 & 0 & 0 & 1 \\ 0 & 0 & 1 & 0 \end{bmatrix}$

8. $A = \begin{bmatrix} -1 & 2 & 3 \\ 0 & -2 & -4 \\ 0 & 0 & -3 \end{bmatrix}$

9. $A = \begin{bmatrix} 0 & 1 & 0 & 1 \\ -1 & 0 & 0 & 0 \\ 0 & 0 & 0 & 1 \\ 0 & 0 & -1 & 0 \end{bmatrix}$

10. $A = \begin{bmatrix} 0 & 1 & 0 & 0 \\ -1 & 0 & 0 & 0 \\ 0 & 0 & 0 & 1 \\ 0 & 0 & -4 & 0 \end{bmatrix}$

11. If A is a 2×2 real matrix whose characteristic roots lie on the imaginary axis, show that each solution of $\mathbf{u}' = A\mathbf{u}$ is periodic.

12. Show that if A is an $n \times n$ real matrix whose characteristic roots are all located on the imaginary axis, and all of these characteristic roots are distinct, then every solution of $\mathbf{u}' = A\mathbf{u}$ is bounded. Is every solution necessarily periodic? Prove or find a counterexample.

13. Suppose that A is an $n \times n$ matrix that has one characteristic root in the right half-plane, and $(n - 1)$ characteristic roots in the left half-plane. A random vector $\mathbf{c}$ is chosen in $\mathbf{R}^n$. How likely is it that $e^{At}\mathbf{c}$ will be bounded as $t \to \infty$?

14. Show that $e^{At}\mathbf{c}$ is bounded on $\mathbf{R}$ if and only if $\mathbf{c}$ can be expressed as a linear combination of characteristic vectors belonging to pure imaginary characteristic roots of A.

8.3 Stationary Points of Nonlinear Systems

Let us consider a system of ODEs, $\mathbf{v}' = \mathbf{f}(\mathbf{v}, t)$, where

$$\mathbf{v}(t) = \begin{bmatrix} x_1(t) \\ x_2(t) \\ \vdots \\ x_n(t) \end{bmatrix} \quad \text{and} \quad \mathbf{f}(\mathbf{v}, t) = \begin{bmatrix} f_1(x_1, x_2, \ldots, x_n, t) \\ f_2(x_1, x_2, \ldots, x_n, t) \\ \vdots \\ f_n(x_1, x_2, \ldots, x_n, t) \end{bmatrix}$$

is a vector function. If $\mathbf{f}(\mathbf{v}, t)$ does not have the form $A(t)\mathbf{v} + \mathbf{b}(t)$, where $A(t)$ is an $n \times n$ matrix, we say that the system is nonlinear.

A system of the form $\mathbf{v}' = \mathbf{f}(\mathbf{v})$, where $\mathbf{f}$ does not depend on t, is called an **autonomous system**. A **stationary point** $\mathbf{v}_1$ is a point where $\mathbf{f}(\mathbf{v}_1) = \mathbf{0}$ [or, if the system is not autonomous, $\mathbf{f}(\mathbf{v}_1, t) = \mathbf{0}$ for all t]. The stationary point $\mathbf{v}_1$ is **isolated** if there is a number $r > 0$ such that for every stationary point $\mathbf{v}_2 \neq \mathbf{v}_1$, the distance

$$\|\mathbf{v}_1 - \mathbf{v}_2\| \geq r.$$

The origin is an isolated stationary point of a linear system if and only if the system is nondegenerate, because a degenerate system would have a *stationary line* through the origin, in the characteristic direction belonging to the characteristic root 0. Stationary points arbitrarily close to the origin can be found on the stationary line.

It does not make sense to say that a nonlinear system is *stable* or *unstable* as a whole, as we did with linear systems in Section 8.2. We must analyze a nonlinear system locally and make statements that apply to small pieces of phase space. The one-dimensional case was covered in Section 2.5, where the focus was on isolated stationary points: Each stationary point could be individually classified as stable or unstable. If the two adjacent orbits were directed toward it, the stationary point was *stable;* and it was *unstable* if at least one of these orbits was directed away from it.

In *n* dimensions, with $n > 1$, more than two categories are necessary to classify stationary points. In Section 8.2, linear systems were classified as either *asymptotically stable, neutrally stable,* or *unstable,* and we will use the same terms to classify isolated stationary points of nonlinear systems.

All of the results considered in this section will be set in the context of an autonomous system of two differential equations

$$\left. \begin{array}{l} x' = f(x, y) \\ y' = g(x, y), \end{array} \right\} \tag{8.22}$$

where the functions f and g have continuous first partial derivatives on an open domain $\mathcal{D} \subset \mathbf{R}^2$. The assumption that there are only two equations is merely for convenience, and except when noted to the contrary, the results hold for systems with any number of equations.

We will say that the orbit defined by a solution $(x, y) = (\phi(t), \psi(t))$ **converges** to a point (x^*, y^*) if $(\phi(t), \psi(t))$ is defined for all $t > 0$ and

$$\lim_{t \to \infty} (\phi(t), \psi(t)) = (x^*, y^*).$$

DEFINITION A stationary point $(x_1, y_1) \in \mathcal{D}$ of the system (8.22) is **asymptotically stable** if there is a number $r > 0$ such that every orbit that enters the circle with radius r and center (x_1, y_1) converges to (x_1, y_1).

If A is an $n \times n$ matrix, the system

$$\mathbf{v}' = A\mathbf{v}$$

was defined to be asymptotically stable if every orbit converges to the origin. Thus, the origin is an asymptotically stable stationary point of an asymptotically stable system.

Before turning to a detailed analysis of stability for stationary points of nonlinear systems, let's look at a couple of pictures.

Example 8.3.1   Figure 8.11 is the phase portrait of

$$\left. \begin{array}{l} x' = -4y + x(1 - \sqrt{x^2 + y^2}) \\[2mm] y' = 4x + y(1 - \sqrt{x^2 + y^2}). \end{array} \right\} \tag{8.23}$$

Find a stationary point of the system, and draw a conclusion about its stability. In what sense does the stability (or lack thereof) appear to be local?

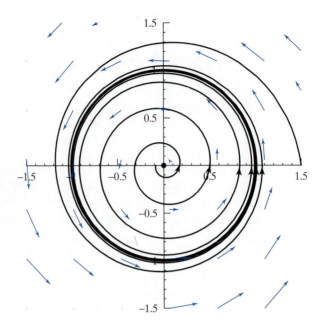

FIGURE 8.11   Phase portrait of the nonlinear
system (8.23). (See Example 8.3.1.)

Solution. You can see that the origin is a stationary point, and the unit circle
$x^2 + y^2 = 1$ is an orbit [in fact, it is easy to verify that $(x, y) = (0, 0)$, and
$(x, y) = (\cos(4t), \sin(4t))$ are solutions]. Since (8.23) is nonlinear, the formula

$$\begin{bmatrix} x \\ y \end{bmatrix} = C \begin{bmatrix} \cos(4t) \\ \sin(4t) \end{bmatrix}$$

only yields a solution when $|C| = 1$. The orbits inside the unit circle appear to
spiral away from the origin, and the orbits outside the circle spiral inward toward it.
Therefore, the phase portrait indicates that the origin is not stable, but the circular
orbit is stable. Of course, these statements are speculative: We can't use a computer-
generated sketch as a proof, and we haven't defined the notion of a stable orbit.
We can look at the figure as a scientist would view an experiment, and seek an
explanation in the form of a rigorous discussion.   ❑

Example 8.3.2   Locate all stationary points of the system

$$\left. \begin{array}{l} x' = y \\ y' = 2x - 3x^2/2, \end{array} \right\} \tag{8.24}$$

whose phase portrait is shown in Figure 8.12, and discuss their stability.

Solution. If (x_1, y_1) is a stationary point, then $x' = 0$ implies $y_1 = 0$. Thus all
stationary points are located on the x-axis. Also, $y' = 0$ implies

$$2x_1 - \frac{3}{2}x_1^2 = 0.$$

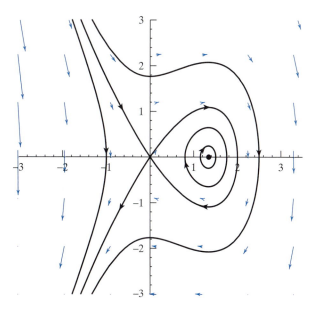

FIGURE 8.12   The origin is a saddle point of the system (8.24).

This quadratic has two solutions: $x_1 = 0$ and $x_1 = \frac{4}{3}$. We conclude that there are two stationary points: $(0, 0)$ and $\left(\frac{4}{3}, 0\right)$. The origin looks like a saddle point of a linear system and is not stable. The orbits appear to circulate around $\left(\frac{4}{3}, 0\right)$ as they do in a linear system that is a center (characteristic roots are located on the imaginary axis). We infer that it too is not asymptotically stable, although here we are stretching. The computer-drawn orbits circulating around $\left(\frac{4}{3}, 0\right)$ appear to be closed ovals, but possibly they spiral inward (or outward) very gradually.   ❑

Linearly stable equilibria

Near the stationary points in the examples that we just considered, the phase portraits all appear to resemble phase portraits of linear systems. To relate a stationary point (x_1, y_1) of

$$\left. \begin{array}{l} x' = f(x, y) \\ y' = g(x, y) \end{array} \right\}$$

to a specific linear system, we will use the matrix of partial derivatives of f and g, all evaluated at (x_1, y_1). We will be generalizing the following stability test, stated as Proposition 2.5.3 on page 85, for stationary points of a one-dimensional system $y' = f(y)$: A stationary point y_1 is stable if $f'(y_1) < 0$, unstable if $f'(y_1) > 0$. The test makes no conclusion if $f'(y_1) = 0$.

Let $f_x(x_1, y_1)$, $f_y(x_1, y_1)$, $g_x(x_1, y_1)$, and $g_y(x_1, y_1)$ denote the partial derivatives of f and g. The **derivative** of the vector function $\mathbf{F}(x, y) = \begin{bmatrix} f(x, y) \\ g(x, y) \end{bmatrix}$ at the point (x_1, y_1) is defined to be the matrix

$$A(x_1, y_1) = \begin{bmatrix} f_x(x_1, y_1) & f_y(x_1, y_1) \\ g_x(x_1, y_1) & g_y(x_1, y_1) \end{bmatrix}.$$

The following proposition indicates that the derivative of a vector function is the best linear approximation of the function.

Proposition 8.3.1

Let (x_1, y_1) be a point such that $f(x_1, y_1) = g(x_1, y_1) = 0$. Suppose that the second partial derivatives of $f(x, y)$ and $g(x, y)$ are continuous on a rectangle

$$\mathcal{R} = \{(x, y) : (x_1 - a) \le x \le (x_1 + a), (y_1 - b) \le y \le (y_1 + b)\},$$

where a and b are positive. Then

$$\mathbf{F}(x, y) = A(x_1, y_1)\begin{bmatrix} x - x_1 \\ y - y_1 \end{bmatrix} + \mathbf{O}((x - x_1)^2 + (y - y_1)^2) \tag{8.25}$$

as $(x, y) \to (x_1, y_1)$.

Equation (8.25) uses "big $\mathbf{O}$" notation, which can be explained as follows: Let $\mathbf{F}(x, y)$ and $\mathbf{G}(x, y)$ be vector functions, and let $h(x, y)$ be a real-valued function. The statement

$$\mathbf{F}(x, y) = \mathbf{G}(x, y) + \mathbf{O}(h(x, y)) \text{ as } (x, y) \to (x_1, y_1)$$

means there is a rectangle $\mathcal{R}$ centered at (x_1, y_1) and a constant K such that

$$|\mathbf{F}(x, y) - \mathbf{G}(x, y)| \le K\, h(x, y)$$

for all $(x, y) \in \mathcal{R}$ Thus we can interpret (8.25) to say that there is a constant K and a rectangle $\mathcal{R}$ such that

$$\left| \mathbf{F}(x, y) - A(x_1, y_1)\begin{bmatrix} x - x_1 \\ y - y_1 \end{bmatrix} \right| \le K((x - x_1)^2 + (y - y_1)^2) \tag{8.26}$$

for all $(x, y) \in \mathcal{R}$.

Proof of Proposition 8.3.1   Put $\bar{x} = x - x_1$ and $\bar{y} = y - y_1$, and define a function

$$p(t) = f(x_1 + t\bar{x}, y_1 + t\bar{y}).$$

Since f has continuous second partial derivatives, it is easy to see that p has continuous second derivatives on the interval $[0, 1]$. By Taylor's theorem there is a number $c_1 \in (0, 1)$ such that

$$p(1) = p(0) + p'(0)(1 - 0) + \frac{1}{2}p''(c_1)(1 - 0)^2.$$

But $p(1) = f(x_1 + \bar{x}, y_1 + \bar{y}) = f(x, y)$, and $p(0) = f(x_1, y_1) = 0$. Furthermore, by the chain rule for partial derivatives,

$$p'(0) = \frac{\partial f}{\partial x}(x_1, y_1)\bar{x} + \frac{\partial f}{\partial y}(x_1, y_1)\bar{y}$$

and

$$p''(c_1) = B_{11}\bar{x}^2 + 2B_{12}\bar{x}\bar{y} + B_{13}\bar{y}^2,$$

where

$$B_{11} = \frac{\partial^2 f}{\partial x^2}(x_1 + c_1\bar{x}, y_1 + c_1\bar{y}),$$

$$B_{12} = \frac{\partial^2 f}{\partial x \partial y}(x_1 + c_1\bar{x}, y_1 + c_1\bar{y}), \text{ and}$$

$$B_{13} = \frac{\partial^2 f}{\partial y^2}(x_1 + c_1\bar{x}, y_1 + c_1\bar{y}).$$

Thus

$$f(x, y) - [f_x(x_1, y_1)\bar{x} + f_y(x_1, y_1)\bar{y}] = B_{11}\bar{x}^2 + 2B_{12}\bar{x}\bar{y} + B_{13}\bar{y}^2. \tag{8.27}$$

Similarly,

$$g(x, y) - [g_x(x_1, y_1)\bar{x} + g_y(x_1, y_1)\bar{y}] = B_{21}\bar{x}^2 + 2B_{22}\bar{x}\bar{y} + B_{23}\bar{y}^2, \tag{8.28}$$

where B_{21}, B_{22}, and B_{23} represent the second partial derivatives of g, evaluated at a point $(x_1 + c_2\bar{x}, y_1 + c_2\bar{y})$. The B_{ij} are not constants: All depend on $(\bar{x}, \bar{y})$. Since any function that is continuous on $\mathcal{R}$ must be bounded on $\mathcal{R}$, there is a constant M such that each second partial derivative of either f or g has absolute value less than M on $\mathcal{R}$. Since each B_{ij} is equal to a second partial derivative of either f or g, the bounds $|B_{ij}| < M$ hold.

By combining (8.27) and (8.28), we obtain

$$\left| \mathbf{F}(x, y) - A(x_1, y_1) \begin{bmatrix} \bar{x} \\ \bar{y} \end{bmatrix} \right| = \left| \begin{bmatrix} B_{11}\bar{x}^2 + 2B_{12}\bar{x}\bar{y} + B_{13}\bar{y}^2 \\ B_{21}\bar{x}^2 + 2B_{22}\bar{x}\bar{y} + B_{23}\bar{y}^2 \end{bmatrix} \right|. \tag{8.29}$$

Using the fact that $(\bar{x} \pm \bar{y})^2 \geq 0$, you can show that

$$2|\bar{x}\bar{y}| \leq (\bar{x}^2 + \bar{y}^2).$$

You can use this inequality, the bounds $|B_{ij}| < M$, and (8.29) to show that

$$\left| \mathbf{F}(x, y) - A(x_1, y_1) \begin{bmatrix} \bar{x} \\ \bar{y} \end{bmatrix} \right| < 2M(\bar{x}^2 + \bar{y}^2) \left| \begin{bmatrix} 1 \\ 1 \end{bmatrix} \right| = 2\sqrt{2}M(\bar{x}^2 + \bar{y}^2).$$

Thus (8.26) holds, with $K = 2\sqrt{2}M$. ■

Proposition 8.3.1 provides the motivation to explore similarities of the nonlinear system

$$\begin{bmatrix} x' \\ y' \end{bmatrix} = \mathbf{F}(x, y) \tag{8.30}$$

and the linear system

$$\begin{bmatrix} \bar{x}' \\ \bar{y}' \end{bmatrix} = A(x_1, y_1) \cdot \begin{bmatrix} \bar{x} \\ \bar{y} \end{bmatrix}. \tag{8.31}$$

Any comparison would only be valid when (x, y) is close to the stationary point (x_1, y_1). It turns out that the two systems have much in common, unless the matrix $A(x_1, y_1)$ has a characteristic root that is either zero or pure imaginary.

The system (8.30) is said to be **linearly stable** at the stationary point (x_1, y_1) if all characteristic roots of $A(x_1, y_1)$ lie to the left of the imaginary axis of the complex plane. The following theorem, which was published in 1929 by the German mathematician Oskar Perron, is a generalization of Proposition 2.5.3 on page 85.

Theorem 8.3

Let (x_1, y_1) be a linearly stable stationary point of the system (8.30) of differential equations, where the components f and g of F have continuous second partial derivatives in a rectangular region centered at (x_1, y_1). Then (x_1, y_1) is asymptotically stable.

You will be asked to prove this theorem in Exercise 25 on page 426.

If $\mathbf{v}_1 = (x_1, y_1)$ is a stationary point of a system, we will say that a point $\mathbf{v}^* = (x^*, y^*)$ is **attracted to** $\mathbf{v}_1$ if the solution $\mathbf{v}(t)$ with initial value $\mathbf{v}(0) = \mathbf{v}^*$ has the property that $\lim_{t \to \infty} \mathbf{v}(t) = \mathbf{v}_1$. The set of all points that are attracted to $\mathbf{v}_1$ is called the **attracted set** of $\mathbf{v}_1$. For example, consider a linear system $\mathbf{v}' = A\mathbf{v}$, where A is a constant matrix. The attracted set of the origin is

- the origin alone if A has no characteristic roots in the left half-plane,
- the entire plane if both characteristic roots of A are in the left half-plane, and
- the line through the origin whose direction is given by characteristic vector belonging to the negative characteristic root, if the characteristic roots of A are real and of opposite sign.

For another example, consider an asymptotically stable stationary point $\mathbf{v}_1$ of a nonlinear system. By definition, there is a disk $\mathcal{U}$ of radius r, centered at $\mathbf{v}_1$ ($\mathcal{U}$ may be very small), such that all points $\mathcal{U}$ are attracted to $\mathbf{v}_1$. Thus we can say that the attracted set of $\mathbf{v}_1$ contains a disk centered at $\mathbf{v}_1$.

A point $\mathbf{v}^*$ is **repelled by** $\mathbf{v}_1$ if the solution $\mathbf{v}(t)$ with initial value $\mathbf{v}(0) = \mathbf{v}^*$ has the property that $\lim_{t \to -\infty} \mathbf{v}(t) = \mathbf{v}_1$, and the **repelled set** is the set of points repelled by $\mathbf{v}_1$. For example, in Figure 8.11 on page 396, the repelled set of the origin appears to be the set of points bounded by the unit circle, and the attracted set is the origin alone.

Advanced texts go beyond Theorem 8.3 and show that the phase portrait of the system (8.30) and its "linearization" at $\mathbf{v}_1$, the system (8.31), share the following properties:

i. If the phase portrait of the linear system (8.31) is an unstable node, then the repelled set of $\mathbf{v}_1$ in the nonlinear system (8.30) contains a disk $\mathcal{U}$ centered at $\mathbf{v}_1$.

ii.   If the phase portrait of the linear system (8.31) is a saddle, then, in the non-linear system (8.30), the repelled set of $\mathbf{v}_1$ is a curve, whose tangent direction at $\mathbf{v}_1$ is parallel to the characteristic vector of $A(x_1, y_1)$ belonging to the positive characteristic root. The attracted set is a curve with tangent direction at $\mathbf{v}_1$ parallel to the characteristic vector belonging to the negative characteristic root. In Figure 8.12, the origin is a saddle for the linearized system. The nonlinear system is more complicated (there is a center on the positive x-axis), but the stationary point at the origin is of the saddle type. The stable and repelled sets are no longer straight lines; they consist of the orbits directed toward and away from the origin. Notice that in the right half-plane, the attracted and repelled sets coincide.

iii.   If the phase portrait of the linearized system is a spiral node, there will be a spiral node near at the stationary point $\mathbf{v}_1$ of the nonlinear system (8.30). This is illustrated by Figure 8.11, in which the node is unstable, and by Figure 8.13, where the node is stable, although orbits not starting inside the circular orbit will not converge to the origin.

iv.   If the phase portrait of the linear system (8.31) is a center, the orbits near $\mathbf{v}_1$ of the system (8.30) will swirl around $\mathbf{v}_1$, but they may not be closed. The second-order term that was dropped when passing from the nonlinear system to the linear system is small, but not too small to disrupt a family of closed orbits. This is illustrated in Figure 8.14.

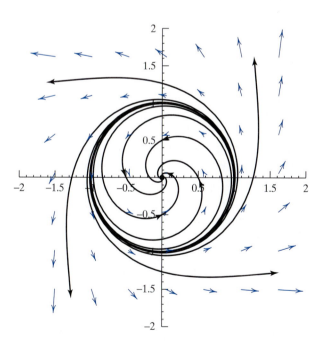

FIGURE 8.13   The origin is a stable stationary point of the system
$x' = -y - x(1 - \sqrt{x^2 + y^2})$, $y' = x - y(1 - \sqrt{x^2 + y^2})$.

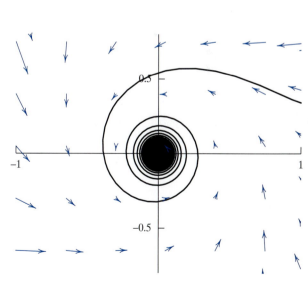

FIGURE 8.14   The system $x' = -y - x(x^2 + y^2)$, $y' = x - y(x^2 + y^2)$ has an asymptotically stable stationary point at the origin. The characteristic roots of the linearization at the origin are $\pm i$, so the linearization is a center. One orbit is shown.

▢ **Example 8.3.3** Find the stationary points of the system

$$\left. \begin{array}{l} x' = x(2x + 3y - 7) \\ y' = y(3x - 4y - 2) \end{array} \right\} \tag{8.32}$$

and determine which are stable.

Solution. The stationary points are the solutions of the equations

$$x(2x + 3y - 7) = 0$$
$$y(3x - 4y - 2) = 0.$$

Recall from Section 3.4 (page 112) that the graphs of these equations are the *null-clines* of the system. When drawing a phase portrait, it is often useful to include them. The nullclines intersect at the points $(0, 0)$, $\left(0, -\frac{1}{2}\right)$, $\left(\frac{7}{2}, 0\right)$, and $(2, 1)$. These are the stationary points.

The derivative matrix is

$$A(x, y) = \begin{bmatrix} \frac{\partial}{\partial x}x(2x + 3y - 7) & \frac{\partial}{\partial y}x(2x + 3y - 7) \\ \frac{\partial}{\partial x}y(3x - 4y - 2) & \frac{\partial}{\partial y}y(3x - 4y - 2) \end{bmatrix}$$

$$= \begin{bmatrix} 4x + 3y - 7 & 3x \\ 3y & 3x - 8y - 2 \end{bmatrix}.$$

This matrix is only of interest when (x, y) is one of the four stationary points of the system (8.32).

At the stationary point $(0, 0)$,

$$A(0, 0) = \begin{bmatrix} -7 & 0 \\ 0 & -2 \end{bmatrix},$$

which has characteristic roots -2 and -7. Since both are negative, $(0, 0)$ is a linearly stable stationary point, and by Theorem 8.3, asymptotically stable.

The characteristic roots of

$$A\left(0, -\frac{1}{2}\right) = \begin{bmatrix} -\frac{17}{2} & 0 \\ -\frac{3}{2} & 2 \end{bmatrix}$$

are $-\frac{17}{2}$ and 2. Since these have opposite signs, there is a saddle point at $\left(0, -\frac{1}{2}\right)$, with attracted set tangent to the characteristic vector $\begin{bmatrix} 7 \\ 1 \end{bmatrix}$ belonging to $-\frac{17}{2}$ and with repelled set tangent to the characteristic vector $\begin{bmatrix} 0 \\ 1 \end{bmatrix}$ belonging to 2.

The characteristic roots of $A\left(\frac{7}{2}, 0\right)$ are both positive, and it follows that there is an unstable node at the stationary point $\left(\frac{7}{2}, 0\right)$. Finally,

$$A(2, 1) = \begin{bmatrix} 4 & 6 \\ 3 & -4 \end{bmatrix},$$

with characteristic roots $\pm\sqrt{34}$ and corresponding characteristic vectors

$$\begin{bmatrix} 6 \\ \pm\sqrt{34} - 4 \end{bmatrix}.$$

This indicates that there is a saddle point at $(2, 1)$; the repelled set has slope $\frac{1}{6}(\sqrt{34}-4) \approx 0.3$ as it crosses the stationary point, and the attracted set has slope $-\frac{1}{6}(\sqrt{34}+4) \approx -1.6$ there. The phase portrait is shown in Figure 8.15.   ❑

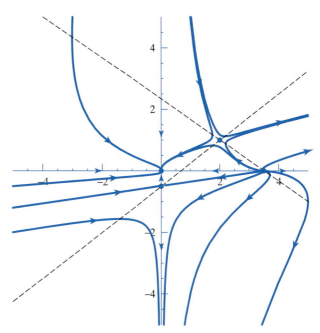

FIGURE 8.15   Phase portrait of the nonlinear system in Example 8.3.3. The dashed lines are nullclines.

Invariant sets

A subset S of the x, y-plane is an **invariant set** for a system (8.22) of differential equations if every orbit of the system that contains a point of S is a subset of S. An equivalent way of stating this definition is that S is invariant if and only if S is the union of a collection of orbits of the system.

Question   It is obvious that the union of two or more invariant sets is also an invariant set. What about the intersection?

Recall that an integral for a system of differential equations is a function $F(x, y)$ such that for any solution $x = \phi(t)$, $y = \psi(t)$, the function $F(\phi(t), \psi(t))$ is constant. It is easy to see that the level curves of F are invariant sets.

Example 8.3.4   Show that the x- and y-axes and the four quadrants of the coordinate plane are invariant sets for the system in Example 8.3.3.

404 Chapter 8 Stability Theory

Solution. The points on the y-axis have x-coordinate equal to 0. If we set $x = 0$, the system reduces to

$$x' = 0$$
$$y' = y(-4y - 2).$$

In particular, $x' = 0$ assures x is constant. Thus, if an orbit meets the y-axis, then $x \equiv 0$ on that orbit, so that it is a subset of the y-axis. Furthermore, on the y-axis, the phase portrait is just the phase diagram of $y' = -y(4y + 2)$. Similarly, the x-axis is invariant, and the the phase portrait reduces to the phase diagram of $x' = x(2x - 7)$. No orbit can cross either of the coordinate axis, and thus every orbit is confined to the quadrant of its initial point. Thus, each quadrant is invariant. $\square$

When drawing phase portraits, it is important to draw the attracted and repelled sets of each saddle point. Each of these will be a smooth curve, called a **separatrix**, passing through the saddle point, and they divide the region around the saddle point into four quadrants. In each of the quadrants, the behavior of the orbits is distinctive. To use an ODE solver to draw a separatrix, follow these steps:

Step 1 Find the characteristic vectors of the linearized system at the saddle point. Draw axes through the saddle point with these characteristic directions. These axes are tangent to the separatrices.

Step 2 Select four points $\mathbf{v}_1, \ldots, \mathbf{v}_4$ on the axes, on opposite sides of the saddle point. These points should be close to the saddle point.

Step 3 Use the ODE solver to draw the orbits passing through the points $\mathbf{v}_1, \ldots, \mathbf{v}_4$. Each orbit $\mathbf{v}_i(t)$ should have initial point $\mathbf{v}_i(0) = \mathbf{v}_i$ and should be drawn for $a \le t \le b$, where $a < 0$ and $b > 0$.

Step 4 If the preceding steps have been done correctly, the four orbits drawn will seem to merge to form the two separatrices. This method was employed in drawing the separatrices for the two saddle points in Figure 8.15.

Example 8.3.5 Find the attracted set of the origin for the system in Example 8.3.3.

Solution. In Example 8.3.4 it was shown that each quadrant of the plane is an invariant set. We will determine the portion to the attracted set of the origin in each quadrant, starting with the second quadrant and proceeding counterclockwise.

It is easy to see that at every point in the second quadrant, $y' < 0$. Thus all orbits will eventually cross the x-nullcline, $y = (7 - 2x)/3$, and after that crossing, $x' > 0$. It follows that all orbits in the second quadrant are attracted toward the origin. In other words, the attracted set of the origin contains the second quadrant.

In the third quadrant, $x' > 0$, so all orbits in that quadrant are directed to the right. One of these orbits is part of a separatrix of the stationary point $\left(0, -\frac{1}{2}\right)$. It separates the other orbits in that quadrant into two classes: Those above it are attracted to the origin, and those below it are asymptotic to the negative y-axis. The portion of the third quadrant that lies above this separatrix is also in the attracted set of the origin,

The entire fourth quadrant is contained in the repelled set of the stationary point $\left(\frac{7}{2}, 0\right)$ on the positive x-axis. It also contains part of the separatrix of the stationary point $\left(0, -\frac{1}{2}\right)$. This is an orbit that goes from $\left(\frac{7}{2}, 0\right)$ to $\left(0, -\frac{1}{2}\right)$, and the wedge-shaped region in the fourth quadrant that lies above this orbit is the portion of the attracted set of the origin in the fourth quadrant. Outside this wedge orbits are asymptotic to the negative y-axis.

Finally, in the first quadrant, the attracted set of the saddle at $(2, 1)$ forms a boundary of the attracted set of the origin. Figure 8.16 shows the attracted set.   ❏

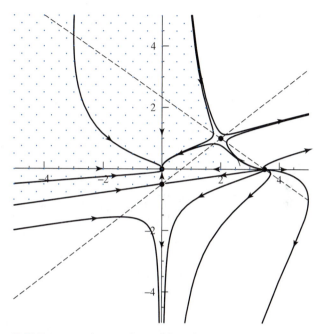

FIGURE 8.16   Attracted set of the origin for the system in Examples 8.3.3–8.3.5.

Neutrally stable stationary points

If the phase portrait of a linear system is a center, then the origin is not *asymptotically* stable, because no orbit, other than the origin itself, actually converges to the origin. However, it is desirable to have a definition of stability that encompasses centers. For example, it is in the context of such a definition that we can say that the solar system, as described by Newton's laws, is stable.[1]

> **DEFINITION**   A stationary point $(x_1, y_1) \in \mathcal{D}$ of the system (8.22) of differential equations is said to be **stable** if, given any number $R > 0$, there is a number $r > 0$ such that for any solution $(x, y) = (\phi(t), \psi(t))$ with $(\phi(0), \psi(0))$ at a distance less than r from (x_1, y_1), $(\phi(t), \psi(t))$ remains inside the circle centered at (x_1, y_1), with radius R for all $t > 0$.
>
>   A stationary point that is not stable is said to be **unstable.**

---

[1]This has never been proved, but we have several billion years of experimental evidence.

Examples of stable stationary points that are not asymptotically stable are not hard to find: See Example 8.3.6. Surprisingly, there are also examples of unstable stationary points that are asymptotically stable (see Exercise 20 at the end of this section).

Example 8.3.6 Show that the stationary point $(0, 0)$ of the system

$$x' = 36y$$
$$y' = -x$$

is stable.

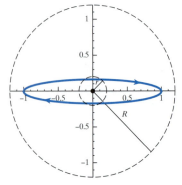

Figure 8.17. A stable stationary point. See Example 8.3.6.

Solution. The phase portrait of this system is a center; the orbits are ellipses; in fact, you can show that each orbit has equation $36x^2 + y^2 = C$, by following the procedure, established in Example 8.1.4 on page 382. The major axis of each ellipse is horizontal, and 6 times as long as the minor axis. Figure 8.17 displays one of the orbits.

Let $R > 0$ be given, and put $r = \frac{1}{6}R$. Any orbit starting inside the circle of radius r will be an ellipse whose minor semiaxis has length less than r; its major semiaxis is therefore less than R units long. Hence the entire orbit is inside the circle of radius R. See Figure 8.17. ❑

Example 8.3.7 Show that $(0, 0)$ is not a stable stationary point of the system

$$x' = y$$
$$y' = x.$$

Solution. We will show that if $R = 1$, then no matter how small r is taken to be, there are orbits with initial points inside the circle of radius r that leave the circle of radius R. In fact, the phase portrait is a saddle, so all orbits starting at a point inside the circle of radius r, except for the orbits lying on the attracted set (the line $y = -x$), are unbounded as $t \to \infty$, and eventually leave the circle of radius 1. ❑

EXERCISES

1. Find the repelled set to the stationary point of $\left(\frac{7}{2}, 0\right)$ of the system in Example 8.3.3.

In Exercises 2–13, find the stationary points, and sketch a phase portrait. In each case, is the stationary point stable? Linearly stable? Asymptotically stable? Draw the separatrices of each saddle point, and determine the attracted and repelled sets of the stable and unstable stationary points, respectively.

2. $\begin{cases} x' = x - y \\ y' = 3x - 2y \end{cases}$

3. $\begin{cases} x' = x - y \\ y' = 5x - 3y \end{cases}$

4. $\begin{cases} x' = x(1 - x) \\ y' = y(2 - y) \end{cases}$

5. $\begin{cases} x' = x + 4y \\ y' = -9x - y \end{cases}$

6. $\begin{cases} x' = x + 2y \\ y' = 2x - y \end{cases}$

7. $\begin{cases} x' = y(1 - x^2 - y^2) \\ y' = -x(1 - x^2 - y^2) \end{cases}$

8. $\begin{cases} x' = y(1 - x^2) \\ y' = -x(1 - x^2) \end{cases}$

9. $\begin{cases} x' = x(2 - x - y) \\ y' = y(x - y) \end{cases}$

10. $\begin{cases} x' = x(1 + x^2 + y^2) \\ y' = y(2 + x - y) \end{cases}$

11. $\begin{cases} x' = x(x^2 + y^2 - 10) \\ y' = y(xy - 3) \end{cases}$

12. $\begin{cases} x' = x(x+y+6) \\ y' = y(5x+y-10) \end{cases}$ **13.** $\begin{cases} x' = x(1-y) \\ y' = y(x-1) \end{cases}$

14. Show that the origin is the only stationary point of (8.23).

15. Show that *every* stationary point of the system

$$x' = y$$
$$y' = -y$$

is stable.

16. Consider the system $\begin{cases} x' = -y + kx(x^2+y^2) \\ y' = x + ky(x^2+y^2), \end{cases}$ where k is a constant (see Figure 8.14).

a. Show that Theorem 8.3 does not determine whether or not the origin is a stable stationary point.

b. Show that if we set $r^2 = x^2 + y^2$, then r satisfies the differential equation $r' = kr^3$.

c. Conclude that the phase portrait is a stable spiral node if $k < 0$, a center if $k = 0$, and an unstable spiral node if $k > 0$.

17. Analyze the system

$$x' = -y + kx(1 - x^2 - y^2)$$
$$y' = x + ky(1 - x^2 - y^2),$$

following steps similar to those taken in Exercise 16. Show that the stability of the stationary at the origin depends on k. Also show that the circle $x^2 + y^2 = 1$ is always a closed orbit. Draw typical phase portraits for the system corresponding to $k > 0$, $k = 0$, and $k < 0$.

18. Modify the proof of Proposition 2.5.2 on page 83 to show that (x^*, y^*) is a stationary point if a solution of the system (8.22) converges to it.

19. Suppose that A is a real 4×4 matrix with two double characteristic roots: $\pm i$. Show that the origin is a stable stationary point of $\mathbf{v}' = A\mathbf{v}$ if and only if there are two independent (complex) characteristic vectors belonging to i.

20. Challenging problem: Asymptotic stability does not imply stability. Show that all nonconstant orbits of

$$x' = -y + x(1 - x^2 - y^2) + \frac{xy}{\sqrt{x^2 + y^2}}$$

$$y' = x + y(1 - x^2 - y^2) - \frac{x^2}{\sqrt{x^2 + y^2}}$$

converge to the stationary point $(1, 0)$ as $t \to \infty$. Furthermore, show that $(1, 0)$ is unstable (see Figure 8.18).

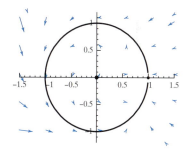

FIGURE 8.18   The stationary point at $(1, 0)$ is unstable, although all orbits except $x \equiv 0$, $y \equiv 0$ converge to it. See Exercise 20.

*8.4   Competing Species

When we considered the population dynamics of two-species systems in Section 3.5, we had not yet developed the concept of stability. With the definitions of stable and asymptotically stable stationary points in place, we will reexamine the competing species equations

$$\left. \begin{array}{l} x' = x(k - ax - by) \\ y' = y(l - cx - dy) \end{array} \right\} \tag{8.33}$$

that were derived in Section 3.5.

The four stationary points of (8.33) are the origin, where both species are extinct, $(k/a, 0)$, where the species represented by y is extinct, $(0, l/d)$, where the x-species is extinct, and a point (x_1, y_1) at the intersection of the nullclines $ax + by = k$ and $cx + dy = l$. The latter stationary point has no biological significance unless it is located in the first quadrant, but then it represents a situation where the two species share the resources. Since neither nullcline enters the third quadrant, (x_1, y_1) must lie in the first, second, or fourth quadrant. Figure 3.19 on page 128 displays the possible configurations of the nullclines. As in Figure 3.19, let $L = l/d$ be the coordinate of the stationary point on the y-axis, and $C = k/a$ be the coordinate of the stationary point on the x-axis. Also put $p = k/b$ (this is the point where the x-nullcline intersects the y-axis) and let $q = l/c$ be the point where the y-nullcline crosses the y-axis.

The derivative matrix for the system (8.33) is

$$A(x, y) = \begin{bmatrix} k - 2ax - by & -bx \\ -cy & l - cx - 2dy \end{bmatrix}.$$

Thus, $A(0, 0) = \begin{bmatrix} k & 0 \\ 0 & l \end{bmatrix}$ has two positive characteristic roots; as expected, there is an unstable node at the origin.

At the stationary point on the y-axis,

$$A(0, L) = \begin{bmatrix} k - bL & 0 \\ -cL & -l \end{bmatrix}.$$

The characteristic roots of $A(0, L)$ are $-l$ and $k - bL = b(p - L)$. The second of these characteristic roots will be negative if $p < L$, and positive if $L < p$. It follows that the stationary point at $(0, L)$ is stable if $p < L$, and is a saddle when $L < p$. Similar reasoning shows that the stationary point $(K, 0)$ is stable if $q < C$ and a saddle if $C < q$.

The fourth stationary point (x_1, y_1) lies in the first quadrant if and only if either

Case I. $p < L$ and $q < C$, or
Case II. $L < p$ and $C < q$.

Furthermore, you can calculate that

$$A(x_1, y_1) = \begin{bmatrix} -ax_1 & -bx_1 \\ -cy_1 & -dy_1 \end{bmatrix}.$$

Since $\det A(x_1, y_1) = x_1 y_1 (ad - bc)$, the characteristic roots are real with opposite sign if $ad < bc$ and (x_1, y_1) is a saddle. In Case I,

$$\frac{l}{d} > \frac{k}{b} \quad \text{and} \quad \frac{k}{a} > \frac{l}{c}.$$

Since each of the parameters a, b, c, d, k, and l is positive, we can multiply the inequalities to obtain

$$\frac{kl}{ad} > \frac{kl}{bc} \tag{8.34}$$

from which it follows that $bc > ad$. Hence, in Case I the stationary point at (x_1, y_1) is a saddle.

In Case II the inequality in (8.34) is reversed, so $\det(A(x_1, y_1)) > 0$. Hence the characteristic roots are either real with the same sign, or complex conjugate. The trace of $A(x_1, y_1)$ is negative, and thus the roots, if real, are negative, and if they are complex, the real parts are negative. The system (8.33) is therefore linearly stable in Case II. By the theorem of Perron (Theorem 8.3 on page 400), the (x_1, y_1) is an asymptotically stable stationary point in Case II.

What are the biological implications? We will start with the situation in Case I, where there is a saddle at (x_1, y_1). If the two populations started at their *exact* equilibrium values of x_1 and x_2, respectively, that equilibrium would be maintained (unless disturbed by external influences). Since the stationary point at (x_1, y_1) is unstable, coexistence is extremely improbable in this situation. Usually *competitive exclusion* occurs: The species with the initial population advantage overwhelms the other. The separatrix of (x_1, y_1) serves as a boundary between the attracted sets of $(0, L)$ and $(C, 0)$, and the destiny of the two competing species is determined by which side of the separatrix the initial population vector lies. Figure 8.19 is a phase portrait displaying the phenomenon of competitive exclusion.

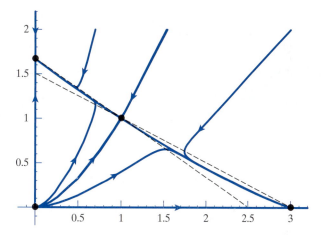

FIGURE 8.19   **Competitive exclusion.** The attracted set of the stationary point (x_1, y_1) is the separatrix. If the initial population vector lies below the separatrix, the x-species dominates and the y-species becomes extinct. If the initial population is above the separatrix, the y-species is dominant. The parameter configuration is $p < L$ and $q < C$. The nullclines are shown as dashed lines. In drawing this phase portrait, the parameter values $(k, a, b, l, c, d) = (3, 1, 2, 5, 2, 3)$ were used.

If the stationary point (x_1, y_1) does not lie in the first quadrant, the two species cannot coexist: One species will dominate and eventually exclude the other without regard to the initial condition, unless all members of the dominant species are removed from the environment. To see why this is so, refer to Figure 8.20, which is a phase portrait of the system for case where $C > q$ and $L < p$. We have noted that when the parameters are configured in this way, $(0, L)$ is a saddle and $(C, 0)$ is a stable node. Furthermore, the attracted set for $(0, L)$ is the positive y-axis, and we can infer that the repelled set is a curve extending into the first quadrant. We saw in Section 3.5 that the quadrilateral $LqCp$ is a trap, in the sense that every orbit in the first quadrant enters $LqCp$ unless it converges directly to $(C, 0)$, and no orbit can leave the quadrilateral. An orbit that enters the quadrilateral is easily seen to converge to $(C, 0)$. Since every orbit in the first quadrant must enter $LqCp$, the attracted set of $(C, 0)$ contains the entire first quadrant.

When $L < p$ and $C < q$, both stationary points $(C, 0)$ and $(0, L)$ are saddles. As long as both species are represented in the initial population vector, the populations will approach the stationary point (x_1, y_1), which is asymptotically stable: See Figure 8.21. In this case, the attracted set of (x_1, y_1) is the first quadrant.

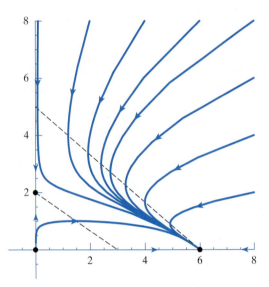

FIGURE 8.20 **Dominance.** The species with population x will outcompete its competitor unless it is entirely removed from the environment. The parameter configuration is $C > q$ and $L < p$. The nullclines are represented by dashed lines. In drawing the phase portrait, the parameter values $(k, a, b, l, c, d) = (30, 5, 6, 6, 2, 3)$ were used.

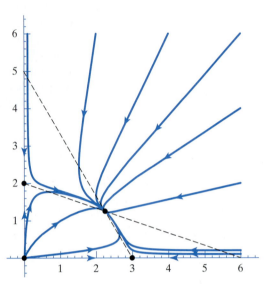

FIGURE 8.21 **Coexistence.** When $L < p$ and $C < q$ the species will coexist at the stationary point in the first quadrant. The initial population vector has no effect on the outcome, as long as both species are represented. The nullclines are represented by dashed lines. In drawing this phase portrait, the parameter values $(k, a, b, c, l, c, d) = (15, 5, 3, 12, 2, 6)$ were used.

EXERCISES

1. In Exercise 5 on page 132, we saw that the system

$$x' = ax(K - x + By)$$
$$y' = dy(L + Cx - y)$$

could serve as a model describing the population dynamics for two species in **symbiosis**. Draw a phase portrait for the system, considering the cases where $BC > 1$ and $BC < 1$ separately.

2. Draw a phase portrait for the following differential equation model for a predator-prey system, in which the prey population would have logistic growth in the absence of the predator.

$$\left. \begin{array}{l} x' = x[A(C - x) - by] \\ y' = cy(x - d) \end{array} \right\} \qquad (8.35)$$

Consider the following three cases, and draw conclusions about the fate of the predator in each.

a. $C - d < 0$

b. $0 < C - d < \dfrac{Ad}{4c}$ c. $C - d > \dfrac{Ad}{4c}$

3. In the system (8.35) (see Exercise 2), let $a = AC$. Then, as $C \to \infty$, with $A = aC^{-1}$ so that a is constant, the right side of the system (8.35) approaches the right side

of the Lotka-Volterra equations (3.23) on page 125. However, show that for any finite $C > d$, the stationary point $(x_1, y_1) = \left(d, \frac{a - Ad}{b}\right)$ of the system (8.35) is asymptotically stable, even though the corresponding stationary point of the Lotka-Volterra equations is not. We express this failure of the phase portrait of the Lotka-Volterra system to resemble the phase portrait of the system (8.35), no matter how large the carrying capacity is taken to be, by saying that the Lotka-Volterra system is not *structurally stable*.

4. Suppose that the differential equations model of a predator-prey system has a limit cycle. If the prey is a pest, how would you devise a pest control strategy?

8.5   Energy Integrals

Suppose that a particle with unit mass is in motion with one degree of freedom. This is the case if the particle is constrained to move on an axis. If the force on the particle is dependent only on its position, not on its velocity (this rules out friction), then Newton's second law of motion takes the form

$$\frac{d^2 x}{dt^2} = g(x), \tag{8.36}$$

where $g(x)$ denotes the force at position x. For example, the particle might be attached to a nonlinear spring with restoring force $g(x)$.

Solutions of equation (8.36) can be represented on the phase plane as orbits of the system

$$\left. \begin{aligned} x' &= y \\ y' &= g(x). \end{aligned} \right\} \tag{8.37}$$

Proposition 8.5.1

Orbits of the system (8.37) are directed to the right at points in the first and second quadrants, and to the left in the third and fourth quadrants. An orbit has a vertical tangent only at points where it crosses the x-axis.

Proof In the first and second quadrants, $y > 0$. Since $y = x'$, it follows that x is increasing in these quadrants; in other words, every orbit is directed to the right as it passes through the quadrants. Similarly, in the third and fourth quadrants, $x' < 0$, and it follows that an orbit will be directed to the left while in these quadrants. A vertical tangent can only appear where the orbit crosses the x-nullcline, which is the x-axis. ∎

Referring to the box on page 116, we can find an integral for the system (8.37) by solving the differential equation

$$y \, dy - g(x) \, dx = 0.$$

The result is

$$\frac{1}{2} y^2 - G(x) = C, \tag{8.38}$$

where $G(x)$ denotes an antiderivative of $g(x)$. It follows that

$$F(x, y) = \frac{1}{2}y^2 - G(x)$$

is an integral of the system (8.37).

The variable y the system (8.37) in represents the velocity of the object, and $\frac{1}{2}y^2$ is the *kinetic energy*. The *potential energy* of the particle is defined to be $U(x) = -G(x)$. Since $G(x)$ is an antiderivative, the potential energy is determined only up to the addition of a constant. The integral $F(x, y)$ represents the total mechanical energy of the particle. In deriving this integral we have established a special case of the *law of conservation of energy*.

To draw a phase portrait of the system (8.37), it is helpful to sketch the graph of the potential energy function first. Since on any orbit, the sum of the kinetic and potential energy is constant, when the potential energy increases, the kinetic energy must decrease, and vice versa. All stationary points of the system lie at points $(x_1, 0)$ on the x-axis where $g(x_1) = 0$; that is, $U'(x_1) = 0$, so that x_1 is a *critical point* of the potential energy. We will say that x_1 is an *isolated* critical point if there is an interval $(x_1 - r, x_1 + s)$ containing no other critical points.

Proposition 8.5.2

Suppose that the potential energy $U(x)$ has an isolated critical point at $x = x_1$. If $U(x_1)$ is a relative minimum, then $(x_1, 0)$ is a stable stationary point of the system (8.37). Furthermore, there are numbers $r, s > 0$ such that every orbit that crosses the x-axis in the interval $(x_1 - r, x_1 + s)$ is a closed orbit, traversed clockwise.

Proof For convenience, assume that $x_1 = 0$. Let $r > 0$ and $s > 0$ be chosen so that $U(x)$ has no other critical points in the interval $\mathcal{I} = (-r, s)$. The potential energy is then decreasing on $\mathcal{I}_- = (-r, 0)$ and increasing on $\mathcal{I}_+ = (0, s)$. We will assume that r and s are chosen so that $U(-r) = U(s)$. Suppose that an orbit crosses the x-axis at a point $x_0 \in \mathcal{I}_-$. Since the kinetic energy at the crossing is 0, the total energy of this orbit is then $U(x_0)$. As shown in Figure 8.22, the orbit cannot proceed to the left, because then both the kinetic and potential energy would increase, and the total energy would not be constant.

The orbit must therefore proceed to the right. Since a right-directed orbit must go above the x-axis, our orbit enters the upper half-plane. It continues to move to the right, reaching a maximum kinetic energy as it crosses the y-axis, where the potential energy is minimum. Then the kinetic energy (and y) will decrease, until the x-axis is crossed again. This crossing will occur at the point $x \in \mathcal{I}_+$ where $U(x) = U(x_0)$. Now the orbit cannot go farther to the right, because the potential energy cannot exceed the value $U(x_0)$. In going to the left, it enters the lower half-plane. To describe the rest of the orbit we can appeal to symmetry. The orbit is the locus of solutions of (8.38), which is obviously symmetric with respect to the x-axis. Therefore, the orbit in the lower half-plane is just the reflection of the portion already traversed in the upper half-plane, and is therefore a closed curve.

We have verified the second assertion of the proposition. It still must be proved that the stationary point at the origin is stable. This means that for any number $R > 0$, it must be possible to find a number $\epsilon > 0$ such that every orbit that starts

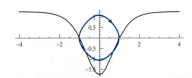

Figure 8.22 A potential well: The orbit cannot escape.

inside the circle of radius ϵ centered at the origin remains inside the circle of radius R.

We can assume that $R < r, s$—otherwise we can replace R by a smaller number that satisfies the assumption. Let

$$C_R = \{(x, y) : x^2 + y^2 = R\}$$

denote the circle of radius R, centered at the origin. Put $m = \min\{F(x, y) : (x, y) \in C_R\}$, the minimum value of the energy integral on C_R. Since there are no stationary points on C_R, $m > F(0, 0)$. Appealing to the continuity of F, there is a number $\epsilon > 0$ such that if (x, y) is inside the circle S_ϵ with radius ϵ, then $F(x, y) < m$. Any orbit that starts inside S_ϵ has total energy less than m, and since every point of C_R has total energy at least m, the orbit cannot cross C_R. ∎

Proposition 8.5.3

Let x_1 be an isolated critical point of the potential $U(x)$. If the potential does not have a relative minimum at x_1, then $(x_1, 0)$ is a not a stable stationary point of the system (8.37).

Proof As in the proof of Proposition 8.5.2, assume that $x_1 = 0$, and let $\mathcal{I} = (-r, s)$ denote an open interval containing no other critical points of $U(x)$. Either $U'(x) > 0$ on $\mathcal{I}_-$ or $U'(x) < 0$ on $\mathcal{I}_+$ since otherwise, by the first derivative test, $U(0)$ would be a relative minimum. To be definite, we will suppose that $U'(x) > 0$ on $\mathcal{I}_-$.

Let C_r denote the circle with center at the origin and radius r. We will show that for any $x_0 \in (-r, 0)$, the orbit of the solution $(x, y) = (\phi(t), \psi(t))$ of the system (8.37) with initial condition $(\phi(0), \psi(0)) = (x_0, 0)$ crosses the circle C_r. As shown in Figure 8.23, there will be two crossing points, since the orbit is symmetric with respect to the x-axis; the one in the lower half-plane is directed to the left and thus outward from the circle.

Since the initial point $(x_0, 0)$ can be made arbitrarily close to the origin, this will demonstrate instability.

The orbit through the point $(x_0, 0)$ is on the level curve of the total energy given by the equation

$$U(x) + \frac{1}{2}y^2 = U(x_0). \tag{8.39}$$

The circle S_r has equation

$$x^2 + y^2 = r^2. \tag{8.40}$$

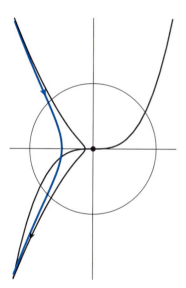

Figure 8.23 An unstable stationary point. The critical point of $U(x)$ at the origin is not a relative minimum.

We will show that there is a point (x^*, y^*), with $-r < x^* < 0$, satisfying equations (8.39) and (8.40) simultaneously. Multiply equation (8.39) by 2 and subtract it from equation (8.40) to eliminate the variable y and get

$$x^2 - 2U(x) = r^2 - 2U(x_0). \tag{8.41}$$

Let $P(x)$ denote the expression on the left side of equation (8.41). Notice that

$$P(x_0) = x_0^2 - 2U(x_0) < r^2 - 2U(x_0).$$

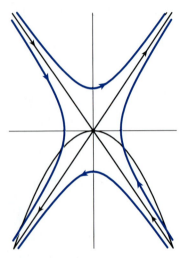

Figure 8.24. A separatrix: $U(x)$ has a relative maximum at x_1.

Also, since $U(-r) < U(x_0)$,

$$P(-r) = r^2 - 2U(-r) > r^2 - 2U(x_0).$$

Appealing to the continuity of $P(x)$ and the intermediate value theorem, there is a solution $x^* \in (-r, x_0)$ of equation (8.41); the orbit through the point $(x_0, 0)$ crosses S_r at $(x^*, \pm y^*)$, where $y^* = \sqrt{2[U(x_0) - U(x^*)]}$. ∎

If x_1 is an isolated critical point of $U(x)$, not a relative minimum, then $(x_1, 0)$ lies on a level energy curve

$$\frac{1}{2}y^2 + U(x) = U(x_1).$$

This level curve is the *separatrix* of the stationary point $(x_1, 0)$. If $U(x)$ has a relative maximum at x_1, then $(x_1, 0)$, is a saddle point, as shown in Figure 8.24. If x_1 is an inflection point of $U(x)$, the separatrix has only one branch; this is illustrated by Figure 8.23.

Example 8.5.1   Draw a phase portrait for the differential equation

$$x'' = 6x(1 - x).$$

Solution. We will draw the phase portrait of the system

$$x' = y$$
$$y' = 6x(1 - x).$$

The potential is $U(x) = -\int_0^x 6z(1 - z)\,dz = 2x^3 - 3x^2$, which has a relative maximum at $x = 0$ and a minimum at $x = 1$. Hence the stationary point at $(0, 0)$ is unstable, while the stationary point $(1, 0)$ is stable. The separatrix corresponding to the stationary point at the origin is the cubic curve

$$\frac{1}{2}y^2 + 2x^3 - 3x^2 = 0.$$

This curve consists of three orbits in addition to the origin itself. The portion of the curve in the second quadrant is an orbit directed toward the origin, and forms part of the attracted set. The intersection of the curve with the third quadrant is an orbit directed away from the origin: It is part of the repelled set. The situation in the right half-plane seems paradoxical but actually represents a relatively common occurrence: The portion of the separatrix in the right half-plane is a single orbit, which converges to the origin as $t \to \infty$ and as $t \to -\infty$; it therefore lies in the intersection of the attracted and repelled sets.

The orbits inside the separatrix in the right half-plane are closed; all other orbits were drawn by reference to the potential function. The phase portrait is shown in Figure 8.12 on page 397. ❑

The pendulum

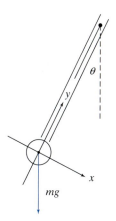

Figure 8.25 shows a pendulum. L denotes the length of the pendulum arm, θ the angle made by the arm with the vertical axis, and m the mass.

The force of gravity on the bob is $\mathbf{F} = mg$, directed downward. We will use a moving coordinate system in which the origin is the center of the bob, with the y-axis is aligned along the pendulum arm, and the x-axis tangent to the trajectory of the pendulum. In this coordinate system,

$$\mathbf{F} = -mg(\sin(\theta)\mathbf{i} + \cos(\theta)\mathbf{j}).$$

The net force on the pendulum bob is $\mathbf{F} + T\mathbf{j}$, where $T\mathbf{j}$, the support provided by the pendulum arm, is exactly canceled by the $\mathbf{j}$-component of $\mathbf{F}$. Thus, the net force is equal to $-mg\sin(\theta)\mathbf{i}$.

By Newton's second law of motion, the net force on the pendulum bob is equal to its mass multiplied by its acceleration. The acceleration is the product $L\theta''$ of the angular acceleration θ'' of the pendulum arm and its length L. Therefore, the motion of the pendulum is governed by the differential equation $mL\theta'' = -mg\sin\theta$, or

$$\theta'' = -\frac{g}{L}\sin\theta. \tag{8.42}$$

The potential energy of the pendulum, normalized so that $U(0) = 0$, is

$$U(\theta) = \int_0^\theta \frac{g}{L}\sin(z)\,dz = \frac{g}{L}(1 - \cos\theta).$$

As we would expect, the stationary points at $\theta = 2n\pi$, when the pendulum is at the bottom of its trajectory, correspond to minima of the potential energy and are stable. The stationary points at odd multiples of π (where the pendulum is inverted) correspond to maxima of the potential energy and are unstable. The separatrix is the curve

$$\frac{1}{2}\omega^2 - \frac{g}{L}\cos\theta = \frac{g}{L},$$

in the phase plane (the variable $\omega = \theta'$ represents angular velocity). In this case, the separatrix separates the closed (periodic) orbits from those that are not periodic. See Figure 8.26.

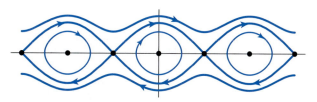

FIGURE 8.26 Phase portrait of the pendulum equation. The variables θ and ω represent angular displacement and angular velocity, respectively.

Figure 8.25 A simple pendulum.

EXERCISES

1. Let x_1 denote an isolated critical point of $U(x)$. Is it possible for the stationary point $(x_1, 0)$ of the system (8.37) to be asymptotically stable?

2. Suppose that $U(x)$ has a *nondegenerate* critical point at $(x_1, 0)$; that is, $U'(x_1) = 0$ and $U''(x_1) \neq 0$. Let A denote the Jacobian matrix of the system (8.37) at $(x_1, 0)$. Show that if $U(x)$ has a relative maximum at x_1, then the characteristic roots of A have opposite sign, while if $U(x)$ has a relative minimum at x_1, then the characteristic roots are pure imaginary.

3. Show that if $g(x_1) = 0$ and $g'(x_1) < 0$, then $(x_1, 0)$ is a stable stationary point of the system (8.37).

In Exercises 4–19, find a potential energy function and sketch its graph. Then draw a phase portrait for the differential equation.

4. $x'' = ax$, where a is a positive constant

5. $x'' = ax$, where a is a negative constant

6. $x'' = 0$ 7. $x'' = 1$

8. $x'' = x^2$ 9. $x'' = -x^2$

10. $x'' = 4x^3$ 11. $x'' = -4x^3$

12. $x'' = 4x - x^3$. Show that the attracted and repelled sets of the origin are the same.

13. $x'' = 3x|x|$ 14. $x'' = e^{-x}(\cos x + \sin x)$

15. $x'' = \dfrac{x^2 - 1}{(x^2 + 1)^2}$. To save time in computing the potential energy, note that $\dfrac{x^2 - 1}{(x^2 + 1)^2} = \text{Re}\left[\dfrac{1}{(x + i)^2}\right]$.

16. $x'' = x^2 \sin x - 2x \cos x$ 17. $x'' = \sin x - \frac{1}{2}$

18. $x'' = \sin x - 1$ 19. $x'' = \sin x - \left(\frac{2}{\pi}\right)x$

20. Suppose that the potential energy function $U(x)$ for the differential equation (8.36) has a relative maximum at $x = x_1$. Show that the when the two branches of the separatrix cross the x-axis of the phase plane at $x = x_1$, their slopes are $\pm\sqrt{g'(x_1)}$.

8.6 Lyapunov Stability Tests

Let θ denote the angular displacement of a pendulum arm from the vertical position. If the pendulum is subject to linear friction, then θ satisfies the differential equation

$$\frac{d^2\theta}{dt^2} = -b\frac{d\theta}{dt} - k\sin\theta, \tag{8.43}$$

where b is the friction constant, and $k = g/L$.

Let ω denote the angular velocity of the pendulum. The phase portrait of the differential equation (8.43) is that of the system

$$\left. \begin{aligned} \frac{d\theta}{dt} &= \omega \\[2mm] \frac{d\omega}{dt} &= -b\omega - k\sin\theta. \end{aligned} \right\} \tag{8.44}$$

Mechanical energy is not conserved when this pendulum is in motion, because the friction dissipates kinetic energy as heat. The total energy,

$$F(\theta, \omega) = \frac{1}{2}\omega^2 + k(1 - \cos\theta),$$

is an integral for the motion of the frictionless pendulum, but it cannot be expected to serve as an integral when there is friction.

We can validate this physical intuition by differentiating $F(\theta, \omega)$.

$$\frac{d}{dt}F(\theta, \omega) = \omega \frac{d\omega}{dt} + k \sin\theta \frac{d\theta}{dt}$$
$$= \omega[-b\omega - k\sin\theta] + k(\sin\theta)\omega$$
$$= -b\omega^2 \qquad (8.45)$$

It follows that $F(\theta, \omega)$ decreases at a rate proportional to ω^2 and is not an integral.

When $(\theta, \omega) = (2n\pi, 0)$ the pendulum is motionless at the bottom of its trajectory. Physical intuition suggests that if the pendulum starts with a modest initial velocity near this equilibrium position, it will eventually stop there. In other words, we expect the stationary point $(2n\pi, 0)$ to be asymptotically stable. It is possible to prove this by using Theorem 8.3, but we will explore a second method, due to Lyapunov, which focuses on the mechanical energy $F(\theta, \omega)$.

Lyapunov functions

Let $L(x, y)$ be a function with $L(0, 0) = 0$. If there is a number $R > 0$ with $L(x, y) > 0.$ for all points (x, y) except the origin inside the circle $x^2 + y^2 = R^2$, then $L(x, y)$ is said to be **positive definite** near the origin. If the inequality is not strict, and we only have $L(x, y) \geq 0$ for $x^2 + y^2 < R^2$, then $L(x, y)$ is **positive semidefinite** near the origin.

We also define a function $N(x, y)$, with $N(0, 0) = 0$, to be negative definite or semidefinite if $-N(x, y)$ is positive definite of semidefinite, respectively. A function $L(x, y)$ with $L(x_1, y_1) = 0$ is positive or negative (semi)definite near (x_1, y_1) if the function $F(x, y) = L(x + x_1, y + y_1)$ has the corresponding property near the origin.

Consider a system of differential equations,

$$\left. \begin{array}{l} x' = f(x, y) \\ y' = g(x, y). \end{array} \right\} \qquad (8.46)$$

Let $L(x, y)$ be a function whose partial derivatives $\frac{\partial L}{\partial x}$ and $\frac{\partial L}{\partial y}$ are continuous, and define the function

$$L'(x, y) = \frac{\partial L}{\partial x}f(x, y) + \frac{\partial L}{\partial y}g(x, y).$$

Proposition 8.6.1

Let $(x, y) = (\phi(t), \psi(t))$ be a solution of (8.46). Then

$$\frac{d}{dt}L(\phi(t), \psi(t)) = L'(\phi(t), \psi(t)).$$

Proof By the chain rule,

$$\frac{dL}{dt}(x, y) = \frac{\partial L}{\partial x}\frac{dx}{dt} + \frac{\partial L}{\partial y}\frac{dy}{dt}$$

$$= \frac{\partial L}{\partial x}f(x, y) + \frac{\partial L}{\partial y}g(x, y)$$

$$= L'(x, y).$$

Now set $x = \phi(t)$, $y = \psi(t)$, and the proof is complete. ∎

DEFINITION A *Lyapunov function* for a stationary point (x_1, y_1) of an autonomous system of differential equations is a function $L(x, y)$ that is positive definite near (x_1, y_1), while $L'(x, y)$ is either positive or negative semidefinite.

Example 8.6.1 The total energy, $F(\theta, \omega) = \frac{1}{2}\omega^2 + k(1 - \cos\theta)$, is a Lyapunov function for the pendulum system (8.44), for each of the stationary points $(2n\pi, 0)$.

Solution. We will confine our attention to the stationary point at the origin. Let $R = 2\pi$. Then for $0 < |\theta| < R$, we have $1 - \cos\theta > 0$, and hence $F(\theta, \omega) > 0$ for $0 < \theta^2 + \omega^2 < R^2$.

Since $\frac{\partial F}{\partial \theta} = k\sin(\theta)$, and $\frac{\partial F}{\partial \omega} = \omega$, we have

$$F'(\theta, \omega) = (k\sin\theta)(\omega) + (\omega)(-b\omega - k\sin\theta)$$

$$= -b\omega^2.$$

Thus F' is negative semidefinite. Because it vanishes on the θ-axis, F' is not negative definite. □

The first stability test of Lyapunov is as follows.

Proposition 8.6.2

Suppose that $L(x, y)$ is a Lyapunov function for a stationary point (x_1, y_1) of the system (8.46), where $L'(x, y)$ is negative semidefinite near (x_1, y_1). Then (x_1, y_1) is stable.

For simplicity, we will assume in the proof of this proposition, and of subsequent results in this section, that the point (x_1, y_1) is the origin. The set of points in the phase plane that lie inside the circle with radius R, centered at the origin, will be denoted $\mathcal{B}_R$. The boundary circle is $\mathcal{C}_R$.

Proof of Proposition 8.6.2. Choose $R > 0$ so that for all $(x, y) \in \mathcal{B}_R \cup \mathcal{C}_R$ with $(x, y) \neq (0, 0)$, $L(x, y) > 0$ and $L'(x, y) \leq 0$. Let m denote the minimum value of $L(x, y)$ for $(x, y) \in \mathcal{C}_R$. Since the origin does not lie on $\mathcal{C}_R$, this number m is positive. Furthermore, since L is continuous and $L(0, 0) = 0$, there is a number r, with $0 < r \leq R$, such that for all points $(x, y) \in \mathcal{B}_r$, $L(x, y) < m$. Let

$(x, y) = (\phi(t), \psi(t))$ be a solution of the system (8.46) with $(\phi(0), \psi(0)) \in \mathcal{B}_r$. By Proposition 8.6.1,

$$\frac{d}{dt} L(\phi(t), \psi(t)) = L'(\phi(t), \psi(t)) \leq 0,$$

provided that $(\phi(t), \psi(t)) \in \mathcal{B}_R$. Therefore,[2] $L(\phi(t), \psi(t))$ cannot increase unless the orbit of $(\phi(t), \psi(t))$ crosses $\mathcal{C}_R$, and the circle $\mathcal{C}_R$ cannot be reached unless $L(\phi(t), \psi(t))$ increases to at least m. ∎

Proposition 8.6.2 and Example 8.6.1 together imply that the lower equilibrium position of the pendulum is stable. Since we expect asymptotic stability, this is disappointing, but we will see that the method of Lyapunov can be refined to obtain the desired result.

Example 8.6.2   Show that $L(x, y) = x^2 + y^2$ is a Lyapunov function for the origin as a stationary point of the system

$$x' = y$$
$$y' = -x$$

and conclude that the origin is stable.

Solution. Since $L(x, y) > 0$ for all $(x, y) \neq 0$, L is positive definite. Further, $L'(x, y) = (2x) \cdot y + (2y) \cdot (-x) \equiv 0$. Therefore, $L'(x, y) \leq 0$ for all (x, y), and, by Proposition 8.6.2, the origin is stable. ❑

The result of Example 8.6.2 is not surprising, because the phase portrait is a center, and the orbits are circles—level curves of $L(x, y)$.

Limit sets

A point $(\bar{x}, \bar{y})$ is said to be a **forward limit point** of a solution $(x, y) = (\phi(t), \psi(t))$ of the system (8.46) if there is a sequence $t_n \to \infty$ such that

$$\lim_{n \to \infty} (\phi(t_n), \psi(t_n)) = (\bar{x}, \bar{y}).$$

The set of all forward limit points is denoted $\lim_+(\phi, \psi)$.

Here are some examples to consider. Proofs are omitted.

- If a stationary point is asymptotically stable, it is the only forward limit point of each orbit whose initial point is sufficiently near to it.
- A saddle point of a linear system is the forward limit point of any solution in its attracted set. Orbits that do not lie on the attracted set may not have any forward limit points.
- Some solutions of nonlinear systems are are only defined on an interval (a, b) with $b < \infty$; these automatically have empty forward limit sets.

---

[2]Readers of *Catch-22*, by Joseph Heller, will recognize this line of thought.

- It is possible for the forward limit set of an orbit to contain more than one point. For example, the forward limit set of a periodic orbit is the entire orbit.
- If an orbit converges to a limit cycle, as in Figure 8.11 on page 396, each point of the limit cycle is a forward limit point.

The following two propositions are presented without proof. The first says that an orbit passing close enough to a stable stationary point must have forward limit points. These forward limit points are not necessarily stationary points themselves. For example, consider the case where the equilibrium is *neutrally stable*: The limit points may lie in a periodic orbit. Of course, if the stationary point is asymptotically stable, all nearby orbits converge to it.

Proposition 8.6.3

Suppose that (x_1, y_1) is a stable stationary point of the system (8.46). Then there is a circle C centered at (x_1, y_1) such that if $(x, y) = (\phi(t), \psi(t))$ is a solution of the system (8.46) starting inside C, then

$$\lim{}_+(\phi, \psi) \neq \varnothing.$$

The forward limit set of an orbit may contain more than one point, but according to the next proposition, it must be an invariant set. In other words, if a point x_1 belongs to the forward limit set of an orbit, then the entire orbit determined by x_1 is either equal to, or a subset of, the forward limit set. The best example to visualize for this proposition is the limit cycle. Referring to the phase portrait of Figure 8.11 on page 396, suppose x_0 is a point, other than the origin, and *not* on the limit cycle. Then the orbit of x_0 has the limit cycle as its forward limit set. There are examples in which the forward limit set of an orbit contains an infinite number of distinct orbits, but these are beyond the scope of this text.

Proposition 8.6.4

Assume that the functions $f(x, y)$ and $g(x, y)$ on the right side of the system (8.46) have continuous partial derivatives everywhere in the x, y-plane. Then the forward limit set $\lim_+(\phi, \psi)$ of any solution $(\phi(t), \psi(t))$ is an invariant set.

The test for asymptotic stability

We can now formulate a criterion for asymptotic stability by combining the concepts of limit sets and Lyapunov functions.

Proposition 8.6.5

Let (x_1, y_1) be a stationary point of the system (8.46). Suppose that there is a Lyapunov function $L(x, y)$ for (x_1, y_1) such that $L'(x, y)$ is negative semidefinite near (x_1, y_1), and that there is a radius $R > 0$ such that the only nonempty invariant subset of

$$A_R = \mathcal{B}_R \cap \{(x, y) : L'(x, y) = 0\}$$

is $\{(x_1, y_1)\}$ itself ($\mathcal{B}_R$ denotes the set of points inside the circle with center (x_1, y_1) and radius R). Then (x_1, y_1) is asymptotically stable.

The proof depends on the following lemma.

Lemma 8.6.6

Let $g(t)$ be a continuous, nonincreasing function, defined for all $t > 0$. Suppose that $g(t) \geq 0$ for all $t > 0$. Then $\lim_{t \to \infty} g(t)$ exists.

The proof of Lemma 8.6.6 is omitted. It is based on a fundamental property of the real number system: Every set that has a lower bound has a greatest lower bound.

Proof of Proposition 8.6.5.   Again, assume that $(x_1, y_1) = (0, 0)$. Let R be a radius such that A_R, the set of points $(x, y) \in \mathcal{B}_R$ where $L'(x, y) = 0$, contains no invariant set except $(0, 0)$. By Proposition 8.6.2, this stationary point is stable; hence there is an $r > 0$ such that if $(x, y) = (\phi(t), \psi(t))$ is a solution of the system (8.46) with

$$(\phi(0), \psi(0)) \in \mathcal{B}_r \tag{8.47}$$

then

$$(\phi(t), \psi(t)) \in \mathcal{B}_R \tag{8.48}$$

for all $t > 0$.

We will show that any solution $(x, y) = (\phi(t), \psi(t))$ satisfying the condition (8.47) converges to the origin. Since condition (8.48) then holds for all $t > 0$, $L(\phi(t), \psi(t)) \geq 0$ and $L'(\phi(t), \psi(t)) \leq 0$. It follows that $L(\phi(t), \psi(t))$ is a nonincreasing nonnegative function. Therefore, $\lim_{t \to \infty} L(\phi(t), \psi(t))$ exists, by Lemma 8.6.6. Denote this limit by l.

If $(\bar{x}, \bar{y})$ is a forward limit point of the solution $(\phi(t), \psi(t))$, then there is a sequence $t_n \to \infty$ such that $(\phi(t_n), \psi(t_n)) \to (\bar{x}, \bar{y})$. Therefore, for any $(\bar{x}, \bar{y}) \in \lim_+(\phi, \psi)$,

$$L(\bar{x}, \bar{y}) = \lim_{n \to \infty} L(\phi(t_n), \psi(t_n)) = l.$$

Let $(x, y) = (\bar{\phi}(t), \bar{\psi}(t))$ be the solution of the system (8.46) with initial condition $(\bar{\phi}(0), \bar{\psi}(0)) = (\bar{x}, \bar{y})$. Since, by Proposition 8.6.4, $\lim_+(\phi, \psi)$ is invariant, $(\bar{\phi}(t), \bar{\psi}(t)) \in \lim_+(\phi, \psi)$ for all t. It follows that $L(\bar{\phi}(t), \bar{\psi}(t)) \equiv l$ for all t; differentiating, we have $L'(\bar{\phi}(t), \bar{\psi}(t)) \equiv 0$. It follows that $L'(\bar{x}, \bar{y}) = 0$; thus $(\bar{x}, \bar{y}) \in A_R$.

We have shown that the forward limit set of any orbit starting in $\mathcal{B}_r$ is an invariant subset of A_R. By Proposition 8.6.3, this forward limit set cannot be empty; therefore, it must be $\{(0, 0)\}$. It follows that the origin is asymptotically stable.   ∎

Example 8.6.3   Show that the stationary points $(\theta, \omega) = (2n\pi, 0)$ of the system (8.44), which describes the motion of a linearly damped pendulum, are asymptotically stable.

Solution.   We have seen in Example 8.6.1 that the total energy $F(\theta, \omega) = \frac{1}{2}\omega^2 + k(1 - \cos\theta)$ is a Lyapunov function for each of these stationary points, and $F'(\theta, \omega) = -b\omega^2$ is negative semidefinite. To apply Proposition 8.6.5, it is necessary to find a number R such that the only nonempty invariant subset of A_R is the stationary point $(2n\pi, 0)$.

The function F' vanishes only on the θ-axis. The invariant subsets of this axis are composed of the equilibria $(n\pi, 0)$. Hence, if $R < \pi$, the set A_R will only contain one stationary point, and thus only one nonempty invariant subset. The equilibrium point, $(2n\pi, 0)$ is therefore asymptotically stable. Figure 8.27 is a phase portrait for equation (8.44). ❑

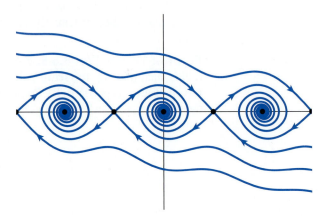

FIGURE 8.27 A phase portrait of the linearly damped pendulum equation (8.44).

There is a Lyapunov test for instability:

Proposition 8.6.7

Suppose that $L(x, y)$ is a Lyapunov function for the stationary point (x_1, y_1) of the system (8.46), such that $L'(x, y)$ is positive definite near (x_1, y_1). Then (x_1, y_1) is not stable.

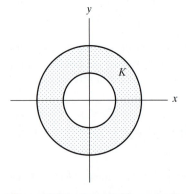

Figure 8.28. L and L' are both positive definite. In the annulus K, $L'(x, y) \geq k$, and $L(x, y) \leq M$.

Proof As usual, we will take (x_1, y_1) to be the origin. Choose $R > 0$ such that $L(x, y) > 0$ and $L'(x, y) > 0$ for all $(x, y) \in \mathcal{B}_R$ or on its boundary circle $\mathcal{C}_R$ [except for $(x, y) = (0, 0)$]. We will show that every orbit that starts at a point inside $\mathcal{B}_R$ other than the origin eventually crosses the circle $\mathcal{C}_R$.

Let M be the maximum value of $L(x, y)$ for $(x, y) \in \mathcal{B}_R \cup \mathcal{C}_R$. Suppose that $(x, y) = (\phi(t), \psi(t))$ is a solution of the system (8.46) with $(\phi(0), \psi(0)) = (x_0, y_0) \in \mathcal{B}_R$. Assume that $(x_0, y_0) \neq (0, 0)$, and put $m = L(x_0, y_0)$. Since L is positive definite, $m > 0$. A postive number r can be chosen such that $L(x, y) < \frac{m}{2}$ for all $(x, y) \in \mathcal{B}_r$, because L is continuous and $L(0, 0) > 0$.

Let K denote the annulus $\{(x, y) : r \leq \sqrt{x^2 + y^2} \leq R\}$ (see Figure 8.28). We will reach a contradiction by assuming that $(\phi(t), \psi(t)) \in \mathcal{B}_R$ for all $t > 0$. Since $L'(x, y) > 0$ for $(x, y) \in \mathcal{B}_R$ it follows that $L(\phi(t), \psi(t))$ is an increasing function that starts with $L(\phi(0), \psi(0)) = m$. The trajectory cannot cross the inner boundary of K without entering $\mathcal{B}_r$, where $L(x, y) < m$ and hence $L(\phi(t), \psi(t)) \in K$ for all $t > 0$.

Let k be the minimum value of $L'(x, y)$ for $(x, y) \in K$. Then $k > 0$, because $L'(x, y)$ vanishes only at the origin. Put $T = \frac{M-m+1}{k}$. Then

$$L(\phi(T), \psi(T)) = m + \int_0^T L'(\phi(t), \psi(t))\, dt$$
$$\geq m + kT = M + 1.$$

This is impossible, because $L(x, y) \leq M$ for all $(x, y) \in K$.   ■

Example 8.6.4   Let $q(x, y)$ be a function with continuous partial derivatives. Show that $L(x, y) = x^2 + y^2$ is a Lyapunov function for the origin as a stationary point of the system

$$\left. \begin{array}{l} x' = -y + xq(x, y) \\ y' = x + yq(x, y) \end{array} \right\} \qquad (8.49)$$

if either

a. q is positive definite, or
b. q is negative definite.

Use this Lyapunov function to determine whether or not the origin is stable, asymptotically stable, or unstable.

Solution. The function $L(x, y) = x^2 + y^2$ is positive definite, and

$$L'(x, y) = 2x(y + xq(x, y)) + 2y(-x + yq(x, y))$$
$$= 2(x^2 + y^2)q(x, y)$$

is positive (negative) definite if $q(x, y)$ is positive (negative) definite. Therefore, $L(x, y)$ is a Lyapunov function in either case. By Lyapunov's stability tests, the origin is an unstable stationary point if q is positive definite, and it is asymptotically stable if q is negative definite.   ❑

The origin is not a linearly stable stationary point of the system (8.49). A brief calculation shows that the derivative of $\mathbf{F}(x, y) = \begin{bmatrix} -y + xq(x, y) \\ x + yq(x, y) \end{bmatrix}$ is

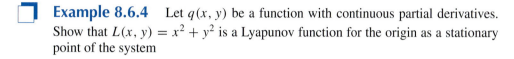

$$A(x, y) = \begin{bmatrix} xq_x(x, y) + q(x, y) & 1 + xq_y(x, y) \\ -1 + yq_x(x, y) & yq_y(x, y) + q(x, y) \end{bmatrix},$$

where q_x and q_y denote the partial derivatives of q. Since $q(0, 0) = 0$, $A(0, 0) = \begin{bmatrix} 0 & 1 \\ -1 & 0 \end{bmatrix}$, with characteristic roots $\pm i$. Since the characteristic roots are not in the left half-plane, this does not imply linear stability. Indeed, $A(0, 0)$ does not involve q at all, whereas we have seen that the stability of the origin depends on q.

Example 8.6.5 Show that $L(x, y) = x^2 + y^2$ is a Lyapunov function for the system

$$\left. \begin{array}{l} x' = -3kx + y + kx^3 \\ y' = -x \end{array} \right\} \tag{8.50}$$

and use it to draw conclusions about the stability of the equilibrium at the origin.

Solution. $L(x, y)$ is obviously positive definite, and

$$L'(x, y) = 2x(-3kx + y + kx^3) - 2xy$$
$$= 2kx^2(-3 + x^2).$$

If $k > 0$, $L'(x, y) \le 0$ for $|x| < \sqrt{3}$, with equality only if $x = 0$. Thus the origin is stable by Proposition 8.6.2. To apply Proposition 8.6.5, the invariant subsets of the y-axis, where $L'(x, y)$ vanishes, must be determined. If $x = 0$, the first equation reduces to $x' = y$. If $y \ne 0$, an orbit starting at $(0, y)$ cannot be a subset of the y-axis, since any orbit on the y-axis must have $x' = 0$. Therefore, the origin is the only invariant subset of the y-axis, and we conclude that if $k > 0$, then the origin is asymptotically stable.

If $k < 0$, the origin is not stable. Unfortunately, L' is not positive definite near the origin (it is only positive semidefinite), so we cannot apply Proposition 8.6.7 to reach this conclusion. Instead, we will assume stability and reach a contradiction. By Proposition 8.6.4, there is a number $r > 0$ such that if $(x, y) = (\phi(t), \psi(t))$ is a solution of the system with $(\phi(0), \psi(0)) \in B_r(0, 0)$, then $\lim_+(\phi, \psi) \ne \varnothing$. Reasoning similar to the proof of Proposition 8.6.5 shows that $\lim_+(\phi, \psi)$ must be an invariant subset of the y-axis. The only such subset is the origin. However, if $(\phi(0), \psi(0)) \ne (0, 0)$, then $L(\phi(t), \psi(t)) \ge L(\phi(0), \psi(0)) > 0$, and it follows that $\lim_+(\phi, \psi) \ne \{(0, 0)\}$, a contradiction. ∎

Notice that if $k = 0$, the system (8.50) is the linear system $x' = y$, $y' = -x$, whose orbits are circles. This system is neutrally stable. The derivative matrix of the system (8.50) is

$$A(x, y) = \begin{bmatrix} 3k(x^2 - 1) & 1 \\ -1 & 0 \end{bmatrix}.$$

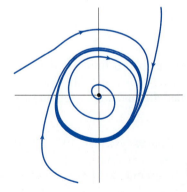

Figure 8.29. Phase portrait of the system (8.50). The parameter $k = -0.2$.

The characteristic polynomial of $A(0, 0)$ is $p(s) = s^2 + 3ks + 1$. The characteristic roots are complex for $|k| < \frac{2}{3}$, double for $k = \pm\frac{2}{3}$, and a pair of real numbers for $|k| > \frac{2}{3}$. The roots have negative real parts for $k > 0$ and positive real parts for $k < 0$. Therefore, Theorem 8.3 is applicable in this case, and the phase portrait will be an unstable node if $k < 0$ and a stable node if $k > 0$, of spiral type if $|k| < \frac{2}{3}$. See Figure 8.29.

Finding Lyapunov functions

The principal difficulty in using the Lyapunov tests for stability is in finding a Lyapunov function. There is no general procedure, but systems that arise in applications are frequently accompanied by natural Lyapunov functions. For example, the total mechanical energy of the damped pendulum is a Lyapunov function.

If the functions $f(x, y)$ and $g(x, y)$ on the right side of the system (8.46) are polynomials in x and y, there may be a Lyapunov function of the form $L(x, y) = a(x-x_1)^2 + b(x-x_1)(y-y_1) + c(y-y_1)^2$. You can show that this function is definite near (x_1, y_1) if and only if $b^2 - 4ac < 0$. Of course, L' will not be quadratic (unless we are dealing with a linear system). The following proposition makes it possible to use this test for definiteness for some non-quadratic functions. A function $P(x, y)$ is said to *vanish to order 2* at the origin if $P(0, 0) = 0$, and

$$\frac{\partial P}{\partial x}(0, 0) = \frac{\partial P}{\partial y}(0, 0) = \frac{\partial^2 P}{\partial^2 x}(0, 0) = \frac{\partial^2 P}{\partial x \partial y}(0, 0) = \frac{\partial^2 P}{\partial^2 x}(0, 0) = 0.$$

Proposition 8.6.8

Let $L_2(x, y) = ax^2 + bxy + cy^2$ and let

$$L(x, y) = L_2(x, y) + P(x, y),$$

where $P(x, y)$ vanishes to order 2 at the origin. Then

- If L_2 is positive or negative definite, then L has the same property near the origin.
- If L_2 is positive semidefinite, L is not necessarily positive semidefinite near the origin.

The proof of the proposition is left to you; see Exercise 24 at the end of this section for hints. To see that there can be no conclusion when L_2 is only semidefinite, consider $L(x, y) = x^3 + y^3$, which is neither positive nor negative semidefinite. If we put $L_2(x, y) = 0$ (this is both positive and negative semidefinite), then $L(x, y) = L_2(x, y) + P(x, y)$, where $P(x, y) = x^3 + y^3$ vanishes to order 2 at the origin.

Example 8.6.6   Find a Lyapunov function of the form $L(x, y) = ax^2 + cy^2$ for the system

$$x' = -x + 4y + x^3 + xy^2$$
$$y' = -x - 2y + 2y^3 + 2x^2y$$

and determine the stability of the origin.

Solution. Since the function $L(x, y) = ax^2 + cy^2$ is required to be positive definite, a and c must be positive. We will now calculate L'.

$$L'(x, y) = 2ax(-x + 4y + x^3 + xy^2) + 2by(-x - 2y + 2y^3 + 2x^2y)$$
$$= -2ax^2 + (8a - 2c)xy - 4cy^2 + 2ax^4 + (2a + 4c)x^2y^2 + 4cy^4$$

By Proposition 8.6.8, $L'(x, y)$ will be positive or negative definite if the expression obtained by deleting the third and higher-order terms is also positive or negative definite. We will therefore try to adjust the parameters a and c so that the expression

$$L'_2(x, y) = -2ax^2 + (8a - 2c)xy - 4cy^2$$

is negative definite (positive definite is out of the question, because the coefficients of x^2 and y^2 are negative). Observe that if we put $a = 1$ and $c = 4$, the coefficient of xy will vanish, giving the expression $L_2'(x, y) = -2x^2 - 16y^2$, which is negative definite. Therefore, $L'(x, y)$ is also negative definite, and hence $L(x, y) = x^2 + 4y^2$ is a Lyapunov function that shows the origin to be asymptotically stable.   ❑

EXERCISES

1. Show that $L(x, y) = ax^2 + bxy + cy^2$ is positive definite near the origin if and only if $b^2 - 4ac < 0$ and $a > 0$.

2. Suppose that $L(x, y)$ is a Lyapunov function for an stationary point (x_1, y_1) of a system of differential equations. Show that if $L' \equiv 0$, then L is an integral, and that (x_1, y_1) is stable but not asymptotically stable.

In each of Exercises 3–8, show that $L(x, y) = x^2 + y^2$ is a Lyapunov function for the given system at the origin, and draw a conclusion about stability.

3. $\begin{cases} x' = -y + x^3 \\ y' = x + y^3 \end{cases}$ 4. $\begin{cases} x' = -5x + 2y \\ y' = 4x - 3y \end{cases}$

5. $\begin{cases} x' = y - \sin x \\ y' = -x - \sin y \end{cases}$ 6. $\begin{cases} x' = 2xy^2 - x^3 + 4y^3 \\ y' = 4x^3 - 8x^2y - y^3 \end{cases}$

7. $\begin{cases} x' = x^3 + x + y \\ y' = -2x + y \end{cases}$ 8. $\begin{cases} x' = -2y + y^2 \\ y' = 2x + y - xy - y^3 \end{cases}$

In Exercises 9–17, try to find a Lyapunov function of the form $L(x, y) = ax^2 + cy^2$ and determine the stability of the origin as a stationary point. It will occasionally be necessary to take $L(x, y) = ax^2 + bxy + cy^2$.

9. $\begin{cases} x' = -x + y \\ y' = -x - y \end{cases}$ 10. $\begin{cases} x' = 100y \\ y' = -x \end{cases}$

11. $\begin{cases} x' = 2x - y \\ y' = 40x + 2y \end{cases}$ 12. $\begin{cases} x' = -x - 5y \\ y' = 5x + 5y \end{cases}$

13. $\begin{cases} x' = x - y \\ y' = 50x - y \end{cases}$ 14. $\begin{cases} x' = -x^3 - y \\ y' = x - 4y^3 \end{cases}$

15. $\begin{cases} x' = 2x + 5(y^2 - y) \\ y' = 5(x - x^2) - 4y \end{cases}$ 16. $\begin{cases} x' = 4y - xy^2 \\ y' = -x - x^2y \end{cases}$

17. $\begin{cases} x' = 3y - x^3 - xy^2 \\ y' = -4x - x^2y - y^3 \end{cases}$

18. Prove *Lyapunov's test for asymptotic stability*: A stationary point is asymptotically stable if there is a Lyapunov function L for it with the property that L' is negative definite.

19. What happens if, in the system (8.49) in Example 8.6.4, $q(x, y)$ is negative semidefinite?

20. The Lotka-Volterra equations

$$\left. \begin{array}{l} x' = x(a - by) \\ y' = cy(x - d) \end{array} \right\}$$

are a model for describing the dynamics of a predator-prey relationship, with x representing the prey population, and y the predator population. The system was explored in some detail in Section 3.5, where we found that there are two stationary points: the origin $\mathbf{0}$, and $\mathbf{p} = (d, \frac{b}{a})$.

Show that $\mathbf{p}$ is stable, but not asymptotically stable.

21. Put $q(x, y) = (x^2 + y^2)(1 - x^2 - y^2)$ in the system (8.49) in Example 8.6.4.

a. Show that the origin is unstable.

b. Show that $x = \cos t$, $y = \sin t$ is a solution of the system (8.49), whose orbit is the unit circle, traversed counterclockwise.

c. Show that the orbit found in part (b) is a limit cycle.

d. **Computer laboratory problem.** Draw the phase portrait of this system.

22. A certain nonlinear, damped mass spring apparatus is governed by the differential equation

$$x'' + b|x'|x' + k(x - x^3) = 0.$$

Show that the total mechanical energy is a Lyapunov function for this spring, and that the solution $x = x' \equiv 0$ is asymptotically stable.

23. **Computer laboratory problem.** Figure 8.29 indicates that the system (8.50) has a limit cycle. Investigate what happens for other values of k. Does the limit cycle persist?

24. Prove Proposition 8.6.8.

◆ *Hints:* (1) The origin is a critical point of L. (2) Use the second derivative test.

25. The following is a proof, in outline form, of Theorem 8.3 on page 400. Suppose that $\mathbf{x}_1$ is a linearly stable stationary point of the system

$$\mathbf{x}' = \mathbf{f}(\mathbf{x}); \qquad (8.51)$$

that is, $\mathbf{f}(\mathbf{x}_1) = \mathbf{0}$, and all characteristic roots of $A(\mathbf{x}_1)$, the derivative matrix of $\mathbf{f}$ at $\mathbf{x}_1$, are either negative real numbers or complex numbers with negative real parts. We will refer to this matrix simply as A.

A symmetric matrix S will be called *positive definite* if the function $L(\mathbf{x}) = \mathbf{x}^T \cdot S \cdot \mathbf{x}$ is positive definite. Here, we view $\mathbf{x}$ as a column matrix; $\mathbf{x}^T$ denotes its transpose, a row matrix, and $L(\mathbf{x})$, which is the product of the three matrices, is a 1×1 matrix.

a. Show that if there is a nonsingular matrix P such that $S = P \cdot P^T$, then S is positive definite.

b. Show that $L(\mathbf{x}) = \mathbf{x}^T \cdot S \cdot \mathbf{x}$ is a Lyapunov function for the origin as a stationary point of

$$\mathbf{x}' = A\mathbf{x}$$

if S is positive definite, and

$$A^T \cdot S + S \cdot A \text{ is negative definite.} \qquad (8.52)$$

c. Use Exercise 18 to conclude that if there exists a positive definite symmetric matrix S that satisfies condition (8.52), then $\mathbf{x}_1$ is an asymptotically stable stationary point of the system (8.51).

To prove Theorem 8.3, we will show how to find a positive definite symmetric matrix S that satisfies condition (8.52), provided that all characteristic roots of A have negative real parts.

d. Let $\mathcal{X}(t)$ denote a fundamental matrix solution of the *transposed* linear system

$$\mathbf{x}' = A^T \mathbf{x},$$

and put $\mathcal{Y}(t) = \mathcal{X}(t) \cdot (\mathcal{X}(t))^T$. Show that if the characteristic roots of A are either negative, or have negative real parts, then

$$\int_0^\infty \mathcal{Y}(t)\, dt$$

converges, and let S denote the value of this integral.

e. Show that S, as defined in part (d), is positive definite.

f. Show that $A^T \cdot S + S \cdot A = -\mathcal{Y}(0)$, and complete the proof.

◆ **Hint:** Differentiate $\mathcal{Y}(t)$, and note that

$$-\mathcal{Y}(0) = \int_0^\infty [d\mathcal{Y}(t)/dt]\, dt.$$

26. Take $A = \begin{bmatrix} -1 & 1 \\ 0 & -1 \end{bmatrix}$. Calculate S by the method outlined in Exercise 25 and show that S is positive definite and satisfies condition (8.52).

8.7 Glossary

In this glossary, definitions apply to a system of n ODEs except when stated to the contrary. Thus we refer to phase space *rather than the phase plane. To fix notation, the system will be denoted*

$$\mathbf{v}' = \mathbf{F}(\mathbf{v}),$$

if it is autonomous, and $\mathbf{v}' = \mathbf{F}(\mathbf{v}, t)$ *if it is not. Here,*

$$\mathbf{v} = \begin{bmatrix} x_1 \\ x_2 \\ \vdots \\ x_n \end{bmatrix}$$

is the vector of dependent variables, and

$$\mathbf{F}(\mathbf{v}, t) = \begin{bmatrix} f_1(\mathbf{v}, t) \\ f_2(\mathbf{v}, t) \\ \vdots \\ f_n(\mathbf{v}, t) \end{bmatrix}$$

denotes a vector function defined on a subset of $\mathbf{R}^n \times \mathbf{R}$. *Finally, if* $\mathbf{v}^* \in \mathbf{R}^n$ *and* $r > 0$, *let* $\mathcal{B}_r(\mathbf{v}^*)$ *denote the set of points in* $\mathbf{R}^n$ *at a distance less than r from* $\mathbf{v}^*$. *The set* $\mathcal{B}_r(\mathbf{v}^*)$ *is called the ball of radius r centered at* $\mathbf{v}^*$.

Asymptotically stable (stationary point) A stationary point $\mathbf{v}^*$ such that there is a ball $\mathcal{B}_r(\mathbf{v}^*)$ and every solution $\mathbf{v}(t)$ with $\mathbf{v}(t_1) \in \mathcal{B}_r(\mathbf{v}^*)$ at some time t_1 has the property

$$\lim_{t \to \infty} \mathbf{v}(t) = \mathbf{v}^*.$$

Attracted A point $\mathbf{v}_1$ is attracted to a stationary point $\mathbf{v}^*$ if any solution with $\mathbf{v}(t_1) = \mathbf{v}_1$ for some t_1 converges to $\mathbf{v}^*$. The set of points that are attracted to $\mathbf{v}^*$ is called the attracted set of $\mathbf{v}^*$. More generally, if $\mathcal{S}$ is an invariant set, then $\mathbf{v}_1$ is attracted to $\mathcal{S}$ if the forward limit set of every solution with $\mathbf{v}(t_1) = \mathbf{v}_1$ for some t_1 is contained in $\mathcal{S}$.

Center A system of two homogeneous linear ODEs in which the characteristic roots are imaginary numbers. The orbits are ellipses, centered at the origin.

Degenerate system A system of ODEs $\mathbf{v}' = A\mathbf{v}$ in which the matrix A is singular.

Forward limit point See **Limit point**.

Improper node A system of linear ODEs in which not all orbits are half-lines, but all orbits are directed toward the origin as $t \to \infty$ (the stable case), or all orbits are directed toward the origin as $t \to -\infty$ (the unstable case).

Invariant set (of an autonomous system) A subset S of phase space with the property that every orbit of the system that contains a point of S is a subset of S.

lim$_+$ See **Limit point**.

Limit cycle A closed orbit that is the forward limit set of a nearby orbit.

Limit point (of a solution $\mathbf{v}(t)$). There are forward and backward limit points. $\mathbf{v}^*$ is a **forward** limit point if there is an increasing, unbounded sequence of real numbers (we use the suggestive notation $t_m \uparrow \infty$) such that

$$\lim_{m \to \infty} \mathbf{v}(t_m) = \mathbf{v}^*. \qquad (8.53)$$

Similarly, $\mathbf{v}^*$ is a **backward** limit point if there is a sequence $t_m \downarrow -\infty$ such that (8.53) holds. Limit points can also be ascribed to the orbit represented by $\mathbf{v}(t)$. The set of forward limit points of $\mathbf{v}(t)$ is denoted lim$_+(\mathbf{v})$ and the set of backward limit points is lim$_-(\mathbf{v})$. These sets can be empty, consist of one point to which the orbit converges, or can be infinite. Some texts refer to the backward and forward limit sets as the alpha and omega sets, respectively, of the orbit. In these texts, the notations $\alpha(\mathbf{v})$ and $\omega(\mathbf{v})$ are used in place of lim$_-(\mathbf{v})$ and lim$_+(\mathbf{v})$, respectively.

Limit set See **Limit point**.

Linearly stable (stationary point of an autonomous system) A stationary point $\mathbf{v}^*$ such that all of the characteristic roots of the derivative $A(\mathbf{v}^*)$ of $\mathbf{F}$ at $\mathbf{v}^*$ are situated to the left of the imaginary axis of the complex plane. By Perron's theorem (page 400), linearly stable implies asymptotically stable, provided that the second partial derivatives of the components of $\mathbf{F}$ are continuous.

Node A system of linear ODEs in which all orbits are directed toward the origin as $t \to \infty$ (the stable case), or all orbits are directed toward the origin as $t \to -\infty$ (the unstable case).

Positive definite A function $L(x_1, x_2, \ldots, x_n)$ such that $L(0, 0, \ldots, 0) = 0$ and $L(x_1, x_2, \ldots, x_n) > 0$ for $(x_1, x_2, \ldots, x_n) \neq L(0, 0, \ldots, 0)$. The function L is said to be **positive semidefinite** if $L(x_1, x_2, \ldots, x_n) \geq 0$. If one of the preceding inequalities holds only for all

$(x_1, x_2, \ldots, x_n) \in \mathcal{B}_r(\mathbf{0})$ for some $r > 0$, then we say that L is positive definite or semidefinite **near the origin**.

Proper node A linear system of ODEs in which the orbits are all half-lines. The half-lines will all be directed toward the origin (the stable case) or all directed away from the origin (the unstable case).

Repelled A point $\mathbf{v}_1$ is repelled by stationary point $\mathbf{v}^*$ if any solution with $\mathbf{v}(t_1) = \mathbf{v}_1$ for some t_1 converges to $\mathbf{v}^*$ as $t \to -\infty$. The set of points that are repelled by $\mathbf{v}^*$ is called the repelled set of $\mathbf{v}^*$. More generally, if S is an invariant set, then $\mathbf{v}_1$ is repelled by S if the backward limit set of every solution with $\mathbf{v}(t_1) = \mathbf{v}_1$ for some t_1 is contained in S.

Saddle A linear system of two ODEs that has two real characteristic roots of opposite sign, or a stationary point $\mathbf{v}^*$ of a nonlinear autonomous system of two equations $\mathbf{v}' = \mathbf{F}(\mathbf{v})$ such that the characteristic roots of the derivative $A(\mathbf{v}^*)$ of $\mathbf{F}$ at $\mathbf{v}^*$ are real and of opposite sign. Near the stationary point, the phase portrait looks like a linear saddle.

Separatrix (applies to an autonomous system of two equations) An orbit that converges to a saddle. Orbits on either side of a separatrix have different limiting behavior.

Spiral node A linear system of two ODEs in which the characteristic roots are neither real nor imaginary. The orbits swirl around the origin in spirals.

Stable (stationary point $\mathbf{v}^*$ of an autonomous system) For any number $R > 0$, there is a number $r > 0$ such that every orbit that passes within a distance r from $\mathbf{v}^*$ will be entirely within the set of points at distance less than R from $\mathbf{v}^*$.

Stable line A straight line through the origin that is the union of the origin, and two orbits directed toward the origin. A linear system of two ODEs that is a saddle has a unique stable line. In the case of a stable proper node, all lines through the origin are stable.

Stable node A system of linear ODEs in which all orbits approach the origin as $t \to \infty$.

Stationary point (of an autonomous system) A point $\mathbf{v}^*$ such that $\mathbf{F}(\mathbf{v}^*) = \mathbf{0}$. A stationary point represents a constant solution of the system.

Unstable line A straight line through the origin that is the union of the origin, and two orbits directed away from the origin. A linear system of two ODEs that is a saddle has a unique unstable line. In the case of an unstable proper node, all lines through the origin are unstable.

8.8 Suggestions for Further Readings

Many advanced texts give complete proofs of the relation between the phase portrait of a system of differential equations near an equilibrium point and the phase portrait of the linearized system. The books *Ordinary Differential Equations*, by Philip Hartman,[3] and *Ordinary Differential Equations*, by Jack Hale,[4] are graduate-level texts that contain rather complete treatments of this topic.

Applications of differential equations to biology are the subject of a great deal of current research. An outstanding text in this area is *Mathematical Biology*, by J. D. Murray.[5] Murray's book is comprehensive, has few prerequisites, and is well written. It has the additional advantage, unusual for a mathematics text, of being suitable for browsing.

---

[3] Second edition, New York: Wiley, 1973.

[4] Second edition, Malabar, Florida: Krieger, 1980.

[5] New York: Springer-Verlag, 1989.

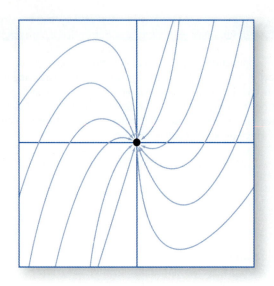

CHAPTER 9

Dynamical Systems and Chaos

9.1 Flows and Dynamical Systems

Consider a system of ODEs,

$$\left. \begin{array}{l} x' = f(x, y, t) \\ y' = g(x, y, t), \end{array} \right\} \tag{9.1}$$

where the functions f and g are continuous and have continuous partial derivatives with respect to x and y.

For each $(x_0, y_0) \in \mathbf{R}^2$ and $t_0 \in \mathbf{R}$, let

$$x = \phi(t; t_0, x_0, y_0)$$
$$y = \psi(t; t_0, x_0, y_0)$$

be the solution satisfying the initial condition

$$(x(t_0), y(t_0)) = (x_0, y_0).$$

Let's consider the mapping defined on a subset of $\mathbf{R}^2$ that takes a point (x, y) to the point $(\phi(t_1; t_0, x, y), \psi(t_1; t_0, x, y))$, where t_0 and t_1 are fixed numbers. A point (x, y) is in the domain of this mapping if and only if the solution

$$(\phi(t; t_0, x, y), \psi(t; t_0, x, y))$$

of (9.1) is defined for $t = t_1$. We will denote this mapping as F_{t_0, t_1}. Thus

$$F_{t_0, t_1}(x, y) = (\phi(t_1; t_0, x, y), \psi(t_1; t_0, x, y)).$$

Proposition 9.1.1

If $F_{t_0,t_2}(x, y)$ is defined and t_1 is a number between t_0 and t_2, then

$$F_{t_0,t_2}(x, y) = F_{t_1,t_2} \circ F_{t_0,t_1}(x, y).$$

Furthermore, for any t_0 we have $F_{t_0,t_0}(x, y) \equiv (x, y)$.

The proof of this proposition is based on the uniqueness theorem for solutions of IVPs and is left to you. A family of mappings with the properties described in Proposition 9.1.1 is called a **flow**.

Example 9.1.1   Describe the flow associated with the system

$$x' = y$$
$$y' = -x.$$

Solution. This linear system of ODEs can be put into matrix form as $\mathbf{v}' = A\mathbf{v}$, where

$$A = \begin{bmatrix} 0 & 1 \\ -1 & 0 \end{bmatrix}.$$

The general solution, $\mathbf{v} = e^{At}\mathbf{c}$, can be used to determine the flow: The solution with $\mathbf{v}(t_0) = \begin{bmatrix} x \\ y \end{bmatrix}$ is $\mathbf{v}(t) = e^{A(t-t_0)}\begin{bmatrix} x \\ y \end{bmatrix}$. Thus,

$$F_{t_0,t_1}(x, y) = e^{A(t_1-t_0)}\begin{bmatrix} x \\ y \end{bmatrix}.$$

The characteristic roots of A are $\pm i$, and the characteristic vectors are $\begin{bmatrix} 1 \\ \pm i \end{bmatrix}$. It follows that

$$e^{At} = \begin{bmatrix} \text{Re}(e^{it}) & \text{Im}(e^{it}) \\ \text{Re}(ie^{it}) & \text{Im}(ie^{it}) \end{bmatrix} = \begin{bmatrix} \cos(t) & \sin(t) \\ -\sin(t) & \cos(t) \end{bmatrix}.$$

Thus

$$F_{t_0,t_1}(x, y) = \begin{bmatrix} \cos(t_1 - t_0) & \sin(t_1 - t_0) \\ -\sin(t_1 - t_0) & \cos(t_1 - t_0) \end{bmatrix}\begin{bmatrix} x \\ y \end{bmatrix}$$
$$= (\cos(t_1 - t_0)x + \sin(t_1 - t_0)y, -\sin(t_1 - t_0)x + \cos(t_1 - t_0)y).$$

This flow can be described as a clockwise rotation of the plane through an angle of $(t_1 - t_0)$ radians.   ❑

A **dynamical system** is a one-to-one mapping P from a set of points A to another set of points B. If the sets A and B overlap—that is, $A \cap B \neq \emptyset$—then for each $x \in A$ such that $P(x) \in A$ we can form $P^2(x) = P(P(x))$. If $P^2(x) \in A$ as well, then $P^3(x)$ is also defined, and so on. Since we are assuming that P is one to one we can also find $P^{-1}(x)$ if $x \in B$; if $P^{-1}(x) \in B$, then we can find $P^{-2}(x)$, and so on.

For any dynamical system P, where the domain A and the range B are subsets of the plane $\mathbf{R}^2$, the **orbit** of a point (x_0, y_0) is a sequence of points extending in both directions,

$$\ldots, (x_{-1}, y_{-1}), (x_0, y_0), (x_1, y_1), (x_2, y_2), \ldots,$$

where for $n > 0$, $(x_n, y_n) = P^n(x_0, y_0)$. If the orbit stays within $A \cap B$, it forms a doubly infinite sequence (x_n, y_n), $-\infty < n < \infty$. Notice that if $P = T_{t_0, t_1}$ is a mapping defined by the flow associated with a system (9.1) of ODEs, then the orbits of P are subsets of corresponding orbits of the ODEs.

If P is not one to one, it is still possible to define its orbits, by iterating P. However, orbits only go forward from their starting point, since there is no P^{-1}. We cannot call P a dynamical system if it isn't one to one, so we will just use the term **mapping.** Thus, every dynamical system is a mapping, but only one to one mappings are dynamical systems.

An orbit can consist of just one point (x^*, y^*), if $P(x^*, y^*) = (x^*, y^*)$. In this case, (x^*, y^*) is called a **fixed point** of P. The following example has been chosen for computational simplicity. The mapping P is not one to one; indeed, $P(x, y) = P(-x, -y)$ for all (x, y).

Example 9.1.2   Find the fixed points of $P(x, y) = (x^2 - y^2, 2xy)$.

Solution. To find the fixed points, we must solve $P(x, y) = (x, y)$ or, equivalently, the system of equations

$$x^2 - y^2 = x$$
$$2xy = y.$$

The second equation holds if either $x = \frac{1}{2}$ or $y = 0$. When $x = \frac{1}{2}$, the first equation reads

$$\frac{1}{4} - y^2 = \frac{1}{2},$$

which has no real solution. Hence all solutions have $y = 0$, and the first equation is simply $x^2 = x$. It follows that $x = 0$ or $x = 1$. The fixed points of P are therefore $(0, 0)$ and $(1, 0)$.   ❑

A point (x, y) is called an **N-periodic point** (N is a positive integer) if

$$\underbrace{P \circ P \circ \cdots \circ P}_{N \text{ times}}(x, y) = (x, y).$$

Let P denote the mapping defined in Example 9.1.2. Then

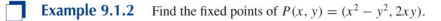

$$P\left(-\frac{1}{2}, \frac{\sqrt{3}}{2}\right) = \left(-\frac{1}{2}, -\frac{\sqrt{3}}{2}\right) \quad \text{and} \quad P\left(-\frac{1}{2}, -\frac{\sqrt{3}}{2}\right) = \left(-\frac{1}{2}, \frac{\sqrt{3}}{2}\right).$$

Thus, the two points $\left(-\frac{1}{2}, \pm\frac{\sqrt{3}}{2}\right)$ are 2-periodic. Periodic points and fixed points are closely related, and in fact an N-periodic point is a fixed point of the mapping P^N obtained by iterating P N times.

We will say that an orbit (x_n, y_n) **converges** to a point (x^*, y^*) as $n \to \infty$ if (x_n, y_n) is defined for $0 < n < \infty$ and

$$\lim_{n\to\infty} x_n = x^*, \quad \lim_{n\to\infty} y_n = y^*.$$

If P is one to one, there is a similar definition for an orbit of a dynamical system to converge to a point (x^*, y^*) as $n \to -\infty$. It will be seen in Exercise 22 that if an orbit converges to a point (x^*, y^*), then (x^*, y^*) is a fixed point.

The notion of convergence permits us to define asymptotic stability for fixed points and periodic points. A fixed point (x^*, y^*) is **asymptotically stable** if there is a circle $\mathcal{C}$ centered at (x^*, y^*) such that every orbit (x_n, y_n) with (x_0, y_0) inside $\mathcal{C}$ converges to (x^*, y^*) as $n \to \infty$. If (x^*, y^*) is an N-periodic point, then it is a fixed point of P^N. We will say that (x^*, y^*) is asymptotically stable as a periodic point of P if it is an asymptotically stable fixed point of P^N.

A fixed point (x^*, y^*) of the mapping P is **unstable** if there is a circle $\mathcal{C}$ centered at (x^*, y^*) such that, given any positive number r, there is a point (x_0, y_0) with

$$\sqrt{(x_0 - x^*)^2 + (y_0 - y^*)^2} < r$$

and such that $P^n(x_0, y_0)$ lies outside $\mathcal{C}$ for some $n > 0$. Informally, this means that points whose orbits lead outside $\mathcal{C}$ can be found arbitrarily close to (x^*, y^*).

The derivative of a mapping can be used to test the stability of a fixed point. We will start with a result about one-dimensional mappings.

Proposition 9.1.2

Let x^* be a fixed point of a mapping f of some subset of the real line. If $f'(x)$ is continuous at x^*, then $|f'(x^*)| < 1$ implies x^* is asymptotically stable, and $|f'(x^*)| > 1$ implies x^* is unstable.

Proof Assume that $|f'(x^*)| < 1$, and let p be a number such that $|f'(x^*)| < p < 1$. Since $f'(x)$ is continuous at x^*, there is a number $r > 0$ such that for all x satisfying $|x - x^*| < r$, $f'(x) < p$. Suppose that the initial point x_0 is chosen so that $|x_0 - x^*| < r$. Since $f(x^*) = x^*$, the mean value theorem implies

$$|x_1 - x^*| = |f(x_0) - f(x^*)|$$
$$= |f'(\xi)| \cdot |x_0 - x^*|,$$

where ξ is between x_0 and x^*. Thus $|\xi - x^*| < r$. Thus $|f'(\xi)| < p$, and it follows that

$$|x_1 - x^*| \le p|x_0 - x^*|. < pr < r.$$

Since $|x_1 - x^*| < r$, the next iteration, $x_2 = f(x_1)$, satisfies $|x_2 - x^*| \le p^2|x_0 - x^*|$. Further iteration of f yields a sequence x_n such that $|x_n - x^*| \le p^n|x_0 - x^*|$. Since

$p^n \to 0$ as $n \to \infty$, it follows that $\lim_{n \to \infty} x_n = x^*$. We have shown that any orbit with $x_0 \in (x^* - r, x^* + r)$ converges to x^* as $n \to \infty$, and hence that x^* is asymptotically stable.

You will be asked to prove that if $|f'(x^*)| > 1$ then x^* is unstable in Exercise 10 at the end of this section. ∎

Example 9.1.3 Let $f(x) = 2x(1-x)$. Find the fixed points of f and test them for stability.

Solution. Let x^* be a fixed point. Then $f(x^*) = x^*$, so x^* is a solution of the quadratic equation $x^* = 2x^*(1 - x^*)$, or $2(x^*)^2 - x^* = 0$. The two fixed points are 0 and $\frac{1}{2}$. Since $f'(x) = 2 - 4x$, $f'(0) = 2$ and $f'\left(\frac{1}{2}\right) = 0$. It follows that 0 is unstable and $\frac{1}{2}$ is stable. ❑

We would like to find a version of Proposition 9.1.2 that would be valid for two-dimensional mappings. To do so, we have to study linear dynamical systems first.

Linear dynamical systems in the plane

A linear dynamical system is generated by an invertible linear transformation of the plane. Thus, let A be a nonsingular 2×2 matrix, and consider the dynamical system

$$\begin{bmatrix} x_n \\ y_n \end{bmatrix} = A^n \begin{bmatrix} x_0 \\ y_0 \end{bmatrix}.$$

We will show that $\mathbf{0}$ is an asymptotically stable fixed point of this dynamical system if and only if both characteristic roots of A have absolute value less than 1. This stability condition is that the characteristic roots are located inside the unit circle of the complex plane. This contrasts with the stability condition for the system of ODEs $\mathbf{v}' = A\mathbf{v}$, which is that the characteristic roots lie to the left of the imaginary axis.

Let us say that a matrix L is a **logarithm** of the matrix A if $e^L = A$. If all the entries of L are real numbers, then L is called a **real logarithm** of A. The following example shows how to verify that a given matrix L is a logarithm of a matrix A. It also shows that logarithms are not unique. While it is obvious that the zero matrix is a logarithm of the identity matrix, we will see that the zero matrix is only one member of an infinite family of logarithms!

Example 9.1.4 Show that 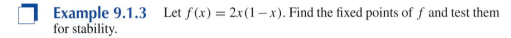 $\begin{bmatrix} 0 & 1 \\ 0 & 0 \end{bmatrix}$ is a logarithm of $\begin{bmatrix} 1 & 1 \\ 0 & 1 \end{bmatrix}$, and that for any integer n, $\begin{bmatrix} 0 & 2n\pi \\ -2n\pi & 0 \end{bmatrix}$ is a logarithm of the identity matrix.

Solution. To show that L is a logarithm of A, we have to calculate e^L and verify that the result is equal to A. If $L = \begin{bmatrix} 0 & 1 \\ 0 & 0 \end{bmatrix}$, then $L^2 = 0$. Hence $e^L = I + L = \begin{bmatrix} 1 & 1 \\ 0 & 1 \end{bmatrix}$.

Now let $L_p = \begin{bmatrix} 0 & p \\ -p & 0 \end{bmatrix}$. The characteristic roots of L_p are $\pm pi$ and the characteristic vector belonging to pi is $\begin{bmatrix} 1 \\ i \end{bmatrix}$. Since e^{L_p} has the same characteristic vectors, and characteristic roots $e^{\pm pi}$, it follows that

$$e^{L_p} \begin{bmatrix} 1 \\ i \end{bmatrix} = \begin{bmatrix} e^{pi} \\ i e^{pi} \end{bmatrix}. \tag{9.2}$$

Taking real and imaginary parts of both sides of (9.2),

$$e^{L_p} \begin{bmatrix} 1 \\ 0 \end{bmatrix} = \begin{bmatrix} \cos(p) \\ -\sin(p) \end{bmatrix} \text{ and } e^{L_p} \begin{bmatrix} 0 \\ 1 \end{bmatrix} = \begin{bmatrix} \sin(p) \\ \cos(p) \end{bmatrix}.$$

Hence

$$e^{L_p} = \begin{bmatrix} \cos(p) & \sin(p) \\ -\sin(p) & \cos(p) \end{bmatrix},$$

and when $p = 2n\pi$,

$$\exp\left(\begin{bmatrix} 0 & 2n\pi \\ -2n\pi & 0 \end{bmatrix} \right) = I. \qquad \square$$

Let

$$\mathbf{x}_n = A^n \mathbf{x}_0, \quad -\infty < n < \infty \tag{9.3}$$

denote an orbit in a linear dynamical system. If A has a logarithm L, then $A^n \mathbf{x}_0 = [e^L]^n \mathbf{x}_0 = e^{nL} \mathbf{x}_0$. The flow associated with the linear system

$$\mathbf{x}' = L\mathbf{x} \tag{9.4}$$

is $F_{0,\tau}(\mathbf{x}_0) = e^{\tau L} \mathbf{x}_0$. Thus $\mathbf{x}_n = F_{0,n}(\mathbf{x}_0)$. The orbit $\mathbf{x}_n$ of the linear dynamical system is a subset of the orbit of the system (9.4) of differential equations.

Thus, to embed a linear dynamical system in a flow we need to find a logarithm of the matrix A, and this induces us to ask which matrices have logarithms. A full discussion of this topic would be beyond the scope of this text, but the answer is simple enough. Since e^L must be nonsingular—it has an inverse, e^{-L}, and all invertible matrices are nonsingular—A must be nonsingular if it is to have a logarithm.

A matrix A with real entries is said to have a real square root if there is a real matrix B such that $B^2 = A$. If L is real, then e^L has a real square root since $(e^{L/2})^2 = e^L$. Thus, a nonsingular matrix A can have a real logarithm only if it has a real square root. The following theorem says that this condition is sufficient: Every nonsingular real matrix that has a real square root has a real logarithm. If this surprises you, consider the one-dimensional case: A nonzero real number x has a real logarithm if and only if it has a real square root, since both conclusions are equivalent to $x > 0$.

Theorem 9.1

A square matrix A has a logarithm if and only if A is nonsingular. If there is a nonsingular matrix B such that $A = B^2$, then A has a real logarithm.

The proof of Theorem 9.1 is omitted.

Proposition 9.1.3

Let s_1 and s_2 be the characteristic roots of a matrix A. If $|s_1| < 1$ and $|s_2| < 1$, then the origin is an asymptotically stable fixed point of the dynamical system defined by A. If $|s_1| > 1$ or $|s_2| > 1$, then the origin is unstable.

Proof We will give two proofs. First, let's assume that A has a real logarithm L. Express each characteristic root of A in polar form as $s_j = r_j e^{i\theta_j}$, where $r_j = |s_j|$ and $\theta_j = \arg(s_j)$. Then the characteristic roots of L are the logarithms $l_j = \ln(s_j) = \ln(r_j) + i\theta_j$, $j = 1, 2$. Since $r_j < 1$, $\text{Re}(l_j) = \ln(r_j) < 0$. By Theorem 8.1 on page 391, if $\mathbf{x}(t)$ is a solution of $\mathbf{x}' = L\mathbf{x}$, $\lim_{t \to \infty} \mathbf{x}(t) = \mathbf{0}$. Therefore,

$$\lim_{n \to \infty} \mathbf{x}_n = \lim_{n \to \infty} \mathbf{x}(n) = \mathbf{0}.$$

The second proof does not depend on A having a real logarithm. If A has independent characteristic vectors $\mathbf{v}_1$ and $\mathbf{v}_2$, we can express $\mathbf{x}_0$ as a linear combination of these:

$$\mathbf{x}_0 = c_1 \mathbf{v}_1 + c_2 \mathbf{v}_2.$$

Hence

$$\mathbf{x}_n = A^n \mathbf{x}_0 = c_1 s_1^n \mathbf{v}_1 + c_2 s_2^n \mathbf{v}_2.$$

Since $|s_j| < 1$, $s_j^n \to 0$ as $n \to \infty$, and we conclude that $\mathbf{x}_n \to \mathbf{0}$.

If A does not have two independent characteristic vectors, then choose $\mathbf{v}_1$ to be a vector that is not characteristic, and put $\mathbf{v}_2 = (A - s_1 I)\mathbf{v}_1$. By the Cayley-Hamilton theorem (page 154), $(A - s_1 I)^2 = 0$. Hence $\mathbf{v}_2$ is a characteristic vector, and

$$A\mathbf{v}_1 = s_1 \mathbf{v}_1 + \mathbf{v}_2. \tag{9.5}$$

We claim that for all $n > 0$,

$$A^n \mathbf{v}_1 = s_1^n \mathbf{v}_1 + n s_1^{n-1} \mathbf{v}_2. \tag{9.6}$$

The proof is by induction, with equation (9.5) establishing the case $n = 1$. Assume that equation (9.6) holds for $n = 1, \ldots, N$. Then

$$\begin{aligned}
A^{N+1}\mathbf{v}_1 &= A(s_1^N \mathbf{v}_1 + N s_1^{N-1} \mathbf{v}_2) \\
&= s_1^N A\mathbf{v}_1 + N s_1^{N-1} A\mathbf{v}_2 \\
&= s_1^N (s_1 \mathbf{v}_1 + \mathbf{v}_2) + N s_1^{N-1}(s_1 \mathbf{v}_2) \\
&= s_1^{N+1}\mathbf{v}_1 + (N+1)s_1^N \mathbf{v}_2.
\end{aligned}$$

We now proceed as before: Express $\mathbf{x}_0$ as a linear combination of $\mathbf{v}_1$ and $\mathbf{v}_2$ and calculate,

$$
\begin{aligned}
A^n \mathbf{x}_0 &= A^n (c_1 \mathbf{v}_1 + c_2 \mathbf{v}_2) \\
&= c_1 (s_1^n \mathbf{v}_1 + n s_1^{n-1} \mathbf{v}_2) + c_2 s_1^n \mathbf{v}_2 \\
&= c_1 s_1^n \mathbf{v}_1 + (c_1 n s_1^{n-1} + c_2 s_1^n) \mathbf{v}_2.
\end{aligned}
$$

Since $|s_j| < 1$, both $s_1^n \to 0$ and $n s_1^{n-1} \to 0$ as $n \to \infty$, it follows that $A^n \mathbf{x}_0 \to \mathbf{0}$ as $n \to \infty$.

The proof that the origin is unstable if either of both characteristic roots have magnitude greater than 1 is an exercise. ∎

Now consider a nonlinear mapping $P(x, y) = (f(x, y), g(x, y))$ taking an open subset of the plane to itself. If $\mathbf{x}^* = (x^*, y^*)$ is an fixed point of P, and the partial derivatives of f and g are continuous at $\mathbf{x}^*$, then P can be approximated near $\mathbf{x}^*$ by the linear mapping associated with the derivative matrix

$$
A = \begin{bmatrix} f_x(x^*, y^*) & f_y(x^*, y^*) \\ g_x(x^*, y^*) & g_y(x^*, y^*) \end{bmatrix}
$$

of P. This follows from Proposition 8.3.1 on page 398, which states that

$$
P(\mathbf{x}) = P(\mathbf{x}^*) + A(\mathbf{x} - \mathbf{x}^*) + \mathbf{O}(|\mathbf{x} - \mathbf{x}^*|^2).
$$

Theorem 9.2

Suppose $\mathbf{x}^*$ is a fixed point of a mapping P, and let A denote the derivative matrix of P at $\mathbf{x}^*$. Then

i. If all of the characteristic roots of A lie inside the unit circle of the complex plane, then $\mathbf{x}^*$ is asymptotically stable.
ii. If the one or more of the characteristic roots of A lie outside the unit circle of the complex plane, then $\mathbf{x}^*$ is unstable.

The proof of Theorem 9.2 is omitted.

Example 9.1.5   Test the fixed points $(0, 0)$ and $(1, 0)$, and the periodic point $\left(-\frac{1}{2}, \frac{\sqrt{3}}{2}\right)$ of $P(x, y) = (x^2 - y^2, 2xy)$ for stability.

Solution.   The derivative matrix of P at the point (x, y) is

$$
A(x, y) = \begin{bmatrix} 2x & -2y \\ 2y & 2x \end{bmatrix}.
$$

Since $A(0, 0)$ is the zero matrix, $(0, 0)$ is stable. $A(1, 0) = 2I$ has a double characteristic root, 2, so it is unstable. Turning to the periodic point, we have

$$
A\left(-\frac{1}{2}, \frac{\sqrt{3}}{2}\right) = \begin{bmatrix} -1 & -\sqrt{3} \\ \sqrt{3} & -1 \end{bmatrix}.
$$

Since this point is a fixed point of P^2, we should examine the characteristic roots of the derivative of P^2, which is A^2. These are the squares of the characteristic roots of A. Since trace $(A) = 2$ and $\det(A) = 4$ the characteristic roots of A are $-1 \pm \sqrt{3}i$, and each has absolute value 2. The characteristic roots of A^2 have absolute value 4, and hence the periodic point is also unstable. ❑

EXERCISES

1. **Staircase representation for orbits of one-dimensional mappings.** Let $f : \mathcal{I} \to \mathcal{I}$ be a mapping of an interval to itself, and let $x_0 \in \mathcal{I}$. Draw the graphs of $y = f(x)$ and $y = x$ on the x, y-plane. The staircase representation of the orbit $x_n = f^n(x_0)$ is generated as follows. Draw a vertical line from $Q_0 : (x_0, x_0)$ on the line $y = x$ to the graph of $y = f(x)$, meeting it at the point $P_1 : (x_0, x_1)$. Draw a horizontal line from P_1 to the graph of $y = x$, meeting it at $Q_1 : (x_1, x_1)$. Next draw a vertical line from Q_1 to the graph of $y = f(x)$, meeting it at $P_2 : (x_1, x_2)$. Next, draw a horizontal line from P_2 to $Q_2 : (x_2, x_2)$ on the line $y = x$. See Figure 9.1.

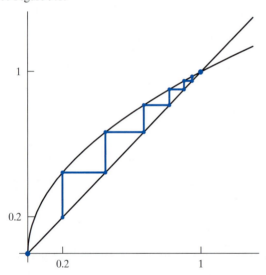

1

0.2

0.2   1

FIGURE 9.1   Staircase representation for the orbit resulting from iterating the mapping $f(x) = \sqrt{x}$ starting with $x_0 = 0.2$.

Show that this process may be continued forever, and that the sequence of points x_n obtained is the orbit of x_0.

2. If f is a decreasing mapping, show that the staircase representation of each orbit of f is a spiral.

In each of Exercises 3–8, draw the staircase representation of the orbit of the given mapping, starting at the initial point x_0. The drawings may be made by hand or by using a computer.

3. $f(x) = x + \frac{1}{2}\sin(\pi x)$; $x_0 = \frac{1}{10}$

4. $f(x) = x + \frac{1}{2}\sin(\pi x)$; $x_0 = \frac{19}{10}$
5. $f(x) = \frac{1}{2}(x^2 - 2x - 1)$; $x_0 = 1$
6. $f(x) = \frac{1}{2}(x^2 - 2x - 1)$; $x_0 = 2$
7. $f(x) = \frac{1}{2}(x^2 - 2x - 1)$; $x_0 = -2$
8. $f(x) = 3.83x(1 - x)$; $x_0 = 0.5$
9. The mapping $f(x) = x^2$ has two fixed points. Which is stable?
10. Complete the proof of Proposition 9.1.2 by showing that a fixed point x^* of a mapping f is unstable if $|f'(x^*)| > 1$.

    ◆ *Hint:* Show that there are numbers $r > 1$ and $\epsilon > 0$ such that for all $x \in (x^* - \epsilon, x^* + \epsilon)$, the following inequality holds: $|f'(x^*)| > r$. Hence show that if $x_0 \neq x^*$ and $f^N(x_0) \in (x^* - \epsilon, x^* + \epsilon)$ for all $N \geq 0$, a contradiction results.

11. Let f be a one-dimensional map with a 2-periodic orbit $\{x, y\}$; that is, $y = f(x)$ and $x = f(y)$. Show that this orbit is stable if

    $$|f'(x)| \cdot |f'(y)| < 1.$$

    Develop a criterion for the stability of 3-periodic orbits, and so on.

12. The mapping $f(x) = \cos(x)$, taking the interval $\left[-\frac{\pi}{2}, \frac{\pi}{2}\right]$ into itself, has a fixed point. Is this fixed point stable?

    In Exercises 13–15, find a logarithm of the matrix A and then determine a flow $P_\tau(x, y)$ such that P_1 is the linear dynamical system defined by A.

13. $A = \begin{bmatrix} 2 & 1 \\ 0 & 2 \end{bmatrix}$ 14. $A = \begin{bmatrix} 2 & 1 \\ 1 & 2 \end{bmatrix}$

15. $A = \begin{bmatrix} 0 & 1 \\ -1 & 0 \end{bmatrix}$

16. Show that if A is a matrix that has a characteristic root s_1 with $|s_1| > 1$, then the dynamical system $\mathbf{x}_n = A^n \mathbf{x}_0$ has an unstable fixed point at the origin.

17. Show that if a nonsingular matrix A has a real square root, then $\det(A) > 0$.

18. Let A be a 2×2 matrix, with determinant $q > 0$ and trace p. Show that A has a real square root if and only if either $A = cI$ for some number c, or $p > -2\sqrt{q}$.

The following three exercises show that monotone one dimensional dynamical systems are relatively simple.

19. Suppose $F : \mathbf{R} \longrightarrow \mathbf{R}$ is an increasing mapping [that is, for all $x < y$, $F(x) < F(y)$]. Let $x_0 \in \mathbf{R}$, and $x_n = F^n(x_0)$ be the orbit of x_0. Show that x_n is a monotone sequence (either increasing, decreasing, or constant).

20. Show that if $F : \mathbf{R} \longrightarrow \mathbf{R}$ is a strictly decreasing mapping, then F^2 is strictly increasing.

21. Suppose F is a one to one mapping of $\mathbf{R}$ to $\mathbf{R}$. Show that every orbit x_n of F is either a fixed point, so that $x_0 = x_1 = \cdots$; 2-periodic, so that $x_0 = x_2 = \cdots$ and $x_1 = x_3 = \cdots$, an infinite increasing sequence that converges to a fixed point, denoted $x_n \uparrow x^*$; an infinite decreasing sequence converging to a fixed point, denoted $x_n \downarrow x^*$; an infinite increasing or decreasing divergent sequence, denoted $x_n \uparrow \infty$ or $x_n \downarrow \infty$, respectively; or one of the following four types of oscillating sequences: $x_{2n} \uparrow x^*$, $x_{2n+1} \downarrow x^*$; $x_{2n} \downarrow x^*$, $x_{2n+1} \uparrow x^*$; $x_{2n} \uparrow \infty$, $x_{2n+1} \downarrow \infty$; or $x_{2n} \downarrow \infty$, $x_{2n+1} \uparrow \infty$.

22. Let $P : \mathbf{R}^2 \to \mathbf{R}^2$ be a continuous mapping. Show that if an orbit $\{(x_n, y_n)\}$ of P converges to (x^*, y^*) as $n \to \infty$ or $-\infty$, then (x^*, y^*) is a fixed point of P.

In Exercises 23–26, find the fixed points of the mapping and identify those that are stable.

23. $F\begin{bmatrix} x \\ y \end{bmatrix} = \begin{bmatrix} \frac{1}{2}(x - y) \\ \frac{1}{2}(x - y) - 2y^2 \end{bmatrix}$

24. $F\begin{bmatrix} x \\ y \end{bmatrix} = \begin{bmatrix} x^2 + y^2 \\ xy \end{bmatrix}$

25. $F\begin{bmatrix} x \\ y \end{bmatrix} = \begin{bmatrix} x^2 + y \\ x - y \end{bmatrix}$

26. $F\begin{bmatrix} x \\ y \end{bmatrix} = \begin{bmatrix} \frac{1}{2}(x^2 + y) \\ \frac{1}{2}(3y^2 - x) \end{bmatrix}$

27. Let P_τ be the flow associated with an autonomous system $\mathbf{x}' = \mathbf{f}(\mathbf{x})$ of differential equations. Show that

a. P_0 is the identity mapping.

b. $P_\tau^{-1} = P_{-\tau}$

c. $P_{\tau_1} \circ P_{\tau_2} = P_{\tau_1 + \tau_2}$

28. Let $F : \mathbf{R}^2 \longrightarrow \mathbf{R}^2$ be a differentiable mapping, and let $J(x, y)$ denote the derivative matrix of F at the point (x, y). We will say that F is an **expanding mapping** is if there exists a number $\rho > 1$ such that for all (x, y), $|\det(J(x, y))| \geq \rho$. Show that an expanding mapping cannot have a stable fixed point.

29. Show that an expanding mapping cannot have a stable periodic point.

◆ **Hint:** The chain rule for partial derivatives can be used to prove that any composition of expanding mappings is expanding.

9.2   Autonomous Systems in the Plane

One-dimensional autonomous ODEs were the subject of Section 2.5. If an autonomous first-order ODE has isolated stationary points, we found that the non-stationary orbits must be open intervals, whose endpoints are consecutive stationary points, or $\pm\infty$.

The transition from one- to two-dimensional autonomous systems allows for orbits that trace curves that are unbounded in one or both directions, bounded curves that connect a stationary point to itself or another stationary point, and closed orbits, or **cycles**. Cycles are the orbits of periodic solutions. We have seen examples where the forward limit set of an orbit is a point (a stable node has this property) or a cycle (in this case the cycle is called a **limit cycle**).

The possibilities for two-dimensional systems are constrained by the topology of the plane. The following theorem is the fundamental result on two-dimensional autonomous systems of ODEs, and it specifies an important property that is not shared by higher-dimensional systems. We might informally paraphrase the theorem to say that two-dimensional autonomous systems of ODEs are not permitted to have chaotic solutions.

Theorem 9.3   **Poincaré-Bendixson**

Let $\mathbf{v}(t) = (\phi(t), \psi(t))$ be a solution of a two-dimensional system of autonomous ODEs,

$$x' = f(x, y), \quad y' = g(x, y),$$

whose forward limit set $L = \lim_{+}(\mathbf{v})$ is nonempty and contains no stationary points. Then one of the following statements is true:

- $\mathbf{v}(t)$ is periodic—in this case L is the orbit of $\mathbf{v}$, and is hence a cycle.
- L is a limit cycle for $\mathbf{v}$.

The proof of this theorem depends on the Jordan curve theorem, a fundamental result concerning the topology of the plane.

Connected sets

Topology is the mathematics of open sets. If $\mathbf{x} \in \mathbf{R}^n$ and $\epsilon > 0$, let

$$B_\epsilon(\mathbf{x}) = \{\mathbf{y} \in \mathbf{R}^n : \|\mathbf{y} - \mathbf{x}\| < \epsilon\}.$$

This set is called the ϵ-ball centered at $\mathbf{x}$. For example, in the one-dimensional case,

$$B_\epsilon(x) = (x - \epsilon, x + \epsilon) \subset \mathbf{R}$$

is an interval. In two dimensions, $B_\epsilon(\mathbf{x}) \subset \mathbf{R}^2$ is the interior of a circle, with center $\mathbf{x}$ and radius ϵ.

A set $U \subset \mathbf{R}^n$ is said to be **open** if for every $\mathbf{x} \in U$ there is $\epsilon > 0$ such that $B_\epsilon(\mathbf{x}) \subset U$.

Example 9.2.1   Show that the following are open subsets of the plane.

- The rectangle $R = \{(x, y) : a < x < b, \text{ and } c < y < d\}$
- The disk $D = B_r((p, q))$

Solution.

Rectangle: If $(x, y) \in R$, let ϵ be the least of the following positive numbers: $x - a, b - x, y - c, d - y$. You can easily verify that $B_\epsilon(x, y) \subset R$. In the drawing to the left, $\epsilon = b - x$.

Disk: If $(x, y) \in D$, let $\epsilon = r - \sqrt{(x - p)^2 + (y - q)^2}$. Then $B_\epsilon(x, y) \subset D$.   ☐

A set $A \subset \mathbf{R}^n$ is **disconnected** if there are *disjoint* open sets $U_1, U_2 \subset \mathbf{R}^n$ such that

$$A \subset U_1 \cup U_2, \quad U_1 \cap A \neq \emptyset, \quad \text{and} \quad U_2 \cap A \neq \emptyset.$$

We say in this case that U_1 and U_2 **separate** A.

Example 9.2.2   Find separations for the following subsets of the plane:

- $\{\mathbf{p}, \mathbf{q}\}$, where $\mathbf{p}$ and $\mathbf{q}$ are distinct points
- The complementary set $\mathbf{R}^2 - \mathcal{C}$ of a circle $\mathcal{C}$

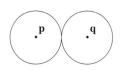

Solution. We will use the result of Example 9.2.1 that a disk is open.

Two points: Let the distance between $\mathbf{p}$ and $\mathbf{q}$ be denoted $2r$. Then $U_1 = B_r(\mathbf{p})$ and $U_2 = B_r(\mathbf{q})$ form a separation of $\{\mathbf{p}, \mathbf{q}\}$.

Complement of circle: Let the center C be $\mathbf{p}$, and let the radius be r. Then $U_1 = B_r(\mathbf{p})$ is an open set, and we will let

$$U_2 = \{\mathbf{x} \in \mathbf{R}^2 : \|\mathbf{x} - \mathbf{p}\| > r\}.$$

Thus U_1 is the set of points inside C, and U_2 is the set of points outside C. These sets are disjoint, and their union is the complement of C. By Example 9.2.1, U_1 is open; therefore, it is only necessary to show that U_2 is open. Let $\mathbf{x} \in U_2$, and put

$$s = \|\mathbf{x} - \mathbf{p}\|.$$

Then $s - r > 0$; we put $\epsilon = s - r$. Since all points in $B_\epsilon(\mathbf{x})$ lie outside C, it follows that $B_\epsilon(\mathbf{x}) \subset U_2$. ❑

A set that is not disconnected; that is, a set for which no separation exists is said to be **connected**. A lawyer will tell you it is always harder to prove a negative: You have to work harder to prove that a set is connected.

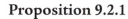

Proposition 9.2.1

An interval $[a, b] \subset \mathbf{R}$ is connected.

Proof Suppose that we have a separation U_1, U_2 of $[a, b]$. We can assume that $a \in U_1$. Let c denote the least upper bound of the set of numbers $K = \{x \in [a, b] : [a, x] \subset U_1\}$.

We can ask, Does c belong to K? If so, then $[a, c] \subset U_1$; since U_1 is open there is $\epsilon > 0$ such that

$$B_\epsilon(c) = (c - \epsilon, c + \epsilon) \subset U_1.$$

Then $[a, c] \cup (c - \epsilon, c + \epsilon/2] \subset U_1$. Since

$$[a, c] \cup (c - \epsilon, c + \epsilon/2] = [a, c + \epsilon/2],$$

there is a number bigger than c, namely $c + \epsilon/2$, that belongs to K. This is impossible, for c is the least upper bound of K.

Thus, $c \notin K$, and hence $[a, c] \not\subset U_1$. However, it is not hard to see that $[a, c) \subset U_1$ (this gap is left to you to fill). Thus, we find that $c \notin U_1$. Hence, $c \in U_2$. Now U_2 is open, so there is $\delta > 0$ such that $(c - \delta, c + \delta) \subset U_2$. The intervals $[a, c)$ and $(c - \delta, c + \delta)$ have a nonempty intersection, namely $(c - \delta, c)$. Thus U_1 and U_2 are not disjoint, and hence do not form a separation. ∎

Proposition 9.2.1 is a basic result about the topology of the real line. For example, it is the main ingredient of the proof of the intermediate value theorem. Here is another corollary.

Proposition 9.2.2

Let $A \subset \mathbf{R}^n$ be a set with the property that for any two points $\mathbf{x}, \mathbf{y} \in A$, there is a continuous function

$$\mathbf{f} : [0, 1] \longrightarrow A$$

such that $\mathbf{f}(0) = \mathbf{x}$ and $\mathbf{f}(1) = \mathbf{y}$. Then A is connected.

Proof Suppose that U_1, U_2 form a separation of A. Let $\mathbf{x} \in U_1 \cap A$ and $\mathbf{y} \in U_2 \cap A$. There is a function $\mathbf{f} : [0, 1] \longrightarrow A$ such that $\mathbf{f}(0) = \mathbf{x}$ and $\mathbf{f}(1) = \mathbf{y}$. Let

$$W_1 = \{t \in [0, 1] : \mathbf{f}(t) \in U_1\} \quad \text{and} \quad W_2 = \{t \in [0, 1] : \mathbf{f}(t) \in U_2\}.$$

For each $t \in [0, 1]$, the point $\mathbf{f}(t)$ falls in either U_1 or U_2; hence $t \in W_1$ or $t \in W_2$. In other words, $[0, 1] \subset W_1 \cup W_2$. Furthermore, $0 \in W_1$ and $1 \in W_2$. We will now show that the W_i are open: If $t \in W_i$, let $\epsilon > 0$ be chosen so that $B_\epsilon(\mathbf{f}(t)) \subset U_i$. This can be done because U_i is open and $\mathbf{f}(t) \in U_i$. Since $\mathbf{f}$ is continuous there is a number $\delta > 0$ such that

$$\|\mathbf{f}(s) - \mathbf{f}(t)\| < \epsilon \text{ when } |s - t| < \delta.$$

Thus $(t - \delta, t + \delta) \subset W_i$. This shows each W_i is open, and thus W_1, W_2 form a separation of $[0,1]$, impossible by Proposition 9.2.1. ∎

If a set A is disconnected, we will refer to its **components**. Points $\mathbf{x}, \mathbf{y} \in A$ are said to be in the same component if there is a connected subset of A that contains $\{\mathbf{x}, \mathbf{y}\}$.

The Jordan curve theorem

A circle is an example of a **Jordan curve**. More generally, a Jordan curve is a plane curve given by parametric equations

$$x = f(t), y = g(t) \text{ for } a \le t \le b,$$

such that

- f and g are continuous on $[a, b]$,
- If $a \le t_1 < t_2 < b$, then the points $(f(t_1), g(t_1))$ and $(f(t_2), g(t_2))$ are distinct (this means the curve does not cross itself), and
- $(f(a), g(a)) = (f(b), g(b))$ (this means the curve is closed).

Jordan curves are also called simple closed curves.

The following theorem was stated, with an incorrect proof, by the French mathematician Camille Jordan in 1887. A correct proof, by the American mathematician Oswald Veblen, appeared in 1905.

Theorem 9.4   Jordan curve theorem

Let $\mathcal{C}$ be a Jordan curve in the plane. Then its complement, $\mathbf{R}^2 - \mathcal{C}$, is disconnected: It has two components, U_1 and U_2, called the inside and outside of $\mathcal{C}$, respectively. They are distinguished as follows: There is a number M such that $\mathcal{C} \subset B_M((0, 0))$. Then $U_1 \subset B_M((0, 0))$, while $U_2 \not\subset B_M((0, 0))$.

Proving the Jordan theorem would lead us on an excursion far from the topic of this text.

Proof of the Poincaré-Bendixson theorem

Consider a system

$$x' = f(x, y), \quad y' = g(x, y), \tag{9.7}$$

where the partial derivatives of f and g are assumed to be continuous.

Recall that we have defined the forward limit point of an orbit $\mathbf{v}(t)$ of (9.7) to be a point that can be expressed as a limit of a sequence $\mathbf{v}(t_n)$, where t_n is an increasing, unbounded sequence of numbers. The forward limit set, denoted $\lim_+ \mathbf{v}$, is the set of all forward limit points.

A subset F of the plane is said to be **closed** if its complement $\mathbf{R}^2 - F$ is open. Thus, you can show that F is closed if and only if $B_\epsilon(\mathbf{x}) \cap F \neq \emptyset$ for every $\epsilon > 0$ implies that $\mathbf{x} \in F$. Recall that a set F is *invariant* if every orbit that intersects F is a subset of F.

Proposition 9.2.3

$\lim_+ \mathbf{v}$ is a closed, invariant set.

Proof We have already mentioned the forward limit set of a solution is invariant (Proposition 8.6.4 on page 420); thus we need only show that $\lim_+ \mathbf{v}$ is closed. Suppose that $B_\epsilon(\mathbf{x}) \cap \lim_+(\mathbf{v}) \neq \emptyset$ for every $\epsilon > 0$. Choose $\mathbf{x}_n$ to be a point in $\lim_+(\mathbf{v}) \cap B_{\frac{1}{n}}(\mathbf{x})$, for $n = 1, 2, 3, \ldots$. Choose t_1 with $\|\mathbf{x}_1 - \mathbf{v}(t_1)\| < 1$, $t_2 > t_1$ such that $\|\mathbf{x}_2 - \mathbf{v}(t_2)\| < \frac{1}{2}$, $t_3 > t_2$ such that $\|\mathbf{x}_3 - \mathbf{v}(t_3)\| < \frac{1}{3}$, and so on. Then by the triangle inequality,

$$\|\mathbf{x} - \mathbf{v}(t_n)\| \leq \|\mathbf{x} - \mathbf{x}_n\| + \|\mathbf{x}_n - \mathbf{v}(t_n)\| < \frac{2}{n},$$

which shows that $\mathbf{x} \in \lim_+(\mathbf{v})$. ∎

Proposition 9.2.4

Suppose that there is a number M such that $\mathbf{v}(t) \in B_M(\mathbf{0})$ for all $t \geq 0$. Then $\lim_+ \mathbf{v}$ is a nonempty, connected set.

Proof The sequence of points $\{\mathbf{v}(n) : n = 0, 1, 2, 3, \ldots\}$ is a **bounded sequence**, since $\|\mathbf{v}(n)\| < M$ for $n \geq 0$. The **Bolzano-Weierstrass theorem** (from advanced calculus) tells us that there is an increasing sequence of integers n_k such that $\mathbf{v}(n_k)$ converges. The limit of this sequence will be a forward limit point.

Suppose that U_1, U_2 form a separation of $\lim_+(\mathbf{v})$, and choose points $\mathbf{x}_1 \in U_1 \cap \lim_+(\mathbf{v})$, and $\mathbf{x}_2 \in U_2 \cap \lim_+(\mathbf{v})$. Let ϵ be a number such that

$$B_\epsilon(\mathbf{x}_i) \subset U_i, \quad i = 1, 2.$$

We can find an increasing sequence $t_n \to \infty$ such that when n is odd, $\|\mathbf{v}(t_n) - \mathbf{x}_1\| < \epsilon$, and when n is even, $\|\mathbf{v}(t_n) - \mathbf{x}_2\| < \epsilon$.

We can consider $\mathbf{v}(t)$ as a continuous function taking each interval $[t_{2n-1}, t_{2n}]$ (for $n \geq 1$) to the plane. Since the sets U_1 and U_2 are disjoint and open, and since $\mathbf{v}(t_{2n-1}) \in U_1$, while $\mathbf{v}(t_{2n}) \in U_2$, it follows from Proposition 9.2.1 that there is a t_n^*, with $t_{2n-1} < t_n^* < t_{2n}$, such that $t_n^* \notin U_1 \cup U_2$. Referring to the Bolzano-Weierstrass theorem again, we find that there is an increasing sequence $t_{n_k}^*$ such that $\mathbf{v}(t_{n_k}^*)$ converges. Since the complement of $U_1 \cup U_2$ is closed and contains each point of the sequence $\mathbf{v}(t_{n_k}^*)$, it must also contain the limit of the sequence, which belongs to $\lim_+(\mathbf{v})$. This contradicts the assumption that U_1, U_2 is a separation of $\lim_+(\mathbf{v})$. ∎

Before turning to the proof of the Poincaré-Bendixson theorem, we must consider one more definition. A line segment T in the plane is called a **transversal** for the system (9.7) if it contains no stationary point of the system, and if, for each $(x, y) \in T$, the vector $f(x, y)\mathbf{i} + g(x, y)\mathbf{j}$, which we call the **phase vector**, is *not* tangent to T. For example, in Section 3.4 we considered the van der Pol system (see page 121). The positive x-axis served as a transversal, shown in Figure 3.15.

We will always treat a transversal as a directed line segment, so that if $\mathbf{q}_1 = (x_1, y_1)$ and $\mathbf{q}_2 = (x_2, y_2) \in T$, we will say $\mathbf{q}_1 < \mathbf{q}_2$ if $x_1 < x_2$. This works unless the transversal is vertical, because then $x_1 = x_2$. In that case, $\mathbf{q}_1 < \mathbf{q}_2$ if $y_1 < y_2$.

The following proposition outlines the important properties of transversals.

Proposition 9.2.5

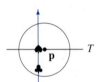

a. All orbits cross a transversal in the same direction.
b. If $\mathbf{p}$ be a point on the plane that is not a stationary point of (9.7), then there is a transversal T with $\mathbf{p} \in T$.
c. If $\mathbf{p}$ is a point on a transversal T, then there is $\epsilon > 0$ such that $B_\epsilon(\mathbf{p})$ has the following property: For every solution $\mathbf{v}(t)$ of (9.7) and t_1 such that $\mathbf{v}(t_1) \in B_\epsilon(\mathbf{p})$, there is a unique t_1' such that $\mathbf{v}(t_1') \in T$ and $\mathbf{v}(t) \in B_\epsilon(\mathbf{p})$ for all t between t_1 and t_1'. [In the drawing to the left, $\mathbf{v}(t_1)$ is marked ♣ and $\mathbf{v}(t_1')$ is marked ♠.]

Proof

a. If T is a transversal and $\mathbf{p} \in T$, the angle $\theta(\mathbf{p})$ made by T and the vector $f(\mathbf{p})\mathbf{i} + g(\mathbf{p})\mathbf{j}$, measured counterclockwise, is not a multiple of π. Let us assume that $0 < \theta(\mathbf{p}) < \pi$. This function θ is continuous on T. Therefore, if $-\pi < \theta(\mathbf{q}) < 0$, where $\mathbf{q}$ is another point on T, the intermediate value theorem tells us that θ has a zero somewhere between $\mathbf{q}$ and $\mathbf{p}$ on T. But then the orbit through that point would be tangent to T, and T would not be a transversal. It follows that $0 < \theta(\mathbf{q}) < \pi$ for all $\mathbf{q} \in T$. ∎
b. Let T' be a line through $\mathbf{p}$ that is not tangent to $f(\mathbf{p})\mathbf{i} + g(\mathbf{p})\mathbf{j}$. We can assume that the measure of the angle $\theta(\mathbf{p})$ between T' and $\mathbf{v}(\mathbf{p})$ is in the interval $(0, \pi)$. Again, since θ is continuous on T', there is an open interval $T \subset T'$ such that $0 < \theta(\mathbf{q}) < \pi$ for all $\mathbf{q} \in T$. ∎
c. To keep the notation simple, assume that T is an interval on the x-axis. Let $F_{t_1, t_2}(x, y)$ be the flow associated with the system (9.7), and define a pair of functions

$$(h(u, t), k(u, t)) = F_{0,t}(u, 0).$$

Thus, if $\mathbf{w}(t)$ is the solution of (9.7) with initial point $\mathbf{w}(0) = (u, 0)$ on the

x-axis, then $(h(u, t), k(u, t)) = \mathbf{w}(t)$. Then $h(u, 0) \equiv u$ and $k(u, 0) \equiv 0$, so

$$\frac{\partial h}{\partial u}(u, 0) = 1 \quad \text{and} \quad \frac{\partial k}{\partial u}(u, 0) = 0.$$

Furthermore, when u is held constant and t varies, $(h(u, t), k(u, t))$ traces an orbit of (9.7); hence

$$\frac{\partial h}{\partial t}(u, 0) = f(u, 0) \quad \text{and} \quad \frac{\partial k}{\partial t}(u, 0) = g(u, 0).$$

We know that $g(u, 0) \neq 0$ when $(u, 0) \in T$ because the vector $\mathbf{i}$ giving the direction of the transversal cannot be parallel to $f(u, 0)\mathbf{i} + g(u, 0)\mathbf{j}$. Therefore, when $(u, 0) \in T$, the matrix

$$D(u, 0) = \begin{bmatrix} \frac{\partial h}{\partial u}(u, 0) & \frac{\partial k}{\partial u}(u, 0) \\ \frac{\partial h}{\partial t}(u, 0) & \frac{\partial k}{\partial t}(u, 0) \end{bmatrix} = \begin{bmatrix} 1 & 0 \\ f(u, 0) & g(u, 0) \end{bmatrix}$$

is nonsingular. We now refer to the **inverse function theorem** (stated in the Chapter Glossary), which implies that there is $\epsilon > 0$ such that for every point $\mathbf{q} \in B_\epsilon(\mathbf{p})$ there is a unique (u, t) such that $\mathbf{q} = (h(u, t), g(u, t))$.

Now suppose that $\mathbf{v}(t)$ is a solution of (9.7) with $\mathbf{v}(t_1) \in B_\epsilon(\mathbf{p})$. Then there is a unique (u_2, t_2) such that

$$\mathbf{v}(t_1) = (h(u_2, t_2), g(u_2, t_2)).$$

Since $\mathbf{z}(t) = \mathbf{v}(t + t_1 - t_2)$ and $(h(u_2, t), g(u_2, t))$ are solutions such that $\mathbf{z}(t_2) = (h(u_2, t_2), g(u_2, t_2))$, it follows from the uniqueness theorem that they are the same. Hence

$$\mathbf{z}(0) = (h(u_2, 0), g(u_2, 0)) = (u_2, 0) \in T;$$

in other words, $\mathbf{v}(t_1 - t_2) \in T$. ∎

If $\mathbf{q}$ is a point on a transversal T, the solution $\mathbf{v}(t)$ of (9.7) with $\mathbf{v}(0) = \mathbf{q}$ may return to T; that is, for some $t_1 > 0$ it may occur that $\mathbf{v}(t_1) \in T$. For points $\mathbf{q} \in T$ with this property, let $P(\mathbf{q})$ denote the point $\mathbf{v}(t_1)$, where t_1 is the least $t > 0$ for which $\mathbf{v}(t_1) \in T$. This P is a mapping that takes a subset of T to T. We call P the **first return mapping.** We have actually used the first return mapping to establish the fact that the van der Pol equation has a limit cycle (see page 123).

The following lemma is the step in the proof of the Poincaré-Bendixson theorem where the Jordan curve theorem makes its appearance.

Lemma 9.2.6

The orbit of a point $\mathbf{q}$ on a transversal T is a cycle if and only if $\mathbf{q}$ is a fixed point of the first return mapping: that is, $P(\mathbf{q}) = \mathbf{q}$.

If $P(\mathbf{q}) > \mathbf{q}$ (see Figure 9.2), then $P(\mathbf{q}_1) > P(\mathbf{q})$ for every point $\mathbf{q}_1$ with $\mathbf{q} \leq \mathbf{q}_1$ and such that $P(\mathbf{q}_1)$ is defined.

If $P(\mathbf{q}) < \mathbf{q}$, then $P(\mathbf{q}_1) < P(\mathbf{q})$ for every point $\mathbf{q}_1$ with $\mathbf{q}_1 \leq \mathbf{q}$ and such that $P(\mathbf{q}_1)$ is defined.

Proof If $P(\mathbf{q}) = \mathbf{q}$, then the solution $\mathbf{v}(t)$ such that $\mathbf{v}(0) = \mathbf{q}$ is periodic—it returns to its starting point. Thus the orbit is a cycle.

Now suppose that $P(\mathbf{q}) > \mathbf{q}$, and that $P(\mathbf{q}_1)$ exists, where $\mathbf{q}_1 > \mathbf{q}$. For convenience, we will assume that the configuration is as shown in Figure 9.2, with T located on the x-axis and with orbits crossing T in the upward direction. The portion of the orbit connecting $\mathbf{q}$ with $P(\mathbf{q})$ and the segment $T_1 \subset T$ with endpoints $\mathbf{q}$ and $P(\mathbf{q})$ form a Jordan curve $\mathcal{C}$. No orbit can cross $\mathcal{C}$ except at a point in T', since orbits of (9.7) cannot cross each other, and orbits that cross T' must do so in the upward (and outward) direction. The orbit through $\mathbf{q}_1$ is directed out of $\mathcal{C}$, if $\mathbf{q}_1 < P(\mathbf{q})$, or starts outside $\mathcal{C}$ if $\mathbf{q}_1 \geq P(\mathbf{q})$. It can therefore never enter the inside component of $\mathbf{R}^2 - \mathcal{C}$ to reach a point of T to the left of $P(\mathbf{q})$.

The proof of the statement when $P(\mathbf{q}) < \mathbf{q}$ is left to you: See Figure 9.3. ■

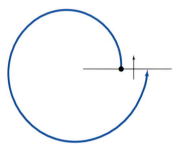

FIGURE 9.2   An orbit starting at $\mathbf{q}$ returns to a transversal at a point $P(\mathbf{q}) > \mathbf{q}$.

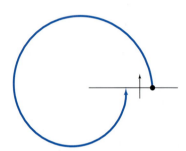

FIGURE 9.3   An orbit starting at $\mathbf{q}$ returns to a transversal at a point $P(\mathbf{q}) < \mathbf{q}$.

The following is an immediate corollary of Lemma 9.2.6.

Corollary 9.2.7

Let $\mathbf{q}$ be a point on a transversal T. The orbit

$$\mathbf{q}, \; P(\mathbf{q}), \; P^2(\mathbf{q}), \; P^3(\mathbf{q}), \; \ldots$$

of $\mathbf{q}$ under the first return mapping P is then either a constant sequence, or a monotone sequence, which may be increasing or decreasing on T.

The final preparation for the proof of the of the Poincaré-Bendixson theorem is given by the following proposition and its corollary:

Proposition 9.2.8

Let $\mathbf{v}(t)$ be an orbit of (9.7), and let T be a transversal. Then $\lim_+(\mathbf{v}) \cap T$ can contain at most one point.

Proof Suppose that $\mathbf{p} \in \lim_+(\mathbf{v}) \cap T$. By Proposition 9.2.5 there is $\epsilon > 0$ such that every orbit that enters $B_\epsilon(\mathbf{p})$ crosses T at exactly one point before leaving $B_\epsilon(\mathbf{p})$. There is an increasing sequence t_n such that $\mathbf{v}(t_n) \to \mathbf{p}$. Thus, there is N such that $\mathbf{v}(t_n) \in B_\epsilon(\mathbf{p})$ for $n > N$, and we can assume that $\mathbf{v}(t_n) \in T$ for each $n > N$. It may not be true that $\mathbf{v}(t_{n+1}) = P(\mathbf{v}(t_n))$, because the orbit traced by $\mathbf{v}(t)$ may cross

T several times between these points. However, if there are k such crossings, then $\mathbf{v}(t_{n+1}) = P^{k+1}(\mathbf{v}(t_n))$, so the sequence

$$\mathbf{v}(t_{N+1}), \mathbf{v}(t_{N+2}), \mathbf{v}(t_{N+3}), \ldots$$

is a *subsequence* of the orbit

$$\mathbf{v}(t_{N+1}), P(\mathbf{v}(t_{N+1})), P^2(\mathbf{v}(t_{N+1})), \ldots, \tag{9.8}$$

which is monotone by Corollary 9.2.7. We conclude that the sequence (9.8) converges to $\mathbf{p}$, because the limit of a monotone sequence must be the same as the limit of any of its subsequences.

If $\mathbf{q}$ were another point in $T \cap \lim_+(\mathbf{v})$, then a sequence of points where the orbit of $\mathbf{v}$ crosses T would converge to $\mathbf{q}$. This sequence would be a subsequence of (9.8), which converges to $\mathbf{p}$. Since a convergent sequence can have only one limit, $\mathbf{q} = \mathbf{p}$.   ∎

Corollary 9.2.9

If $\mathbf{v}(t)$ is an orbit of (9.7) such that $L = \lim_+(\mathbf{v})$ contains a cycle C, then $L = C$.

Proof The complement of a cycle is an open set. If C is a proper subset of L, then every open set that contains C must also contain points of $L - C$, for otherwise L would be disconnected, contrary to Proposition 9.2.4. It follows that there is a sequence of points $\mathbf{p}_n \in L - C$ that converges to a point $\mathbf{p} \in C$. If T is a transversal with $\mathbf{p} \in T$, then there is a number N such that for $n > N$ the orbit of p_n crosses T. Since $L - C$ is invariant, these crossings also belong to $L - C$. This means that T would meet L at an infinite set of points, in contradiction to Proposition 9.2.8.   ∎

Proof of the Poincaré-Bendixson Theorem.   We are given a solution $\mathbf{v}(t)$ such that $L = \lim_+(\mathbf{v})$ is not empty and contains no stationary points. Let $\mathbf{p} \in L$, and let $\mathbf{w}(t)$ be the solution with $\mathbf{w}(0) = \mathbf{p}$. Since L is invariant, the orbit of $\mathbf{w}$ is a subset of L. It follows that $M = \lim_+(\mathbf{w}) \subset L$. M is nonempty because the orbit of $\mathbf{w}(t)$ is bounded. Let $\mathbf{p}' \in M$. By Proposition 9.2.5 there is a transversal $T \ni \mathbf{p}'$. The intersection of the orbit of $\mathbf{w}(t)$ with T can be only at $\mathbf{p}'$, because every point of that orbit is a limit point of $\mathbf{v}(t)$, and Proposition 9.2.8 guarantees that $\mathbf{v}(t)$ has only one limit point on T. It follows that M contains a cycle, and by Corollary 9.2.9, M is equal to this cycle. Since $M \subset L$ we can refer to Corollary 9.2.9 again and conclude that $L = M$.

If $P(\mathbf{p}) = \mathbf{p}$, then the orbit of $\mathbf{v}$ is a cycle. If $P(\mathbf{p}) \neq \mathbf{p}$, then $\mathbf{v}$ converges to a limit cycle.   ∎

The Poincaré-Bendixson theorem is a beautiful piece of pure mathematics that is unfortunately rather difficult to apply to actual ODEs. To show that there is an orbit whose forward limit set does not contain any stationary points, the standard approach is to look for a **trap** that contains no stationary points. (Recall that a trap is a region in the phase plane such that every orbit that crosses its boundary does so by going from outside the region to the inside.)

The following example concerns the van der Pol equation, which we encountered in Chapter 3 (see page 121). We were able to show that when $c = 0.5$, the van der Pol equation had a cycle in the phase plane. The approach used there involved numerical calculations and was not suitable for treating the equation when no particular value of the parameter is specified.

Example 9.2.3   Show that for any positive value of the parameter c, the van der Pol equation,

$$\frac{d^2x}{dt^2} + c(x^2 - 1)\frac{dx}{dt} + x = 0,$$

has a cycle in the phase plane.

Solution. In the phase plane, the van der Pol equation takes the form

$$\left. \begin{aligned} x' &= y \\ y' &= -x + c(1 - x^2)y. \end{aligned} \right\} \tag{9.9}$$

Our analysis is based on three observations. First, the system is symmetric, in the sense that $(x, y) = (\phi(t), \psi(t))$ is a solution implies that $(x, y) = (-\phi(t), -\psi(t))$ is also a solution. Proving this is left to you. Second, let $L(x, y) = x^2 + y^2$, and notice that

$$\begin{aligned} L' &= 2x \cdot x' + 2y \cdot y' \\ &= 2x\,y + 2y(-x + c(1 - x^2)y) \\ &= 2c(1 - x^2)y^2. \end{aligned}$$

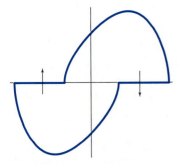

Finally, the x-nullcline is the x-axis and the only stationary point is the origin. Orbits cross the positive x-axis going downward, and they cross the negative x-axis in the upward direction; hence they are clockwise.

Consider an orbit that starts at a point $(a, 0)$ on the positive x-axis, and next crosses the negative x-axis at $(-b, 0)$. The function L measures the square of the distance to the origin, and if $0 < a, b < 1$, then $L' = 2c(1 - x^2)y^2$ is positive. This indicates that the endpoint $(-b, 0)$ is farther from the origin than the starting point $(a, 0)$: $b > a$. If we are only given that $0 < a < 1$, it may happen that $b \geq 1$, but in this case $b > a$ as well. Thus, $b > a$ for all $a < 1$.

Symmetry indicates that a second orbit, starting at $(-a, 0)$, would be the reflection of the first orbit through the origin, and it would therefore cross the positive x-axis at $(b, 0)$. Let $\mathcal{J}$ be the Jordan curve that starts at $(a, 0)$ and follows the orbit until it gets to $(-b, 0)$. From this point, follow the x-axis to the right until $(-a, 0)$ is reached, and then follow the orbit from there to $(b, 0)$. Close the curve by following the x-axis again to the left until $(a, 0)$ is reached. Orbits can cross the curve $\mathcal{J}$ only through the intervals $(-b, -a)$ and (a, b) on the x-axis, and when crossing, they will be directed downward on (a, b) and upward on $(-b, -a)$. Thus no orbit can enter the region bounded by $\mathcal{J}$ from the outside. Figure 9.4 shows the curve $\mathcal{J}$. The trap that we are building will be the region confined between two Jordan curves, $\mathcal{J}$, and a curve $\mathcal{K}$ that we will now construct.

Figure 9.5 shows the graph of the equation

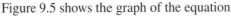

$$y = \frac{x}{c(1 - x^2)},$$

which is the y-nullcline. The nullcline has three branches: A is a curve in the second quadrant with asymptotes $x = -1$ and $y = 0$; C in the fourth quadrant is the reflection of A through the origin; and B passes from the third to the first quadrant through the origin and has asymptotes $x = \pm 1$.

Figure 9.4. The Jordan curve $\mathcal{J}$. An orbit cannot cross the boundary directed inward.

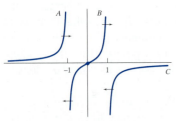

Figure 9.5. The y-nullcline of the van der Pol equation consists of three branches, A, B, and C.

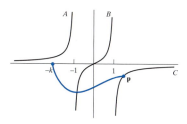

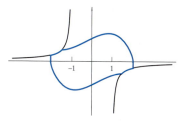

Figure 9.6 The segment of the orbit $\mathcal{O}$ extending from the point **p** on the C-branch of the nullcline to $(-k, 0)$ on the x-axis.

An orbit that crosses the x-axis above the branch C will not subsequently cross C, because to cross C moving from above and to the left, the orbit would have to have a positive slope to cross this nullcline—and because it is a nullcline, the slope would have to be 0.

Consider an orbit $\mathcal{O}$ that has its initial point at **p** $\in C$. This orbit will cross the negative x-axis at a point $(-k, 0)$ (see Figure 9.6). To construct the Jordan curve $\mathcal{K}$, start by following the orbit $\mathcal{O}$ to the x-axis at $(-k, 0)$. Next follow a vertical segment to a point on the branch A of the y-isocline, and from there follow A until reaching the point $-\mathbf{p}$ that is the reflection of **p** through the origin. This give us half of $\mathcal{K}$; to complete this Jordan curve, simply reflect the portion already drawn through the origin. Thus, $\mathcal{K}$ continues by following the orbit $\mathcal{O}'$ that is opposite to $\mathcal{O}$, then drops vertically to the C-branch of the isocline, then follows the isocline back to **p**. This Jordan curve is displayed in Figure 9.7.

We have seen that orbits cannot cross the isocline segments in the outward direction, and of course they cannot cross the portions of $\mathcal{K}$ that are on orbits at all. Since all orbits follow a clockwise orientation, they will be directed into the region surrounded by $\mathcal{K}$ when crossing the vertical segments that form part of $\mathcal{K}$.

The region surrounded by $\mathcal{K}$ is thus a trap. However, it does contain a stationary point, so we cannot apply the Poincaré-Bendixson theorem.

The region $\mathcal{R}$ that is surrounded by $\mathcal{K}$ but lies outside the region surrounded $\mathcal{J}$ is a trap that does not contain the origin. It follows from the Poincaré-Bendixson theorem that there is a cycle in $\mathcal{R}$. Figure 9.8 displays the region $\mathcal{R}$ and its cycle. ❑

Figure 9.7 The Jordan curve $\mathcal{K}$ is outlined in blue. Orbits can enter the region surrounded by $\mathcal{K}$, but they cannot leave.

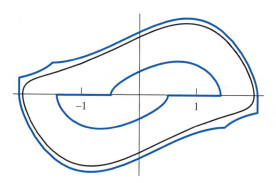

FIGURE 9.8 The trap $\mathcal{R}$ is the region confined by the two blue Jordan curves; the periodic orbit that is contained within $\mathcal{R}$ is shown in black.

EXERCISES

1. Let $\mathcal{C}$ be the circle in $\mathbf{R}^3$ with the equations

$$x^2 + y^2 = 1$$
$$z = 0.$$

Show that $\mathbf{R}^3 - \mathcal{C}$ is connected.

2. Finish the proof of Lemma 9.2.6.

3. The complications involved in proving the Jordan curve theorem are both local and global. On the small scale, it isn't so clear that the complement of a fractal curve like the Koch snowflake[1] has two components. Since we have only

---

[1] If you're not familiar with Koch snowflake, you can find information on it by using a Web search engine, such as http://www.google.com.

used piecewise differentiable Jordan curves, this complication does not affect us. To understand the global complications, consider a Jordan curve that is drawn on a surface.

a. Give an example (if possible) of a Jordan curve on a
torus (surface of an inner tube) with a disconnected
complement.

b. Give an example (if possible) of a Jordan curve on a
torus with a connected complement.

c. Repeat (a) and (b), where the surface is a Möbius band
(a strip of paper that is given a half-twist and then taped
together).

4. Points on a torus can be located by two angular coordinates. To see this, note that one angular coordinate θ is
all that is needed to locate a point in the unit circle in the
xy-plane. As usual, $x = \cos(\theta)$ and $y = \sin(\theta)$. A torus
is formed by rotating the unit circle about the line $x = 2$
(we are picturing the z-axis as vertical and pointed upward
when the xy-plane is on the surface of a table). A rotation
through an angle ϕ about $x = 2$ takes a point (x, y) to
$((x - 2)\cos(\phi) + 2, y, (x - 2)\sin(\phi))$ in space. Thus the
pair (θ, ϕ) corresponds to a point on the torus with xyz-
coordinates

$$((\cos(\theta) - 2)\cos(\phi) + 2, \sin(\theta), (\cos(\theta) - 2)\sin(\phi)).$$

An autonomous system of differential equations on the
torus can thus take the form

$$\theta' = f(\theta, \phi) \qquad \phi' = g(\theta, \phi).$$

The complication is that for a torus, coordinates are only
defined modulo 2π. Thus, the solution of the system
$\theta' = \phi' = 1$ with initial conditions $\theta(0) = \phi(0) = 0$ is
$\theta = \phi = t$. The orbit is a closed curve that winds once
around the torus, since $\theta(2\pi) = \phi(2\pi) = 0 \mod 2\pi$. Describe the orbits of the solution of the following systems,
all with initial conditions $\theta(0) = \phi(0) = 0$.

a. $\theta' = 2, \phi' = 1$ b. $\theta' = 2, \phi' = 3$
c. $\theta' = 141, \phi' = 100$ d. $\theta' = 1.41, \phi' = 1$
e. $\theta' = \sqrt{2}, \phi' = 1$

5. In Example 9.2.3 it was shown that when the parameter c is
positive, the van der Pol equation has a cycle in the phase
plane. When $c = 0$, the equation reduces to $y'' + y = 0$,
and the orbit of every solution is a cycle. What happens if
$c < 0$?

6. Verify the symmetry of the van der Pol equation: If $x = \phi(t)$ is a solution, then $x = -\phi(t)$ is a solution, too. Remember, the van der Pol equation isn't linear.

7. The van der Pol equation with parameter $c > 0$ has exactly one cycle in the phase plane, and all orbits except the
stationary orbit converge to it. Perform computer experiments to determine how the shape of the cycle evolves as c
increases from near 0 to large values. How does the period
of a periodic solution of the van der Pol equation vary with
c?

8. Let C be a cycle for an autonomous system of two ODEs,
and let U be the set of points that are on orbits, other than
C itself, that converge to C. Show that U is an open set.

◆ **Hint:** Show that if the solution $\mathbf{v}(t)$ with $\mathbf{v}(0) = \mathbf{p}$
crosses a transversal T, then there is $\epsilon > 0$ such that every
solution $\mathbf{w}(t)$ with $\mathbf{w}(0) \in B_\epsilon(\mathbf{p})$ crosses T as well.

9. Suppose that an autonomous system of ODEs in the plane
has exactly two cycles, one surrounding the other, and that
no stationary points are located between the cycles. Show
that one of the cycles is the forward limit set of all orbits
whose initial points lie between the cycles, and that the
other cycle is the backward limit set. You may use the fact
that the region between the cycles is connected.

10. Let $r(x, y) = |(x + iy)^2 + 1|$. This function satisfies a
Lipschitz condition on $\mathbf{R}^2$, and $r(x, y) \geq 0$, with equality
if and only if $(x, y) = (0, \pm 1)$. This exercise concerns a
system

$$\begin{cases} x' = r(x, y)[y + x(1 - \sqrt{x^2 + y^2})] \\ y' = r(x, y)[-x + y(1 - \sqrt{x^2 + y^2})]. \end{cases} \quad (9.10)$$

a. Let $L(x, y) = x^2 + y^2$. Calculate L' and hence show
that the circle C of points such that $L(x, y) = 1$ is invariant.

b. Show that as subsets of the plane, the orbits that are
not subsets of C are identical to the orbits of the system (8.23) on page 395. Hence show that C is the forward limit set of every orbit of (9.10) except the orbits
that are subsets of C.

c. Show that C is not a cycle for (9.10) [although it is a
cycle for (8.23)]. Does this observation contradict the
Poincaré-Bendixson theorem?

d. Construct a system that has an orbit whose forward limit
set contains exactly three stationary points.

e. Suppose a system has an orbit whose forward limit set
L includes exactly one stationary point. Is it possible
that L contains more than one point? Explain.

9.3   **Three-Dimensional Systems**

Systems of differential equations serve as models for many phenomena of physics, chemistry, and biology, and these models often involve more than two equations. It is then necessary to use a higher-dimensional *phase space*. Thus, for example, the phase portrait of a system of three autonomous equations would be three dimensional. Although the results that we have established for systems of two equations in Chapter 8 hold for larger systems, the results in Section 9.2 do not. Thus, in moving from two-dimensional phase space to three dimensions, the degree of complexity that is possible jumps dramatically.

A system of differential equations is said to be sensitive to initial conditions if small changes in initial conditions can cause dramatic changes in the solutions. By this definition, the linear growth equation $y' = y$ is sensitive to initial conditions because nearby solutions diverge from one another exponentially as t increases. We have not yet encountered solutions of differential equations that are *bounded* and still diverge from one another. However, this can happen with dynamical systems—even in the two-dimensional case.

The weather seems to be inherently unpredictable. A mathematical model for the weather would involve a system of partial differential equations based on the physics of fluid motion and heat convection. In 1963, E. N. Lorenz, an American meteorologist, published[2] the results of a numerical study of the system

$$\left. \begin{array}{l} x' = \sigma(y - x) \\ y' = x(r - z) - y \\ z' = xy - bz, \end{array} \right\} \tag{9.11}$$

which was derived from a partial differential equation describing heat convection in a gas by making many simplifying assumptions. Lorenz reasoned that if we can't solve this system, we won't have much hope of handling the partial differential equation. The title of his paper, "Deterministic non-periodic flows," describes what he found. In the system (9.11), σ, r, and b are thermodynamic parameters, each representing a positive quantity.

With any set of initial conditions, the system (9.11) has a unique solution. However, the solution is *extremely* sensitive to changes in the initial conditions, changes that inevitably result when data are recorded with finite precision. Small truncation errors can also create problems in numerical approximation. In other words, the solution of the Lorenz equations is not predictable, even though it is determined by the data! The Lorenz equations are the subject of numerous research papers and at least one book.[3] Figure 9.9 displays a projection of *one* orbit of the Lorenz equations. You are encouraged to draw orbits of the Lorenz equations in the laboratory. The truly chaotic nature of the solution is best appreciated by watching it develop on the computer display.

---

[2] "Deterministic non-periodic flows," *Journal of Atmospheric Science* **20** (1963), pp. 130–141.

[3] *The Lorenz Equations: Bifurcations, Chaos, and Strange Attractors*, by Colin Sparrow. New York: Springer-Verlag, 1982.

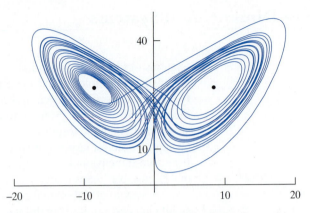

FIGURE 9.9   The projection on the x, z-plane of part of one orbit of the Lorenz equations, with parameters $\sigma = 10$, $r = 28$, and $b = 8/3$. The projections of the equilibrium points $Q_\pm$ are marked.

To find the stationary points of the Lorenz equations, we must find the solutions of

$$\sigma(y - x) = 0$$
$$x(r - z) - y = 0$$
$$xy - bz = 0.$$

It is clear that one solution is $(x, y, z) = (0, 0, 0)$. To find other solutions, notice that the first equation holds if and only if $x = y$. Setting $y = x$ in the second and third equations, we have

$$x(r - z - 1) = 0 \tag{9.12}$$
$$x^2 - bz = 0. \tag{9.13}$$

If $x = 0$, these equations tell us that $y = z = 0$ as well. Hence we can factor x from equation (9.12), so that $z = r - 1$. Turning to equation (9.13), we have $x^2 = b(r - 1)$. If $r < 1$, there is no solution, but if $r \geq 1$, we have found the stationary points

$$Q_\pm = \left(\pm\sqrt{b(r - 1)}, \pm\sqrt{b(r - 1)}, r - 1 \right).$$

To determine the stability of $Q_\pm$, we will apply Theorem 8.3 on page 400 (the theorem is stated for systems of two equations, but it is valid for any number of equations). For the Lorenz equations, the Jacobian at (x, y, z) is

$$J(x, y, z) = \begin{bmatrix} -\sigma & \sigma & 0 \\ r - z & -1 & -x \\ y & x & -b \end{bmatrix}.$$

It is easy to see that the characteristic roots of $J(0, 0, 0)$ are $-b$, and

$$\frac{1}{2}\left[-(\sigma + 1) \pm \sqrt{(\sigma + 1)^2 - 4(1 - r)\sigma} \right].$$

It follows that if $r < 1$, all three characteristic roots are negative, and the origin is asymptotically stable. For $r > 1$, there are still three real characteristic roots, but one of them is positive. Therefore, for $r > 1$, the origin has a one-dimensional unstable manifold.

When $r > 1$, we must consider the points $Q_\pm$. The characteristic equation of the Jacobian at each of these points turns out to be

$$s^3 + (\sigma + b + 1)s^2 + b(r + \sigma)s + 2b\sigma(r - 1) = 0. \tag{9.14}$$

Solving a cubic equation is troublesome, so we turn to the following lemma.

Lemma 9.3.1

Let a, b, and c be positive numbers. Then all real roots of the equation

$$s^3 + as^2 + bs + c = 0$$

are negative. If $ab > c$, there will either be three negative real roots, or one negative real root and a pair of conjugate complex roots with negative real parts. If $ab = c$, the roots are $s = -a$ and $s = \pm\sqrt{b}i$. If $ab < c$, then there will be one real root and a pair of conjugate complex roots with positive real parts.

The proof of Lemma 9.3.1 is elementary, but it is omitted because it has nothing to do with differential equations. The lemma is a special case of the Routh-Hurwitz conditions, which are necessary and sufficient conditions, based on the coefficients of a polynomial, for all of the roots of the polynomial to lie in the left half of the complex plane.

We can apply the lemma to determine the stability of $Q_\pm$. As long as $r > 1$, each coefficient of the characteristic equation (9.14) is positive. Therefore, all real characteristic roots are negative. Provided that

$$(\sigma + b + 1) \cdot b(r + \sigma) > 2b\sigma(r - 1),$$

all complex characteristic roots will also lie in the left half-plane, and both stationary points will be stable. We can divide through by b and rearrange the inequality in the form

$$(\sigma - b - 1)r < \sigma(\sigma + b + 3).$$

Assuming that $\sigma > b + 1$, we conclude that $Q_\pm$ are asymptotically stable if

$$r < \frac{\sigma(\sigma + b + 3)}{(\sigma - b - 1)}. \tag{9.15}$$

If the inequality (9.15) is replaced by equality, then by Lemma 9.3.1 there is a pair of pure imaginary characteristic roots, and the stability of $Q_\pm$ is not determined. Finally, when

$$r > \frac{\sigma(\sigma + b + 3)}{(\sigma - b - 1)},$$

there is a pair of complex characteristic roots with positive real parts, and a negative real root. In Lorenz's study, $\sigma = 10$, and $b = 8/3$. Thus, $\sigma > b + 1$, and by inequality (9.15), the stationary points $Q_\pm$ are stable for $r < 470/19 \approx 24.7$. When r exceeds that critical value, there are no longer any stable stationary points. Using $r = 28$, Lorenz discovered that while instability was evident, the orbits seemed to be bounded and to alternate unpredictably between two spirals.

Nonautonomous systems

Consider systems of two nonautonomous first-order equations

$$\left. \begin{array}{l} x' = f(x, y, t) \\ y' = g(x, y, t). \end{array} \right\} \tag{9.16}$$

We will assume that the functions f and g on the right side of the system (9.16) are continuous and possess continuous partial derivatives with respect to x and y. These assumptions will enable us to invoke the uniqueness theorem. The three-dimensional character of this system is obvious, because f and g depend on three variables. We can replace (9.16) with an autonomous system of three ODEs by introducing a new independent variable τ:

$$\left. \begin{array}{l} \dfrac{dx}{d\tau} = f(x, y, t) \\[2mm] \dfrac{dy}{d\tau} = g(x, y, t) \\[2mm] \dfrac{dt}{d\tau} = 1. \end{array} \right\} \tag{9.17}$$

The third equation in (9.17), coupled with an initial condition $t(0) = 0$, ensures that as far as the first two equations are concerned, $t \equiv \tau$. Thus any information that we obtain about (9.17) is applicable to (9.16). For example, if some sort of chaos occurs in (9.17), we can expect the same in (9.16).

Suppose that that f and g in (9.16) have the following periodicity property: For some number T, and all t, $f(x, y, T + t) = f(x, y, t)$ and $g(x, y, T + t) = g(x, y, t)$. We express this property by saying that the right side of the system (9.16) is T-periodic in t. Points (x, y, t) and (x, y, t_1) in the phase space of (9.17) should then be considered equivalent if $t - t_1$ is an integer multiple of T.

The **first return mapping**, introduced in Section 9.2 in the context of a two-dimensional system of ODEs, can be used in three dimensions as well. The role of a transversal, which is to serve as the domain and range of the first return mapping, will be played by the plane $t = 0$ in the xyt-phase space of the system (9.17). The phase vector, $f(x, y, t)\mathbf{i} + g(x, y, t)\mathbf{i} + \mathbf{k}$, cannot be tangent to this plane, because all tangent vectors to $t = 0$ have zero $\mathbf{k}$ components.

The first return mapping associated with the T-periodic system (9.16) is a mapping P_T of the xy-plane to itself. P_T is defined, in terms of the flow F associated with the system (9.17), by advancing the independent variable τ through T time units:

$$F_{T,0}(x, y, 0) = (P_T(x, y), T).$$

We can call this a first return mapping because the T-periodicity of the system means that the planes $t = 0$ and $t = T$ in phase space are the same. Thus, when the variable t has advanced from $t = 0$ to $t = T$, the orbit has "returned" to the plane $t = 0$.

Proposition 9.3.2

Let $(x, y) = (\phi(t), \psi(t))$ be a solution of the T-periodic system (9.16). Then for all integers n and $t \in [0, T)$ such that $(\phi(nT + t), \psi(nT + t))$ is defined

$$(\phi(nT + t), \psi(nT + t)) = P_T^n(\phi(t), \psi(t)).$$

Proof We assume $n \geq 0$, and proceed by induction on n. For $n = 0$, the statement is obviously true. Assume that $m \geq 1$ and that the statement holds for all $n < m$. Let $(x, y) = (\bar{\phi}(t), \bar{\psi}(t))$ be the solution of the system (9.16) with initial point

$$(\bar{\phi}(0), \bar{\psi}(0)) = P_T(\phi(0), \psi(0)).$$

We will show that

$$(\bar{\phi}(t), \bar{\psi}(t)) = (\phi(t + T), \psi(t + T)). \tag{9.18}$$

Since $\phi'(t) = f(\phi(t), \psi(t), t)$, it follows that

$$\phi'(t + T) = f(\phi(t + T), \psi(t + T), t + T)$$
$$= f(\phi(t + T), \psi(t + T), t)$$

because f is T-periodic in t. Similarly,

$$\psi'(t + T) = g(\phi(t + T), \psi(t + T), t).$$

Thus $(x, y) = (\phi(t+T), \psi(t+T))$ is a solution of the system (9.16). Its initial point is $P_T(\phi(0), \psi(0))$, the same as the initial point of $(\bar{\phi}(t), \bar{\psi}(t))$. By the uniqueness theorem, the two solutions are identical, and this proves that equation (9.18) holds. Hence

$$(\bar{\phi}((m - 1)T + t), \bar{\psi}((m - 1)T + t)) = (\phi(mT + t), \psi(mT + t)).$$

By the induction hypothesis,

$$(\bar{\phi}((m - 1)T + t), \bar{\psi}((m - 1)T + t)) = P_T^{m-1}(\bar{\phi}(t), \bar{\psi}(t))$$
$$= P_T^m(\phi(t), \psi(t)),$$

and hence $(\phi(mT + t), \psi(mT + t)) = P_T^m(\phi(t), \psi(t))$. This advances the induction and completes the proof. ∎

The first return mapping is one to one (see Exercise 1), and therefore defines a dynamical system on the plane.

Proposition 9.3.3

If (x^*, y^*) is a fixed point of P_T, then the solution of the system (9.16) with initial values $(x(0), y(0)) = (x^*, y^*)$ is T-periodic solution. More generally, a solution whose initial values are at an n-periodic point (x^*, y^*) is nT-periodic.

If (x^*, y^*) is an asymptotically stable fixed or periodic point of P_T, then the corresponding periodic solution of the system (9.16) is also stable.

Proof Suppose that (x^*, y^*) is a fixed point of P_T, and $(x(t), y(t))$ is the solution of the system with initial values $(x(0), y(0)) = (x^*, y^*)$. Then

$$(x(T), y(T)) = P_T(x^*, y^*) = (x^*, y^*) = (x(0), y(0)).$$

It follows that $(x(t), y(t))$ and $(x(t + T), y(t + T))$ are solutions of the same initial value problem; by the uniqueness theorem $(x(t), y(t)) \equiv (x(t + T), y(t + T))$.

If (x^*, y^*) is an n-periodic point of P_T, then it is a fixed point of $P_T^n = P_{nT}$. It follows from what we have just proved that the solution with initial point at (x^*, y^*) is nT-periodic.

Now suppose that (x^*, y^*) is an asymptotically stable fixed point of P_T, and let $(x(t), y(t))$ denote the solution with initial point (x^*, y^*). Let S be a circle, centered at (x^*, y^*) such that for all points (x_0, y_0) inside S, $\lim_{m \to \infty} P_T^m(x_0, y_0) = (x^*, y^*)$. Let $(x, y) = (\phi(t), \psi(t))$ be a solution with $(\phi(0), \psi(0))$ inside S. Then

$$\lim_{m \to \infty} (\phi(mT), \psi(mT)) = \lim_{m \to \infty} P_T^m(\phi(0), \psi(0)) = (x^*, y^*).$$

Since the solution $(x(t), y(t))$ is T-periodic, $(x(mT), y(mT)) = (x^*, y^*)$ for all m. It follows that $\lim_{m \to \infty} \|(\phi(mT), \psi(mT)) - (x(mT), y(mT))\| = 0$.

This proof is not quite complete, because we have not shown that

$$\lim_{t \to \infty} \|(\phi(t), \psi(t)) - (x(t), y(t))\| = 0,$$

having handled only the case where t runs through multiples of T. The remaining steps would cause an excessive digression and are omitted. ∎

Floquet theory

The first return mapping was defined by H. Poincaré in 1893. Ten years earlier, another French mathematician, Gustave Floquet, had considered a special case: systems of linear differential equations with periodic coefficients. Let $A(t)$ be a matrix whose entries are T-periodic functions of t, and let P_T denote the first return mapping associated with the linear system

$$\mathbf{x}' = A(t)\mathbf{x}. \tag{9.19}$$

If the coefficient matrix $A(t)$ is continuous on $\mathbf{R}$, then by Theorem 4.11 on page 167, the system (9.19) has a fundamental matrix solution $\mathcal{X}(t)$, also defined for all $t \in \mathbf{R}$. Of course, analytic expressions for the entries of $\mathcal{X}(t)$ will not usually be available, but for any value of t the entries of $\mathcal{X}(t)$ can be calculated with great precision by using numerical methods.

Given an initial point $\mathbf{a} \in \mathbf{R}^n$, the solution of the system (9.19) satisfying $\mathbf{x}(0) = \mathbf{a}$ is $\mathbf{x}(t) = \mathcal{X}(t)\mathcal{X}(0)^{-1} \cdot \mathbf{a}$. By definition of P_T, $P_T(\mathbf{a}) = \mathbf{x}(T)$. Thus

$$P_T\mathbf{a} = \mathcal{X}(T) \cdot [\mathcal{X}(0)]^{-1} \cdot \mathbf{a}. \tag{9.20}$$

This establishes the first result of Floquet theory:

Proposition 9.3.4

Let $A(t)$ be an $n \times n$ matrix whose entries are continuous, T-periodic functions. Then the first return mapping P_T associated with the system $\mathbf{x}' = A(t)\mathbf{x}$ is a linear dynamical system.

The characteristic roots of the first return mapping P_T are called the **Floquet multipliers** of the system.

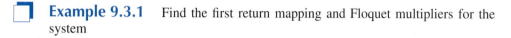 **Example 9.3.1**   Find the first return mapping and Floquet multipliers for the system

$$\begin{bmatrix} x' \\ y' \end{bmatrix} = \begin{bmatrix} 0 & 1 \\ -[\pi^2 + 2\cos(2\pi t)]/4 & 0 \end{bmatrix} \begin{bmatrix} x \\ y \end{bmatrix}. \tag{9.21}$$

Solution. The coefficient matrix is 1-periodic, so we will compute P_1. Let $\mathbf{v}_1(t)$ and $\mathbf{v}_2(t)$ denote the solutions with initial conditions

$$\mathbf{v}_1(0) = \begin{bmatrix} 1 \\ 0 \end{bmatrix} \quad \text{and} \quad \mathbf{v}_2(0) = \begin{bmatrix} 0 \\ 1 \end{bmatrix}.$$

Using a program based on the fourth-order Runge-Kutta algorithm with step size 0.01, it was found that

$$\mathbf{v}_1(1) = \begin{bmatrix} 0 \\ -1.5196 \end{bmatrix} \quad \text{and} \quad \mathbf{v}_2(1) = \begin{bmatrix} 0.658 \\ 0 \end{bmatrix}.$$

Hence,

$$P_1 = \begin{bmatrix} 0 & 0.658 \\ -1.520 & 0 \end{bmatrix}.$$

Since trace $(P_1) = 0$ and $\det(P_1) = 1$, the characteristic equation of P_1 is $s^2 + 1 = 0$, and the Floquet multipliers are therefore $\pm i$.   ❑

The following theorem, which states the relationship between Floquet multipliers and stability of solutions of systems with periodic coefficients, may be compared with the constant coefficient case (Theorem 8.1 on page 391).

Theorem 9.5

Let $A(t)$ denote an $n \times n$ matrix whose entries are T-periodic functions of t, and let $\mathcal{F} = \{\mu_1, \mu_2, \dots, \mu_k\}$ denote the set of Floquet multipliers of the system

$$\mathbf{x}' = A(t)\mathbf{x}. \tag{9.22}$$

i. Every solution $\mathbf{x}(t)$ of the system (9.22) converges to $\mathbf{0}$ as $t \to \infty$ if and only if $|\mu_j| < 1$ for all $\mu_j \in \mathcal{F}$.

ii. The system (9.22) has a T-periodic solution if and only if $1 \in \mathcal{F}$.

iii. The system (9.22) has a $2T$-periodic solution that is not T-periodic if and only if $-1 \in \mathcal{F}$, and a $4T$-periodic solution that is not $2T$-periodic if and only if $i \in \mathcal{F}$

iv. The system (9.22) has a solution that is unbounded as $t \to \infty$ if $\mathcal{F}$ contains a multiplier μ with $|\mu| > 1$.

Let us apply this theorem to the system in Example 9.3.1. The entries of the coefficient matrix are 1-periodic, and the Floquet multipliers are $\pm i$. Therefore, the system has a 4-periodic solution; in fact, all solutions are 4-periodic. This system is a special case of a class of ODEs called Hill's equation. For more information, see Exercise 7 at the end of this section.

The proof of Theorem 9.5 is based on the following lemma.

Lemma 9.3.5

Let L be a logarithm of the matrix P_T representing the first return mapping for the system (9.22). If $\mathbf{x}(t)$ is a solution of the system, then $\mathbf{y}(t) = e^{-(t/T)L}\mathbf{x}(t)$ is T-periodic.

Proof By the definition of logarithm, $e^L = P_T$. Hence for any t,

$$\mathbf{y}(t + T) = e^{-((t+T)/T)L}x(t + T)$$
$$= e^{-(t/T)L-L} P_T \mathbf{x}(t)$$
$$= e^{-(t/T)L}e^{-L}e^L\mathbf{x}(t) = \mathbf{y}(t)$$

after canceling the $e^{-L}e^L$ factor.   ■

Proof of Theorem 9.5.   To prove statement (i), consider a solution $\mathbf{x}(t)$ of the system (9.22). By Theorem 9.1 (page 436) the square of every nonsingular matrix has a real logarithm. Hence $P_{2T} = P_T^2$ has a real logarithm, which we will denote $2L$. By Lemma 9.3.5, there is a $2T$-periodic[4] function $\mathbf{y}(t)$ such that

$$\mathbf{x}(t) = e^{(t/T)L}\mathbf{y}(t).$$

The characteristic roots of L are the logarithms of the Floquet multipliers. Therefore, the multipliers lie inside the unit disc of the complex plane if and only if the characteristic roots of L have negative real parts, and this occurs if and only if

$$\lim_{t \to \infty} e^{(t/T)L} = 0.$$

It follows that the characteristic roots of P_T lie inside the unit disk of the complex plane if and only if $\lim_{t \to \infty} \mathbf{x}(t) = \mathbf{0}$.   ■

We now turn to part (ii). If there is a T-periodic solution $\mathbf{x}_p(t)$, then $\mathbf{x}_p(T) = \mathbf{x}_p(0)$; in other words,

$$P_T\mathbf{x}_p(0) = \mathbf{x}_p(0).$$

This means that $\mathbf{x}_p(0)$ is a characteristic vector of P_T with characteristic root 1. The Floquet multipliers are the characteristic roots of P_T, and therefore 1 is a Floquet

---

[4]$\mathbf{y}(t)$ is T-periodic if P_T has a real logarithm.

multiplier. Conversely, if $1 \in \mathcal{F}$, the corresponding characteristic vector is a fixed point of P_T, and by Proposition 9.3.3 determines a T-periodic solution.   ∎

The proof of part (iii) is similar to that of part (ii) and is left to you.

To prove part (iv), suppose first that P_T has a characteristic root μ that lies outside the unit circle of the complex plane, and let $\mathbf{f}$ denote a characteristic vector belonging to μ. Let $\mathbf{x}(t)$ be the solution of the system (9.22) with initial condition $\mathbf{x}(0) = \mathbf{f}$. Then

$$\|\mathbf{x}(nT)\| = \|P_T^n \mathbf{f}\| = |\mu|^n \|\mathbf{f}\|.$$

Since $|\mu| > 1$, $\lim_{n \to \infty} |\mu|^n = \infty$, and therefore $\mathbf{x}(t)$ is an unbounded solution.   ∎

General first return mappings

Consider a system

$$\left. \begin{array}{l} x' = f(x, y, z) \\ y' = g(x, y, z) \\ z' = h(x, y, z). \end{array} \right\} \tag{9.23}$$

The orbits of the system (9.23) will be curves in three-dimensional space. Let Π be a segment of a plane in phase space. We will say that Π is a **transversal** if the phase vector

$$f(x, y, z)\mathbf{i} + g(x, y, z)\mathbf{j} + h(x, y, z)\mathbf{k}$$

is *not* parallel to Π at *any* point $(x, y, z) \in \Pi$.

Let Π be a transversal to the system (9.23) and $\mathbf{e} \in \Pi$, and let $\mathbf{v}(t)$ be the solution of the system (9.23) with $\mathbf{v}(0) = \mathbf{e}$. Let $t_1 > 0$ denote the least positive number such that $\mathbf{v}(t_1) \in \Pi$, if such a number exists, and define $P(\mathbf{e}) = \mathbf{v}(t_1)$. Of course, it is possible that the orbit defined by $\mathbf{v}(t)$ never intersects Π for $t > 0$; in that case, $P(\mathbf{e})$ would be undefined.

The mapping P is the first return mapping of the plane Π defined by the system (9.23). Exercise 2 asks the reader to show that P is one to one. That result implies that P generates a dynamical system on a subset of Π. Fixed points of P correspond to periodic orbits (recall that by Theorem 3.3 on page 120, an orbit that returns to its starting point must be periodic), and the stability of the periodic orbit is determined by the stability of the fixed point. Unlike the two-dimensional case, in three dimensions a single periodic orbit may intersect a transversal at several different points; if there are exactly n points of intersection, then these points form an n-periodic orbit of the dynamical system defined by P.

EXERCISES

1. Show that the first return mapping P_T associated with a system of differential equations with T-periodic right side is one to one.

2. Show that the first return mapping of any plane $\Pi \subset \mathbf{R}^3$, defined by a system of three autonomous differential equations, is one to one.

3. Prove part (iii) of Theorem 9.5.

4. Can 0 be a Floquet multiplier of a system $\mathbf{x}' = A(t)\mathbf{x}$ of differential equations with T-periodic coefficients?

5. Calculate the first return mapping for the system

$$\left. \begin{array}{l} x' = y \\ y' = -2(x + y) + 5\sin t. \end{array} \right\} \tag{9.24}$$

6. Use the result of Exercise 5 to show that the system (9.24) has a unique stable periodic orbit.

7. Hill's equation. The differential equation

$$x'' + [\lambda + f(t)]x = 0,$$

where λ is a parameter, treated as a constant, and f is periodic, is called Hill's equation. It is named for an American astronomer, George William Hill, who used it as a model for the orbit of the moon in an 1877 paper that profoundly influenced the future of celestial mechanics. The problem with the moon is that it is influenced by the gravitation of both the sun and the earth. Before Hill, research on the moon's orbit had focused on the three-body problem, which is notoriously intractable. Hill's equation approximates by assuming that the earth and the moon have negligible effects on the motion of the sun.

a. Let $\{x_1(t), x_2(t)\}$ be a fundamental set of solutions of Hill's equation. Show that the Wronskian $W[x_1, x_2](t)$ is constant.

  ◆ **Hint:** Differentiate the Wronskian and simplify the result.

b. Let T be the period of $f(t)$. Show that the determinant of the first return mapping P_T for Hill's equation is equal to 1.

c. Show that the Floquet multipliers lie on the unit circle if

$$\text{tr}(P_T) \le 2$$

and that in this case, P_T is a rotation of the plane.

d. Show that if $\text{trace}(P_T) > 2$, then the zero solution of Hill's equation is unstable.

e. Let $f(t) = \epsilon \cos(2\pi t)$. Calculate $\text{trace}(P_1)$ when $\epsilon = 0$.

f. What will happen if ϵ is small and positive in part (e)?

8. This problem introduces a differential equation that has been the subject of recent research: the forced Duffing equation

$$x'' + \epsilon x' - (x - x^3) = \lambda \sin(\omega t), \qquad (9.25)$$

where the parameters ϵ, λ, and ω represent real numbers that may be positive or 0.

a. Draw a phase portrait for equation (9.25) when $\epsilon = \lambda = 0$.

  ◆ **Hint:** Find an energy integral.

b. Assume that $\epsilon > 0$, but $\lambda = 0$ in equation (9.25), so that the equation is still autonomous. Show that the origin is a saddle point and that there are two stable equilibria.

  ◆ **Hint:** The total energy will be a Lyapunov function.

c. Draw a phase portrait for the case where ϵ is positive, and $\lambda = 0$.

d. If $T = 2\pi/\omega$ is the period of the right side of equation (9.25), we can consider the first return mapping P_T. Show that for $\lambda = 0$, P_T has three fixed points.

e. This part is optional and open ended. It is reasonable to suspect that P_T will still have three fixed points if λ is small but positive. Do a computer study to verify that this is the case. What happens as λ increases?

9. Show that if $\mathcal{X}(t)$ is a fundamental matrix solution of the system $\mathbf{x}' = A(t)\mathbf{x}$ with T-periodic coefficients, then there is a $2T$-periodic matrix function $\mathcal{Z}(t)$ and a real matrix M such that $\mathcal{X}(t) = e^{Mt}\mathcal{Z}(t)$.

9.4 Glossary

$B_\epsilon(\mathbf{p})$ The set of points $\mathbf{q}$ whose distance from $\mathbf{p}$, $\|\mathbf{q} - \mathbf{p}\|$, is less than ϵ.

lim$_+$ Forward limit set.

T-periodic in t   (T is a positive real number) A function $f(x, y, t)$ with the property

$$f(x, y, t + T) = f(x, y, t).$$

Asymptotically stable   (fixed point of a mapping P) A fixed point $\mathbf{p}^*$ such that there is $\epsilon > 0$ such that the orbit of each point in ϵ $B_\epsilon(\mathbf{p}^*)$ converges to $\mathbf{p}^*$.

Bolzano-Weierstrass theorem   Let

$$\mathbf{x}_1, \mathbf{x}_2, \mathbf{x}_3, \ldots$$

be a bounded sequence in $\mathbf{R}^n$. Then there are integers

$$0 < k_1 < k_2 < k_3 < \cdots$$

and a point $\mathbf{x}^* \in \mathbf{R}^n$ such that

$$\mathbf{x}_{k_1}, \mathbf{x}_{k_2}, \mathbf{x}_{k_3}, \ldots$$

converges to $\mathbf{x}^*$. For example, the sequence of real numbers $x_k = \sin(k)$ is a bounded. The Bolzano-Weierstrass theorem tells us that there is an increasing sequence of integers k_n such that $x_{k_n} = \sin(k_n)$ converges to some number. Here's a proof of this fact (the proof of the Bolzano-Weierstrass theorem uses the same idea). We know that for all k, $x_k \in [-1, 1]$. One of the subintervals, $[-1, 0]$ or $[0, 1]$, must contain an infinite number of the x_k (in this particular case, both halves do); choose k_1 so that x_{k_1} is in that half. Bisect the subinterval again; thus if $x_{k_1} \in [0, 1]$ we note that one of the intervals,

[0, .5] or [.5, 1] has an infinite number of elements of our sequence. Choose $k_2 > k_1$ such that x_{k_2} is in that subinterval. The process can be continued, and the resulting subsequence is convergent.

Bounded sequence   A sequence $\mathbf{x}_k$ in $\mathbf{R}^n$ such that there is a constant C with $\|\mathbf{x}_k\| \leq C$ for all k.

Closed set   A set whose complement is open.

Component   A connected subset of a set A that is not properly contained in any connected subset of A.

Connected set   A subset of $\mathbf{R}^n$ for which no separation exists.

Convergence   (of an orbit of a mapping P) The orbit of $\mathbf{p}$ converges to $\mathbf{p}^*$ if $\lim_{n \to \infty} P^n(\mathbf{p}) = \mathbf{p}^*$.

Cycle   An orbit of an autonomous system of ODEs that is closed. A solution that traces a cycle is periodic.

Disconnected set   A subset of $\mathbf{R}^n$ that has a separation.

Dynamical system   A one-to-one mapping of a set to itself.

First return mapping   A mapping $P : T \longrightarrow T'$, where $T \subset T'$ and T is a transversal in phase space. $P(\mathbf{x})$ is defined to be the next intersection point with T' of an orbit that crosses T at $\mathbf{x}$, and T is the set of points in T' whose orbits return to T'.

Given a system of ODEs,

$$x' = f(x, y, t), \quad y' = g(x, y, t), \qquad (9.26)$$

where f and g are T-periodic in t, the first return mapping takes a point $(x, y) = (u, v)$ to $(x, y) = F_{0,T}(u, v)$, where F is the flow associated with (9.26).

Fixed point   (of a mapping P) A point $\mathbf{p}$ such that $P(\mathbf{p}) = \mathbf{p}$.

Floquet multiplier   (of a linear system of ODEs with periodic coefficients). A characteristic root of the first return mapping.

Flow   A family of mappings $F_{s,t}(x, y)$ with the following properties:

- $F_{s,s}(x, y) = (x, y)$.
- $F_{u,t}(F_{s,u}(x, y)) = F_{s,t}(x, y)$.

Forward limit set   The set of forward limit points of an orbit.

Inverse function theorem   Let $h(u, v)$ and $k(u, v)$ be functions with continuous partial derivatives in an open set $U \subset$

$\mathbf{R}^2$ and let $\mathbf{p} = (u_0, v_0)$ be a point such that the matrix

$$D(u_0, v_0) = \begin{bmatrix} \frac{\partial h}{\partial u}(u_0, v_0) & \frac{\partial k}{\partial u}(u_0, v_0) \\ \frac{\partial h}{\partial t}(u_0, v_0) & \frac{\partial k}{\partial t}(u_0, v_0) \end{bmatrix}$$

is nonsingular. This matrix D is called the derivative matrix of (h, k) at $\mathbf{p}_0$ Put $\mathbf{q}_0(x_0, y_0) = (h(u_0, v_0), k(u_0, v_0))$. Then there is $\epsilon > 0$ and functions $a(x, y)$ and $b(x, y)$, both defined and with continuous partial derivatives on $B_\epsilon((x_0, y_0))$, such that for all $(x, y) \in B_\epsilon((x_0, y_0))$,

$$x = h(a(x, y), b(x, y)) \quad \text{and} \quad y = k(a(x, y), b(x, y)).$$

There is also $\epsilon' > 0$ such that for all $(u, v) \in B_{\epsilon'}((u_0, v_0))$,

$$u = a(h(u, v), k(u, v)) \quad \text{and} \quad v = b(h(u, v), k(u, v)).$$

In particular, every point in $B_\epsilon((x_0, y_0))$ can be expressed in the form $(h(u, v), k(u, v))$.

Jordan curve   A curve that can be expressed as the graph of a pair of parametric equations $(x, y) = (f(t), g(t))$, for $a \leq t \leq b$ where f and g are continuous, $(f(t_1), g(t_1)) \neq (f(t_2), g(t_2))$ for $a \leq t_1 < t_2 < b$, and $(f(a), g(a)) = (f(b), g(b))$.

Logarithm   (of a matrix A) A matrix L such that $e^L = A$. Logarithms are not unique: a nonsingular matrix A will have an infinite family of logarithms, but a singular matrix will have none.

Mapping   A function whose domain and range are subsets of the line, or of the plane, and so on. We always assume our mappings to be differentiable.

Open set   A set $U \subset \mathbf{R}^n$ such that for each $\mathbf{x} \in U$ there is $\epsilon > 0$ for which every $\mathbf{y}$ with $\|\mathbf{y} = \mathbf{x}\| < \epsilon$ belongs to U.

Orbit   (of a point $\mathbf{p}$ in a dynamical system or mapping P) The sequence, infinite in both directions in the case of a dynamical system, and starting with $\mathbf{p}$ if P is a noninvertible mapping,

$$\dots, P^{-2}(\mathbf{p}), P^{-1}(\mathbf{p}), \mathbf{p}, P(\mathbf{p}), P^2(\mathbf{p}), P^3(\mathbf{p}), \dots.$$

Periodic point   (of a dynamical system P) A point $\mathbf{p}$ such that for some integer $N > 0$, $P^N(\mathbf{p}) = \mathbf{p}$. The number N is the period.

Phase vector   (of an autonomous system of ODEs) The right side of the system, considered as a vector field. Thus, the phase vector of

$$x' = f(x, y) \quad y' = g(x, y)$$

is $f(x, y)\mathbf{i} + g(x, y)\mathbf{j}$.

Real logarithm   A logarithm of a matrix whose entries are real.

Separation   (of a subset $A \subset \mathbf{R}^n$) A pair U_1, U_2 of *disjoint* open sets, both with *nonempty* intersections with A, such that $A \subset U_1 \cup U_2$.

Transversal   In the phase plane: A line segment that contains no stationary point and is tangent to no phase vector. Transversals are ordered: If the transversal is not vertical, the positive direction is to the right; the positive direction on a vertical transversal is up.

In phase space: A plane segment that contains no stationary point, and is tangent to no phase vector.

Trap   A region in the plane, bounded by a polygon, or more generally a Jordan curve, such that orbits that cross its boundary are always directed toward its inside.

Unstable   (fixed point $\mathbf{p}^*$ of a mapping P) There is $\epsilon > 0$ such that there are points $\mathbf{p}$, arbitrarily close to $\mathbf{p}^*$, with $P^n(\mathbf{p}) \notin B_\epsilon(\mathbf{p}^*)$, for some $n > 0$.

9.5 Suggestions for Further Readings

Research on chaos focuses on a paradox that a deterministic process such as the solution of an initial value problem or a dynamical system can be unpredictable. This fact has been a source of frustration for those who would, for example, make long-range weather predictions. At the turn of the twentieth century, Poincaré was drawn to the subject as he studied the many body problem, which is to predict the future behavior of a planetary system subject to the gravitational force attracting the bodies in the system toward each other.

Activity in chaos research accelerated in the 1980's and 90's. The popular book, *Chaos: Making a New Science*[5] by James Gleick is an overview of the chaos phenomenon. *Chaos: An Introduction to Dynamical Systems*[6] by K. T. Alligood, T. D. Sauer, and J. A. Yorke, is an undergraduate-level textbook (with exercises at the end of each chapter) suitable for anyone who has completed a differential equations course.

Chaos impinges on many scientific disciplines. *The Nature of Chaos*[7], edited by Tom Mullen, is a well-organized collection of articles whose authors include a biologist, a chemist, earth scientists, an engineer, and of course a mathematician.

How many limit cycles can a system

$$\frac{dx}{dt} = p(x, y)$$

$$\frac{dy}{dt} = q(x, y)$$

have? Here is an example (in polar coordinates) in which each circle of integer radius and centered at the origin is a limit cycle: $r' = -\sin(2\pi r)$, $\theta' = 1$. When $p(x, y)$ and $q(x, y)$ are required to be polynomials, finding a bound for the number of limit cycles is a problem that has been the subject of more than one hundred years of research, without a complete answer. This is the sixteenth of a famous set of 23 problems set by David Hilbert in 1900 at the Second International Congress of Mathematicians, which met in Paris. "Centennial History of Hilbert's 16th Problem"[8] by Yu. Ilyashenko gives an account of this research.

---

[5]New York: Viking Penguin, 1987

[6]New York: Springer-Verlag, 1996

[7]New York: Oxford University Press, 1993

[8]*Bulletin (New Series) of the American Mathematical Society*, vol 39 (2002), pages 301–354

PART IV

BOUNDARY PROBLEMS AND PDEs

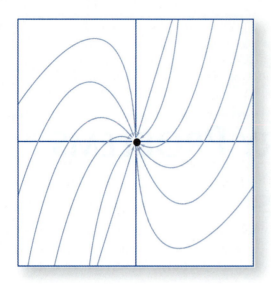

CHAPTER 10

Boundary Value Problems

10.1 Inhomogeneous BVPs

Consider a mechanical system involving a weight of mass m, a dashpot with damping b, and a spring of stiffness k, with an external force of magnitude $f(t)$. The following observations are made: At $t = t_1$, the weight is 1 unit above the equilibrium position, and at a later time, $t = t_2$, the weight is at the equilibrium position. We would like to determine the displacement function. The given information can be restated as an ODE,

$$m\,y'' + b\,y' + k\,y = f(t) \tag{10.1}$$

(where y represents displacement); an initial condition, $y(t_1) = 1$; and a final condition, $y(t_2) = 0$. The existence and uniqueness theorem does not apply, since this is not an IVP. It is a **boundary value problem**, or **BVP**.

In this chapter we will study BVPs involving an ODE of the form

$$a_2(t)\,y'' + a_1(t)\,y' + a_0(t)\,y = b(t). \tag{10.2}$$

The general solution of (10.2) is

$$y = y_p(t) + C_1\,y_1(t) + C_2\,y_2(t), \tag{10.3}$$

where $y_p(t)$ is a particular solution of (10.2), $\{y_1(t),\,y_2(t)\}$ is a fundamental set of solutions of the associated homogeneous equation, and C_1, C_2 are constants. To determine a particular solution, we need data to calculate C_1 and C_2.

Instead of initial conditions that specify all the data at an initial time, a BVP imposes **boundary conditions**, where the data are specified at two boundary points t_1 and t_2. It is always required that equation (10.2) has no singular points in the interval $[t_1, t_2]$ between the boundary points. For the aforementioned mechanical system, the boundary conditions would be $y(t_1) = 1$, $y(t_2) = 0$.

In principle, BVPs can be solved in the same way that IVPs are solved. For example, the BVP consisting of equation (10.2) with boundary conditions $y(t_1) = K$, $y(t_2) = L$ is solved by first determining the general solution (10.3) and then substituting the boundary conditions to obtain a pair of equations

$$\left. \begin{array}{l} y_p(t_1) + C_1\, y_1(t_1) + C_2\, y_2(t_1) = K \\ y_p(t_2) + C_1\, y_1(t_2) + C_2\, y_2(t_2) = L, \end{array} \right\} \tag{10.4}$$

which determine C_1 and C_2. There is a chance that equations (10.4) will be either inconsistent (then the BVP will have no solution) or redundant (there will be multiple solutions in this case).

Example 10.1.1   A mechanical system has $m = 1$, $b = 2$, and $k = 2$. A force $f(t) = e^{-t}$ is applied. At $t = 0$, the displacement of the weight is $y = 1$, and at $t = \pi/2$ the displacement is $y = 0$. Determine the displacement as a function of time.

Solution. We are to solve the BVP,

$$y'' + 2\,y' + 2\,y = e^{-t}; \quad y(0) = 1,\ y\left(\tfrac{\pi}{2}\right) = 0.$$

Let $\mathcal{L}(y) = y'' + 2\,y' + 2\,y$. The characteristic equation of the homogeneous ODE $\mathcal{L}(y) = 0$ is $s^2 + 2s + 2 = 0$, and by using the quadratic formula we can see that the characteristic roots are $r_1, r_2 = -1 \pm i$. Therefore, the general solution of the $\mathcal{L}(y) = 0$ is $y_h = e^{-t}\,[C_1\,\cos(t) + C_2\,\sin(t)]$. To find a particular solution of $\mathcal{L}(y) = e^{-t}$ we will use the method of undetermined coefficients, with $y_p = Ae^{-t}$. Thus,

$$\mathcal{L}(Ae^{-t}) = Ae^{-t} - 2Ae^{-t} + 2Ae^{-t} = Ae^{-t}.$$

Thus, $A = 1$ gives $\mathcal{L}(y_p) = e^{-t}$. The general solution is

$$y = y_p + y_h = e^{-t} + e^{-t}\,[C_1\,\cos(t) + C_2\,\sin(t)].$$

Substituting the boundary data, we obtain, for $t = 0$,

$$1 + C_1 = 1, \tag{10.5}$$

and for $t = \tfrac{\pi}{2}$,

$$e^{-\pi/2} + e^{-\pi/2}C_2 = 0. \tag{10.6}$$

By (10.5), $C_1 = 0$, and (10.6) reduces to $C_2 = -1$. Therefore, the BVP has the unique solution,

$$y = e^{-t}(1 - \sin(t)).$$

Figure 10.1 is a graph of this displacement function.

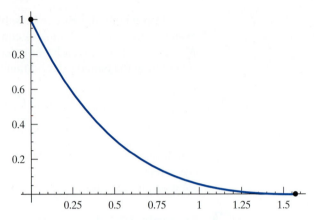

FIGURE 10.1   The displacement of the weight in
Example 10.1.1. The graph of the solution is required to
pass through the boundary points as marked.

In the following example, we will explore a family of BVPs. Some of the
problems in the family turn out not to have solutions.

Example 10.1.2   For which values of λ does the BVP

$$y'' + \lambda y = 1; \; y(0) = y(\pi) = 0 \tag{10.7}$$

have a unique solution? Find the solution in these cases.

Solution. The nature of this BVP depends on whether λ is negative, zero, or positive.

Case 1: $\lambda < 0$. Put $\lambda = -k^2$. The general solution of $y'' - k^2 y = 1$ is

$$y = -\frac{1}{k^2} + C_1 e^{kt} + C_2 e^{-kt},$$

where C_1 and C_2 are arbitrary parameters. Then

$$y(0) = -\frac{1}{k^2} + C_1 + C_2, \quad \text{and} \quad y(\pi) = -\frac{1}{k^2} + C_1 e^{k\pi} + C_2 e^{-k\pi},$$

so the boundary conditions are equivalent to the system of equations

$$C_1 + C_2 = \frac{1}{k^2}$$

$$C_1 e^{k\pi} + C_2 e^{-k\pi} = \frac{1}{k^2}.$$

Calculation shows that the solution of this system is

$$C_1 = \frac{1}{k^2(e^{k\pi} + 1)}; \; C_2 = \frac{1}{k^2(e^{-k\pi} + 1)}.$$

Thus, there is a unique solution to the BVP:

$$y = -\frac{1}{k^2} + \frac{e^{kt}}{k^2(e^{k\pi} + 1)} + \frac{e^{-kt}}{k^2(e^{-k\pi} + 1)}.$$

Case 2: $\lambda = 0$. The general solution of $y'' + 0 \cdot y = 1$ is $y = \frac{1}{2}t^2 + C_1 t + C_2$. Since $y(0) = C_2$, the boundary condition $y(0) = 0$ implies $C_2 = 0$. The boundary condition $y(\pi) = 0$ combines with

$$y(\pi) = \frac{1}{2}\pi^2 + C_1\pi + C_2 = \left(\frac{\pi}{2} + C_1\right)\pi$$

to imply $C_1 = -\pi/2$. Therefore, there is a unique solution of the BVP:

$$y = \frac{1}{2}t(t - \pi).$$

Case 3: $\lambda > 0$. Put $\lambda = k^2$. The general solution of $y'' + k^2 y = 1$ is

$$y = \frac{1}{k^2} + C_1\cos(kt) + C_2\sin(kt).$$

Now $y(0) = (1/k^2) + C_1$, so the boundary condition at $t = 0$ implies that $C_1 = -1/k^2$. Since

$$y(\pi) = \frac{1}{k^2} + C_1\cos(k\pi) + C_2\sin(k\pi),$$

the boundary condition at $t = \pi$ is equivalent to

$$C_1\cos(k\pi) + C_2\sin(k\pi) = -\frac{1}{k^2}, \tag{10.8}$$

so $C_2 = (\cos(k\pi) - 1)/(k^2\sin(k\pi))$, provided that $\sin(k\pi) \neq 0$. Thus, unless k is an integer, the BVP (10.7) has the unique solution

$$y = \frac{1}{k^2}\left(1 - \cos(kt) + \frac{\cos(k\pi) - 1}{\sin(k\pi)}\sin(kt)\right).$$

What happens if k is an integer? When k is odd, $\cos(k\pi) = -1$ and $\sin(k\pi) = 0$. Substitution of these values in equation (10.8) leads to

$$C_1 = \frac{1}{k^2},$$

which is impossible since the boundary condition at $t = 0$ requires $C_1 = -\frac{1}{k^2}$. On the other hand, if $k = 2m$ is even, $\cos(2m\pi) = 1$, and we still have $\sin(2m\pi) = 0$. Hence in this case equation (10.8) is equivalent to the boundary condition at $t = 0$, so $C_1 = -\frac{1}{4m^2}$, and the parameter C_2 is undetermined. It follows that when k is an odd integer, the BVP has *no* solution, and when $k = 2m$ is even, there is a family of solutions

$$y = \frac{1}{4m^2}[1 - \cos(2mt)] + C_2\sin(2mt).$$

In summary, the BVP has a solution for all λ except $\lambda = (2m + 1)^2$, with m an integer. The solution is unique unless λ is of the form $(2m)^2$. Figure 10.2 shows solutions of the BVP for several values of λ. ❑

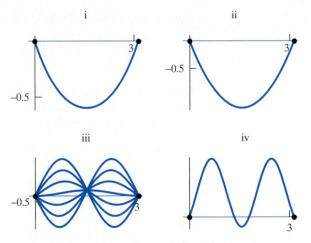

FIGURE 10.2   Solutions of the BVP in Example 10.1.2 for (i) $\lambda = -1$, (ii) $\lambda = 0$, (iii) $\lambda = 4$, and (iv) $\lambda = (\frac{7}{2})^2$.

The boundary conditions in Example 10.1.2 are **homogeneous**, in the sense that they are satisfied by $y(t) \equiv 0$. If a BVP has homogeneous boundary conditions, and involves a homogeneous ODE, then $y(t) \equiv 0$ is a solution. We call this solution the **trivial solution**.

When $\lambda = k^2$ is the square of an integer, the BVP

$$y'' + \lambda y = 0; \; y(0) = y(\pi) = 0 \tag{10.9}$$

has a nontrivial solution $y = \sin(k\,t)$. The BVP in Example 10.1.2 fails to have a unique solution exactly when the homogeneous BVP (10.9) has a nontrivial solution. This observation is a special case of the following theorem.

Theorem 10.1

Let $\mathcal{L}(y) = y'' + p(t)y' + q(t)y$, and let $[t_1, t_2]$ be an interval containing no singular points of $\mathcal{L}$. Suppose also that the function $f(t)$ is continuous on the interval $[t_1, t_2]$. Then the BVP

$$\mathcal{L}(y) = f(t); \; y(t_1) = y(t_2) = 0 \tag{10.10}$$

has a unique solution if and only if the associated homogeneous problem

$$\mathcal{L}(y) = 0; \; y(t_1) = y(t_2) = 0 \tag{10.11}$$

admits only the trivial solution $y \equiv 0$.

Proof of the uniqueness statement of Theorem 10.1.   Suppose that the BVP (10.10) admits two solutions $y_1(t)$ and $y_2(t)$, so that $\mathcal{L}(y_1) = \mathcal{L}(y_2) = f(t)$. Since $\mathcal{L}$ is linear differential operator, $\mathcal{L}(y_1 - y_2) = 0$, and $v(t) = y_1(t) - y_2(t)$ satisfies the boundary conditions. It follows that $v(t)$ is a solution of the homogeneous BVP (10.11). This shows that if there are distinct solutions of the problem (10.10), then there is a nontrivial solution of the associated homogeneous problem. Conversely, suppose that $v(t)$ is a nontrivial solution of the homogeneous BVP (10.11). We must show that the inhomogeneous BVP (10.10) has either no solution or multiple solutions. If there is a solution $y(t)$ of the BVP, then for any constant A,

$$\mathcal{L}(y(t) + Av(t)) = \mathcal{L}(y(t)) + A\mathcal{L}(v(t))$$
$$= f(t) + 0.$$

Hence $y(t) + Av(t)$ is a one parameter family of solutions, and the BVP does not have a *unique* solution.   ■

Proof of the existence statement of Theorem 10.1.   Suppose that the homogeneous BVP (10.11) has only the trivial solution. We will use variation of constants to construct a solution of the inhomogeneous BVP (10.10).

We will need a fundamental set of solutions of the associated homogeneous equation $\mathcal{L}(y) = 0$. Let $\phi_1(t)$ be the solution of the IVP

$$\mathcal{L}(y) = 0; \quad y(t_1) = 0, \ y'(t_1) = 1.$$

Notice that $\phi_1(t)$ will satisfy the boundary condition $y(t_1) = 0$. Similarly, let $\phi_2(t)$ be the solution of the IVP

$$\mathcal{L}(y) = 0; \quad y(t_2) = 0, \ y'(t_2) = 1,$$

which will satisfy the boundary condition $y(t_2) = 0$. The existence and uniqueness theorem for linear IVPs (Theorem 5.1 on page 186) ensures that both functions $\phi_i(t)$ are defined for all $t \in [t_1, t_2]$.

We will establish that $\{\phi_1, \phi_2\}$ is a fundamental set of solutions of $\mathcal{L}(y) = 0$ by showing that it is linearly independent. It will be sufficient to show that $\phi_2(t)$ is not a constant multiple of $\phi_1(t)$. Suppose that, to the contrary, $\phi_2(t) = C\phi_1(t)$. Then

$$\phi_2(t_1) = C\phi_1(t_1) = 0.$$

It follows that the homogeneous BVP (10.11) has the nontrivial solution $\phi_2(t)$: a contradiction.

Question   How do we know that $\phi_2(t) \neq 0$?

The method of variation of constants yields a particular solution of $\mathcal{L}(y) = f(t)$ of the form

$$y_p(t) = v_1(t)\phi_1(t) + v_2(t)\phi_2(t),$$

where v_1 and v_2 are determined by solving the equations

$$\left. \begin{array}{c} v_1'(t)\phi_1(t) + v_2'(t)\phi_2(t) = 0 \\ v_1'(t)\phi_1'(t) + v_2'(t)\phi_2'(t) = f(t) \end{array} \right\} \tag{10.12}$$

and integrating. The constants of integration can be chosen so that $v_1(t_2) = 0$ and $v_2(t_1) = 0$. Then

$$y_p(t_1) = v_1(t_1)\phi_1(t_1) + v_2(t_1)\phi_2(t_1)$$
$$= v_1(t_1) \cdot 0 + 0 \cdot \phi_2(t_1)$$
$$= 0,$$

and similarly $y_p(t_2) = 0$. It follows that $y_p(t)$ satisfies the BVP (10.10, 10.11).   ■

Let us calculate $v_1(t)$ and $v_2(t)$ explicitly, and thus derive a formula for the solution of the BVP (10.10, 10.11). Let

$$W(t) = \phi_1(t)\phi_2'(t) - \phi_1'(t)\phi_2(t)$$

be the Wronskian of the solutions ϕ_1, ϕ_2. To calculate v_1', multiply the first equation of the system (10.12) by $\phi_2'(t)$, multiply the second equation by $\phi_2(t)$, and subtract the equations. This yields

$$v_1'(t) = -\frac{f(t)\phi_2(t)}{W(t)}.$$

We can ensure that $v_1(t_2) = 0$ by taking

$$v_1(t) = -\int_{t_2}^{t} \frac{f(u)\phi_2(u)}{W(u)}\,du$$

$$= \int_{t}^{t_2} \frac{f(u)\phi_2(u)}{W(u)}\,du.$$

A similar calculation leads to

$$v_2(t) = \int_{t_1}^{t} \frac{f(u)\phi_1(u)}{W(u)}\,du;$$

note that as required, $v_2(t_1) = 0$. Therefore,

$$y_p(t) = \phi_1(t)\int_{t}^{t_2} \frac{f(u)\phi_2(u)}{W(u)}\,du + \phi_2(t)\int_{t_1}^{t} \frac{f(u)\phi_1(u)}{W(u)}\,du \qquad (10.13)$$

is the solution of the BVP (10.10, 10.11).

We can write the solution (10.13) in a more compact form, as follows. Define a function $G(t, u)$ by the formula

$$G(t, u) = \begin{cases} \dfrac{\phi_1(u)\phi_2(t)}{W(u)} & \text{if } u \leq t, \text{ and} \\[3mm] \dfrac{\phi_1(t)\phi_2(u)}{W(u)} & \text{if } u > t, \end{cases}$$

where the functions $\phi_i(t)$ are defined in the proof of Theorem 10.1, and $W(t)$ denotes their Wronskian. This function $G(t, u)$ is called *Green's function.* It is named for the nineteenth-century English mathematician George Green. Green's function depends on the linear differential operator $\mathcal{L}$ and the boundary conditions. It does not depend on the function $f(t)$.

Proposition 10.1.1

Let $G(t, u)$ denote Green's function for the linear differential operator $\mathcal{L}$ with boundary conditions $y(t_1) = y(t_2) = 0$ on the interval $[t_1, t_2]$. Then the solution of the BVP (10.10, 10.11) is

$$y(t) = \int_{t_1}^{t_2} G(t, u) f(u)\,du.$$

Proof   We will use the same notation as the proof of Theorem 10.1:

$$\int_{t_1}^{t_2} f(u)G(t,u)\,du = \int_{t_1}^{t} f(u)G(t,u)\,du + \int_{t}^{t_2} f(u)G(t,u)\,du$$

$$= \int_{t_1}^{t} \frac{f(u)\phi_1(u)\phi_2(t)}{W(u)}\,du + \int_{t}^{t_2} \frac{f(u)\phi_1(t)\phi_2(u)}{W(u)}\,du$$

$$= y_p(t)$$

by formula (10.13).   ∎

Example 10.1.3   Determine Green's function for $\mathcal{L}(y) = y'' + k^2 y$ with boundary conditions $y(0) = y(\pi) = 0$, where k is a nonzero real number.

Solution.   The function $\phi_1(t)$ is the solution of the IVP $y'' + k^2 y = 0$; $y(0) = 0$, $y'(0) = 1$. Thus, $\phi_1(t) = \frac{1}{k}\sin(kt)$. Similarly, $\phi_2(t)$ is the solution of the IVP $y'' + k^2 y = 0$; $y(\pi) = 0$, $y'(\pi) = 1$. Therefore, $\phi_2(t) = \frac{1}{k}\sin(k(t - \pi))$. The Wronskian is

$$W(k,t) = \det \begin{bmatrix} \frac{1}{k}\sin(kt) & \frac{1}{k}\sin(k(t-\pi)) \\ \cos(kt) & \cos(k(t-\pi)) \end{bmatrix}$$

$$= \frac{1}{k}[\sin(kt)\cos(k(t-\pi)) - \cos(kt)\sin(k(t-\pi))]$$

$$= \frac{1}{k}\sin(k\pi).$$

(Here we have used the difference formula for the sine function.)

Green's function provides a formula for the solution of the inhomogeneous BVP, so it exists only when the BVP has a unique solution. We know from Example 10.1.2 that BVPs involving $\mathcal{L}$ with boundary conditions $y(0) = y(\pi) = 0$ don't have unique solutions when k is an integer, so it is interesting to see what happens to Green's function in this case. The answer is that when k is an integer, $W(k,t) \equiv 0$, so Green's function, which has $W(k,t)$ in its denominator, is undefined.

When k is not an integer, we have

$$G(t,u) = \begin{cases} \dfrac{\sin(ku)\sin(k(t-\pi))}{k\sin(k\pi)} & \text{if } u \le t, \text{ and} \\[3mm] \dfrac{\sin(kt)\sin(k(u-\pi))}{k\sin(k\pi)} & \text{if } u > t. \end{cases}$$

The formula for the solution of the BVP that is given in Proposition 10.1.1 is in terms of an integral operator

$$\mathcal{G}(f) = \int_{t_1}^{t_2} G(t,u)f(u)\,du.$$

This operator $\mathcal{G}$ is an *inverse operator* to the operator $\mathcal{L}$, in the following sense. Let $\mathcal{V}(t_1, t_2)$ denote the vector space of all functions $y(t)$ that possess continuous second derivatives on the interval $[t_1, t_2]$ and that satisfy the boundary conditions $y(t_1) = y(t_2) = 0$. Denote the vector space of functions that are continuous on $[t_1, t_2]$ by $C[t_1, t_2]$. Then we have $\mathcal{L} : V(t_1, t_2) \longrightarrow C[t_1, t_2]$ and $\mathcal{G} : C[t_1, t_2] \longrightarrow V(t_1, t_2)$, and we have shown that $\mathcal{L}(\mathcal{G}(f)) = f$. That is, if $f(t)$ is any function that is continuous on $[t_1, t_2]$ then $\mathcal{G}(f)$ is the solution of $\mathcal{L}(y) = f(t)$ that satisfies the boundary conditions.

It is also true that $\mathcal{G}(\mathcal{L}(y)) = y$ for any $y(t) \in V(t_1, t_2)$. The proof is very simple: Start with any $y(t) \in V(t_1, t_2)$, and put $f(t) = \mathcal{L}(y)$. Then $f \in C[t_1, t_2]$, so $\mathcal{L}(\mathcal{G}(f)) = f$. Therefore, $\mathcal{L}(\mathcal{G}(f) - y) = 0$, and thus our assumption that the homogeneous BVP has only the trivial solution implies $\mathcal{G}(f) = y$. Since $f = \mathcal{L}(y)$, we have proved that $\mathcal{G}(\mathcal{L}(y)) = y$.

Other types of boundary conditions

Our discussion has so far been limited to boundary conditions of the form $y(t_1) = y(t_2) = 0$, but there are other possibilities. The most general homogeneous boundary conditions have the form

$$A_1 y(t_1) + A_2 y'(t_1) + A_3 y(t_2) + A_4 y'(t_2) = 0 \qquad (10.14)$$
$$B_1 y(t_2) + B_2 y'(t_2) + B_3 y(t_1) + B_4 y'(t_1) = 0. \qquad (10.15)$$

(Inhomogeneous boundary conditions are discussed in Exercise 18.) The procedure for solving a BVP with more general boundary conditions is similar to that used in Example 10.1.1.

Example 10.1.4 Solve the BVP

$$y'' - y = 2e^t; \quad y(0) = y'(1) = 0.$$

Solution. Let $\mathcal{L}(y) = y'' - y$. The characteristic roots of the associated homogeneous equation, $\mathcal{L}(y) = 0$, are $r_1, r_2 = \pm 1$. Therefore, the homogeneous solution is $y_h = C_1 e^t + C_2 e^{-t}$. To find a particular solution of $\mathcal{L}(y) = 2e^t$, we will use undetermined coefficients, with $y_p(t) = Ate^t$ (why?). Then $y_p'' = Ate^t + 2Ae^t$, so $\mathcal{L}(y_p) = 2Ae^t$. It follows that $A = 1$. The general solution of the ODE is thus

$$y = te^t + C_1 e^t + C_2 e^{-t},$$

and

$$y' = (t + 1)e^t + C_1 e^t - C_2 e^{-t}.$$

The boundary condition $y(0) = 0$ holds if

$$C_1 + C_2 = 0,$$

and the boundary condition $y'(1) = 0$ holds if

$$C_1 e - C_2 e^{-1} = -2e.$$

Thus, $C_1 = \frac{-2e}{e + e^{-1}}$ and $C_2 = \frac{2e}{e + e^{-1}}$. The solution is

$$y = te^t + \frac{2(e^{1-t} - e^{1+t})}{e + e^{-1}}.$$

A graph of this solution is shown in Figure 10.3.

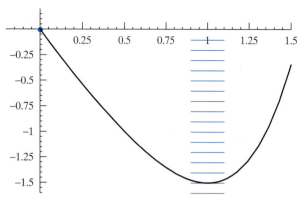

FIGURE 10.3 Solution of the BVP of Example 10.1.4. The boundary conditions are shown as a heavy dot at $(0, 0)$ and line segments at $t = 1$ indicating the horizontal direction.

The boundary conditions (10.14) and (10.15) are **separated** if $A_3 = A_4 = B_3 = B_4 = 0$. In this case, boundary condition (10.14) involves only the endpoint t_1, and boundary condition (10.15) involves only the endpoint t_2. The following theorem extends Theorem 10.1 to cover BVPs with separated boundary conditions.

Theorem 10.2

Let $\mathcal{L}(y) = a_2(t)\, y'' + a_1(t)\, y' + a_0(t)\, y$ be a linear differential operator with no singular points in $[t_1, t_2]$, and suppose that $f(t)$ is continuous on $[t_1, t_2]$. Assume also that $(A_1, A_2) \neq (0, 0)$ and $(B_1, B_2) \neq (0, 0)$. Then the BVP

$$\mathcal{L}(y) = f(t); \; A_1\, y(t_1) + A_2\, y'(t_1) = 0, \; B_1\, y(t_2) + B_2\, y'(t_2) = 0$$

has a unique solution if and only if the associated homogeneous problem $\mathcal{L}(y) = 0$ with the same boundary conditions has only the trivial solution.

The proof of this theorem is left to you. You will need to construct a Green's function in the proof, and the following example illustrates the procedure for doing so.

Example 10.1.5 Find Green's function for the $\mathcal{L}(y) = y'' + 2y' + y$ with boundary conditions $y(0) + y'(0) = 0$, $y(1) - y'(1) = 0$. Use this Green's function to solve the BVP $\mathcal{L}(y) = -2e^{-t}$ with these boundary conditions.

Solution. The general solution of $\mathcal{L}(y) = 0$ is $y = e^{-t}(C_1 + C_2 t)$, and its derivative is $y' = -e^{-t}(C_1 + C_2(t - 1))$. The function $\phi_1(t)$ will be chosen to satisfy $\mathcal{L}(\phi_1) = 0$ and the boundary condition at $t = 0$: $\phi_1(0) + \phi_1'(0) = 0$. Thus, $\phi_1(t) = e^{-t}(C_1 + C_2 t)$, where

$$C_1 - C_1 + C_2 = 0.$$

Hence $C_2 = 0$, and we will arbitrarily say $C_1 = 1$, so $\phi_1(t) = e^{-t}$. Now put $\phi_2(t) = e^{-t}(D_1 + D_2 t)$. We will choose the coefficients D_i so that ϕ_2 satisfies the right boundary condition, $y(1) - y'(1) = 0$. Thus

$$e^{-1}(D_1 + D_2) + e^{-1}D_1 = 0.$$

Thus we can take $D_1 = -1$, $D_2 = 2$, and $\phi_2(t) = e^{-t}(2t - 1)$.
To solve $\mathcal{L}(y) = f(t)$ by variation of constants, set

$$y = v_1(t)\,\phi_1(t) + v_2(t)\,\phi_2(t),$$

where v_1' and v_2' satisfy the system

$$
\begin{aligned}
v_1'(t)\,\phi_1(t) + v_2'(t)\,\phi_2(t) &= 0 \\
v_1'(t)\,\phi_1'(t) + v_2'(t)\,\phi_2'(t) &= f(t)
\end{aligned}
\tag{10.16}
$$

with constants of integration chosen so that $v_2(0) = v_1(1) = 0$. Let us check the left boundary condition:

$$y(0) + y'(0) = v_1(0)\,\underbrace{(\phi_1(0) + \phi_1'(0))}_{=\,0}$$

$$+ \underbrace{v_2(0)}_{=\,0}(\phi_2(0) + \phi_2'(0)) + \underbrace{v_1'(0)\,\phi_1(0) + v_2'(0)\,\phi_2(0)}_{=\,0}.$$

Here, the first term is zero because of the choice of initial conditions for ϕ_1, the second is zero because $v_2(0) = 0$, and the third term is zero by equation (10.16). The right boundary condition can be verified in the same way, and you should carry this out before reading further.

Now let us determine Green's function.

$$W[\phi_1, \phi_2](t) = \det\begin{bmatrix} e^{-t} & e^{-t}(2t-1) \\ -e^{-t} & -e^{-t}(2t-3) \end{bmatrix}$$

$$= 2e^{-2t}$$

Thus,

$$G(t, u) = \begin{cases} \dfrac{e^{-u}e^{-t}(2t-1)}{2e^{-2u}} & \text{for } u \le t, \text{ and} \\[3mm] \dfrac{e^{-t}e^{-u}(2u-1)}{2e^{-2u}} & \text{for } u > t. \end{cases}$$

This can be simplified as

$$G(t, u) = \begin{cases} e^{u-t}\left(t - \tfrac{1}{2}\right) & \text{for } u \le t, \text{ and} \\[2mm] e^{u-t}\left(u - \tfrac{1}{2}\right) & \text{for } u > t \end{cases}$$

$$= e^{u-t}\left(\max\{t, u\} - \frac{1}{2}\right).$$

The solution of the BVP with $f(t) = -2e^{-t}$ is

$$\int_0^1 G(t, u) f(u) \, du$$

$$= \int_0^t e^{u-t} \left(t - \frac{1}{2} \right) (-2e^{-u}) \, du + \int_t^1 e^{u-t} \left(u - \frac{1}{2} \right) (-2e^{-u}) \, du$$

$$= (-2t + 1) e^{-t} \int_0^t du + e^{-t} \int_t^1 (-2u + 1) \, du$$

$$= (-2t + 1) e^{-t} (t) + e^{-t}(-u^2 + u)|_t^1$$

$$= -t^2 e^{-t}. \qquad \qquad \square$$

Figure 10.4 shows a graph of the solution of the BVP in Example 10.1.5. It displays the boundary conditions as **boundary fields**, which are similar to direction fields. If the boundary points of a BVP are t_1 and t_2, the elements of the boundary field are drawn at selected points (t_1, y_j) and (t_2, y_j) in the t, y-plane. Let the boundary conditions be $A_1 y(t_1) + A_2 y'(t_1) = 0$ and $B_1 y(t_2) + B_2 y'(t_2) = 0$. Then the boundary field element drawn at (t_1, y_j) is a line segment of slope

$$m_{1,j} = -\frac{A_1}{A_2} y_j$$

and it gives the slope of the solution if its graph contains the point (t_1, y_j). Similarly, the slope of the element at (t_2, y_j) is $m_{2,j} = -(B_1/B_2)y_j$. The boundary field only exists for boundary conditions that involve y'. Thus, if $A_2 = 0$, there is no boundary field at t_1, and the boundary condition, which would be $y(t_1) = 0$, should be shown as a dot at the point $(t_1, 0)$ instead.

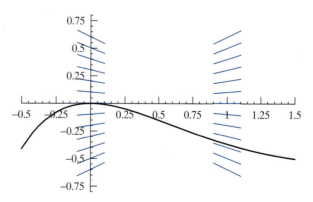

FIGURE 10.4   Graph of the solution of the BVP in Example 10.1.5.

In Figure 10.4 the boundary field corresponding to $y(0) + y'(0) = 0$ is $m_{1,j} = -y_j$, and the boundary field corresponding to $y(1) - y'(1) = 0$ is $m_{2,j} = y_j$.

EXERCISES

In Exercises 1–4, BVPs involving a parameter λ are given. In each case, determine the values of λ for which the problem has a unique solution, and draw a boundary field where appropriate. You are not required to solve the BVP.

1. $y'' + \lambda y = \sin t$; $y(0) = y(\pi) = 0$
2. $y'' + \lambda y = 1$; $y(0) = y'(\pi) = 0$
3. $y'' + 2y' + \lambda y = t$; $y'(0) = y'(1) = 0$
4. $y'' + \lambda y = 0$; $y(0) + y'(0) = 0$, $y(\pi) + y'(\pi) = 0$
5. Show that Green's function for $\mathcal{L}(y) = y''$, with boundary conditions $y(0) = y(\pi) = 0$ is

$$G(t, u) = \frac{t\,u}{\pi} - \min\{t, u\}.$$

In each of Exercises 6–13, find Green's function for the given linear differential operator and boundary conditions. Illustrate the boundary conditions with boundary fields or dots at the boundary points, and sketch the graphs of several functions that satisfy the boundary conditions.

6. $\mathcal{L}(y) = y'' - y$; $y(0) = y(\ln 2) = 0$
7. $\mathcal{L}(y) = y'' - 3y' + 2y$; $y(0) - y'(0) = 0$, $2y(1) - y'(1) = 0$
8. $\mathcal{L}(y) = y'' - k^2 y$; $y(0) = y'(1) = 0$ (Green's function will depend on the parameter k.)
9. $\mathcal{L}(y) = y'' + y$; $y(0) = y'(\pi) = 0$
10. $\mathcal{L}(y) = y'' + k^2 y$; $y(0) = y'(2\pi) = 0$ (Green's function will be undefined for certain values of k; which are they?)
11. $\mathcal{L}(y) = y''$; $y(0) = 0$, $y(1) + y'(1) = 0$
12. $\mathcal{L}(y) = (e^t y')'$; $y(0) = y'(1) = 0$
13. $\mathcal{L}(y) = [(1 + t^2) y']'$; $y(0) - y'(0) = y(1) = 0$
14. Find Green's function for $\mathcal{L}(y) = y''$ with boundary conditions $A_1 y(0) + A_2 y'(0) = 0$, $B_1 y(1) + B_2 y'(1) = 0$. Of course, Green's function will depend on the parameters A_i and B_i. Is it defined for all possible values of these parameters?
15. Suppose that $f(t)$ is continuous on the interval $[0, 1]$. Find all values of A_i and B_i for which the BVP $\mathcal{L}(y) = f(t)$; $A_1 y(0) + A_2 y'(0) = 0$, $B_1 y(1) + B_2 y'(1) = 0$ has a unique solution, where
    a. $\mathcal{L}(y) = y'' + \pi^2 y$ b. $\mathcal{L}(y) = y'' - y$
    c. $\mathcal{L}(y) = y'' + 2y' + y$ d. $\mathcal{L}(y) = (t^2 + 1)y'' + 2ty'$
16. Complete the proof of Theorem 10.1 by showing that $y_p(t_2) = 0$.
17. Prove Theorem 10.2. The uniqueness part is proved just as in the proof of Theorem 10.1. For existence, you need to construct a Green's function. Follow the steps in Example 10.1.5, and show that the resulting $G(t, u)$ does produce a solution of the BVP.

18. Let $\mathcal{L}(y) = a_2(t) y'' + a_1(t) y' + a_0(t) y$, and consider the BVP with inhomogeneous boundary conditions,

$$\mathcal{L}(y) = f(t); \qquad (10.17)$$

$$A_1 y(t_1) + A_2 y'(t_1) = M,$$
$$B_1 y(t_2) + B_2 y'(t_2) = N, \qquad (10.18)$$

where $(A_1, A_2) \neq (0, 0)$ and $(B_1, B_2) \neq (0, 0)$. Assume that $\mathcal{L}$ has no singular points in the interval $[t_1, t_2]$, and that $y \equiv 0$ is the only solution of $\mathcal{L}(y) = 0$ that satisfies the homogeneous versions of the boundary conditions:

$$A_1 y(t_1) + A_2 y'(t_1) = 0,$$
$$B_1 y(t_2) + B_2 y'(t_2) = 0. \qquad (10.19)$$

a. Show that there is a function $l(t)$, with continuous second derivative on $[t_1, t_2]$, that satisfies the boundary conditions (10.18).

   ◆ **Hint:** You may use the following lemma (without proof, if you wish):

   Lemma 10.1.2   Suppose that $t_1 < t_2$ and that p, q, r, and s are arbitrarily given numbers. Then there is a polynomial $l(t)$ of degree ≤ 3 with the following properties:

   $$l(t_1) = p, \quad l'(t_1) = q, \quad l(t_2) = r, \quad \text{and } l'(t_2) = s.$$

b. Put $g(t) = \mathcal{L}(l(t))$, and let $G(t, u)$ denote Green's function for the linear differential operator $\mathcal{L}$ with the homogeneous boundary conditions (10.19). Show that the function

$$y(t) = l(t) + \int_{t_1}^{t_2} [f(u) - g(u)] G(t, u)\, du$$

is a solution of the BVP (10.17), (10.18).

c. Is the solution unique?

19. For each of the following BVPs, draw boundary fields where possible, and find a solution if one exists. If there is a solution, determine whether or not it is unique, and sketch its graph. If there are multiple solutions, sketch three of them.

    a. $y'' + y = 0$; $y(0) = y'(\pi) = 1$
    b. $y'' = t$; $y(0) + y'(0) = 2$, $y(1) - y'(1) = 1$
    c. $y'' - k^2 y = 0$; $y(0) = 1$, $y'(1) - y(1) = 0$
    d. $y'' - 3y' + 2y = e^t$; $y(0) = 1$, $y(1) = e$
    e. $y'' + y = 0$; $y(0) = 1$, $y(\pi) = 0$

20. Show that if $a_2(t)a_0(t) < 0$ on $[t_1, t_2]$, then the homogeneous BVP

$$a_2(t)y'' + a_1(t)y' + a_0(t)y = 0; \quad y(t_1) = y(t_2) = 0$$

has no nontrivial solutions. What does this result imply about inhomogeneous BVPs?

◆ **Hint:** Suppose that $y = \phi(t)$ is a solution of the ODE. Show that $\phi(t)$ cannot have a positive maximum value or a negative minimum value on the interval (t_1, t_2).

21. **Solving BVPs with a CAS**

a. If your computer algebra system can find the general solution of a given ODE, then it can determine the solution of a BVP involving that ODE (provided, of course, that the BVP has a unique solution). You just have to enter the ODE and the boundary conditions. For example, the *Maple* command

```
>dsolve({diff(y(t),t$2)-y(t)=0,
y(0)=1,D(y)(2)=-1},y(t));
```

solves the BVP $y'' - y = 0$, $y(0) = 1$, $y'(2) = -1$. Use your CAS to solve this BVP, and experiment with other boundary conditions.

b. Use a CAS to solve the BVP

$$y'' + 2y' + 5y = e^{-t}\sin(2t);$$
$$y(0) - 2y'(0) = 0, \ y(\pi) + 3y'(\pi) = 0$$

and plot a graph of the solution.

c. Try to produce a graph of the solution of the BVP,

$$y'' + t^2y = 1; \ y(0) = y'(5) = 0.$$

◆ **Suggestion:** Use your CAS to determine solutions of the homogeneous equation $y'' + t^2y = 0$ with initial conditions $y(0) = 0$ and $y'(5) = 0$. These will be expressed in terms of modified Bessel functions (Bessel functions were introduced in Chapter 7). You can then use the variation of constants formula to find a solution of the BVP. Your CAS will probably prefer to do the integral numerically. Unless you have a supercomputer, allow approximately one-half hour to plot the solution.

22. Exercise 21c points out a weakness of symbolic ODE solvers. As a general rule, symbolic methods should be

avoided unless the solution they produce is a simple and useful formula, or there is some reason that a formula for the solution is needed, even if it may be complicated. If we only need a graph of a solution, a numerical method is usually more efficient. Our computer is only equipped with IVP solvers, though, so how can we solve a BVP? An IVP solver can be adapted to the purpose of solving BVPs by a procedure known as **shooting**. Suppose that we need to solve a BVP,

$$\mathcal{L}(y) = f(t);$$
$$A_1 \, y(t_1) + A_2 \, y'(t_1) = M, \qquad (10.20)$$
$$B_1 \, y(t_2) + B_2 \, y'(t_2) = N,$$

where $\mathcal{L}$ is a linear differential operator with no singular points in $[t_1, t_2]$.

Let $a = M \, A_1/(A_1^2 + A_2^2)$ and $b = M \, A_2/(A_1^2 + A_2^2)$, and let $\phi(t)$ be the solution of the IVP

$$\mathcal{L}(y) = f(t); \ y(t_1) = a, \ y'(t_1) = b.$$

Also let $\psi(t)$ be a solution of the IVP

$$\mathcal{L}(y) = 0; \ y(t_1) = -A_2, \ y'(t_1) = A_1.$$

a. Show that if the only solution of

$$\mathcal{L}(y) = 0;$$
$$A_1 \, y(t_1) + A_2 \, y'(t_1) = 0, \ B_1 \, y(t_2) + B_2 \, y'(t_2) = 0$$

is the trivial solution, then there is a constant K such that $y = \phi(t) + K \, \psi(t)$ is the solution of (10.20).

b. Explain how to use an IVP solver to determine K.

c. Explain how to use an IVP solver to produce a graph of the solution of (10.20).

d. Use the shooting method to produce a graph of the solution of the BVP given in Exercise 21c.

e. Use the shooting method to graph the solution of the BVP

$$y'' + \left(1 + \frac{1}{2}\sin(t)\right)y = 0;$$
$$y(0) - y'(0) = 1; \ y(60) + 3\,y'(60) = 2.$$

10.2 Eigenvalue Problems

Let $\mathcal{L}(y) = p(x)y'' + q(x)y' + r(x)y$ be a linear differential operator with no singular points on an interval $[x_1, x_2]$, and let $\mathcal{V}$ be the set of functions that have a continuous second derivative on $[x_1, x_2]$ and satisfy homogeneous, separated boundary conditions

$$\left. \begin{array}{r} A_1\, y(x_1) + A_2\, y'(x_1) = 0 \\ B_1\, y(x_2) + B_2\, y'(x_2) = 0. \end{array} \right\} \qquad (10.21)$$

We are using x rather than t as the independent variable here because in the applications to follow, the independent variable will represent position, rather than time.

> **DEFINITION**   A real or complex number λ is called an **eigenvalue** of $\mathcal{L}$ on $\mathcal{V}$ if the BVP
>
> $$-\mathcal{L}(y) + \lambda\, y = 0, \qquad (10.22)$$
>
> with boundary conditions (10.21) has a nontrivial solution. A nontrivial solution $y = \phi(x)$ is called an **eigenfunction** of $\mathcal{L}$ belonging to the eigenvalue λ. Finally, the problem of finding nontrivial solutions of the BVP (10.22), (10.21) is called a one-dimensional **eigenvalue problem**, or **EVP**.

You may ask why we should be interested in EVPs, since their solutions only represent situations in which BVPs fail to have unique solutions. We will see in the next section that solving EVPs has important applications. Another reason to suspect that EVPs are important is hinted by their name.

The words *eigenvalue* and *eigenfunction* are hybrids, formed by attaching the German word *eigen* (meaning "characteristic") as a prefix to English words.[1] In linear algebra courses, *eigenvalue* is synonymous with *characteristic root*, and these terms are central to the study of square matrices. The EVP (10.22), (10.21) has a nontrivial solution $y = \phi(x)$ if and only if λ is a characteristic root of $\mathcal{L}$, because (10.22) implies that $\mathcal{L}(\phi) = \lambda\phi$.

As a first example of an EVP, let us consider one involving the linear differential operator $\mathcal{L}(y) = -y''$:

$$y'' + \lambda y = 0; \;\; y(0) = y(L) = 0. \qquad (10.23)$$

Proposition 10.2.1

Let $\omega = \pi/L$. The eigenvalues of the EVP (10.23) are $\lambda_m = (m\omega)^2$, for $m = 1, 2, 3, \ldots$. The eigenfunctions belonging to λ_m are all of the form

$$y = C \sin(m\omega x),$$

where C is a nonzero constant.

---

[1] The German words for eigenvalue and eigenfunction, *der Eigenwert* and *die Eigenfunktion*, are completely German.

Proof Suppose first that λ is positive, and put $\nu = \sqrt{\lambda}$. The general solution of

$$y'' + \nu^2 y = 0$$

is $y = C \sin(\nu x) + D \cos(\nu x)$. The left boundary condition is $y(0) = C \sin 0 + D \cos 0 = 0$, which simplifies to $D = 0$. Thus $y = C \sin(\nu x)$. Now we turn to the right boundary condition: $y(L) = C \sin(\nu L) = 0$. This will hold only if $\nu L = m\pi$ for some integer m; hence

$$\nu = m \frac{\pi}{L} = m\omega.$$

We have thus shown that all of the positive eigenvalues have the form $\lambda = (m\omega)^2$.

We will complete the proof by showing that there are no other eigenvalues. $\lambda = 0$ is not an eigenvalue because the general solution of $y'' = 0$ is the linear function $y = Cx + D$. If $y(0) = y(L) = 0$, the graph (a straight line) crosses the x-axis twice; hence $y \equiv 0$. Now consider the case $\lambda < 0$. Put $\lambda = -\nu^2$. The general solution of

$$y'' - \nu^2 y = 0$$

is $y = C \sinh(\nu x) + D \cosh(\nu x)$. Again, $y(0) = D = 0$, so $y = C \sinh(\nu x)$. But then $y(L) = C \sinh(\nu L) \neq 0$, so $\lambda = -\nu^2$ cannot be an eigenvalue. ∎

The method used in the proof of Proposition 10.2.1 can be applied to many other EVPs, such as the one in the following example, where $\mathcal{L}(y) = -y'' - 2y' - y$.

Example 10.2.1   Solve the EVP

$$y'' + 2y' + (1 + \lambda)y = 0; \quad y(0) = y(\pi) + y'(\pi) = 0.$$

Solution. The characteristic equation for the ODE is

$$s^2 + 2s + 1 + \lambda = 0,$$

or $(s + 1)^2 + \lambda = 0$. The characteristic roots are therefore $s = -1 \pm \sqrt{-\lambda}$; these are real if $\lambda \leq 0$ and complex if $\lambda > 0$.

Case 1: $\lambda < 0$. Set $\lambda = -\nu^2$; then the characteristic roots are $s = -1 \pm \nu$, so the general solution is

$$y = C_1 e^{(-1+\nu)x} + C_2 e^{(-1-\nu)x}.$$

Thus, $y(0) = C_1 + C_2$, $y(\pi) = C_1 e^{(-1+\nu)\pi} + C_2 e^{(-1-\nu)\pi}$, and

$$y'(\pi) = C_1(-1 + \nu)e^{(-1+\nu)\pi} + C_2(-1 - \nu)e^{(-1-\nu)\pi}.$$

Hence,

$$y(\pi) + y'(\pi) = C_1 \nu e^{(-\nu+1)\pi} - C_2 \nu e^{(-\nu-1)\pi},$$

and therefore the coefficients C_1 and C_2 satisfy

$$C_1 + C_2 = 0$$
$$C_1 v e^{(-1+v)\pi} - C_2 v e^{(-1-v)\pi} = 0.$$

This system of equations has only the trivial solution $C_1 = C_2 = 0$; hence λ is not an eigenvalue.

Case 2: $\lambda = 0$. The general solution of $y'' + 2y' + y = 0$ is $y = e^{-x}(C_1 + C_2 x)$. To meet the left boundary condition, we require that $C_1 = 0$, and this leaves $y = C_2 x e^{-x}$. Therefore,

$$y(\pi) + y'(\pi) = C_2 \pi e^{-\pi} + C_2 e^{-\pi} - C_2 \pi e^{-\pi} = C_2 e^{-\pi}.$$

To satisfy the right boundary condition, the coefficient $C_2 = 0$ as well. It follows that $\lambda = 0$ is not an eigenvalue.

Case 3: $\lambda > 0$. Put $\lambda = v^2$. The characteristic roots are $-1 \pm vi$, so the general solution is

$$y = e^{-x}[C_1 \cos(vx) + C_2 \sin(vx)].$$

The left boundary condition is then

$$y(0) = C_1 = 0,$$

so $y = C_2 e^{-x} \sin(vx)$. Hence,

$$y(\pi) + y'(\pi) = C_2 e^{-\pi} \sin(v\pi) - C_2 e^{-\pi} \sin(v\pi) + C_2 e^{-\pi} \cos(v\pi)$$
$$= C_2 e^{-\pi} \cos(v\pi).$$

The right boundary condition is thus satisfied if $\cos(v\pi) = 0$; that is, for $v = \frac{2m-1}{2}$ for some integer $m \geq 1$. The eigenvalues are thus $\lambda_m = \left(\frac{2m-1}{2}\right)^2$, for $m = 1, 2, 3, \ldots$, and the corresponding eigenfunctions are $\phi_m(x) = e^{-x} \sin\left(\frac{2m-1}{2}x\right)$. ∎

We know that a matrix can have multiple characteristic roots, and corresponding to such a characteristic root there can be several linearly independent characteristic vectors. The following proposition states that no eigenvalue of an EVP admits two independent eigenfunctions.

Proposition 10.2.2

Let $\phi_1(x)$ and $\phi_2(x)$ be eigenfunctions belonging to the same eigenvalue λ of the EVP (10.22), (10.21). Then $\phi_2(x) = C\phi_1(x)$ for some constant C.

Proof The eigenfunctions $\phi_1(x)$ and $\phi_2(x)$ are solutions of the homogeneous linear ODE (10.22). By Theorem 5.5 on page 195, they are linearly dependent if their Wronskian vanishes at some point in the interval $[a, b]$. Since $\phi_1(x)$ and $\phi_2(x)$ satisfy the boundary conditions (10.21), this is a consequence of the following lemma.

Lemma 10.2.3

Suppose that $\phi_1(x)$ and $\phi_2(x)$ are functions that are differentiable at $x = a$, and such that there are constants C and D, not both zero, with

$$C\phi_1(a) + D\phi_1'(a) = C\phi_2(a) + D\phi_2'(a) = 0.$$

Then $W[\phi_1, \phi_2](a) = 0$.

Proof of Lemma 10.2.3. Assume that $C \neq 0$ and put $k = -D/C$. Then $\phi_1'(a) = k\phi_1(a)$ and $\phi_2'(a) = k\phi_2(a)$. Therefore,

$$\begin{aligned} W[\phi_1, \phi_2](a) &= \phi_1(a)\,\phi_2'(a) - \phi_2(a)\,\phi_1'(a) \\ &= \phi_1(a)\,(k\,\phi_2(a)) - \phi_2(a)\,(k\,\phi_1(a)) \\ &= 0. \end{aligned}$$

If $C = 0$, then $D \neq 0$, and the proof is the same. ∎

Notice that Lemma 10.2.3 is valid for any pair of functions ϕ_1, ϕ_2, as long as they satisfy the boundary conditions. They do not have to be eigenfunctions.

EXERCISES

1. Solve the following EVPs:

  a. $y'' + \lambda y = 0$; $y'(0) = y'(L) = 0$

  b. $y'' + \lambda y = 0$; $y'(0) = y(L) = 0$

  c. $y'' + 4y' + \lambda y = 0$; $y(0) = y(L) = 0$

2. Solve the EVP

$$t^2\,y'' + t\,y' + \lambda\,y = 0; \quad y(1) = y(e^\pi) = 0.$$

3. Let $\mathcal{L}$ be a second-order linear operator with no singular points on the interval $[a, b]$. Show that if the boundary conditions are omitted, every number λ is an eigenvalue of

$$\mathcal{L}(y) + \lambda y = 0,$$

with two independent eigenfunctions.

4. Show that the eigenvalues of

$$y'' + \lambda y = 0; \quad y(0) = 0 = y(\pi) + y'(\pi) = 0$$

are the numbers v_m^2 such that $v_m + \tan(v_m \pi) = 0$.

5. Verify that the numbers v_m defined in Exercise 4 form an unbounded, increasing sequence of positive numbers.

  ◆ **Hint:** Sketch the graphs of $s = -t$ and $s = \tan(t\pi)$, and show that the graphs intersect at points where the t-coordinate is v_m.

6. Find boundary conditions such that $y'' + \lambda y = 0$ has exactly one negative eigenvalue.

7. Show that 0 is an eigenvalue of

$$y'' + \lambda y = 0; \quad y(0) = y(1) - y'(1) = 0$$

and that there are no negative eigenvalues. Describe the set of positive eigenvalues.

8. Periodic boundary conditions. Let $p(x)$ and $r(x)$ be L-periodic functions. The EVP

$$[p(x)y']' + [r(x) + \lambda]y = 0;$$
$$y(0) = y(L), \ y'(0) = y'(L)$$

is said to have periodic boundary conditions. Show that every eigenfunction is L-periodic.

9. Show that Proposition 10.2.2 does not hold for periodic boundary conditions. This is not a contradiction because periodic boundary conditions are not separated.

  ◆ **Hint:** Consider the EVP $y'' + \lambda y = 0$; $y(0) = y(\pi)$, $y'(0) = y'(\pi)$.

10. There are several ways to prove that the eigenvalues of

$$y'' + \lambda y = 0; \quad y(0) = y(L) = 0$$

are positive, in addition to the one used in the proof of Proposition 10.2.1.

  a. Use the result of Exercise 20 on page 477 to show that there are no negative eigenvalues of $\mathcal{L}(y) = -y''$.

b. If $y = \phi(x)$ satisfies the boundary conditions, then you can integrate by parts and obtain

$$\int_0^L \phi(x)\phi''(x)\,dx = -\int_0^L [\phi'(x)]^2\,dx < 0.$$

You can then multiply the equation $\phi''(x) + \lambda\phi(x) = 0$ by $\phi(x)$, integrate, and complete the proof.

Can you state and prove a more general theorem based on either of these methods?

11. Show that all of the eigenvalues of

$$(1 + x^2)\,y'' + \lambda y = 0; \quad y(0) = y(1) = 0$$

are positive.

12. Let $G(x, t)$ denote Green's function for the linear differential operator

$$\mathcal{L}(y) = -[p(x)y'' + q(x)y' + r(x)y]$$

and boundary conditions $A_1y(a) + A_2y'(a) = B_1y(b) + B_2y'(b) = 0$. Recall that Green's function does not exist unless $y \equiv 0$ is the only solution of $\mathcal{L}(y) = 0$ that satisfies the boundary conditions. Thus, assume that $\lambda = 0$ is not an eigenvalue of the EVP

$$-\mathcal{L}(y) + \lambda y = 0;$$
$$A_1y(a) + A_2y'(a) = B_1y(b) + B_2y'(b) = 0. \qquad (10.24)$$

Then Green's operator is

$$\mathcal{G}(u)(x) = \int_a^b G(x, t)u(t)\,dt.$$

We will say that $\phi(x)$ is an eigenfunction of $\mathcal{G}$ with eigenvalue μ if $\mathcal{G}(\phi) = \mu\phi$. Show that the eigenfunctions of $\mathcal{G}$ are the same as the eigenfunctions of the EVP (10.24), but the eigenvalues of $\mathcal{G}$ are reciprocals of the corresponding eigenvalues of (10.24).

13. Let $\mathcal{G}$ denote Green's operator (see Exercise 12). Show that if M is the maximum value of $|G(x, t)|$ for $a \leq x, t \leq b$, then for every eigenvalue μ of $\mathcal{G}$, $|\mu| \leq M(b - a)$.

◆ **Hint:** Show that if ϕ is an eigenfunction belonging to μ, then $\int_a^b \mathcal{G}(\phi)(x)\phi(x)\,dx = \mu\int_a^b [\phi(x)]^2\,dx$, and use Schwarz's inequality: For any functions $f(x)$, $g(x)$ that are continuous on $[a, b]$,

$$\left(\int_a^b f(x)\,g(x)\,dx\right)^2$$
$$\leq \left(\int_a^b [f(x)]^2\,dx\right)\left(\int_a^b [g(x)]^2\,dx\right).$$

14. Find a number $A > 0$ such that for all eigenvalues λ of the EVP (10.24), $|\lambda| > A$.

◆ **Hint:** Use the result of Exercise 13.

10.3 Diffusion

If the tip of an iron rod is heated until it is red hot and then removed from the fire, we expect that the tip will cool until it is no longer red and that the rest of the rod will become warmer. The process by which heat flows in the rod is called **diffusion**, and it was first explored by the French mathematician Joseph Fourier in a book entitled *The Analytic Theory of Heat,*[2] published in 1822. Fourier's book had an immediate and lasting influence on physics and on mathematics. If it had not been written (and it might not have been: Fourier was sentenced to death on two separate occasions in the 1790s), other scientists would have made the same discoveries. However, Fourier's genius as an expositor would have been lost. The notation $\int_a^b$ for the definite integral was invented by Fourier, and first appeared in print in his book.

Newton's law of cooling (see Section 1.4), formulated in the 1670s, is an observation about temperature change, not heat flow. Prior to 1761, no distinction between heat and temperature was made. In that year a Scottish scientist, Joseph Black, discovered that when ice was heated the temperature did not change until the melting was complete. This led him to make further experiments to elucidate the distinction between temperature and heat. Black defined the *specific heat* of a

---

[2]*Théorie analytique de la chaleur.* An English translation, prepared in 1878 by Alexander Freeman, is available from Dover Publications.

material to be the quantity of heat required to warm one unit mass of that material by one temperature unit. For example, the specific heat of water, in centimeter-gram-second units, is 1 calorie per gram per degree centigrade (this varies a little with temperature and is exact only when the temperature of the water is 15°C). By comparison, the specific heats of aluminum, iron and silver are approximately 0.2, 0.1, and 0.05, respectively.

Consider a thin rod that is not in thermal equilibrium. The rod is assumed to be insulated, so that the heat is not transmitted between the rod and the environment, *except through the ends of the rod.* Let $u(x, t)$ denote the temperature at a point x on the rod, at time t. Fourier showed in his book how to determine the function $u(x, t)$, given the initial temperature $u(x, 0)$, and information about the amount of heat conducted through the ends of the rod. We will let C denote the specific heat of the material of which the rod is composed and ρ denote its density. The density per unit length of the rod is ρA, where A is the area of the rod's cross section. If a segment of the rod between $x = a$ and $x = b$ should be brought from a uniform temperature of 0 to the temperature $u(x, t)$, the amount of heat necessary would be

$$Q_a^b(t) = \int_a^b C\rho A\, u(x, t)\, dx. \tag{10.25}$$

Heat flux is the rate at which heat moves in the rod, divided by the cross-sectional area of the rod. Let $f(x, t)$ denote the flux at the point x on the rod, at time t. The rate at which heat is absorbed by the segment from the left is $Af(a, t)$. Similarly, $Af(b, t)$ is the rate at which heat is lost through the right end of the segment. See Figure 10.5. Of course, $f(a, t)$ or $f(b, t)$ could be negative, if heat is being lost at the left end or gained at the right end of the segment. Since we are assuming that heat is conducted in and out of any segment of the rod only through the ends of the segment,

$$\frac{d}{dt} Q_a^b = A(f(a, t) - f(b, t)). \tag{10.26}$$

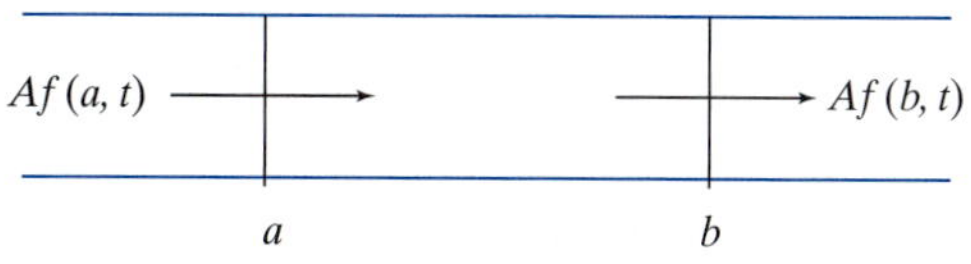

FIGURE 10.5   Heat flow in a rod with uniform cross-sectional area A. The heat flux is $f(x, t)$.

A new version of Newton's law of cooling, **Fourier's law**, states that *the heat flux is proportional to minus the derivative of the temperature with respect to x.* Thus,

$$f(x, t) = -K \frac{\partial u}{\partial x}(x, t).$$

The constant K is called the *thermal conductivity.* Typical values are approximately 0.08, .56, .21, and 1, for water, aluminum, iron, and silver, respectively (in centimeter-gram-second units). Water, being a liquid, must be assumed not to circulate. A good thermal insulator will have a low thermal conductivity; for example, fiberglass has a thermal conductivity of 0.001.

Question   How is Fourier's law similar to Newton's law of cooling?

If we combine Fourier's law with equation (10.26), we can eliminate the flux:

$$\frac{d}{dt}Q_a^b = -AK\left(\frac{\partial u}{\partial x}(a, t) - \frac{\partial u}{\partial x}(b, t)\right). \tag{10.27}$$

It is also possible to determine $\frac{d}{dt}Q_a^b(t)$ by differentiating formula (10.25). Thus,

$$\frac{d}{dt}Q_a^b(t) = \frac{d}{dt}\int_a^b C\rho A\, u(x, t)\, dx$$

$$= \int_a^b C\rho A\frac{\partial u}{\partial t}\, dx.$$

Using the fundamental theorem of calculus, the right side of equation (10.27) can be written as an integral. Since $\frac{\partial u}{\partial x}$ is an antiderivative of $\frac{\partial^2 u}{\partial x^2}$ with respect to x,

$$\frac{\partial u}{\partial x}(a, t) - \frac{\partial u}{\partial x}(b, t) = -\int_a^b \frac{\partial^2 u}{\partial x^2}(x, t)\, dx.$$

Therefore, equation (10.27) can be rewritten as

$$\int_a^b C\rho A\frac{\partial u}{\partial t}(x, t)\, dx = AK\int_a^b \frac{\partial^2 u}{\partial x^2}(x, t)\, dx.$$

Since this equation holds for any choice of the limits a and b of integration, it follows that $u(x, t)$ satisfies the partial differential equation, or **PDE**

$$\frac{\partial u}{\partial t} = c\frac{\partial^2 u}{\partial x^2}, \tag{10.28}$$

where $c = \frac{K}{C\rho}$ is called the **diffusion coefficient**. Equation (10.28) is the one-dimensional diffusion equation.

The initial-boundary value problem

Suppose that the temperature at each end of the rod is maintained at $0°$. Suppose also that the initial temperature is a given function $\phi(x)$, for $0 \le x \le L$. We pose the following mathematical problem as a model for the temperature of the rod.

Find a function $u(x, t)$ such that

$$\frac{\partial u}{\partial t} = c\frac{\partial^2 u}{\partial x^2}, \tag{10.29}$$

satisfying the *initial condition*

$$u(x, 0) = \phi(x) \text{ for all } x \in [0, L], \tag{10.30}$$

and satisfying the *boundary conditions*

$$u(0, t) = u(L, t) = 0 \text{ for all } t \ge 0. \tag{10.31}$$

This is an **initial-boundary value problem**, or **IBVP** for the diffusion equation. For it to make sense, we must require $\phi(0) = \phi(L) = 0$.

It is shown in partial differential equations texts that the IBVP (10.29), (10.30), (10.31) has a unique solution provided that the function $\phi(x)$ satisfies some very reasonable conditions.[3] We will confine our attention to finding a solution by the method of **separation of variables.** To begin, we look for solutions of the diffusion equation with the special form

$$u(x, t) = h(x)k(t). \tag{10.32}$$

We will insist that these solutions satisfy the boundary conditions, by requiring that

$$h(0) = h(L) = 0. \tag{10.33}$$

The initial condition will be set aside for the moment.

Substituting the expression (10.32) into the diffusion equation, we have

$$h(x)k'(t) = c \cdot h''(x)k(t).$$

Divide this equation by the expression $c \cdot h(x)k(t)$ to get

$$\frac{k'(t)}{c \cdot k(t)} = \frac{h''(x)}{h(x)}. \tag{10.34}$$

The left side of equation (10.34) does not depend on x, and the right side does not depend on t. Since the two expressions are equal, neither depends on x or t. It follows that each expression is constant, and we denote this constant by $-\lambda$. The minus sign is only for convenience, because it will turn out that the expressions in (10.34) are negative. From $h''(x)/h(x) = -\lambda$ we obtain a second-order linear ODE,

$$h''(x) + \lambda h(x) = 0. \tag{10.35}$$

Similarly, $k'(t)/(c\,k(t)) = -\lambda$ translates to a first-order ODE,

$$k'(t) = -c \cdot \lambda\, k(t). \tag{10.36}$$

Let us focus on equation (10.35), with boundary conditions (10.33). This is an EVP. If $u(x, t) = h(x)k(t)$ is a solution of the diffusion equation satisfying $u(0, t) = u(L, t) = 0$, then the parameter λ must be an eigenvalue of the EVP (10.35), (10.33), and $h(x)$ must be an eigenfunction belonging to λ. By Proposition 10.2.1 on page 478, $\lambda = m^2\omega^2$, where $\omega = \pi/L$ and m is an integer, while $h(x) = C \sin(m\omega x)$, where C is constant. Put $\lambda = m^2\omega^2$ in equation (10.36); it then follows that

$$k(t) = e^{-cm^2\omega^2 t}.$$

We have constructed an infinite sequence

$$\{u_1(x, t), u_2(x, t), \dots\} \tag{10.37}$$

of solutions of the diffusion equation, where

$$u_m(x, t) = e^{-cm^2\omega^2 t} \sin(m\omega x).$$

You should verify, by differentiating both sides, that

$$\frac{\partial}{\partial t}u_m(x, t) = c\frac{\partial^2}{\partial x^2}u_m(x, t).$$

---

[3]For a proof of uniqueness when $\phi(x)$ is continuous, see Exercise 10 at the end of this section.

Superposition

We can put the solutions $u_m(x, t)$ of the diffusion equation together to form a solution of the IBVP when the initial condition is of a special type of function, called a **trigonometric polynomial**:

$$u(x, 0) = \phi(x) = b_1 \sin(\omega x) + b_2 \sin(2\omega x) + \cdots + b_n \sin(n \omega x). \quad (10.38)$$

Here, ω and $b_1 \ldots b_n$ are constants. In general, trigonometric polynomials may have a constant term and cosine terms as well, but our boundary conditions preclude these.

The solution $u_m(x, t)$ satisfies the IBVP with initial condition

$$u_m(x, 0) = \sin(m\omega x).$$

The **principle of superposition** applies to the diffusion equation: Any linear combination of solutions is again a solution. Thus, when the initial condition is of the form (10.38), the solution of the corresponding IBVP is a linear combination of the solutions $u_m(x, t)$, where the coefficients are the same as the b_i in (10.38):

$$u(x, t) = b_1 e^{-c\omega^2 t} \sin(\omega x) + b_2 e^{-c4\omega^2 t} \sin(2\omega x) + \cdots + b_n e^{-cn^2\omega^2 t} \sin(n\omega x).$$

Example 10.3.1   An insulated uniform rod of length π is composed of material with thermal diffusion coefficient $c = 0.1$. At $t = 0$, the temperature of the rod is

$$u(x, 0) = \phi(x) = \sin x - 2 \sin 4x,$$

and the ends of the rod are held at the constant temperature $u(0, t) = u(\pi, t) = 0$. Determine the temperature of the rod as a function of position x and time t.

Solution. We are to solve the IBVP,

$$\frac{\partial u}{\partial t} = 0.1 \frac{\partial^2 u}{\partial x^2}; \quad u(0, t) = u(\pi, t) = 0 \text{ for } t > 0, \quad u(x, 0) = \phi(x).$$

By inspection, $b_1 = 1$ and $b_4 = -2$, and $c = 0.1$. Therefore,

$$u(x, t) = e^{-.1t} \sin x - 2e^{-1.6t} \sin 4x. \qquad \square$$

Figure 10.6 displays the initial temperature of points on the rod of Example 10.3.1, and the temperature after 1 second. This reveals the typical features of the diffusion process: The temperature approaches a constant steady state, and the short wavelength variations [coming from the $\sin(4x)$ term in the initial temperature] are smoothed the most rapidly.

The initial temperature cannot always be expressed as a trigonometric polynomial. To make his theory of diffusion work more generally, Fourier defined infinite trigonometric series, now called Fourier series. The idea is similar to the step taken when going from ordinary polynomials to Taylor series, but Fourier series present more complicated convergence problems, which were not completely resolved by Fourier.

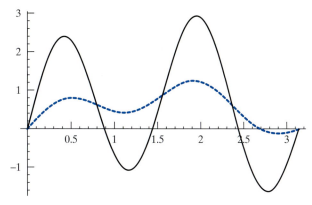

FIGURE 10.6   Temperature of the rod in Example 10.3.1 at $t = 0$ (solid curve) and $t = 1$ (dashed curve).

Any phenomenon that is modelled by the diffusion equation is called a diffusion process. For example, consider a glass tube that is filled with smoke. The particles of smoke, when watched individually under magnification, appear to move randomly, a process known as *Brownian motion*. On the macroscopic scale, let $u(x, t)$ denote the density of the smoke at a point x and time t. It can be shown that $u(x, t)$ obeys the diffusion equation. An IBVP that is completely analogous to Example 10.3.1 results if the initial density of the smoke in the tube is specified as the function $u(x, 0) = \phi(x)$, and the ends of the tube are open so that $u(x, 0) = u(x, L) = 0$.

Temperature in caves

Visitors to underground caverns notice that the temperature is much less variable underground than it is on the surface. This phenomenon is explained by the diffusion equation. Thus, let $u(x, t)$ denote the temperature at depth x and time t. Although the earth is not like the insulated, thin rod for which the diffusion equation was derived, it is nevertheless true that

$$\frac{\partial u}{\partial t} = c \frac{\partial^2 u}{\partial x^2},$$

where c is the thermal diffusion coefficient for the soil, limestone, or whatever material lies above the cave.

The temperature of the soil on the surface above the cave is subject to daily and seasonal fluctuation. For simplicity, the daily fluctuations will be ignored, and it will be assumed that surface temperature is given by a sine function. We will choose a temperature scale so that the average temperature on the surface is 0, the maximum temperature is 1, and the minimum temperature is -1. It is convenient to measure t in years, with $t = 0$ marking the hottest time of year at the surface. We can then approximate the surface temperature as $u(0, t) = \cos(2\pi t)$. It will be convenient to use the complex exponential function here, so we will put $u(0, t) = \operatorname{Re}(e^{2\pi i t})$.

It is reasonable to assume that the temperature at depth x varies periodically, with a period of one year, but its amplitude and phase may not agree with the surface temperature. We will therefore assume that $u(x, t) = \operatorname{Re}[f(x)e^{2\pi i t}]$, where $f(x)$ is a complex-valued function. If $f(x)$ is written in polar form as $f(x) = r(x)e^{i\theta(x)}$, then $r(x)$ gives the amplitude of the temperature fluctuation at depth x, while $\theta(x)$ gives the phase.

Put $w(x, t) = f(x)e^{2\pi it}$, so that $u(x, t) = \text{Re}[w(x, t)]$. Substituting $w(x, t)$ into the diffusion equation yields

$$2\pi i f(x)e^{2\pi it} = cf''(x)e^{2\pi it}.$$

After canceling the exponential factor, we have an ordinary differential equation with complex coefficients,

$$f''(x) - 2\pi i c^{-1} f(x) = 0. \tag{10.39}$$

The characteristic roots of this second-order linear differential equation are $\pm[2\pi c^{-1}i]^{1/2}$. Recall that

$$\pm\sqrt{i} = \pm e^{\pi i/4} = \pm\frac{1+i}{\sqrt{2}}.$$

The two characteristic roots are therefore $\pm\lambda$, where

$$\lambda = \sqrt{\frac{2\pi}{c}} \cdot \frac{1+i}{\sqrt{2}} = \sqrt{\frac{\pi}{c}}(1+i),$$

and the general solution of equation (10.39) is

$$f(x) = C_1 e^{\lambda x} + C_2 e^{-\lambda x}.$$

The solution $e^{\lambda x}$ is not bounded as $x \to \infty$, but $e^{-\lambda x}$ is bounded as $x \to \infty$. Since it is reasonable to assume that $w(x, t)$ is bounded, we must take $C_1 = 0$. Furthermore, the assumption that $w(0, t) = e^{2\pi it}$ implies that $f(0) = 1$, and hence $C_2 = 1$. We conclude that

$$w(x, t) = e^{-\lambda x}e^{2\pi it}$$
$$= e^{-x\sqrt{\pi/c}}e^{i(2\pi t - x\sqrt{\pi/c})},$$

and

$$u(x, t) = \text{Re}[w(x, t)]$$
$$= e^{-x\sqrt{\pi/c}}\cos\left(2\pi t - x\sqrt{\frac{\pi}{c}}\right)$$
$$= e^{-x\sqrt{\pi/c}}\cos\left(2\pi(t - \frac{x}{2\sqrt{c\pi}})\right).$$

The amplitude of oscillation of the temperature $u(x, t)$ is $e^{-x\sqrt{\pi/c}}$, which decays exponentially as one descends into a cave. Below the surface, the hottest day of the year does not occur when $t = 0$. At depth x, it is hottest when $t = \frac{x}{2\sqrt{c\pi}}$.

Example 10.3.2   At what depth in a cave does the hottest day occur at the same time as the coldest day on the surface, and how hot is it down there on that day?

Solution. On the surface, it is coldest when $t = \frac{1}{2}$; then the temperature is -1. It will be hottest when $t = \frac{1}{2}$ at depth x_r, where

$$\frac{1}{2} = \frac{x_r}{2\sqrt{c\pi}};$$

hence $x_r = \sqrt{c\pi}$. At depth x_r, the seasons are reversed, and the temperature there will be maximum on the coldest day of winter. Then the temperature will be $e^{-\sqrt{\pi/c}\cdot\sqrt{c\pi}} = e^{-\pi} \approx 0.04$ in our special temperature scale. In a locale where the surface temperature fluctuates annually between a minimum of $-15°C$ and a maximum of $35°C$, with a mean temperature of $10°C$, the temperature at depth x_r will be about $11°C$ in midwinter, and about $9°C$ in midsummer.

The thermal diffusion coefficient of limestone is approximately 27 square meters per year. The seasons will therefore reverse at a depth of $x_r \approx \sqrt{27\pi} \approx 9$ meters. ◻

EXERCISES

1. Solve the IBVP

$$\frac{\partial u}{\partial t} = c\frac{\partial^2 u}{\partial x^2}; \; u(x,0) = \phi(x), \, u(0,t) = u(\pi,t) = 0,$$

where

a. $c = 1$, and $\phi(x) = \sin(3x)$

b. $c = 3$, and $\phi(x) = \sin(x)$

c. $c = 1$, and $\phi(x) = \sin(100x)$

d. $c = 2$, and $\phi(x) = \sin(x) + 3\sin(3x) + 100\sin(100x)$

2. A glass tube is open at both ends and is filled with smoke. The tube is 1 meter long; the density of the smoke x meters from an end of the tube is initially $\sin(\pi x)$. One minute later, the density at the middle of the tube is 0.7. Determine the diffusion coefficient for the smoke in this tube.

3. The temperature of an insulated piece of wire is initially a trigonometric polynomial

$$u(x,0) = \phi(x)$$
$$= b_1\sin(\omega x) + b_2\sin(2\omega x)$$
$$+\cdots+ b_n\sin(n\omega x),$$

where the length of the wire is π/ω. Show that if the ends of the wire are held at a temperature of 0, then at each point of the wire, the temperature will converge to 0 as $t \to \infty$.

4. The temperature of a piece of wire with length $L = \pi$, thermal diffusion coefficient $c = 0.1$, and initial temperature $u(x,0) = 100\sin(100x)$ converges to 0 much faster than it would if the initial temperature were $u(x,0) = 100\sin(x)$.

a. How long will it take for the maximum temperature at any point of the wire to decrease to 1 degree in each case?

b. How would you explain this phenomenon to someone who has no knowledge of calculus?

5. What are the units of thermal conductivity, in the centigrade-calorie-centimeter-gram-second (cccgs) system?

6. Determine the thermal diffusion coefficients of aluminum, iron, and silver. You will need the following densities (in grams per cubic centimeter, at $20°C$): aluminum, 2.7, iron, 7.9, and silver, 10.5.

7. Derive the cccgs units for the thermal diffusion coefficient from the units for the thermal conductivity (Exercise 5) and the units for heat capacity and density. Then check the heat equation for dimensional consistency by verifying that the right side is denominated in units that reduce to the units of the left side, centigrade degrees per second.

8. Show that variables can be rescaled, with distance $X = ax$, time $T = bt$, and temperature $U = du$ so that the IBVP (10.29), (10.30), (10.31) takes the form

$$\frac{\partial U}{\partial T} = \frac{\partial^2 U}{\partial X^2};$$

$$U(0,T) = U(\pi,T) = 0; \; U(X,0) = \Phi(X),$$

where $\max\{|\Phi(X)| : 0 \le X \le \pi\} = 1$. Determine the scaling factors a, b, and d in terms of the parameters of the diffusion problem.

9. Suppose that a rod does not have a uniform cross-section, and let $A(x)$ denote the cross-sectional area at x. Suppose also the composition of the rod is not homogeneous, so that the specific heat, density, and thermal conductivity depend on x. Show that the temperature $u(x, t)$ satisfies the linear partial differential equation with variable coefficients,

$$C(x)\rho(x)A(x)\frac{\partial u}{\partial t} = \frac{\partial}{\partial x}\left(K(x)A(x)\frac{\partial u}{\partial x}\right).$$

10. Assume a rod lies on the x-axis between $x = 0$ and $x = L$. As in Exercise 9, the parameters influencing the rate of diffusion are variable. Put $\alpha(x) = C(x)\rho(x)A(x)$ and $\beta(x) = K(x)A(x)$, so that the temperature $u(x, t)$ satisfies the partial differential equation

$$\frac{\partial u}{\partial t} = \frac{1}{\alpha(x)}\frac{\partial}{\partial x}\left(\beta(x)\frac{\partial u}{\partial x}\right). \tag{10.40}$$

The boundary conditions will either be constant temperature,

$$\frac{\partial u}{\partial t}(0, t) = \frac{\partial u}{\partial t}(L, t) = 0;$$

or zero heat flux,

$$\frac{\partial u}{\partial x}(0, t) = \frac{\partial u}{\partial x}(L, x) = 0.$$

This problem will use a function

$$E(t) = \int_0^L \frac{1}{2}\beta(x)\left(\frac{\partial u}{\partial x}(x, t)\right)^2 dx$$

to show that solutions of IBVPs for the equation (10.40) are unique. As the integral of $\beta(x)$ times the square of the temperature gradient, $E(t)$ measures the degree to which the temperature of the rod is not a constant function of position at time t. Our proof will be based on showing that $E(t)$ is a decreasing function of time. If the rod is uniform in composition and cross section, equation (10.40) is just the diffusion equation; therefore, we will prove uniqueness for solutions of the IBVP (10.29), (10.30), (10.31).

a. Show that $E(t) \ge 0$, with equality holding if and only if $u(x, t)$ is constant.

b. Use integration by parts to show that

$$\frac{dE}{dt} = -\int_0^L \frac{\partial}{\partial x}\left(\beta(x)\frac{\partial u}{\partial x}(x, t)\right)\frac{\partial u}{\partial t} dx.$$

c. Show that $\frac{dE}{dt} \le 0$, with equality holding if and only if

$$\frac{\partial}{\partial x}\left(\beta(x)\frac{\partial u}{\partial x}\right) \equiv 0.$$

d. Show that if $u_1(x, t)$ and $u_2(x, t)$ are solutions of equation (10.40) that satisfy the same initial and boundary conditions, then

$$u_1(x, t) \equiv u_2(x, t).$$

◆ **Hint:** $v(x, t) = u_1(x, t) - u_2(x, t)$ will satisfy equation (10.40) with initial condition $v(x, 0) = 0$ and boundary conditions $v(0, t) = v(L, t) = 0$ (if u_1 and u_2 satisfy constant temperature boundary conditions), or $\partial v/\partial x(0, t) = \partial v/\partial x(L, t) = 0$ (if the boundary conditions for u_1 and u_2 are zero flux).

11. In the discussion of the temperature in caves, the daily temperature fluctuation was neglected. Explain how a surface temperature of

$$\psi(t) = A\,\text{Re}(e^{2\pi it}) + D\,\text{Re}(e^{730\pi it})$$

would take both the daily and annual temperature fluctuations into account. Represent the temperature at depth x as

$$u(x, t) = A\,\text{Re}[f(x)e^{2\pi it}] + D\,\text{Re}[g(x)e^{730\pi it}],$$

where $f(x)$ and $g(x)$ are complex-valued functions such that $f(0) = g(0) = 1$, $|f(x)|$ and $|g(x)|$ are bounded as $x \to \infty$, and $w(x, t) = f(x)e^{2\pi it}$ and $v(x, t) = g(x)e^{730\pi it}$ satisfy the diffusion equation.

a. Determine $g(x)$.

b. Is it reasonable to consider the daily temperature fluctuations to be negligible?

12. The *frost line* is the depth below the surface at which the temperature never falls below the freezing point of water. The importance of the frost line is obvious to plumbers, because water pipes that are buried above it may freeze. Determine the frost line for a locale where the temperature varies between $-15°C$ and $35°C$, with an average temperature of $10°C$. Assume the thermal diffusion coefficient of the soil in this locale is 3 square meters per year.

10.4   Fourier series

In this brief introduction to Fourier series, we will see how to present a solution of a diffusion equation IBVP when the initial function is not a trigonometric polynomial.

A **Fourier sine series** on an interval $[0, L]$ is an infinite series of the form

$$\psi(x) = \sum_{n=1}^{\infty} C_n \sin(n\omega x), \tag{10.41}$$

where $0 \leq x \leq L$ and $\omega = \pi/L$. Each partial sum

$$\psi_m(x) = C_1 \sin(\omega x) + C_2 \sin(2\omega x) + \cdots + C_m \sin(m\omega x)$$

is a trigonometric polynomial. Thus, if $\psi(x)$ is equal to the sum of a convergent Fourier sine series, we might say that it can be approximated by trigonometric polynomials. However, this would blur an important distinction about convergence of series of functions. When we say that the series (10.41) converges, we mean that for any particular number x,

$$\lim_{m \to \infty} \psi_m(x) = \psi(x).$$

To be precise, we should say that the series **converges pointwise**. On the other hand, the assertion that $\psi(x)$ can be approximated by trigonometric polynomials means that for any specified tolerance $\epsilon > 0$, there is a trigonometric polynomial $\phi(x)$ that approximates $\psi(x)$ within ϵ; that is,

$$|\psi(x) - \phi(x)| \leq \epsilon \quad \text{for all } x \in [0, L].$$

If for any $\epsilon > 0$ there is a number M such that each of the trigonometric polynomials $\psi_M(x), \psi_{M+1}(x), \psi_{M+2}(x), \ldots$ approximates $\psi(x)$ within ϵ, then (10.41) is said to **converge uniformly on the interval** $[0, L]$. There are examples of functions that have pointwise convergent, but not uniformly convergent Fourier series, and cannot be approximated by trigonometric polynomials.

Proposition 10.4.1

If the series $\sum_{n=1}^{\infty} |C_n|$ converges, then the Fourier series (10.41) converges uniformly on $[0, L]$.

Proof Let ϵ be a given positive number, and let $S = \sum_{n=1}^{\infty} |C_n|$. Denote the partial sum $\sum_{n=1}^{M} |C_n|$ by S_M. Since it is given that $\sum_{n=1}^{\infty} |C_n|$ converges, there is a number M such that $0 \leq S - S_M < \epsilon$, because it is given that $\sum_{n=1}^{\infty} |C_n|$ converges. Since the terms $|C_n|$ are not negative, for any integer $k \geq 1$ and any value of $N > M + k$,

$$\sum_{n=M+k}^{N} |C_n| \leq \sum_{n=M+1}^{\infty} |C_n| = S - S_M < \epsilon.$$

Since $|\sin(n\omega x)| \le 1$ for all $x \in [0, L]$,

$$\sum_{n=M+k}^{N} |C_n \sin(n\omega x)| \le \sum_{n=M+k}^{N} |C_n| < \epsilon.$$

By the triangle inequality,

$$\left| \sum_{n=M+k}^{N} C_n \sin(n\omega x) \right| \le \sum_{n=M+k}^{N} |C_n \sin(n\omega x)| < \epsilon.$$

According to the definition of an infinite series,

$$\left| \sum_{n=M+k}^{\infty} C_n \sin(n\omega x) \right| = \lim_{N \to \infty} \left| \sum_{n=M+k}^{N} C_n \sin(n\omega x) \right|.$$

The limit of a sequence of numbers, all less than ϵ, cannot exceed ϵ. Therefore,

$$|\psi(x) - \psi_{M+k-1}(x)| = \left| \sum_{n=M+k}^{\infty} C_n \sin(n\omega x) \right| \le \epsilon.$$

In other words, each of the trigonometric polynomials corresponding to $k = 1, 2, 3, \dots$; that is,

$$\psi_M(x), \psi_{M+1}(x), \psi_{M+2}(x), \dots$$

approximates $\psi(x)$ within ϵ.   ■

The result of Section 10.3 was that when the initial function of the IBVP,

$$\frac{\partial u}{\partial t} = c \frac{\partial^2 u}{\partial x^2}; \quad u(0, t) = u(L, t) = 0, \tag{10.42}$$

is a trigonometric polynomial $u(x, 0) = \psi_m(x)$, then the solution is

$$u(x, t) = C_1 e^{-c\omega^2 t} \sin(\omega x) + C_2 e^{-4c\omega^2 t} \sin(2\omega x) + \cdots + C_m e^{-m^2 c\omega^2 t} \sin(m\omega x).$$

We will extend that result by stating the following theorem:

Theorem 10.3

Suppose that the initial condition for the IBVP (10.42) is $u(x, 0) = \psi(x)$, where $\psi(x)$ is equal to the sum of a pointwise convergent Fourier sine series, as in (10.41). Then the solution of the IBVP is

$$u(x, t) = \sum_{n=1}^{\infty} C_n e^{-n^2 c\omega^2 t} \sin(n\omega x),$$

where the coefficients C_n are the coefficients of the Fourier sine series representing $\psi(x)$.

Given a one-dimensional diffusion problem where an initial function $u(x, 0) = \psi(x)$ is specified, and $u(x, t) = 0$ at the boundary points $x = 0$ and $x = L$, the first thing to do is to represent $\psi(x)$ as a Fourier sine series. Theorem 10.3 can then be used to present the solution. Since the initial function is not usually given in Fourier series form, the Fourier coefficients C_n must be calculated. This involves *orthogonal functions*.

Orthogonal functions

Let $C[0, L]$ be the set of all real-valued continuous functions defined on the interval $[0, L]$. On $C[0, L]$ we define an **inner product**,

$$\langle f, g \rangle = \int_0^L f(x)g(x)\,dx.$$

A real inner product must satisfy the following three axioms.

Linearity $\langle \psi_1, C\psi_2 + D\psi_3 \rangle = C\langle \psi_1, \psi_2 \rangle + D\langle \psi_1, \psi_3 \rangle$

Symmetry $\langle \psi_1, \psi_2 \rangle = \langle \psi_2, \psi_1 \rangle$

Positivity $\langle \psi, \psi \rangle \geq 0$ with equality only if $\psi(t) = 0$ for all $t \in [0, L]$

You can verify these axioms hold by referring to the following properties of integrals:

$$\int_0^L (A_1\, f_1(x) + A_2\, f_2(x))\,dx = A_1 \int_0^L f_1(x)\,dx + A_2 \int_0^L f_2(x))\,dx,$$

and for any nonnegative function $f(x)$ that is continuous on $[0, L]$,

$$\int_0^L f(x)\,dx > 0$$

unless $f(x) \equiv 0$.

A set of functions $\{\phi_1, \phi_2, \dots\} \subset C[0, L]$ is **orthogonal** if none of the ϕ_i is identically 0, and $\langle \phi_m, \phi_n \rangle = 0$ when $m \neq n$. Suppose that

$$\psi(t) = C_1\phi_1(t) + C_2\phi_2(t) + \cdots + C_n\phi_n(t)$$

is a linear combination of orthogonal functions. The inner product of $\psi(x)$ with $\phi_m(x)$ is

$$\langle \psi, \phi_m \rangle = C_1\langle \phi_1, \phi_m \rangle + \cdots + C_m\langle \phi_m, \phi_m \rangle + \cdots + C_n\langle \phi_n, \phi_m \rangle$$
$$= C_1\, 0 + \cdots + C_m\langle \phi_m, \phi_m \rangle + \cdots + C_n\, 0.$$

It follows that

$$C_m = \frac{\langle \psi, \phi_m \rangle}{\langle \phi_m, \phi_m \rangle}.$$

Proposition 10.4.2

___

Let $\omega = \pi/L$, and $\phi_m(x) = \sin(m\omega x)$ for $m = 1, 2, 3\dots$. Then the set $\{\phi_1, \phi_2, \phi_3, \dots\}$ is orthogonal on $[0, L]$, and for each m,

$$\langle \phi_m, \phi_m \rangle = \frac{L}{2}.$$

Proof Suppose $n \neq m$. Then

$$\langle \sin(n\omega x), \sin(m\omega x) \rangle = \int_0^L \sin(n\omega x)\, \sin(m\omega x)\, dx = 0,$$

as you can verify by doing the integral. Also,

$$\langle \sin(n\omega x), \sin(n\omega x) \rangle = \int_0^L \sin^2(n\omega x)\, dx = \frac{L}{2}. \qquad \blacksquare$$

Calculating the Fourier coefficients is done by using the inner product. Assuming that it is permissable to form the inner product of a Fourier sine series with a function term by term, we have

$$\langle \psi(x), \sin(n\omega x) \rangle = \sum_{m=1}^{\infty} C_m \, \langle \sin(m\omega x), \sin(n\omega x) \rangle = C_n \frac{L}{2}.$$

It follows that

$$C_n = \frac{2}{L} \langle \psi(x), \sin(n\omega x) \rangle = \frac{2}{L} \int_0^L \psi(x)\, \sin(n\omega x)\, dx.$$

Example 10.4.1   Find the Fourier sine series for the tent function

$$\psi(x) = \begin{cases} x & \text{if } 0 \le x \le 1 \\ 2 - x & \text{if } 1 < x \le 2 \end{cases}$$

on the interval $[0, 2]$.

Solution. The length of the interval $[0, 2]$ is $L = 2$, so we will put $\omega = \frac{\pi}{L} = \frac{\pi}{2}$. The Fourier coefficients are

$$C_n = \frac{2}{2} \int_0^2 \psi(x)\, \sin(n\omega x)\, dx$$

$$= \int_0^1 x \, \sin\left(\frac{n\pi}{2}x\right) dx + \int_1^2 (2 - x)\, \sin\left(\frac{n\pi}{2}x\right) dx.$$

These integrals are evaluated by parts. Thus

$$\int_0^1 x \, \sin\left(\frac{n\pi}{2}x\right) dx = -\frac{2}{n\pi} x \cos\left(\frac{n\pi}{2}x\right)\Big|_0^1$$

$$+ \left(\frac{2}{n\pi}\right)^2 \sin\left(\frac{n\pi}{2}x\right)\Big|_0^1$$

$$= -\frac{2}{n\pi}\cos\left(\frac{n\pi}{2}\right) + \left(\frac{2}{n\pi}\right)^2 \sin\left(\frac{n\pi}{2}\right).$$

Similarly, it can be shown that

$$\int_1^2 (2-x) \sin\left(\frac{n\pi}{2}x\right) dx = \left(\frac{2}{n\pi}\right) \cos\left(\frac{n\pi}{2}\right) + \left(\frac{2}{n\pi}\right)^2 \sin\left(\frac{n\pi}{2}\right).$$

Adding these, we find that

$$C_n = 2\left(\frac{2}{n\pi}\right)^2 \sin\left(\frac{n\pi}{2}\right)$$

$$= \begin{cases} 0 & \text{if } n \text{ is even} \\ (-1)^k \dfrac{8}{(2k+1)^2\pi^2} & \text{if } n = 2k+1. \end{cases}$$

It follows that the Fourier series of $\psi(t)$ is

$$\frac{8}{\pi^2}\left(\sin\left(\frac{\pi}{2}x\right) - \frac{1}{9}\sin\left(3\frac{\pi}{2}x\right) + \frac{1}{25}\sin\left(5\frac{\pi}{2}x\right) + \cdots\right)$$

$$= \frac{8}{\pi^2}\sum_{k=0}^{\infty} \frac{(-1)^k}{2k+1}\sin\left((2k+1)\frac{\pi}{2}x\right). \qquad \square$$

Figure 10.7 displays the graphs to the tent function $\psi(x)$ of Example 10.4.1 and the partial sum

$$\psi_5(x) = \frac{8}{\pi^2}\left(\sin(\pi x/2) - \frac{1}{9}\sin(3\pi x/2) + \frac{1}{25}\sin(5\pi x/2)\right),$$

drawn on the interval $[0, 2]$. You will notice that except for points near $x = 1$, the approximation is good. We have to expect some imperfection at the midpoint, since the at peak of the tent, ψ is not differentiable, while any linear combination of sine functions is differentiable. Nevertheless, the Fourier series does converge at $x = 1$.

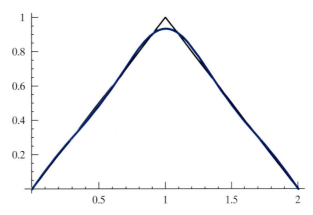

FIGURE 10.7   The tent function of Example 10.4.1 and a trigonometric polynomial approximation.

If we substitute $x = 1$ in each term of the series and note that $\sin((2k + 1)\pi/2) = (-1)^k$, the following identity results:

$$\frac{\pi^2}{8} = 1 + \frac{1}{9} + \frac{1}{25} + \cdots = \sum_{k=0}^{\infty} \frac{1}{(2k + 1)^2}. \tag{10.43}$$

Figure 10.8 shows the graph of $\psi_{11}(x)$ on the interval $[-4, 4]$. Although $\psi(x)$ is defined only for $0 \le x \le 2$, its Fourier sine series is defined and converges for all real x. The sum of the series inherits two properties from the sine function: It is an odd function, and it is periodic (the period is 4).

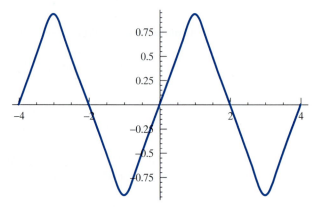

FIGURE 10.8   A Fourier series extends the tent function as a periodic, odd function.

A function $\psi(x)$ is **piecewise differentiable** on $[0, L]$ if there is a partition $0 = x_0 \le x_1 \le x_2 \le \dots \le x_n = L$ of the interval such that $\psi(t)$ is differentiable on each subinterval $[x_i, x_{i+1}]$. For example, the tent function of Example 10.4.1 is piecewise differentiable. In this case, the partition is $\{x_0, x_1, x_2\} = \{0, 1, 2\}$. On $[0, 1]$, $\psi(x) = x$, and on $[1, 2]$, $\psi(x) = 2 - x$.

Piecewise differentiable functions do not have to be continuous. For example, $\psi(x) = u(x - 1)$, where u is the unit step function, is piecewise differentiable on the interval $[0, 2]$, and to show this, the same partition that was used with the tent function should be chosen.

In his work on the diffusion equation, Fourier made the incorrect assumption that any continuous function can be expressed as the sum of a convergent Fourier sine series. In fact, there are continuous functions that cannot be expressed this way, and discontinuous functions that can.

The following theorem, due to the German mathematician G. L. Dirichlet, gives an answer to the question of convergence of Fourier series. Its proof may be found in any Fourier series text.[4]

---

[4]I recommend *Fourier Series* by G. P. Tolstov, Dover, 1976, page 75.

Theorem 10.4

Let $\psi(x)$ be a piecewise differentiable function defined on the interval $[0, L]$. For each integer $m \geq 1$, put

$$C_m = \frac{2}{L} \int_0^L \psi(x) \sin(m\omega x)\, dx, \qquad (10.44)$$

where $\omega = \pi/L$. Then the Fourier sine series $\sum_{m=1}^{\infty} C_m \sin(m\omega x)$ converges for each real value of x. Furthermore, for each $x \in (0, L)$ such that ψ is continuous at x,

$$\sum_{m=1}^{\infty} C_m \sin(n\omega x) = \psi(x).$$

Heat diffusion

When $\psi(x)$ is piecewise differentiable on $[0, L]$, the solution of the IBVP

$$\frac{\partial u}{\partial t} = c\, \frac{\partial^2 u}{\partial x^2}; \quad u(0, t) = u(L, t), \; u(x, 0) = \psi(x) \qquad (10.45)$$

can be represented by the Fourier series,

$$u(x, t) = \sum_{m=1}^{\infty} C_m e^{-cm^2 \omega^2 t} \sin(m\omega x), \qquad (10.46)$$

where the coefficients are given by (10.44). The factors $e^{-cm^2\omega^2 t}$ decay extremely rapidly when m is sufficiently large. As we will now see, it is acceptable to use a partial sum of (10.46) to approximate the solution for any positive value of t.

Proposition 10.4.3

Suppose[5] that the initial function $\psi(x) = u(x, 0)$ is bounded. This means we are assuming that there is a constant K such that $|\psi(x)| \leq K$ for $0 \leq x \leq L$. Then for a fixed $t > 0$, then the series (10.46) converges uniformly for $0 \leq x \leq L$.

Proof We will use Proposition 10.4.1, so we need to show that for any fixed, positive t, the series

$$\sum_{m=1}^{\infty} |C_m| e^{-cm^2 \omega^2 t}. \qquad (10.47)$$

converges.

The coefficient C_m satisfies the following inequality:

$$|C_m| = \frac{2}{L} \left| \int_0^L \psi(x) \sin(m\omega x)\, dx \right| \leq \frac{2}{L} \int_0^L |\psi(x)|\, dx \leq 2K.$$

---

[5]If you know Schwarz's inequality, you can prove this result with the weaker hypothesis that $\int_0^L (\psi(x))^2\, dx$ is finite.

Therefore, each term of the series (10.47) is less than or equal to the corresponding term of

$$2K \sum_{m=1}^{\infty} e^{-cm^2\omega^2 t}. \tag{10.48}$$

Let $v = e^{-c\omega^2 t}$ and notice that $0 < v < 1$. The series (10.48) is equal to $2K \sum_{m=1}^{\infty} v^{m^2}$, and each term of this series appears as the term corresponding to $k = m^2$ in the convergent geometric series $2K \sum_{k=1}^{\infty} v^k$. It follows that (10.48) converges, and therefore the series it dominates, (10.47), converges too. ■

Example 10.4.2 Let $\psi(x)$ be the tent function defined in Example 10.4.1. We suppose that the initial temperature of an iron rod, two meters in length, is $u(x, 0) = 100\psi(x)$ for $0 \le x \le 2$. The rod is insulated, except at the ends, which are held at a constant temperature of 0. Find the temperature of the rod as a function of position and time, and plot its graph. (The thermal diffusion coefficient of iron is $c = 3 \times 10^{-4}$, in meter-kilogram-second units.)

Solution. The Fourier series,

$$\psi(x) = \frac{8}{\pi^2} \sum_{m=0}^{\infty} \frac{(-1)^m}{(2m+1)^2} \sin\left(\frac{(2m+1)\pi}{2}x\right),$$

was determined in Example 10.4.1. As an infinite series, the solution of our IBVP is determined by multiplying the mth term of the Fourier series expansion of $u(x, 0) = 100\psi(x)$ by $e^{-c(2m+1)^2(\pi/2)^2 t}$, for $m = 0, 1, 2 \ldots$. Thus,

$$u(x, t) = \frac{800}{\pi^2} \sum_{m=0}^{\infty} \frac{(-1)^m}{(2m+1)^2} e^{-c(2m+1)^2(\pi/2)^2 t} \sin\left(\frac{(2m+1)\pi}{2}x\right). \tag{10.49}$$

Let us use the following one-digit approximations (more precision is not justified since we were only given the diffusion coefficient to one-digit accuracy): $800/\pi^2 \approx 80$, $c(\pi/2)^2 \approx 7 \times 10^{-4}$, $\pi/2 \approx 1.6$, $800/(3\pi)^2 \approx 9$, $c(3\pi/2)^2 \approx 7 \times 10^{-3}$, $3\pi/2 \approx 4.7$, $800/(5\pi)^2 \approx 3$, $c(5\pi/2)^2 \approx 2 \times 10^{-2}$, and $5\pi/2 \approx 7.9$. Taking just the first three terms of the solution (10.49) yields

$$u(x, t) \approx 80\, e^{-0.0007\,t}\, \sin(1.6\,x) - 9\, e^{-.007\,t}\, \sin(4.7\,x) + 3\, e^{-.02\,t}\, \sin(7.9\,x).$$

The fourth term, which we are neglecting, has absolute value less than

$$\frac{80}{7^2} e^{-(7\pi/2)^2\, c\, t} \approx 1.6 e^{-0.04\,t}.$$

After 30 seconds, this term decreases to less than $1.6\, e^{-1.2} \approx 0.5°$, so it can be considered negligible.

A graph of $u(x, t)$ is shown in Figure 10.9. ❑

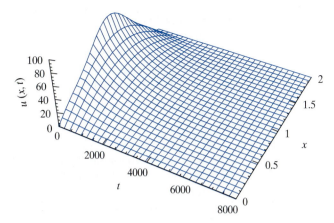

FIGURE 10.9   Solution of a heat diffusion problem: See
Example 10.4.2.

Diffusion with an insulated boundary

A thin rod is insulated so that no heat is transmitted to the environment except
through the ends. The ends are insulated too, but the insulation may be imperfect
there, so that some heat transmission to or from the environment can occur.

Let $u(x, t)$ denote the temperature of the rod at position x and time t, and let
$f(x, t)$ denote the heat flux. At the ends of the rod, $x = 0$ and $x = L$, Fourier's law
tells us that

$$f(0, t) = -K\frac{\partial u}{\partial x}(0, t) \quad \text{and} \quad f(L, t) = -K\frac{\partial u}{\partial x}(L, t), \tag{10.50}$$

where K is the thermal conductivity coefficient for the rod. For example, if the
insulation is perfect, allowing no heat to flow, then $f(0, t) = f(L, t) = 0$. It
follows from (10.50) that the boundary conditions for the diffusion problem are
$\frac{\partial u}{\partial x}(0, t) = \frac{\partial u}{\partial x}(L, t) = 0$.

If the insulation at the ends is imperfect, then the heat flux at each end of
the rod is proportional to the difference in temperature between the rod and the
environment. Thus, the flux at the ends is

$$f(0, t) = -k_l(u(0, t) - T_l) \quad \text{and} \quad f(L, t) = -k_r(T_r - u(L, t)), \tag{10.51}$$

where k_l and k_r are thermal conductivity coefficients for the insulation at the left
and right ends of the rod, and T_l and T_r are the ambient temperatures at the ends.
By combining (10.50) and (10.51), we obtain the following boundary conditions for
the diffusion equation:

$$k_l u(0, t) - K\frac{\partial u}{\partial x}(0, t) = k_l T_l \tag{10.52}$$

and

$$k_r u(L, t) + K\frac{\partial u}{\partial x}(L, t) = k_r T_r. \tag{10.53}$$

Since we need to work with homogeneous boundary conditions, assume that $T_l = T_r = 0$. (If the ambient temperature is the same at each end, but nonzero, we can change the temperature scale to make the ambient temperature 0. The case where $T_l \neq T_r$ is handled in Exercise 11 at the end of this section.)

To solve the IBVP,

$$\frac{\partial u}{\partial t} = c \frac{\partial^2 u}{\partial x^2}, \tag{10.54}$$

with initial condition $u(x, 0) = \psi(x)$ and boundary conditions (10.52), (10.53) with $T_l = T_r = 0$, we separate variables to find all solutions of (10.54) that have the form $u(x, t) = h(x)k(t)$ and satisfy the boundary conditions. In Exercise 9 you are asked to show that for each solution of this form, $h(x)$ must be an eigenfunction of $\mathcal{L}(y) = -y''$ that satisfies the boundary conditions

$$k_l\, y(0) - K\, y'(0) = 0, \quad k_r\, y(L) + K\, y'(L) = 0 \tag{10.55}$$

and $k(t) = e^{-c\lambda t}$, where λ is the eigenvalue belonging to $h(x)$.

Let $\phi_1(x), \phi_2(x), \dots$ be the eigenfunctions of the EVP

$$y'' + \lambda\, y = 0, \tag{10.56}$$

with boundary conditions (10.55). If it is possible to express the initial function as a Fourier series,

$$\psi(x) = \sum_{m=1}^{\infty} C_m\, \phi_m(x), \tag{10.57}$$

then the solution of the IBVP is

$$u(x, t) = \sum_{m=1}^{\infty} C_m\, e^{-c\lambda_m t}\, \phi_m(x).$$

Thus the following problems remain:

I.   Solve the EVP (10.56), (10.55).
II.   Show that $\psi(x)$ can be represented as a Fourier series of type (10.57).
III.   Determine the Fourier coefficients.

Task II will be set aside, as it is best handled in a Fourier series text. Let us start with task I.

Perfect insulation. If the insulation at the ends of the rod is perfect, then $k_l = k_r = 0$, and we have to solve (10.56) with boundary conditions

$$y'(0) = y'(L) = 0.$$

In Exercise 1 at the end of this section, you will show that the eigenfunctions of $y'' + \lambda y = 0$ with these boundary conditions are the constant function $\phi_0(x) = 1$ and the cosine functions $\phi_n(x) = \cos(n\omega x)$, where $\omega = \pi/L$. The eigenvalue belonging to ϕ_n is $\lambda_n = (n\omega)^2$. Furthermore, $\langle \phi_n, \phi_m \rangle = 0$ if $n \neq m$, while $\langle \phi_0, \phi_0 \rangle = L$ and

$\langle \phi_n, \phi_n \rangle = L/2$ for $n = 1, 2, 3, \ldots$. Thus, to solve a heat diffusion problem in the case of a perfectly insulated rod, the initial temperature $\psi(x)$ should be expressed as the sum of a **Fourier cosine series**,

$$\psi(x) = C_0 + \sum_{m=1}^{\infty} C_m \cos(m\omega x) = \sum_{m=0}^{\infty} C_m \phi_m(x),$$

where $C_m = \langle \psi, \phi_m \rangle / \langle \phi_m, \phi_m \rangle$, for $m = 0, 1, 2, 3, \ldots$. Thus

$$C_0 = \frac{1}{L} \int_0^L \phi(x)\, dx$$

is the average value of ψ on $[0, L]$, and

$$C_m = \frac{2}{L} \int_0^L \phi(x) \cos(m\omega x)\, dx$$

for $m = 1, 2, 3, \ldots$.

The solution of the diffusion IBVP in the case of perfect insulation is therefore

$$u(x, t) = C_0 + \sum_{m=1}^{\infty} C_m e^{-cm^2\omega^2 t} \cos(m\omega x).$$

This confirms physical intuition, because as $t \to \infty$, the $u(x, t) \to C_0$. In other words, the temperature approaches a constant value, and that constant is the average value of the initial temperature.

This discussion is applicable in other contexts that involve diffusion. For example, if smoke is injected into a glass tube and then the ends are sealed, the smoke density can be represented by a function, $u(x, t)$, that is a solution of the same IBVP.

Imperfect insulation. Now consider a rod with insulation at each end. We will assume that the conductivity of the insulating material at each end of the rod is the same; that is, $k_l = k_r$. If you wish, you may also consider a glass tube filled with smoke whose density at position x and time t is denoted $u(x, t)$. The initial density is a given function $u(x, 0) = \psi(x)$, and the ends of the tube are closed with a semiporous material.

Let $r = k_l/K = k_r/K$ be the ratio of the conductivity of the insulator to the conductivity of the rod. The solution of the IBVP can then be expressed as a Fourier series

$$u(x, t) = \sum_{m=1}^{\infty} C_m e^{-c\lambda_m t} \phi_m(x),$$

where $\phi_m(x)$ is an eigenvalue of

$$y'' + \lambda y = 0; \quad r\, y(0) - y'(0) = r\, y(L) + y'(L) = 0. \tag{10.58}$$

You will be asked, in Exercise 10, to show that all eigenvalues of (10.58) are positive. If $\lambda = v^2 > 0$, then the general solution of the ODE in (10.58) can be presented as

$$y = A \sin(v\, x + \theta),$$

where A and θ are constants. The boundary conditions are

$$A(r \, \sin(\theta) - v \, \cos(\theta)) = 0 \tag{10.59}$$

and

$$A(r \, \sin(v \, L + \theta) + v \, \cos(v \, L + \theta)) = 0. \tag{10.60}$$

By equation (10.59),

$$\tan(\theta) = \frac{v}{r}$$

and it follows that

$$\sin(\theta) = \frac{v}{\sqrt{v^2 + r^2}} \quad \text{and} \quad \cos(\theta) = \frac{r}{\sqrt{v^2 + r^2}}. \tag{10.61}$$

Using the addition formulas for the sine and cosine, equation (10.60) can be rearranged as

$$r \, (\sin(v \, L) \cos(\theta) + \cos(v \, L) \sin(\theta)) + v \, (\cos(v \, L) \cos(\theta) - \sin(v \, L) \sin(\theta)) = 0.$$

Using (10.61) and simplifying, we obtain

$$\sin(v \, L)(r^2 - v^2) + \cos(v \, L)(2 \, v \, r) = 0$$

and it follows that

$$\tan(vL) = \frac{2 \, v \, r}{v^2 - r^2}. \tag{10.62}$$

To find the eigenvalues, we have to solve equation (10.62).

Figure 10.10 displays the graphs of two functions: $L(v) = \tan(vL)$ is the left side of (10.62), and $R(v) = 2 \, v \, r/(v^2 - r^2)$ is the right side. For this figure, we have

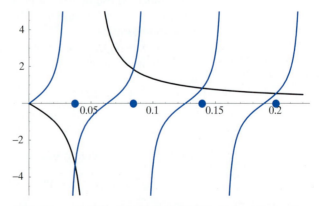

FIGURE 10.10   The solutions of (10.62) are the values of v corresponding to intersections of the graphs of $2vr/(v^2 - r^2)$ (in black) and $\tan(vL)$ (in blue), where $r = 0.05$ and $L = 50$. The dots on the v-axis show $v_1, \ldots, v_4$.

taken $L = 50$ and $r = 0.05$. It is clear that there is an infinite sequence of numbers v_n such that $L(v_n) = R(v_n)$, and these are the solutions of (10.62).
The eigenfunction belonging to $\lambda_n = v_n^2$ is

$$\phi_n(x) = \sin(v_n x + \theta_n),$$

where $\theta_n = \tan^{-1}(v_n/r)$. In Exercise 10, you are asked to show that these eigenfunctions are orthogonal. Armed with this result, we can expand an initial temperature function $\psi(x)$ as a Fourier series

$$\psi(x) = \sum_{n=1}^{\infty} C_n \sin(v_n x + \theta_n).$$

The Fourier coefficients are given by the formula $C_n = \langle \psi, \phi_n \rangle / \langle \phi_n, \phi_n \rangle$.
When the Fourier coefficients of ψ are determined (this will be a job for a CAS), the solution of the IBVP is

$$u(x, t) = \sum_{n=1}^{\infty} C_n e^{-cv_n^2 t} \sin(v_n x + \theta_n).$$

EXERCISES

1. Find the eigenvalues and eigenfunctions of the BVP,

   $$y'' + \lambda y = 0; \quad y'(0) = y'(L) = 0.$$

2. Does the Fourier series representing the tent function (see Example 10.4.1) converge uniformly?

3. Determine the Fourier sine series representing the function $f(x) = 1$ for $0 \le x \le \pi$.

   a. Does the series converge uniformly?

   b. Use a computer to plot a graph of the partial sum $\sum_{n=1}^{15} C_n \sin(nx)$ for $-2\pi \le x \le 2\pi$. Explain the appearance of this graph, and describe the features of the sum of the Fourier series.

   c. Substitute $x = \pi/2$ in the series and use it to evaluate the sum

   $$\sum_{k=0}^{\infty} \frac{(-1)^k}{2k + 1}.$$

   d. The ends of an insulated rod (length π) are held at a temperature of 0, while the initial temperature of the rod is $u(x, 0) = 1$ for $0 < x < \pi$. The diffusivity coefficient is $c = 1$. Find an expression for $u(x, t)$ for $t > 0$.

4. Determine the Fourier sine series representing the function $f(x) = \pi x - x^2$ for $0 \le x \le \pi$.

   a. Does the series converge uniformly?

   b. Use a computer to plot a graph of the partial sum $\sum_{n=1}^{15} C_n \sin(nx)$ for $-2\pi \le x \le 2\pi$. Explain the appearance of this graph, and describe the features of the sum of the Fourier series.

   c. Substitute $x = \pi/2$ in the series and use it to evaluate the sum

   $$\sum_{k=0}^{\infty} \frac{(-1)^k}{(2k + 1)^3}.$$

   d. The ends of an insulated rod (length π) are held at a temperature of 0, while the initial temperature of the rod is $u(x, 0) = \pi x - x^2$ for $0 < x < \pi$. The diffusivity coefficient is $c = 1$. Find an expression for $u(x, t)$ for $t > 0$.

5. Find the Fourier cosine series representing the tent function

   $$f(x) = \begin{cases} x & \text{if } 0 \le x < \frac{\pi}{2} \\ \pi - x & \text{if } \frac{\pi}{2} \le x \le \pi. \end{cases}$$

   a. Use a computer to plot a graph of the partial sum

   $$f_{15}(x) = C_0 + \sum_{n=1}^{15} C_n \cos(nx)$$

for $-2\pi \le x \le 2\pi$. Explain the appearance of this graph, and describe the features of the sum of the Fourier series.

b. Substitute $x = \pi/2$ in the series and use it to evaluate the sum

$$\sum_{k=0}^{\infty} \frac{1}{(2k+1)^2}.$$

c. The ends of an insulated rod (length π) conduct no heat, while the initial temperature of the rod is $u(x, 0) = f(x)$ (the tent function) for $0 < x < \pi$. The diffusivity coefficient is $c = 1$. Find an expression for $u(x, t)$ for $t > 0$.

6. An iron rod (thermal diffusion coefficient $c = 0.3$, thermal conductivity $K = 0.2$) of length π is situated on the x-axis with one end at the origin. Find the temperature as a function of x and time, given that

a. The ends of the rod are kept at temperature of $0°C$, and the initial temperature is $\psi(x) = 40x$ for $0 \le x < \pi - 1$ and $\psi(x) = -40(\pi - 1)(x - \pi)$ for $\pi - 1 \le x \le \pi$.

b. The ends of the rod are perfectly insulated, and the initial temperature is $\psi(x) = \cos(x) - \cos(10x)$.

c. The ends of the rod are insulated with thermal conductivity at the ends of the rod $k = 0.02$, and the ambient temperature is $0°C$. The initial temperature of the rod is a constant $100°C$.

◆ **Warning:** This problem requires the use of a CAS to approximate the eigenvalues and Fourier coefficients.

d. The left end of the rod is held at $0°C$; the right end is perfectly insulated, and at time 0, the temperature is $\psi(x) = \sin(x/2) - 3\sin(3x/2)$.

e. The left end of the rod is held at $0°C$; the right end is insulated with thermal conductivity $k = 0.02$. The initial temperature is as in (d).

◆ **Warning:** You will need a CAS to approximate the eigenvalues and Fourier coefficients.

In each case, plot the temperature as a function of x for times $t = 0$, $t = 1$, and $t = 2$.

7. Find the first four eigenvalues of

$$y'' + \lambda y = 0;$$
$$.05\, y(0) - y'(0) = 0, \ .05\, y(50) + y'(50) = 0$$

accurately to six decimal places. Display graphs of the corresponding eigenfunctions. Give a rough estimate of the remaining eigenvalues.

8. Show that $\lambda = 1$ is an eigenvalue of

$$y'' + \lambda y = 0;$$
$$y(0) - y'(0) = 0, \ y(\pi/2) + y'(\pi/2) = 0$$

and find the corresponding eigenfunction. Are there any smaller eigenvalues? Explain how this EVP would occur in a diffusion problem.

9. Show that if $u(x, t) = h(x) k(t)$ is a solution of the diffusion equation that satisfies the boundary conditions (10.52) and (10.53), then $h(x)$ is a solution of the EVP

$$y'' + \lambda\, y = 0;$$
$$k_l\, y(0) - K y'(0) = k_r\, y(L) + K y'(L) = 0$$

with eigenvalue λ and $k(t) = e^{-c\lambda t}$.

10. Prove that all of the eigenvalues of the EVP (10.58) are positive and that the eigenfunctions are orthogonal.

◆ **Hint:** Let ϕ_n and ϕ_m be eigenfunctions belonging to distinct eigenvalues. Notice that

$$v_n^2 \int_0^L \phi_n(x)\phi_m(x)\, dx = -\int_0^L \phi_n''(x)\phi_m(x)\, dx$$

and

$$v_m^2 \int_0^L \phi_n(x)\phi_m(x)\, dx = -\int_0^L \phi_n(x)\phi_m''(x)\, dx.$$

Use integration by parts to prove that the integrals on the right sides of these equations are equal.

11. Consider the IBVP

$$\frac{\partial u}{\partial t} = c\frac{\partial^2 u}{\partial x^2};$$
$$\alpha u(0, t) - \frac{\partial u}{\partial x}(0, t) = A, \ \beta u(L, t) + \frac{\partial u}{\partial x}(L, t) = B,$$

with initial condition $u(x, 0) = \phi(x)$. Assume that α, β and c are positive. An equilibrium solution is an initial function $\phi(x)$ with the property that $u(x, t) = \phi(x)$ satisfies the IBVP. For example, if $A = B = 0$ in the boundary conditions, then $\phi(x) \equiv 0$ is an equilibrium solution.

a. Show that if $\phi(x)$ is an equilibrium solution, then $\phi(x) = mx + d$, where m and d are constants.

b. Determine m and d in terms of the parameters α, β, A, and B of the boundary conditions.

c. Show that if $u(x, t)$ is any solution of the IBVP and $\phi_\infty(x)$ is an equilibrium solution, then

$$v(x, t) = u(x, t) - \phi_\infty(x)$$

satisfies an IBVP with homogeneous boundary conditions. Specify this new IBVP.

d. Show that $v(x, t) \to 0$ as $t \to \infty$, and hence that $u(x, t) \to \phi_\infty(x)$ as $t \to \infty$.

*10.5   EVPs and the Phase Plane

This section is about an EVP consisting of an ODE

$$y'' + (\lambda + q(x))\, y = 0, \tag{10.63}$$

where $q(x)$ is continuous on an interval $[0, L]$, and boundary conditions

$$y(0) = 0,\ y(L) = 0. \tag{10.64}$$

We will not develop methods for solving this EVP, because when $q(x)$ is not constant, there is a good chance that the general solution of (10.63) cannot be expressed in terms of elementary functions. Instead, our purpose is to identify circumstances in which we can be sure that the EVP (10.63), (10.64) has an infinite sequence of eigenvalues, and in which the corresponding eigenfunctions form an orthogonal set.

In the case where $q(x) = q$ is constant, we already know how to solve the EVP. Assuming that $\lambda > -q$, set $k = \sqrt{\lambda + q}$. The general solution of (10.63) is a family of oscillating functions,

$$y = A\, \cos(kx) + B\, \sin(kx).$$

The constant A is determined by the left boundary condition, $y(0) = 0$, since substituting $x = 0$ gives $A = 0$. The right boundary condition, $y(L) = 0$, translates to $\sin(kL) = 0$. Hence $kL = n\pi$ for some integer n, so $\lambda = k^2 - q = (n\pi/L)^2 - q$. The eigenvalues form an infinite sequence, $\lambda_1 = (\pi/L)^2 - q$, $\lambda_2 = (2\pi/L)^2 - q, \ldots$.

Our objective is to show that if $q(x)$ is not constant, the eigenvalues can still be arranged in an infinite, increasing sequence. We will do so by analyzing the ODE (10.63) in the phase plane. The system

$$\left. \begin{aligned} y' &= v \\ v' &= -(\lambda + q(x))\, y \end{aligned} \right\} \tag{10.65}$$

is equivalent to (10.63). Let $r(y, v) = (y^2 + v^2)^{1/2}$ and $\theta = \arg(y + i\, v)$ be polar coordinates for the phase plane, as shown in Figure 10.11. We will be especially interested in the **phase angle** θ because it is the key to understanding the oscillations of solutions of (10.63). By differentiating the relation

$$\tan(\theta) = \frac{v}{y} = y^{-1} v$$

with respect to x, we find that

$$\sec^2(\theta) \frac{d\theta}{dx} = y^{-1} \frac{dv}{dx} - y^{-2} v \frac{dy}{dx}. \tag{10.66}$$

By (10.65), $y^{-1} \frac{dv}{dx} = -(\lambda + q(x))$, and $\frac{dy}{dx} = v$. Using these facts, rewrite (10.66) as

$$\sec^2(\theta) \frac{d\theta}{dx} = -(\lambda + q(x)) - \frac{v^2}{y^2}$$

$$= -(\lambda + q(x)) - \tan^2(\theta).$$

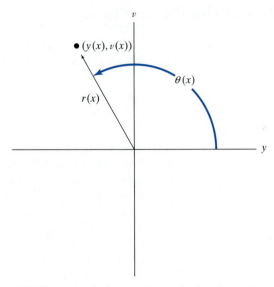

FIGURE 10.11 Polar coordinates in the phase plane.

Now let us multiply through by $\cos^2(\theta)$ to obtain

$$\frac{d\theta}{dx} = -(\lambda + q(x)) \cos^2(\theta) - \sin^2(\theta). \qquad (10.67)$$

Equation (10.67) is a first-order ODE that—when combined with an initial condition—determines the phase angle θ as a function of x.

If λ is chosen so that $\lambda + q(x) > 0$ for all x, (10.67) implies that $\theta(x)$ is a decreasing function. If we were able to show that $\theta(x)$ decreases without bound, then the corresponding phase curve would wind clockwise around the origin an infinite number of times as $x \to \infty$. Every crossing of the v-axis corresponds to a point where $y = 0$, and at every crossing of the y-axis, $y' = v = 0$. In other words, the solution whose phase angle is $\theta(x)$ would be shown to be oscillating.

Example 10.5.1 Make phase plots of the solutions of the IVP

$$y'' + (\lambda + \cos(\pi x))y = 0; \quad y(0) = 0, \; y'(0) = 1$$

for $\lambda = 1$ and $\lambda = 4$. In each case, also plot the function $\theta(x)$.

Solution. This is a problem for an IVP solver. The IVP in the phase plane is

$$\left. \begin{array}{lll} y' = & v; & y(0) = 0 \\ v' = & -(\lambda + \cos(\pi x))y; & v(0) = 1. \end{array} \right\} \qquad (10.68)$$

The initial phase angle is $\theta(0) = \arg(0 + 1 \cdot i) = \frac{\pi}{2}$. This initial condition and the ODE

$$\theta' = -(\lambda + \cos(\pi x)) \cos^2(\theta) - \sin^2(\theta) \qquad (10.69)$$

completely determine $\theta(x)$. See Figure 10.12.

FIGURE 10.12 Solution of the IVP of Example 10.5.1 with $\lambda = 1$. The phase plot is at the left, the graph of the phase angle is the middle graph, and the graph of the solution as a function of x is at the right.

We can determine the number of times the solution winds around the origin for $0 \leq x \leq 20$ by dividing the difference $\theta(20) - \theta(0)$ by 2π. According to the IVP solver, $\theta(20) \approx -19$, so the solution makes approximately -3.3 rotations (the negative sign indicates the direction is clockwise). By examining the phase plot, you can see that this is about right.

Figure 10.13 shows the phase plot, the phase angle variable, and the graph of $y(x)$ for $\lambda = 4$. In this case, $\theta(20) \approx -38.8$, so there are approximately 6.4 clockwise revolutions. ❑

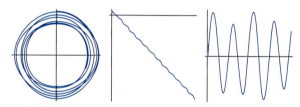

FIGURE 10.13 Solution of the IVP of Example 10.5.1 with $\lambda = 2$. The arrangement of the display is as in Figure 10.12. Notice that the phase curve intersects itself. This often happens with systems that are not autonomous, but never with autonomous systems.

Figure 10.14 shows the graphs of the phase angle θ in the IVP (10.69) for several values of the parameter λ. It shows clearly that as λ increases, the graph of $\theta(x)$ becomes more negatively sloped. This means that the solution $y(x)$ of the original ODE oscillates increasingly rapidly with increasing λ, and consequently that its zeros are more closely spaced as λ increases.

Let us return to the general eigenvalue problem (10.63), (10.64). Denote the solution of the IVP,

$$\theta' = -(\lambda + q(x))\cos^2(\theta) - \sin^2(\theta); \quad \theta(0) = \tfrac{\pi}{2}, \tag{10.70}$$

as a function of the independent variable x *and* the parameter λ, by $\theta(x, \lambda)$. Thus $\theta(x, \lambda)$ is the phase angle of a solution of (10.63) with $y(0) = 0$. If λ is an eigenvalue of (10.63), (10.64), then $y(L) = 0$ as well, and in terms of the phase angle, $\theta(L, \lambda) = n\pi/2$, where n is odd. Therefore, we need to prove that there is a sequence $\lambda_1, \lambda_2, \ldots$ such that for each integer $k \geq 1$,

$$\theta(L, \lambda_k) = -(2k - 1)\frac{\pi}{2}. \tag{10.71}$$

These λ_k would be the eigenvalues.

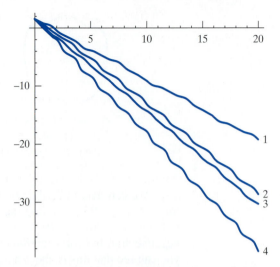

FIGURE 10.14   Graphs of the phase angle $\theta(x)$ for
the IVP of Example 10.5.1, using the parameter
values $\lambda = 1, 2, 3$, and 4. The value of the
parameter used is shown to the right of each graph.

We will first rule out the possibility of an infinite sequence of eigenvalues
$\to -\infty$.

Proposition 10.5.1

Let M be the maximum value of $q(x)$ on $[0, L]$. If $\lambda < -M$, then $0 < \theta(L, \lambda) < \frac{\pi}{2}$.
Thus all eigenvalues of (10.63), (10.64) satisfy $\lambda_k \geq -M$.

FIGURE 10.15   Direction field of
$\theta' = -(\lambda + q(x)) \cos^2(\theta) - \sin^2(\theta)$ when $\lambda < -M$.

Proof Figure 10.15 shows part of the direction field for the ODE in (10.70), un-
der the assumption that $\lambda < -M$. For $\theta = \frac{\pi}{2}$, notice that $\theta' = -1$ (this is true
independently of the value of λ). For $\theta = 0$, we have

$$\theta' = -(\lambda + q(x)) > -(-M + q(x)) \geq 0$$

It is clear from this direction field that the graph of the solution of (10.70) must
remain in the strip $0 < \theta < \frac{\pi}{2}$ for all $x \in [0, L]$. ∎

To show that there is an infinite sequence

$$\lambda_1 < \lambda_2 < \lambda_3 < \cdots$$

such that (10.71) holds, we will use the following comparison lemma.

Lemma 10.5.2

Suppose that $\alpha(x)$ is a solution of the IVP $y' = f(x, y)$; $y(x_0) = y_0$ on an interval $[x_0, x_1]$, and that $\beta(x)$ is a solution of another IVP, $y' = g(x, y)$, with the same initial condition $y(x_0) = y_0$. Assume also that for all (x, y) with $x \in [x_0, x_1]$ we have $g(x, y) \leq f(x, y)$, and that the function f satisfies the following Lipschitz condition[6]: For any interval $\mathcal{I}$ there is a constant M such that for all $x \in [x_0, x_1]$ and $y \in \mathcal{I}$,

$$|f(x, y_1) - f(x, y_2)| \leq M|y_1 - y_2|.$$

Then for all $x \in [x_0, x_1]$, $\beta(x) \leq \alpha(x)$.

Proof Suppose that the statement is false. Then there is a number $x^* \in [x_0, x_1]$ such that $\beta(x^*) > \alpha(x^*)$. Let $s \in [x_0, x_1]$ be the least number with the property that $\beta(x) \geq \alpha(x)$ for all $x \in [s, x^*]$, and put $\delta(x) = \beta(x) - \alpha(x)$. Then $\delta(x) \geq 0$ on $[s, x^*]$, and by continuity, $\delta(s) = 0$. Furthermore,

$$\begin{aligned}
\delta'(x) &= g(x, \beta(x)) - f(x, \alpha(x)) \\
&\leq f(x, \beta(x)) - f(x, \alpha(x)) \\
&\leq M|\beta(x) - \alpha(x)| \\
&= M\,\delta(x).
\end{aligned}$$

Thus, we have shown that $\delta'(x) - M\,\delta(x) \leq 0$ on $[s, x^*]$. Multiply this inequality by e^{-Mx} to obtain

$$\frac{d}{dx}[e^{-Mx}\delta(x)] \leq 0.$$

This implies that $e^{-Mx^*}\delta(x^*) \leq e^{-Ms}\delta(s) = 0$ and contradicts our assumption that $\beta(x^*) > \alpha(x^*)$. ∎

We also need to know that solutions of the IVP (10.70) depend continuously on the parameter λ as well as the variable x. That is the purpose of the following lemma.

Lemma 10.5.3

Let $y = \phi(x, \lambda)$ be the solution of an IVP

$$y' = f(x, y, \lambda); \quad y(x_0) = y_0, \tag{10.72}$$

where $f(x, y, \lambda)$ has continuous partial derivatives with respect to y and λ on a rectilinear domain

$$\mathcal{D} = \{(x, y, \lambda) : A \leq x \leq B, C \leq y \leq D, E \leq \lambda \leq F\}$$

containing the point (x_0, y_0, λ_0). If $x_1 \in (A, B)$ is a point such that $\phi(x_1, \lambda_0)$ is defined, then $\phi(x, \lambda)$ is jointly continuous in x and λ at (x_1, λ_0).

---

[6]If the partial derivative $\partial f/\partial y$ is continuous, then the Lipschitz condition is automatically satisfied.

Proof The IVP (10.72) is equivalent to the system

$$\left. \begin{array}{ll} y' = f(x, y, \lambda) & y(x_0) = y_0 \\ \lambda' = 0 & \lambda(x_0) = \lambda_0. \end{array} \right\}$$

The second equation of the system ensures that λ maintains a constant value, as a parameter should. The first equation is the original ODE. Theorem 3.1 on page 119 says that the solution of an IVP depends continuously on the variable and the initial conditions (in this case, (x_0, y_0, λ_0)). Therefore, it is continuous as a function of (x, λ). ∎

The two lemmas imply the following theorem:

Theorem 10.5

Let $q(x)$ be a function that is continuous on an interval $[0, L]$ and such that $q''(x)$ is continuous on $(0, L)$. Then the eigenvalues of

$$y'' + (\lambda + q(x)) y' = 0; \ \ y(0) = y(L) = 0$$

form an infinite, increasing, and unbounded sequence,

$$\lambda_1 < \lambda_2 < \cdots .$$

Furthermore, the eigenfunctions ϕ_i satisfy the orthogonality relation

$$\int_0^L \phi_i(x)\phi_j(x)\, dx = 0$$

when $i \neq j$.

Proof We have identified the eigenvalues λ_k as the numbers that satisfy equation (10.71): $\theta(L, \lambda_k) = -(2k-1)\frac{\pi}{2}$.

By Lemma 10.5.3, $\theta(L, \lambda)$ is a continuous function of λ, and we will use Lemma 10.5.2 to show that

$$\lim_{\lambda \to \infty} \theta(L, \lambda) = -\infty.$$

Recall that $\theta(x, \lambda)$ is determined by the IVP (10.70). Let m be the minimum value of $q(x)$ on $[0, L]$ and let $\phi(x, \lambda)$ be the solution of the IVP

$$\phi' = -(\lambda + m)\cos^2(\phi) - \sin^2(\phi); \ \ \phi(0) = \frac{\pi}{2}. \tag{10.73}$$

Lemma 10.5.2 tells us that $\theta(x, \lambda) \leq \phi(x, \lambda)$ provided that $x \in [0, L]$. On the other hand, we see that $\phi(x, \lambda)$ is the phase angle for the solution of the IVP

$$y'' + (\lambda + m)y = 0; \ \ y(0) = 0, y'(0) = 1. \tag{10.74}$$

The solution of (10.74) is

$$y = \frac{\sin(\sqrt{\lambda + m}\, x)}{\sqrt{\lambda + m}}.$$

The orbit of $(y(x), y'(x))$ in the phase plane is a clockwise ellipse that returns to its starting point at $x = 0, 2\pi/\sqrt{\lambda + m}, 4\pi/\sqrt{\lambda + m}, \ldots$. Thus the phase angle $\phi(x, \lambda)$ is decreasing (since the direction is clockwise), and $\phi(2m\pi/\sqrt{\lambda + m}, \lambda) = -2m\pi$. By solving

$$\frac{2m\pi}{\sqrt{\lambda + m}} = L$$

for λ, we find that

$$\phi\left(L, \frac{4m^2\pi^2}{L^2} - m\right) = -(2m - 1)\pi.$$

Since $\theta(L, \lambda) \le \phi(L, \lambda)$, it follows that $\theta(L, \lambda)$ decreases without bound as $\lambda \to -\infty$. Noting that θ is continuous, we conclude that there is an infinite sequence of numbers λ_m with $\theta(L, \lambda_m) = (2m - 1)\pi/2$. These λ_m are eigenvalues.

To prove orthogonality, suppose that $\phi_i(x)$ and $\phi_j(x)$ are eigenfunctions belonging to distinct eigenvalues λ_i and λ_j. Let $A = \int_0^L \phi_i(x)\phi_j(x)\, dx$, so we must show $A = 0$. Also, let $B = \int_0^L \phi_i(x)\phi_j(x)q(x)\, dx$, and $C = \int_0^L \phi_i'(x)\phi_j'(x)\, dx$. Since $\phi_i(x)$ satisfies the ODE,

$$\lambda_i A + B = \int_0^L [(\lambda_i + q(x))\phi_i(x)]\,\phi_j(x)\, dx$$

$$= -\int_0^L \phi_i''(x)\,\phi_j(x)\, dx.$$

Integrating by parts,

$$-\int_0^L \phi_i''(x)\,\phi_j(x)\, dx = \int_0^L \phi_i'(x)\,\phi_j'(x)\, dx - \phi_i'(x)\,\phi_j(x)\Big|_0^L$$

$$= C.$$

Putting this together, we see that $\lambda_i A = C - B$. In the same way, we could show that $\lambda_j A = C - B$. Hence

$$\lambda_i A = \lambda_j A.$$

Since $\lambda_i \ne \lambda_j$, it follows that $A = 0$. ∎

Eigenfunction expansions

An **eigenfunction expansion** is a generalization of Fourier series in which a function $f(x)$ is expanded as the sum of a series

$$f(x) = \sum_{n=1}^{\infty} C_n \phi_n(x), \tag{10.75}$$

where each function ϕ_n is an eigenfunction of a given eigenvalue problem on an interval $[0, L]$. A Fourier sine series is an eigenfunction expansion where the eigenvalue problem is

$$y'' + \lambda y = 0; \quad y(0) = y(L) = 0.$$

Theorem 10.5 assures us that there is an infinite sequence of eigenfunctions to use in an expansion—if the EVP is of the restricted form

$$y'' + (\lambda + q(x)) \, y' = 0; \quad y(0) = y(L) = 0. \tag{10.76}$$

By Theorem 10.5, the eigenfunctions ϕ_n form an orthogonal set, and this enables us to compute the coefficients in the eigenfunction expansion in the same way that we computed Fourier coefficients:

$$C_n = \frac{\langle f, \phi_n \rangle}{\langle \phi_n, \phi_n \rangle},$$

where $\langle f, \phi_n \rangle = \int_0^L f(x) \, \phi_n(x) \, dx$. If f is continuous at each point of $[0, L]$, these coefficients can be calculated whether or not f is the sum of an eigenfunction expansion such as (10.75). We would like to know how to tell if a function f can be represented as the sum of a convergent eigenfunction expansion and the following theorem, a generalization of Theorem 10.4 on page 497, identifies a large class of functions that have convergent eigenfunction expansions:

Theorem 10.6

Let $\{\lambda_n\}$ be the sequence of eigenvalues of the EVP (10.76) and for each n, let $\phi_n(x)$ be an eigenfunction belonging to the eigenvalue λ_n.

If the function f is piecewise differentiable on $[0, L]$, with $f(0) = f(L) = 0$, then the eigenfunction expansion displayed on the right side of (10.75) is convergent at each point of $[0, L]$. Furthermore, at each point x where f is continuous, the equality in (10.75) holds. At a point p where f is not continuous,

$$\sum n = 1^{\infty} C_n \phi_n(x) = \frac{f(p^+) + f(p^-)}{2},$$

where $f(p^{\pm})$ denotes the one-sided limit,

$$f(p^+) = \lim_{x \to p^+} f(x) \quad \text{and} \quad f(p^-h) = \lim_{x \to p^-} f(x).$$

The convergence of the eigenfunction expansion is in the pointwise sense. If f is continuous on $[0, L]$, then the convergence is uniform.

The proof of the convergence statement in Theorem 10.6 is beyond the scope of this text. We will use the theorem only as a model statement, since the eigenfunction expansions that we need to explore involve EVPs that are not of the form (10.76).

Legendre polynomials

Recall from Section 7.2 (see page 330) that when $\lambda = n(n+1)$, where n is an integer, Legendre's equation,

$$(1 - x^2)\, y'' - 2x\, y' + \lambda\, y = 0, \tag{10.77}$$

has a solution $P_n(x)$ that is a polynomial of degree n. This $P_n(x)$ is called the Legendre polynomial of degree n, and by convention it is normalized so that $P_n(1) = 1$. Notice that Legendre's equation has singular points at $x = \pm 1$.

Proposition 10.5.4

If $\lambda \neq n(n+1)$ for all nonnegative integers n, then the only solution of Legendre's equation (10.77) that is defined and finite at both singular points is $y \equiv 0$.

If $\lambda = n(n+1)$ for some positive integer n, then every solution of (10.77) that is defined and finite at both singular points is equal to a constant multiple of $P_n(x)$.

An outline of the proof of this proposition is given as Exercise 9 at the end of this section.

Consider the following boundary conditions for an eigenvalue problem that involves Legendre's equation:

$$|y(-1)| < \infty \quad \text{and} \quad |y(1)| < \infty. \tag{10.78}$$

Theorem 10.6 cannot be applied to the EVP (10.77), (10.78), since neither the ODE nor the boundary conditions are of the required form. Nevertheless, we will see that its conclusion is true. According to Proposition 10.5.4, the eigenvalues are the numbers $\lambda_n = n(n+1) = 0, 2, 6, 12, 20, \ldots$. The eigenfunctions can be chosen to be the Legendre polynomials.

The following proposition asserts that the Legendre polynomials form an orthogonal set. Let the inner product of functions f and g that are continuous on $[-1, 1]$ be

$$\langle f, g \rangle = \int_{-1}^{1} f(x)g(x)\, dx.$$

Proposition 10.5.5

If $n \neq m$, then $\langle P_n, P_m \rangle = 0$. Furthermore,

$$\langle P_n, P_n \rangle = \frac{2}{2n + 1}. \tag{10.79}$$

Proof Let us rewrite (10.77) in the form

$$\frac{d}{dx}\left((1 - x^2)\frac{dy}{dx} \right) = -\lambda y.$$

Thus,

$$-\lambda_n \langle P_n, P_m \rangle = \int_{-1}^{1} \frac{d}{dx}\left((1-x^2)\frac{dP_n(x)}{dx}\right) P_m(x)\,dx$$

$$= (1-x^2)P_n'(x)P_m(x)\big|_{-1}^{1} - \int_{-1}^{1}(1-x^2)P_n'(x)P_m'(x)\,dx$$

$$= -\int_{-1}^{1}(1-x^2)P_n'(x)P_m'(x)\,dx,$$

where we have integrated by parts. The same calculation shows that

$$\lambda_m \langle P_n, P_m \rangle = \int_{-1}^{1}(1-x^2)P_n'(x)P_m(x)\,dx,$$

so we can conclude that $\lambda_n \langle P_n, P_m \rangle = \lambda_m \langle P_n, P_m \rangle$. If $n \neq m$, we conclude that $\langle P_n, P_m \rangle = \lambda_m = 0$.

The proof of (10.79) would involve a considerable digression and is omitted. ∎

To complete the development of eigenfunction expansions in terms of Legendre polynomials, we have to address the issue of convergence. This is done by Theorem 10.7, which is stated without proof.

Theorem 10.7

Let $f(x)$ be a function that is continuous on $[-1, 1]$, and put

$$C_n = \left(n + \frac{1}{2}\right)\langle f, P_n \rangle, \quad \text{for } n = 0, 1, 2, \ldots.$$

Then the series

$$\sum_{n=0}^{\infty} C_n P_n(x)$$

converges uniformly to f on $[-1, 1]$.

We conclude the discussion of Legendre polynomials with two useful facts (the proof are exercses). The first, called **Rodriguez's formula**, is a formula for $P_n(x)$:

$$P_n(x) = \frac{1}{2^n\,n!}\frac{d^n}{dx^n}(x^2-1)^n.$$

Exercise 10 at the end of this section outlines a proof of Rodriguez's formula.

The second fact is a recursion formula that allows the computation of Legendre polynomials. The following identity holds, for all $n \geq 2$:

$$n\,P_n(x) = (2n-1)x\,P_{n-1}(x) - (n-1)\,P_{n-2}(x). \tag{10.80}$$

Example 10.5.2   Use the recursion formula (10.80) to make a list of the first five Legendre polynomials, starting with $P_0(x) = 1$ and $P_1(x) = x$.

Solution.

$$P_2(x) = \frac{1}{2}[3x\, P_1(x) - P_0(x)]$$

$$= \frac{3}{2}x^2 - \frac{1}{2}$$

$$P_3(x) = \frac{1}{3}[5x\, P_2(x) - 2P_1(x)]$$

$$= \frac{5}{2}x^3 - \frac{3}{2}x$$

$$P_4(x) = \frac{1}{4}[7x\, P_3(x) - 3P_2(x)]$$

$$= \frac{35}{8}x^4 - \frac{15}{4}x^2 + \frac{3}{8}$$

$$P_5(x) = \frac{1}{5}[9x\, P_4(x) - 4P_3(x)]$$

$$= \frac{63}{8}x^5 - \frac{35}{4}x^3 + \frac{15}{8}x$$ ❑

Example 10.5.3   Find the coefficients in the eigenfunction expansion

$$|x| = \sum_{n=0}^{\infty} C_n P_n(x).$$

Solution.   We begin by noting that $|x|$ is an even function and is thus orthogonal to every odd function. Since $P_n(x)$ is odd when n is odd, $C_n = 0$ for all odd n.
If $n = 2k$ is even, then

$$C_{2k} = \left(2k + \frac{1}{2}\right) \int_{-1}^{1} |x|\, P_{2k}(x)\, dx.$$

Since $|x|\, P_{2k}(x)$ is an even function,

$$\int_{-1}^{1} |x|\, P_{2k}(x)\, dx = 2 \int_{0}^{1} |x|\, P_{2k}(x)\, dx = 2 \int_{0}^{1} x\, P_{2k}(x)\, dx.$$

Let $I_k = \int_0^1 x\, P_{2k}(x)\, dx$. Since $P_{2k}(x)$ satisfies Legendre's equation with parameter $\lambda_k = 2k(2k + 1)$,

$$-\lambda_k I_k = \int_{0}^{1} x[(1 - x^2)P'_{2k}(x)]'\, dx$$

Integrating by parts twice yields the identity

$$-\lambda_k I_k = P_{2k}(0) - 2I_k,$$

and it follows that

$$I_k = \frac{P_{2k}(0)}{2 - \lambda_k},$$

and hence

$$C_{2k} = \left(2k + \frac{1}{2}\right) \cdot 2I_k = -\frac{4k+1}{4k^2 + 2k - 2} P_{2k}(0).$$

Substituting $x = 0$ in the recursion formula (10.80), we find that

$$P_{2k}(0) = -\frac{2k-1}{2k} P_{2k-2}(0).$$

Since $P_0(0) = 1$, we can conclude that $P_{2k}(0) = \frac{(2k-1)!!}{(2k)!!}$, and hence

$$C_{2k} = (-1)^{k+1} \frac{4k+1}{4k^2 + 2k - 2} \frac{(2k-1)!!}{(2k)!!}$$

The Legendre expansion appears as follows:

$$|x| = \frac{1}{2} P_0(x) + \frac{5}{8} P_2(x) - \frac{3}{16} P_4(x) - \frac{13}{128} P_6(x) + \cdots$$

A graph, showing the sum of the first three terms of this expansion, is shown in Figure 10.16.   ❑

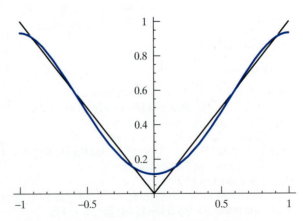

FIGURE 10.16   Graph of a truncated Legendre polynomial expansion of $f(x) = |x|$, including terms of degree up to 4.

EXERCISES

Exercises 1–4 are about the EVP,

$$y'' + (\lambda + \cos(x) - x^2)\, y = 0; \quad y(0) = y(3) = 0. \quad (10.81)$$

1. Let $\theta(x)$ denote the phase angle variable in the phase plane. Thus, if $y(x)$ is a solution of the ODE in (10.81), the corresponding phase angle is $\theta(x) = \arg(y(x) + i y'(x))$. Find the first-order ODE satisfied by $\theta(x)$.

2. Use an IVP solver to plot the phase angle of solutions of the ODE in (10.81) that satisfy the left boundary condition, $y(0) = 0$, but possibly not the right boundary condition. Use integer values of the parameter starting with $\lambda = -1$ and increasing until $\lambda = 15$.

3. Using the data from Exercise 2, estimate the first three eigenvalues of (10.81).

4. Refine your estimates of the first three eigenvalues to a precision of $\pm\frac{1}{8}$.

5. Consider the EVP

$$y'' + (\lambda\, w(x) + q(x))\, y = 0; \quad y(0) = y(L) = 0,$$

where $w(x) > 0$ on $(0, L)$.

a. Let $\phi_i(x)$ and $\phi_j(x)$ be eigenfunctions belonging to eigenvalues $\lambda_i \neq \lambda_j$. Prove that the following orthogonality relation holds:

$$\int_0^L \phi_i(x)\,\phi_j(x)\,w(x)\,dx = 0. \quad (10.82)$$

b. Find a first-order ODE satisfied by the phase angle and use it to show that the eigenvalues form an infinite increasing sequence.

6. A **Sturm-Liouville problem** is an EVP of the form

$$(p(x)\, y')' + (\lambda\, w(x) + q(x))\, y = 0 \quad (10.83)$$

with separated boundary conditions,

$$A\, y(0) + B\, y'(0) = 0, \quad C\, y(L) + D\, y'(L) = 0,$$

where $A^2 + B^2 \neq 0$ and $C^2 + D^2 \neq 0$. The **phase angle** for a solution $\phi(x)$ of the ODE (10.83) is $\theta(x) = \arg(\phi(x) + i\, p(x)\, \phi'(x))$. Find a first-order ODE satisfied by the phase angle of any solution of (10.83). Use this ODE to prove that if $p(x) > 0$ and $w(x) > 0$ for all $x \in [0, L]$, then the eigenvalues of any Sturm-Liouville problem involving the ODE (10.83) form an infinite, increasing sequence and that the eigenfunctions belonging to distinct eigenvalues satisfy the orthogonality relation (10.82).

7. An EVP involving an ODE

$$a(x)\, y'' + b(x)\, y' + \lambda\, c(x)\, y = 0 \quad (10.84)$$

with separated boundary conditions on an interval $[0, L]$ can be replaced by an equivalent Sturm-Liouville problem, obtained by multiplying (10.84) by $m(x) = [a(x)]^{-1} e^{P(x)}$, where $P(x)$ is an antiderivative of $b(x)/a(x)$.

   Convert each of the following eigenvalue problems to Sturm-Liouville problems by this procedure, and specify the orthogonality relation satisfied by the eigenfunctions.

a. $\cos(x)\, y'' + \sin(x)\, y' + (\lambda + \cos^3(x))\, y = 0$; $y(0) = y\left(\frac{\pi}{3}\right) = 0$

b. $y'' - x\, y' + \lambda\, y = 0$; $y(0) = y'(1) = 0$

c. $x\, y'' + 2\, y' + \lambda\, y = 0$; $y(1) = y(2) = 0$

d. $(x^2 + 1)\, y'' + x\, y' + \lambda y = 0$; $y'(-1) = y'(1) = 0$

8. The Airy wave functions[7] are the solutions of the ODE $y'' = x\, y$. Let $\alpha(x)$ be the Airy wave function that satisfies $\alpha(0) = 0$ and $\alpha'(0) = 1$. Show that the zeros of $\alpha(x)$ form an infinite sequence, $-a_1, -a_2, -a_3, \ldots$, with $0 < a_1 < a_2 < \cdots$ and $\lim_{n \to \infty} a_n = \infty$. Furthermore, express the eigenvalues and eigenfunctions of the EVP

$$y'' + \lambda x\, y = 0; \quad y(0) = y(1) = 0$$

in terms of the function $\alpha(x)$ and the numbers a_n.

9. Fill in the details of this proof of Proposition 10.5.4.

a. The singular points of the Legendre equation are regular and have double exponent equal to 0.

b. For any value of λ, the set of solutions that are finite at $x = 1$ is one dimensional.

c. If $y(x)$ is a solution represented by a power series based at $x_0 = 1$, and if that series is not a polynomial, then the radius of convergence of that power series is 2.

d. Suppose that $\lambda \neq n(n + 1)$ for any integer n, so that there is no solution that has the form of a polynomial. Let $y_1(x)$ be a nontrivial solution represented by a power series based at $x_0 = 1$ and let $y_{-1}(x)$ be a nontrivial solution represented by a power series based at $x_0 = -1$. Then $\{y_1(x), y_{-1}(x)\}$ is linearly independent.

e. Conclude that the series representing $y_1(x)$ must diverge at $x = -1$ and that the series representing $y_{-1}(x)$ diverges at $x = 1$.

---

[7]The special functions Ai (x) and Bi(x) are linearly independent Airy wave functions. The Airy wave function that we will use is not one of these, but like any Airy wave function, it can be expressed as a linear combination of Ai (x) and Bi(x).

f. If $\lambda = n(n + 1)$, show that every solution that is finite at $x = 1$ or $x = -1$ is a constant multiple of $P_n(x)$.

10. To prove Rodriguez's formula, let $v_n(x) = \dfrac{d^n}{dx^n}(x^2 - 1)^n$.

a. Show that for $k < n$, $\dfrac{d^k}{dx^k}(x^2 - 1)^n$ vanishes at $x = \pm 1$.

b. Use repeated integration by parts to show that for any function $f(x)$ that is sufficiently differentiable on $[-1, 1]$,

$$\int_{-1}^{1} v_n(x)\, f(x)\, dx$$

$$= (-1)^n \int_{-1}^{1} (x^2 - 1)^n\, f^{(n)}(x)\, dx.$$

c. Let $v_0 = 1$. Show that if $n \neq m$, then $\langle v_n, v_m \rangle = 0$.

d. Show that $v_n(x)$ is a polynomial of degree n.

e. Show that $P_n(x)$ is equal to a constant multiple of $v_n(x)$.

f. Show that $v_n(1) = n! 2^n$.

11. Prove the recursion formula for Legendre polynomials by taking the following steps:

a. Since $x P_{n-1}(x)$ is a polynomial of degree n, it can be expressed as a linear combination of the Legendre polynomials $P_0(x), P_1(x), \ldots, P_n(x)$:

$$x P_{n-1}(x) = A_0 P_n(x) + A_1 P_{n-1}(x)$$
$$+ A_2 P_{n-2}(x) + \cdots + A_n P_0(x).$$

b. Noting that $\langle x P_{n-1}, P_k \rangle = \langle P_{n-1}, x P_k \rangle$, show that $A_k = 0$ for $k > 2$.

c. Show that $A_1 = 0$.

◆ **Hint:** $x P_{n-1}^2$ is odd.

d. Show that $A_0 + A_2 = 1$.

e. Use the coefficients of x^n given by Rodriguez's formula to calculate A_0.

12. Show that the Laguerre polynomials (see Exercise 12 on page 348) satisfy the following orthogonality relationship:

$$\int_0^\infty e^{-t} L_N(t) L_M(t)\, dt = 0$$

for $N \neq M$.

◆ **Hint:** First convert the Laguerre equation into Sturm-Liouville form, by using the result of Exercise 7. Then show that

$$N \int_0^\infty e^{-t} L_N(t) L_M(t)\, dt = \int_0^\infty t e^{-t} L_N'(t) L_M'(t)\, dt$$

$$= M \int_0^\infty e^{-t} L_N(t) L_M(t)\, dt.$$

10.6 Glossary

$\langle \psi_1, \psi_2 \rangle$ (on $[a, b]$) The inner product of the functions ψ_1 and ψ_2 on the interval $[a, b]$, which has the value

$$\int_a^b \psi_1(x)\, \psi_2(x)\, dx.$$

BVP   Boundary value problem.

Boundary condition   An equation, involving the values of a function or its derivatives at one or both endpoints of an interval or on the boundary curve of a plane region.

Boundary field   A pair of direction fields in the plane that is associated with the *separated* boundary conditions of a two-point BVP. If we denote the endpoints by a and b, then the boundary field is only defined at points (x, y), where $x = a$ or $x = b$. The boundary field element at a point gives the slope of a solution of a BVP that happens to pass through the point.

Boundary value problem   A two-point BVP consists of an ODE of order 2, with no singular points on an interval $[a, b]$,

provided with two boundary conditions involving the solution and its derivatives evaluated at the boundary points, a and b.

Diffusion   A process by which heat propagates, a cloud of smoke is distributed through a room, and so on. The governing principle is Fourier's law.

Diffusion coefficient   The coefficient c in the diffusion equation,

$$\frac{\partial u}{\partial t} = c \frac{\partial^2 u}{\partial x^2}$$

(in one-dimensional form).

Diffusion equation   (one dimensional). The PDE that governs heat flow in a rod. Let c be the diffusion coefficient of the rod, and let the temperature of the rod at the point x and tine t be denoted $u(x, t)$. Then

$$\frac{\partial u}{\partial t} = c \frac{\partial^2 u}{\partial x^2}.$$

EVP Eigenvalue problem: to find the eigenvalues and eigenfunctions of a linear differential operator $\mathcal{L}$ with specified boundary conditions.

Eigenfunction (of a linear differential operator, $\mathcal{L}$, with specified boundary conditions) A nontrivial solution of $\mathcal{L}(y) = \lambda\, y$ that satisfies the boundary conditions.

Eigenvalue (of a linear differential operator, $\mathcal{L}$, with specified boundary conditions) A number λ such that there is a nontrivial solution of $\mathcal{L}(y) = \lambda\, y$ that satisfies the boundary conditions.

Flux The rate of heat flow in a rod, divided by the cross-sectional area of the rod. The flux is a function of the position x on the rod and of the time t.

Fourier cosine series A series of the form

$$\psi(x) = C_0 + \sum_{n=1}^{\infty} C_n \cos(n\omega x).$$

Fourier sine coefficient (of a function $\psi(x)$ on an interval $[0, L]$) The ratio

$$C_m = \frac{\langle \psi(x), \sin(m\omega\, x)\rangle}{\langle \sin(m\omega\, x), \sin(m\omega\, x)\rangle},$$

where $\omega = \pi/L$ and the inner product is taken on the interval $[0.L]$.

Fourier sine series (of a function $\psi(x)$ on an interval $[0, L]$) The series

$$\sum_{m=1}^{\infty} C_m \sin(m\omega x),$$

where C_m are the Fourier sine coefficients of $\psi(x)$.

Fourier's law In the flow of heat in a rod, let $f(x, t)$ be the flux and $u(x, t)$ be the temperature ant position x and time t. Then there is a positive constant K such that

$$f(x, t) = -K\frac{\partial u}{\partial x}.$$

In words, heat flows from hotter regions to colder regions, at a rate proportional to the temperature gradient.

Homogeneous boundary conditions Boundary conditions that are satisfied by the zero function.

IBVP Initial boundary value problem. In the context of diffusion of heat in a rod, the initial condition is a function that specifies the initial temperature at each point of the rod, and the boundary conditions specify the temperature—or the temperature gradient, or some combination of the two—at the ends of the rod.

Inner product A bilinear, symmetric, and positive definite pairing (these terms are defined on page 493). The inner product of functions f and g that are defined on an interval $[0, L]$ is denoted $\langle f, g\rangle$ and is given by the formula

$$\langle f, g\rangle = \int_0^L f(x)\, g(x)\, dx.$$

Legendre polynomials A sequence of polynomials $P_n(x)$ that can be defined in a number of equivalent ways. Among them are the following:

- $P_n(x)$ is of degree n, normalized so that $P_n(1) = 1$, and for $n \neq m$ the orthogonality relation

$$\int_{-1}^{1} P_n(x)\, P_m(x)\, dx = 0.$$

- (Rodriguez's formula) $P_n(x) = \dfrac{1}{2^n\, n!}\dfrac{d^n}{dx^n}(x^2 - 1)^n$
- (Recursion formula) $P_0(x) \equiv 1$, $P_1(x) = x$, and for $n > 1$,

$$P_n(x) = \left(2 - \frac{1}{n}\right) x\, P_{n-1}(x) - \left(1 - \frac{1}{n}\right) P_{n-2}(x)$$

- $y = P_n(x)$ satisfies the normalization requirement $P_n(1) = 1$ and the Legendre ODE

$$(1 - x^2)\, y'' - 2x\, y' + n(n + 1)y = 0.$$

Table 10.1 is a list of the first eight Legendre polynomials.

Table 10.1 The first eight Legendre polynomials.

ν	$P_\nu(\zeta)$
0	1
1	ζ
2	$\frac{1}{2}(-1 + 3\zeta^2)$
3	$\frac{1}{2}(-3\zeta + 5\zeta^3)$
4	$\frac{1}{8}(3 - 30\zeta^2 + 35\zeta^4)$
5	$\frac{1}{8}(15\zeta - 70\zeta^3 + 63\zeta^5)$
6	$\frac{1}{16}(-5 + 105\zeta^2 - 315\zeta^4 + 231\zeta^6)$
7	$\frac{1}{16}(-35\zeta + 315\zeta^3 - 693\zeta^5 + 429\zeta^7)$

Mixed boundary conditions Boundary conditions of more than one kind in the same BVP: For example, a two-point BVP might have a constant temperature boundary condition at one end [e.g. $u(0, t) = 0$] and an insulated condition at the other end [e.g. $\frac{\partial u}{\partial x}(L, t) = 0$].

Orthogonal functions   Functions ψ_1 and ψ_2 such that $\langle \psi_1, \psi_2 \rangle = 0$.

PDE   Partial differential equation. A differential equation that includes partial derivatives of an unknown function with respect to two or more independent variables.

Periodic boundary conditions   (of a second order linear differential operator on an interval $[a, b]$) The boundary conditions $y(a) = y(b)$, $y'(a) = y'(b)$. Boundary conditions of this type are not separated.

Piecewise differentiable function   (on an interval $[a, b]$) A function $\psi(x)$ that is differentiable on each subinterval $[x_{i-1}, x_i]$ of a partition $a = x_0 < x_1 < \cdots < x_n = b$ of $[a, b]$.

Phase angle   The angular polar coordinate in the phase plane. Thus, for a solution $y = \psi(x)$ of a second-order ODE, the phase angle $\theta(x) = \arg(\psi(x) + i \, \psi'(x))$.

Rodriguez's formula   (See Legendre polynomials.)

Separated boundary conditions   A set of boundary conditions for a two-point BVP in which each involves the value of the solution and derivatives at only *one* of the endpoints of the interval.

Separation of variables   Finding all solutions of a PDE that have the special form of a product of one-variable functions. For example, if the independent variables are x and t, then the objective is to find all solutions that have the form $u(x, t) = g(x) h(t)$.

Shooting   A way of adapting any IVP solver to solve a BVP. See Exercise 22 on page 477.

Sturm-Liouville problem   An EVP with separated boundary conditions on an interval $[a, b]$, where the ODE takes the form

$$(p(x) \, y')' + (\lambda \, w(x) + q(x)) \, y = 0,$$

with $p(x)$ positive and differentiable on $[a, b]$, $w(x)$ positive and continuous on (a, b), and $q(x)$ continuous on (a, b).

Superposition, principle   If $u_1, \ldots, u_n$ are solutions of a homogeneous linear ODE or PDE and $c_1, \ldots, c_n$ are constants, then

$$c_1 u_1 + \cdots + c_n u_n$$

is also a solution.

Temperature gradient   In a one dimensional setting, the partial derivative of the temperature function with respect to the position x.

Trigonometric polynomial   A function that can be expressed as either the real part or the imaginary part of $P(e^{i k x})$, where $P(X) = a_0 + a_1 X + a_2 X^2 + \cdots + a_n X^n$ is a polynomial with real or complex coefficients, and k is a real constant. For example, if $P(X) = 1 + X + 2X^3$, then

$$\mathbf{Re}(P(e^{i x})) = 1 + \cos(x) + 2 \, \cos(3x).$$

Trivial solution   The function $y \equiv 0$, which satisfies any homogeneous linear ODE with homogeneous boundary conditions.

10.7 Chapter Review

1. Solve the BVP $y'' + y = \sin(x)$; $y(0) = y(\pi/2) = 0$.

2. Find Green's function for each of the following LDOs with the given boundary conditions. If there is no Green's function, explain why.

   a. $\mathcal{L}(y) = y'' + y$; $y(0) = y'(\pi/2) = 0$

   b. $\mathcal{L}(y) = y'' + y$; $y(0) = y'(\pi) = 0$

   c. $\mathcal{L}(y) = y''$; $y(0) = y'(0)$, $y(1) = 2y'(1)$

   d. $\mathcal{L}(y) = y''$; $y(0) = y'(0)$, $y(1) = -y'(1)$

3. Find Green's function for the LDO $\mathcal{L}(y) = y'' - y$ with boundary conditions $y(-1) = y(1) = 0$ and use it to solve the BVP $\mathcal{L}(y) = \cosh(t)$ with the given boundary conditions.

4. Find separated homogeneous boundary conditions on $[-1, 1]$ for $\mathcal{L}(y) = y'' - y$ such that Green's function does not exist.

5. Solve the EVP

$$y'' + \lambda \, y = 0; \quad y(0) = y'(\pi) = 0.$$

6. Specify a homogeneous boundary condition at $x = 1$ so that if the boundary condition at $x = 0$ is $y(0) = 0$, then $\lambda = -1$ is an eigenvalue of $y'' + \lambda \, y = 0$.

7. Solve the EVP

$$y'' + 2 \, y' + \lambda \, y = 0; \quad y(0) = y'(\pi) + y(\pi) = 0.$$

8. Solve the IBVP

$$\frac{\partial u}{\partial t} = 0.15 \frac{\partial^2 u}{\partial x^2};$$

$$u(0, x) = 5 \sin(\pi \, x/2) - \sin(3\pi \, x/2),$$

$$u(0, t) = u'(1, t) = 0.$$

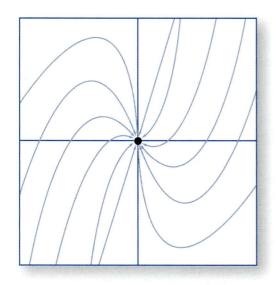

CHAPTER 11

Partial Differential Equations

11.1 The Wave Equation

Figure 11.1 A vibrating string profile, in which the string is frozen at time t_1. The arrow indicates the displacement, $u(x_1, t_1)$.

Figure 11.1 shows a profile of the vibrating string, frozen at an instant $t = t_1$ of time. The ends of the string are fastened to the points $x = 0$ and $x = L$ of the x-axis, and as the string vibrates, each point of the string moves in a vertical path. The displacement from the axis at a time t of a point x on the string is denoted $u(x, t)$.

We will see that this displacement function $u(x, t)$ is a solution of the PDE,

$$\frac{\partial^2 u}{\partial t^2} = c^2 \frac{\partial^2 u}{\partial x^2}. \tag{11.1}$$

Equation (11.1) is the **wave equation**. The parameter c in the wave equation is a constant that gives the speed that waves propagate on the string. It is related to the tension of the string, its density, and its thickness. The **tension** is a vector $\mathbf{T}(x, t)$, tangent to the string profile, that represents the force holding the part of the string on the left of x to the part on the right of x.

If the string is in its equilibrium position along the x-axis, then the tension is constant: $\mathbf{T}(x, t) \equiv T_0 \mathbf{i}$. Let ρ denote the density of the string per unit volume, and let A denote its cross-sectional area. Then

$$c^2 = \frac{T_0}{\rho A}.$$

You should verify that the dimension of c is that of speed, $[\text{distance}] \cdot [\text{time}]^{-1}$.

Although the wave equation appears to be similar to the heat equation, it involves a second derivative with respect to time, while the heat equation only involves a first derivative. It turns out that the solutions of the two equations behave differently.

Derivation[1]

We will suppose that the string is stretched horizontally between two points, and each point of the string vibrates in the vertical direction. The horizontal component of the tension $\mathbf{T}(x, t)$ must be constant, because if it were not, the string would undergo horizontal motion. The magnitude of this horizontal component is T_0, so

$$\mathbf{T}(x, t) = T_0 \mathbf{i} + f(x, t)\mathbf{j},$$

where $f(x, t)$ is the magnitude of the vertical component.

The slope of the profile $u(x, t_1)$ at (x_1, t_1) is $\partial u / \partial x (x_1, t_1)$. Since the direction of $\mathbf{T}(x_1, t_1)$ is tangent to the string's profile and its slope is equal to $f(x_1, t_1)/T_0$,

$$\frac{\partial u}{\partial x}(x, t) = \frac{f(x, t)}{T_0}.$$

We can therefore express the tension vector as

$$\mathbf{T}(x, t) = T_0 \mathbf{i} + T_0 \frac{\partial u}{\partial x}(x, t)\mathbf{j}.$$

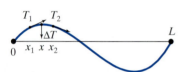

Figure 11.2 Tension forces on a vibrating string.

Consider a short segment of string, extending between points x_1 and x_2. The net force on the segment is

$$\Delta \mathbf{T} = \mathbf{T}(x_2, t) - \mathbf{T}(x_1, t)$$

$$= T_0 \left(\frac{\partial u}{\partial x}(x_2, t) - \frac{\partial u}{\partial x}(x_1, t) \right) \mathbf{j}$$

(see Figure 11.2). Assuming that $u(x, t)$ is sufficiently differentiable, the mean value theorem applies, and we conclude that

$$\Delta \mathbf{T} = T_0(x_2 - x_1)\frac{\partial^2 u}{\partial x^2}(x^*, t)\mathbf{j}, \tag{11.2}$$

where x^* is some point located between x_1 and x_2.

Make an approximation by assuming that every point in the segment has acceleration equal to $\frac{\partial^2 u}{\partial t^2}$. The volume of the segment is $A(x_2 - x_1)$, and its mass is therefore $\rho A(x_2 - x_1)$. By Newton's second law of motion,

$$\rho A(x_2 - x_1)\frac{\partial^2 u}{\partial t^2}\mathbf{j} = \Delta \mathbf{T}.$$

Referring back to equation (11.2), we see that

$$\rho A(x_2 - x_1)\frac{\partial^2 u}{\partial t^2} = T_0(x_2 - x_1)\frac{\partial^2 u}{\partial x^2}.$$

Dividing both sides of this equation by $\rho A(x_2 - x_1)$ yields

$$\frac{\partial^2 u}{\partial t^2} = \frac{T_0}{\rho A}\frac{\partial^2 u}{\partial x^2},$$

and the derivation is complete.

---

[1]This material can be skipped without loss of continuity.

Separation of variables

The boundary conditions for the vibrating string are

$$u(0, t) = u(L, t) = 0 \quad \text{for all } t \geq 0. \tag{11.3}$$

These are justified by noting that the ends of the string are held fixed.

We will find all solutions of the wave equation having the form $u(x, t) = h(x)k(t)$, and satisfying the boundary conditions (11.3). Substitution of this $u(x, t)$ into the wave equation yields

$$h(x)k''(t) = c^2 h''(x)k(t).$$

Now divide by $c^2 h(x)k(t)$:

$$\frac{k''(t)}{c^2 k(t)} = \frac{h''(x)}{h(x)}. \tag{11.4}$$

The left side of equation (11.4) does not depend on x, and the right side is independent of t. Therefore, both sides are constant, and we denote that constant by $-\lambda$. This leads to a pair of ordinary differential equations,

$$h''(x) + \lambda h(x) = 0, \tag{11.5}$$

$$k''(t) + \lambda c^2 k(t) = 0. \tag{11.6}$$

In terms of the functions $h(x)$ and $k(t)$, the boundary conditions (11.3) are

$$h(0)k(t) = h(L)k(t) = 0 \quad \text{for all } t \geq 0.$$

It can be assumed that $k(t) \neq 0$, and it then follows that $h(0) = h(L) = 0$. Thus $u(x, t) = h(x)k(t)$ can satisfy the wave equation and the boundary conditions (11.3) only if λ is an eigenvalue of the *eigenvalue problem*

$$y'' + \lambda y = 0; \quad y(0) = y(L) = 0, \tag{11.7}$$

and $y = h(x)$ is a corresponding eigenfunction. By Proposition 10.2.1 on page 478, the eigenvalues are $\lambda_m = (m\pi/L)^2$, and the eigenfunction belonging to λ_m is $h_m(x) = \sin(m\pi x/L)$.

Substitution of $\lambda = (m\pi/L)^2$ in equation (11.6) yields

$$k''(t) + \left(\frac{cm\pi}{L}\right)^2 k(t) = 0.$$

The general solution is

$$k(t) = C_m \cos\left(\frac{cm\pi t}{L}\right) + C_m' \sin\left(\frac{cm\pi t}{L}\right),$$

where C_m and C_m' are parameters. We have thus determined all solutions of the wave equation, satisfying the boundary conditions, and having the form $u(x, t) = h(x)k(t)$. They are

$$u_m(x, t) = \sin\left(\frac{m\pi x}{L}\right)\left[C_m \cos\left(\frac{cm\pi t}{L}\right) + C_m' \sin\left(\frac{cm\pi t}{L}\right)\right].$$

More general solutions of the wave equation, still satisfying the boundary conditions (11.3), take the form of sums, either infinite or finite, of the solutions $u_m(x, t)$. These solutions can be written out as

$$u(x, t) = \sum_{m=1}^{\infty} \sin\left(\frac{m\pi x}{L}\right) \left[C_m \cos\left(\frac{cm\pi t}{L}\right) + C_m' \sin\left(\frac{cm\pi t}{L}\right) \right], \quad (11.8)$$

where the coefficients C_m, C_m' must be chosen so that the series converges. Notice that

$$u(x, 0) = \sum_{m=1}^{\infty} C_m \sin\left(\frac{m\pi x}{L}\right), \quad (11.9)$$

and

$$\frac{\partial u}{\partial t}(x, 0) = \frac{c\pi}{L} \sum_{m=1}^{\infty} m C_m' \sin\left(\frac{m\pi x}{L}\right). \quad (11.10)$$

The coefficients C_m, C_m' are thus determined by the Fourier sine coefficients of the initial displacement profile, $u(x, 0)$, and the initial velocity profile, $\frac{\partial u}{\partial t}(x, 0)$.

Theorem 11.1

Suppose that the initial displacement and velocity profiles are given to be continuous, piecewise differentiable functions $\phi(x)$ and $\psi(x)$, respectively. We assume that both are defined on the interval $[0, L]$, with $\phi(0) = \phi(L) = \psi(0) = \psi(L) = 0$. Denote the Fourier sine series of these functions by

$$\phi(x) = \sum_{m=1}^{\infty} b_m \sin\left(\frac{m\pi x}{L}\right) \quad \text{and} \quad \psi(x) = \sum_{m=1}^{\infty} b_m' \sin\left(\frac{m\pi x}{L}\right).$$

Then the initial boundary value problem for the wave equation,

$$\left. \begin{array}{c} \dfrac{\partial^2 u}{\partial t^2} = c^2 \dfrac{\partial^2 u}{\partial x^2} \\[2mm] u(x, 0) = \phi(x), \quad \dfrac{\partial u}{\partial t}(x, 0) = \psi(x) \\[2mm] u(0, t) = u(L, t) = 0, \end{array} \right\} \quad (11.11)$$

has a unique solution, given by formula (11.8) with $C_m = b_m$ and $C_m' = \frac{L}{cm\pi} b_m'$.

The proof of the uniqueness portion of Theorem 11.1 is outlined in Exercise 14 at the end of this section. The formulas for C_m and C_m' are obtained by equating the Fourier coefficients of $\phi(x)$ and $\psi(x)$ with those of $u(x, 0)$ and $\frac{\partial u}{\partial t}(x, 0)$ as given by equations (11.9) and (11.10), respectively.

Example 11.1.1   Solve the initial boundary value problem

$$\frac{\partial^2 u}{\partial t^2} = 360{,}000 \frac{\partial^2 u}{\partial x^2};$$

$$u(0) = u(6) = 0, \quad u(x,0) = 0.01 \sin(\pi x), \quad \frac{\partial u}{\partial t}(x,0) = 10\pi \sin\left(\frac{\pi x}{3}\right).$$

Solution. Since $L = 6$ and $c = \sqrt{360{,}000} = 600$, the solution is

$$u(x,t) = \sum_{m=1}^{\infty} \sin\left(\frac{m\pi x}{6}\right) [C_m \cos(100 m\pi t) + C_m' \sin(100 m\pi t)],$$

where C_m and C_m' are to be determined by the initial conditions. We have

$$u(x,0) = \sum_{m=1}^{\infty} C_m \sin\left(\frac{m\pi x}{6}\right) = 0.01 \sin\left(\frac{6\pi x}{6}\right);$$

thus $C_6 = 0.01$ and $C_m = 0$ for $m \neq 6$. Also,

$$\frac{\partial u}{\partial t}(x,0) = \sum_{m=1}^{\infty} 100 m\pi C_m' \sin\left(\frac{m\pi x}{6}\right) = 10\pi \sin\left(\frac{2\pi x}{6}\right).$$

Therefore, $200\pi C_2' = 10\pi$, and $C_m' = 0$ for $m \neq 2$. The solution is

$$u(x,t) = 0.05 \sin\left(\frac{\pi x}{3}\right) \sin(200\pi t) + 0.01 \sin(\pi x) \cos(600\pi t). \qquad \square$$

Example 11.1.2   A 25-gauge steel wire has density 8 grams per cubic centimeter and cross-sectional area of 0.002 square centimeters. It is stretched between two pins 30 centimeters apart, with a tension of 5×10^7 dynes. The wire is plucked at a point 7.5 centimeters from one end, by displacing that point 0.1 centimeter (initially, the segments of wire to the left and to the right form straight lines to the pins). Determine the displacement function of the wire, assuming zero initial velocity.

Solution.   In centimeter-gram-second units, $c^2 = T_0/(\rho A)$, with $T_0 = 5 \times 10^7$, $\rho = 8$, and $A = 0.002$. We therefore calculate that $c \approx 56000$ centimeters per second. The initial data are $u(x,0) = \phi(x)$ and $\frac{\partial u}{\partial t}(x,0) = 0$, where

$$\phi(x) = \begin{cases} .1x/7.5 & \text{if } 0 \le x \le 7.5, \text{ and} \\ .1 - (x - 7.5)/22.5 & \text{if } 7.5 \le x \le 30. \end{cases}$$

The Fourier sine coefficients for $\phi(x)$ are

$$b_m = \frac{2}{30} \left[\int_0^{7.5} \frac{x}{75} \sin\left(\frac{m\pi x}{30}\right) dx + \int_{7.5}^{30} \frac{30 - x}{225} \sin\left(\frac{m\pi x}{30}\right) dx \right]$$

$$= \frac{32}{30\pi^2 m^2} \sin\left(\frac{m\pi}{4}\right).$$

Since $C_m = b_m$ and $C_m' = 0$, $u(x,t) = \sum_{m=1}^{\infty} b_m \sin\left(\frac{m\pi x}{30}\right) \cos\left(\frac{56000 m\pi t}{30}\right)$, or, in approximate terms,

$$u(x,t) = 0.108 \sum_{m=1}^{\infty} \frac{1}{m^2} \sin\left(\frac{m\pi}{4}\right) \sin(.105 mx) \cos(5900 mt). \qquad \square$$

Application to acoustics

The wire that was the subject of Example 11.1.2 vibrates at frequencies that are multiples of 5900 radians per second, or 933 hertz (1 hertz = 2π radians per second). An audible sound will be heard, roughly corresponding to the second B♭ above middle C on the piano. The coefficients C_m give the amplitude (which is interpreted as the loudness of the sound by the ear) of the mth frequency. The most prominent frequency corresponds to $m = 1$: $C_1 = 0.108 \sin(\pi/4) \approx 0.076$ centimeters.

In general, the frequencies of vibration for the string are $cm\pi/L$ radians per second, or $f_m = (cm)/(2L)$ hertz. Since $c = \sqrt{T_0/(\rho A)}$, we have

$$f_m = \frac{m\sqrt{T_0}}{2L\sqrt{\rho A}}.$$

The lowest of these frequencies, f_1, is called the fundamental frequency of the string. In a stringed musical instrument, the fundamental frequency is associated with the musical note that the string produces. Thus, a long, heavy string produces a low note, while a short, fine string produces a high note. If the string in question is "middle C" on a piano, $f_1 = 262$ hertz. By comparison, the A string of a violin has $f_1 = 220$ hertz.

Example 11.1.3   The A string of a violin is held to the fingerboard at its midpoint and is plucked. What fundamental tone is heard?

Solution. The effective length of the string is half the normal length. Since the frequency is inversely proportional to the length, the note heard will be $2 \times 220 = 440$ hertz.   ❑

When two notes have a frequency ratio of 2:1, they are separated by an *octave*. Thus, all eight of the C keys on the piano have fundamental frequencies that are related to middle C by powers of 2, the lowest C having $f_1 = 32.7$ hertz, and the highest having $f_1 = 4186$ hertz.

For a given string, the higher frequencies of vibration, $f_2, f_3, \ldots$, are called the *overtones*. A string that produces a fundamental frequency f_1 will also produce these overtones, whose frequencies are integer multiples of f_1. Suppose that two strings have fundamental frequencies $f_1 = \omega$ and $f_1' = 2\omega$. In terms familiar to musicians, they are pitched an octave apart. The sequence of frequencies of the first string is

$$f_1 = \omega, \; f_2 = 2\omega, \; f_3 = 3\omega, \ldots,$$

while the sequence of frequencies of the second are

$$f_1' = 2\omega, \; f_2' = 4\omega, \; f_3' = 6\omega, \ldots.$$

Notice that the frequencies of the second string form a subset of the frequencies of the first. If the strings are in close proximity and the higher pitched one is vibrating, the lower pitched one will vibrate *sympathetically*, imitating the sound of the higher one, because just the even overtones will be heard. On the other hand, if the lower pitched string is vibrating, the higher pitched one will vibrate sympathetically with the even overtones, and its natural sound will be heard.

If a properly tuned piano is available, try the following experiment. First, hold the middle C key down without causing the note to sound (this frees the string to vibrate). Briefly strike the C key one octave higher so that it makes a sound that is immediately extinguished by the damping mechanism of the piano. The sound that you will hear next comes from sympathetic vibrations of the middle C strings. Describe this sound, and repeat the experiment with the other six C keys of the piano. Test each of the other keys within an octave of middle C. Do any of them cause middle C to vibrate sympathetically? If you find any, devise an explanation, and determine each fundamental frequency (one such key is G, with $f_1 = 392$ hertz).

Overtones determine whether two musical notes are *consonant* or *dissonant*. Consonant notes have common overtones, while dissonant ones do not. Overtones produce the distinctive sound of an instrument, enabling the ear to easily distinguish a note played by a harp, where the higher overtones are not prominent in the sound, from the same note played by a violin, where the higher overtones are prominent.

EXERCISES

1. A string has length π, and units are chosen so that $c = 1$. Initially, the string is in its equilibrium position, with velocity $\frac{\partial u}{\partial t}(x, 0) = \sin(2x) - \frac{1}{2}\sin(4x)$. Determine the displacement function $u(x, t)$.

2. A cable 40 meters long has a density per unit length of $\rho A = 1$ kilogram per meter. It is suspended between two brackets with a tension of 1000 newtons. Initially, the cable is deflected from equilibrium, with each point x deflected $0.1 \sin\left(\frac{\pi x}{40}\right)$ meters, and is motionless. Determine the subsequent deflection $u(x, t)$, for $t \geq 0$.

3. The string in Exercise 1 is given an initial velocity $\frac{\partial u}{\partial t}(x, 0) = x(\pi - x)$. The initial displacement is $u(x, 0) \equiv 0$. Find the displacement function $u(x, t)$.

4. The cable in Exercise 2 is given an initial displacement of $u(x, 0) = 1 - 0.05|x - 20|$, and an initial velocity of zero. Determine the displacement function $u(x, t)$.

5. Calculate the first six nonzero amplitudes for the wire in Example 11.1.2.

6. *Longitudinal vibrations.* Suppose that a metal rod is struck on its end by a hammer. The rod will then vibrate, with the motion parallel to the axis of the rod. The displacement function $u(x, t)$ still satisfies the wave equation, with c being the speed of sound for longitudinal waves in the metal. The boundary conditions are $u(0, t) = 0$, assuming that the end of the rod at $x = 0$ is fixed and cannot vibrate, and $\frac{\partial u}{\partial x}(L, t) = 0$, assuming that the end of the rod at $x = L$ is free to vibrate.

   Find all solutions with the form $u(x, t) = h(x)k(t)$ of the initial boundary value problem for longitudinal vibrations of a rod with one free end.

7. Assume that the rod in Exercise 6 is made of stainless steel, with sound speed 5790 meters per second and length 0.5 meter. Find the longitudinal displacement if

   a. the initial displacement is $0.001 \sin(\pi x)$ meters, and the initial velocity is zero.

   b. the initial velocity is $500 \sin(3\pi x)$ meters per second, and the initial displacement is zero

   c. the initial displacement is $0.004(x - x^2)$ and the initial velocity is zero.

8. A bicycle spoke is 0.3 meters long and has a mass of 0.005 kilograms. The correct tension for the spoke, when installed in the wheel, is 900 newtons. What is the fundamental frequency of vibration for the spoke?

   ✔ *Answer:* 387 hertz—approximately the G tone above middle C on the piano.

9. In the equal tempered chromatic musical scale, each octave contains twelve semitones: A, B♭, B, C, C♯, D, E♭, E, F, F♯, G, and G♯. The ratio of frequencies between two consecutive semitones is $\sqrt[12]{2}$, so that, in musical terminology, the semitones divide the octave into twelve equal intervals.

   a. Show that the only musical notes that are exactly consonant are those that are separated by an octave.

   b. Which semitone is closest to being consonant with A?

   ✔ *Answer:* E, which forms with A a "perfect fifth."

An energy integral   The derivation of the wave equation ignores friction, and it is thus reasonable to assume that in this model for the vibrating string, the total mechanical energy is constant. As usual, the mechanical energy is the sum of the potential and kinetic energy of the string. Write the wave equation *as*

$$\rho A \frac{\partial^2 u}{\partial t^2} = T_0 \frac{\partial^2 u}{\partial x^2}.$$

Multiply the wave equation by $\frac{\partial u}{\partial t}$ and integrate to get

$$\int_0^{t_1} \int_0^L \rho A \frac{\partial^2 u}{\partial t^2} \frac{\partial u}{\partial t} \, dx \, dt = \int_0^{t_1} \int_0^L T_0 \frac{\partial^2 u}{\partial x^2} \frac{\partial u}{\partial t} \, dx \, dt,$$

where $t_1 > 0$ is a fixed time.

10. By integrating by parts, show that if the boundary conditions $u(0, t) = u(L, t) = 0$ hold, then

$$\int_0^{t_1} \int_0^L T_0 \frac{\partial^2 u}{\partial x^2} \frac{\partial u}{\partial t} \, dx \, dt = -\int_0^{t_1} \int_0^L T_0 \frac{\partial u}{\partial x} \frac{\partial^2 u}{\partial x \partial t} \, dx \, dt.$$

11. Show that

$$-\int_0^{t_1} \int_0^L T_0 \frac{\partial u}{\partial x} \frac{\partial^2 u}{\partial x \partial t} \, dx \, dt$$
$$= -\int_0^L \frac{1}{2} T_0 \left[\left(\frac{\partial u}{\partial x}(x, t_1) \right)^2 - \left(\frac{\partial u}{\partial x}(x, 0) \right)^2 \right] dx.$$

12. Show that

$$\int_0^{t_1} \int_0^L \rho A \frac{\partial^2 u}{\partial t^2} \frac{\partial u}{\partial t} \, dx \, dt$$
$$= \int_0^L \frac{1}{2} \rho A \left[\left(\frac{\partial u}{\partial t}(x, t_1) \right)^2 - \left(\frac{\partial u}{\partial t}(x, 0) \right)^2 \right] dx.$$

*We define the **potential energy** as*

$$PE(t) = \int_0^L \frac{1}{2} T_0 \left(\frac{\partial u}{\partial x} \right)^2 dx$$

*and the **kinetic energy** as*

$$KE(t) = \int_0^L \frac{1}{2} \rho A \left(\frac{\partial u}{\partial t} \right)^2 dx.$$

13. Show that for any $t_1 > 0$, $PE(t_1) + KE(t_1) = PE(0) + KE(0)$. Thus the total mechanical energy,

$$TE = \frac{1}{2} \int_0^L \left[T_0 \left(\frac{\partial u}{\partial x} \right)^2 + \rho A \left(\frac{\partial u}{\partial t} \right)^2 \right] dx,$$

is constant. This is the energy integral for the wave equation.

14. Use the energy integral to prove that solutions of the initial boundary value problem (11.11) for the wave equation are unique.

◆ **Hint:** First show uniqueness for the case of homogeneous initial conditions $\phi(x) \equiv \psi(x) \equiv 0$.

15. A linear model for a vibrating string with friction would be given by the partial differential equation

$$\rho A \frac{\partial^2 u}{\partial t^2} + \epsilon \frac{\partial u}{\partial t} - T_0 \frac{\partial^2 u}{\partial x^2} = 0, \qquad (11.12)$$

where ϵ is a friction parameter. Show that if the ends of the string at $x = 0$ and $x = L$ are held fixed, then

$$\frac{d}{dt} TE(t) = -\epsilon \int_0^L \left(\frac{\partial u}{\partial t} \right)^2 dx,$$

where $TE(t)$ is the total mechanical energy. Conclude that the total mechanical energy is nonincreasing.

16. Use the result of Exercise 15 to show that solutions of initial boundary value problems for equation (11.12) are unique.

11.2 The Laplace Equation

The Laplace equation in two dimensions is the PDE

$$\frac{\partial^2 u}{\partial x^2} + \frac{\partial^2 u}{\partial y^2} = 0.$$

It has a prominent role in many applications to physics and engineering, including studies of thermal equilibria, electric and gravitational fields, and the flow of incompressible fluids. The Laplace equation is named for the French mathematician Pierre Simon Laplace, who used its three-dimensional version in 1785 to determine the potential energy of an object in a gravitational field induced by an ellipsoidal mass. A function $u(x, y)$ that satisfies Laplace's equation is called a **harmonic function.**

Let $\mathcal{D}$ be a region in the plane bounded by a curve $\mathcal{S}$. The curve $\mathcal{S}$ is assumed to be differentiable, except at a finite number of "corners." The unit outward normal vector to $\mathcal{S}$ will be denoted **n**. In this section, we will assume that $\mathcal{D}$ is either the rectangle

$$\mathcal{R}_{L,M} = \{(x, y) : 0 < x < L, 0 < y < M\},$$

or the interior of a circle of radius R:

$$\mathcal{C}_R = \{(x, y) : 0 \le x^2 + y^2 < R^2\}.$$

In partial differential equations courses, it is shown that the correctly posed problems involving the Laplace equation are *boundary value problems* of various sorts. These problems ask for a harmonic function that is defined on $\mathcal{D}$ and satisfies certain *boundary conditions* on $\mathcal{S}$. The boundary conditions, in turn, always involve a function ϕ that is piecewise differentiable on $\mathcal{S}$, a term that we will now define.

> **DEFINITION**   A function $\phi(x, y)$, defined for $(x, y) \in \mathcal{S}$, is **piecewise differentiable** if
>
> i.   in case $\mathcal{D} = \mathcal{R}_{L,M}$, the functions $\phi(0, y)$ and $\phi(L, y)$ are piecewise differentiable functions of y on $[0, M]$, and the functions $\phi(x, 0)$ and $\phi(x, M)$ are piecewise differentiable functions of x on $[0, L]$.
> ii.   in case $\mathcal{D} = \mathcal{C}_R$, $\phi(R \cos \theta, R \sin \theta)$ is a piecewise differentiable function of θ on $[0, 2\pi]$.

There are three standard boundary value problems for the Laplace equation on the region $\mathcal{D}$.

First boundary value problem
A function $\phi(x, y)$, piecewise differentiable on $\mathcal{S}$, is given. Find a harmonic function $u(x, y)$, defined at all points of $\mathcal{D}$, such that for all $(x, y) \in \mathcal{S}$, $u(x, y) = \phi(x, y)$.

Second boundary value problem
A function $\phi(x, y)$, piecewise differentiable on $\mathcal{S}$, is given. Denote by **n** the unit vector, normal to $\mathcal{S}$, directed outward. The problem is to find a harmonic function $u(x, y)$, defined at all points of $\mathcal{D}$, such that for all $(x, y) \in \mathcal{S}$, $\frac{\partial u}{\partial \mathbf{n}}(x, y) = \phi(x, y)$, where $\frac{\partial u}{\partial \mathbf{n}}(x, y)$ denotes the directional derivative of u.

Third boundary value problem
A function $\phi(x, y)$, piecewise differentiable on $\mathcal{S}$, and a constant $\kappa > 0$ are given. Find a harmonic function $u(x, y)$, defined at all points of $\mathcal{D}$, such that for all $(x, y) \in \mathcal{S}$,

$$\frac{\partial u}{\partial \mathbf{n}}(x, y) + \kappa u(x, y) = \phi(x, y).$$

These boundary value problems are also known as the **Dirichlet problem**, the **Neumann problem**, and the **Robin problem**, in honor of mathematicians who made early contributions to the theory. Dirichlet and Neumann were Germans, and Robin (whose name rhymes with that of the French sculptor Rodin) was French.

Existence and Uniqueness

It is shown in partial differential equations texts that, provided that ϕ is piece-wise differetiable on S, the first and third boundary valuse problems have solutions.

The second problem has a solution if and only if the average value of ϕ on S is equal to 0, and uniqueness is only up to an additive constant.

The approach to the boundary value problems depends on the shape of the domain $\mathcal{D}$. If $\mathcal{D}$ is a rectangle or a circle, each of the three boundary value problems can be solved by separation of variables. Some other highly symmetric domains (ellipses, for example) can also be handled by the method of separation of variables, but this topic is beyond the scope of this text. The method of separation of variables cannot be adapted to solve boundary value problems on domains without sufficient symmetry. It is necessary to resort to numerical analysis in these cases.

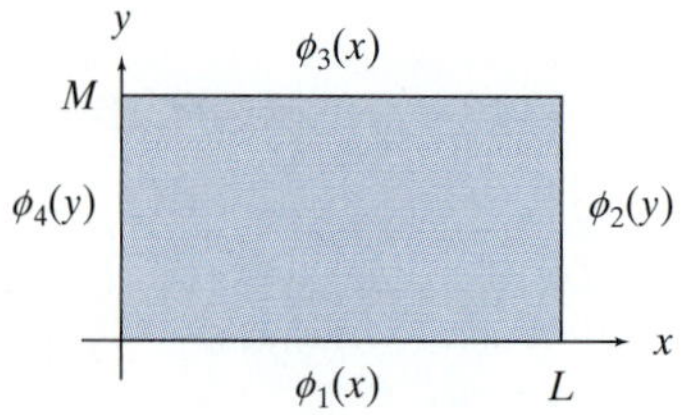

Figure 11.3. Boundary function on the rectangular domain $\mathcal{R}_{L,M}$.

Rectangular domains

Suppose that $\mathcal{D} = \mathcal{R}_{L,M}$. The function $\phi(x, y)$ can be specified by defining four functions, $\phi(x, 0) = \phi_1(x)$, $\phi(L, y) = \phi_2(y)$, $\phi(x, M) = \phi_3(x)$, and $\phi(0, y) = \phi_4(y)$, on the sides of the rectangle, as shown in Figure 11.3.

The method of separation of variables for Laplace's equation on a rectangular domain is based on finding solutions of the form $u(x, y) = h(x)k(y)$.

Proposition 11.2.1

Let $u(x, y) = h(x)k(y)$ denote a harmonic function defined on the rectangle $\mathcal{R}_{L,M}$. Then there is a constant λ such that $h(x)$ and $k(y)$ satisfy the differential equations

$$h''(x) - \lambda h(x) = 0 \tag{11.13}$$
$$k''(y) + \lambda k(y) = 0, \tag{11.14}$$

respectively.

Proof Substituting $u(x, y) = h(x)k(y)$ into Laplace's equation, we have

$$h''(x)k(y) + h(x)k''(y) = 0;$$

hence

$$\frac{h''(x)}{h(x)} = -\frac{k''(y)}{k(y)}. \tag{11.15}$$

The left side of equation (11.15) depends only on x, and the right side depends only on y. Therefore, both are constant. Denote this constant by λ. Thus $-\frac{k''(y)}{k(y)} = \lambda$, and $\frac{h''(x)}{h(x)} = \lambda$. Equations (11.13) and (11.14) now follow immediately. ∎

Our experience with the heat and wave equations suggests that we should formulate an eigenvalue problem to determine possible values for λ. This is only possible if the boundary conditions on two opposite sides of the rectangle are *homogeneous:* that is, either $\phi_1(x) \equiv \phi_3(x) \equiv 0$, or $\phi_2(y) \equiv \phi_4(y) \equiv 0$. In the following proposition, it will be assumed that $\phi_1(x) \equiv \phi_3(x) \equiv 0$. A similar proposition can be stated for $\phi_2(y) \equiv \phi_4(y) \equiv 0$.

Proposition 11.2.2

Let $u(x, y) = h(x)k(y)$ be a solution of one of the three boundary value problems for the Laplace equation on the rectangle $\mathcal{R}_{L,M}$, with boundary conditions specified by functions $\phi_1(x) \equiv \phi_3(x) \equiv 0$ on the top and bottom of the rectangle, and $\phi_2(y)$, $\phi_4(y)$ on the right and left sides, respectively. Then $k(y)$ is a solution of the eigenvalue problem given by equation (11.14) with boundary values

$$k(0) = k(M) = 0 \tag{11.16}$$

in the case of the first boundary value problem,

$$k'(0) = k'(M) = 0 \tag{11.17}$$

in the second boundary value problem, and

$$-k'(0) + \kappa k(0) = k'(M) + \kappa k(M) = 0 \tag{11.18}$$

in the case of the third.

Proof In terms of $h(x)$ and $k(y)$, the boundary conditions for the first boundary value problem are $h(x)k(0) = \phi_1(x)$, $h(L)k(y) = \phi_2(y)$, $h(x)k(M) = \phi_3(x)$, and $h(0)k(y) = \phi_4(y)$. It follows from our assumption that the top and bottom boundary conditions are homogeneous that $h(x)k(0) = h(x)k(M) = 0$ for all $x \in [0, L]$, and hence the condition (11.16) holds.

The normal derivative of $u(x, y)$ at the bottom of the rectangle is

$$\frac{\partial u}{\partial \mathbf{n}}(x, 0) = -\frac{\partial u}{\partial y}(x, 0) = -h(x)k'(0).$$

It follows that $k(y)$ must satisfy the boundary condition $k'(0) = 0$ for the second problem, and $-k'(0) + \kappa k(0) = 0$ in the case of the third. Similar reasoning establishes the boundary conditions at $y = L$ for these problems. ∎

Propositions 11.2.1 and 11.2.2 allow us to apply Fourier series to solve the boundary value problems for the Laplace equation. Let us consider the first boundary value problem on $\mathcal{R}_{L,M}$ with homogeneous boundary conditions on the top and bottom sides. The propositions imply that if $u(x, y) = h(x)k(y)$ is a solution, then $k(y)$ is a solution of the eigenvalue problem

$$k''(y) + \lambda k(y) = 0; \quad k(0) = k(M) = 0. \tag{11.19}$$

Thus, by Proposition 10.2.1 on page 478, the eigenvalues are $\lambda_m = \nu_m^2$, where $\nu_m = m\pi/M$ for $m = 1, 2, 3, \ldots$, and the eigenfunctions are $k_m(y) = \sin(\nu_m x)$.

To determine $h(x)$, we refer again to Proposition 11.2.1. Substituting v_m^2 for λ in equation (11.13), we see that the characteristic roots of that differential equation are $\pm v_m$. Hence the general solution of equation (11.13) is

$$h_m(x) = C_m e^{v_m x} + C'_m e^{-v_m x}.$$

Suppose that we are given that $\phi_4(y) \equiv 0$. Then $h(0) = 0$, so $C_m + C'_m = 0$, and we have

$$h_m(x) = C_m \left(e^{v_m x} - e^{-v_m x} \right) = 2C_m \sinh(v_m x).$$

We have thus constructed an infinite family of harmonic functions,

$$u_m(x, y) = \sinh(v_m x) \sin(v_m y),$$

that vanish on the top, bottom, and left sides of $\mathcal{R}_{L,M}$. We also have

$$u_m(L, y) = 2C_m \sinh(v_m L) \sin(v_m y). \tag{11.20}$$

If we expand the remaining boundary function $\phi_2(y)$ as a Fourier sine series,

$$\phi_2(y) = \sum_{m=1}^{\infty} b_m \sin(v_m y),$$

then it is possible to choose the coefficients C_m in equation (11.20) so that

$$2C_m \sinh(v_m L) = b_m.$$

It then follows that the solution of the first boundary value problem is

$$u(x, y) = \sum_{m=1}^{\infty} u_m(x, y)$$

$$= \sum_{m=1}^{\infty} \frac{b_m}{2 \sinh(v_m L)} \sinh(v_m x) \sin(v_m y).$$

This solution of the first boundary value problem makes the same use of superposition that aided the solution of the initial boundary value problem for the heat equation in Section 10.3.

In a realistic problem, the boundary values would not be zero on three sides of the rectangle. Such problems are handled by more superposition! Let $u^{(1)}(x, y)$ be the harmonic function that vanishes on the top, left, and right sides of $\mathcal{R}_{L,M}$, and with $u^{(1)}(x, 0) = \phi_1(x)$. Similarly, $u^{(2)}(x, y)$ would be the harmonic function that satisfies the right boundary condition and vanishes on the other three sides, and so on. It is easily seen that

$$u(x, y) = u^{(1)}(x, y) + u^{(2)}(x, y) + u^{(3)}(x, y) + u^{(4)}(x, y)$$

is then a harmonic function satisfying the boundary condition on all four sides of the rectangle.

The solution of the second and third boundary value problems on a rectangular domain proceeds in a manner similar to the solution of the first boundary value problem.

Example 11.2.1   Solve the first boundary value problem on the rectangle $0 \leq x \leq 1, 0 \leq y \leq \pi$, with $\phi_4(y) = \sin y$, and $\phi_j \equiv 0$ for $j = 1, 2$, and 3.

Solution. Set $u(x, y) = h(x)k(y)$. Since $\phi_1(x) = \phi_3(x) = 0$, it follows from Proposition 11.2.2 that $k(y)$ is a solution of the eigenvalue problem (11.19) with $M = \pi$, and thus is a multiple of one of the eigenfunctions $k_m(y) = \sin(my)$. The corresponding eigenvalues are numbers $\lambda_m = m^2$. The general solution of equation (11.13), with $\lambda = m^2$, is $h_m(x) = C_m e^{mx} + C'_m e^{-mx}$. Since $u(1, y) = \phi_2(y) = 0$, we have $h_m(1)k_m(y) \equiv 0$, so $h_m(1) = C_m e^m + C'_m e^{-m} = 0$. Put $K_m = C_m e^m = -C'_m e^{-m}$, so that $C_m = K_m e^{-m}$ and $C'_m = -K_m e^m$. Then $h_m(x) = K_m(e^{m(x-1)} - e^{-m(x-1)})$. In terms of hyperbolic functions, $h_m(x) = 2K_m \sinh(m(x - 1))$. Thus, for each positive integer m and constant K_m, the function $u_m(x, y) = 2K_m \sinh(m(x - 1)) \sin(my)$ is a harmonic function that vanishes on the top, bottom, and right sides of the rectangle $\mathcal{R}_{L,M}$, while on the left side we have $u_m(0, y) = -2K_m \sinh(m) \sin(my)$. We now observe that when $m = 1$, and $K_1 = \frac{-1}{2 \sinh 1}$, we have $u_1(0, y) = \phi_4(y)$. The solution is therefore

$$u_1(x, y) = \frac{-1}{\sinh 1} \sinh(x - 1) \sin y. \qquad \square$$

Example 11.2.2   Solve the second boundary value problem on $\mathcal{R}_{1,1}$ with boundary conditions $\phi_1(x) = \phi_2(y) = \phi_4(y) = 0$, and $\phi_3(x) = \sin(2\pi x)$ for $0 \leq x \leq 1$.

Solution. Since this is the second boundary value problem, we must first verify that the average boundary value is zero; otherwise there is no solution. The average boundary value is

$$\int_0^1 \phi_3(x)\, dx = \int_0^1 \sin(2\pi x)\, dx = 0,$$

so we can proceed.

The boundary conditions are homogeneous on the vertical sides of the square. Therefore, if $u(x, y) = h(x)k(y)$ is a harmonic function satisfying these homogeneous boundary conditions, $h(x)$ satisfies the eigenvalue problem

$$h''(x) + \lambda h(x) = 0; \quad h'(0) = h'(1) = 0 \qquad (11.21)$$

[we have reversed the sign of the parameter λ in equation (11.13) and we will do the same in equation (11.14)]. You can verify that there are no negative eigenvalues, so let's put $\lambda = \nu^2$, where ν is a real number.

Corresponding to $\nu = 0$, we have the eigenfunction $h_0(x) \equiv 1$. For $\nu > 0$ the general solution of (11.21) is

$$h = B_1 \cos(\nu x) + B_2 \sin(\nu x).$$

Thus, $h'(x) = -B_1 \nu \sin(\nu x) + B_2 \nu \cos(\nu x)$, and hence $h'(0) = B_2 \nu = 0$. It follows that $B_2 = 0$. Now $h'(1) = -B_1 \nu \sin(\nu) = 0$, which implies that $\nu = n\pi$ for some integer n. Thus, the eigenfunctions are $h_0(x) = 1$ and $h_n(x) = \cos(n\pi x)$, and the corresponding eigenvalues are $\lambda_n = (n\pi)^2$.

Put $u_n(x, y) = h_n(x)k_n(y)$. Then $k_n(y)$ satisfies the ODE

$$k_n''(y) - \lambda_n k_n(y) = 0. \tag{11.22}$$

For $n > 0$ the general solution of (11.22) is

$$k_n(y) = C \cosh(n\pi y) + C' \sinh(n\pi y).$$

Applying the remaining homogeneous boundary condition, $h_n(x)k_n'(0) = \phi_1(x) = 0$, we see that $k_n'(0) = 0$. But $k_n'(y) = n\pi(C_n \sinh(n\pi y) + C_n' \cosh(n\pi y))$. Hence $k_n'(0) = C_n' = 0$, and $k_n(y) = C_n \cosh(n\pi y)$.

If $n = 0$, the general solution of (11.22) is $k_0(y) = C + C'y$, and $k_0'(y) = C'$. Since $k_0'(0) = 0$, we have $C' = 0$. We have thus determined an infinite family of harmonic functions

$$u_n(x, y) = \cos(n\pi x) \cosh(n\pi y)$$

that satisfy the homogeneous boundary conditions on three edges of the square. [Note that $u_0(x, y) \equiv 1$ can be included.] If we write

$$u(x, y) = \sum_{n=0}^{\infty} C_n u_n(x, y),$$

then the remaining boundary condition is $\partial u/\partial \mathbf{n}(x, 1) = \phi_3(x)$. Since

$$\frac{\partial u}{\partial \mathbf{n}}(x, 1) = \frac{\partial u}{\partial y}(x, 1)$$

$$= \sum_{n=1}^{\infty} C_n (n\pi) \sinh(n\pi) \cos(n\pi x), \tag{11.23}$$

we can determine the coefficients C_n by calculating the coefficients A_n in the Fourier *cosine* series for $\phi_3(x) = \sin(2\pi x)$:

$$\sin(2\pi x) = A_0 + \sum_{n=1}^{\infty} A_n \cos(n\pi x).$$

As we have seen, $A_0 = \int_0^1 \sin(2\pi x)\, dx = 0$. For $n \geq 1$,

$$A_n = 2 \int_0^1 \sin(2\pi x) \cos(n\pi x)\, dx$$

$$= \int_0^1 (\sin((n+2)\pi x) - \sin((n-2)\pi x))\, dx,$$

where we have used the trigonometric identity

$$2 \sin(\alpha) \cos(\beta) = \sin(\beta + \alpha) - \sin(\beta - \alpha).$$

Thus, $A_n = 0$ when n is even, and when $n = 2k + 1$ is odd,

$$A_{2k+1} = \frac{1}{\pi} \left[-\frac{\cos((2k+3)\pi x)}{2k+3} + \frac{\cos((2k-1)\pi x)}{2k-1} \right]\Big|_0^1$$

$$= \frac{2}{\pi} \left[\frac{1}{2k+3} - \frac{1}{2k-1} \right]$$

$$= \frac{8}{(3 - 4k - 4k^2)\pi}.$$

By (11.23),

$$C_n (n\pi) \sinh (n\pi) = A_n.$$

Thus, when n is even, $C_n = 0$, and for $k = 0, 1, 2, \ldots,$

$$C_{2k+1}(2k + 1)\pi \sinh((2k + 1)\pi) = A_{2k+1} = \frac{8}{(3 - 4k - 4k^2)\pi}.$$

It follows that

$$C_{2k+1} = \frac{8}{(3 - 4k - 4k^2)(2k + 1)\pi^2 \sinh((2k + 1)\pi)}.$$

The solution of the boundary value problem is therefore

$$u(x, y) = \frac{8}{\pi^2} \sum_{k=0}^{\infty} \frac{\cos((2k + 1)\pi x) \cosh((2k + 1)\pi y)}{(3 - 4k - 4k^2)(2k + 1) \sinh((2k + 1)\pi)}. \qquad \Box$$

EXERCISES

1. Find a harmonic function $u(x, y)$, defined on the square $0 \le x, y \le \pi$, such that on the boundary of the square, $u(x, y) = x \sin y$.

2. Find a harmonic function $u(x, y)$ that vanishes on the left, right, and bottom of the rectangle $0 \le x \le \pi, 0 \le y \le 1$, such that $u(x, 1) = \cos(2x) - 1$.

3. Solve the second boundary value problem on the rectangle $0 \le x \le 1, 0 \le y \le 2\pi$, where the boundary function vanishes on the top, bottom, and left edges of the rectangle, and $\frac{\partial u}{\partial x}(1, y) = \sin y$.

4. Solve the first boundary value problem on the rectangle $0 \le x \le \pi, 0 \le y \le 1$, with $u(0, y) = u(\pi, y) = 0$, $u(x, 0) = \sin x$, and $u(x, 1) = \sin(2x)$.

5. Solve the third boundary value problem on the unit square, with boundary conditions $-\frac{\partial u}{\partial y}(x, 0) + u(x, 0) = 0$, $\frac{\partial u}{\partial x}(1, y) + u(1, y) = \phi_2(y)$, $\frac{\partial u}{\partial y}(x, 1) + u(x, 1) = 0$, and $-\frac{\partial u}{\partial x}(0, y) + u(0, y) = 0$. Use $\phi_2(y) = k \cos(ky) + \sin(ky)$,

where k is a number defined by the equation $(k^2 - 1) \tan k - 2k = 0$, and $1 < k < \frac{\pi}{2}$.

6. Complete the proof of Proposition 11.2.2.

7. **Mixed boundary conditions.** Consider the problem of finding a harmonic function $u(x, y)$, defined on a rectangle $0 \le x \le L, 0 \le y \le M$, such that for $0 \le y \le L$,

$$-\frac{\partial u}{\partial x}(0, y) = u(L, y) = 0.$$

In other words, a boundary condition of the second kind holds on the left edge of the rectangle, while the boundary condition on the right edge is of the first kind. On the top and bottom edges, we will assume boundary conditions of the first kind: $u(x, 0) = 0$, $u(x, M) = \phi_3(x)$. Devise a strategy for solving this problem by separation of variables.

8. Show that $u(x, y) = xy$ is a harmonic function, and use this fact to solve the first boundary value problem on the

square $\mathcal{R}_{1,1}$ with $\phi_1(x) \equiv \phi_4(y) \equiv 0$, and $\phi_2(y) = y$, $\phi_3(x) = x$.

9. Solve the first boundary value problem stated in Exercise 8 by using Fourier series. You are not expected to reconcile the resulting series solution with the solution of Exercise 8.

10. Show that each of the functions $u(x, y)$ listed is a harmonic function. Find $\frac{\partial u}{\partial n}$ on the boundary of the square $0 \le x, y \le 1$, and verify that its average value is zero.

  a. $u(x, y) = ax + by + c$

  b. $u(x, y) = x^2 - y^2$

  c. $u(x, y) = 3x^2 y - y^3$

  d. $u(x, y) = \cos(4\pi x) \cosh(4\pi y)$

  e. $u(x, y) = \frac{1}{2}a_0$
     $\quad + \sum_{m=1}^{\infty} (a_m e^{m\pi x} + b_m e^{-m\pi x}) \cos(m\pi y)$

11. Solve the following second boundary value problem on $\mathcal{R}_{\pi,\pi}$:

$$\frac{\partial^2 u}{\partial x^2} + \frac{\partial^2 u}{\partial y^2} = 0;$$

$$\frac{\partial u}{\partial n}(0, y) = \frac{\partial u}{\partial n}(x, 0) = \frac{\partial u}{\partial n}(x, \pi) = 0,$$

$$\frac{\partial u}{\partial n}(\pi, y) = \cos y.$$

12. Solve the second boundary value problem on $\mathcal{R}_{\pi,\pi}$, with boundary conditions

$$\frac{\partial u}{\partial n}(0, y) = 0, \; \frac{\partial u}{\partial n}(\pi, y) = 2\pi + \cos y$$

and

$$\frac{\partial u}{\partial n}(x, 0) = 0, \; \frac{\partial u}{\partial n}(x, \pi) = -2\pi.$$

◆ **Hint:** See Exercises 10b and 11.

13. Solve the second boundary value problem on the square $\mathcal{R}_{\pi,\pi}$ with boundary conditions as follows:

$$\frac{\partial u}{\partial n}(0, y) = \frac{\partial u}{\partial n}(\pi, y) = \frac{\partial u}{\partial n}(x, 0) = 0$$

and

$$\frac{\partial u}{\partial n}(x, \pi) = x - \frac{\pi}{2}.$$

Is the solution $u(x, y)$ continuous on $\mathcal{R}_{\pi,\pi}$?

14. Consider the second boundary value problem on a rectangle $\mathcal{R}_{L,M}$, where only the boundary function ϕ_2 is nonzero. Show that if

$$\int_0^M \phi_2(y) \, dy \neq 0,$$

then this problem cannot be solved by the method of separation of variables. It is not necessary to use the result of Exercise 15.

15. This exercise outlines the proof that for the second boundary value problem to have a solution, it is necessary that $\int_S \phi(s) ds = 0$. Exercise 14 does not establish this fact, since it only indicates that in a special case, the condition is necessary for the method of separation of variables to work. A knowledge of Green's theorem is assumed.

Let $u(x, y)$ denote a harmonic function on a region $\mathcal{D}$ whose boundary is a curve S, and for $(x, y) \in S$, put $\phi(x, y) = \frac{\partial u}{\partial n}(x, y)$.

  a. Show that

$$\int_S \phi(s) \, ds = \int_S \left[\frac{\partial u}{\partial x} \, dy - \frac{\partial u}{\partial y} \, dx \right].$$

  b. Use Green's theorem to conclude that

$$\int_S \phi(s) \, ds = \int_{\mathcal{D}} \left[\frac{\partial^2 u}{\partial x^2} + \frac{\partial^2 u}{\partial y^2} \right] dx \, dy = 0.$$

16. This problem outlines the proof of uniqueness for the boundary value problems.

  a. Let $u_1(x, y)$ and $u_2(x, y)$ denote solutions of one of the three boundary value problems, satisfying identical boundary conditions. Show that $v(x, y) = u_1(x, y) - u_2(x, y)$ is a harmonic function satisfying *homogeneous* boundary conditions.

Henceforth, $v(x, y)$ will denote a harmonic function defined of a region $\mathcal{D}$ with boundary S, satisfying homogeneous boundary conditions on S. The objective will be to show that $v \equiv 0$ in the case of the first and third boundary value problems, and $v \equiv$ constant in the second problem.

  b. Show that

$$\int_S v(x, y) \left(\frac{\partial v}{\partial x}(x, y) \, dy - \frac{\partial v}{\partial y}(x, y) \, dx \right) \le 0$$

in the first, second, or third boundary value problems. (Equality will hold for the first and second.)

  c. Using Green's theorem and the fact that v is a harmonic function, conclude that

$$\int_{\mathcal{D}} \left[\left(\frac{\partial v}{\partial x} \right)^2 + \left(\frac{\partial v}{\partial y} \right)^2 \right] dx \, dy \le 0.$$

  d. Deduce that v is constant.

  e. In the case of the first and third problems, further deduce that the constant is 0.

*11.3   BVPs on Circular and Spherical Domains

We will consider the classical three boundary value problems for the Laplace equation on a circular domain. There is a well-established notation Δ for the Laplace operator. Thus, in rectangular coordinates,

$$\Delta u = \frac{\partial^2 u}{\partial x^2} + \frac{\partial^2 u}{\partial y^2}.$$

Lemma 11.3.1

Let $u(x, y)$ be a twice differentiable function, and put $v(r, \theta) = u(r \cos \theta, r \sin \theta)$. Then

$$\Delta u(r \cos \theta, r \sin \theta) = \frac{\partial^2 v}{\partial r^2} + \frac{1}{r} \frac{\partial v}{\partial r} + \frac{1}{r^2} \frac{\partial^2 v}{\partial \theta^2}.$$

Proof By the chain rule for partial derivatives,

$$\frac{\partial u}{\partial x} = \frac{\partial v}{\partial r} \frac{\partial r}{\partial x} + \frac{\partial v}{\partial \theta} \frac{\partial \theta}{\partial x},$$

and

$$\frac{\partial^2 u}{\partial x^2} = \frac{\partial^2 v}{\partial r^2} \left(\frac{\partial r}{\partial x} \right)^2 + \frac{\partial v}{\partial r} \left(\frac{\partial^2 r}{\partial x^2} \right) + 2 \frac{\partial^2 v}{\partial r \partial \theta} \left(\frac{\partial r}{\partial x} \right) \left(\frac{\partial \theta}{\partial x} \right)$$

$$+ \frac{\partial^2 v}{\partial \theta^2} \left(\frac{\partial \theta}{\partial x} \right)^2 + \frac{\partial v}{\partial \theta} \frac{\partial^2 \theta}{\partial x^2}.$$

A similar formula holds for $\frac{\partial^2 u}{\partial y^2}$, and hence

$$\Delta u = \frac{\partial^2 v}{\partial r^2} \left[\left(\frac{\partial r}{\partial x} \right)^2 + \left(\frac{\partial r}{\partial y} \right)^2 \right] + \frac{\partial v}{\partial r} \left[\frac{\partial^2 r}{\partial x^2} + \frac{\partial^2 r}{\partial y^2} \right]$$

$$+ 2 \frac{\partial^2 v}{\partial r \partial \theta} \left[\frac{\partial r}{\partial x} \frac{\partial \theta}{\partial x} + \frac{\partial r}{\partial y} \frac{\partial \theta}{\partial y} \right]$$

$$+ \frac{\partial^2 v}{\partial \theta^2} \left[\left(\frac{\partial \theta}{\partial x} \right)^2 + \left(\frac{\partial \theta}{\partial y} \right)^2 \right] + \frac{\partial v}{\partial \theta} \left[\frac{\partial^2 \theta}{\partial x^2} + \frac{\partial^2 \theta}{\partial y^2} \right]. \qquad (11.24)$$

Since $r^2 = x^2 + y^2$, $2r \frac{\partial r}{\partial x} = 2x$ and $2r \frac{\partial r}{\partial y} = 2y$. Hence $\frac{\partial r}{\partial x} = \frac{x}{r}$, $\frac{\partial r}{\partial y} = \frac{y}{r}$, and

$$\left[\left(\frac{\partial r}{\partial x} \right)^2 + \left(\frac{\partial r}{\partial y} \right)^2 \right] = \frac{x^2 + y^2}{r^2} = 1.$$

Furthermore, by the quotient rule for differentiation,

$$\frac{\partial^2 r}{\partial x^2} = \frac{r - x \cdot (x/r)}{r^2} = \frac{r^2 - x^2}{r^3}$$

$$= \frac{y^2}{r^3};$$

similarly, $\frac{\partial^2 r}{\partial y^2} = \frac{x^2}{r^3}$, so

$$\frac{\partial^2 r}{\partial x^2} + \frac{\partial^2 r}{\partial y^2} = \frac{y^2 + x^2}{r^3} = \frac{1}{r}.$$

Now let us calculate the partial derivatives of θ. Since $\theta = \tan^{-1}(y/x)$,

$$\frac{\partial \theta}{\partial x} = \frac{1}{1 + (y/x)^2} \left(-\frac{y}{x^2} \right) = -\frac{y}{r^2}.$$

A similar calculation shows that $\frac{\partial \theta}{\partial y} = \frac{x}{r^2}$. It follows that

$$\frac{\partial r}{\partial x} \frac{\partial \theta}{\partial x} + \frac{\partial r}{\partial y} \frac{\partial \theta}{\partial y} = 0.$$

Furthermore,

$$\left(\frac{\partial \theta}{\partial x} \right)^2 + \left(\frac{\partial \theta}{\partial y} \right)^2 = \frac{y^2 + x^2}{r^4} = \frac{1}{r^2}.$$

Finally, θ is a *harmonic* function of x and y:

$$\frac{\partial^2 \theta}{\partial x^2} = -\frac{\partial}{\partial x} \left[\frac{y}{r^2} \right]$$

$$= 2\frac{y}{r^3} \frac{\partial r}{\partial x} = \frac{2xy}{r^4},$$

and similarly $\frac{\partial^2 \theta}{\partial y^2} = -\frac{2xy}{r^4}$. Therefore,

$$\Delta \theta = \frac{\partial^2 \theta}{\partial x^2} + \frac{\partial^2 \theta}{\partial y^2} = 0.$$

To complete the proof, we have only to substitute these formulas for the partial derivatives of r and θ into equation (11.24). ∎

Suppose that a harmonic function has the special form $u = v(r, \theta) = f(r)g(\theta)$. Since, by definition, $\Delta u = 0$, we have

$$f''(r)g(\theta) + \frac{1}{r} f'(r)g(\theta) + \frac{1}{r^2} f(r)g''(\theta) = 0.$$

Now multiply this equation by r^2, and divide through by $f(r)g(\theta)$. After a slight rearrangement, the following equation results:

$$\frac{r^2 f''(r) + rf'(r)}{f(r)} = -\frac{g''(\theta)}{g(\theta)}.$$

The left side of this equation is independent of θ, and the right side does not depend on r. Hence, both sides are constant. Denoting this constant by λ, the following two equations result:

$$r^2 f''(r) + rf'(r) - \lambda f(r) = 0 \qquad (11.25)$$

and

$$g''(\theta) + \lambda g(\theta) = 0. \qquad (11.26)$$

To determine solutions, we must consider boundary conditions. Since (r, θ) and $(r, \theta + 2\pi)$ represent the same point, $v(r, \theta) = v(r, \theta + 2\pi)$ for all (r, θ). In other words, $g(\theta)$ has *periodic boundary conditions* on $[0, 2\pi]$ (see Exercise 7 on page 481 for a discussion of this kind of boundary conditions). You can show that the eigenvalues of equation (11.26) with periodic boundary conditions on $[0, 2\pi]$ are $\lambda_m = m^2$, for $m = 0, 1, 2, \ldots$, and the eigenfunctions belonging to λ_m are $\cos(m\theta)$ and $\sin(m\theta)$.

Equation (11.25) is a Cauchy-Euler equation. Taking $\lambda = m^2$, its general solution is $f(r) = cr^m + dr^{-m}$ when $m > 0$, and $f(r) = c + d \ln |r|$ when $m = 0$. In either case, $d = 0$, since otherwise $f(r)$ would be undefined at $r = 0$, the center of the circle. Hence $u = r^m[a_m \cos(m\theta) + b_m \sin(m\theta)]$ for some integer $m \geq 0$. Using superposition, we can get a large family of harmonic functions of the form

$$u = v(r, \theta) = \frac{1}{2}a_0 + \sum_{m=1}^{\infty} r^m[a_m \cos(m\theta) + b_m \sin(m\theta)]. \qquad (11.27)$$

Example 11.3.1   Solve the first boundary value problem for the interior of the unit circle $\mathcal{D} = \{(x, y) : x^2 + y^2 < 1\}$, with boundary condition $\phi(x, y) = 2x^2$ when $x^2 + y^2 = 1$.

Solution. In terms of polar coordinates, we are to find a harmonic function $u = v(r, \theta)$ such that $v(1, \theta) = 2x^2 = 2\cos^2 \theta$. By equation (11.27),

$$v(1, \theta) = \frac{1}{2}a_0 + \sum_{m=1}^{\infty}[a_m \cos(m\theta) + b_m \sin(m\theta)].$$

Since $\cos^2 \theta = \frac{1}{2} + \frac{1}{2}\cos(2\theta)$, $\frac{1}{2}a_0 = a_2 = 1$, and all other coefficients are 0. Thus

$$v(r, \theta) = 1 + r^2 \cos(2\theta).$$

It is left as an exercise for you to show that in cartesian coordinates, the solution is $u(x, y) = 1 + x^2 - y^2$.   ❑

Now let us turn to the second boundary value problem on the interior of the unit circle. Given a 2π-periodic function $\phi(\theta)$, we must determine a harmonic function $u = v(r, \theta)$, such that $\frac{\partial v}{\partial \mathbf{n}}(1, \theta) = \phi(\theta)$, where $\mathbf{n}$ is the outward unit normal vector of the circle. We know that this will be impossible unless the average value of ϕ is 0, so we will assume that

$$\int_0^{2\pi} \phi(\theta) \, d\theta = 0.$$

The unit normal to the circle $r = 1$ is $\mathbf{n} = (x\mathbf{i} + y\mathbf{j})$. Therefore,

$$\frac{\partial u}{\partial \mathbf{n}} = x \frac{\partial u}{\partial x} + y \frac{\partial u}{\partial y}.$$

Referring to the proof of Lemma 11.3.1, it follows that

$$\frac{\partial u}{\partial \mathbf{n}} = \frac{\partial v}{\partial r}.$$

Differentiate the expression (11.27) with respect to r and set $r = 1$ to put the boundary condition into the form

$$\sum_{m=1}^{\infty} m[a_m \cos(m\theta) + b_m \sin(m\theta)] = \phi(\theta). \tag{11.28}$$

To complete the calculations, it is only necessary to find the Fourier coefficients of $\phi(\theta)$ and substitute their values into equation (11.27). Since the left side of equation (11.28) has no constant term, the Fourier coefficient a_0 of $\phi(\theta)$ must be zero. We have provided for this requirement by insisting that the average value of $\phi(\theta)$ must be zero.

Example 11.3.2   Solve the following third boundary value problem. Find a harmonic function $u = v(r, \theta)$, defined in the circular domain $\mathcal{D} = \{(r, \theta) : r < 2\}$, such that on the boundary $r = 2$,

$$v(2, \theta) + 2 \frac{\partial v}{\partial \mathbf{n}}(2, \theta) = 1 + \sum_{m=1}^{\infty} \frac{1}{m^2} \cos(m\theta).$$

Solution. We have noted that $\frac{\partial v}{\partial \mathbf{n}} = \frac{\partial v}{\partial r}$. If $v(r, \theta)$ has the form of equation (11.27), then

$$v(2, \theta) + 2\frac{\partial v}{\partial r}(2, \theta) = a_0 + \sum_{m=1}^{\infty} 2^m[a_m \cos(m\theta) + b_m \sin(m\theta)]$$

$$+ 2\sum_{m=1}^{\infty} m \cdot 2^{m-1}[a_m \cos(m\theta) + b_m \sin(m\theta)]$$

$$= a_0 + \sum_{m=1}^{\infty} 2^m(1 + m)[a_m \cos(m\theta) + b_m \sin(m\theta)].$$

Equating Fourier coefficients in this expression with those of

$$\phi(\theta) = 1 + \sum_{m=1}^{\infty} \frac{1}{m^2} \cos(m\theta),$$

we have $a_0 = 1$, $a_m = \frac{1}{m^2 2^m (1+m)}$, and $b_m = 0$. The solution is

$$v(r, \theta) = 1 + \sum_{m=1}^{\infty} \frac{r^m}{m^2 2^m (1 + m)} \cos(m\theta). \qquad \Box$$

Spherical coordinates

To solve boundary value problems with spherical domains, it is natural to use spherical coordinates, $\rho = \sqrt{x^2 + y^2 + z^2}$, $\theta = \arg(x + iy)$, and $\phi = \cos^{-1}(z/\rho)$. Thus, a sphere of radius R, centered at the origin, is described by the equation $\rho = R$. The coordinate θ is the same in the spherical and the cylindrical coordinate systems, and, as with cylindrical coordinates, points on the z-axis are singular, in the sense that the θ-coordinate is undefined there. The "north pole" of the sphere, with rectangular coordinates $(x, y, z) = (0, 0, R)$, has $\phi = 0$, and the "south pole," $(x, y, z) = (0, 0, -R)$, has $\phi = \pi$. The great circles that pass through the poles (geographers call them "meridians") are given by the equations $\theta = $ constant. The circles $\phi = $ constant correspond to what are called "parallels," or "lines of latitude" in geography. The only parallel that is a great circle is the equator, with latitude $\phi = \cos^{-1}(0) = \frac{\pi}{2}$. In this we differ from the geographers, for whom the latitude varies from $90° = \frac{\pi}{2}$ radians at the north pole to $-90°$ at the south pole.

Our first task is to present the Laplace operator in terms of spherical coordinates. Recall that the conversion from spherical to rectangular coordinates is given by

$$x = \rho \cos(\theta) \sin(\phi), \quad y = \rho \sin(\theta) \sin(\phi), \quad z = \rho \cos(\phi).$$

Lemma 11.3.2

Let $u(x, y, z)$ be a twice-differentiable function, and put

$$v(\rho, \theta, \phi) = u(\rho \cos(\theta) \sin(\phi), \rho \sin(\theta) \sin(\phi), \rho \cos(\phi)).$$

Then

$$\Delta[u(\rho \cos(\theta) \sin(\phi), \rho \sin(\theta) \sin(\phi), \rho \cos(\phi))] = \Delta_S[v(\rho, \theta, \phi)],$$

where Δ_S is the spherical Laplace operator,

$$\Delta_S[v(\rho, \theta, \phi)] = \frac{\partial^2 v}{\partial \rho^2} + \frac{2}{\rho} \frac{\partial v}{\partial \rho} + \frac{1}{\rho^2} \left(\frac{\partial^2 v}{\partial \phi^2} + \cot(\phi) \frac{\partial v}{\partial \phi} + \csc^2(\phi) \frac{\partial^2 v}{\partial \theta^2} \right).$$

The proof of Lemma 11.3.2 covers much the same ground as the proof of Lemma 11.3.1 and is left as an exercise.

To solve boundary value problems for the Laplace equation $\Delta u = 0$ on a spherical domain by separation of variables, we must determine all harmonic functions of the form

$$u = v(\rho, \theta, \phi) = X(\rho)Y(\theta, \phi). \tag{11.29}$$

A function $Y(\theta, \phi)$ such that the expression in (11.29) satisfies Laplace's equation is called a **spherical harmonic**. Applying the Laplace operator, we have

$$\Delta_S(X(\rho)Y(\theta, \phi)) = \left(X'' + \frac{2}{\rho}X'\right)Y + \frac{1}{\rho^2}X\left(\frac{\partial^2 Y}{\partial \phi^2} + \cot(\phi)\frac{\partial Y}{\partial \phi} + \csc^2(\phi)\frac{\partial^2 Y}{\partial \theta^2}\right)$$

$$= 0.$$

To separate the variable ρ from θ and ϕ, we will divide through by $\rho^{-2}XY$ to obtain

$$\frac{\rho^2 X'' + 2\rho X'}{X} = -\frac{1}{Y}\left(\frac{\partial^2 Y}{\partial \phi^2} + \cot(\phi)\frac{\partial Y}{\partial \phi} + \csc^2(\phi)\frac{\partial^2 Y}{\partial \theta^2}\right). \tag{11.30}$$

The left side of (11.30) depends only on ρ, and the right side is independent of ρ. Therefore, both sides are constant; we will denote this constant by λ. Then (11.30) spawns two equations:

$$\rho^2 X'' + 2\rho X' - \lambda X = 0 \tag{11.31}$$

$$\sin^2(\phi)\frac{\partial^2 Y}{\partial \phi^2} + \sin(\phi)\cos(\phi)\frac{\partial Y}{\partial \phi} + \frac{\partial^2 Y}{\partial \theta^2} + \lambda \sin^2(\phi)Y = 0. \tag{11.32}$$

Equation (11.32), which is the defining equation for spherical harmonics, was obtained by equating the right side of (11.30) to λ, multiplying through by $Y \sin^2(\phi)$, and simplifying. Although it is tempting to solve the Cauchy-Euler equation (11.31) right away, we will work on (11.32) first, since it will determine which values of the parameter λ can be used.

You will have the opportunity to separate variables in (11.32) and look for solutions with the form $Y = g(\theta)h(\phi)$ in Exercise 12 at the end of this section. We will now look for the spherical harmonics that depend only on the latitude ϕ. These are called the axial symmetric harmonics.

With our assumption that Y depends only on ϕ, (11.32) reduces to the following ODE:

$$\sin^2(\phi)\frac{d^2 Y}{d\phi^2} + \sin(\phi)\cos(\phi)\frac{dY}{d\phi} + \lambda \sin^2(\phi)Y = 0. \tag{11.33}$$

Let us substitute $\zeta = \cos(\phi)$ in this equation and put $Y(\phi) = P(\cos(\phi))$. By the chain rule,

$$\frac{dY}{d\phi} = \frac{dP}{d\zeta}\frac{d\zeta}{d\phi} = -\sin(\phi)\frac{dP}{d\zeta},$$

and similarly

$$\frac{d^2 Y}{d\phi^2} = \sin^2(\phi)\frac{d^2 P}{d\zeta^2} - \cos(\phi)\frac{dP}{d\zeta}.$$

Since $\sin^2(\phi) = 1 - \zeta^2$, our substitution results in the ODE

$$(1 - \zeta^2)\left((1 - \zeta^2)\frac{d^2 P}{d\zeta^2} - \zeta\frac{dP}{d\zeta}\right) - (1 - \zeta^2)\zeta\frac{dP}{d\zeta} + \lambda(1 - \zeta^2)P = 0.$$

Let $P' = \frac{dP}{d\zeta}$. We can divide the above ODE through by $(1 - \zeta^2)$ and combine the two $\zeta P'$ terms to obtain

$$(1 - \zeta^2)P'' - 2\zeta P' + \lambda P = 0. \tag{11.34}$$

We recognize (11.34) as Legendre's equation, with the parameter λ. To be of any use, a solution must be defined and finite on the interval $[-1, 1]$, because that is the range of the variable $\zeta = \cos(\phi)$. Thus, λ must be an eigenvalue of (11.34) with boundary conditions $P(\pm 1) < \infty$. Referring to Section 10.5 (see page 513), we see that the eigenvalues are $\lambda_\nu = \nu(\nu + 1)$, where $\nu = 0, 1, 2, \ldots$. The corresponding eigenfunctions can be taken to be the Legendre polynomials $P_\nu(\zeta)$.

If we put $\lambda = \nu(\nu + 1)$ in equation (11.31), the result is the following Cauchy-Euler equation:

$$\rho^2 X'' + 2\rho X' - \nu(\nu + 1) X = 0.$$

The indicial equation is

$$s^2 + s - \nu(\nu + 1) = 0,$$

or $(s - \nu)(s + \nu + 1) = 0$. The exponents are therefore ν and $-(\nu + 1)$. We will eliminate the negative exponent because $\rho^{-(\nu+1)}$ is not defined at the origin. Therefore, X is a constant multiple of ρ^ν. We have thus found a sequence of axial symmetric spherical harmonics, $v_\nu(\rho, \phi) = \rho^\nu P_\nu(\cos(\phi))$, for $\nu = 0, 1, 2, \ldots$.

To solve a boundary value problem on a spherical domain, it will be necessary to express the boundary condition in terms of the functions $P_\nu(\cos(\phi))$. This will be impossible unless the function that specifies the boundary condition is axially symmetric; that is, it depends only on the spherical coordinate ϕ. To express an axial symmetric function $h(\cos(\phi))$ in terms of axial symmetric spherical harmonics, we have to produce an eigenfunction expansion of $h(\zeta)$ in terms of the Legendre polynomials, as in Example 10.5.3 on page 515.

Example 11.3.3 Find a function u that is harmonic on the ball of radius 3, $x^2 + y^2 + z^2 < 9$ and such that on the boundary sphere $x^2 + y^2 + z^2 = 9$ of the ball, $u(x, y, z) = z^2$.

Solution. In spherical coordinates, the boundary condition $u(x, y, z) = z^2$ becomes $v(3, \phi) = (3\cos(\phi))^2$, because when $\rho = 3$, we have $z = 3\cos(\phi)$. We are thus to find a function $v(\rho, \phi)$ satisfying this boundary condition, and such that $\Delta_S(v) = 0$ for all (ρ, ϕ) such that $\rho < 3$.

Let us substitute $\zeta = \cos(\phi)$. Our boundary function in terms of the new ζ is $h(\zeta) = (3\zeta)^2 = 9\zeta^2$. We will express $h(\zeta)$ as the sum of a Legendre series,

$$h(\zeta) = b_0 P_0 + b_1 P_1(\zeta) + b_2 P_2(\zeta) + \cdots.$$

In fact, $b_n = 0$ for $n > 2$ since $h(\zeta)$ is a polynomial of degree 2, and so can be expressed as a linear combination of P_0, $P_1(\zeta)$, and $P_2(\zeta)$.

The three coefficients that are potentially nonzero are

$$b_0 = \frac{\langle P_0, h \rangle}{\langle P_0, P_0 \rangle} = \frac{\int_{-1}^{1} 1 \cdot (9\zeta^2) \, d\zeta}{\int_{-1}^{1} 1 \cdot 1 \, d\zeta} = 3,$$

$$b_1 = \frac{\langle P_1, h \rangle}{\langle P_1, P_1 \rangle} = \frac{\int_{-1}^{1} \zeta \cdot (9\zeta^2) \, d\zeta}{\int_{-1}^{1} \zeta \cdot \zeta \, d\zeta} = 0,$$

and

$$b_2 = \frac{\langle P_2, h \rangle}{\langle P_2, P_2 \rangle} = \frac{\int_{-1}^{1} \frac{1}{2}(-1 + 3\zeta^2) \cdot (9\zeta^2) \, d\zeta}{\int_{-1}^{1} \frac{1}{2}(-1 + 3\zeta^2) \cdot \frac{1}{2}(-1 + 3\zeta^2) \, d\zeta} = 6.$$

Thus, $h(\zeta) = 3P_0 + 6P_2(\zeta)$ and hence $v(3, \phi) = 3P_0 + 6P_2(\cos(\phi))$. The solution $v(\rho, \phi)$ of $\Delta_S v = 0$ will be a linear combination of the solutions P_0 and $\rho^2 P_2(\cos(\phi))$:

$$v(\rho, \phi) = a_0 P_0 + a_2 \rho^2 P_2(\cos(\phi)).$$

Thus, $v(3, \phi) = a_0 P_0 + 9a_2 P_2(\cos(\phi))$. It follows that $a_0 = b_0 = 3$ and $9a_2 = b_2 = 6$, and therefore

$$v(\rho, \phi) = 3 + \frac{2}{3}\rho^2 P_2(\cos(\phi))$$

$$= 3 + \frac{1}{3}\rho^2(-1 + 3\cos^2(\phi)).$$

In terms of the cartesian coordinates x, y, and $z = \rho \cos(\phi)$, we have

$$u(x, y, z) = 3 - \frac{1}{3}(x^2 + y^2 + z^2) + z^2$$

$$= 3 + \frac{2}{3}z^2 - \frac{1}{3}x^2 - \frac{1}{3}y^2. \qquad \square$$

EXERCISES

1. Convert the solution of Example 11.3.1 to cartesian coordinates and show that $u(x, y) = 1 + x^2 - y^2$.

2. A **Newton potential** on $\mathbf{R}^n$ is a harmonic function of the form $u = v(r)$, where r is the distance to the origin. A Newton potential is not required to be defined at the origin but $v(r)$ must be defined for all $r > 0$. Find all Newton potentials on $\mathbf{R}^2$, and show that only the constant Newton potential is defined at the origin.

3. Find all Newton potentials on $\mathbf{R}^3$.

4. Solve the first boundary value problem on a circular domain, with the radius R and the boundary function $\phi(\theta)$ given as follows.

   a. $R = 1$; $\phi(\theta) = \sin \theta \cos \theta$ [Incidentally, in cartesian coordinates $\phi(x, y) = xy$. Notice that $u(x, y) = xy$ is a harmonic function.]

   b. $R = 2$; $\phi(\theta) = \sin^3 \theta$

c. $R = 1$; $\phi(\theta) = |\theta|$ for $-\pi < \theta \leq \pi$

d. $R = 3$; $\phi(\theta) \begin{cases} 1 & \text{if } 0 \leq \theta \leq \pi \\ 0 & \text{if } -\pi < \theta < 0 \end{cases}$

5. It is shown in complex analysis courses that if u is a harmonic function defined on an open domain $\mathcal{D}$, then u is infinitely differentiable on $\mathcal{D}$. Verify that the solution $v(r, \theta)$ of Example 11.3.2 is infinitely differentiable for $r < 2$. Is $v(r, \theta)$ defined for $r > 2$?

6. Let $\mathcal{D}$ denote the domain whose boundary is the unit circle. Find a harmonic function $u = v(r, \theta)$ defined on $\mathcal{D}$ such that when $r = 1$,

    a. $\frac{\partial v}{\partial n} = \theta$ $(-\pi < \theta < \pi)$ b. $u + \frac{\partial v}{\partial n} = \sin^2(4\theta)$

    c. $\frac{\partial v}{\partial n} = S(\theta/\pi)$, where $S(t) = (-1)^{\lfloor t \rfloor}$ is the square wave

7. Let $\mathcal{W}$ denote a sector of a circle with radius R and angle α:

    $$\mathcal{W} = \{(r, \theta) : 0 < r < R, 0 < \theta < \alpha\}.$$

    Suppose a piecewise differentiable function $\phi(\theta)$, defined for $0 \leq \theta \leq \alpha$, is given. Devise a way of finding a harmonic function $u = v(r, \theta)$ on $\mathcal{W}$ such that for all r, $v(r, 0) = v(r, \alpha) = 0$, and $v(R, \theta) = \phi(\theta)$.

8. Find a harmonic function $u = v(r, \theta)$ that is defined on the region bounded by the concentric circles $r = 1$ and $r = 2$, with $v(1, \theta) \equiv 0$ and $v(2, \theta) = 2 \sin \theta$.

9. Find a harmonic function $u = v(r, \theta)$ defined on the semicircular domain $\{(r, \theta) : r < 1, 0 < \theta < \pi\}$, with $v(1, \theta) = \theta(\pi - \theta)$, and $v(r, 0) = v(r, \pi) = 0$ for $0 \leq r < 1$.

10. Show that every polynomial can be expressed as a linear combination of Legendre polynomials.

    ◆ **Hint:** It is sufficient to show for each $n \geq 0$, that ζ^n can be expressed as a linear combination of Legendre polynomials. Show how to express 1, ζ, ζ^2, and ζ^3 in this way, and then complete the proof by using mathematical induction.

11. Prove Lemma 11.3.2.

    ◆ **Hint:** The first thing to do is to show that, in cylindrical

coordinates,

$$\Delta u(r \cos \theta, r \sin \theta, z)$$
$$= \frac{\partial^2 w}{\partial r^2} + \frac{1}{r}\frac{\partial w}{\partial r} + \frac{1}{r^2}\frac{\partial^2 w}{\partial \theta^2} + \frac{\partial^2 w}{\partial z^2},$$

where $w(r, \theta, z) = u(r \cos \theta, r \sin \theta, z)$. Since θ is the same in the spherical and cylindrical coordinate systems, you just have to express the partial derivatives of w with respect to r and z in terms of the partial derivatives of v with respect to ρ and ϕ. To accomplish this, mimic the proof of Lemma 11.3.1.

12. Separate variables in equation (11.32); that is, substitute $Y(\theta, \phi) = g(\theta)h(\phi)$ to obtain

    $$g(\theta)(\sin^2(\phi)h''(\phi) + \sin(\phi)\cos(\phi)h'(\phi))$$
    $$= -g''(\theta)h(\phi) - \lambda g(\theta)h(\phi)$$

    and divide through by $g(\theta)h(\phi)$. Show that $g(\theta)$ must satisfy $y'' + \mu y = 0$ with periodic boundary conditions on $[0, \pi]$, and give an ODE whose independent variable is $\zeta = \cos(\phi)$ that determines h.

13. Solve the first boundary value problem on the unit sphere with boundary condition $u(x, y, z) = 15z - 70z^3 + 63z^5$ when $x^2 + y^2 + z^2 = 1$.

14. Show that if u is a function differentiable on the ball $x^2 + y^2 + z^2 \leq R^2$ and $\mathbf{n}$ is the unit outward normal vector of the boundary sphere of this ball, then

    $$\frac{\partial u}{\partial \mathbf{n}} = \frac{\partial v}{\partial \rho},$$

    where $u(x, y, z) = v(\rho, \theta, \phi)$ in spherical coordinates.

15. Solve the second boundary value problem on the unit sphere with boundary condition

    $$\frac{\partial u}{\partial \mathbf{n}} = 3z^2 - 1.$$

16. A harmonic function $u(x, y, z)$ is defined in the region between the spheres $\rho = 1$ and $\rho = 2$. At every point of the inner sphere, $u = 0$, and at every point of the outer the outer sphere, $u = 2$. Find u.

*11.4   Heat Diffusion in Solids

The diffusion equation is derived from a principle, Fourier's law, that expresses the way that heat flows from hotter regions toward cooler regions. We have used this principle in the one-dimensional setting in Section 10.3, and it applies equally well in three dimensions. Let $u(x, y, z, t)$ denote the temperature at a point (x, y, z) in a solid S at time t. Its gradient,

$$\nabla u = \frac{\partial u}{\partial x}\mathbf{i} + \frac{\partial u}{\partial y}\mathbf{j} + \frac{\partial u}{\partial z}\mathbf{k},$$

is a vector field that indicates, at each point in the solid, the direction of greatest increase of temperature. Fourier's law tells us that heat flows in the direction opposite to the gradient of the temperature.

Heat flow in three dimensions can be confusing, and we have to think carefully about it. What is happening is that heat energy is moving from place to place in the solid, and we need to define a vector field, called the **flux**, that indicates the rate and direction of heat flow. The magnitude of the flux is the rate at which heat crosses a minuscule disk, situated perpendicularly to the flux direction, divided by the area of that disk.

The definition of the flux just given is adequate for intuitive purposes, but it needs a more careful statement if it is to be used in deriving the diffusion equation. Given any unit vector $\mathbf{v}$, let $D_r(\mathbf{v}, x, y, z)$ denote a disk with radius r, centered at (x, y, z), in the plane whose normal vector is $\mathbf{v}$. Let

$$H_r(\mathbf{v}, x, y, z, t) = \text{ the rate that heat flows through } D_r(\mathbf{v}, x, y, z).$$

We take H_r to be negative if the direction of heat flow is opposite to $\mathbf{v}$, and positive if heat flows through $D_r(\mathbf{v}, x, y, z)$ in the $\mathbf{v}$-direction. Then put

$$F(\mathbf{v}, x, y, z, t) = \lim_{r \to 0} \frac{H_r(\mathbf{v}, x, y, z, t)}{\pi r^2}.$$

The **heat flux** at (x, y, z) and time t is the unique vector $\mathbf{f}(x, y, z, t)$ such that for all unit vectors $\mathbf{v}$,

$$F(\mathbf{v}, x, y, z, t) = \mathbf{f}(x, y, z, t) \cdot \mathbf{v}.$$

Fourier's law states that the heat flux vector is proportional to the gradient of the temperature, but pointed in the opposite direction:

$$\mathbf{f}(x, y, z, t) = -K\, \nabla u.$$

The constant K is again the thermal conductivity of the solid in question. Fourier's law leads to the **three-dimensional diffusion equation**,

$$\frac{\partial u}{\partial t} = c\Delta(u), \tag{11.35}$$

where Δ is the three-dimensional Laplace operator. The derivation of (11.35) is outlined in Exercise 9 at the end of this section.

Diffusion in a cylinder

When considering heat diffusion in a solid that has cylindrical shape, it makes sense to use cylindrical coordinates. We therefore make a change of variables $x = r \cos(\theta)$, $y = r \sin(\theta)$ and put

$$v(r, \theta, z, t) = u(r \cos(\theta), r \sin(\theta), z, t)$$

in the diffusion equation (11.35). As we saw in Exercise 11.3.2 on page 541,

$$\Delta[u(r \cos(\theta), r \sin(\theta), z, t)] = \Delta_C(v),$$

where

$$\Delta_C(v)(r, \theta, z) = \frac{\partial^2 v}{\partial r^2} + \frac{1}{r}\frac{\partial v}{\partial r} + \frac{1}{r^2}\frac{\partial^2 v}{\partial \theta^2} + \frac{\partial^2 v}{\partial z^2}.$$

Thus, in cylindrical coordinates, the diffusion equation takes the form

$$\frac{\partial v}{\partial t} = c\Delta_C(v). \tag{11.36}$$

Suppose that the temperature of a cylindrical solid depends only on the radial coordinate r. In other words, the temperature does not vary as the longitudinal coordinate (z) or the angular coordinate (θ) changes. We can then simplify equation (11.36) by dropping the derivatives with respect to θ and z to obtain

$$\frac{\partial v}{\partial t} = c\left(\frac{\partial^2 v}{\partial r^2} + \frac{1}{r}\frac{\partial v}{\partial r}\right). \tag{11.37}$$

Let us look for solutions of this equation with the special form $v(r, t) = h(t)k(r)$. Substituting this for v in equation (11.37) yields

$$h'(t)\,k(r) = c\,h(t)\left(k''(r) + \frac{1}{r}k'(r)\right),$$

which we can divide through by $c\,h(t)\,k(r)$ to obtain

$$\frac{h'(t)}{c\,h(t)} = \frac{1}{k(r)}\left(k''(r) + \frac{1}{r}k'(r)\right). \tag{11.38}$$

The left side of equation (11.38) is independent of r, and the right side is independent of t. This implies that both sides are constant, and we will let $-\lambda$ denote this constant. Then

$$h'(t) = -c\,\lambda h(t) \tag{11.39}$$

and

$$r\,k''(r) + k'(r) + \lambda\,r\,k(r) = 0. \tag{11.40}$$

We will need to obtain the general solution of equation (11.40). If $\lambda = 0$, we can multiply through by r to obtain a Cauchy-Euler equation, $r^2 k'' + r k' = 0$. The indicial equation, $s^2 = 0$, tells us that 0 is a double exponent; therefore, the general solution is $k(r) = C_1 + C_2 \ln(r)$. If λ is positive, we can put $\lambda = \omega^2$. Again we multiply through by r to obtain

$$r^2 k''(r) + r k'(r) + (\omega r)^2 k(r) = 0. \tag{11.41}$$

Let us rescale this equation by changing the independent variable. Put $R = \omega r$, so that

$$\frac{dk}{dr} = \omega \frac{dk}{dR} \quad \text{and} \quad \frac{d^2 k}{dr^2} = \omega^2 \frac{d^2 k}{dR^2}.$$

With this substitution, equation (11.41) becomes

$$\left(\frac{R}{\omega}\right)^2 \omega^2 \frac{d^2 k}{dR^2} + \left(\frac{R}{\omega}\right) \omega \frac{dk}{dR} + R^2 k = 0,$$

which is easily simplified to the form

$$R^2 \frac{d^2 k}{dR^2} + R \frac{dk}{dR} + R^2 k = 0. \tag{11.42}$$

We recognize equation (11.42) as Bessel's equation with parameter 0. Hence its general solution is

$$k = C_1 J_0(R) + C_2 Y_0(R).$$

When we undo the rescaling, we can express the solution as

$$k(r) = C_1 J_0(\omega r) + C_2 Y_0(\omega r).$$

The case where $\lambda < 0$ is reserved for you to work out in Exercise 10.

Boundary conditions for cylindrical domains

Equation (11.40) has a singular point at $r = 0$, which corresponds to the central axis of the cylinder. To avoid it, we will consider a cylindrical shell; that is, the solid whose boundary is a pair of concentric cylinders, with inner radius r_1 and outer radius r_2. Assuming that the temperature at each boundary cylinder is 0—that is, $v(r_1, t) = v(r_2, t) = 0$—any solution of equation (11.37) that has the form $v(r, t) = h(t) k(r)$ will satisfy the boundary conditions if and only if $k(r_1) = k(r_2) = 0$. We will seek solutions of equation (11.40) that satisfy these boundary conditions.

Let $\mathcal{L}(k) = -r^{-1}(r k'' + k')$. A nontrivial solution of equation (11.40) with parameter λ is an eigenfunction of $\mathcal{L}$, and λ is the corresponding eigenvalue. Therefore, our diffusion problem has been reduced to finding the eigenvalues and eigenfunctions of $\mathcal{L}$.

First, if $\lambda = 0$, we have $k(r) = C_1 + C_2 \ln(r)$. Therefore, $k'(r) = C_2/r$ is nonzero unless $C_2 = 0$. Since Rolle's theorem, combined with $k(r_1) = k(r_2) = 0$, guarantees $k'(r) = 0$ somewhere in the interval (r_1, r_2), we conclude $C_2 = 0$. We are left with $k(r) = C_1$, and then the boundary conditions make it obvious that $C_1 = 0$. We conclude that 0 is not an eigenvalue. Rolle's theorem is also the best way to rule out negative eigenvalues. If $k = \phi(r)$ is an eigenfunction belonging to a negative eigenvalue λ, then for some $c \in (r_1, r_2)$, $\phi'(c) = 0$. Multiplying ϕ by -1 if necessary, we can assume $\phi(c) > 0$, and that $\phi(c)$ is the maximum of ϕ on $[r_1, r_2]$. By equation (11.40),

$$c\,\phi''(c) = -\phi'(c) - \lambda\,c\,\phi(c),$$

and it follows that $\phi''(c) > 0$. By the second derivative test, ϕ has a relative *minimum* at $r = c$. Since we know that ϕ has a relative maximum there instead, this is a contradiction and we must conclude that it is impossible for $\mathcal{L}$ to have a negative eigenvalue.

To find positive eigenvalues, we must consider the equations

$$\left.\begin{array}{c} C_1 J_0(\omega\, r_1) + C_2 Y_0(\omega\, r_1) = 0 \\ C_1 J_0(\omega\, r_2) + C_2 Y_0(\omega\, r_2) = 0. \end{array}\right\} \qquad (11.43)$$

Let

$$f(\omega) = \det \begin{bmatrix} J_0(\omega\, r_1) & Y_0(\omega\, r_1) \\ J_0(\omega\, r_2) & Y_0(\omega\, r_2) \end{bmatrix}$$

denote the determinant of the coefficient matrix of the system (11.43). The system has a nontrivial solution (either $C_1 \neq 0$ or $C_2 \neq 0$) if and only if $f(\omega) = 0$, and hence $\lambda = \omega^2$ is an eigenvalue if and only if $f(\omega) = 0$. Finding the zeros of $f(\omega)$ is accomplished by graphing f to approximate the zeros and then using Newton's method to refine the approximation. There will always be an infinite family of zeros since this is a Sturm-Liouville problem (see Section 10.5), but for practical purposes three or four eigenvalues at most are sufficient to produce an accurate solution of the diffusion problem.

Example 11.4.1   Find the first three eigenvalues and corresponding eigenfunctions of

$$\mathcal{L}(k) = -r^{-1}(r\,k'' + k')$$

with boundary conditions $k(1) = k(2) = 0$.

Solution. All of the eigenvalues are positive. $\lambda = \omega^2$ is an eigenvalue if and only if

$$f(\omega) = J_0(\omega)\, Y_0(2\,\omega) - Y_0(\omega)\, J_0(2\,\omega) = 0.$$

We will use a CAS to locate the first three zeros of $f(\omega)$. Figure 11.4 shows a graph of $f(\omega)$, indicating that the first three zeros of f are $\omega \approx 3.2$, 6.3, and 9.4. We will therefore ask the CAS to find solutions in the intervals $(3, 4)$, $(6, 7)$, and $(9, 10)$.

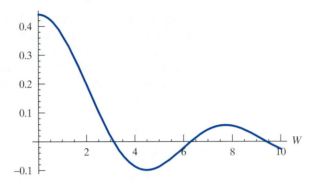

FIGURE 11.4   A graph of $f(\omega)$: See Example 11.4.1.

The CAS determines the first zero is $\omega_1 = 3.123030920$; the second and third are $\omega_2 = 6.273435714$ and $\omega_3 = 9.418207542$. The eigenvalues are

$$\lambda_1 = \omega_1^2 = 9.753322127,$$
$$\lambda_2 = \omega_2^2 = 39.35599566,$$
$$\lambda_3 = \omega_3^2 = 88.70263330.$$

To find the corresponding eigenfunctions, it is sufficient to know that the system (11.43) has a nontrivial solution, because that means the second equation is just a multiple of the first one and can be ignored. To satisfy the first equation, we can put $C_1 = -Y_0(\omega_i)$ and $C_2 = J_0(\omega_i)$. The eigenfunction belonging to λ_i is therefore

$$\phi_i(r) = -Y_0(\omega_i)\, J_0(\omega_i\, r) + J_0(\omega_i)\, Y_0(\omega_i\, r).$$

Figure 11.5 shows graphs of ϕ_1, ϕ_2, and ϕ_3. The graphs can be distinguished by their number of oscillations: $\phi_1(r)$ has no zeros in $(1, 2)$; ϕ_2 has one; and ϕ_3 has two. ❑

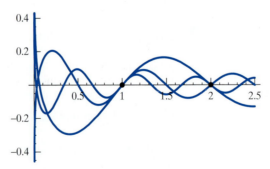

FIGURE 11.5   The first three eigenfunctions of the eigenvalue problem in Example 11.4.1.

To state a diffusion problem in a cylindrical shell properly, we need an initial condition, specifying the initial temperature in the cylindrical shell as a function of r, as well as boundary conditions. Let us specify the initial temperature as $u(0, r) = \psi(r)$, where ψ is continuous on $[r_1, r_2]$ and satisfies the boundary conditions $\psi(r_1) = \psi(r_2) = 0$.

Let $\phi_1(r)$, $\phi_2(r), \ldots$ be the eigenfunctions of $\mathcal{L}$ with boundary conditions $\phi_i(r_1) = \phi_i(r_2) = 0$, and let $\lambda_1, \lambda_2 \ldots$ be the corresponding eigenvalues. Each eigenfunction leads to a solution of (11.37) of the form

$$v_i(r, t) = h_i(t)\phi_i(r),$$

where $h_i(t)$ is a solution of the ODE (11.39) with $\lambda = \lambda_i$:

$$h_i'(t) = -c\,\lambda_i\,h_i(t).$$

Thus

$$h_i(t) = C_i\,e^{-c\lambda_i t},$$

where C_i is a constant. These solutions of the diffusion equation in cylindrical coordinates all satisfy the boundary conditions $v(r_1, t) = v(r_2, t) = 0$.

Any finite sum

$$v(r, t) = C_1\,e^{-c\lambda_1 t}\,\phi_1(r) + C_2\,e^{-c\lambda_2 t}\,\phi_2(r) + \cdots + C_n\,e^{-c\lambda_n t}\,\phi_n(r) \qquad (11.44)$$

is also a solution. We can find a good approximation to the solution of the diffusion problem if we can find constants $C_1, C_2, \ldots, C_n$ such that the initial temperature is approximated as

$$\psi(r) \approx C_1\,\phi_1(r) + C_2\,\phi_2(r) + \cdots + C_n\,\phi_n(r).$$

The following lemma will help to determine these constants.

Lemma 11.4.1

Let λ_i and λ_j be eigenvalues of

$$\mathcal{L}(k) = -r^{-1}\,(r\,k'' + k')$$

with boundary conditions $k(r_1) = k(r_2) = 0$, and let $\phi_i(r)$ and $\phi_j(r)$ be corresponding eigenfunctions. Then

a. If $\lambda_i = \lambda_j$ then $\phi_j(r)$ is a constant multiple of $\phi_i(r)$.
b. If $\lambda_i \neq \lambda_j$ then ϕ_i and ϕ_j are orthogonal, in the following sense:

$$\int_{r_1}^{r_2} \phi_i(r)\,\phi_j(r)\,r\,dr = 0. \qquad (11.45)$$

Proof The proof of statement (a) uses the uniqueness theorem. Let $A_i = \phi_i'(r_1)$ and $A_j = \phi_j'(r_1)$, and put $M = A_j/A_i$. Since $\lambda_j = \lambda_i$, $\phi_j(r)$ satisfies the IVP, $\mathcal{L}(k) = \lambda_i\,k$; $k(r_1) = 0$, $k'(r_1) = M\,A_i$. Since $M\,\phi_i(r)$ obviously satisfies the same IVP, $\phi_j(r) = M\phi_i(r)$ for all $r > 0$.

Statement (b) is proved by integration by parts. Observe that the ODE $\mathcal{L}(k) = \lambda k$ can be written in the form

$$-\frac{d}{dr}\left(r\,\frac{dk}{dr}\right) = \lambda\,r\,k.$$

Thus

$$\lambda_j \int_{r_1}^{r_2} \phi_i(r)\,\phi_j(r)\,r\,dr = -\int_{r_1}^{r_2} \phi_i(r)\,\frac{d}{dr}\left(r\,\frac{d}{dr}\phi_j(r)\right)\,dr.$$

Now integrate by parts to see that

$$\lambda_j \int_{r_1}^{r_2} \phi_i(r)\,\phi_j(r)\,r\,dr = -\phi_i(r)\,r\phi_j'(r)\Big|_{r_1}^{r_2} + \int_{r_1}^{r_2} \phi_i'(r)\,\phi_j'(r)\,r\,dr.$$

Since the boundary conditions say that $\phi_i(r_1) = \phi_i(r_2) = 0$, it follows that

$$\lambda_j \int_{r_1}^{r_2} \phi_i(r)\,\phi_j(r)\,r\,dr = \int_{r_1}^{r_2} \phi_i'(r)\,\phi_j'(r)\,r\,dr.$$

The same calculation with the roles of i and j reversed would show that

$$\lambda_i \int_{r_1}^{r_2} \phi_i(r)\,\phi_j(r)\,r\,dr = \int_{r_1}^{r_2} \phi_j'(r)\,\phi_i'(r)\,r\,dr.$$

Since $\lambda_i \neq \lambda_j$, the orthogonality (11.45) holds.   ∎

Now suppose that the initial temperature function $\psi(r)$ can be expressed as a linear combination of a finite number of eigenfunctions ϕ_j:

$$\psi(r) = C_1\,\phi_1(r) + C_2\,\phi_2(r) + \cdots + C_n\,\phi_n(r).$$

Then for any m in the range $1 \le m \le n$,

$$\int_{r_1}^{r_2} \psi(r)\,\phi_m(r)\,r\,dr = C_1 \int_{r_1}^{r_2} \phi_1(r)\,\phi_m(r)\,r\,dr + \cdots + C_m \int_{r_1}^{r_2} [\phi_m(r)]^2\,r\,dr$$

$$+ \cdots + C_n \int_{r_1}^{r_2} \phi_n(r)\,\phi_m(r)\,r\,dr. \tag{11.46}$$

By Lemma 11.4.1, the only nonzero integral on the right side of (11.46) is

$$\kappa_m = \int_{r_1}^{r_2} [\phi_m(r)]^2\,r\,dr.$$

Furthermore, κ_m is definitely nonzero, since $\phi_m(r)$ is not trivial. Therefore, we can solve (11.46) to obtain

$$C_m = \frac{1}{\kappa_m} \int_{r_1}^{r_2} \psi(r)\,\phi_m(r)\,r\,dr \tag{11.47}$$

If the initial function $\psi(r)$ cannot be expressed as a finite linear combination of eigenfunctions, we have to rely on a result about generalized Fourier series that states that nice functions can be closely approximated by finite linear combinations of eigenfunctions. This statement is intentionally vague, and you are encouraged to consult a text[2] on Fourier series for a more precise one. Thus, we can compute the coefficients C_m by using equation (11.47), for $1 \leq m \leq n$. Larger values n will give greater precision, but $n = 3$ or 4 is more than adequate for most purposes. The solution of the diffusion problem is then given by formula (11.44). Since the λ_i form a rapidly increasing sequence, the exponential factor in

$$C_m \, e^{-c\lambda_m t} \phi_m(r)$$

decays rapidly when m is large. That is why only a few terms are necessary to approximate $u(r, t)$ accurately for $t > 0$.

Example 11.4.2   A cylindrical shell has inner radius 1 and outer radius 2. The material constituting the shell has thermal diffusion coefficient $c = 0.1$. At $t = 0$, the temperature of the shell is $\psi(r) = 100(r-1)(r-2)$. Determine the temperature function $u(r, t)$.

Solution. We have determined the first three eigenfunctions in Example 11.4.1. All we need to do is to calculate the coefficients C_m such that

$$100 \, (1 - r) \, (2 - r) \approx C_1 \, \phi_1(r) + C_2 \, \phi_2(r) + C_3 \, \phi_3(r). \tag{11.48}$$

The solution will then be

$$u(r, t) \approx C_1 \, e^{-0.1\lambda_1 t} \phi_1(r) + C_2 \, e^{-0.1\lambda_2 t} \phi_2(r) + C_3 \, e^{-0.1\lambda_3 t} \phi_3(r). \tag{11.49}$$

The details for computing C_1 will be given, and the same procedure can be repeated to determine C_2 and C_3. We know that

$$C_1 = \frac{1}{\kappa_1} \left(100 \int_1^2 (r-1) \, (r-2) \, \phi_1(r) \, r \, dr \right),$$

where

$$\kappa_1 = \int_1^2 [\phi_1(r)]^2 \, r \, dr.$$

These integrals were evaluated two graphs of the initial function $\psi(r) = 100 \, (r - 1) \, (r - 2)$ and the approximation given by equation (11.48) with the values of $C_1, C_2,$ and C_3 that the CAS computed. You will notice that the graphs are almost identical.

Figure 11.6 displays the graph of the solution of the diffusion equation given by equation (11.49), as a function of (r, t) on the domain $1 \leq r \leq 2, 0 \leq t \leq 4$. The accuracy of this graph is worst for $t = 0$, since the inaccuracy stems from neglected coefficients $C_4, C_5,$ and so on, and the exponential factors associated with these coefficients decay increasingly rapidly. Figure 11.7 shows that the accuracy at $t = 0$ is not very bad.   ❑

---

[2]See *Fourier Series*, by G. P Tolstov, translated by Richard Silverman, Dover Publications, 1962, page 221.

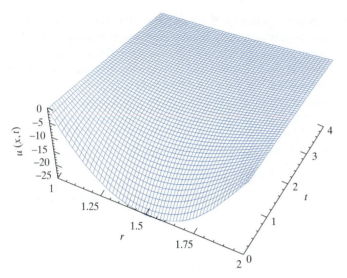

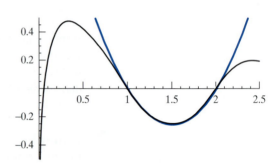

FIGURE 11.6 Solution of the diffusion problem in Example 11.4.2. The spatial scale is on the right; the time scale is on the left. The graph shows that by $t = 4$ the temperature is practically constant.

FIGURE 11.7 The initial condition in Example 11.4.2, $\psi(r) = 100(r-1)(r-2)$, and its approximation by a linear combination of ϕ_1, ϕ_2, and ϕ_3. Although the approximation is extremely good for $1 \le r \le 2$, it fails completely outside this interval.

A solid cylinder

The diffusion problem on a solid cylinder is theoretically more complicated but computationally simpler than the problem on a cylindrical shell. The theoretical complication is due to the singular point that Bessel's equation has at $r = 0$, corresponding to the central axis of the cylinder. Since the singular point is incompatible with Bessel functions of the second kind, it simplifies computation.

The IBVP for the cylinder can be stated as follows. Find a solution $u(r, t)$ of the diffusion equation in cylindrical coordinates,

$$\frac{\partial u}{\partial t} = c \left(\frac{\partial^2 u}{\partial r^2} + \frac{1}{r} \frac{\partial u}{\partial r} \right),$$

that is defined for all (r, t) such that $t \ge 0$, and $0 \le r \le R$, where R is the radius of the cylinder, and that satisfies the boundary condition $u(R, t) = 0$ for all $t \ge 0$ and the initial condition $u(r, 0) = \psi(r)$, where $\psi(r)$ is given and $\psi(R) = 0$.

The procedure is the same as for a cylindrical shell. We look for solutions of the diffusion equation of the form $u(r, t) = h(t) k(r)$. Separating variables shows that there is a constant $\lambda = \omega^2$ such that

$$h(t) = C e^{-c\lambda t}$$

and $k(r)$ is a solution of

$$r k'' + k' + \omega^2 r k = 0 \qquad (11.50)$$

that satisfies $k(R) = 0$. There is apparently only one boundary condition, but since the general solution of (11.50) is $k(r) = C_1 J_0(\omega r) + C_2 Y_0(\omega r)$, and $Y_0(\omega r)$ is unbounded as $r \to 0$, there is a second, implicit boundary condition that can be stated as $k(0) < \infty$. This boundary condition implies that $C_2 = 0$, and hence we can take

$$k(r) = J_0(\omega r),$$

where ω is chosen so that $J_0(\omega R) = 0$. Figure 11.8 displays a graph of the Bessel function J_0. You will notice that it is an oscillating functions, with an infinite sequence $j_{0,1}, j_{0,2}, \ldots$ of zeros. These zeros are important mathematical constants and can be found in published tables. They can also be computed by any CAS.

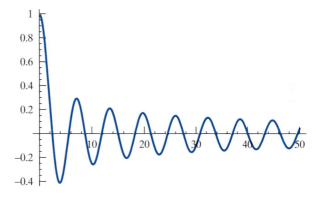

FIGURE 11.8 The Bessel function $J_0(x)$.

The eigenvalues of the operator $\mathcal{L}(k) = -r^{-1}(r\,k'' + k)$ with boundary conditions $\phi(0) < \infty$ and $\phi_i(R) = 0$ are

$$\lambda_i = \left(\frac{j_{0,i}}{R}\right)^2$$

and the corresponding eigenfunctions are $\phi_i(r) = J_0(j_{0,i}\, r/R)$. Hence the solution of the diffusion equation on the solid cylinder, with the temperature at the surface of the cylinder held at 0, and assuming that at all times the temperature can be expressed as a function $u(r, t)$, is

$$u(r, t) = C_1 e^{-c\lambda_1 t} J_0\left(\frac{j_{0,1}}{R}r\right) + C_2 e^{-c\lambda_2 t} J_0\left(\frac{j_{0,2}}{R}r\right) + \cdots.$$

EXERCISES

All computations in these exercises require a CAS.

1. Let $\mathcal{L}(y) = -r^{-1}(r\,y'' + y')$.

    a. Find a solution $y = \eta(r)$ of the BVP $\mathcal{L}(y) = 0$;

$y(r_1) = A$, $y(r_2) = 0$. Is your solution the only one?

    b. The interior wall $r = r_1$ of a cylindrical shell is held at a constant temperature $v(r_1, t) = A$, and the exterior wall $r = r_2$ is held at temperature $v(r_2, t) = 0$.

At $t = 0$ the temperature is $\psi(r) = A\frac{r-r_2}{r_1-r_2}$. Show that $u(r, t) = v(r, t) - \eta(r)$ satisfies the IBVP

$$\frac{\partial u}{\partial t} = c\left(\frac{\partial^2 u}{\partial r^2} + \frac{1}{r}\frac{\partial u}{\partial r}\right);$$

$$u(r_1, t) = u(r_2, t) = 0, \quad u(r, 0) = \psi(r) - \eta(r).$$

c. Put $r_1 = 1$, $r_2 = 2$, $c = 0.03$, and $A = 100$. Plot the function $u(r, t)$ for $0 \le t \le 6$ and $1 \le r \le 2$. How many terms of the generalized Fourier series are nedded for an accurate plot?

2. Use a CAS to compute $j_{0,1}$, $j_{0,2}$, and $j_{0,3}$. Use the result to find the first three eigenvalues of

$$r\, y'' + y' + \lambda r\, y = 0; \quad y(0) < \infty, \quad y(1) = 0$$

and the corresponding eigenfunctions. Graph the eigenfunctions.

3. Show that the eigenfunctions $\phi_n(r)$ of $\mathcal{L}(y) = r^{-1}(r\,y'' + y')$ with boundary conditions $y(0) < \infty$, $y(1) = 0$ satisfy the orthogonality condition

$$\int_0^1 \phi_m(r)\,\phi_n(r)\,r\,dr = 0 \quad \text{for } m \ne n.$$

4. The outer surface of a solid cylinder with radius 1 is held at a temperature $u(1, t) = 0$. Initially, the temperature at distance r from the axis is $u(r, 0) = 100(1 - t)$. The cylinder is made of a substance whose thermal diffusion coefficient is $c = 0.1$. Determine, approximately, the temperature function $u(r, t)$, and produce a three-dimensional plot of its graph.

5. Solve the EVP

$$r\, y'' + y' + \lambda r\, y = 0; \quad y(0) < \infty, \quad y'(1) = 0.$$

Derive an orthogonality relation for the eigenfunctions.

6. Use Fourier's law to derive an appropriate boundary condition for an insulated solid cylinder (heat does not flow through the boundary). Solve the IBVP for the insulated cylinder. Your answer should involve the EVP that was solved in Exercise 5.

7. An insulated cylinder with radius 1 has initial temperature $u(r, 0) = 100(1 - r)$. Assuming that the thermal diffusion coefficient is $c = 0.1$, find, approximately, the temperature function $u(r, t)$, and display a three-dimensional graph of this function.

8. Find the first three eigenvalues and corresponding eigenfunctions of the EVP

$$y'' + \lambda\, t^2\, y = 0; \quad y(0) = y(1) = 0$$

◆ **Hint:** Use Proposition 7.5.7 on page 356.

9. **Derivation of the three-dimensional diffusion equation.**[3] Let S be a solid where the temperature is $u(x, y, z, t)$, and let $\mathbf{f}(x, y, z, t)$ be the heat flux vector.

a. If $\mathcal{R} \subset S$ is a region whose boundary is a smooth surface Σ, show that the rate at which $\mathcal{R}$ loses heat is

$$\int_\Sigma \mathbf{f}(x, y, z, t) \cdot \mathbf{n}(x, y, z)\, dA(x, y, z),$$

where $\mathbf{n}(x, y, z)$ is the unit outward normal vector to the surface Σ at the point (x, y, z) and $dA(x, y, z)$ is the element of area on the surface.

b. Show that the rate at which $\mathcal{R}$ loses heat can also be expressed as

$$-\int_\mathcal{R} C\rho\, \frac{\partial u}{\partial t}\, dV,$$

where C is the specific heat and ρ is the density.

c. Use Fourier's law and the divergence theorem to complete the derivation.

10. Show that when $\lambda = -\omega^2$, the general solution of equation (11.40) is

$$k(r) = C_1\, I_0(\omega r) + C_2\, K_0(\omega r).$$

11.5 Glossary

Boundary value problem There are three prominent boundary value problems that are studied in the context of the Laplace equation: They can be found in this glossary as the Dirichlet problem (first BVP), the Neumann problem (second BVP), and the Robin problem (third BVP).

Diffusion A process by which heat propagates, a cloud of smoke is distributed through a room, and so on. The governing principle is Fourier's law.

Diffusion equation (three dimensional). The PDE that governs heat flow in solids. For a solid with diffusion co-

---

[3]This Exercise is appropriate for you if you have had an advanced calculus course.

efficient c, and temperature $u(t, x, y, z)$,

$$\frac{\partial u}{\partial t} = c \left(\frac{\partial^2 u}{\partial x^2} + \frac{\partial^2 u}{\partial y^2} + \frac{\partial^2 u}{\partial z^2} \right).$$

Dirichlet problem (two-dimensional version) The first boundary problem for the Laplace equation: *Given a plane region $\mathcal{R}$ with boundary curve $\mathcal{C}$, and a function f that is continuous on $\mathcal{C}$, to find a function $u(x, y)$ that is harmonic on $\mathcal{R}$ and continuous on $\mathcal{R} \cup \mathcal{C}$, such that $u(x, y) = f(x, y)$ for all $(x, y) \in \mathcal{C}$.*

Flux A time-dependent vector field $\mathbf{f}(t, x, y, z)$ that indicates the rate and direction of flow of heat, smoke, and so on. If an infinitesimal disk D with area dA located at (x, y, z), and its unit normal vector is $\mathbf{n}$, then at time t, the rate of heat (or smoke) flow through D in the $\mathbf{n}$-direction is $\mathbf{n} \cdot \mathbf{f}(t, x, y, z) \, dA$.

Fourier's Law In the flow of heat, the flux vector is proportional to the gradient of the temperature and directed oppositely to it.

Harmonic function A function $u(x, y)$ that satisfies Laplace's equation.

IBVP Initial-boundary value problem. In the context of the diffusion equation, the initial conditions specify the initial temperature distribution and the boundary conditions specify the temperature or its gradient on the boundary of the region of heat diffusion. In the context of the wave equation, the initial position and velocity profile are specified as the initial conditions, and the boundary conditions specify the motion and position at the endpoints of the vibrating string.

Kinetic energy The energy required to convert a mechanical system (such as a vibrating string) into its present state of motion, starting from the stationary state.

Laplace equation The PDE

$$\frac{\partial^2 u}{\partial x^2} + \frac{\partial^2 u}{\partial y^2} = 0.$$

Mixed boundary conditions Boundary conditions of more than one kind in the same BVP. For example, consider the Laplace equation with Dirichlet boundary conditions on the top and bottom edges of a square, and Neumann boundary conditions on the left and right sides.

Neumann problem The second boundary problem for the Laplace equation: *Given a region $\mathcal{R}$ with a piecewise smooth differentiable boundary curve $\mathcal{C}$ whose outward unit normal*

vector is $\mathbf{n}$, and a function f that is continuous on $\mathcal{C}$, to find a function $u(x, y)$ that is harmonic on $\mathcal{R}$ and has continuous partial derivatives on $\mathcal{R} \cup \mathcal{C}$, such that $\frac{\partial u}{\partial \mathbf{n}}(x, y) = f(x, y)$ for all $(x, y) \in \mathcal{C}$, where $\frac{\partial u}{\partial \mathbf{n}}(x, y)$ is defined.

Newton potential A harmonic function u that depends only on the distance to the origin.

PDE Partial differential equation. A differential equation that includes partial derivatives of an unknown function with respect to two or more independent variables.

Potential energy The energy that must be expended to put a mechanical system (such as a string that is vibrating) into its present configuration.

Robin Problem The third boundary problem for the Laplace equation: *Given a region $\mathcal{R}$ with piecewise differentiable boundary curve $\mathcal{C}$ whose outward unit normal vector is $\mathbf{n}$, and functions f and κ that are continuous on $\mathcal{C}$, and $\kappa > 0$ on $\mathcal{C}$, to find a function $u(x, y)$ that is harmonic on $\mathcal{R}$ and has continuous partial derivatives on $\mathcal{R} \cup \mathcal{C}$, such that $\frac{\partial u}{\partial \mathbf{n}}(x, y) + \kappa(x, y)u(x, y) = f(x, y)$ for all $(x, y) \in \mathcal{C}$, where $\frac{\partial u}{\partial \mathbf{n}}(x, y)$ is defined.*

Separation of variables Finding all solutions of a partial differential equation that have the special form of a product of one-variable functions. For example, if the independent variables are x and t, then the objective is to find all solutions that have the form $u(x, t) = g(x) h(t)$.

Spherical harmonic A function $Y(\theta, \phi)$ such that for some function $X(\rho)$,

$$u(\rho \sin(\phi) \cos(\theta), \rho \sin(\phi) \sin(\theta), \rho \cos(\phi))$$
$$= X(\rho)Y(\theta, \phi)$$

is harmonic.

Tension The force at a point x on a taut string that holds the part of the string on the left of x to the part on the right of x.

Wave equation A partial differential equation that is used as a model of the displacement function for points on a vibrating string (such as a piano string). If $u(x, t)$ is the displacement of the point x on the string at time t, then

$$\frac{\partial^2 u}{\partial t^2} = c^2 \frac{\partial^2 u}{\partial x^2}.$$

The parameter c in the wave equation represents the speed at which disturbances on the string propagate.

11.6 Chapter Review

1. Solve the IBVP,

$$\frac{\partial u}{\partial t} = \frac{\partial^2 u}{\partial x^2} + 2\frac{\partial u}{\partial x};$$

$$u(0, x) = e^{-x}\sin(x/2),$$

$$u(x,0) = u(\pi, t) + u'(\pi, t) = 0.$$

2. An iron rod is 1 meter long. At time 0, the temperature x meters from one end of the rod is $u(x, 0) = 100\sin(2\pi x)$ degrees. The rod is insulated so that heat flows only through the ends and each of the ends is held at a temperature of $0°$. Given that the thermal diffusivity of iron is 2.7×10^{-4} in meter-kilogram-second units, how long will it take until the temperature of the iron ranges from -10 degrees to 10 degrees?

3. A string is stretched between the points 0 and L on the x-axis. It is initially in the equilibrium configuration and is struck at its midpoint, imparting an initial velocity profile of $\delta(x - L/2)$. Show that the velocity profile for $0 < t < L/(2c)$ is

$$\frac{\partial u}{\partial t}(x, t) = \frac{1}{2}\left(\delta\left(x - ct - \frac{L}{2}\right) + \delta\left(x + ct - \frac{L}{2}\right)\right)$$

and describe this profile in words. What is the displacement function? See Section 6.6 for information about the delta function.

4. Show that the real and imaginary parts $u(x, y)$ and $v(x, y)$ of $(x + iy)^n$ are harmonic functions, and then explain why it is that if the power series $f(z) = \sum_{n=0}^{\infty} a_n(z - z_0)^n$ is convergent for $|z - z_0| < R$, where the coefficients a_n are complex numbers, the real and imaginary parts of $f(x+iy)$ are harmonic in the region $\|(x, y) - (x_0, y_0)\| < R$.

Answers

Chapter 1

1.1

1. a. Order 2   b. ODE

3. a. Order 1   b. ODE

5. a. Order 1   b. ODE

7. a. Order 1   b. ODE

9. a. Order 1   b. ODE

11. a. Order 2   b. PDE

13. a. Order 3   b. ODE

15. a. $y = C\sin(t)$ b. $y = C\sin(t)$
     c. $y = 2\sin(t)$ d. $y = -\sin(t)+\cos(t)$

17. $y = 3e^{t^2}$ **19.** Use implicit differentiation.

21. $\frac{d}{dt}\left(\frac{y}{t}\right) = \frac{1}{t}$. Use the equal derivatives theorem.

1.2

3. By 27.726 per thousand

5. The relative growth rate of **B** is 2% per year.

9. $y = \frac{1}{2^{12}}t^{12}$

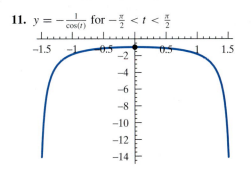

11. $y = -\frac{1}{\cos(t)}$ for $-\frac{\pi}{2} < t < \frac{\pi}{2}$

13. $y = \frac{1}{t}$ for $t > 0$

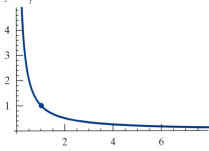

15. $y = 0$ **17.** 0.675 ppb   **19.** 0.556 ppb

1.3

1. $y = \frac{3}{4}t - \frac{3}{16} + \frac{3}{16}e^{-4t}$

3. $y = \frac{1}{2}\ln t e^{-t/2} - \frac{1}{2}\ln 2 e^{-t/2}$ **5.** $y = t e^{-t^2}$

7. a. $y = \frac{1}{10} + \frac{9}{10}e^{-10t}$ b. $y = -\frac{1}{10} + \frac{11}{10}e^{10t}$

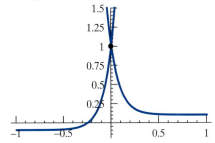

9. a. $y = -\cos(2t)e^{-4t} + e^{-4t}$
     b. $y = -\frac{1}{17}(4e^{-4t}\sin(2t) + \cos(2t)e^{-4t}) + \frac{1}{17}e^{4t}$

11. $y = 1 - \sec(t)$

15. 81°C. The assumptions are that the transmission coefficient in the oven is the same as that outside the oven, and that there are no other sources of heat.

19. 35°C

21. We use the model $T'(t) = -k[T(t) - A(t)] + mH(t)$, with $A(t) = 0$, $k = 0.04$, and

$$mH(t) = \begin{cases} 1 & \text{if } 0 \le t < 4 \\ 0 & \text{if } 4 \le t < 5 \\ 1 & \text{if } 5 \le t < 9 \\ \vdots \end{cases}$$

In simpler terms, $T' + 0.04T = mH(t)$. Since the furnace is on $\frac{4}{5}$ of the time, an intuitive approach would be to use a different model, where $mH(t) = \frac{4}{5} = 0.8$.

The homogeneous solution is $T_h = e^{-0.04t}$. Substitute $T = ve^{-0.04t}$; after canceling this results in $e^{-0.04t} v' = mH(t)$, or $v' = mH(t)e^{0.04t}$. With $mH(t) = 0.8$, we can integrate to get $v = 20e^{0.04t} + C$. Hence $T = e^{-0.04t}v = 20 + Ce^{-0.04t}$. As $t \to \infty$, this solution converges to $20°$C.

A rigorous approach to this problem is as follows. The heat cycles on and off, so the temperature in the house will never be constant. The ODE $T' + 0.04T = mH(t)$ does have a periodic solution $T_p(t)$. If $T(t)$ is any other solution, then $y = T - T_1$ satisfies the ODE $y' = -.04y$; hence $y = Ce^{-0.04t} \to 0$. If we find $T_p(t)$, we have essentially solved the problem.

Let $f_1(t)$ be the solution of the IVP

$$y' = -.04y + 1; \quad y(0) = a,$$

and let $f_2(t)$ be the solution of the IVP

$$y' = -.04y; \quad y(4) = f_1(4).$$

If the initial temperature is $T(0) = a$, then

$$T(t) = \begin{cases} f_1(t) & \text{if } 0 \le t < 4 \\ f_2(t) & \text{if } 4 \le t \le 5. \end{cases}$$

Thus, $T(t)$ will be periodic if $T(0) = T(5)$; that is, $f_2(5) = a$. Using a CAS, I expressed $f_2(5)$ as a function of a, and solved for a. I found $a = 19.592°$C. This is the temperature when the furnace cycles on. The average temperature is

$$\frac{1}{5} \int_0^5 T_p(t)\, dt = \frac{1}{5} \left[\int_0^4 f_1(t)\, dt + \int_4^5 f_2(t)\, dt \right].$$

Turning to the CAS to evaluate these integrals, I found that the average temperature was indeed equal to $20°$C. The drawing shows the graph of the periodic temperature.

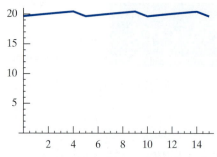

22. a. The temperature obeys the IVP $T'(t) = -k[T(t) - A(t)] + H(T)$; $H(0) = 20$. The ambient temperature $A(t) = 3 + 10\sin\left(\frac{2\pi t}{24}\right)$, and $H(T)$ represents the heat source.

b. $H(T)$ depends on the dependent variable T.

c.

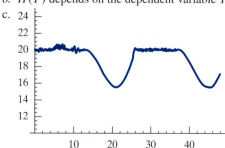

The horizontal segments appear to be jagged because the IVP solver mirrors the actual function of the thermostat, turning on and off repeatedly as the temperature varies above and below $20°$. Different IVP solvers will produce different results because, just as the thermostat cannot go on and off in an infinitesimal time period, the IVP solver takes discrete steps. The IVP solver is, in this respect, a better model than the actual IVP being solved, for this IVP does not actually have a solution! Discrete steps of the kind that the IVP solver takes are necessary.

If you are using a sophisticated IVP solver, it may not function well with this problem, since it will detect that the problem has no solution. If possible, use a solver that employs Euler's method.

23. a. $\frac{5}{29}(5\cos 2t + 2\sin 2t)$; stable

b. $y = -\frac{7}{17}(\cos 4t - 4\sin 4t)$; not stable

c. $y = \cos t - \sin t$; stable

d. $y = -\frac{1}{26}(23\cos t + 11\sin t)$; not stable

e. There is no periodic solution

25. a. $\dfrac{C}{e^{(0.1)t}}$

$$+ \dfrac{(-0.498753)e^{0.1t}\cos(2.t) + (0.0249377)e^{0.1t}\sin(2.t)}{e^{(0.1)t}}$$

b. The solution is

$$y = e^{-t^2} \int e^{t^2}\, dt + Ce^{-t^2}.$$

The CAS found

$$\frac{C}{e^{t^2}} + \frac{\sqrt{\pi}\,\mathrm{Erfi}(t)}{2e^{t^2}}$$

and seems to be using a special function of the form

$$\mathrm{Erfi}(t) = \frac{2}{\sqrt{\pi}} \int e^{t^2}\, dt.$$

c. $\dfrac{C}{t^{12}} + \dfrac{\frac{-t^{12}}{3456} + \frac{t^{12}\ln(t)}{288} - \frac{t^{12}\ln(t)^2}{48} + \frac{t^{12}\ln(t)^3}{12}}{t^{12}}.$

d. $\dfrac{1}{2} - \sqrt{t} + t + \dfrac{C}{e^{2\sqrt{t}}}$

26. a. $y = 1 + Ce^{-t}$

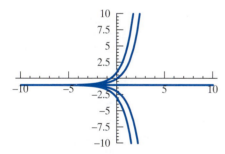

b. $y = -1 + Ce^{t}$

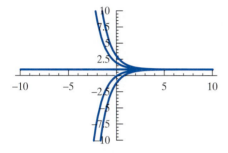

c. $y = t - 1 + Ce^{-t}$

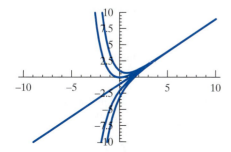

d. $y = t + 1 + Ce^{t}$

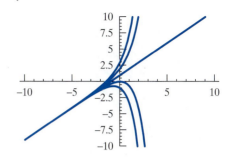

e. $y = \dfrac{C}{e^{t}} + \dfrac{\frac{-(e^{t}\cos(t))}{2} + \frac{e^{t}\sin(t)}{2}}{e^{t}}$

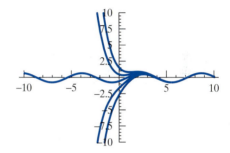

f. $y = \dfrac{C}{e^{(0.05)\,t}}$
$\quad + \dfrac{(-0.997506)\,e^{0.05t}\cos(t) + (0.0498753)\,e^{0.05t}\sin(t)}{e^{(0.05)\,t}}$

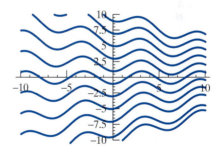

g. $y = \csc(\pi t)(\pi t + C)$

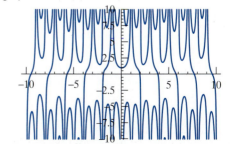

h. $y = C \cos\left(\frac{\pi}{20}t\right) + \frac{20}{\pi}\sin\left(\frac{\pi}{20}t\right)$

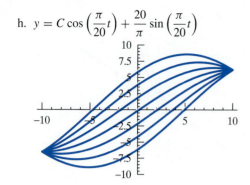

27.

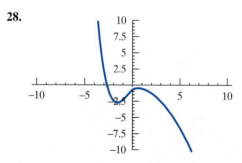

28.

29.

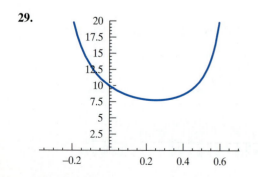

30.

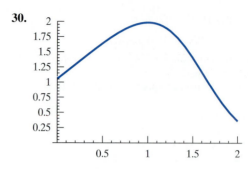

1.4

1. 35, 69, and 138 seconds, respectively. Notice that the time doubles for each increment—can you say why?

5. No more than 13 hours ago.

9. The concentration in tank **A** is $0.09 + 0.03e^{-\frac{4}{15}t}$, and the concentration in tank **B** is $0.09 - 0.03e^{-\frac{4}{15}t}$.

1.5

1. $y = (1 + Ct^{-3})^{1/3}$

3. $y = \pm\dfrac{\sqrt{37}}{\sqrt{2\cos t + 12\sin t + Ce^{6t}}}$

5. $y = \sqrt[4]{4(2t+1)^4 + C(2t+1)^{-2}}$

9. $y = \left(-\dfrac{1}{3} + Ce^{3t}\right)^{-1} + 1$

11. $y = \left(-\dfrac{1}{2t} + \dfrac{C}{t}e^{\frac{2}{3}t^3}\right)^{-1} + t$

1.7

1. $y = -1 + Ce^{t^2/2+t}$.

2. $y = (2 + 2t + t^2) + Ce^t$

3. $y = (t + C)e^t$

4. $y = (\sin t + C)\cos t$

5. $y = \frac{1}{2}e^t + Ce^{-t}$

6. $y = -\frac{1}{5t}e^t(2\cos(2t) - \sin(2t)) + C/t$

7. $y = \frac{5}{14}t^2 + \frac{3}{13}t - \frac{1}{6} + Ct^{-12}$

8. $y = (\ln|\sec(t) + \tan(t)| + C)\cos t$

9. $y = \left(\int e^{-t^2/2}\,dt + C\right)e^{t^2/2}$

10. $y = Ct^5$

11. $y = (t + C)e^{-t}$

12. $y = Ce^{\sin t}$

13. $y = t\ln t + 2t$

14. $y = te^{-t}$

15. $y = t^2$

16. $y = -\cos(3t)e^{-4t} + e^{-4t}$

17. $y = 5t^3\ln t + t^3$

18. $y = (2/3t + 1/3)^{3/2}$

19. $y = -2 + 2e^{t^2/2+3t}$

20. $y = -2$

21. $y = 2\frac{2+t}{\sqrt{(2-t)}} + \sqrt{2\frac{2+t}{2-t}}$

22. $y = t^2 + e^{-\frac{t^2}{2}}$ **23.** $30\frac{\ln 2}{\ln 3}$ years

24. 53 years   **25.** 1844 years

26. $y = \frac{1}{17}(\cos(4t) + 4\sin(4t)) + Ce^{-t}$; the solution with $C = 0$ is the stable periodic solution.

27. 62.5 grams per liter

Chapter 2

2.1

1. a. Separable   b. Separable
   c. Not separable   d. Separable
   e. Separable   f. Not separable

3. a. $-\ln(0.8) = 0.223$
   b. $\frac{\ln(0.5)}{\ln(0.8)} = 3.106$ seconds
   c. $-\frac{0.2}{\ln(0.8)} = 0.8963$ meters
   d. $-\frac{1}{\ln(0.8)} = 4.4814$ meters

5. Put the right side of $mv' = -bv + mg$ equal to 0: $-bv + mg = 0$. The constant solution $v = v_\infty = \frac{mg}{b}$ results. If $v(t) > v_\infty$, then $v' < 0$, and if $0 < v(t) < v_\infty$, then $v' > 0$; hence $v(t) \to v_\infty$ as $t \to \infty$.

9. $y = \pm C\sqrt{t}$; $y = 2\sqrt{t}$ **11.** $y = \pm Ce^{e^t}$; $y = e^{e^t - 1}$

13. $y = -\frac{1}{t+c}$; $y = -\frac{1}{t-1}$ **15.** $\frac{y-1}{y+1} = \pm Ce^{t^2}$; $y = \frac{e^{t^2}+1}{1+e^{t^2}}$

19. a. $\frac{v_\infty^2}{g} \ln \frac{\sqrt{v_0^2 + v_\infty^2}}{v_\infty}$ b. $T_1 = \frac{v_\infty}{g} \arctan \frac{|v_0|}{v_\infty}$
   c. Let T_2 be the time taken to fall from the maximum height to the ground. Then

$$T_2 = \frac{v_\infty}{g} \ln \frac{\sqrt{v_0^2 + v_\infty^2} + |v_0|}{v_\infty}.$$

If the drag force is negligible, then

   a. the maximum height attained by the ball is $\frac{v_0^2}{2g}$.
   b. the time taken to reach that height is $T_1 = \frac{v_0}{g}$.
   c. the time taken to fall from the maximum height to the ground is $T_2 = \frac{v_0}{g}$.
   d. the velocity when the ball hits the ground is v_0

2.2

1. $F(x, y) = x^2 + 5xy + 3x - 2y^2 + 2y$.

3. The exactness condition does not hold.

5. $F(x, y) = e^{3x}y(\ln y - 1) + \frac{y^2}{4} - \frac{y^2 \ln y}{2}$

7. $F(x, y) = \frac{x^4}{4} + \frac{x^2 y^2}{2} - \frac{x^2}{2} + \frac{y^4}{4} + \frac{y^2}{2}$

9. $F(x, y) = x - \frac{y^2}{x+y}$

11. $F(x, y) = x^3 + 3yx^2 + 9y^2x + 17y^3$

13. The exactness condition does not hold.

15. $m = e^{\frac{x^2}{2}}$ is an integrating factor, and $F(x, y) = y^2 e^{\frac{x^2}{2}} + xe^{\frac{x^2}{2}}$ is an integral.

17. $m = e^x$ is an integrating factor, and $F(x, y) = 2xye^x + 3y^2e^x + x^2e^x$ is an integral.

19. $m = y^2$ is an integrating factor, and $F(x, y) = x^2y^3 + \frac{y^4}{2}$ is an integral.

23. $m = e^{\int p(x)\, dx}$ is an integrating factor.

2.3

1.

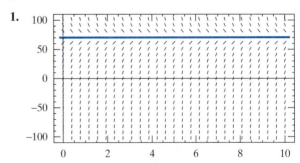

3.

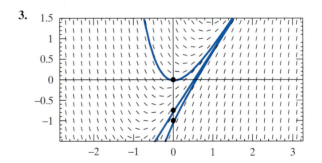

5.

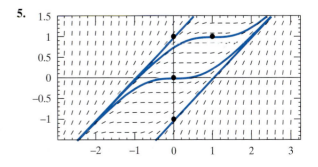

7.

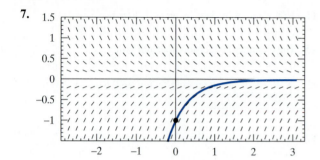

9.

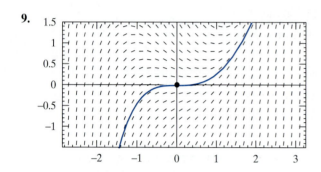

11.

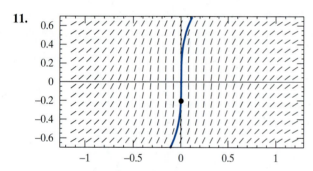

13. At a crossing point, $y' = 0$ and $y'' = 1 > 0$; hence there is a relative minimum by the second derivative test.

15. a. $0, 1, 2, 3, 4, \ldots; \; y_m = m - 1$

b. $1, 2, 4, 8, \ldots; \; y_m = 2^{m-1}$

c. $C, Ck + 1, C((k+1)^2, C(k+1)^3, \ldots;$
$y_m = C(k+1)^{m-1}$

d. $0, 1, 3, 7, \ldots; \; y_m = 2^{m-1} - 1$

17. c. $y_m = Cm$, where C is a constant

d. $y_m = m(m-1)$

19.

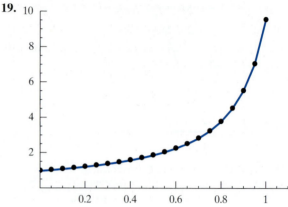

21.

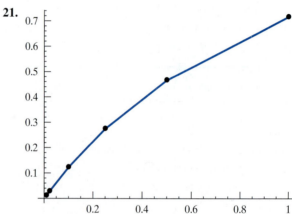

23. Local errors: $0.101691, 0.0771727, 0.013463,$
-0.0506145
Accumulated errors: $0.130573, 0.266751, 0.359802$

25. a. No   b. $0 < h < 0.02$

27. a. In the forward version, using $y_{m+1} = y_m + \sqrt{1 - y_m^2}$, y_m eventually increases to a value bigger than 1, and then $\sqrt{1 - y_m^2}$ is undefined.

b. The plus sign should be used.

c. Forward version: $-0.000166583, 0.199499, 0.297489,$
$0.392961, 0.484917, 0.572373, 0.654372, 0.729989,$
$0.798335, 0.858556, 0.909828, 0.951327, 0.982145,$
1.00096 (terminates)
Backward version: $0.0995037, 0.00113599, 0.29314,$
$0.385415, 0.473494, 0.55657, 0.633914, 0.70485,$
$0.768799, 0.825273, 0.873886, 0.914373, 0.946611,$
$0.970657, 0.986832, 0.99589, 0.999388, 0.999982, 1,$
$1, \ldots$

d. The backward version is the more accurate.

29. $y = \pm 3$ (singular solutions); $y = 3\sin(t + C)$, where C is a constant, is a solution for all values of t such that $y' = 3\cos(t + C)$ is positive.

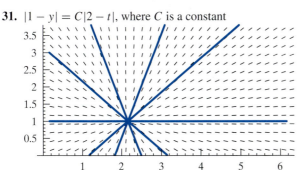

31. $|1 - y| = C|2 - t|$, where C is a constant

2.4

1. $y = -2$, defined on $(-\infty, +\infty)$

3. $y = -\frac{1}{29}(5\cos(5t) - 2\sin(5t) - 5e^{-2t})$ defined on $(-\infty, +\infty)$

5. $y = t^3 - t^2$, defined on $(-\infty, +\infty)$

7. $y = -\frac{1}{2} + e^{t^2}$ defined on $(-\infty, +\infty)$

9. If the graphs intersect, then there would be two solutions to the IVP with initial condition at the intersection point.

11. $y \equiv 1$

13. i. t_0, y_0 could be any real numbers.
    ii. t_0, y_0 could be any real numbers.

15. i. $t_0 \neq k\pi$, where k is an integer
    ii. $t_0 \neq k\pi$, where k is an integer

17. i. $y_0^2 > 4t_0$, and $t_0 \neq 0$ ii. Ø

19. i. $t_0 y_0 > 0$ b. $t_0 y_0 > 0$ and $y_0 \neq 1$

21. i. t_0, y_0 could be any real numbers.   ii. $t_0 - y_0 \neq 0$

2.5

3. $y = 200/(19e^{-4t} + 1)$

5. Let $y(t)$ denote the population, in millions, t years after 1790. Then

$$y(t) = \frac{258.5}{1 + 63.6(0.97017)^t}.$$

11. The function $f(y) = ky\ln\left(\frac{M}{y}\right)$ is defined for $y > 0$ only, but $\lim_{y\to 0} f(y) = 0$. Thus we can say 0 is stationary. The other stationary point at M is verified because $\ln(1) = 0$. If $0 < y < M$, then $\ln(M/y) > 0$, indicating that the population is increasing; for $y > M$ we have $\ln(M/y) < 0$ and the population is decreasing. Thus M is a stable stationary point and 0 is unstable.

15. $\frac{dy}{dt} = ku(t)y(t)$. In a closed system, the number of total molecules is a constant C. Therefore, $u(t) + y(t) = C$, and $\frac{dy}{dt} = ky(t)(C - y(t))$. It's a logistic equation.

17. (a) and (d)

19. Stationary point: 0

21. No stationary points

23. a. This is the solution of first-order autonomous ODE with no stationary points.
    b. This is not a solution of a first-order autonomous ODE.
    c. This is the solution of first-order autonomous ODE with a stationary point $y = 0$.
    d. This is the solution of first-order autonomous ODE with no stationary points.
    e. This is not a solution of a first-order autonomous ODE.

29. a. $H = 0.25$ b. $P_2 = 0.816$
    c.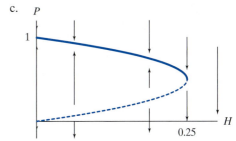

31.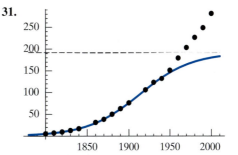

2.7

1.

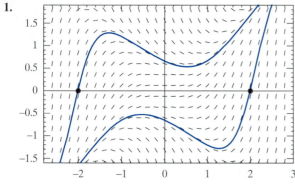

2.

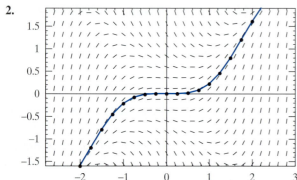

3.

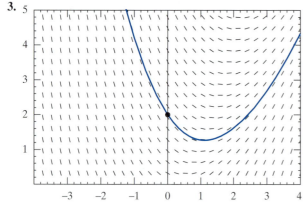

5. a. $y' = 0.03(5t - y^2)$: Graph IV
b. $y' = \sin(y)$: Graph III
c. $y' = \sin(t + y)$: Graph II
d. $y' = -0.01y$: Graph VI
e. $y' = 0.05y(\pi - y)$: Graph V
f. $y' = 0.02(t^2 + y^2)$: Graph I

6. $y = \frac{2}{C - t^2}$, with a singular solution, $y \equiv 0$

7. $y = -\ln(-t - C)$

8. $y = \frac{Ce^t}{1 + Ce^t}$, with a singular solution, $y \equiv 1$

9. $y = \sin(\arcsin t + C)$, with two singular solutions: $y = \pm 1$

10. $y = \cosh(\cosh^{-1}(t) + C)$, with two singular solutions:
$y = \pm 1$

11. $y = \tan(\ln \sqrt{1 + t^2} + C)$

12. $y = (\sqrt{t} + C)^2$, with a singular solution $Y \equiv 0$

13. $y = -\frac{1}{4} \ln(\ln(C \cos(3t))^{4/3}))$

14. a. 166.4 meters   **b.** 72 seconds
c. 236 meters; 68.4 seconds

15. a. For year A, 4.5%. For year B, 4%   **b.** 1,000,000

16.

$y = -\frac{1}{2}$ is stable. $y(t) \to -\frac{1}{2}$.

17.

$y = 0$ is stable. If $y(0) = 30$, then $y(t) \to 9\pi$.

18.

No stationary points; $y(t) \to +\infty$

19.

$y \equiv \frac{5}{3}$ is stable, and $y(t) \to \frac{5}{3}$.

20.

$y \equiv 1$ is stable, and $y(t) \to 1$.

21.

There are no stationary points in $(-\infty, 0)$; $y(t) \to -\infty$.

22. If $y' = \frac{1}{y}$, then $y'' = -\frac{y'}{y^2}$. Thus the signs of y' and y'' are opposite.

23. Notice that $y'' = 1 + y' = 1 + t + y$, and use the second derivative to determine concavity.

24. For $y' = e^y$, we have $y'' = e^{2y}$. Both are positive. Solutions of $y' = e^{-y}$ are increasing and concave down.

25. a.

h	$Y(h)$
0.1	0.423
0.05	0.440
0.025	0.448

b. $Z(h_1) = 2Y_{h_2} - Y_{h_1}$

c. $Z_{0.1} = 0.456638$, and $Z_{0.05} = 0.455814$

25. There will be a solution unless $y_0 = 1$, and a unique solution unless $y_0 = \pm 1$.

26. The right side of the differential equation satisfies a Lipschitz condition with respect to y with Lipschitz constant equal to 2.

Chapter 3

3.1

1. $x = A e^{-t}, y = -A e^{-t}$

3. $x = 2A \sin(3t), y = A \sin(3t) - 3A \cos(3t)$

5. $x = e^{2t}(t + 1 + C), y = e^{2t}(t - 1 + C)$

7. $y' = v, v' = -3v - 4y + t^2$

9. $u' = v, v' = \frac{1+t^2}{uv}$

11. $x = \dfrac{-1}{t + C}, y = -\dfrac{1}{2}t^2 - Ct + D$, where C, D are constants

13. This system is not uncoupled.

15. $x = \pm\sqrt{t^2 + C}, y = De^{\pm\frac{1}{3}(t^2+C)^{3/2}}$, where C and D are constants.

17. This system is not uncoupled.

3.2

1.

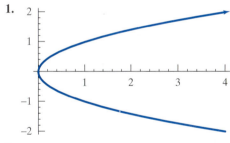

3.

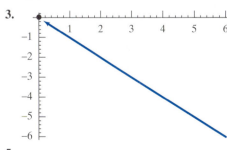

5.

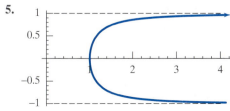

7.

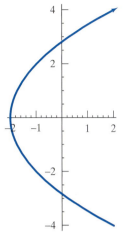

9. Stationary point: $(0, 0)$

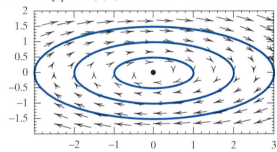

11. Stationary point $(0, 0)$

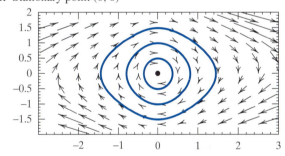

13. a.

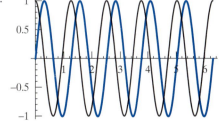

Graphs of $x(t) = \sin(5t)$ and $y(t) = \cos(5t)$

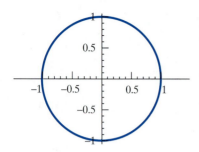

Phase portrait

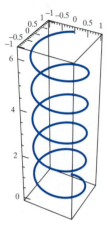

Three-dimensional graph

Phase portrait

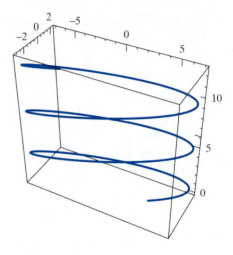

Three-dimensional graph

b.

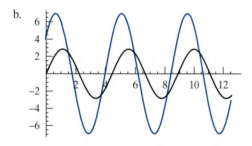

Graphs of $x(t) = 4(\cos(\sqrt{2}t) + \sqrt{2}\sin(\sqrt{2}t))$ and $y(t) = 2\sqrt{2}\sin[(\sqrt{2}t)]$

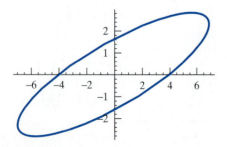

c.

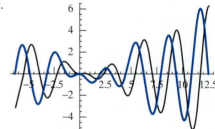

Graphs of $x(t) = \frac{1}{12}(4\cos(t) + (-4 + 6t)\cos(2t) - 3\sin(2t))$ and $y(t) = -\frac{1}{3}((2 + (-2 + 3t)\cos(t))\sin(t))$

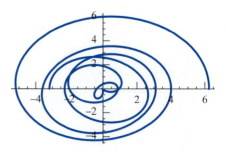

Phase portrait

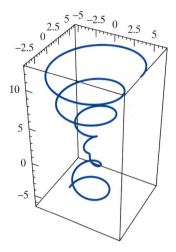

Three-dimensional graph.

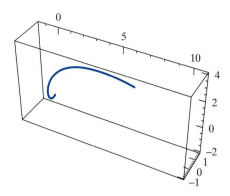

Three-dimensional graph

3.3

1. $y(1) \approx 0.841471076$

3. The phase diagram depends on ϵ:

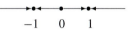

Phase diagram for $\epsilon < 0$ Phase diagram for $\epsilon > 0$

d.

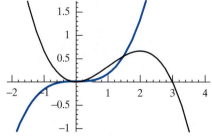

Graphs of $x(t) = \frac{1}{6}t^3$ and $y(t) = \frac{1}{6}(3t^2 - t^3)$

a.

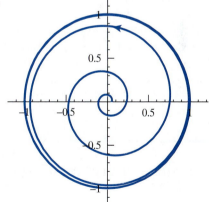

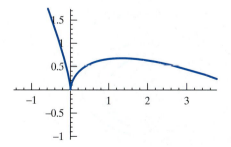

Phase portrait

b. i.

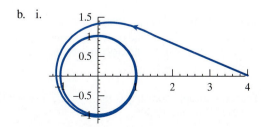

ii.

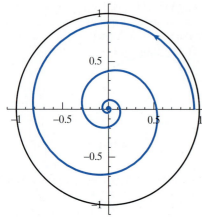

iii.

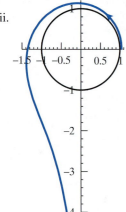

5. The blue curve in the following graph is the actual solution, and the black curve is the Euler approximation using step size 0.05.

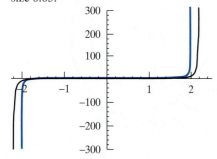

7. a. Solve the ODE $y' = f(y)/1000$ with initial conditions $y(0) = 0, -3, -2, -1$, and 0, over the interval $[-5, 5]$. The zeros of f will be the ordinates of the horizontal asymptotes of these solutions.

b. The roots are approximately 0.193, 1.027, 2.568, 4.900, 8.182, 12.734, and 19.396.

9. a.

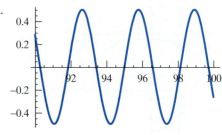

b.

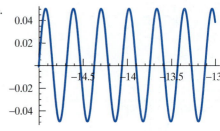

c.

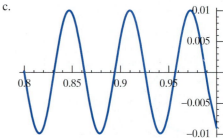

3.4

1. x-nullcline: $(0, 0)$
y-nullclines: $y = \pm x$
Stationary point: $(0, 0)$

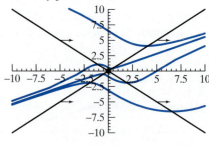

3. x-nullcline: the y-axis
y-nullclines: $y = 4$ and $y = 0$
Stationary points: $(0, 0)$ and $(0, 4)$

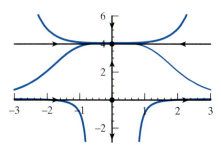

5. x-nullcline: the y-axis
y-nullcline: the x-axis
Stationary point: $(0, 0)$

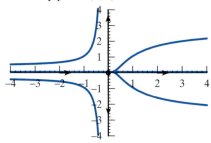

7. x-nullcline: $y = 2 - x$
y-nullcline: $y = x$
Stationary point: $(1, 1)$

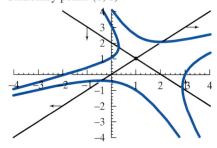

9. $F(x, y) = -2x^{-1/2}y + 2x^{1/2}$

11. $F(x, y) = 5 \ln x - 3 \ln y$

13. $F(x, y) = x^2 y + \frac{y^3}{3}$

15. The system is uncoupled. The solution of the first equation, $x' = 1$, with $x(0) = x_0$ is $x = t + x_0$. Thus the second equation can be written $\frac{dy}{dx} = f(x, y)$; the solution of the system that passes through (x_0, y_0) when $t = 0$ would satisfy the initial condition $y(0) = y_0$, and since $x = t + x_0$ the solution follows the graph of the solution of the IVP

$$\frac{dy}{dx} = f(x, y); \quad y(x_0) = y_0.$$

17. Set $y = \frac{dx}{dt}$ and $y' = \frac{d^2x}{dt^2}$. Then the system

$$x' = y; \quad y(t_0) = y_0$$
$$y' = f(t, x, y); \quad x(t_0) = x_0$$

replaces the given ODE. Let $g(t, x, y) = y$; then $f(t, x, y)$ and $g(t, x, y)$ satisfy Lipschitz conditions. By the existence and uniqueness theorems there is a unique solution of the system. Thus there is a unique solution of the IVP.

19. The system is equivalent to

$$x' = \frac{1}{ad - bc}(ag(t, x, y) - cf(t, x, y))$$

$$y' = \frac{1}{ad - bc}(df(t, x, y) - bg(t, x, y)).$$

If the functions $f(t, x, y)$, $g(t, x, y)$ are continuous and satisfy a Lipschitz condition, then $\frac{1}{ad-bc}(ag(t, x, y) - cf(t, x, y))$ and $\frac{1}{ad-bc}(df(t, x, y) - bg(t, x, y))$ are continuous and satisfy a Lipschitz condition. Thus by the existence and uniqueness theorems the system has a unique solution with the initial conditions.

3.5

1. a. $\frac{a}{b}$

b. The strategy is ineffective.

c. Wait until there are $\approx d$ pests.

3. The fish populations are governed by the system

$$x' = k_1 x(4000 - 4x - y)$$
$$y' = k_2 y(12000 - 7x - 8y).$$

The phase portrait has the configuration (d) of Figure 3.19. The triangles ABC and CDE in the following figure are trapping regions, and orbits within these triangles converge to the point $C = (800, 800)$.

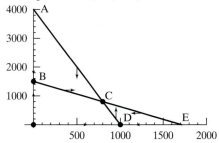

5. The x-nullclines are the lines $x = 0$ and $K - x + By = 0$. The y-nullclines are the lines $y = 0$ and $L + Cx - y = 0$. The system has at most four stationary points. Three of these occur when one or both of the species are extinct. The fourth stationary point, (x_1, y_1) exists if and only if $BC \neq 1$. Then the stationary point is

$$(x_1, y_1) = \left(\frac{BL + K}{1 - BC}, \frac{CK + L}{1 - BC} \right).$$

If $BC > 1$, the fourth stationary point is in the third quadrant. There is no trap, and the species will increase in numbers without bound. If $BC < 1$, we have the fourth stationary point in the first quadrant, and it is stable. All orbits in the first quadrant converge to this point.

The drawing shows the configuration when $BC < 1$. The quadrilateral whose vertices are the four stationary points is a trap.

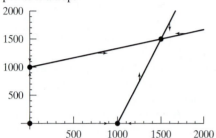

7. The system with fishing can be written as Lotka-Volterra equations:

$$x' = x[(a - R) - by]$$

$$y' = cy\left[x - \left(d + \frac{R}{c}\right)\right].$$

By the result of Exercise 6, the average populations will be $\bar{x} = d + \frac{R}{c}$ and $\bar{y} = \frac{a-R}{b}$. Thus the average prey population increases proportionally to the catch rate, and the average predator population decreases.

3.7

1. $x = te^t$, $y = \frac{1}{e^t(1-t)}$

2. General solution: $x = Ce^{-t}$, $y = De^t$
Integral: $F(x, y) = xy$
Phase portrait:

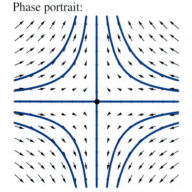

3. General solution: $x = Ce^t$, $y = De^{2t}$
Integral: $F(x, y) = x^2/y$
Phase portrait:

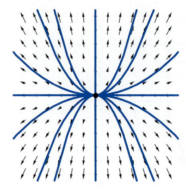

4. $F(y, v) = \frac{v^2}{2} + \int f(y)\,dy$ is an integral for the system $y' = v$, $v' = f(y)$.

a. $F(y, v) = \frac{v^2}{2} - \alpha^2 \cos(y)$

b. $F(y, v) = \frac{v^2}{2} + \frac{\alpha^2 y^2}{2}$ c. $F(y, v) = \frac{v^2}{2} - \alpha^2 y^{-1}$

5. a. $y' = v$, $v' = -y$ b. $y' = v$, $v' = v - t^2 \sin(y)$

6.

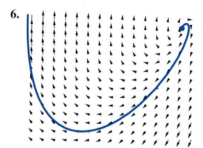

7. $F(x, y) = yx^2 + 2y^2 x$

8. a. $(1, 1)$ b. $(0, 0)$, $(-1, -1)$, $(1, 1)$

c. $(0, 0)$, $(-4, 0)$, $(-5, 1)$

9. a. $x' = x + y - 2$, $y' = x - 3y + 2$

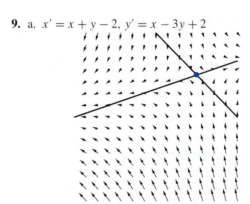

b. $x' = y - x$, $y' = y - x^3$.

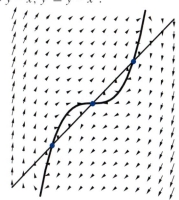

c. $x' = x(x + y + 4)$, $y' = y(x + 5y)$

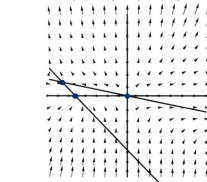

10. In the drawing, the trap is shown in blue.

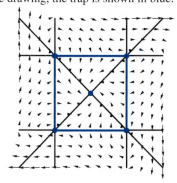

11. a.

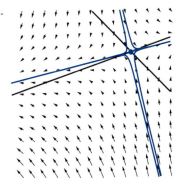

b.

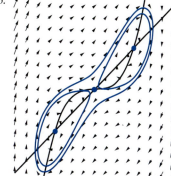

c.

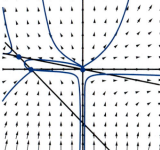

12. Both the linear and nonlinear versions are shown: The linear version is in black; the nonlinear version in blue. The bottom figure is an enlargement of the region inside the rectangle in the top figure, and shows more detail.

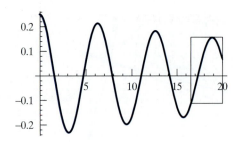

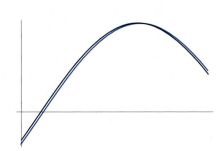

Chapter 4

4.1

1. $\begin{bmatrix} x \\ y \end{bmatrix} = C_1 e^{3t} \begin{bmatrix} 1 \\ 1 \end{bmatrix} + C_2 e^{-t} \begin{bmatrix} 1 \\ -1 \end{bmatrix}$, where C_1, C_2 are constants

3. $x = e^{2t}$, $y = 2e^{3t}$ **5.** $x = e^{3t} + e^{-t}$, $y = e^{3t} - e^{-t}$

7. $x \equiv 1$, $y \equiv 1$ **9.** $e^{-t} \begin{bmatrix} 1 \\ -1 \end{bmatrix}$

11. $\begin{bmatrix} -t^2 + 3t + 1 \\ -2t^2 + 8t + 1 \end{bmatrix}$ **13.** $t \begin{bmatrix} \cosh(t) \\ \sinh(t) \end{bmatrix} + 2 \begin{bmatrix} e^t \\ e^t \end{bmatrix}$

15. $\begin{bmatrix} y' \\ v' \end{bmatrix} = \begin{bmatrix} 0 & 1 \\ -q(t) & -p(t) \end{bmatrix} \begin{bmatrix} y \\ v \end{bmatrix} + \begin{bmatrix} 0 \\ r(t) \end{bmatrix}$

4.2

1. Characteristic roots: $s_1 = -4$ and $s_2 = 1$. Characteristic vectors: $\mathbf{b}_1 = \begin{bmatrix} 0 \\ 1 \end{bmatrix}$ and $\begin{bmatrix} 1 \\ 0 \end{bmatrix}$

3. Characteristic roots: $s_1 = -4$ and $s_2 = 1$; characteristic vectors: $\mathbf{b}_1 = \begin{bmatrix} 1 \\ -5 \end{bmatrix}$ and $\mathbf{b}_2 = \begin{bmatrix} 1 \\ 0 \end{bmatrix}$

5. This matrix has no real characteristic roots.

7. (c) Every nonzero vector is a characteristic vector in this case.

9. $c_1 e^t \begin{bmatrix} -1 \\ 1 \end{bmatrix} + c_2 e^{4t} \begin{bmatrix} 2 \\ 1 \end{bmatrix}$ **11.** $c_1 \begin{bmatrix} 1 \\ 0 \end{bmatrix} + c_2 \begin{bmatrix} t \\ 1 \end{bmatrix}$

13. $c_1 e^{3t} \begin{bmatrix} 1 \\ 1 \end{bmatrix} + c_2 e^{-t} \begin{bmatrix} 1 \\ -1 \end{bmatrix}$ **15.** $e^{-2t} \begin{bmatrix} t \\ 1 - t \end{bmatrix}$

17. $e^{3t} \begin{bmatrix} 2t - 1 \\ 2t - 2 \end{bmatrix}$

4.3

1. $\dfrac{1}{5}(2 - i)$

3. $1, \dfrac{1}{2} + i\dfrac{\sqrt{3}}{2}, -\dfrac{1}{2} + i\dfrac{\sqrt{3}}{2}, -1, -\dfrac{1}{2} - i\dfrac{\sqrt{3}}{2},$ and $\dfrac{1}{2} + i\dfrac{\sqrt{3}}{2}$

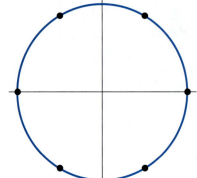

5. $|e^{\lambda + i\omega}| = e^\lambda |e^{i\omega}|$
$= e^\lambda |\cos(\omega) + i \sin(\omega)|$
$= e^\lambda \sqrt{\cos^2(\omega) + \sin^2 \omega} = e^\lambda$

7. $e^{-t/2} \begin{bmatrix} -c_1 \sin(t/2) + c_2 \cos(t/2) \\ c_1 \cos(t/2) + c_2 \sin(t/2) \end{bmatrix}$

9. $\begin{bmatrix} (c_1 + c_2) \cos(4t) + (c_2 - c_1) \sin(4t) \\ c_1 \cos(4t) + c_2 \sin(4t) \end{bmatrix}$

4.4

1. $\mathcal{X}(t) = \begin{bmatrix} e^t & e^{2t} \\ -e^t & e^{2t} \end{bmatrix}$

3. $\mathcal{X}(t) = \begin{bmatrix} 2\cos(4t) & 2\sin(4t) \\ \cos(4t) + 2\sin(4t) & \sin(4t) - 2\cos(4t) \end{bmatrix}$

5. a. $\mathbf{v}(t) = \begin{bmatrix} e^t & e^{2t} \\ -e^t & e^{2t} \end{bmatrix} \mathbf{c} + \begin{bmatrix} -e^t \\ -e^t \end{bmatrix}$

  b. $\mathbf{v}(t) = \begin{bmatrix} e^t & e^{2t} \\ -e^t & e^{2t} \end{bmatrix} \mathbf{c} + \begin{bmatrix} te^t \\ -te^t \end{bmatrix}$

  c. $\mathbf{v}(t) = \begin{bmatrix} e^t & e^{2t} \\ -e^t & e^{2t} \end{bmatrix} \mathbf{c} + \begin{bmatrix} (t^2/4 - t/2 - 1/2)e^t \\ (-t^2/4 - t/2 - 1/2)e^t \end{bmatrix}$

7. $\mathbf{v}(t) = \begin{bmatrix} 2\cos(4t) & 2\sin(4t) \\ \cos(4t) + 2\sin(4t) & \sin(4t) - 2\cos(4t) \end{bmatrix} \mathbf{c}$
$+ \begin{bmatrix} -16t \cos(4t) - 8t \sin(4t) - \cos(4t) - 2\sin(4t) \\ -20t \sin(4t) - \frac{5}{2} \cos(4t) \end{bmatrix}$

9. a. $\begin{bmatrix} \frac{-7}{2} t \\ \frac{1}{2} t \end{bmatrix}$ b. $\begin{bmatrix} t^3 \ln|t| \\ -t^3 \ln|t| \end{bmatrix}$

11. $(\mathcal{X}C)' = \mathcal{X}'C = (A(t)\mathcal{X})C = A(t)(\mathcal{X}C)$.

4.5

1. $e^{At} = \begin{bmatrix} 1 & t & \frac{1}{2}t^2 & \frac{1}{6}t^3 \\ 0 & 1 & t & \frac{1}{2}t^2 \\ 0 & 0 & 1 & t \\ 0 & 0 & 0 & 1 \end{bmatrix}$

7. $A^2 = \begin{bmatrix} 0 & 0 & ac \\ 0 & 0 & 0 \\ 0 & 0 & 0 \end{bmatrix}$, and $A^k = 0$ for $k \geq 3$

a. $e^{At} = \begin{bmatrix} 1 & at & bt + \frac{1}{2}act^2 \\ 0 & 1 & ct \\ 0 & 0 & 1 \end{bmatrix}$

b. $e^{(A+\lambda I)t} = \begin{bmatrix} e^{\lambda t} & ate^{\lambda t} & (bt + \frac{1}{2}act^2)e^{\lambda t} \\ 0 & e^{\lambda t} & cte^{\lambda t} \\ 0 & 0 & e^{\lambda t} \end{bmatrix}$

$x = e^{-2t}(3 + t + 6t^2)$, $y = -4te^{-2t}$, and $z = e^{-2t}$

9. $\begin{bmatrix} \cos(t) + \sin(t) & \sin(t) \\ -2\sin(t) & \cos(t) - \sin(t) \end{bmatrix}$

11. $\frac{1}{\sqrt{2}}e^t \begin{bmatrix} \sqrt{2}\cos(\sqrt{2}t) & 2\sin(\sqrt{2}t) \\ -2\sin(\sqrt{2}t) & \\ -3\sin(\sqrt{2}t) & 2\sin(\sqrt{2}t) \\ & +\sqrt{2}\cos(\sqrt{2}t) \end{bmatrix}$

13. $e^{-t} \begin{bmatrix} -2t + 1 & -2t \\ 2t & 2t + 1 \end{bmatrix}$

15. $\begin{bmatrix} (t+1)e^t & te^t & 0 \\ -te^t & e^t - te^t & 0 \\ 0 & 0 & e^{-t} \end{bmatrix}$

17. $(e^{t\mathcal{D}}f)(x) = f(x + t)$

4.7

1. $x = \frac{1}{2}e^{3t} + \frac{1}{2}e^t$, $y = \frac{1}{2}e^{3t} - \frac{1}{2}e^t$

2. a. $\begin{bmatrix} \cos(3t) - 3\sin(3t) & \sin(3t) + 3\cos(3t) \\ 5\cos(3t) & 5\sin(3t) \end{bmatrix}$

b. $\begin{bmatrix} -e^t\sin(t) & e^t\cos(t) \\ e^t\cos(t) & e^t\sin(t) \end{bmatrix}$

c. $\begin{bmatrix} 2e^t & 2te^t \\ 0 & e^t \end{bmatrix}$ d. $\begin{bmatrix} e^t & 2e^{2t} & \frac{9}{2}e^{3t} \\ 0 & e^{2t} & 3e^{3t} \\ 0 & 0 & e^{3t} \end{bmatrix}$

3. $x = -\frac{1}{2} + \left(\frac{1}{2}\sin(2t) - \frac{1}{4}\cos(2t)\right)\ln\left(\frac{1 + \sin(2t)}{\cos(2t)}\right)$

$\qquad + C_1[2\sin(2t) - \cos(2t)] + C_2[\sin(2t) + 2\cos(2t)]$

$y = \frac{1}{4}\cos(2t)\ln\left(\frac{1 + \sin(2t)}{\cos(2t)}\right) + C_1\cos(2t) + C_2\sin(2t)$

4. $e^{At} = V \cdot E(t) \cdot V^{-1}$

Chapter 5

5.1

1. (a), (b), and (d) are linear.

2. a. The singular points are $t = \pm 1$.

c. Divide through by the coefficient of y'' to obtain $y'' - \frac{t}{\cos(t)}y' + \frac{t^2 - 1}{\cos(t)}y = \frac{1}{\cos(t)}$. The singular points are $t = (k + \frac{1}{2})\pi$, where k is a integer.

3. a. Substitute $t = 0$, $y = \phi(0)$, $y' = \phi'(0)$. and $y'' = \phi''(0)$ in the ODE:

$$0\phi''(0) + \phi'(0) + 0\phi(0) = 0$$

b. Differentiating the ODE results in

$$ty''' + 2y'' + ty' + y = 0$$

Therefore, $2y''(0) + y(0) = 0$. If $y(0) > 0$ then $y''(0) < 0$; hence there is a relative maximum at $t = 0$ by the second derivative test.

4. a. $(-1, 1)$ **6.** $(\frac{1}{2}\pi, \frac{3}{2}\pi)$ **8.** $(\frac{1}{2}\pi, 2)$

6. If $y = t^n$ with $n \geq 2$, then $y(0) = y'(0) = 0$. If 0 is an ordinary point then the only solution with $y(0) = y'(0) = 0$ is $y \equiv 0$.

5.2

1. Let $f(t)$ be a polynomial of degree no more than 3. Then $\frac{d^n}{dt^n}f(t) \equiv 0$ for $n > 3$, so by Taylor's theorem,

$$f(t) = f(1) \cdot 1 + f'(1) \cdot (t - 1) + \frac{1}{2}f''(1) \cdot (t - 1)^2$$
$$+ \frac{1}{6}f'''(1) \cdot (t - 1)^3.$$

3. $\cos\left(t - \frac{\pi}{4}\right) = \frac{\sqrt{2}}{2}\cos(t) + \frac{\sqrt{2}}{2}\sin(t)$

5. $t = -\frac{3}{4}t^2 + (t + 1)^2 - \frac{1}{4}(t + 2)^2$

7. $S = \{1, t\}$ **9.** $y = C_1e^{t/4} + C_2e^{-t}$

11. $y = 5t^3 - 18t$

13. The two functions in S have the form $f(t) = \sin(4t - k)$. Every function of this form is a solution of the ODE, since $f''(t) = -4f(t)$. A fundamental set of solutions of a homogeneous linear second-order ODE consists of two solutions that are linearly independent. Neither of these functions is a constant multiple of the other, so they are independent.

15. All solutions of the ODE are linear polynomials. Thus the functions in S are solutions. A fundamental set of solutions of a homogeneous linear second-order ODE consists of two solutions that are linearly independent. Neither of these functions is a constant multiple of the other, so they are independent.

17. If $y = 1$, then $y' = 0$ and $(ty')' = 0$. If $y = \ln(t)$, then $y' = 1/t$ and $(ty')' = (1)' = 0$. Therefore, both functions in S are solutions. A fundamental set of solutions of a homogeneous linear second-order ODE consists of two solutions that are linearly independent. Neither of these functions is a constant multiple of the other, so they are independent.

19. If $y = t$, then $y' = 1$ and $y'' = 0$. Thus $t^2 y'' - ty' + y = 0 - t + t = 0$. If $y = t \ln(-t)$, then $y' = \ln(-t) + 1$ and $y'' = 1/t$. Thus $t^2 y'' - ty' + y = t - t \ln(-t) - t + t \ln(-t) = 0$. Therefore, both functions in S are solutions. A fundamental set of solutions of a homogeneous linear second-order ODE consists of two solutions that are linearly independent. Neither of these functions is a constant multiple of the other, so they are independent.

20. a. $t_0 = 0$ and $a + b \neq 0$ c. $t_0 = \pi/2$, $-a + b \neq 0$

21. If there were a constant c such that $t^3 = c|t|^3$, then by taking $t = 1$, we find $c = 1$, while we would find $c = -1$ if we took $t = -1$. Thus there is no such constant, and the functions are independent. The derivatives of the two functions are $3t^2$ and $3t|t|$, respectively, so $W[t^3, |t|^3] = t^3 \cdot 3t|t| - |t|^3 \cdot 3t^2$. If $t \geq 0$, then $W[t^3, |t|^3] = 3t^5 - 3t^5 = 0$, and if $t < 0$, then $W[t^3, |t|^3] = -3t^5 - (-3t^5) = 0$; hence $W[t^3, |t|^3] \equiv 0$. This contradicts nothing; it merely implies that t^3 and $|t|^3$ cannot both be solutions of the same homogeneous linear second-order ODE.

23. a. If ϕ_1 and ϕ_2 both vanish at the same point $t_0 \in (c, d)$, then $W[\phi_1(t_0), \phi_2(t_0)] = 0$. Hence ϕ_1 and ϕ_2 are linearly dependent, and cannot be a fundamental set of solutions.

    c. If ϕ_1 and ϕ_2 both have an inflection point at $t_0 \in (c, d)$, then $\phi_1''(t_0) = \phi_2''(t_0) = 0$. Let $A = p(t_0)$ and $B = q(t_0)$. Substituting $t = t_0$, $y = \phi_1(t_0)$, $y' = \phi_1'(t_0)$, and $y'' = 0$ in the ODE, we have

$$A\phi_1'(t_0) + B\phi_1(t_0) = 0.$$

    Similarly,

$$A\phi_2'(t_0) + B\phi_2(t_0) = 0.$$

$(A, B) = (0, 0)$ is the only pair of numbers that satisfies both equations, because $W[\phi_1, \phi_2](t_0) \neq 0$.

25. Since $W(t) = \phi_1(t)\phi_2'(t) - \phi_2(t)\phi_1'(t)$, differentiation of both sides and simplifying yields

$$W'(t) = \phi_1(t)\phi_2''(t) - \phi_2(t)\phi_1''(t).$$

Now substitute $\phi_1''(t) = -p(t)\phi_1'(t) - q(t)\phi_1(t)$, and $\phi_2''(t) = -p(t)\phi_2'(t) - q(t)\phi_2(t)$. After canceling, the result is

$$W'(t) = -\phi_1(t)p(t)\phi_2'(t) + \phi_2(t)p(t)\phi_1'(t)$$
$$= -p(t)W(t).$$

27. $W = \dfrac{7}{1 - t^2}$

5.3

1. Characteristic roots: ± 2
   General solution: $y = C_1 e^{2t} + C_2 e^{-2t}$

3. Double characteristic root: 0
   General solution: $y = C_1 t + C_2$

5. Characteristic roots: $\pm 5i$
   General solution: $y = C_1 \cos(5t) + C_2 \sin(5t)$

7. Double characteristic root: -2
   General solution: $y = e^{-2t}(C_1 t + C_2)$

9. Characteristic roots: $-1 \pm \sqrt{3}i$
   General solution: $y = e^{-t}(C_1 \cos(\sqrt{3}t) + C_2 \sin(\sqrt{3}t))$

11. Double characteristic root: 7
    General solution: $y = e^{7t}(C_1 + C_2 t)$

13. Characteristic roots: $\dfrac{1}{2}, -1$
    General solution: $y = C_1 e^{t/2} + C_2 e^{-t}$

15. Characteristic roots: $-\dfrac{1}{2}, -\dfrac{1}{6}$
    General solution: $y = C_1 e^{-t/2} + C_2 e^{-t/6}$

17. Let $A = \begin{bmatrix} 0 & 1 \\ -q & -p \end{bmatrix}$. Then $\mathrm{tr}(A) = -p$, $\det(A) = q$. Thus the characteristic equation of the equivalent system is $s^2 + ps + q$. It's the same as the characteristic equation of the ODE.

19. a. $W[e^{r_1 t}, e^{r_2 t}] = \det \begin{bmatrix} e^{r_1 t} & e^{r_2 t} \\ r_1 e^{r_1 t} & r_2 e^{r_2 t} \end{bmatrix}$
    $= (r_2 - r_1)e^{(r_1 + r_2)t} \neq 0$
    Thus $e^{r_1 t}$ and $e^{r_2 t}$ are linearly independent.

    c. $W[e^{r_1 t}, te^{r_1 t}] = \det \begin{bmatrix} e^{r_1 t} & te^{r_1 t} \\ r_1 e^{r_1 t} & e^{r_1 t} + r_1 te^{r_1 t} \end{bmatrix}$
    $= e^{2r_1 t} \neq 0$
    Thus $e^{r_1 t}$ and $te^{r_1 t}$ are linearly independent.

21. $\phi(t)$ is a solution of the ODE, $y'' - 2p\,y' + (p^2 + q^2)y = 0$. If $\phi(t_0) = \phi'(t_0) = 0$ for some t_0, then, by Corollary 5.1.1 on page 188, $\phi(t) \equiv 0$. Thus $\phi(0) = C_1 = 0$ and $\phi'(0) = pC_1 - qC_2 = 0$. It follows that $C_1 = C_2 = 0$, because $q \neq 0$.

5.4

1. a. $y = \dfrac{40}{3} - \dfrac{40}{3}e^{-1.5t}$ The displacement never reaches a maximum, but $\lim_{t \to \infty} y(t) = \frac{40}{3}$ meters.

  c. $y = 20te^{-3t/4}$. The maximum displacement $y = \frac{80}{3e} \approx$ 9.81012 meters is reached when $t = 4/3$.

2. a. $x = 0.1\cos(50t)$

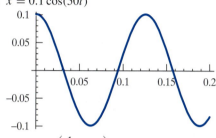

  c. $x = e^{-50t}\left(\dfrac{1}{10} + 5t\right)$

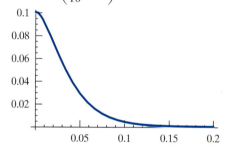

3. (a), underdamped; (c), critically damped

5. $\bar{\psi}(t) - \psi(t)$ is a solution of the associated homogeneous ODE $a_2 y'' + a_1 y' + a_0 y = 0$. Because it is stable, $\lim_{t \to \infty}(\bar{\psi}(t) - \psi(t)) = 0$.

7. $x = 100e^{-t}\sin(t)$

9. Suppose $\phi'(t_1) > 0$. Then there is a number t^* such that $\phi_1(t^*) < 0$ because $\lim_{t \to \infty} \phi(t) = 0$, due to the damping. By the intermediate value theorem, there is $\phi'(t_2) = 0$ for some $t_2 \in (t_1, t^*)$. A similar proof applies if $\phi'(t_1) < 0$; and, if $\phi'(t_1) = 0$, then $\phi(t) \equiv 0$ by the uniqueness theorem, and there is nothing to prove.

  Turning to the second derivative, notice that differentiating both sides of

$$m\phi''(t) + b\phi'(t) + k\phi(t) = 0$$

shows that $\phi'(t)$ is also a solution. Since we know $\phi'(t_2) = 0$, it follows by what we have just proved that $\phi''(t_3) = 0$ for some $t_3 > t_2$.

10. a. $t = 0.2$ seconds

  c. $y(1.2) = -e^{-1.2} \approx 0.301194$ meters

11. b. Never

12. a. $t = \arctan(0.2) \approx 0.197396$

  c. $y(\arctan(3/2)) \approx -0.269886$ meters

13. 1 newton $=$ 1 kilogram meter (per second)^2. Thus 1 damping unit is equal to $\dfrac{1\ \text{newton}}{1\ \text{meter per second}}$, or

$$\dfrac{1\ \text{kilogram meter (per second)}^2}{1\ \text{meter per second}} = 1\ \text{kilogram per second}.$$

15. $\dfrac{d^2 y}{dT^2} + \dfrac{dy}{dT} + \dfrac{mk}{b^2}y = 0$

5.5

1. The method applies to equations (a) and (c) only.

3. $y_p = \dfrac{1}{12}e^{2t}$ **5.** $y_p = -\dfrac{1}{2}te^{-t}$

7. $y_p = -\dfrac{1}{5}(\sin(2t) + 2\cos(2t))$ **9.** $y_p = \dfrac{1}{2}t^2 e^{-t}$

11. $y = -\dfrac{1}{2}e^{2t} + C_1 e^{(2+\sqrt{2})t} + C_2 e^{(2-\sqrt{2})t}$

13. $y = 1 + \dfrac{1}{2}t^2 e^t + e^t(C_1 + C_2 t)$

15. $y = -\dfrac{1}{2}e^{2t} + \dfrac{1}{4}e^{(2+\sqrt{2})t} + \dfrac{1}{4}e^{(2-\sqrt{2})t}$

17. $y = \dfrac{1}{6}t\sin(3t) + \dfrac{2-\pi}{72}\cos(3t) - \dfrac{1}{36}\sin(3t)$

19. $y = \left(\dfrac{1}{2}(t - \ln(2)) + \dfrac{15}{32}\right)\cosh(t) - \dfrac{25}{32}\sinh(t)$

21. $y = -4\cos(t) + 2\sin(t) + e^{-t}(4\cos(t) - 4\sin(t))$

23. a. If $y = t^p e^{kt}$ then $y' = e^{kt}(kt^p + pt^{p-1})$ and $y'' = e^{kt}(k^2 t^p + 2kpt^{p-1} + p(p-1)t^{p-2})$. Thus

$$\mathcal{L}(t^p e^{kt}) = a_2 e^{kt}(k^2 t^p + 2kpt^{p-1} + p(p-1)t^{p-2})$$
$$+ a_1 e^{kt}(kt^p + pt^{p-1}) + a_0 t^p e^{kt}$$
$$= t^p e^{kt}(a_2 k^2 + a_1 k + a_0)$$
$$+ pt^{p-1}e^{kt}(2ka_2 + a_1)$$
$$+ a_2 p(p-1)t^{p-2}e^{kt}$$
$$= (t^p f(k) + pt^{p-1} f'(k)$$
$$+ \dfrac{p(p-1)}{2}t^{p-2}f''(k))e^{kt}$$

  b. $y = e^{st}(C_1 + C_2 t) + \displaystyle\sum_{i=0}^{n} \dfrac{c_i t^{i+2}}{a_2(i+1)(i+2)}e^{kt}$

24. b. In the following drawing, the parameters are (a), $\omega = 1$, (b), $\omega = 2.5$, (c), $\omega = 2.9$, and (d), $\omega = 3$.

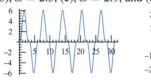

(a)   (b)

(c)   (d)

25.

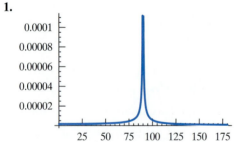

(a)   (b)

(c)   (d)

5.6

1.

$$
0.0001 \\
0.00008 \\
0.00006 \\
0.00004 \\
0.00002
$$

25   50   75   100   125   150   175

3. a. The resonant frequency decreases.
   b. The resonant frequency increases.

5. a. The displacement is $y = 0.3\cos(40t) - 0.25\sin(40t)$, indicating simple harmonic motion.
   b. $y = \dfrac{1}{6}\sin(20t) - \dfrac{1}{12}\sin(40t)$
   c. The motion is periodic, with a minimum periodic of $\dfrac{\pi}{10}$.

7. $y = -499.5\sin(0.001t)\cos(t) + 0.4995\sin(t)\cos(0.001t) + \cos(1.001t)$. The second and third terms are negligible.

9. The displacement function is

$$
y = \frac{1}{9 - \pi^2}\left[\left(1 - \frac{\pi}{3}\right)\sin(\theta t)\cos(\delta t) + \left(1 + \frac{\pi}{3}\right)\sin(\delta t)\cos(\theta t)\right],
$$

where $\theta = \dfrac{\pi + 3}{4}$ and $\delta = \dfrac{\pi - 3}{4}$.

The beat term is

$$
\frac{1}{3(3 - \pi)}\sin\left(\frac{\pi - 3}{4}t\right)\cos\left(\frac{\pi + 3}{4}t\right).
$$

The beat amplitude is $\dfrac{1}{3(3 - \pi)}$ and the beat frequency is $\dfrac{\pi - 3}{4}$.

11. a.

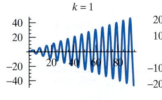
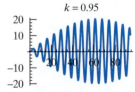
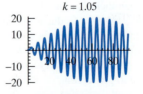
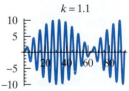
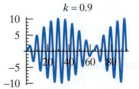
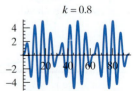

b.

$k = 1$

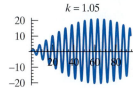

$k = 0.95$

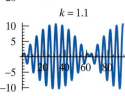

$k = 1.05$

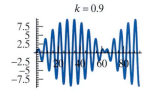

$k = 1.1$

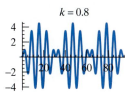

$k = 0.9$

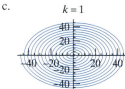

$k = 0.8$

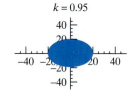

c.

$k = 1$

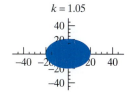

$k = 0.95$

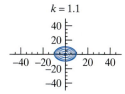

$k = 1.05$

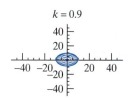

$k = 1.1$

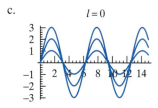

$k = 0.9$

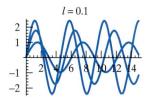

$k = 0.8$

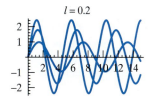

13. $y(t) = 0.078 \sin(t - 0.0078)$

15.

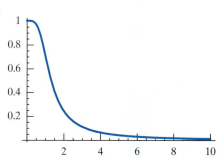

17. a. $r(\omega) = 40$ and $\theta(\omega) = \dfrac{\pi}{2}$

  c. $r(\omega) = 16010$ and $\theta(\omega) = 0.05$

19. a. $\omega = \sqrt{1 - \dfrac{1}{2} b^2}$

  c. The resistance is an increasing function of frequency.

  e. $b_0 = \sqrt{2}$

21. a. $G_S(x) = -\dfrac{1}{x} F_S(x) = -(k + lx^2 + \cdots)$, Thus $G_S''(0) = 2l$. If $G_S''(0) < 0$, $l < 0$, S is a soft spring. If $G_S''(0) > 0$, $l > 0$, S is hard.

  c.

$l = 0$

$l = 0.1$

$l = 0.2$

d.

$l = 0$

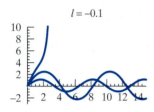

$l = -0.1$

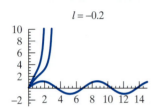

$l = -0.2$

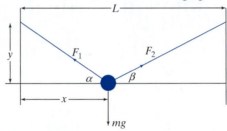

5.7

1. The fundamental frequency is 7.

3. Every solution of the system is a linear combination of a vector function $\mathbf{v}_1(t)$ with period 2π and a vector function $\mathbf{v}_2(t)$ of period $\frac{2}{3}\pi$. Since 2π is a multiple of both periods, the linear combination must be 2π-periodic.

5. a. The fundamental frequencies are 1 and $\dfrac{1}{2}(\sqrt{3 \pm \sqrt{5}})$.

  b. 1

  c. Suppose that the initial positions of the left and right objects are b units to the right of their respective equilibrium positions. In the mode with frequency $\frac{1}{2}(\sqrt{3 + \sqrt{5}})$, the middle object would have initial position at $b\dfrac{1 + \sqrt{5}}{2}$ units to the right of its equilibrium position, while in the mode with frequency $\frac{1}{2}(\sqrt{3 - \sqrt{5}})$, the middle object would have initial position $b\dfrac{-1+\sqrt{5}}{2}$ to the left of its equilibrium position.

7. The first fundamental frequency is $\sqrt{2 - \sqrt{2}}$, where the weights vibrate in phase, and the upper weight vibrates with amplitude $\sqrt{2} - 1 \approx 41\%$ times the amplitude of the lower weight. The second fundamental frequency is $\sqrt{2 + \sqrt{2}}$. In this mode of vibration, the weights vibrate in opposite phase, and the lower weight vibrates with amplitude $\sqrt{2} - 1 \approx 41\%$ times the amplitude of the upper weight.

9. Let the springs have natural lengths J and K, and let (x, y) be the position of the weight as measured in a coordinate system where the origin is the point of attachment on the left wall, and the y-axis points down. The motion is governed by the following nonlinear system of ODEs:

$$\begin{cases} mx'' = F_2 \cos(\beta) - F_1 \cos(\alpha) \\ my'' = mg - F_2 \sin(\beta) - F_1 \sin(\alpha), \end{cases}$$

where $F_1 = k_1(\sqrt{x^2 + y^2} - J)$ and $F_2 = k_2(\sqrt{(L - x)^2 + y^2} - K)$ are the magnitudes of the restoring forces for each of the springs, and $\alpha = \arctan(\frac{y}{x})$ and $\beta = \arctan(\frac{y}{L-x})$ are the angles made by the spring axes with the horizontal. See the following figure.

The system is nonlinear.

5.8

1. $y_p = \frac{1}{12}e^{2t}$ **3.** $y_p = \frac{1}{130}(4\cos(2t) + 7\sin(2t))$

5. $y_p = \frac{1}{2}\ln t - \frac{3}{4}$ **7.** $y_p = -\frac{1}{4}t\cos(2t)$

9. $y_p = te^{3t}(\ln t - 1)$

11. $y_p = \frac{1}{16}e^{-\frac{1}{2}t}(-3t^2 + 2t^2\ln(t))$

15. $y = \frac{1}{t}(-\ln(t) + \ln(t)\ln(\ln(t)) + C_1 + C_2\ln(t))$

17. $y = e^{-t}\displaystyle\int_0^t [-sb(s)e^s + tb(s)e^s]ds$

19. $y = e^{-t^2}\left(\frac{1}{12}t^4 + \frac{1}{4}t^2 + C_1 + C_2t\right)$

5.9

1. $y(t) = -C_1(2t^{-3} + 2t^{-2} + t^{-1})e^{-t} + C_2t^{-3}$

3. $y(t) = t/4 - C_1(2t^{-3} + 2t^{-2} + t^{-1})e^{-t} + C_2t^{-3}$

5. $y = \ln\left|\dfrac{1 - \sin t}{\cos t}\right|\cos t + C_1\sin(t) + C_2\cos(t)$

7. $y = \dfrac{1}{12}e^{-2t}(3t - 3t^2 + 2t^3) - \dfrac{1}{2}C_1e^{-4t} + C_2e^{-2t}$

9. $y(t) = C_1(1 + t\tan(t)) + C_2\sec(t)$

11. $y(t) = C_1e^{t - t^2/2} + C_2e^{-t^2/2} - te^{-t^2/2}$

5.11

1. $y = C_1e^t + C_2e^{5t/2}$

2. $y = e^{-t/2}(C_1 \cos(t/2) + C_2 \sin(t/2))$

3. $y = e^{3t}(C_1 + C_2 t)$

4. $y = t^3 - 6t + C_1 \cos(t) + C_2 \sin(t)$

5. $y = -\frac{1}{2}t \cos(t) + C_1 \cos(t) + C_2 \sin(t)$

6. $y = (\frac{1}{16}t + \frac{1}{32})e^{-2t} + e^{2t}(C_1 + tC_2)$

7. $y = \frac{1}{2}t^2 e^{2t} + e^{2t}(C_1 + tC_2)$ **8.** $y = 2te^{0.1t}$

9. $y = (\frac{1}{5}t + \frac{4}{25})e^{-2t} - \frac{4}{25}\cos(t) + \frac{3}{25}\sin(t)$

10. $y = \frac{1}{12}e^t - \frac{1}{2}te^{-t} + \frac{1}{4}e^{-t} - \frac{1}{3}e^{-2t}$

11. $y = \cos(t) - 2\sin(t)$

12. $y = \frac{1}{6}e^{-t} + \frac{1}{2}e^{-3t} - \frac{2}{3}e^{-4t}$

13. a. $\{e^{-t}, e^{-2t}\}$
  b. $\{e^{-t/2}\cos(\sqrt{7}t/2), e^{-t/2}\sin(\sqrt{7}t/2)\}$
  c. $\{\sec(t), t\sec(t)\}$

14. a. $y_p = e^t(t-2)$
  b. $y_p = \frac{1}{5}\ln(t)t^2 - \frac{1}{25}t^2$
  c. $y_p = e^{-5t}(\frac{1}{2}t^2\ln(t) - \frac{3}{4}t^2)$
  d. $y_p = -\frac{1}{4}\ln(\cosh(2t))\cosh(2t) + \frac{1}{2}t\sinh(2t)$

15. $\mathcal{M}(b)(t) = \int_0^t (t-s)e^{s-t}b(s)\,ds$

16. a. $y = (C_1 t + C_2)e^{-3t}$
  b. $y(t) = t^{-1}(C_1 \ln(t) + C_2)$
  c. $y(t) = t^{-1}(C_1 \ln(t) + C_2) + \ln(t) - 2$
  d. $y(t) = C_1\left(\frac{1}{4} - \frac{1}{2}t + \frac{1}{2}t^2\right) + C_2 e^{-2t}$

17. The mass matrix is $M = \begin{bmatrix} 5 & 0 \\ 0 & 2 \end{bmatrix}$, and the coupling matrix is $B = \begin{bmatrix} -12 & 10 \\ 10 & -15 \end{bmatrix}$. The system of ODEs $M\mathbf{x}'' = B\mathbf{x}$ is a model for the mechanical system. The fundamental frequencies are $\sqrt{\frac{99\mp\sqrt{6601}}{20}}$. In the mode of vibration with frequency $\sqrt{\frac{99-\sqrt{6601}}{20}}$, the weights move in phase with one another, and the left weight has an amplitude of $\frac{40}{\sqrt{6601}-51} \approx$ 1.32 times the amplitude of the right weight. In the mode of vibration with frequency $\sqrt{\frac{99+\sqrt{6601}}{20}}$, the upper and lower weights move in opposite phase, and the left weight has an amplitude of $\frac{40}{\sqrt{6601}+51} \approx 0.30\%$ times the amplitude of the right weight.

Chapter 6

6.1

1. $L(t-2) = -\frac{2}{s} + \frac{1}{s^2}$

3. $L(f(t)) = \frac{1}{s^2}(2e^{-4s} + 4s - 1)$

5. $L(f(t)) = \dfrac{e^{-s} + e^{-2s}}{s}$

7. (a), (b), (c), (d), (f), and (g)

9. $\int_0^B f'(t)\,dt = t^{-1/2}\big|_0^B = -\infty$. Hence $f(t)$ is not locally integrable.

11. Solve $s^2 Y(s) - s + k^2 Y(s) = 0$ to obtain $Y(s) = \dfrac{s}{s^2 + k^2}$.

13. By Proposition 6.1.3, $L(f(t)) = sL(g(t)) - g(0)$. Noting that $g(0) = \int_0^0 f(u)\,du$, we conclude that $L(g(t)) = \dfrac{1}{s}L(f(t))$.

15. $y = e^t$

6.2

1. a. $I(t) = \dfrac{5}{6}(e^{-4t} - e^{-16t})$ **c.** $I(t) = 2\sin(5t)e^{-10t}$

2. a. $\dfrac{5}{32}$ **c.** $\dfrac{2}{25}$

3. $C = \dfrac{10}{401}$

5. a. The current $I(t)$ satisfies the IVP

$$0.1I' + 7.5I + 100\int_0^t I(\tau)\,d\tau = 1000;\ I(0) = 0$$

The solution is

$$I(t) = \frac{400\sqrt{65}}{13}(e^{(-\frac{75}{2}+\frac{5}{2}\sqrt{65})t} - e^{(-\frac{75}{2}-\frac{5}{2}\sqrt{65})t}).$$

Eventually the current will decay to 0 and the capacitor will hold a charge of 10 coulombs. Here is a graph of the current as a function of time:

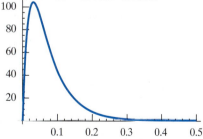

c. The current $I(t)$ satisfies the IVP

$$0.1I' + 0.75I + 100\int_0^t I(\tau)\,d\tau = 1000\sin(32t);$$

$$I(0) = 0.$$

The solution is

$$I(t) = -\frac{40000}{303}\cos(32t) + \frac{400000}{303}\sin(32t)$$
$$+ \frac{40000}{303}e^{-\frac{15}{4}t}\cos\left(\frac{5}{4}\sqrt{631}t\right)$$
$$- \frac{10120000}{191193}\sqrt{631}e^{-\frac{15}{4}t}\sin\left(\frac{5}{4}\sqrt{631}t\right).$$

Here is a graph of the current as a function of time:

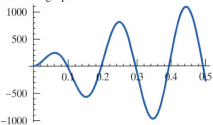

6. a. Here are graphs of $x(t)$ over intervals $0 \le t \le 10$ and $100 \le t \le 110$. The transient, which has a significant role in the the first graph, is not noticeable in the second.

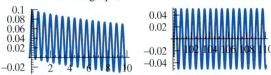

The transient component of $y_1(t)$ decays more rapidly. The graph of y_1 over the interval $0 \le t \le 20$ is shown below.

7. a. $\begin{cases} x' = \frac{3}{5}I - 6x + 3.6 \\ I' = -2I - 30x + 18 \end{cases}$

8. a. $\begin{cases} x = -3/5\,e^{-4t}\cos(\sqrt{14}\,t) + 3/5 \\ \qquad + \frac{3}{35}\sqrt{14}e^{-4t}\sin(\sqrt{14}\,t) \\ I = \frac{9}{7}\sqrt{14}e^{-4t}\sin(\sqrt{14}\,t) \end{cases}$

The graphs of $x(t)$ and $I(t)$, assuming homogeneous initial conditions, are shown below.

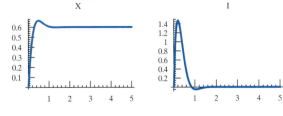

6.3

1. $\dfrac{6}{s^4} + \dfrac{10}{s^3} + \dfrac{2}{s^2} - \dfrac{1}{s}$ **3.** $\dfrac{1}{(s-1)^2}$

5. $\dfrac{1}{s^2 - 1}$ **7.** $\dfrac{2}{(s-1)^2 + 4}$

9. $\dfrac{4(s-1)}{((s-1)^2 + 4)^2}$

11. $\dfrac{2}{(s+3)^3} - \dfrac{3}{(s+3)^2} + \dfrac{5}{s+3}$

13. $\dfrac{6}{(s-3)^4}$ **15.** $\dfrac{100!}{(s-7)^{101}}$

17. $\dfrac{1}{2}\left(\dfrac{1}{(s-1)^2 + 1} - \dfrac{1}{(s+1)^2 + 1}\right)$

19. $\dfrac{(s-3)^2 - 25}{s((s-3)^2 + 25)^2}$ **21.** $\dfrac{2}{s^3} + \dfrac{4s}{(s^2+1)^2} + \dfrac{2}{s(s^2+4)}$

23. $\dfrac{2s}{(s-3)^3} + \dfrac{3s}{(s-3)^2} - \dfrac{6s}{s-3} + 6$

27. $\dfrac{\sqrt{\pi}}{2s^{\frac{3}{2}}}$ **29.** $\dfrac{\sqrt{2\pi}}{2}\sqrt{\dfrac{\sqrt{s^2+1} - s}{s^2 + 1}}$

31. $y = te^{-2t}$ **33.** $y = e^{-t}\cos(t)$

35. $y = e^{-4t}\cos(3t) + \frac{7}{3}e^{-4t}\sin(3t)$

37. $y = \cos(t)e^{-t} - 3\sin(t)e^{-t}$

39. $x = y = e^t$ **41.** $x = 3e^t + 6te^t$, $y = 6te^t$

43. $y = \cos(t) + \sin(t)$

6.4

1. $-6 + 6t + 6e^{-t}$ **3.** $\cos(t)e^{-3t} + 2\sin(t)e^{-3t}$

5. $-1 - \frac{1}{2}t^2 + e^t$ **7.** $-1 + \cos(t)e^{-t} + 2\sin(t)e^{-t}$

9. $y(t) = 4t - 2\sin(2t)$ **11.** $y = -\sin(2t) + \sin(t)$

13. $y = \cos(t)$

15. $y = te^{-t} + \frac{1}{5}\cos(t)e^{-t} - \frac{2}{5}\sin(t)e^{-t} - \frac{1}{5}e^{-3t}$

17. $y = \frac{1}{3}e^t - \frac{1}{3}\cos(\frac{\sqrt{3}}{2}t)e^{-t/2} + \frac{\sqrt{3}}{3}\sin(\frac{\sqrt{3}}{2}t)e^{-t/2}$

19. $x = 1 + t - \frac{1}{6}t^3$, $y = -\frac{1}{2}t^2 - t + \frac{1}{6}t^3$

21. $x = e^{-2t} + e^t$, $y = -\frac{1}{2}(e^{-2t} + 4e^t)$

23. $y = 4t^4e^{-2t}$ **25.** $y = \cosh(2t) - \cos(2t)$

26. a. $y_p = \frac{1}{5}\sin t - \frac{2}{5}\cos t$ **c.** $y_p = -\dfrac{1 + 3i}{10}e^{it}$

27. The natural unit of time is $T = \frac{t}{\sqrt{LC}}$. The natural unit of current is $Y = y\frac{1}{A}\sqrt{\frac{L}{C}}$.

29. b. $\omega = \frac{1}{CL}$

31. a. The circuit is underdamped.

b. $= -25\sin(4t)e^{-8t}$

c. $x(t) = -\dfrac{396}{697}\cos(6t) - \dfrac{864}{697}\sin(6t)$

$\qquad + \dfrac{396}{697}\cos(4t)e^{-8t} + \dfrac{4622.5}{697}\sin(4t)e^{-8t}$

d. $P(\omega) = 8 + i\left(0.5\omega - \dfrac{40}{\omega}\right)$

e. $\omega = 4\sqrt{5}$

33. $\lim_{s\to\infty} sF(s) = \lim_{s\to\infty}(sF(s) - f(0)) + f(0)$

$\qquad = \lim_{s\to\infty}(L(f'(t))) + f(0)$

$\qquad = f(0)$

6.5

1. $f(t) = 2u(t - 15)$

3. $f(t) = u(t) - 2u(t - 2) + u(t - 4)$

5. $f(t) = -3t + 7 + 6c(t - 7/3)$

7. $f(t) = c(t) - 2c(t - 1) + 2u(t - 3) - c(t - 4)$

9. $L(c(t - 5)) = e^{-5s}L(t) = \dfrac{e^{-5s}}{s^2}$

11. $L(e^{3t}e^{-3}u(t - 3)) = e^{-3}L(e^{3t}u(t - 3)) = \dfrac{e^{-3(s-3)}e^{-3}}{s - 3}$

13. $L(t^2u(t - 1)) = \dfrac{2e^{-s}}{s^3} + \dfrac{2e^{-s}}{s^2} + \dfrac{e^{-s}}{s}$

15. $L(f(t)) = \dfrac{2e^{-10s}}{s^3} + \dfrac{20e^{-10s}}{s^2}$

17. $L^{-1}(G(s)) = 1 - u(t - 1)$

19. $L^{-1}(G(s)) = \sin(t)(1 - u(t - 1))$

21. $y = \begin{cases} 1 - \cos(t) & \text{if } t < 2\pi, \\ -[1 - \cos(t)] & \text{if } 2\pi \le t < 4\pi \\ 0 & \text{if } t \ge 4\pi \end{cases}$

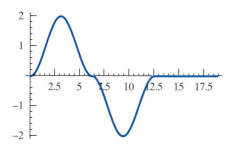

23. $y = \begin{cases} -\frac{1}{2}t\cos(2t) + \frac{1}{4}\sin(2t) & \text{if } t \le \pi, \\ -\frac{1}{2}\pi\cos(2t) & \text{if } t > \pi \end{cases}$

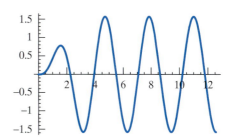

25. $y = \begin{cases} 1 - te^{-t} - e^{-t} & \text{if } t \le 1, \\ 2e^{1-t}t - 1 - te^{-t} - e^{-t} & \text{if } 1 < t \le 2 \\ 2e^{1-t}t + e^{2-t} - e^{2-t}t - te^{-t} - e^{-t} & \text{if } t > 2 \end{cases}$

27. $y = u(t - 1)e^{t-1}$

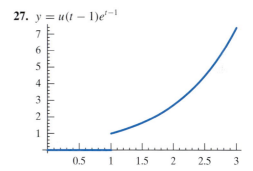

29. $y = \begin{cases} \cos(\pi t) + \dfrac{\sin(\pi t)}{\pi} & \text{if } t \le 1, \\ \cos(\pi t) + 2\dfrac{\sin(\pi t)}{\pi} & \text{if } 1 < t \le 2 \\ \cos(\pi t) + 3\dfrac{\sin(\pi t)}{\pi} & \text{if } t \ge 2 \end{cases}$

31. $y(t) = \begin{cases} 0 & \text{if } t \le \pi \\ -60\sin(100t) & \text{if } t > \pi \end{cases}$

6.6

1. $e^{-s} + e^{-2s} + 3e^{-3s}$ **3.** $4e^{-2s}$

5. $y = e^t + u(t-1)e^{t-1}$

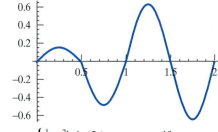

7. $y = \begin{cases} \dfrac{\sin(2\pi t)}{2\pi} & \text{if } t \le 1/2, \\[2mm] \dfrac{3\sin(2\pi t)}{2\pi} & \text{if } 1/2 < t \le 1 \\[2mm] \dfrac{2\sin(2\pi t)}{\pi} & \text{if } t > 1 \end{cases}$

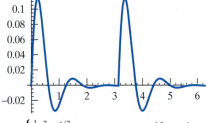

9. $y = \begin{cases} \frac{1}{5}e^{-2t}\sin(5t) & \text{if } t \le \pi, \\[2mm] \frac{1}{5}e^{-2t}\sin(5t)(1-e^{2\pi}) & \text{if } t > \pi \end{cases}$

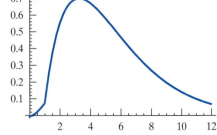

11. $y = \begin{cases} \frac{1}{8}t^2 e^{-t/2} & \text{if } t \le 1, \\[2mm] \frac{1}{8}t^2 e^{-t/2} + (t-1)e^{-t/2} & \text{if } t > 1 \end{cases}$

13. $y = \frac{1}{6} - \frac{1}{4}e^t + \frac{1}{10}e^{2t} - \frac{1}{60}e^{-3t} - \frac{1}{6}u(t-1) + \frac{1}{4}u(t-1)e^{t-1}$
$\qquad - \frac{1}{10}u(t-1)e^{2(t-1)} + \frac{1}{60}u(t-1)e^{-3(t-1)}$

15. $y = \frac{1}{5}\delta(t) - 2000000e^{-10000000t}$

17. $y(t) \approx \frac{1}{2}e^{-.01t}\sin(100t)$
$\qquad + \frac{1}{2}u\left(t - \frac{\pi}{100}\right)e^{-.01\left(t - \frac{\pi}{100}\right)}\sin\left(100\left(t - \frac{\pi}{100}\right)\right)$

6.7

1. $L(1*f(t)) = L(1)L(f(t))$
$$= \frac{L(f(t))}{s} = L\left(\int_0^t f(u)\,du\right)$$

3. $t - \sin(t)$ **5.** $\sinh(t)$ **7.** $u(t - 2\pi)\sin(t)$

9. $\dfrac{1}{(n-1)!}t^{n-1}$ **11.** $\frac{1}{2}(t\cos(t) + \sin(t))$

13. $f(t) = \frac{1}{2}(8!)^{1/3}t^2$

15. Let $L(f) = F(s)$ and $L(g) = G(s)$. Then $L(f*g) = F(s)G(s)$, so

$$L(e^{at}[f*g](t)) = F(s-a)G(s-a)$$
$$= L([e^{at}f(t)] * [e^{at}g(t)]).$$

17. $y(t) = \displaystyle\int_0^t e^{-4(t-u)}g(u)\,du$

19. $y(t) = \dfrac{1}{2}\displaystyle\int_0^t \sin\left(\dfrac{t-u}{2}\right)g(u)\,du$

21. $y(t) = -\dfrac{1}{2}\displaystyle\int_0^t (\cos(t-u) + (t-u)\sin(t-u))g(u)\,du$

23. $y(t) = \displaystyle\int_0^t e^{(t-u)/2}\left(\cosh\left(\dfrac{\sqrt5}{2}(t-u)\right)\right.$
$\qquad\qquad \left. + \dfrac{1}{\sqrt5}\sinh\left(\dfrac{\sqrt5}{2}(t-u)\right)\right)g(u)\,du$

6.8

3. a. $\Lambda(t) = 1 - 2t - 4\displaystyle\sum_{k=1}^{\infty}(-1)^k c(t-k)$

c. $L(\Lambda(t)) = \dfrac{1}{s} - \dfrac{2}{s^2} - 4\displaystyle\sum_{k=1}^{\infty}(-1)^k \dfrac{e^{-ks}}{s^2}$

5. $y(t) = \lfloor t/(2\pi)\rfloor \sin(t)$

7. If n is an odd integer, $(-1)^m \cos(n\pi(t-m)) = \cos(n\pi t)$. Therefore,

$$y_p = \frac{1}{(n\pi)^2}\left[S(t) - \cos(n\pi t)\left(1 + 2\sum_{m=1}^{\infty}u(t-m)\right)\right]$$
$$= \frac{1}{(n\pi)^2}[S(t) - \cos(n\pi t)(1 + 2\lfloor t\rfloor)].$$

Because of the term $2\lfloor t \rfloor$, y is unbounded. If n is an even integer, $(-1)^m \cos(n\pi(t-m)) = (-1)^m \cos(n\pi t)$. Therefore,

$$y_p = \frac{1}{(n\pi)^2}\left[S(t) - \cos(n\pi t)\left(1 + 2\sum_{m=1}^{\infty}(-1)^m u(t-m)\right)\right]$$

$$= \frac{1}{(n\pi)^2}\left[S(t) - \cos(n\pi t)\left(\frac{1}{2}S(t) + \frac{1}{2}\right)\right].$$

It follows that 2 is a period of y_p. It is thus also a period of the general solution, $y_p(t) + C_1 \cos(n\pi t) + C_2 \sin(n\pi t)$.

9. Resonance occurs when k is an odd multiple of π, and in this case the resonant solution is

$$y(t) = \frac{1}{(m\pi)^2}\left[\Lambda(t) - \cos(m\pi t) + \frac{2\sin(m\pi t)}{m\pi}(1 + 2\lfloor t \rfloor)\right]$$

11. a. $y = \dfrac{e^{-\theta(t)}}{1 - e^{-1}} - \dfrac{e^{-2\theta(t)}}{1 - e^{-2}}$

c. $y = \dfrac{1}{1 + \pi^2}\left[S(t) - \dfrac{2e^{-\theta(t)}}{1 + e^{-1}}\left(\dfrac{1}{\pi}\sin(\pi t) + \cos(\pi t)\right)\right]$

15.

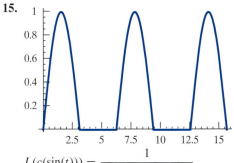

$$L(c(\sin(t))) = \frac{1}{(1 - e^{-s\pi})(s^2 + 1)}$$

6.10

1. $L(f) = \dfrac{2}{s} - \dfrac{1}{s^2}(1 - e^{-2s})$

2. $y(t) = -2e^{-3t} + e^{-t}(2\cos(t) + \sin(t))$

3. $y(t) = e^{-3t}$ **4.** $y(t) = -5e^{-6t} + 9e^t$

5. $y(t) = \dfrac{e}{32}\left(e^{2t}(-1 + 2t) + e^{-2t}(1 + 2t)\right)$

6. $y(t) = 2e^{-t} - e^{-2t}$ **7.** $y(t) = -\dfrac{1}{4}e^{-6t} + te^{2t} + \dfrac{5}{4}e^{2t}$

8. $y(t) = \dfrac{1}{20}\left(e^{2t}(\cos(2t) + 2\sin(2t)) - \cos(2t) - 3\sin(2t)\right)$

9. $y(t) = e^{-t} - e^{3t/5}$ **10.** $y(t) = \dfrac{1}{2}\left(e^{-t} - \cos(t) + \sin(t)\right)$

11. $y(t) = \dfrac{2}{6!}t^6 e^{-t} = \dfrac{1}{360}t^6 e^{-t}$ **12.** $y(t) = e^t$

13. $y(t) = 2\sin(t) + 2\sin(t - \pi)u(t - \pi)$
$= 2\sin(t)[1 - u(t - \pi)]$

14. $y(t) = e^{-t}\sin(t)[1 - 2e^\pi u(t - \pi) + 3e^{2\pi}u(t - 2\pi)]$

15. $y(t) = \dfrac{1}{\pi}e^{-t}\sin(\pi t)[1 - eu(t - 1) + e^2 u(t - 2)]$

16. $y(t) = \dfrac{1}{\pi^2}[1 - u(t - 1) - \cos(\pi t) - \cos(\pi t)u(t - 1)]$

17. $y(t) = \dfrac{1}{4\pi^2}S(t)(1 - \cos(2\pi t))$

18. $y(t) = \dfrac{1}{\pi}\lfloor t \rfloor \sin(\pi t)$

19. $y(t) = \begin{cases} \sin(\pi t) & \text{if } \lfloor t \rfloor \text{ is even,} \\ 0 & \text{if } \lfloor t \rfloor \text{ is odd} \end{cases}$

20. $y(t) = \begin{cases} \frac{1}{2}(\sin(t) - t\cos(t)) & \text{if } 0 \le t < \pi, \\ -\frac{\pi}{2}\cos(t) & \text{if } t \ge \pi \end{cases}$

21. $y(t) = \left(\dfrac{\cosh(2t) - \cos(2t)}{2}\right)$
$\qquad + \left(\dfrac{\cosh(2(t-1)) + \cos(2(t-1))}{2} - 1\right)u(t - 1)$

22. $y(t) = \begin{cases} \frac{1}{3}\sin(t) - \frac{1}{6}\sin(2t) & \text{if } 0 \le t < \pi, \\ \frac{2}{3}\sin(t) & \text{if } \pi \le t < 2\pi, \\ \frac{1}{3}\sin(t) + \frac{1}{6}\sin(2t) & \text{if } t \ge 2\pi \end{cases}$

23. a. Transfer function: $H(s) = \dfrac{1}{s + 1}$
Impulse response: $\phi(t) = e^{-t}$

b. Transfer function: $H(s) = \dfrac{1}{(s + 1/2)^2 + 3/4}$
Impulse response: $\phi(t) = \dfrac{2}{\sqrt{3}}e^{-t/2}\sin(\sqrt{3}t/2)$

c. Transfer function: $H(s) = -\dfrac{1}{s + 1} + \dfrac{2}{2s + 1}$
Impulse response: $\phi(t) = -e^{-t} + e^{-t/2}$

d. Transfer function: $H(s) = \dfrac{1}{2}\left(\dfrac{1}{s + 1} - \dfrac{s - 1}{s^2 + 1}\right)$
Impulse response: $\phi(t) = \dfrac{1}{2}\left(e^{-t} - \cos(t) + \sin(t)\right)$

24. Let $x(t)$ denote the charge of the capacitor. Then

$$\frac{1}{0.005}x + 100(x' - I) = 6$$
$$-100(x' - I) + 0.01I' = -6.$$

Solution: $I(t) \approx -\dfrac{3}{50} + \dfrac{3}{50}e^{-t}\cos(100\sqrt{2}t)$

25. $y = e^{-t/2} - \dfrac{1}{2}te^{-t/2}$ **26.** $y = \dfrac{1}{3}\left(e^{2t} - e^{-t}\right)$

27. a. $f(t) = e^{t^2}$ **b.** $f(t) = u(t - 1)$ **c.** $f(t) = t^{-1/2}$

28. $L\left(\dfrac{1}{\lfloor t\rfloor+1}\right) = \dfrac{1}{s}\left(1 - \displaystyle\sum_{k=1}^{\infty}\dfrac{e^{-sk}}{k(k+1)}\right)$, or, in closed

form, $\dfrac{1-e^{s}}{s}\ln\left(\dfrac{1}{1-e^{-s}}\right)$

29. $I = 12\sin(10t)\dfrac{e^{-10t+\lfloor\frac{10t}{\pi}\rfloor\pi}}{1-e^{-\pi}}$

31. $L(J_0(t)) = \dfrac{1}{\sqrt{1+s^2}}$

$J_0(t) = \displaystyle\sum_{n=0}^{\infty}(-1)^{n}\dfrac{1}{2^{2n}\,(n!)^2}t^{2n}$

32. a. $L\left(\dfrac{\sin t}{t}\right) = \arctan\left(\dfrac{1}{s}\right)$

    b. $L\left(\dfrac{\cos t-1}{t^2}\right) = -\arctan\left(\dfrac{1}{s}\right)+\dfrac{s}{2}\ln\left(1+\dfrac{1}{s^2}\right)$

Chapter 7

7.1

1. $A_0 + \displaystyle\sum_{n=1}^{\infty}(A_n + A_{n-1})t^{n}$

3. $2A_0t + \displaystyle\sum_{n=1}^{\infty}(2A_n - A_{n-1})t^{2n+1}$ **5.** $3t - \displaystyle\sum_{n=2}^{\infty}t^{n}$

7. $A_0,\ (A_0 + A_1)t,\ \left(\frac{1}{2}A_0 + A_1 + A_2\right)t^2,$

    and $\left(\frac{1}{6}A_0 + \frac{1}{2}A_1 + A_2 + A_3\right)t^3$

9. $\displaystyle\sum_{n=0}^{\infty}(n^2 + 1)A_n t^{n}$

11. $2A_2 + \displaystyle\sum_{n=1}^{\infty}[(n+2)(n+1)A_{n+2} - A_{n-1}]t^{n}$

13. $\displaystyle\sum_{n=0}^{\infty}\dfrac{(-1)^{n+1}}{(2n)!}t^{2n}$

15. a. $\displaystyle\sum_{n=0}^{\infty}[4n^2 + 10n + 9]2^{n}t^{n}$

    c. $\displaystyle\sum_{n=0}^{\infty}(n^2 - n + 1)(t-1)^{n}$

    e. $\displaystyle\sum_{n=0}^{\infty}[(n+2)(n+1)A_{n+2}-2nA_n+2(n+1)A_{n+1}+A_n](t+1)^{n}$

7.2

1. $\displaystyle\sum_{n=0}^{\infty}(-1)^{n}t^{n}$ **3.** $\displaystyle\sum_{n=0}^{\infty}(-1)^{n}\dfrac{1}{n!}t^{2n}$

5. $y''(t) - 4y'(t) + 13y(t) = 0$;

    $A_0 = 1,\ A_1 = 2,\ A_2 = -\dfrac{5}{2},\ A_3 = -\dfrac{23}{3},\ A_3 = -\dfrac{119}{24}$

7. $y(t) = 1 + \dfrac{1}{2}t^2 - \displaystyle\sum_{k=2}^{\infty}(-1)^{k}\dfrac{(2k-3)!!}{(2k)!!}t^{2k}$

9. $\left\{1 + \displaystyle\sum_{n=1}^{\infty}\dfrac{1}{(3n)!!!(3n-1)!!!}t^{3n}\right.,$

    $\left. t + \displaystyle\sum_{n=1}^{\infty}\dfrac{1}{(3n+1)!!!(3n)!!!}t^{3n+1}\right\}$

15. $y = 1 + \displaystyle\sum_{k=1}^{\infty}\dfrac{(-1)^{k+1}(3\cdot 7\cdot 11\cdots(4k-5)}{(2k)!}t^{2k}$

17. $y = t - 1$ **19.** $y = \displaystyle\sum_{n=0}^{\infty}(t-1)^{2n+1} = \dfrac{t-1}{2t-t^2}$

21. $y(t) = \displaystyle\sum_{n=0}^{\infty}(2n+1)t^{2n} = \dfrac{1-t^2}{(1+t^2)^2}$

23. $y = \displaystyle\sum_{n=0}^{\infty}\dfrac{(4k-3)(4k-7)\cdots 1}{2^{2k}(2k)!}t^{2k}$

25. $y(t) = \displaystyle\sum_{k=0}^{\infty}(k+1)(2k+1)t^{2k}$

27. $\begin{cases} y_1 = 1 - \frac{1}{2}(t-1)^2 + \frac{1}{6}(t-1)^3 + \frac{1}{24}(t-1)^4\cdots \\ y_2 = (t-1) - \frac{1}{2}(t-1)^2 - \frac{1}{12}(t-1)^3 \\ \qquad + \frac{11}{96}(t-1)^4 + \cdots \end{cases}$

29. $\begin{cases} y_1 = t - \frac{1}{6}t^3 - \frac{1}{12}t^4 - \frac{1}{24}t^5 + \cdots \\ y_2 = 1 - \frac{1}{2}t^2 - \frac{1}{6}t^3 - \frac{1}{24}t^4 + \cdots \end{cases}$

31. $y = 1 + 3t^2 + t^4 + \dfrac{1}{15}t^6$ **33.** $y = (t-1)^2$

35. $\{t, 1 + t^2\}$ is a fundamental set of solutions; thus all solutions are polynomials of the form $y = At^2 + Bt + A$.

37. $v = 1,$ $y = t;$

      $v = 3,$ $y = t - \frac{2}{3}t^3;$

      $v = 5,$ $y = t - \frac{4}{3}t^3 + \frac{4}{15}t^5$

7.3

1. $4s^2 - 4s + 1 = 0$ **3.** $s^2 - 10s + 15 = 0$

5. $y = C_1 t^3 + C_2 t^5$

7. $y = t^{3/2}(C_1 + C_2\ln t)$ (for $t > 0$)

9. $y = t[C_1\cos(\ln t) + C_2\sin(\ln t)]$ (for $t > 0$)

11. $b = -\dfrac{a_0}{a_2},\ c = 1 - \dfrac{a_1}{a_2}$

13. $y = \dfrac{1 + 2\ln|t|}{t^2}$ (for $t < 0$)

16. a. $y = C_1(t+5)^2 + \dfrac{C_2}{t+5}$

c. $y = C_1(t - 2) + \dfrac{C_2}{t - 2}$

e. $y = C_1(t - 1)^{-4}\cos(3\ln|t - 1|) + C_2(t - 1)^{-4}\sin(3\ln|t - 1|)$

19. $y = 2(t - 1)\ln|t - 1|$ (for $t < 1$)

7.4

1. The indicial equation is $s^2 - s = 0$, and the exponents are 1, which is unblocked, and 0, which is blocked.

3. The indicial equation is $s^2 + s + 1 = 0$, and the exponents are $\dfrac{-1 \pm i\sqrt{3}}{2}$, both unblocked.

5. The indicial equation is $s^2 = 0$, and 0 is a double exponent (unblocked).

7. The indicial equation is $s^2 - 1 = 0$, and the exponents are ± 1. The exponent -1 is a blocked.

9. $R = \displaystyle\lim_{n\to\infty}\left|\dfrac{n + 4}{n + 5}\right| = 1$

11. $y = \dfrac{C_1\sin(x) + C_2\cos(x)}{\sqrt{x}}$

13. $L_1(t) = 1 - t$, $L_2(t) = 1 - 2t + \frac{1}{2}t^2$, $L_3(t) = 1 - 3t + \frac{3}{2}t^2 - \frac{1}{6}t^3$

15. $y_1(t) = te^{-t}$, $y_2(t) = \displaystyle\int_1^t \dfrac{t}{s}e^{s-t}\,ds$

17. $y = C_1\displaystyle\sum_{n=0}^{\infty}\dfrac{1}{n!(2n + 1)!!}t^n + C_2\left(1 + \displaystyle\sum_{n=1}^{\infty}\dfrac{1}{n!(2n - 1)!!}t^{n-\frac{1}{2}}\right)$

19. $y = 1 + \frac{1}{3}t$,

$y = \displaystyle\sum_{n=0}^{\infty}(-1)^n\dfrac{10(n + 3)(n + 2)(n + 1)}{3(n + 5)(n + 4)}(t - 1)^{n+5}$

7.5

9. In the following graph, the blue dashed curve represents $y = \sqrt{t}\,J_0(t)$, the solid black curve is $y = \sqrt{t}\,J_{1/2}(t) = -\dfrac{2}{\sqrt{\pi}}\cos(t)$, and the solid blue curve is $y = \sqrt{t}\,J_1(t)$.

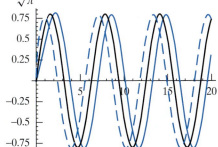

10. In the following graph, the blue dashed curve represents $y = Y_0(t)$, the solid black curve is $y = Y_{1/2}(t) =$

$-\dfrac{2}{\sqrt{\pi t}}\cos(t)$, and the solid blue curve is $Y_1(t)$.

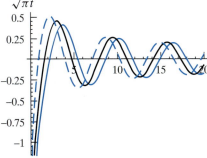

12. Here is a table of zeros of the Bessel functions $J_\nu(t)$ for $\nu = 0$, $\frac{1}{2}$, and 1. The right three columns show differences between consecutive zeros.

	Zeros			Differences		
n	$j_{0,n}$	$j_{1/2,n}$	$j_{1,n}$	j_0	$j_{1/2}$	j_1
1	2.40483	π	3.83171			
2	5.52008	2π	7.01559	3.11525	π	3.18388
3	8.65373	3π	10.1735	3.13365	π	3.15788
4	11.7915	4π	13.3237	3.13781	π	3.15022
5	14.9309	5π	16.4706	3.13938	π	3.14694
6	18.0711	6π	19.6159	3.14015	π	3.14523
7	21.2116	7π	22.7601	3.14057	π	3.14423
8	24.3525	8π	25.9037	3.14083	π	3.14359
9	27.4935	9π	29.0468	3.14101	π	3.14316

13. a. $J_\nu'(t) = t^\nu\displaystyle\sum_{m=0}^{\infty}\dfrac{(-1)^m(2m + \nu)}{2^{2m+\nu}m!\Gamma(m + \nu + 1)}t^{2m-1}$

15. The general solution of Cauchy-Euler equation cannot be expressed in terms of Bessel functions.

17. $w(x) = C_1x^{-\frac{1}{2}}I_0(\sqrt{x}) + C_2x^{-\frac{1}{2}}K_0(\sqrt{x})$

19. $w(x) = C_1x^{1/2}J_{\frac{1}{4}}\left(\frac{1}{2}x^2\right) + C_2x^{1/2}Y_{\frac{1}{4}}\left(\frac{1}{2}x^2\right)$

21. $\sinh(x) = \sqrt{\dfrac{x\pi}{2}}\,I_{\frac{1}{2}}(x)$, $\cosh(x) = \sqrt{\dfrac{x\pi}{2}}\,I_{-\frac{1}{2}}(x)$

22. a. $w(x) = C_1J_0(e^t) + C_2Y_0(e^t)$

23. $y = \dfrac{[C_1t^{1/2}J_{\frac{1}{4}}(t^2) + C_2t^{1/2}Y_{\frac{1}{4}}(t^2)]'}{a[C_1t^{1/2}J_{\frac{1}{4}}(t^2) + C_2t^{1/2}Y_{\frac{1}{4}}(t^2)]}$

7.6

5. The solutions $y(t; 1) = \displaystyle\sum_{n=0}^{\infty}\dfrac{1}{n!}t^{n+1} = te^t$ and $y_2(t) = te^t\ln(t) - \displaystyle\sum_{n=1}^{\infty}\dfrac{\phi(n)}{n!}t^{n+1}$ are independent solutions.

9. The indicial equation, $s^2 - 2s + 1 = 0$, has a double root.

15. The second solution is $y = \sum_{n=0}^{\infty}(n+1)t^n = (1-t)^{-2}$

7.8

1. No singular points; general solution in terms of Bessel functions: $y(t) = \sqrt{t}\left(C_1 J_{\frac{1}{2}}(t) + C_2 Y_{\frac{1}{2}}(t)\right)$

2. Regular singular point at 0; indicial equation, $s^2+12s+5 = 0$. The exponents, $-6 \pm \sqrt{31}$, are not blocked. Solutions of Cauchy-Euler equations cannot be expressed in terms of Bessel functions.

3. Regular singular point at -1; indicial equation, $s^2+1 = 0$. The exponents, $\pm i$, are not blocked. Solutions of Cauchy-Euler equations cannot be expressed in terms of Bessel functions.

4. There are regular singular points at $t = 0, \pi$, and irregular singular points at $t = n\pi$, for n any integer other than 0, 1. For the singular point $t = 0$: The indicial equation is $s^2 - (\pi + 1)s + 2 = 0$; the exponents are $\frac{1}{2}(\pi + 1 \pm \sqrt{\pi^2 + 2\pi - 7})$; and neither is blocked. For the singular point $t = \pi$: The indicial equation is $s^2+(\pi - 1)s+2 = 0$; the exponents are $\frac{1}{2}(1 - \pi \pm \sqrt{2\pi + 7 - \pi^2}i)$; and neither is blocked. There is no reasonable way to express solutions in terms of Bessel functions.

5. There is a regular singular point at 0. Its indicial equation is $s^2 - s - 2 = 0$; the exponents are -1 (blocked) and 2 (unblocked). In terms of Bessel functions, the general solution is

$$y = C_1 t^{1/2} Y_{\frac{1}{2}}\left(\frac{1}{3}t^3\right) + C_2 t^{1/2} J_{\frac{1}{2}}\left(\frac{1}{3}t^3\right)$$
$$= \frac{1}{t}\sqrt{\frac{6}{\pi}}\left[-C_1 \cos\left(\frac{1}{3}t^3\right) + C_2 \sin\left(\frac{1}{3}t^3\right)\right].$$

6. There is one singular point, 0, and it is regular. The indicial equation is $s^2 - 1 = 0$, and the exponents are -1 (blocked) and 1 (unblocked). This is the modified Bessel equation with parameter $\nu = 1$; hence the general solution is $C_1 I_1(t) + C_2 K_1(t)$.

7. There are singular points at 0 and 1; both regular. The indicial equation at 0 is $s^2 + 2s = 0$, and the exponents are -2 (blocked) and 0 (unblocked). The indicial equation at 1 is $s^2 - 3s = 0$; the exponents are -3 (blocked) and 0 (unblocked). We can't use Proposition 7.5.7 to express the general solution in terms of Bessel functions.

8. There is one singular point, 0, and it is irregular.

9. $y = C_1(t-1) + C_2\left(1 - \sum_{k=1}^{\infty}\frac{(t-1)^{2k}}{(2k-1)(2k)!!}\right)$

10. a. $y = C_1 y_1 + C_2 y_2$, where

$$y_1 = \sum_{n=0}^{\infty}\frac{(-1)^n}{(4n)(4n-1)(4n-4)(4n-5)\cdots(4)(3)}x^{4n}$$

$$y_2 = \sum_{n=0}^{\infty}\frac{(-1)^n}{(4n+1)(4n)(4n-3)(4n-4)\cdots(5)(4)}x^{4n+1}$$

b. $y(t) = C_1 t^{1/2} J_{\frac{1}{4}}\left(\frac{1}{2}t^2\right) + C_2 t^{1/2} Y_{\frac{1}{4}}\left(\frac{1}{2}t^2\right)$

11. $\left\{ t^{3/4}\sum_{n=0}^{\infty}\frac{(-1)^n 3 \cdot 7 \cdots (4n-1)}{n! 2 \cdot 6 \cdots (4n-2)}t^n , \right.$

$\left. t^{5/4}\sum_{n=0}^{\infty}\frac{(-1)^n 1 \cdot 5 \cdots (4n+1)}{n! 6 \cdot 10 \cdots (4n+2)}t^n \right\}$

Chapter 8

8.1

1. E   **3.** I   **5.** F   **7.** L   **9.** A   **11.** D

13. Degenerate system:

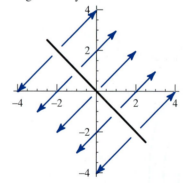

17. The phase portrait is a saddle:

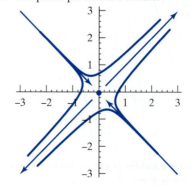

19. The phase portrait is a stable spiral node:

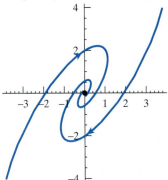

21. This degenerate system has $y = x$ as the stationary line:

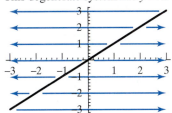

23. The phase portrait is a center, with orbits directed counter-clockwise on ellipses $5y^2 - 4xy + 17x^2 = C$

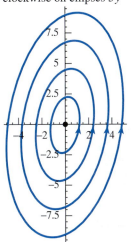

25. This phase portrait is a saddle; the stable line is $y = -5x$, and the unstable line is $y = -x$.

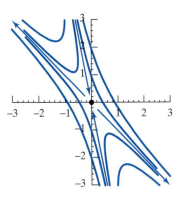

33. D   **35.** B   **37.** A

8.2

1. Stable, but not asymptotically stable: degenerate system

3. Unstable spiral: every nonstationary orbit is unbounded as $t \to \infty$. Example of an unbounded solution:
$$\begin{bmatrix} x \\ y \end{bmatrix} = e^t \begin{bmatrix} \cos(t) \\ -\sin(t) \end{bmatrix}.$$

5. Unstable. Here is an unbounded solution: $\mathbf{v} = e^{3t} \begin{bmatrix} 0 \\ 0 \\ 1 \\ 1 \end{bmatrix}.$

7. Unstable. Here is an unbounded solution: $\mathbf{v} = e^t \begin{bmatrix} 0 \\ 0 \\ 1 \\ 1 \end{bmatrix}.$

9. Unstable. Here is an unbounded solution:
$$\mathbf{v} = \mathbf{Re}(\mathbf{v}_1(t)) = \begin{bmatrix} t\cos(t) \\ \cos(t) + t\sin(t) \\ 0 \\ 0 \end{bmatrix}.$$

11. The probability is equal to 0.

8.3

1. The repelled set consists of the fourth quadrant, and the points of the first quadrant that lie below the separatrices that form the unstable manifold of the stationary point $(2, 1)$.

3. The origin is the only stationary point. It is asymptotically stable.

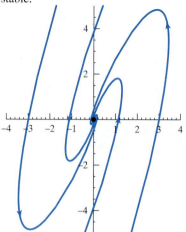

5. The origin is the only stationary point, and is stable, but not asymptotically stable.

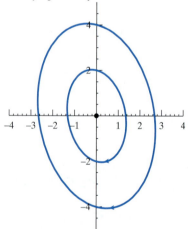

7. The stationary points are the origin, and all points on the circle of radius 1, centered at the origin. The origin is stable, but not asymptotically stable; the points on the circle are not stable.

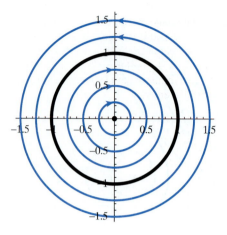

9. The stationary point $(0, 0)$, which is linearly degenerate (characteristic roots, 0 and -2) and unstable; $(2, 0)$, a saddle point with the x-axis as the stable separatrix; and $(1, 1)$, a stable spiral node.

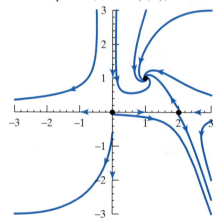

11. The seven stationary points are $(0, 0)$ (stable node), $(\pm\sqrt{10}, 0)$ (saddle points with the x-axis as unstable separatrices), $(\pm3, \pm1)$ (unstable nodes), and $(\pm1, \pm3)$ (saddle points).

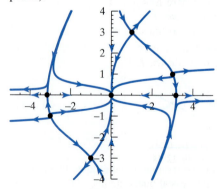

13. There are two stationary points, $(0, 0)$ (a saddle with stable separatrix on the y-axis and unstable separatrix on the y-axis) and $(1, 1)$ (characteristic roots are $\pm i$). There is an integral: $F(x, y) = x + y - \ln|x\,y|$ defined except on the axes; F has a local minimum at $(1,1)$, confirming that this stationary point is a center.

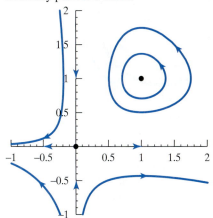

15. The origin is asymptotically stable if $k < 0$ (a counterclockwise inward spiral), stable if $k = 0$ (a linear center, oriented counterclockwise), and unstable if $k > 0$ (a counterclockwise outward spiral).

$k = -0.1$ $k = 0$ $k = 0.1$

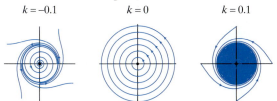

8.4

1. There are four stationary points, including the origin, $(K, 0)$ on the x-axis, and $(0, L)$ on the y-axis. If $BC < 1$, the fourth stationary point is a stable node in the first quadrant, and if $BC > 1$ the fourth stationary point is in the third quadrant, where it is not of interest—since populations can't be negative.

In the phase portraits shown below, $(B, C) = \left(2, \frac{1}{3}\right)$ and $(B, C) = \left(2, \frac{2}{3}\right)$, respectively. The other parameters are the same for both: $(a, d, K, L) = (0.05, 0.05, 2, 1)$.

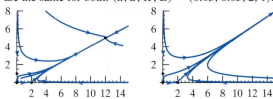

2. b. There are three stationary points: The origin, a sad-

dle where both species are extinct; the point $(C, 0)$, a saddle where the predator is extinct and the prey population is equal to the carrying capacity and is enough to support some predators should they arrive on the scene; and a stable node (d, y^*), where $y^* = \frac{A}{b}(C - d)$. When the predator population is y^*, there are just enough predators to limit the prey population to d, the minimum number needed to support any predators at all. In the phase portrait that follows, the parameters have been set as follows: $(A, C, b, c, d) = (0.5, 1500, 1, 0.125, 1000)$.

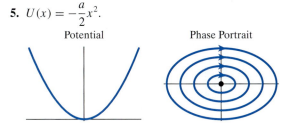

8.5

1. No

5. $U(x) = -\dfrac{a}{2}x^2$.

Potential   Phase Portrait

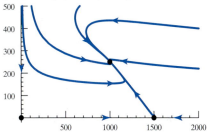

7. $U(x) = -x$

Potential   Phase Portrait

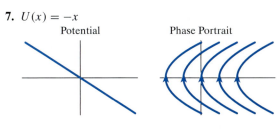

9. $U(x) = \frac{1}{3}x^3$

Potential   Phase Portrait

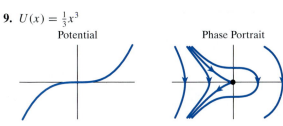

11. $U(x) = x^4$

Potential

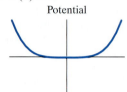

Phase Portrait

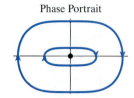

13. $U(x) = -|x^3|$

Potential

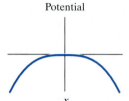

Phase Portrait

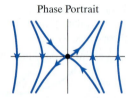

15. $U(x) = \dfrac{x}{x^2 + 1}$

Potential

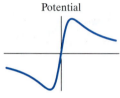

Phase Portrait

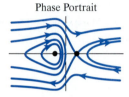

17. $U(x) = \cos(x) + \frac{1}{2}x$

Potential

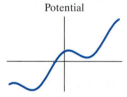

Phase Portrait

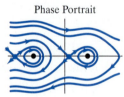

19. $U(x) = \cos(x) + \left(\frac{1}{\pi}\right)x^2$

Potential

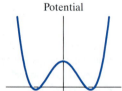

Phase Portrait

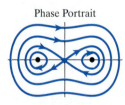

8.6

1. $L(x, y) = ax^2 + bxy + cy^2$
$$= a\left(x + \tfrac{b}{2a}y\right)^2 + \left(c - \tfrac{b^2}{4a}\right)y^2$$
$$= a\left(x + \tfrac{b}{2a}y\right)^2 + \tfrac{1}{4a}(4ac - b^2)y^2$$
Thus if $b^2 - 4ac < 0$, and $a > 0$, then $L(x, y) \geq 0$ and the equality holds if and only if $x = y = 0$. In other words, $L(x, y)$ is positive definite.

  Converse: $L(\epsilon, 0) = a\epsilon^2$; hence $a > 0$ if L is positive definite. Furthermore,

$$L(b\epsilon, -2a\epsilon) = a(4ac - b^2)\epsilon^2,$$

and thus $4ac - b^2 > 0$ as well.

3. The origin is unstable.

5. The origin is asymptotically stable.

7. The origin is unstable.

9. The origin is asymptotically stable. Use $L(x, y) = x^2 + y^2$ as the Lyapunov function.

11. Using the Lyapunov function $L(x, y) = 40x^2 + y^2$, the origin can be shown to be unstable.

13. $L(x, y) = 50x^2 - 2xy + y^2$ is a Lyapunov function that shows the origin to be stable but not asymptotically stable.

15. $L(x, y) = 5x^2 - 5xy + 4y^2$ is a Lyapunov function that shows the origin to be asymptotically stable.

17. $L(x, y) = 4x^2 + 3y^2$ is a Lyapunov function that shows the origin to be asymptotically stable.

19. If $q(x, y)$ is negative semidefinite, $L(x, y)$ is still a Lyapunov function for the system. and the origin is a stable stationary point.

21. a. $L'(x, y) = 2(x^2 + y^2)^2(1 - x^2 - y^2)$ is positive definite for on the set $x^2 + y^2 < 1$. Therefore, the origin is unstable.

d.

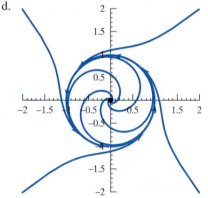

23. The system has a limit cycle for any negative value of k.

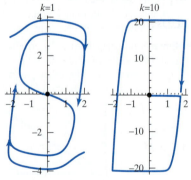

26. $S = \begin{bmatrix} \frac{1}{2} & \frac{1}{4} \\ \frac{1}{4} & \frac{3}{4} \end{bmatrix}$

Chapter 9

9.1

1. If the point x_i is in $\mathcal{I}$, then the following point x_{i+1} is also in $\mathcal{I}$, hence the process may be continued forever.

3.

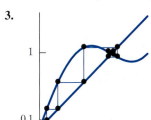

5.

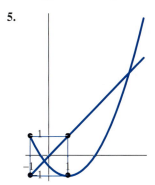

7.

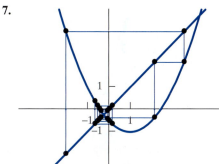

9. The fixed point at 0 is stable.

11. Let f be a one dimensional map with a 3-periodic orbit $\{x, y, z\}$; that is, $y = f(x)$, $z = f(y)$ and $x = f(z)$. Then this orbit is stable if

$$|f'(x)| \cdot |f'(y)| \cdot |f'(z)| < 1.$$

13. The logarithm of A is $M = \begin{bmatrix} \ln 2 & \frac{1}{2} \\ 0 & \ln 2 \end{bmatrix}$, and

$$P_\tau(x, y) = \begin{bmatrix} 2^t & t2^{t-1} \\ 0 & 2^t \end{bmatrix} \begin{bmatrix} x \\ y \end{bmatrix}.$$

15. The logarithm of A is $M = \begin{bmatrix} 0 & \frac{\pi}{2} \\ -\frac{\pi}{2} & 0 \end{bmatrix}$, and

$$P_\tau(x, y) = \begin{bmatrix} \cos\left(\frac{\pi}{2}t\right) & \sin\left(\frac{\pi}{2}t\right) \\ -\sin\left(\frac{\pi}{2}t\right) & \cos\left(\frac{\pi}{2}t\right) \end{bmatrix} \begin{bmatrix} x \\ y \end{bmatrix}.$$

17. Suppose $A = B^2$, then $\det(A) = \det(B^2) = [\det(B)]^2 > 0$.

19. If $x_0 < x_1$, then $f(x_0) < f(x_1)$; hence we get $x_1 < x_2$. If $x_0 > x_1$, then $f(x_0) > f(x_1)$; hence we get $x_1 > x_2$. If $x_0 = x_1$, then $f(x_0) = f(x_1)$; hence we get $x_1 = x_2$. In all three cases, we can continue by mathematical induction.

21. If F is increasing, then by the result of Exercise 19, x_n must be monotone increasing, monotone decreasing, or constant. If F is decreasing, then F^2 is increasing and hence the preceding subsequences x_{2n} and x_{2n+1} are each monotone increasing, monotone decreasing, or constant sequences.

23. The fixed points are $(0, 0)$, (stable), and $(1, -1)$ (unstable).

25. There are two fixed points $(0, 0)$, $(1/2, 1/4)$. Both are unstable.

27. Use the uniqueness theorem.

9.2

1. Use the principle that a set is connected if, for any given two points in the set, a continuous path can be found that starts at one of the points and ends at the other.

3. a. The edge of a patch on the inner tube is a curve whose complement is not connected.

   b. Cut the inner tube along a meridian line. The resulting surface will be a cylinder, which is connected.

   c. A cutting around a patch on the Möbius band will disconnect it. If you cut the band along its midline, you will get a strip with a full twist (try it!).

4. Visualize the orbits by representing the torus as a square with edges of length 2π. Points on the edges whose coordinates are equal mod 2π are considered to be the same.

   a. The orbit is a closed curve. See the drawing below.

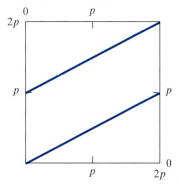

b. The orbit is a closed curve. See the drawing below.

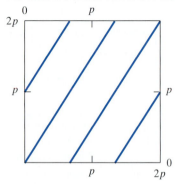

c. The curve is closed and goes around the torus in the θ direction 141 times, as it goes around in the ϕ direction 100 times.

d. The orbit is the same as the orbit in part (c).

e. This orbit is not closed.

5. There is a cycle, and orbits converge to it as $t \to -\infty$.

9. By the Poincaré-Bendixson theorem, every orbit of this system has a limit cycle, which must be one of the two boundary cycles. If we reverse time— that is, we replace the system, which we can denote $(x', y') = (f(x, y), g(x, y))$, with $(x', y') = (-f(x, y), -g(x, y))$—the orbits will be the same (as sets), but they will be oppositely directed. One of the two boundary curves must serve as a limit cycle for the reversed system.

10. b. The system (9.10) was obtained from (8.23) by multiplying the right side of each equation by $r(x, y)$.

c. The points $(0, 1)$ and $(0, -1)$ are stationary.

d. Multiply the right side of each equation of (9.10) by $|x + iy - 1|$.

e. Yes

9.3

1. Suppose that v_1 is in the range of the first return mapping P_T belonging to a system $v' = f(v, t)$, where f is T-periodic in time. The IVP with initial condition $v(T) = v_1$ has a unique solution, $v(t)$. Thus, $P_T(u) = v_1$ if and only if $u = v(0)$.

4. No   5. $P_{2\pi}\begin{bmatrix} x \\ y \end{bmatrix} = \begin{bmatrix} -2 \\ 1 \end{bmatrix} + e^{-2\pi}\begin{bmatrix} x + 2 \\ y - 1 \end{bmatrix}$

7. a. This follows from Proposition 5.2.3 on page 196.

b. The determinant of P_T is equal to $W(\phi_1, \phi_2)(T)/W(\phi_1, \phi_2)(0)$, where $\{\phi_1, \phi_2\}$ is a fundamental set of solutions. Since the Wronskian is constant, this quotient is equal to 1.

c. The characteristic equation for P_T is $s^2 - \text{tr}(P_T)s + \det(P_T) = 0$, and we know that $\det(P_T) = 1$. Hence the characteristic roots are

$$s_1, s_2 = \frac{1}{2}(\text{tr}(P_T) \pm \sqrt{\text{tr}(P_T)^2 - 4}.$$

If $|\text{tr}(P_T)| < 2$ the roots s_1, s_2 form a conjugate pair of complex numbers: Thus $|s_1| = |s_2| = 1$. If $|\text{tr}(P_T)| = 2$ then the roots are equal. It follows that $s_1 = s_2 = \pm 1$, because $\text{tr}(P_T) = s_1 + s_2$.

d. If $\text{trace}(P_T) > 2$, then $|s_1| + |s_2| > 2$, hence either $|s_1| > 1$ or $|s_2| > 1$, thus the origin is unstable.

e. The equation is equal to the system $x' = Ax$, where $A = \begin{bmatrix} 0 & 1 \\ -\lambda & 0 \end{bmatrix}$ The characteristic equation for A is $s^2 + \lambda = 0$. Therefore $\text{trace}(P_1) = 2\cos(\sqrt{\lambda})$ if $\lambda \geq 0$; $\text{trace}(P_1) = 2\cosh(\sqrt{|\lambda|})$ if $\lambda < 0$.

f. Suppose that $\lambda > 0$. If $|\cos(\sqrt{\lambda})| < 1$; that is, if $\lambda \neq n^2\pi^2$, then the zero solution will be stable for ϵ sufficiently small. On the other hand, if $\lambda < 0$, the zero solution is unstable for small ϵ. Finally, if $\lambda = n^2\pi^2$ for some n, the zero solution may be unstable for any positive ϵ.

8. a.

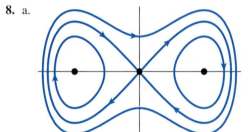

c.

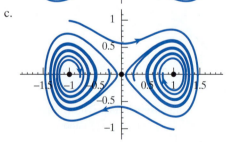

d. The three fixed points are $(1, 0)$, $(0, 0)$, $(-1, 0)$.

9. Let P_T be the first return mapping for the system $x' = A(t)x$, and let L be a logarithm of P_T^2 (which has a real logarithm because it has a square root). Suppose $\mathcal{X}(t) = [x_1(t), x_2(t)]$; then by Lemma 9.3.5, $z_1(t) = e^{-(t/T)L}x_1(t)$ and $z_2(t) = e^{-(t/T)L}x_2(t)$ are $2T$-periodic. Let $\mathcal{Z}(t) = [z_1(t), z_2(t)]$, and $M = L/(2T)$, then

$$\mathcal{X}(t) = [x_1(t), x_2(t)] = e^{(t/T)L}[z_1(t), z_2(t)] = e^{Mt}\mathcal{Z}(t)$$

Chapter 10

10.1

1. The BVP has a unique solution unless λ is of the form k^2, where k is a integer.

3. There is a unique solution unless $\lambda = 0$ or $\lambda = 1 + (n\pi)^2$, where n is an integer.

7. $G(t, u) = \begin{cases} e^{2(t-u)} & \text{if } u \le t, \\ e^{t-u} & \text{if } u > t \end{cases}$

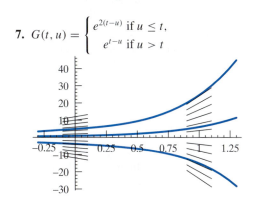

9. $G(t, u) = \begin{cases} -\sin(u)\cos(t) & \text{if } u \le t, \\ -\sin(t)\cos(u) & \text{if } u > t \end{cases}$

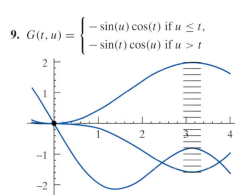

11. $G(t, u) = \begin{cases} \frac{1}{2}u(t-2) & \text{if } u \le t, \\ \frac{1}{2}t(u-2) & \text{if } u > t \end{cases}$

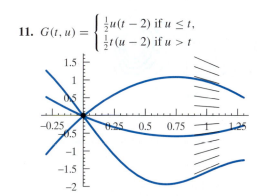

13. $G(t, u) = \begin{cases} \frac{(\arctan(u)-1)(\arctan(t)-\pi/4)}{\pi/4-1} & \text{if } u \le t, \\ \frac{(\arctan(t)-1)(\arctan(u)-\pi/4)}{\pi/4-1} & \text{if } u > t \end{cases}$

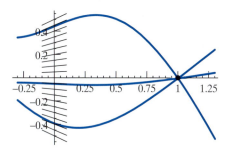

15. a. $A_1 B_2 + B_1 A_2 \ne 0$

  c. $A_1 B_1 + A_2 B_2 - 2A_2 B_1 \ne 0$

19. a. There is a unique solution: $y = \cos(t) - \sin(t)$.

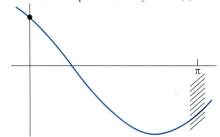

  c. There is a unique solution if $k \ne 0$: $y = \cosh(kt) + C\sinh(kt)$, where $C = \dfrac{k\sinh(k) - \cosh(k)}{\sinh(k) - k\cosh(k)}$; but there is no solution if $k = 0$. In the following graph, the values of k, chosen so that $y(1) = 1, 2$, and 3, were $k = 1.54, 1.15$, and 0.96, respectively.

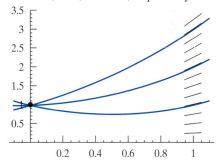

  e. There is no solution.

21. b.

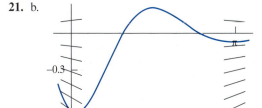

22. d.

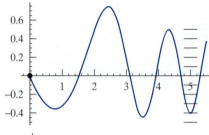

e.

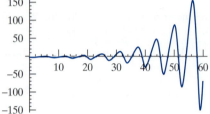

10.2

1. a. The eigenvalues are $\lambda_n = \left(\dfrac{n\pi}{L}\right)^2$, where n is an integer ≥ 0; the eigenfunctions are $\phi_0(x) \equiv 1$, and $\phi_n(x) = \cos\left(\dfrac{n\pi}{L}x\right)$.

c. The eigenvalues are $\lambda_n = 4 + \left(\dfrac{n\pi}{L}\right)^2$, where n is an integer > 0; the eigenfunctions are $\phi_n(x) = e^{-2x}\sin\left(\dfrac{n\pi}{L}x\right)$.

5. The first four intersection points of the graphs of $s = -t$ and $s = \tan(t\pi)$, are displayed in the following graph.

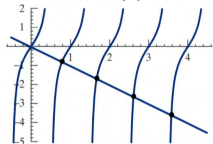

7. The positive eigenvalues are ν_n^2, where the ν_n are the sequence of solutions of the equation $\tan(\nu) = \nu$.

9. The eigenvalues of the EVP $y'' + \lambda y = 0$; $y(0) = y(\pi)$, $y'(0) = y'(\pi)$ are $\lambda = n^2$, where n is a nonnegative integer. To each eigenvalue except 0 there belong two independent eigenfunctions, namely, $y = \cos(n x)$ and $y = \sin(n x)$.

11. Suppose that $\lambda < 0$ is an eigenvalue, and $\phi(x)$ is an eigenfunction belonging to λ. Multiply the equation

$$(1 + x^2)\phi''(x) = -\lambda\phi(x)$$

by $(1 + x^2)^{-1}\phi(x)$ and integrate:

$$\int_0^1 \phi(x)\phi''(x)\,dx = -\lambda\int_0^1 \frac{(\phi(x))^2}{1+x^2}\,dx. \qquad (*)$$

The right side of $(*)$ is positive because λ is negative; the left side is equal to $-\int_0^1 (\phi'(x))^2\,dx$ and is negative, a contradiction. If $\lambda = 0$, it follows from $(*)$ that ϕ is constant, and the boundary conditions force $\phi \equiv 0$. Thus 0 is not an eigenvalue.

13. Let ϕ be an eigenfunction of Green's operator, with eigenvalue μ, and let $G(x, y)$ be Green's function. We will assume $|G(x, y)| \leq M$ in the square $a \leq x, y \leq b$. Thus

$$\mathcal{G}(\phi) = \mu\,\phi,$$

and hence

$$|\mu|\int_a^b [\phi(x)]^2\,dx = \left|\int_a^b \mathcal{G}(\phi)(x)\,\phi(x)\,dx\right|$$

$$= \left|\int_a^b\int_a^b G(x, t)\phi(t)\,\phi(x)\,dt\,dx\right|$$

$$\leq M\left|\int_a^b\int_a^b |\phi(t)|\,|\phi(x)|\,dt\,dx\right|$$

$$= M\left(\int_a^b |\phi(x)|\,dx\right)^2$$

So far, we haven't needed the Schwarz inequality—now we do.

$$M\left(\int_a^b |\phi(x)|\,dx\right)^2 = M\left(\int_a^b 1\cdot|\phi(x)|\,dx\right)^2$$

$$\leq M\int_a^b 1^2\,dx\int_a^b [\phi(x)]^2\,dx$$

$$= M(b-a)\int_a^b [\phi(x)]^2\,dx$$

It follows that $|\mu| \leq M(b-a)$.

10.3

1. a. $u(x, t) = e^{-9t}\sin(3x)$

c. $u(x, t) = e^{-10000t}\sin(100x)$

3. The temperature is given by the formula

$$u(x, t) = b_1 e^{-c\omega^2 t}\sin(\omega x) + b_2 e^{-4c\omega^2 t}\sin(2\omega x)$$
$$+ \cdots + b_n e^{-cn^2\omega^2 t}\sin(n\omega x),$$

where c is the diffusion coefficient. Thus $\lim_{t\to\infty} u(x, t) = 0$ for each x.

5. Calories per centimeter per second per centigrade degree

7. Square centimeters per second

11. a. $g(x) = e^{-\sqrt{\frac{365\pi}{c}}(1+i)x}$ **b.** Yes

10.4

1. The eigenvalues are $\lambda_m = \left(\frac{m\pi}{L}\right)^2$, for $m = 0, 1, 2, 3, \dots$, and the corresponding eigenfunctions are $\phi_0(x) = 1$, and $\phi_m(x) = \cos\left(\frac{m\pi}{L}x\right)$.

3. The Fourier series is

$$F(x) = \sum_{k=0}^{\infty} \frac{4}{(2k+1)\pi} \sin((2k+1)x).$$

  a. The series does not converge uniformly.

  b.

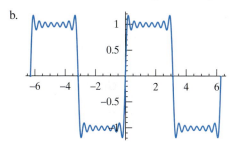

  c. $\displaystyle\sum_{k=0}^{\infty} \frac{(-1)^k}{2k+1} = \frac{\pi}{4}$

  d. $u(x,t) = \displaystyle\sum_{k=0}^{\infty} \frac{4}{(2k+1)\pi} e^{-(2k+1)^2 t} \sin((2k+1)x)$

5. $f(x) = \dfrac{\pi}{4} - \displaystyle\sum_{k=0}^{\infty} \frac{2}{(2k+1)^2 \pi} \cos((4k+2)x)$

  a.

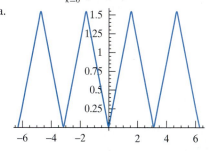

  b. $\displaystyle\sum_{k=0}^{\infty} \frac{1}{(2k+1)^2} = \frac{\pi^2}{8}$

  c. $u(x,t) = \dfrac{\pi}{4} - \displaystyle\sum_{k=0}^{\infty} \frac{2}{(2k+1)^2 \pi} e^{-(4k+2)^2 t} \cos((4k+2)x)$

7. The eigenvalues are numbers of the form $\lambda_n = v_n^2$, where $\tan(50v_n) = -\frac{0.1v_n}{0.05^2 - v^2}$. In the following drawing, the graphs of $\tan(50v)$ and $-\frac{0.1v_n}{0.05^2 - v^2}$ are displayed, and it can be seen that there is an infinite sequence of eigenvalues.

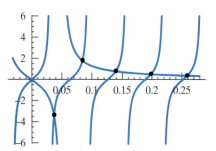

The first four eigenvalues are 0.00138609, 0.00709891, 0.0194424, and 0.0393514. For $n > 4$, $\lambda_n \approx \left(\frac{(n-1)\pi}{50}\right)^2$. $\phi_n(x) = 20v_n \cos(vx) + \sin(vx)$ is an eigenfunction belonging to the eigenvalue $\lambda_n = v_n^2$. The following drawing displays the boundary field and the graphs of $\phi_1, \dots, \phi_4$.

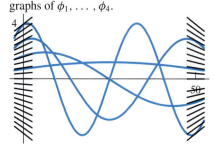

11. b. $m = \dfrac{B\alpha - A\beta}{\alpha + \beta + \alpha\beta L}$; $d = \dfrac{A + B + AL\beta}{\alpha + \beta + \alpha\beta L}$

  c. $v(x,t)$ satisfies the IBVP

$$\frac{\partial v}{\partial t} = c\frac{\partial^2 v}{\partial x^2};$$

$$\alpha v(0,t) - \frac{\partial v}{\partial x}(0,t) = 0, \quad \beta v(L,t) + \frac{\partial v}{\partial x}(L,t) = 0$$

10.5

1. $\dfrac{d\theta}{dx} = -(\lambda + \cos(x) - x^2)\cos^2(\theta) - \sin^2(\theta)$

2.

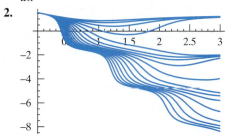

3. There are eigenvalues in the intervals (2,3), (7,8), and (12,13).

4. $2.5 < \lambda_1 < 2.625$, $7.25 < \lambda_2 < 7.375$, and $12.875 < \lambda_3 < 13$.

5. b. Let $\theta(x, \lambda)$ be the solution of the IVP

$$\theta'(x, \lambda) = -(\lambda\, w(x) + q(x)) \cos^2(\theta) - \sin^2(\theta);$$
$$\theta(0) = \frac{\pi}{2}.$$

The eigenvalues λ_m form infinite sequence of numbers such that $\theta(L, \lambda_m) = (2m - 1)\pi/2$.

6. The phase angle satisfies the ODE

$$\frac{d\theta}{dx} = -(\lambda\, w(x) + q(x)) \cos^2(\theta) - \sin^2(\theta),$$

and λ is an eigenvalue if the solution of the IVP with initial condition $\theta(0) = \arg(B - i\, p(0)\, A)$ also satisfies $\theta(L) = \arg(D - i\, p(L)\, C)$.

7. a. The equivalent Sturm-Liouville problem is

$$\left(\sec(x) y'\right)' + (\lambda \sec^2(x) + \cos(x))\, y = 0$$

with boundary conditions $y(0) = y(\pi/3) = 0$. The eigenfunctions $\phi_i(x)$ satisfy the following orthogonality relation:

$$\int_0^{\frac{\pi}{3}} \phi_i(x)\, \phi_j(x)\, \sec^2(x)\, dx = 0.$$

c. The equivalent Sturm-Liouville problem is

$$\left(x^2 y'\right)' + \lambda\, xy = 0,$$

with boundary conditions $y(1) = y(2) = 0$. The eigenfunctions $\phi_i(x)$ satisfy the following orthogonality relation:

$$\int_1^2 \phi_i(x)\, \phi_j(x)\, x\, dx = 0.$$

10.7

1. $y = -\frac{1}{2}x \cos(x)$

2. a. There is no Green's function; the BVP $\mathcal{L}(y) = 0$; $y(0) = y'(\pi/2) = 0$ has a nontrivial solution, $y = \sin(t)$. It follows from Theorem 10.1 (page 468) that Green's function can exist only if the homogeneous BVP has only the solution $y \equiv 0$.

b. $G(t, u) = \begin{cases} -\sin(u)\cos(t) & \text{for } u \le t, \text{ and} \\ -\sin(t)\cos(u) & \text{for } u > t \end{cases}$

c. $\mathcal{L}(y) = 0$ has the solution $y = 1 + t$ that satisfies both boundary conditions; therefore, there is no Green's function.

d. $G(t, u) = \begin{cases} \frac{1}{3}(u + 1)(t - 2) & \text{for } u \le t, \text{ and} \\ \frac{1}{3}(t + 1)(u - 2) & \text{for } u > t. \end{cases}$

3. Green's function:

$$G(t, u) = \begin{cases} \dfrac{1}{\sinh(2)} \sinh(u + 1)\sinh(t - 1) & \text{for } u \le t, \\[2ex] \dfrac{1}{\sinh(2)} \sinh(t + 1)\sinh(u - 1) & \text{for } u > t. \end{cases}$$

Solution of the BVP:

$$\frac{(1 - e^2)\cosh(t) + t(e^2 + 1)\sinh(t)}{2 + 2e^2}.$$

4. Green's function will not exist in any case where there is a nontrivial solution of $\mathcal{L}(y) = 0$ that satisfies the boundary conditions. For example, $\mathcal{L}(e^t) = 0$, and $y = e^t$ satisfies the boundary conditions $y(-1) = y'(-1)$, $y(1) = y'(1)$. Another solution, $y = \cosh(t)$, satisfies the boundary conditions $y'(0) = 0$, $y'(1) - \tanh(1)y(1) = 0$. The operator $\mathcal{L}$, with either of these two sets of boundary conditions, has no Green's function.

5. The eigenvalues are $\lambda_m = \left(\dfrac{2m - 1}{2}\right)^2$, for $m = 1, 2, 3, \dots$, and the corresponding eigenfunctions are $\phi_m(x) = \sin\left(\dfrac{2m - 1}{2}x\right)$.

6. $y(1) - \tanh(1)\, y'(1) = 0$

7. The eigenvalues are $\lambda_m = 1 + \left(\dfrac{2m - 1}{2}\right)^2$, and the corresponding eigenfunctions are $\phi_m(x) = e^{-x}\sin\left(\dfrac{2m - 1}{2}x\right)$.

8. $u(x, t) = 5e^{-\frac{0.15\pi^2}{4}t}\sin\left(\dfrac{\pi x}{2}\right) - e^{-\frac{1.35\pi^2}{4}t}\sin\left(\dfrac{3\pi x}{2}\right)$

Chapter 11

11.1

1. $u(x, t) = \dfrac{1}{2}\sin(2x)\sin(2t) - \dfrac{1}{8}\sin(4x)\sin(4t)$

3. $u(x, t) = \displaystyle\sum_{k=0}^{\infty} \dfrac{8}{(2k + 1)^4\pi}\sin((2k + 1)x)\sin((2k + 1)t)$

5.

m	C_m
1	0.0764212
2	0.027019
3	0.00849125
4	0.00000000
5	-0.00305685
6	-0.00300211
7	-0.00155962

7. a. $u(x, t) = 0.001\sin(\pi x)\cos(5790\pi t)$

c. $u(x, t) = \displaystyle\sum_{m=0}^{\infty} \frac{4}{125(\pi + 2m\pi)^3}$

$$\sin\left(\frac{(2m+1)\pi}{2L}x\right)\cos\left(\frac{c(2m+1)\pi}{2L}t\right)$$

11.2

1. $u(x, y) = \dfrac{\pi}{\sinh(\pi)} \sinh(x)\sin(y)$

3. $u(x, y) =$

$-\displaystyle\sum_{k=0}^{\infty} \frac{16\cosh((2k+1)x/2)\cos((2k+1)x/2)}{(2k+1)(4k^2+4k-3)\pi \sinh((2k+1)/2)}$

5. $u(x, y) = \dfrac{[\sinh(kx) + k\cosh(kx)][k\cos(ky) + \sin(ky)]}{2k\cosh(k) + (k^2+1)\sinh(k)}$

7. Let $\omega_m = \dfrac{(2m-1)\pi}{2L}$, for $m = 1, 2, \ldots$. Express the function $\phi_3(x)$ as a Fourier cosine series:

$$\phi_3(x) = \sum_{m=1}^{\infty} A_m \cos(\omega_m x),$$

where $A_m = \dfrac{2}{L}\displaystyle\int_0^L \phi_3(x)\,\cos(\omega_m x)\,dx$. Then the solution of the mixed BVP is

$$u(x, y) = \sum_{m=1}^{\infty} \frac{A_m}{\sinh(\omega_m M)}\sinh(\omega_m y)\cos(\omega_m x).$$

9. $u(x, y) = \displaystyle\sum_{m=1}^{\infty} \frac{2(-1)^{m-1}}{m\pi \sinh(m\pi)}[\sin(m\pi x)\sinh(m\pi y) + \sin(m\pi y)\sinh(m\pi x)]$

10. In the drawings, normal derivatives are noted on the edges of the squares.

a.

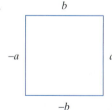

b

$-a$ a

$-b$

c.

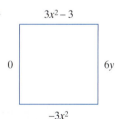

$3x^2 - 3$

0 $6y$

$-3x^2$

e.

$\displaystyle\sum_{m=1}^{\infty} m\pi(a_m e^{m\pi} - b_m e^{-m\pi})\cos(m\pi x)$

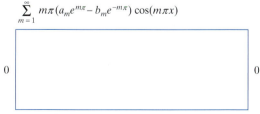

0 0

$\displaystyle\sum_{m=1}^{\infty} m\pi(b_m - a_m)\cos(m\pi x)$

11. $u(x, y) = \dfrac{1}{\sinh(\pi)}\cosh(x)\cos(y)$

13. $u(x, y) = -\dfrac{4}{\pi}\displaystyle\sum_{k=0}^{\infty} \frac{\cos((2k+1)x)\cosh((2k+1)y)}{(2k+1)^3 \sinh((2k+1)\pi)}$

11.3

3. $v(r) = \dfrac{C_1}{r} + C_2$, where C_1, C_2 are constants

4. a. $v(r, \theta) = \dfrac{1}{2}r^2 \sin(2\theta)$

c. $v(r, \theta) = \dfrac{\pi}{2} - \displaystyle\sum_{m=0}^{\infty} \frac{4r^{2m+1}}{(2m+1)^2\pi}\cos((2m+1)\theta)$

5. The solution $v(r, \theta)$ is not defined for $r > 2$.

6. a. $v(r, \theta) = \displaystyle\sum_{m=1}^{\infty} \frac{2(-r)^m}{m\pi}\sin(m\theta)$

c. $v(r, \theta) = \displaystyle\sum_{m=1}^{\infty} \frac{4r^m}{m(2m-1)\pi}\sin((2m-1)\theta)$

7. $v(r, \theta) = \displaystyle\sum_{n=1}^{\infty} C_n r^{n\pi/\alpha}\sin\left(\frac{n\pi}{\alpha}\theta\right)$, where

$$C_n = \frac{2}{\alpha R^{n\pi/\alpha}}\int_0^\alpha \sin\left(\frac{n\pi}{\alpha}\theta\right)\phi(\theta)\,d\theta.$$

9. $v(r, \theta) = \dfrac{8}{\pi}\displaystyle\sum_{k=0}^{\infty} \frac{1}{(2k+1)^3}r^{2k+1}\sin((2k+1)\theta)$

13. $u(x, y, z) = 15z(x^2+y^2+z^2)^2 - 70z^3(x^2+y^2+z^2) + 63z^5$

15. $v(x, y, z) = z^2 - \dfrac{1}{2}x^2 - \dfrac{1}{2}y^2$

11.4

1. a. The solution, $y = A\dfrac{\ln\left(\frac{r_2}{r}\right)}{\ln\left(\frac{r_2}{r_1}\right)}$, is unique.

c. Experimentation shows that only the first term of this series is needed for graphing purposes.

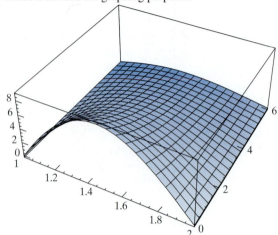

4. The eigenvalues are $\lambda_n = \omega_n^2$, where $r = \omega_1, \omega_2, \ldots$ are the roots of $J_0(r) = 0$. The eigenfunctions are $\phi_n(r) = J_0(\omega_n r)$, and they satisfy the orthogonality relation

$$\int_0^1 \phi_m(r)\,\phi_n(r)\,r\,dr = 0 \text{ for } m \neq n.$$

6. $u(r, t) \approx 33.33 + 57.84\,e^{-1.468t}\,J_0(3.83\,r)$

$\quad - 7.36\,e^{-4.922t}\,J_0(7.02\,r)$

$\quad + 10.86e^{-10.35t}\,J_0(10.17\,r).$

The graph of this approximation follows.

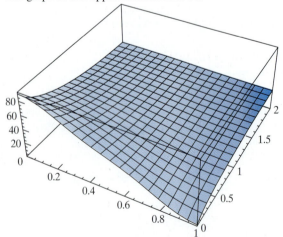

11.6

1. $u(t, x) = e^{-(x+\frac{5}{4}t)}\sin(x/2)$ 2. 216 seconds

3. As a Fourier series, the displacement function is

$$u(x, t) = \frac{1}{c\pi}\sum_{k=0}^{\infty}\frac{(-1)^k}{(2k+1)}$$

$$\left[2\sin\left(\frac{c\,(2k+1)\pi\,t}{L}\right)\sin\left(\frac{(2k+1)\pi\,x}{L}\right)\right].$$

It can be shown that an equivalent expression for $u(x, t)$ is

$$\begin{cases} -1 \text{ if } -ct < x < ct \\ 0 \text{ otherwise.} \end{cases}$$

The velocity profile, as a Fourier series, is

$$\frac{\partial u}{\partial t}(x, t) = \frac{1}{L}\sum_{k=0}^{\infty}(-1)^k$$

$$\left[2\cos\left(\frac{c\,(2k+1)\pi\,t}{L}\right)\sin\left(\frac{(2k+1)\pi\,x}{L}\right)\right]$$

Using the identity $2\sin(X)\cos(Y) = \sin(X+Y)+\sin(X-Y)$ and the Fourier series of the delta function,

$$\delta(x - L/2) = \frac{2}{L}\sum_{k=0}^{\infty}(-1)^k\sin\left(\frac{n\pi x}{L}\right)$$

you may show that

$$\frac{\partial u}{\partial t}(x, t) = \frac{1}{2}[\delta(x - ct - L/2) + \delta(x + ct - L/2)].$$

Disturbances, with half the magnitude of the initial velocity profile, emanate in both directions on the string.

4. The second partial derivatives of $w(x, y) = (x + iy)^n$ are $\frac{\partial^2 w}{\partial x^2} = n(n-1)(x+iy)^{n-2}$, and $\frac{\partial^2 w}{\partial y^2} = -n(n-1)(x+iy)^{n-2}$. Therefore, w is a complex-valued harmonic function; that is,

$$\frac{\partial^2 w}{\partial x^2} + \frac{\partial^2 w}{\partial y^2} = 0 + 0i$$

Thus the real and imaginary parts of w are also harmonic. The real and imaginary parts of the power series $f(z) = \sum_{n=0}^{\infty} a_n(z - z_0)^n$ are sums of harmonic functions, and are thus themselves harmonic.

Index